# Turbulence and Instabilities in Magnetised Plasmas, Volume 1

Fluid drift turbulence

# IOP Series in Plasma Physics

## Series Editors

**Richard Dendy**

*Culham Centre for Fusion Energy and the University of Warwick, UK*

**Uwe Czarnetzki**

*Ruhr-University Bochum, Germany*

### About the series

The IOP Plasma Physics ebook series aims at comprehensive coverage of the physics and applications of natural and laboratory plasmas, across all temperature regimes. Books in the series range from graduate and upper-level undergraduate textbooks, research monographs and reviews.

The conceptual areas of plasma physics addressed in the series include:

- Equilibrium, stability and control
- Waves: fundamental properties, emission, and absorption
- Nonlinear phenomena and turbulence
- Transport theory and phenomenology
- Laser-plasma interactions
- Non-thermal and suprathermal particle populations
- Beams and non-neutral plasmas
- High energy density physics
- Plasma-solid interactions, dusty, complex and non-ideal plasmas
- Diagnostic measurements and techniques for data analysis

The fields of application include:

- Nuclear fusion through magnetic and inertial confinement
- Solar-terrestrial and astrophysical plasma environments and phenomena
- Advanced radiation sources
- Materials processing and functionalisation
- Propulsion, combustion and bulk materials management
- Interaction of plasma with living matter and liquids
- Biological, medical and environmental systems
- Low temperature plasmas, glow discharges and vacuum arcs
- Plasma chemistry and reaction mechanisms
- Plasma production by novel means

# Turbulence and Instabilities in Magnetised Plasmas, Volume 1

Fluid drift turbulence

**Bruce Scott**
*Max-Planck Institute for Plasma Physics, Garching, Germany*

**IOP** Publishing, Bristol, UK

ISBN 978-0-7503-2504-2 (ebook)
ISBN 978-0-7503-2502-8 (print)
ISBN 978-0-7503-2505-9 (myPrint)
ISBN 978-0-7503-2503-5 (mobi)

DOI 10.1088/978-0-7503-2504-2

Version: 20210301

IOP ebooks

British Library Cataloguing-in-Publication Data: A catalogue record for this book is available from the British Library.

Published by IOP Publishing, wholly owned by The Institute of Physics, London

IOP Publishing, Temple Circus, Temple Way, Bristol, BS1 6HG, UK

US Office: IOP Publishing, Inc., 190 North Independence Mall West, Suite 601, Philadelphia, PA 19106, USA

*This work is dedicated to the memory of Professor Dieter Pfirsch, one of the founding directors of the Max-Planck-Institut für Plasmaphysik, and a colleague and friend, 1927–2004.*

# Contents


**Preface** xv

**Author biography** xvii

**1 Overview: magnetised plasma dynamics** **1-1**

1.1 Dynamics in plasmas 1-1

1.2 Magnetised plasmas 1-4

1.3 Outline of the work 1-11

Further Reading 1-12

**2 Introduction to turbulence** **2-1**

2.1 Statistical nonlinearity and cascade dynamics 2-1

2.2 Eddy mitosis and the cascade model 2-4

2.3 The statistical nature of turbulence 2-6

2.4 Quadratic nonlinearity and three-wave coupling 2-8

2.5 Fluid turbulence—energy and enstrophy 2-10

2.6 MHD turbulence 2-15

2.7 Selective decay 2-19

2.8 How the turbulence becomes two-dimensional 2-19

2.9 Plan 2-21

Further Reading 2-21

**3 Turbulence in two-dimensional systems** **3-1**

3.1 Various model systems 3-1

3.2 2D hydrodynamic turbulence 3-2

3.2.1 Bracket notation for the advection nonlinearity 3-2

3.2.2 Dissipation 3-4

3.2.3 The numerical scheme 3-4

3.2.4 Energetics 3-6

3.2.5 Dynamical range 3-8

3.2.6 The base case for fluid decaying turbulence 3-9

3.2.7 Self similarity 3-13

3.2.8 Selective decay 3-14

3.2.9 Structure formation 3-14

3.2.10 Intermittence and structure functions 3-15

3.3 2D MHD turbulence 3-16
3.3.1 MHD energetics 3-17
3.3.2 The base case for MHD decaying turbulence 3-19
3.3.3 Equipartition in the MHD model 3-24
3.4 2D electron MHD turbulence 3-29
3.4.1 Electron MHD energetics 3-32
3.4.2 The base case for EMHD decaying turbulence 3-32
3.5 2D Hall MHD turbulence 3-34
3.5.1 Hall MHD energetics 3-37
3.5.2 Contrast to the MHD cases 3-37
3.6 Compressibility in MHD 3-38
Further Reading 3-41

**4 Driven/dissipative turbulence 4-1**

4.1 Parallel dynamics along the guide field 4-1
4.2 The model system for dissipative ExB turbulence 4-2
4.2.1 The adiabatic response and the drift scale 4-3
4.2.2 The causal direction of the adiabatic response 4-4
4.2.3 Minimal model of adiabatic coupling 4-5
4.3 Turbulence in the adiabatic and hydrodynamic limits 4-6
4.3.1 Decaying turbulence in the hydrodynamic limit 4-7
4.3.2 Decaying turbulence in the adiabatic limit 4-9
4.3.3 Driven turbulence in the hydrodynamic limit 4-12
4.3.4 Driven turbulence in the adiabatic limit 4-14
4.3.5 Dissipative drift wave turbulence 4-17
4.3.6 Energy transfer dynamics 4-20
4.4 Implication of the ion gyroradius 4-25
Further Reading 4-26

**5 Absolute equilibrium ensembles 5-1**

5.1 AEQ and the role of dissipation in turbulence 5-1
5.2 The conserved quantities and equipartition 5-2
5.3 The phase space of degrees of freedom 5-3
5.4 Computational verification 5-4
5.4.1 The fluid case with passive scalar 5-4
5.5 Equipartition among the energies 5-7
5.5.1 The MHD case 5-8
5.5.2 The four-field two-fluid MHD case 5-15

5.6 Reintroduction of dissipation 5-18
Further Reading 5-20

**6 Fluid electrodynamics in a plasma 6-1**

6.1 Introduction 6-1
6.2 Ideal fluid equations and electrodynamics 6-2
6.3 High frequency motion under fluid electrodynamics 6-9
6.4 Quasineutral motion in a neutral plasma 6-13
6.5 Fluid plasma dynamics under quasineutrality 6-15
6.6 E Pluribus Unum—the steps to MHD 6-16
6.7 MHD waves—Alfvén waves 6-18
6.8 Energetics of the ideal fluid dynamical systems 6-21
6.9 Dissipation—corrections to the ideal plasma 6-24
6.10 Chapman–Enskog procedure—dissipation 6-25
6.10.1 Cross-species temperature differences 6-31
6.11 The moment approach—diamagnetic fluxes 6-32
Further Reading 6-36

**7 Fluid drift dynamics in a magnetised plasma 7-1**

7.1 Introduction 7-1
7.2 What the drift approximation is 7-2
7.3 Perpendicular force balance—diamagnetic current 7-3
7.4 Parallel dynamics–shear Alfvén nonlinearity 7-4
7.5 Perpendicular force balance—fluid drifts 7-5
7.6 The polarisation drift 7-6
7.7 Drift ordering and 'delta-f' 7-8
7.8 Derivation of the fluid drift equations 7-10
7.9 Energetics of the fluid drift equations 7-17
7.10 Summary 7-23
7.11 Delta-f versus total-f energetics 7-23
7.12 Quasineutrality in Drift Dynamics 7-25
Further Reading 7-26

**8 Parallel dynamics—Alfvén/sound waves 8-1**

8.1 Introduction 8-1
8.2 The four-field fluid drift model 8-3
8.3 Wave-like motion 8-4

8.4 Energetics, dissipation 8-10
8.5 Transient responses to a disturbance 8-13
8.6 Numerical examples 8-14
8.7 Energetics and decay rates 8-18
8.8 Thermal transport by the current 8-21
8.9 Effects of temperature dynamics 8-24
8.10 Summary 8-24
Further Reading 8-25

**9 Perpendicular dynamics—drift waves 9-1**

9.1 Introduction 9-1
9.2 ExB advection in a gradient—the drift frequency 9-2
9.3 Drift waves—the very simplest model 9-3
9.4 Drift waves—polarisation and dispersion 9-5
9.5 Drift waves—self-consistent dynamics 9-7
9.5.1 Perpendicular forcing, parallel response 9-7
9.5.2 Normalised linear equations 9-8
9.5.3 Limits under conventional ordering 9-10
9.5.4 Self consistency: drift-Alfvén modes 9-11
9.6 Dissipation: phase shifts and energetics 9-13
9.6.1 Excitation of the phase shift 9-13
9.6.2 Self-consistent energetics 9-15
9.7 Alfvénic transients 9-18
9.7.1 The initial value problem for drift waves 9-18
9.8 Numerical examples 9-19
9.9 Drift Alfvén waves—the magnetic flutter effect 9-23
9.10 Reactive instabilities 9-27
9.11 Mode structure 9-29
9.12 Summary 9-30
Further Reading 9-30

**10 Mode structure diagnostics 10-1**

10.1 Introduction 10-1
10.2 Temporal diagnostics 10-2
10.3 Spectral diagnostics 10-4
10.3.1 Normalisation of the Fourier transform 10-5
10.3.2 Spectra in the 2D domain 10-6
10.3.3 In drift wave turbulence 10-8

10.4 Energetics 10-9
10.5 Correlations 10-13
10.6 Linear growth phase versus turbulence 10-15
10.7 Randomness 10-18
10.8 Cross coherence 10-20
10.9 Interscale transfer 10-23
10.10 Three-dimensional diagnostics 10-23
10.11 Summary—mode structure in turbulence 10-23
Further Reading 10-25

**11 Three-dimensional drift wave turbulence 11-1**

11.1 Introduction 11-1
11.2 Drift Alfvén model and energetics 11-1
11.2.1 Normalised nonlinear equations 11-3
11.2.2 Energetics 11-5
11.2.3 Subgrid dissipation 11-6
11.2.4 Computational Courant limit 11-7
11.3 Periodic cases 11-7
11.3.1 Decaying cases 11-9
11.3.2 Driven cases 11-14
11.4 Aspect ratio 11-16
11.5 Bounded cases 11-21
11.6 Cases with magnetic shear 11-25
11.6.1 Mode structure changes with shear 11-28
11.7 On pathology 11-31
11.8 Summary 11-33
Further Reading 11-34

**12 Drift wave turbulence in a sheared magnetic field 12-1**

12.1 Introduction 12-1
12.2 Field line connection and magnetic shear 12-2
12.3 The 2D sheared slab model 12-4
12.4 Linear stability of electrostatic drift waves 12-6
12.5 Magnetic shear in 3D—field-aligned coordinates 12-11
12.6 Self-sustained drift wave turbulence 12-16
12.6.1 Role of nonlinear polarisation 12-19

| | | |
|---|---|---|
| 12.7 | Magnetic shear and drift wave mode structure | 12-24 |
| | 12.7.1 Details of the energetics | 12-26 |
| 12.8 | Electromagnetic effects | 12-34 |
| 12.9 | Contingent role of linear stability | 12-37 |
| 12.10 | Summary | 12-39 |
| | Further Reading | 12-40 |
| **13** | **MHD interchange turbulence** | **13-1** |
| 13.1 | Introduction | 13-1 |
| 13.2 | Magnetic divergences and the interchange model | 13-2 |
| | 13.2.1 Considering the MHD equilibrium | 13-3 |
| | 13.2.2 The curvature terms | 13-4 |
| 13.3 | Interchange energetics | 13-5 |
| 13.4 | The 2D interchange model | 13-6 |
| 13.5 | The ideal interchange mode | 13-7 |
| 13.6 | 2D interchange turbulence | 13-9 |
| 13.7 | Radial flows versus zonal flows | 13-16 |
| | 13.7.1 Aspect ratio effects for interchange flows | 13-19 |
| 13.8 | The mode structure of interchange turbulence | 13-20 |
| 13.9 | A simple model of a toroidal magnetic field | 13-21 |
| 13.10 | The ballooning mode | 13-27 |
| 13.11 | Three dimensions—ballooning mode turbulence | 13-31 |
| | 13.11.1 Resistive ballooning modes and transition to turbulence | 13-32 |
| 13.12 | Curvature forcing and ballooning mode structure | 13-35 |
| | 13.12.1 Details of the energetics | 13-42 |
| 13.13 | Electromagnetic and collisional effects | 13-44 |
| 13.14 | Summary | 13-47 |
| | Further Reading | 13-47 |
| **14** | **Toroidal drift Alfvén turbulence** | **14-1** |
| 14.1 | Introduction | 14-1 |
| 14.2 | The toroidal drift Alfvén model | 14-2 |
| | 14.2.1 Equations | 14-2 |
| | 14.2.2 Energetics | 14-4 |
| 14.3 | Toroidal drift Alfvén turbulence | 14-5 |
| 14.4 | The energetics of toroidal turbulence | 14-6 |

14.5 The mode structure of toroidal turbulence 14-8
14.5.1 Details of the energetics 14-13
14.6 From the linear stage to turbulence 14-16
14.7 Electromagnetic and collisional effects 14-22
14.8 Warm ion effects 14-27
14.9 Comparison to the control cases 14-33
14.9.1 Drift wave and ballooning mode turbulence as controls 14-33
14.9.2 Mode structure and energetics 14-34
14.9.3 Electron force balance and the drift scale 14-35
14.10 Summary 14-35
Further Reading 14-36

**15 Turbulence on open field lines 15-1**

15.1 Introduction—open field line geometry 15-1
15.1.1 Speeds, characteristics, exit fluxes 15-1
15.2 Model characteristics 15-5
15.3 Effects on the turbulence 15-8
15.3.1 The slab and toroidal cases 15-8
15.3.2 Effective adiabatic coupling 15-11
15.3.3 Self-consistent gradient drive 15-13
15.3.4 Cases with a source 15-15
15.4 Turbulence in a dipole magnetic field 15-22
15.4.1 The dipole magnetic field 15-22
15.4.2 Flux-tube geometry in a dipole field 15-23
15.4.3 The turbulence 15-25
15.5 Summary 15-32
Further Reading 15-33

**16 Drift wave turbulence and flows 16-1**

16.1 Introduction—eddies and flows 16-1
16.2 Kelvin–Helmholtz stability 16-3
16.2.1 Aspect ratio effect 16-5
16.3 Sheared flows and decorrelation 16-6
16.4 ExB flow energetics 16-7
16.4.1 In a magnetised plasma 16-9
16.5 Effect of background flow shear 16-12

16.5.1 In toroidal geometry 16-13
16.5.2 In slab geometry 16-13
16.5.3 On open field lines 16-14
16.6 Flow shear in warm-ion toroidal cases 16-16
16.7 Properties of the flux surface average 16-20
16.8 Zonal and equilibrium flows 16-23
16.8.1 Geodesic curvature and sidebands 16-23
16.8.2 Sideband analysis of the dynamics 16-25
16.8.3 Geodesic acoustic oscillations 16-26
16.8.4 Geodesic Alfvén oscillations 16-28
16.8.5 Acoustic/MHD adiabatic coupling 16-30
16.9 Self-generated zonal flows 16-32
16.9.1 Effect of geodesic curvature 16-33
16.9.2 Warm-ion zonal flow energetics 16-34
16.10 Summary 16-38
Further Reading 16-38

**17 Interlude** **17-1**

**Index** **18-1**

# Preface

I wrote this because of the large number of concepts more than 30 years of work have developed on this subject, which too large a fraction of the community hasn't learned about yet.

The 1970s picture of a plasma is that at the macroscopic level you have a conducting fluid interacting with magnetic fields. There is one density, one velocity, and one pressure, and there is the current which is the source for an embedded magnetic field. The electric field itself is submerged in ways that will be explained in detail herein. This is the picture of magnetohydrodynamics, or MHD for short. This serves the conceptual structure in an analogous way to fluid dynamics (the technical term is hydrodynamics), wherein an ideal or perfect fluid does the dynamics and some kinetic mechanism is added on as a series of diffusion or friction coefficients to do the dissipation. The plasma is expected to follow MHD on large scales and then one turns to kinetic mechanisms for the dissipation, with the addition that these mechanisms are often the cause of instabilities. This is the picture as it developed into the late 1970s and it is still the picture most people have today.

Even in dissipation free environments, a much more useful picture allows the electrons to be a separate fluid to the ions, even in the case of a 'single component' plasma (one positive charge carrier). The electrons are no longer a massless species behaving like electricity, they are another fluid with finite mass density and pressure, just with a disparate mass ratio (1845 for protons or normal hydrogen, 3670 for deuterium). Now, we have two densities, two velocities, and two pressures, to go with the electric and magnetic fields. This can also be treated in an ideal (dissipation free) way, together with Maxwell's equations.

The complication now is charge neutrality. It is sort of automatic in MHD, and has to be treated carefully when there are at least in principle two densities. So at low frequencies, the system of a 'quasineutral, two-fluid plasma' is what has to be developed. There are rigorous ways to incorporate this concept, which means you have zero charge density but you can still have an electric field. This is a result of finite polarisability, so it is formulated as one density (electrons) and then instead of another density you have the ion polarisation as the other variable. This also takes care of the electric field. At low frequency, which means slower than electrodynamics but also slower than compressional MHD processes, we know how to treat this and it forms the theoretical substance of this work. There are two approaches to it, one more fluid-like, and one with more of a kinetic basis, and we will do both.

This is the case with magnetised plasma turbulence, not plasma turbulence and not turbulence. So we have to say first what a magnetised plasma is (strong background magnetic field, but what are the conditions of that statement?). Then we have to say what turbulence is. Then we have to build up a look and feel for the turbulence and understand the significance of a self-consistent, statistical process. Then we explore the first version of the underlying theory (more problems, but easier to understand, and with a more obvious capture of MHD as a subset). After that, and only then, do we start addressing waves and instabilities. When the concepts, especially polarisation become familiar, then we can discuss the second version of the theory, gyrokinetic and gyrofluid theory, which is more rigorous and can treat a wider range of situations. At the end, we go back and make some still remaining connections to the standard picture of MHD instabilities.

It is important, very important, to understand this subject as turbulence first, and instabilities second. The reason for that is that linear theory for small fluctuations has a range of applicability which is much narrower than is often realised, and can give an incorrect interpretive physical picture (cartoon) to what the energetics and dynamics actually are.

It takes a lot of effort to do this, and I would not have done so had I not felt a need for it. If everybody was already familiar with this material and could teach it to their students, then this work would not have been necessary.

So I did my best. One can always do better. You are never finished. You are never ready. You are never certain. If you think you are, you are in the wrong field.

# Author biography

B Scott is a research plasma physicist having graduated with a Doctorate from the University of Maryland in 1985, and with the German Habilitation from the Heinrich-Heine-Universität Düsseldorf in 2001. He is a Fellow of the American Physical Society with membership since 1979. He has several tens of first author papers in peer-reviewed journals in the field of theoretical plasma physics.

**IOP** Publishing

Turbulence and Instabilities in Magnetised Plasmas, Volume 1

Fluid drift turbulence

**Bruce Scott**

# Chapter 1

# Overview: magnetised plasma dynamics

## 1.1 Dynamics in plasmas

Plasma physics is the study of large populations of charged particles interacting with electric and magnetic fields. Each particle moves according to the equations of motion for a charged point mass in an electromagnetic field, and the electromagnetic field reacts to what the charges do, according to Maxwell's equations. In this sense it is a classical theory; quantum effects enter only when individual particles interact directly with each other, or in certain instances where packets of electromagnetic energy (photons) interact directly with an atom or one or more of the charged particles. Interactions of the latter type set various reaction rates or atomic ionisation/recombination, or more to our interest, Coulomb collision frequencies and associated dissipation coefficients. In a plasma, most of the interactions between particles are through self-consistent responses of the fields with arbitrarily many particles always in play.

A gas of charged particles is a plasma when the number of particles involved in any interaction of interest is large enough that they act collectively, rather than simply as individuals. One can think of a cloud of electrons made to contract by some external force. The net electrostatic potential in the cloud will then decrease, and the electric field caused by the gradient of that potential will cause the cloud to expand. If there are two clouds, as shown in figure 1.1, each of electrons and ions in equal numbers, then the gas is net neutral, and in this case the electron cloud will oscillate between expansion and contraction, around an equilibrium in which the two clouds would coincide (this neglects the thermal expansion of the neutral cloud, which is much slower). The individual particles are always in motion if the gas is of a finite temperature, but the plasma dynamics proceeds according to what the populations do in a net sense. The simple situation just described is the basic plasma oscillation between the charge density and the electric field. It is fundamentally a collective interaction, and it is with such collective interactions between particle populations and electric and magnetic fields that plasma physics is concerned.

doi:10.1088/978-0-7503-2504-2ch1  

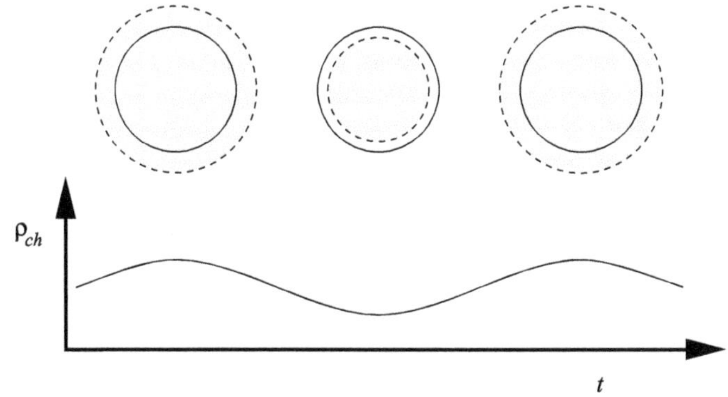


**Figure 1.1.** The basic oscillation of a plasma consisting of clouds of equal numbers of electrons (dashed circle, filled) and ions (solid circle). When the electron cloud expands, it is pulled back to the ion cloud by the electric field created by the nonzero charge density. When the electron contracts, it is pushed back by the same force, now of opposite sign. The ion cloud has a motion opposite to the electrons, but smaller by a factor of the square root of the mass ratio. The sinusoidal curve gives the charge density, $\rho_{ch}$, at the centre of the plasma as a function of time, $t$. The oscillation proceeds at the plasma frequency, $\omega_{pe}$.

There is a very wide range of basic interactions in a plasma. Some involve only a specific subset of the particles, those which for example have a velocity in a certain direction that coincides with the frequency of an electromagnetic wave propagating in that direction. Effects such as these are called kinetic, since the details of which particles are at which velocities, or more generally, which particles are in a given region of phase space, is of key importance to the situation. These effects are usually most important on microscopic scales, or on larger scales as damping mechanisms for otherwise dissipation free fluid-like motions. Examples are the conduction of heat by particles scattering back and forth over the distance of a mean free path between collisions, or the absorption of high frequency waves by resonance with particles moving at a velocity equal to the phase velocity of the wave.

Other interactions, particularly macroscopic ones whose spatial scale is comparable to the overall dimensions of the plasma, can be described with a simpler fluid model. In this case the details of particular velocities are unimportant, and one can describe the situation using only the densities, mean velocities, and temperatures of the constituent gases making up the plasma. This situation is obviously relevant to considerations of global stability. But it is also important to a class of smaller scale, fluid-like motions which occur in the form of turbulence, for which kinetic effects provide dissipation channels. These motions are responsible for the transport of particles and energy out of an inhomogeneous system, whenever microscopic processes do not suffice. The situation is analogous to turbulence in a normal fluid such as the Earth's atmosphere, which transports energy from hot regions to colder regions, and does so much more efficiently than microscopic processes. This type of dynamics is the one with which this work is most concerned.

Plasma physics is important to situations found both in nature and in the laboratory. The most familiar case in nature is the outer regions of the solar atmosphere, called the corona. The gas there is very hot, over a million degrees

Kelvin. We denote temperatures in terms of the average kinetic energy of random motion per particle, expressed in units of energy. So one million degrees Kelvin corresponds roughly to $k_B T = 100$ eV, or an average energy of one hundred electron volts. Here, $k_B = 1.38 \times 10^{-23}$ J K$^{-1}$, or equivalently about one electron volt per about 11 600 K. One electron volt is the quantity of energy acquired by an electron which is accelerated through a potential change of one volt, that is, $1.6 \times 10^{-19}$ J. In plasma physics we normally express temperatures in eV, with the Boltzmann constant understood. With $T = 100$ eV the average energy of collision between any two particles is 100/13.6 or about seven times what is necessary to strip an electron away from a proton, thereby ionising a hydrogen atom. Such a hot gas is fully ionised, and the ions in it are simply individual nuclei. Whether it is a plasma or not depends on the density at a given temperature.

Charged particles in populations screen each other, such that on distances comparable to any relevant scale of motion the individual interactions are negligible. This process is called 'Debye screening,' after its initial author. If one considers a homogeneous gas of electrons and singly charged ions in local thermodynamic equilibrium, and then places a test charge $Ze$, then the local charge density will respond. Assume *a priori* that the ions are static and that the changes to the electron density $n_e$ and temperature $T_e$ are small, with the charge density $\rho_{ch}$ finding equilibrium to the test charge. Force balance on the electrons states that the relation

$$n_e e \frac{\partial \phi}{\partial r} = T_e \frac{\partial n_e}{\partial r} \qquad \text{hence} \quad \phi = -\frac{T_e}{n_e e^2} \rho_{ch}, \tag{1.1}$$

where $r$ is the radial distance from the test charge, holds between the electrostatic and thermal pressure forces. The Poisson equation for the field due to the charges is

$$\nabla_\perp^2 \phi = -\frac{1}{\epsilon_0} \rho_{ch} = \frac{n_e e^2}{\epsilon_0 T_e} \phi, \tag{1.2}$$

for $r > 0$, that is, away from the test charge. The solution to this equation for a point test charge is a Yukawa potential given by

$$\phi = \frac{Ze}{r} \exp[-r/\lambda_D], \tag{1.3}$$

where $\lambda_D$ is the Debye length given by

$$\lambda_D^2 = \frac{\epsilon_0 T_e}{n_e e^2}. \tag{1.4}$$

The Debye length defines the distance over which any individual charge has any influence. The *a priori* assumptions are satisfied if $\lambda_D \nabla \ll 1$ on any quantity of interest to the dynamics.

Outside of a distance given by the Debye length, the influence of an individual particle is no longer felt. The particles respond to each other collectively through their effects on the electromagnetic field. If the number of charged particles in a Debye sphere is large, $n_e \lambda_D^3 \gg 1$, then the gas behaves as a plasma. This parameter for the solar corona, whose density is very roughly $n_e = 10^{12}\ \mathrm{m}^{-3}$, is of the order $10^8$. Differing regions of the corona are threaded by magnetic fields of varying strength, so depending on the situation the coronal plasma is either magnetised or unmagnetised.

## 1.2 Magnetised plasmas

Our interest is not turbulence in plasmas, but turbulence in magnetised plasmas, in which the influence of an equilibrium magnetic field on the dynamics is strong. It is regarded as a 'guide field,' ordering the dynamics into parallel and perpendicular components. A laboratory plasma is the clearest example of a magnetised plasma. The simplest case is a straight magnetic field set up by means of solenoidal coils which carry an electric current. Within this is a vacuum chamber into which a diffuse gas is admitted and then ionised according to some means, or the gas can be introduced already ionised and hot enough to remain so. Individual particles of mass $m$ and charge $Ze$ (electrons have $Z = -1$) in a magnetic field of strength $B$ and unit direction $\mathbf{b}$ represented by the continuous vector field, $\mathbf{B} = B\mathbf{b}$, follow orbits around the lines of force, called magnetic field lines, whose orientation is given by $\mathbf{b}$. This gyromotion occurs in the plane locally perpendicular to the field line. It follows from the magnetic part of the Lorentz force,

$$\mathbf{F}_L = Ze(\mathbf{E} + \mathbf{w} \times \mathbf{B}), \tag{1.5}$$

which acts at right angles to both the instantaneous velocity, $\mathbf{w}$, and the magnetic field, with magnitude $ZewB$. Parallel to the field line the motion is unaffected, but the motion projected onto a plane perpendicular to $\mathbf{B}$ is a circle of radius $\rho_L$ and frequency $\Omega_L$ given by

$$\rho_L = \frac{w}{\Omega} \qquad \Omega_L = \frac{ZeB}{m}. \tag{1.6}$$

These are called the gyroradius and gyrofrequency, respectively. Gyromotion is illustrated in figure 1.2.

Thermal populations of charged particles have characteristic values of these scales. A gas of electrons, for example, has an average velocity, $V_e$, or thermal velocity given in terms of their temperature and mass, since temperature is related to the average kinetic energy per particle in local thermodynamic equilibrium. Substituting this velocity in for $w$ in $\rho_L$ and noting that the magnitude of the charge is just $e$, we find the electron gyroradius and gyrofrequency. Together these are given by

$$\Omega_e = \frac{eB}{m_e} \qquad \rho_e = \frac{V_e}{\Omega_e} \qquad \text{where} \quad V_e^2 = \frac{T_e}{m_e}. \tag{1.7}$$

Note the convention for these is to define them without factors of 3/2 or $\sqrt{2}$. We find these quantities similarly for the ions. For ions of charge $Ze$ and mass $m_i$ they are

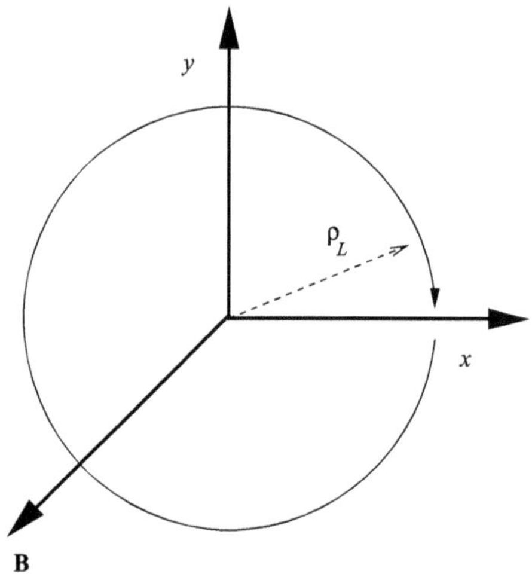


**Figure 1.2.** Gyromotion of an ion in a magnetic field, **B**. The particle orbits the field lines in the *xy*-plane, perpendicular to **B**. The radius of the orbit is $\rho_L$, and the sense of the motion is such that the Lorentz force is centripetal. For positive ions the orbit is clockwise as seen by an observer up the field line looking down on the plane.

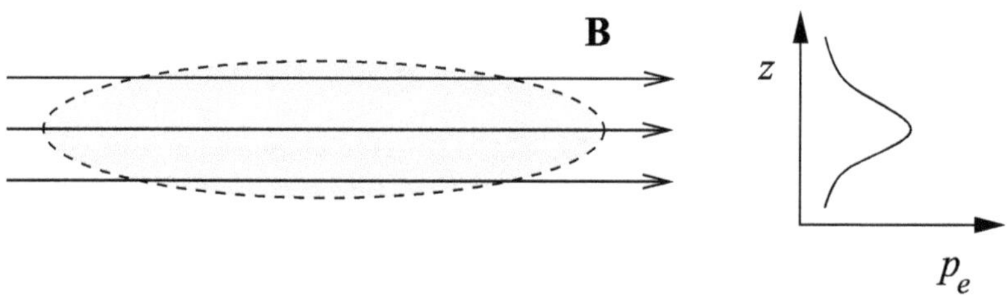


**Figure 1.3.** Sketch of a linear magnetised plasma. The particle gyroradii are small compared to the perpendicular scale length, which is in this case the plasma radius, *a*. If there is an edge layer with steep gradients scaling as $L_\perp \ll a$, then the gyroradii would have to be smaller than $L_\perp$ in order for the dynamics in that layer to be magnetised.

$$\Omega_i = \frac{ZeB}{m_i} \qquad \rho_i = \frac{V_i}{\Omega_i} \qquad \text{where} \quad V_i^2 = \frac{T_i}{m_i}, \tag{1.8}$$

where it is important to note that the masses of electrons and ions are very disparate. We always have

$$\mu_e \equiv \frac{m_e}{m_i} \ll 1, \tag{1.9}$$

since for the lightest ions, protons, $\mu_e = 5.4 \times 10^{-4}$. We usually have

$$\rho_e \ll \rho_i, \tag{1.10}$$

although in some cases the ions are so cold that the mass ratio is overcome. A laboratory plasma is considered magnetised if all of these gyroradii are small compared to its overall dimensions. A sketch is shown in figure 1.3. The limiting case is the ratio between $\rho_i$ and the plasma's scale length perpendicular to **B**. This constraint is well satisfied, particularly for plasmas which are confined by an externally applied magnetic field, which is the usual case of a laboratory plasma.

We will often refer to the concept of low frequency motion. In one sense it refers to the neglect of the finite speed of light in the response of the electromagnetic field to its sources. In this sense $c \to \infty$ reflects $\epsilon_0 \to 0$ in the approximations. Of more significance is the frequency regime

$$\omega \ll \Omega_i \tag{1.11}$$

where the ions are used since their gyromotion is the slowest. Nonlinear macroscopic dynamics resulting from pressure, flow, or current gradients in a magnetised plasma fall into this regime, which can also be taken as another condition for the plasma being magnetised. When we say low frequency herein, we will be referring to this regime, except when discussing Maxwell's equations directly.

One of the most important applications of plasma physics in the laboratory is to the case in which a plasma is confined by means of an appropriately constructed magnetic field, and heated by external means to a temperature hot enough that the individual nuclei undergo fusion reactions. When the energy responsible for this comes from the random thermal motion, this situation is called thermonuclear fusion. The long range attempt to bring this process to bear under controlled conditions is of obvious importance to the future of energy generation in the prevailing industrialised society. The two main lines of research are called magnetic and inertial fusion. In the first, the plasma is confined by a magnetic field and expected to undergo fusion under conditions which are steady for times much longer than those characteristic of dynamical stability. In the second, a small pellet of high density frozen material is very suddenly subjected to intense laser energy, such that it implodes to high density under enough temperature to undergo fusion. As far as fusion applications are concerned, this work will treat only the first situation.

A plasma can be confined by a magnetic field due to the Lorentz forces acting on all its particles. If the plasma is magnetised the individual particles cannot simply slip past the magnetic field. But the magnetic field and the plasma together form a dynamical system, and one must take care that it is stable. The study of macroscopic dynamics of a current carrying plasma and a magnetic field is most often done using a model called 'magnetohydrodynamics,' or MHD. The plasma constituents are assumed to be sufficiently well coupled that it can be treated as a single, charge neutral fluid, which due to its makeup can carry an electric current. In a neutral plasma the electric part of the Lorentz force adds up to zero, and the magnetic part adds up in terms of the individual velocities and charges to what is called the J-cross-B (JxB) force, given by the sum over all particles of the magnetic Lorentz force,

$$\sum_j Z_j e_j \mathbf{w}_j \times \mathbf{B} = \mathbf{J} \times \mathbf{B} \qquad \text{with} \quad \mathbf{J} = \sum_j Z_j e_j \mathbf{w}_j. \tag{1.12}$$

This current is in turn given by the spatial gradients of the magnetic field according to Ampère's law

$$\mu_0 \mathbf{J} = \nabla \times \mathbf{B}, \tag{1.13}$$

which holds at low enough frequencies that we can ignore propagation delay at the finite speed of light. If we substitute this relation into the JxB force, and use the vector identity

$$\nabla \cdot (\mathbf{F} \times \mathbf{G}) = \mathbf{G} \cdot \nabla \times \mathbf{F} - \mathbf{F} \cdot \nabla \times \mathbf{G}, \tag{1.14}$$

(for any two vectors **F**, **G**) to reduce the expression with $(\nabla \times \mathbf{B}) \times \mathbf{B}$, we find the two types of forces that a magnetic field exerts on a current carrying plasma:

$$\mathbf{J} \times \mathbf{B} = -\nabla\left(\frac{B^2}{2\mu_0}\right) + \frac{\mathbf{B} \cdot \nabla \mathbf{B}}{\mu_0} = -\nabla_\perp\left(\frac{B^2}{2\mu_0}\right) + \frac{B^2}{\mu_0}\mathbf{b} \cdot \nabla \mathbf{b}, \tag{1.15}$$

where $\nabla_\perp$ is the component of the gradient acting in the plane perpendicular to **B**. These two terms are referred to as magnetic pressure, or field line compression, and magnetic tension, or field line bending, since $\mathbf{b} \cdot \nabla \mathbf{b}$ is a vector directed towards the centre of the field line curvature with magnitude given by the inverse of the curvature radius.

A simple picture of the MHD model is one of magnetic field lines as wires, bundles of which exert a pressure in the plane perpendicular to their orientation, and which exhibit tension, that is, a restoring force against bending. These two analogies describe the nature of the two terms in the JxB force. Further expanding the wire analogy is the reaction of the magnetic field to the plasma—the wires are simply dragged along by the plasma and bunched together if the plasma flow is compressional. This gives rise to two types of wave motion called Alfvén waves, as shown in figure 1.4, which propagate at a velocity called the Alfvén speed,

$$v_A = \left(\frac{B^2}{\mu_0 n_i m_i}\right)^{1/2}, \tag{1.16}$$

reflecting the fact that the magnetic forces scale as $B^2/\mu_0$ and that $n_i m_i$ is the mass density. Magnetic pressure gives rise to compressional Alfvén waves, and tension leads to shear Alfvén waves, also called torsional Alfvén waves if the perpendicular

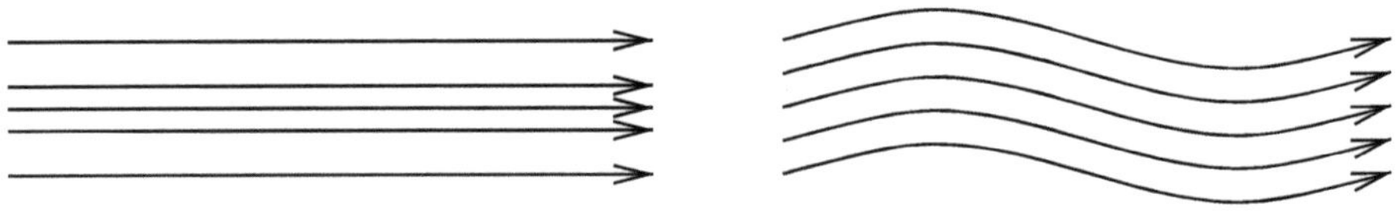

**Figure 1.4.** The two types of magnetic forces. Magnetic pressure (left) acts perpendicular to the field lines in response to compression of the plasma, which carries the field lines with it. This gives rise to compressional Alfvén waves which propagate longitudinally, perpendicular to the magnetic field. Magnetic tension (right) results from shearing, compression free motion, which bends the field lines. The resulting curvature represents a JxB Lorentz force tending to straighten the field lines. This gives rise to shear Alfvén waves, which propagate transverse, parallel to the magnetic field. Both waves propagate at the Alfvén speed, $v_A$, and constitute the basic characteristics of MHD.

flow is in the form of localised vortices, or eddies. Motion at global scales is connected to overall stability: if the net magnetic forces are in balance with the pressure force the system is in equilibrium, and if the net reaction to small disturbances is towards restoration of the equilibrium, then the system is stable.

A stable plasma is still subject to smaller scale motion, however, and it is with this that we are most concerned. Magnetised plasmas, whether in nature or in the laboratory, have transport properties due both to kinetic particle motion and also to small-scale fluid vortical motion perpendicular to the magnetic field. Particles have a probability to collide with one another through very short range (within a Debye length) interactions, which is characterised by a collision frequency. Coulomb collisions are not like those between hard spheres; the interactions are mediated by the two-particle electromagnetic field. The dominant interactions are small-angle scatterings which add up into a characteristic time after which a narrow distribution of velocities becomes thermalised. This is expressed in terms of the collision frequencies for electrons and ions,

$$\nu_e = \frac{4\sqrt{2\pi}n_e\lambda e^4}{3m_e^{1/2}T_e^{3/2}} \qquad \nu_i = \frac{4\sqrt{\pi}n_i\lambda e^4}{3m_i^{1/2}T_i^{3/2}}, \tag{1.17}$$

where it is noted that the characteristic electron–ion collision frequency in a neutral plasma is also $\nu_e$. The factor of $\lambda$ is the Coulomb logarithm, the logarithm of the ratio between the Debye length and the distance of closest approach, usually the DeBroglie wavelength of the two particles in the centre-of-momentum frame. Collisional transport is a random walk process with the gyroradius as the step size and the inverse collision frequency as the step time. For electrons and ions the characteristic classical diffusion coefficients are therefore

$$D_e = \rho_e^2\nu_e \qquad D_i = \rho_i^2\nu_i \sim \left(\frac{T_e}{T_i}\frac{m_i}{m_e}\right)^{1/2} D_e. \tag{1.18}$$

The corresponding diffusion rates for dynamics are given by these coefficients divided by the scale of motion squared. If the plasma is hot enough, however, this diffusion rate is slow enough to be largely negligible. A plasma is considered hot if it is fully ionised and these diffusion rates are slower than any of the intrinsic dynamics under consideration.

Instead of collisional transport, what controls confinement in a hot, magnetised plasma is a class of motions called fluid drift dynamics. These are turbulent flows, whose scale is somewhat larger but not much larger than $\rho_i$, and whose characteristic velocity is well below the sound speed. The standard definition of the sound speed takes into account the possibility of electrons providing pressure and ions providing inertia, through electrostatic coupling between the two fluids. The sound speed $c_s$ in a plasma with singly charged ions is

$$c_s = \left(\frac{T_e}{m_i}\right)^{1/2}, \tag{1.19}$$

which is the analogue of the value in terms of the pressure and mass density familiar from neutral fluids. The fastest frequencies available for incompressible perpendicular motion are those given by the sound speed and the local scale of inhomogeneity, $L_\perp$, so in setting up models we will assume frequencies of this order:

$$\omega \sim \frac{c_s}{L_\perp}, \tag{1.20}$$

which is called the 'drift frequency' and is how the vorticity scales. In keeping with this relation between electron pressure and ion inertia, we have a drift scale, $\rho_s$, which has the same form as the ion gyroradius

$$\rho_s = \frac{c_s}{\Omega_i} = \frac{1}{eB}(m_i T_e)^{1/2}. \tag{1.21}$$

Many laboratory plasmas, especially discharges, have ions at room temperature but electrons at a few to several eV, which gives the motivation for these definitions. If the ions are also hot, their temperature is usually comparable to $T_e$ and so $\rho_s$ and $c_s$ keep their relevance.

Fluid drift dynamics can display a moderately wide range of scales, down to about $\rho_s$ in space and up to about $c_s/L_\perp$ in frequency. The drift concept is illustrated in figure 1.5. It is relevant if the particles undergo many gyro orbits in the time that their gyro centres drift over one gyroradius. This requires dynamical frequencies to be small compared to any of the gyrofrequencies, of which $\Omega_i$ is the slowest. This can only be satisfied if the drift parameter, defined by

$$\delta \equiv \frac{\rho_s}{L_\perp} \ll 1, \tag{1.22}$$

**Figure 1.5.** The ExB drift in a magnetised plasma. The dashed lines are contours of constant electrostatic potential, $\phi$, representing a localised disturbance in the electric field, $\mathbf{E} = -\nabla\phi$ in the $xy$-plane, perpendicular to the magnetic field, **B**. The four thick arrows A–D represent phases of the gyro orbit of a positive ion. During phase A, the particle is accelerated by the electric field, so that during phase B it coasts faster. During phase C it is decelerated, so that during phase D it coasts slower. The centre of the orbit drifts in a direction mutually perpendicular to **E** and **B**, such that the average net Lorentz force vanishes. The drift velocity, $\mathbf{v}_E$, is therefore the same for all charged particles. It is clockwise around a potential maximum, as seen by an observer up the field line looking down on the plane. This ExB drift velocity constitutes the basic characteristic of fluid drift motion.

is small. Another important figure of merit is the ratio of plasma gas pressure to magnetic field pressure,

$$\beta = \frac{2\mu_0 p}{B^2}, \tag{1.23}$$

called the plasma beta. Consideration of fluid drift dynamics while neglecting any compressional effects of the magnetic field typically require that $\beta \ll 1$. A related parameter is the electron dynamical beta, which is the square of the ratio of the sound speed to the Alfvén speed and is given by

$$\beta_e = \frac{\mu_0 p_e}{B^2} = \frac{c_s^2}{v_A^2} \ll 1. \tag{1.24}$$

This parameter must be small in order to separate shear Alfvén from sound wave dynamics. Furthermore, it is $\beta_e$, specifically, that must be small to allow the neglect of compressional dynamics and hence the consideration of perpendicular motions as a quasistatic evolution of force balances.

The important consideration for drift dynamics is that we cannot treat the plasma as a single fluid, but must instead use a two fluid (or several fluid) model. The key assumption behind the MHD model is that the Lorentz force for each fluid species approximately vanishes, leaving a plasma velocity of

$$\mathbf{u} = \mathbf{v}_E \equiv \frac{\mathbf{E} \times \mathbf{B}}{B^2}, \tag{1.25}$$

(neglecting the parallel motion) called the 'E-cross-B' (ExB) velocity. For frequencies as low as $c_s/L_\perp$ this no longer holds; instead, the pressures enter the force balance. It then follows that due to their strongly disparate masses, the electrons and ions will have qualitatively different responses to the electromagnetic field. In particular, the forces on the electrons parallel to the magnetic field can balance, leaving a quasistatic evolution. Ion inertial corrections to the ExB velocity resulting from time dependence in the electric field give rise to the 'polarisation drift'

$$\mathbf{u} = \mathbf{v}_E + \frac{1}{\Omega_i B}\frac{\partial \mathbf{E}}{\partial t}, \tag{1.26}$$

in which the ratio $\Omega_i^{-1}(\partial/\partial t) \ll 1$ is used as an expansion parameter so that $\mathbf{v}_E$ is put into the inertia term. This process controls the response of the ExB vorticity

$$\mathbf{b} \cdot \nabla \times \mathbf{v}_E \approx \frac{1}{B}\nabla_\perp^2 \phi \equiv \Omega_E, \tag{1.27}$$

to the other currents. With $\epsilon_0 \ll \rho/B^2$, meaning that the polarisability of the ions ($\rho$ is the mass density) overcomes that of the vacuum, the charge density itself is neglected and the dynamics is 'quasineutral.' These concepts are central to all of the dynamics we consider. They are treated in detail in Chapters 6 and 7.

Small disturbances exhibiting these situations are called drift waves, and their nonlinear state is called drift wave turbulence, or drift Alfvén turbulence in the more general electromagnetic case. These waves have been directly measured in linear plasma devices, and there is a large body of circumstantial evidence that they are responsible for the turbulent fluctuations among the ExB velocity and thermodynamic quantities that lead to transport of particles and thermal energy out of magnetically confined plasmas. Drift wave turbulence, and more generally fluid drift dynamics, is therefore of obvious importance to laboratory plasmas, and it provides the subject of study for this exposition.

The typical scales of fluid drift turbulence are roughly $10\rho_s$ and $2\pi \times 10L_\perp/c_s$, with $\delta \sim 10^{-2}$. In fusion experiments these correspond to roughly 1 cm and 10 $\mu$s, which is small and fast compared to the transport time scales but large compared to most kinetic processes and slow compared to compressional MHD processes. Due to this last property, the perpendicular flow vortices are largely incompressible. The forces across the magnetic field lines are therefore always in approximate balance, which can be regarded as quasistatic. Part of this is the electrostatic nature of the perpendicular electric field, so that the ExB velocity is given in terms of the electrostatic potential as a stream function. The fluid velocities perpendicular to the magnetic field are still similar to the ExB velocity, and due to some subtle properties which are covered in detail herein, the ExB velocity is still the one according to which thermodynamic quantities are transported, or advected. Shear Alfvén dynamics is still quite important, however, leaving a dynamical system of very rich, electromagnetic character. For those familiar with geophysical fluid dynamics, the situation is not unlike geostrophic flow in character, and there are many analogies between the two models. In a magnetised plasma, small scales across the magnetic field are coupled to long parallel wavelengths, with a parallel/perpendicular scale ratio of the order $10^3$. This is generally called flute mode character and is due to the fact that the shear Alfvén dynamics tends to eliminate gradients parallel to the magnetic field.

## 1.3 Outline of the work

The basic outline of this work is to introduce the turbulence first, and then examine the concept of turbulence in a magnetised plasma through various flavours of two-dimensional fluid dynamics and MHD. This is to get the fundamental understanding of turbulence in a magnetised plasma, before any consideration of instabilities arises to interfere. Then, we go back to the fundamentals to see how the fluid drift model descends from general considerations starting from Maxwell's equations and a set of charged fluids. The shear Alfvén and acoustic dynamics, and then the drift wave instability are introduced. Then, we develop three-dimensional drift wave and drift Alfvén turbulence concept by concept, including the effects of curved and sheared magnetic field lines, the interaction with flows and currents, and open and closed magnetic field lines, or in a confined plasma situation, flux surfaces.

At that point we have enough content to expand the scope to include temperature dynamics and kinetic theory, which takes place in Volume Two. The electron temperature dynamics introduces the heat flux as a dynamical variable, not just a diffusion effect. The ion temperature gradient brings in a new class of dynamics and instabilities, the ITG (ion temperature gradient driven) turbulence. Its nonlinear evolution nevertheless requires the conceptual background provided by the detailed analysis of the simpler isothermal version of drift Alfvén turbulence. Along the way, we encounter the complications presented by the finite gyroradius, which is the same size as the drift scale for equal ion and electron temperatures.

Most of Volume Two is a return to the kinetic theory to do the approximations in a different order. Doing the low frequency approximation at the kinetic level with a field theoretical treatment produces what is called 'gyrokinetic' theory—dynamics well below the gyrofrequencies of all species. The main point will be that the representation changes and we no longer consider the plasma in terms of individual charged particle motion, but as a self-consistent combination of gyrocentres and dynamical field quantities. There is a derivation path from that picture back to different sorts of fluid equations, which constitute the 'gyrofluid' model. This now makes the spectral range at scales comparable to or below the ion gyroradius accessible, so long as the frequency situation remains intact. Some new mode types appear, but we consider them always in the context of self-consistent turbulence, rather than instabilities in isolation. Dynamics due to magnetic particle trapping in axisymmetric systems introduce topological considerations, which are treated in a gyrofluid model with a bounce averaging procedure. A Chapter on MHD instabilities, mostly driven by currents parallel to the magnetic field, within our context but with contact to conventional analysis. We need to see these instabilities from within the context of temperature effects within the gyrofluid model and then make the connection between those instabilities, especially the ones driven by the current, with turbulence. A set of Chapters on topics in gyrokinetic and gyrofluid theory, specifically the gauge transform approach and functional Poisson bracket theory, close the work.

Each Chapter closes with a Section on Further Reading. In the next, we turn to the basic concepts behind turbulence, what it is, what the difference between two- and three-dimensional domains leads to, fluids versus MHD turbulence in a plasma, and what difference the strong guiding magnetic field in a plasma makes.

## Further Reading

These fundamental processes and parameters are treated in basic plasma physics texts such as F F Chen 2016 *Introduction to Plasma Physics and Controlled Fusion* 3rd edn (Berlin: Springer); R Fitzpatrick 2014 *Plasma Physics: An Introduction* (Boca Raton, FL: CRC Press); K Nishikawa and M Wakatani 2013 *Plasma Physics: Basic Theory with Fusion Applications* 3rd edn (Berlin: Springer); R J Goldston and P H Rutherford 1995 *Introduction to Plasma Physics* (Boca Raton, FL: CRC Press); N Krall and A Trivelpiece 1973 *Principles of Plasma Physics* (New York: McGraw-Hill). The standard reference for a long time in the USA was L Spitzer 1956 *Physics of Fully Ionized Gases* (New York: Interscience). Those with Internet access can find the lectures by R Fitzpatrick, *Plasma Physics*, at `http://farside.ph.utexas.edu/teaching/plasma/lectures/Plasmahtml.html`.

The papers on Debye screening and the Yukawa potential are: P Debye and E Hückel 'The theory of electrolytes. I. Lowering of freezing point and related phenomena' *Phys. Z.* **24** (1923) 185; H Yukawa 'On the interaction of elementary particles' *Proc. Phys. Math. Soc. Jpn.* **17** (1935) 48.

Magnetohydrodynamics (MHD) started with H Alfvén 'Existence of electromagnetic-hydrodynamic waves' *Nature* **150** (1942) 405. Further background is cited in Chapter 6. The text with the most relevance to this situation is D Biskamp 1993 *Nonlinear Magnetohydrodynamics* (Cambridge: Cambridge University Press).

Drifts, including the ExB velocity, and MHD were treated in an early monograph by H Alfvén 1950 *Cosmical Electrodynamics* (Oxford: Clarendon Press).

# Chapter 2

## Introduction to turbulence

Includes extracts adapted by permission from Springer: B Scott 2005 *Introduction to Turbulence in Magnetized Plasmas* A Dinklage, Th Klinger, G Marx and L Schweikhard (Eds), copyright Springer-Verlag, Berlin, Heidelberg 2005 (https://doi.org/10.1007/11360360_8).

### 2.1 Statistical nonlinearity and cascade dynamics

In general situations the dynamics of a magnetised plasma is not comprised solely of a single wave or instability. This is due to the general nonlinearity of the system when two or more eigenfunctions are excited; the various waves are coupled together and interact with each other as well as with the background. Turbulence due to a fluid flow like the basic plasma ExB velocity has as its dominant coupling mechanism a quadratic nonlinearity: advection of each dependent variable by the velocity, itself a dependent variable. Quadratic nonlinearities exhibit a phenomenon known as three-wave coupling: in a Fourier decomposition in terms of wavenumbers, each wave $\mathbf{k}$ interacts with the other ones $\mathbf{k}'$ through beat waves $\mathbf{k}''$ satisfying the constraint $\mathbf{k} + \mathbf{k}' + \mathbf{k}'' = 0$. One can consider this as a process of modes $\mathbf{k}'$ and $\mathbf{k}''$ beating together to drive mode $\mathbf{k}$ (in the shorthand in which the word 'mode' is used to signify an individual eigenmode, wave, or Fourier component). We are thinking of a relatively homogeneous system with many degrees of freedom; even if the background has gradients, the turbulence can be thought of as homogeneous if the typical scale of motion is smaller than any gradient scale length of the background.

We can distinguish between quasilinear and nonlinear effects. Quasilinear dynamics refers to a single wave interacting with the background; in other words, mode zero (zero wavenumber, nonzero frequency) and mode $-\mathbf{k}$ driving mode $\mathbf{k}$, as in a linear system, but now together with modes $\mathbf{k}$ and $-\mathbf{k}$ driving mode zero. The waves excite changes in the background which affect the subsequent evolution of the waves. Here, $\mathbf{k}$ and $-\mathbf{k}$ are actually the same mode, since for a real scalar field $\rho(\mathbf{x})$ we have $\rho_{-\mathbf{k}} = \rho_{\mathbf{k}}^*$ in the (complex) Fourier components, where the superscript asterisk denotes the complex conjugate. The quasilinear system therefore involves the same set of modes the linear system does; the only exception is that mode zero is allowed to evolve. When many modes are excited, the dynamics could still be quasilinear if this particular wave/background interaction were still dominant. Such dynamics tends to produce relaxation oscillations in conservative systems: the action of the wave on the background is stabilising but overshoots the point at which the growth rate goes to zero, and then the free energy in the wave all goes back to the background, reestablishing the original unstable state, and so on. The interaction typically has a coherent character, with long term memory of initial conditions.

Turbulence *per se* is very different from this. The same three-wave coupling mechanism works as well between all possible triplets in the full set of waves. Each three-wave interaction works independently of all the others, but many such interactions affect each individual wave. The dynamics involves many degrees of freedom (one for each dependent variable or eigenmode at each $\mathbf{k}$) all exchanging energy, and while each three-wave interaction tends towards coherence, the fact that there are many of them going on simultaneously results in a weak degree of overall coherence between waves at differing $\mathbf{k}$. When these three-wave interactions are as strong as the quasilinear interactions, they result in an incoherent type of dynamics, which must be treated statistically. Moreover, it has a short time memory: dynamics at the same spatial point but separated in time become uncorrelated beyond a time range called the correlation time. The correlation time is somewhat longer than the time required for individual eddies to turn over, but it is shorter than the time required for the collective disturbance free energy to respond to changes in the background parameters. The dynamics therefore quickly loses memory of any initial state, and somewhat more slowly reaches a statistically equilibrated state called 'saturation.' The individual waves fluctuate vigorously in amplitude and phase, but their free energy (or amplitude) level remains statistically stationary over time scales of several correlation times. This type of dynamics is what we call turbulence. It is a nonequilibrium, dissipative system with many degrees of freedom, with short correlation times and lengths. It is characterised by its statistical properties, not so much by the evolutionary fate of individual structures. For those familiar with geophysical fluid dynamics, the distinction between individual eddies, or structures, within the turbulence and the statistical properties of the fluctuating dynamics is analogous to the distinction between weather and climate.

The basic picture of turbulence one starts with is an 'eddy mitosis' model. Large eddies are driven by some 'stirring' process and a balance is formed as they break up into roughly two smaller ones each, on average, then these smaller ones into two smaller ones, each, and then so forth down to scales at which dissipation becomes stronger than the nonlinear dynamics. This energy transfer is called a 'cascade' because it occurs in a chain of interactions, each of which is between eddies at not very disparate scales. In terms of the turbulence 'spectrum', the distribution of energy in terms of wavenumbers, the cascade process is 'local' in the sense that the ratio of wavenumbers involved in most of the transfer process is rarely larger than two.

The separation between the main energy containing scales and the small-scale dissipation range is characterised by the 'Reynolds number', given by the ratio of the product of typical scale length and velocity, and the viscous diffusion coefficient. The turbulence is called 'high Reynolds number turbulence' if the cascade process and not the details of the dissipation process controls the balance between drive and dissipation. In terms of an expansion, the small parameter is the inverse Reynolds number, and the turbulence is thought of as random to lowest order, with the energy transfer and any transport processes entering at next order.

Turbulence in fluids is generally three-dimensional, but under certain circumstances the vorticity of the motion is ordered along a particular axis. This is either the rotation axis in a rapidly spinning system, like a thin accretion disk of material orbiting a central object like a star, or in a thin-atmosphere system with some rotation like geophysical turbulence, or it is the magnetic field in a plasma whose kinetic energy of motion is not sufficient to strongly disturb the magnetic field. Two-dimensional turbulence has the special property that its vorticity is strictly perpendicular to its velocity. As a result, a quantity called the 'enstrophy' is conserved in addition to the energy. The energy density is one half the square of the velocity, and the enstrophy density is one half the square of the vorticity. With both quantities arising from the single velocity variable conserved, the one with the highest wavenumber dependence, namely enstrophy, controls the statistics and affects the direct cascade to small scales, and the other one, namely energy, then cascades to larger scales. This phenomenon is called a 'dual cascade' and the transfer of energy to large scales is called an 'inverse cascade.'

Turbulence in an idealised sense is incompressible, with all the motion of the shearing type. It is important to note this in light of the common concept of one-dimensional nonlinear chaos. Turbulence is not only more statistical than the results of deterministic chaos models, it also has this shearing character. Neutral fluid turbulence is kept incompressible by the pressure, so long as the velocity of the flow eddies is slower than the sound speed. Disturbances in the density can in certain circumstances (e.g. thermal convection) remain, but are merely 'passively' advected, with any coupling back to the flows rather weak in comparison to the time scale represented by the action of the eddies upon each other. A quantity whose dynamics is simple incompressible advection by the turbulence but with no back reaction upon the turbulence is called a 'passive scalar.' In a magnetised plasma, incompressibility is maintained by the magnetic field if the velocity is slower than the Alfvén velocity, in the usual 'low beta' situation in which plasma pressure is much smaller than the effective magnetic field pressure which is part of the magnetohydrodynamic (MHD) force. The pressure evolves freely, and like the density it is under the MHD approximations a passive scalar.

In any discussion of two-dimensional turbulence one must address how it becomes two-dimensional. A background rotation does this through the Coriolis force, following the Proudman–Taylor theorem of fluid mechanics. The influence of a background magnetic field to guide the motion is somewhat similar, with the Lorentz force playing the same role. In this role the background magnetic field is called a 'guide field.' The largest force taken alone leads to the result that the spatial variation along the guiding vorticity or magnetic field vector vanishes, so all other effects provide corrections. One of the definitions of turbulence specifically in a magnetised plasma is that the dynamics takes place within the strong guide field limit.

The above concepts comprise the rest of this Chapter. The following Sections will add substance to this introduction, starting with the eddy mitosis model, continuing with the statistical character, the fluid model, three-wave coupling and the cascades, the contrast between two- and three-dimensional models, and finally MHD turbulence and passive scalar dynamics.

The following Chapters will give some demonstrations of homogeneous turbulence in various two-dimensional fluid and MHD dynamical systems before turning to the special case of gradient-driven turbulence in a magnetised plasma, whose dynamics parallel to the magnetic field invokes dissipative coupling mechanisms between the fluid vorticity and the transported quantities, such that the latter no longer behave as passive scalars. This type of turbulence is very important for two reasons: (1) it is determined by energetic processes not treated by the MHD model, and (2) it is found in many instances to be responsible for the dominant transport processes in a confined plasma.

## 2.2 Eddy mitosis and the cascade model

The basic picture of turbulence in fluids is the one introduced by Kolmogorov in 1941 and independently by von Weizäcker and Heisenberg in a different language in subsequent years. At a certain range of spatial scales, energy is supposed to be introduced into the fluid motions. This is called 'stirring' and beyond the characterisation of a drive rate it is independent of the details. The turbulence is supposed to be homogeneous and isotropic, so that spatial scale and spectral wavenumber magnitude, related through $k \leftrightarrow 2\pi/\Delta$, can be used interchangeably. A collection of eddies at the driven spectral range $k \sim k_0$ finds shearing motion tearing each of the eddies apart at about the same rate at which they are driven, so that there is a statistical equilibrium. This 'nonlinear decorrelation' process conserves energy, so the energy taken out of the driven range is passed to other scales. Since the dominant instability of a sheared flow has a wavelength slightly shorter than the scale of the flow, most of this energy is passed to the next smallest scale in a logarithmic hierarchy, with a scale ratio of about two. With all of the energy transfer taking place between neighbours in the hierarchy, the picture we have in mind is a 'local cascade.' At each scale below the driven range, there is a statistical balance mostly between energy passed in from the next larger scale, and nonlinear decorrelation passing the energy further to the next smaller scale. The shearing rate at each scale is proportional to the vorticity of the flow at that scale, so that the coherent lifetime of a typical eddy is comparable to the 'eddy turnover time', the time the eddy takes to transport the flow energy and momentum across itself.

Following these ideas, at each level $n$ with wavenumber magnitude $k_n$, a balance is struck between incoming energy due to nonlinear decorrelation at level $n - 1$ and stirring at level $n$, and outgoing energy due to nonlinear decorrelation at level $n$. This is expressed as

$$(kv)_{n-1} E_{n-1} + \gamma_n E_n = (kv)_n E_n, \tag{2.1}$$

where $E_n$ and $(kv)_n$ are the energy and vorticity within level $n$, and $\gamma_n$ is the rate of stirring at level $n$. We can take $E_n = v_n^2/2$.

At the low-$k$ end of the spectrum the turbulence is generated by stirring and hence $\gamma > 0$ there. At the high-$k$ end lies the dissipation range, where $\gamma < 0$. Since the separation of scales is supposed to be arbitrarily large, there is a wide spectral range in between for which $\gamma$ is negligible compared to $(kv)$. This is called the 'inertial range' of the turbulence and is the range in which the turbulence is most pure, that is, the conservative nonlinearity is the only part of the dynamics which is active. Within the inertial range, we then have

$$(kv)_{n-1}\, E_{n-1} = (kv)_n E_n = T_0, \tag{2.2}$$

where $T_0$ is a constant giving the energy 'throughput' of the cascade. With $E_n = v_n^2/2$ we find the velocity and energy density spectrum,

$$v_n^3 = 2T_0 k_n^{-1} \qquad E_n = (2T_0)^{2/3} k_n^{-2/3}. \tag{2.3}$$

To find $E_k$, the energy per unit range of $k$, we have to take into account the fact that each level $n$ has a range of $k$ which is proportional to $k$ (i.e. we have a logarithmic spacing). In integral form we have

$$\sum_n E_n = \int d\log k\; E_n = \int dk\; E_k, \tag{2.4}$$

which identifies $E_k$ as $E_n/k$, or $E_n(k_0/k)$ to preserve the dimensions. This yields

$$E_k = E_0\, k^{-5/3}, \tag{2.5}$$

with $E_0 = (2T_0)^{2/3} k_0$. This 5/3 power law is known as the Kolmogorov spectrum. Within the inertial range, this spectrum is 'universal' and the dynamics is 'self similar', indicating that the scales of motion in time and space can be arbitrarily renormalised.

If the dissipation is caused by viscosity or some other diffusive process, we have in the dissipation range $\gamma(k) = -\chi k^2$, where $\chi$ is called the diffusivity. The Reynolds number is defined by comparing the typical velocity ($V_0$) and space ($L$) scales in the *driven range* with the diffusivity,

$$\mathrm{Re} = L V_0/\chi. \tag{2.6}$$

The 'high Reynolds number' regime is the one for which the inertial range involves several levels of the cascade hierarchy, and an important result of that is that the energy balances of both the energy containing range and the inertial range are independent of the strength or form of the dissipation process. For diffusion, the amount of energy consumed by the dissipation should be independent of the diffusion coefficient.

Clearly, the existence of an inertial range, with several levels in the hierarchy for which $\gamma \ll (kv)$, is connected with a large value for Re. It does, however, take a rather large Reynolds number before the high Reynolds number regime is entered. To see this, we may take a simple situation for which Eq. (2.1) can be applied:

assume that $\gamma_0 > 0$ only for the first level in the hierarchy, the driven range $k_0$. For all other scales, let $\gamma_n = -\chi k_n^2$. The amplitude of the driven range is given by the 'mixing condition',

$$\gamma_0 = (kv)_0 \qquad \text{hence} \qquad v_0 = \frac{\gamma_0}{k_0}. \tag{2.7}$$

It is a property in the inertial range that the shearing rate is progressively faster at smaller scale, due to the conservation of energy: with Eq. (2.1) satisfied for $\gamma = 0$ we have

$$\frac{(kv)_{n+1}}{(kv)_n} = \left(\frac{k_{n+1}}{k_n}\right)^{2/3}, \tag{2.8}$$

and hence with $k_{n+1}/k_n = 2$ the ratio is $2^{2/3}$ or about 1.6. Even with a process like convection for which $\gamma$ is roughly constant below the scale of inhomogeneity, one does therefore reach an inertial range if $\chi$ is small enough, that is, if Re is large enough. But the factor of 1.6 is small enough that Re must be very large in order to have a real inertial range. Typically, a flow is not even turbulent for Re lower than a few thousand, and for flow in three dimensions the high Reynolds number regime is not reached until Re is at least $10^6$. In astrophysical situations Re can be in the range $10^{12}$–$10^{15}$. For atmospheric turbulence one finds values some orders of magnitude smaller than that but still large; for air at STP the kinematic viscosity is about $1.5 \times 10^{-5}$ m$^2$s$^{-1}$, so for $L > 15$ m and $V_0 > 1$ ms$^{-1}$ we have Re $> 10^6$. The highest resolution three-dimensional fluid turbulence computations to date are barely into the high Reynolds number regime, although the mere demonstration of a local cascade is possible with a discretised domain of only $64^3$ grid nodes. In two dimensions, as discussed below, the cascade dynamics is different and turbulence may set in for Re values rather lower than this; typically, at least two decades between driven and dissipation range in the spectrum are required ($k_N/k_0 \sim 10^2$). For magnetised plasma turbulence the Reynolds number is not as large as this: dissipative coupling processes inhibit the cascade process sufficiently that only one or two decades are needed to produce converged cases.

## 2.3 The statistical nature of turbulence

One is motivated to ask 'this may be nonlinear dynamics, but what makes it turbulence?' The answer is that the motion involved exhibits short space/time correlation and is 'ultimately nonperiodic' in the sense that no matter how long the time domain in an experiment or computation is allowed to persist, the longest time periods become dominant in a frequency spectrum, the standard deviation of a time series saturates instead of falling with increasing time intervals, and the statistical description of the turbulent flow amplitudes and decorrelation rates in terms of a mean and a standard deviation is independent of how the dynamical system was started. In terms of a computation, the particular initial conditions chosen should have no effect on the statistical result, even though certain initial scenarios can save time by getting the system to statistical equilibrium in a comparatively shorter time than others. Fully developed turbulence is characterised by a 'saturated state' in which any relevant statistical measure is stationary in this sense. This is never certain; one should accept that results are only known within error bars characterised by that standard deviation, and in this case one should always quote the standard deviation of the distribution, not that of the mean (the latter falls with the square root of the number of samples if the former is stationary). In this sense turbulence 'converges' only statistically, not monotonically as a coherent dissipative process would do.

Current research is occupied with the question of rare but strong events: those well down the tails of the 'probability distribution function' (PDF). The PDF is essentially a histogram. The amplitude of a certain variable is sampled and considered as a series. The mean and standard deviation of the series are measured, and the PDF is computed by dividing the abscissa into intervals which are small compared to the standard deviation. The fraction of the events falling into each of these intervals is plotted as the ordinate, normalised in terms of the number of intervals per standard deviation. A purely random process has a Gaussian PDF for any of its measures, but many features of the turbulence are 'intermittent' in the sense that the first one or two standard deviations of the PDF appear Gaussian but the tails are different, usually above the Gaussian curve of the same mean and standard deviation. Nevertheless, the basic statistical character of turbulence can be understood in terms of this near-Gaussian PDF. The PDF curves for turbulence are usually compared to the corresponding Gaussian which has the same mean and standard deviation.

Statistical memory in turbulence can be characterised in terms of autocorrelation functions in the principal fluctuating variables. Both time and space correlations are described, with the $1/e$ half-width of these curves defined as the correlation time or correlation length. The time autocorrelation function of a variable $\phi$ is defined as

$$F(\tau) = \lim_{T\to\infty} \int_{-T/2}^{T/2} \frac{dt}{T}\, \phi(t)\, \phi(t-\tau), \tag{2.9}$$

also averaged over space. For a purely wave-like process (e.g. a linear oscillation), $F(\tau)$ is sinusoidal. For a purely random process, it is Gaussian. If the spatial domain is periodic on a length $L$, it is simple to compute the space autocorrelation function in terms of Fourier components,

$$F(\Delta) = c_0 \sum_{l\neq 0} |\,\phi_l\,|^2 \cos(2\pi l\Delta/L), \tag{2.10}$$

also averaged over time, with the normalising constant $c_0$ chosen such that $F(0) = 1$. The curve for $F(\Delta)$ is itself periodic and therefore is only significant for $-L/2 < \Delta < L/2$. In turbulence within a bounded domain (e.g. convection, or tokamak edge turbulence), the correlation length in the relevant direction (e.g. down the gradient) can be comparable to the domain size, but it should be much smaller than the domain size in the 'horizontal' direction (e.g. east–west for geophysical fluid turbulence, or perpendicular to the magnetic field lines but within the flux surfaces for gradient-driven turbulence within a closed, toroidal MHD equilibrium). Most importantly, the correlation time must be short compared to the time intervals required for the statistical measures to reach saturation. In relevantly scaled units, tokamak turbulence typically has a dynamical scale of order 1, a correlation time of order 10, a saturation time at least of an order of several hundred, and a transport equilibration time at least of an order of several thousand. This level of temporal scale separation is what makes the model of thermal transport by turbulent mixing on a slowly varying background plausible.

## 2.4 Quadratic nonlinearity and three-wave coupling

When treating turbulence as such one usually either assumes a homogeneous background, or one with a gradient that drives the turbulence while nevertheless leading to disturbances which are small compared to background quantities. The flow velocity—as in certain circumstances, the magnetic field—belongs to the disturbances. The pressure and density belong to the background, and since the disturbances are small these two quantities are treated as constant coefficients except where acted upon by the gradient operator. The scale of motion is also expected to be small compared to the scale of the background, and this simplification is used as a model even when in the case of a thin boundary region (e.g. convection in a thin atmosphere, or tokamak edge turbulence) the scales of inhomogeneity and of the turbulence are comparable but both small compared to the overall size of the equilibrium.

In a situation like this the equations governing the dynamics are linearised except for the terms representing advection or, in the case of turbulence on a magnetised plasma background, the disturbed parallel gradient. These terms have the special character that they are quadratic in the dependent variables. The simplest example of such an effect is $\mathbf{v} \cdot \nabla\mathbf{v}$, the self-advection of the velocity, which is the basis of what gives turbulence its character. The thermodynamic state variables are also advected through quadratic nonlinearities in their respective continuity equations. The simplest example of these equations is the one for the mass density $\rho$,[1]

$$\frac{\partial \rho}{\partial t} + \nabla \cdot \rho\mathbf{v} = 0, \tag{2.11}$$

expressing mass conservation. This is rearranged by separating the advection and the divergence terms, and linearising the latter, formally writing

$$\left(\frac{\partial}{\partial t} + \mathbf{v} \cdot \nabla\right)\log\rho + \nabla \cdot \mathbf{v} = 0, \tag{2.12}$$

which we can do with no loss of generality, but then also considering small disturbances so that $\delta \log\rho \rightarrow \delta\rho/\rho$, writing

[1] We use $\rho$ for mass density. In other uses such as for gyroradii it appears with a subscript.

$$\frac{\partial\tilde{p}}{\partial t} + \mathbf{v}\cdot\nabla\tilde{p} + \rho\nabla\cdot\mathbf{v} = 0, \tag{2.13}$$

and treating the factor of $\rho$ multiplying the divergence as a constant parameter while using the tilde symbol to mark the dependent variable. The advection term $\mathbf{v}\cdot\nabla\tilde{p}$ is recognised as a quadratic nonlinearity. The nonlinear part of the divergence term is smaller than this because we are treating situations under the 'small-scale mixing' condition:

$$\tilde{p}\ll\rho \qquad \text{although} \qquad \nabla\tilde{p}\sim\nabla\rho. \tag{2.14}$$

In an incompressible model, of course, the advection term is the only effect controlling time evolution in this equation.

The three-wave coupling condition follows from the properties of Fourier transforms. We treat the background as homogeneous in understanding the properties of the turbulence. The dependent variables are Fourier decomposed according to

$$\tilde{p}(\mathbf{x}) = \sum_{\mathbf{k}}\rho_{\mathbf{k}}\, e^{i\mathbf{k}\cdot\mathbf{x}} \qquad \rho_{\mathbf{k}} = \oint\frac{d^n x}{L^n}\, e^{-i\mathbf{k}\cdot\mathbf{x}}\,\tilde{p}(\mathbf{x}), \tag{2.15}$$

where $n$ is the number of dimensions (usually 2 or 3) and the spatial domain is taken to be periodic on length $L$ in each dimension (this is easily generalised). The transform is orthogonal, which means that the integration picks out a single component in the sum, since the integration yields zero if the arguments of the exponential do not cancel.

The incompressible continuity equation (with $\nabla\cdot\mathbf{v} = 0$) then becomes

$$\frac{\partial\rho_{\mathbf{k}}}{\partial t} = -\oint\frac{d^n x}{L^n}\, e^{-i\mathbf{k}\cdot\mathbf{x}}\sum_{-\mathbf{k}'}\sum_{-\mathbf{k}''}\mathbf{v}_{-\mathbf{k}'}\cdot(-i\mathbf{k}'')\rho_{-\mathbf{k}''}e^{-i\mathbf{k}'\cdot\mathbf{x}}\, e^{-i\mathbf{k}''\cdot\mathbf{x}}, \tag{2.16}$$

where we have used the decomposition in terms of the sum for both dependent variables in the nonlinearity and the integral operation to isolate the single component under the time derivative. Note that the choice of signs in $\mathbf{k}$ is arbitrary due to the condition that $\rho(\mathbf{x})$ is real valued, $\rho_{-\mathbf{k}} = \rho_{\mathbf{k}}^*$, since we sum over all the components. The factors involving $\mathbf{x}$ and the integral can be isolated from the rest of the right-hand side,

$$\oint\frac{d^n x}{L^n}\, e^{-i(\mathbf{k}+\mathbf{k}'+\mathbf{k}'')\cdot\mathbf{x}} = \delta_{(\mathbf{k}+\mathbf{k}'+\mathbf{k}''),0}, \tag{2.17}$$

yielding the Kronecker delta. The two components are said to beat against one another (the parlance for waves) to drive component $\mathbf{k}$. The condition under which there is a finite interaction is then

$$\mathbf{k} + \mathbf{k}' + \mathbf{k}'' = 0, \tag{2.18}$$

which is called the three-wave condition: the three wavenumbers must vectorially add to zero. Interactions of this type are called 'three-wave interactions' and they form the basis for the interactions in turbulence. The three-wave condition is therefore a relatively trivial consequence of the quadratic nonlinearity and the orthogonality of Fourier transforms. Even if the spatial domain is not periodic, the basic character of turbulence has this qualitative form.

Three-wave models are constructed using a single triplet $\{\mathbf{k}_1, \mathbf{k}_2, \mathbf{k}_3\}$, satisfying the three-wave condition, for explanatory purposes. The model consists of an evolution equation for each Fourier component which is a member of the triplet. Turbulence, however, derives its character from the fact that there are very many such triplets and the vast majority of them are mutually incoherent, with the three members having no special relation to each other. To the lowest order of approximation (the small parameter is usually $\mathrm{Re}^{-1}$ or some equivalent) in the mathematical theories of turbulence the various triplets are supposed to be mutually random, with the PDF of cross correlation having Gaussian form with zero mean. All the energy transfer dynamics comes from the corrections, which are computed according to one or another model. The details are very complicated and are left to the references listed at the end. We will return to three-wave interactions and what they say about the transfer dynamics generally, after introducing the incompressible turbulence models.

## 2.5 Fluid turbulence—energy and enstrophy

We start with simple hydrodynamic turbulence, following the ideal equation of motion (same as in MHD but without the magnetic field)

$$\frac{\partial \mathbf{v}}{\partial t} + \mathbf{v} \cdot \nabla \mathbf{v} = -\frac{1}{\rho}\nabla p. \tag{2.19}$$

We recognise the role of the pressure in maintaining incompressibility by taking the divergence and assuming that $\nabla \cdot \mathbf{v} = 0$,

$$\nabla \cdot \frac{1}{\rho}\nabla p = -\nabla \cdot (\mathbf{v} \cdot \nabla \mathbf{v}). \tag{2.20}$$

This condition controls the evolution of the pressure, and so even in an inhomogeneous model the only other equation controlling evolution would be an incompressible advection equation for $\rho$. In a homogeneous model we take $\rho$ to be constant in Eq. (2.19) since $\rho\nabla^2 p \gg \nabla\rho \cdot \nabla p$ due to the small fluctuation approximation (cf. Eq. (2.14)). Taking the curl, and defining the vorticity as

$$\mathbf{W} = \nabla \times \mathbf{v}, \tag{2.21}$$

we then have

$$\frac{\partial \mathbf{W}}{\partial t} + \mathbf{v} \cdot \nabla \mathbf{W} = \mathbf{W} \cdot \nabla \mathbf{v}. \tag{2.22}$$

This equation says that the fluid vorticity $\mathbf{W}$ evolves not only due to advection but also due to a finite component of the velocity gradient along the vorticity. The squared magnitude of the vorticity is related to a quantity called 'enstrophy' in hydrodynamics. Defining the enstrophy and its density as

$$\mathcal{W} = \int d\mathcal{V}\; W \qquad W = \frac{1}{2}\, |\mathbf{W}|^2, \tag{2.23}$$

where the integration is over the spatial domain, we may form the equation for $W$ by contracting (taking the dot product of) Eq. (2.22) with the vorticity to obtain

$$\frac{\partial W}{\partial t} + \nabla \cdot (W\, \mathbf{v}) = \mathbf{W}\mathbf{W}\colon \nabla\mathbf{v}, \tag{2.24}$$

(take care to note that $W$ is not the same as the magnitude of $\mathbf{W}$) using the condition $\nabla \cdot \mathbf{v} = 0$. This equation says that the enstrophy is conserved except for a class of motions which involves stretching the velocity component along the vorticity further in the direction of the vorticity. Since the motion is incompressible, the components of the velocity making up most of the vorticity must have a convergence to make up the divergence involved in the stretching.

To illustrate what this type of motion signifies and what its consequence is, we reconsider the results arising from the turbulent cascade model. Recall that a cascade towards small scale which conserves energy must exhibit a shearing rate which increases towards small scale. Given the energy conservation condition, it is a simple matter to conclude that the enstrophy must grow during this cascade, simply from the extra factors of $k$ in enstrophy compared to energy. The type of motion which we concluded above allows the enstrophy to grow may be displayed in a simple sketch: most of the velocity is in eddies, each of which rotates in a local two-dimensional plane, to which the vorticity is perpendicular. The third dimension has to have a finite velocity component, along the vorticity vector, and it must stretch in this direction. The eddy motion in the local plane must also contract to maintain incompressibility. This basically involves a vortex tube structure which is stretched in the time it maintains its coherence (not more than a few eddy turnover times), before it is sheared apart by the motion both in the local plane and along the local vorticity vector. We therefore conclude simply on the basis of (1) the incompressibility constraint, (2) energy conservation, (3) the cascade to smaller scales, and (4) the growth of the enstrophy, that the turbulent cascade process behind the eddy mitosis model must involve this 'vortex tube stretching' in three dimensions.

The foregoing is significant because in several important applications (e.g. motion at large horizontal scale in a thin atmosphere, 'geostrophic' motion in a rapidly rotating background, or motion perpendicular to the magnetic field in a magnetised plasma) the turbulence is quite strictly two-dimensional (2D). We will see in later Chapters how the dynamics parallel to the guide field causes this for gradient-driven turbulence in a magnetised plasma. In a 2D model, we have the additional constraint that the vorticity is strictly perpendicular to the plane in which the velocity and the gradient operator both lie. In this case we have both $\mathbf{W} \cdot \mathbf{v} = 0$ and $\mathbf{W} \cdot \nabla = 0$, so that the right-hand side of Eq. (2.24) is zero. Therefore, the enstrophy as well as the energy is conserved.

With these two quantities simultaneously conserved, we cannot have the cascade of energy to small scale. In a 2D incompressible flow, the velocity is given in terms of a stream function, which we label $\psi$, and the vorticity is a scalar, $\varpi$, times the unit normal to the plane, which we label $\hat{\mathbf{s}}$, so that

$$\mathbf{v} = \hat{\mathbf{s}} \times \nabla\psi \qquad \mathbf{W} = \varpi\,\hat{\mathbf{s}} \qquad \varpi = \nabla_\perp^2\psi, \tag{2.25}$$

with the perp (perpendicular to the plane) symbol used to denote the fact that only the two dimensions in the plane are involved. Eq. (2.22) becomes

$$\frac{\partial\varpi}{\partial t} + \mathbf{v}\cdot\nabla\varpi = 0, \tag{2.26}$$

which is called the 2D Euler equation. Multiplying by $-\psi$ or by $\varpi$, we find

$$\begin{aligned} &\frac{\partial U}{\partial t} + \nabla\cdot\left(U\,\mathbf{v} - \left[\psi\left\{\frac{\partial}{\partial t} + \mathbf{v}\cdot\nabla\right\}\nabla_\perp\psi\right]\right) = 0 \\ &\frac{\partial W}{\partial t} + \nabla\cdot(W\,\mathbf{v}) = 0, \end{aligned} \tag{2.27}$$

where the energy and enstrophy densities in the 2D version are, respectively, given by

$$U = \frac{1}{2}\,|\,\nabla\psi\,|^2 \qquad W = \frac{1}{2}\,|\,\varpi\,|^2 \tag{2.28}$$

hence showing simultaneous energy and enstrophy conservation[2]. If we integrate these over the spatial domain and neglect surface terms we have

$$\frac{\partial\mathcal{E}}{\partial t} = 0 \qquad \frac{\partial\mathcal{W}}{\partial t} = 0 \qquad \text{where} \quad \mathcal{E} = \int d\mathcal{V}\;U \quad \text{and} \quad \mathcal{W} = \int d\mathcal{V}\;W, \tag{2.29}$$

as the global energy and enstrophy conservation laws.

Considering the case of finite dissipation we add a viscosity term with a constant coefficient to Eq. (2.19) so that

$$\frac{\partial\mathbf{v}}{\partial t} + \mathbf{v}\cdot\nabla\mathbf{v} = -\frac{1}{\rho}\nabla p + \chi\nabla^2\mathbf{v}, \tag{2.30}$$

that is, with the viscosity in the form of a vector Laplacian. Using the vector identity

$$\nabla^2\mathbf{F} = \nabla(\nabla\cdot\mathbf{F}) - \nabla\times(\nabla\times\mathbf{F}), \tag{2.31}$$

(for any vector $\mathbf{F}$) on the vector $\mathbf{v}$, then assuming $\nabla\cdot\mathbf{v} = 0$ incompressible flow so that only the term with $\nabla\times\mathbf{W}$ remains, we contract the equation with $\mathbf{v}$ to obtain

$$\frac{\partial U}{\partial t} + \nabla\cdot\left(U + \frac{p}{\rho}\right)\mathbf{v} = \nabla\cdot(\chi\mathbf{v}\times\mathbf{W}) - \chi|\mathbf{W}|^2, \tag{2.32}$$

having again used $\nabla\cdot\mathbf{v} = 0$, and another vector identity,

$$\nabla\cdot(\mathbf{F}\times\mathbf{G}) = \mathbf{G}\cdot\nabla\times\mathbf{F} - \mathbf{F}\cdot\nabla\times\mathbf{G}, \tag{2.33}$$

[2] The term in square brackets in the equation for $U$ is a quantity related to compression and becomes important only if the density is strongly variable, at which point the assumption of incompressibility itself breaks down.

(for any two vectors **F**, **G**) on the last term. Integrating over the spatial domain and neglecting surface terms we have

$$\frac{\partial \mathscr{E}}{\partial t} = -\chi \int d\mathscr{V}\ |\mathbf{W}|^2, \tag{2.34}$$

which is referred to as a well-known theorem of fluid mechanics by Onsager (1949). In three dimensions, while $\chi \to 0$, the entrophy can be sufficiently generated in the cascade to make the product $\chi W$ remain finite. But in two dimensions small $\chi$ leads to small dissipation, reaching zero in the limit. The same treatment for enstrophy in a 2D incompressible flow leads to

$$\frac{\partial W}{\partial t} = -\chi \int d\mathscr{V}\ |\nabla \varpi|^2, \tag{2.35}$$

which can remain finite in the limit $\chi \to 0$ by an increasingly pronounced small-scale structure (cf. Batchelor 1969). Hence we expect the possibility of a direct cascade in enstrophy but not in energy. To find out how this situation is resolved dynamically we may examine three-wave coupling in 2D turbulence.

Three-wave interactions set up by Fourier decomposition of Eq. (2.26) will also have these conservation properties, therefore representing a conservative mutual transfer of energy and enstrophy from wave to wave. For Eq. (2.26), for example, one can easily verify that

$$\frac{1}{2}\frac{\partial}{\partial t}(\varpi_{\mathbf{k}}^* \varpi_{\mathbf{k}} + \varpi_{\mathbf{k}'}^* \varpi_{\mathbf{k}'} + \varpi_{\mathbf{k}''}^* \varpi_{\mathbf{k}''}) = 0, \tag{2.36}$$

for any particular triplet $\{\mathbf{k}, \mathbf{k}', \mathbf{k}''\}$. The equations satisfied by the members of the triplet are

$$\frac{\partial \varpi_{\mathbf{k}}}{\partial t} = C_{\mathbf{k}\mathbf{k}'}(\varpi_{-\mathbf{k}''}\psi_{-\mathbf{k}'} - \varpi_{-\mathbf{k}'}\psi_{-\mathbf{k}''}), \tag{2.37}$$

$$\frac{\partial \varpi_{\mathbf{k}'}}{\partial t} = C_{\mathbf{k}\mathbf{k}'}(\varpi_{-\mathbf{k}}\psi_{-\mathbf{k}''} - \varpi_{-\mathbf{k}''}\psi_{-\mathbf{k}}), \tag{2.38}$$

$$\frac{\partial \varpi_{\mathbf{k}''}}{\partial t} = C_{\mathbf{k}\mathbf{k}'}(\varpi_{-\mathbf{k}'}\psi_{-\mathbf{k}} - \varpi_{-\mathbf{k}}\psi_{-\mathbf{k}'}), \tag{2.39}$$

which we can form by simple permutation among the members of the triplet. The coupling coefficient is given by

$$C_{\mathbf{k}\mathbf{k}'} = \frac{1}{2}\hat{\mathbf{s}} \cdot (\mathbf{k} \times \mathbf{k}'), \tag{2.40}$$

where we use the relative symmetry in the interactions,

$$\hat{\mathbf{s}} \cdot (\mathbf{k} \times \mathbf{k}') = \hat{\mathbf{s}} \cdot (\mathbf{k}' \times \mathbf{k}'') = \hat{\mathbf{s}} \cdot (\mathbf{k}'' \times \mathbf{k}), \tag{2.41}$$

since the wavenumbers in each triplet satisfy Eq. (2.18).

Using these forms we can build the energy equations for each of the three components (and their complex conjugates) in the triplet, noting that for each one we have the reality condition, e.g. $\psi_{\mathbf{k}} = \psi^*_{-\mathbf{k}}$. For each energy component given by

$$U_{\mathbf{k}} = |\,\mathbf{k}\psi_{\mathbf{k}}\,|^2 = -\psi_{\mathbf{k}}\varpi_{-\mathbf{k}}, \tag{2.42}$$

we obtain

$$\frac{\partial U_{\mathbf{k}}}{\partial t} = 2C_{\mathbf{kk}'}\mathrm{Re}[\psi_{\mathbf{k}}\,\varpi_{\mathbf{k}'}\,\psi_{\mathbf{k}''} - \psi_{\mathbf{k}}\,\psi_{\mathbf{k}'}\,\varpi_{\mathbf{k}''}], \tag{2.43}$$

$$\frac{\partial U_{\mathbf{k}'}}{\partial t} = 2C_{\mathbf{kk}'}\mathrm{Re}[\psi_{\mathbf{k}'}\,\varpi_{\mathbf{k}''}\,\psi_{\mathbf{k}} - \psi_{\mathbf{k}'}\,\psi_{\mathbf{k}''}\,\varpi_{\mathbf{k}}], \tag{2.44}$$

$$\frac{\partial U_{\mathbf{k}''}}{\partial t} = 2C_{\mathbf{kk}'}\mathrm{Re}[\psi_{\mathbf{k}''}\,\varpi_{\mathbf{k}}\,\psi_{\mathbf{k}'} - \psi_{\mathbf{k}''}\,\psi_{\mathbf{k}}\,\varpi_{\mathbf{k}'}]. \tag{2.45}$$

where the Re symbol before a set of square bracket parentheses denotes the real part of the quantity inside. Pairs of equal terms appearing with opposite sign give the energy transfer effects. Considering the equations as a permutation among the three modes in the triplet, the second term in each equation gives the energy transfer from the equation below it, and the first term in each equation gives the energy transfer from the equation above it. For example, the energy transfer from mode $\mathbf{k}'$ to mode $\mathbf{k}$ is given by

$$T_U(\mathbf{k} \leftarrow \mathbf{k}') = 2C_{\mathbf{kk}'}\,\mathrm{Re}[-\psi_{\mathbf{k}}\,\psi_{\mathbf{k}'}\,\varpi_{\mathbf{k}''}] = 2C_{\mathbf{kk}'}\,\mathrm{Re}[(k'')^2\psi_{\mathbf{k}}\,\psi_{\mathbf{k}'}\,\psi_{\mathbf{k}''}]. \tag{2.46}$$

Similarly, for each enstrophy component given by

$$W_{\mathbf{k}} = |\,\varpi_{\mathbf{k}}\,|^2 = k^4\,|\,\psi_{\mathbf{k}}\,|^2, \tag{2.47}$$

we build the enstrophy transfer equations,

$$\frac{\partial W_{\mathbf{k}}}{\partial t} = 2C_{\mathbf{kk}'}\,\mathrm{Re}[\varpi_{\mathbf{k}}\,\psi_{\mathbf{k}'}\,\varpi_{\mathbf{k}''} - \varpi_{\mathbf{k}}\,\varpi_{\mathbf{k}'}\,\psi_{\mathbf{k}''}], \tag{2.48}$$

$$\frac{\partial W_{\mathbf{k}'}}{\partial t} = 2C_{\mathbf{kk}'}\,\mathrm{Re}[\varpi_{\mathbf{k}'}\,\psi_{\mathbf{k}''}\,\varpi_{\mathbf{k}} - \varpi_{\mathbf{k}'}\,\varpi_{\mathbf{k}''}\,\psi_{\mathbf{k}}], \tag{2.49}$$

$$\frac{\partial W_{\mathbf{k}''}}{\partial t} = 2C_{\mathbf{kk}'}\,\mathrm{Re}[\varpi_{\mathbf{k}''}\,\psi_{\mathbf{k}}\,\varpi_{\mathbf{k}'} - \varpi_{\mathbf{k}''}\,\varpi_{\mathbf{k}}\,\psi_{\mathbf{k}'}]. \tag{2.50}$$

Hence, the transfer of enstrophy from mode $\mathbf{k}'$ to mode $\mathbf{k}$ is given by

$$T_W(\mathbf{k} \leftarrow \mathbf{k}') = 2C_{\mathbf{kk}'}\,\mathrm{Re}[-\varpi_{\mathbf{k}}\,\varpi_{\mathbf{k}'}\,\psi_{\mathbf{k}''}] = 2C_{\mathbf{kk}'}\,\mathrm{Re}[-k^2(k')^2\psi_{\mathbf{k}}\,\psi_{\mathbf{k}'}\,\psi_{\mathbf{k}''}]. \tag{2.51}$$

For the three-wave system to have an average spectral transfer, the triple correlation $\langle\psi_{\mathbf{k}}\psi_{\mathbf{k}'}\psi_{\mathbf{k}''}\rangle$ will have a definite sign. If only a single triplet is present, we expect the energy or enstrophy to simply redistribute to reach a statistical equilibrium with a triple correlation of zero. But in turbulence, with many triplets present all with differing relationships between the wavenumbers, we expect an asymmetry to develop since there are more available states towards high $k$ than towards low for any physically realisable instantaneous configuration (i.e. with finite total energy content at $k \to \infty$). The triple correlation and hence the average transfer dynamics will be constrained by the simultaneous conservation of both energy and enstrophy. We can consider the effects of interactions which spread energy and enstrophy towards both ends in the spectrum, out of some intermediate spectral range. Examining Eqs (2.46) and (2.51), noting that the squared factors are positive definite and the coefficient $C_{\mathbf{kk}'}$ is the same in both, we find that for any definite sign of the triple correlation, the signs of the transfer are opposite. If $k < k' < k''$ then the direction of energy transfer from $k''$ to $k'$ and from $k'$ to $k$ is the same, and similarly for the oppositely directed enstrophy transfer. For both energy and enstrophy, then, the intermediate range $k'$ sees a throughput.

The overall process is called a 'dual cascade' since the two quantities arising from the same dependent variable $\psi$ are transferred in opposite directions in the spectrum. The direction of the cascade dynamics will be determined by which one of these transfer effects controls the overall statistics. It is expected to be a local cascade since the coupling coefficients are largest when the magnitudes $\{k, k', k''\}$ are all comparable. The enstrophy and enstrophy transfer contain extra wave-number factors compared to the energy and energy transfer, so the enstrophy can be expected to mix faster. The statistical behaviour of the whole should therefore be controlled by the enstrophy. The enstrophy should be redistributed towards higher $k$ due to the fact that there are more available states there given any physically realisable initial state (i.e. with finite total energy content at $k \to \infty$). Besides, we already found that for the energy cascade to be 'direct' (towards smaller scale) the enstrophy must grow. So, with enstrophy also conserved, we expect the energy cascade to be 'inverse' (towards larger scale). We therefore expect the dual cascade to send enstrophy preferentially to higher $k$ and energy preferentially to lower $k$.

The main points of this Section are that (1) a direct energy cascade requires the vortex tube stretching process in three dimensions since it requires a growing enstrophy, (2) two-dimensional incompressible hydrodynamics conserves enstrophy as well as energy, (3) the directions of spectral transfer for energy and enstrophy are opposite given a definite sign of the triple correlation, and therefore (4) we expect to find a dual cascade in two-dimensional incompressible hydrodynamic turbulence. We can learn all this simply from the properties of the equations the dynamical model satisfies. In computations, direct measurement of the three-wave coupling energetics can be used to show that it actually occurs.

## 2.6 MHD turbulence

Magnetohydrodynamic (MHD) turbulence is similar to its neutral fluid counterpart, with two differences: incompressibility is maintained by the magnetic field strength (magnetic rather than plasma pressure), and the additional magnetic nonlinearities cause special effects through which flow eddies and Alfvén waves interact with each other to allow long-range spectral transfer.

To examine these effects we start with the MHD force equation,

$$\rho\left(\frac{\partial \mathbf{v}}{\partial t} + \mathbf{v} \cdot \nabla \mathbf{v}\right) = \mathbf{J} \times \mathbf{B} - \nabla p. \tag{2.52}$$

Using Ampère's law

$$\nabla \times \mathbf{B} = \mu_0 \mathbf{J}, \tag{2.53}$$

and the vector identity

$$\nabla \frac{B^2}{2} = \mathbf{B} \cdot \nabla \mathbf{B} - (\nabla \times \mathbf{B}) \times \mathbf{B}, \tag{2.54}$$

the force equation can be re-written as

$$\frac{\partial \mathbf{v}}{\partial t} + \mathbf{v} \cdot \nabla \mathbf{v} = -\frac{1}{\rho}\nabla\left(p + \frac{B^2}{2\mu_0}\right) + \frac{\mathbf{B} \cdot \nabla \mathbf{B}}{\mu_0 \rho}. \tag{2.55}$$

In most cases one is interested in the 'low beta regime' characterised by

$$\beta = \frac{2\mu_0 p}{B^2} \ll 1. \tag{2.56}$$

Here, the plasma pressure $p$ is insignificant to the magnetic pressure $B^2/2\mu_0$ in the maintenance of incompressibility, which we can see by taking the divergence of Eq. (2.55) assuming $\nabla \cdot \mathbf{v} = 0$,

$$\nabla \cdot \frac{1}{2\mu_0 \rho}\nabla B^2 = -\nabla \cdot \left(\frac{1}{\rho}\nabla p + \mathbf{v} \cdot \nabla \mathbf{v} - \frac{1}{\mu_0 \rho}\mathbf{B} \cdot \nabla \mathbf{B}\right). \tag{2.57}$$

It is therefore possible for $p$ as well as $\rho$ to enter the dynamics of background disturbances while $B^2$ maintains the dynamical incompressibility as long as $\beta \ll 1$, and the dynamics is slow compared to the compressional Alfvén frequency at the same scale, $\omega \ll k_\perp v_A$, and the velocity is sub-Alfvénic, $v \ll v_A$ (the last two conditions are different if the motion is driven by the magnetic field). The perp symbol is now reckoned against the background magnetic field. The Alfvén velocity $v_A$ is defined by

$$v_A^2 = \frac{B^2}{\mu_0 \rho}. \tag{2.58}$$

Most models satisfy these conditions, with the principal exception being the kinematic dynamo process (thermal convection with $\beta \gg 1$, with the magnetic field passively advected).

The equations of incompressible MHD turbulence also include the kinematics of the magnetic field, which assuming $\nabla \cdot \mathbf{v} = 0$ satisfies

$$\frac{\partial \mathbf{B}}{\partial t} + \mathbf{v} \cdot \nabla \mathbf{B} = \mathbf{B} \cdot \nabla \mathbf{v}. \tag{2.59}$$

Shear Alfvén dynamics basically involves a magnetic field whose direction is fluctuating but whose field strength is constant except for the small changes required to maintain the incompressibility. The dependent variable then becomes the unit vector. We can also scale the velocity against the background Alfvén velocity $v_{A0}$, which is $v_A$ defined in terms of the background field strength $B_0$. The dependent variables are then

$$\mathbf{u} = \frac{\mathbf{v}}{v_{A0}} \qquad \mathbf{b} = \frac{\mathbf{B}}{B_0}, \tag{2.60}$$

in whose terms Eqs (2.55) and (2.59) become

$$\frac{1}{v_{A0}}\frac{\partial \mathbf{u}}{\partial t} + \mathbf{u}\cdot\nabla\mathbf{u} = -\nabla I + \mathbf{b}\cdot\nabla\mathbf{b}, \tag{2.61}$$

$$\frac{1}{v_{A0}}\frac{\partial \mathbf{b}}{\partial t} + \mathbf{u}\cdot\nabla\mathbf{b} = \mathbf{b}\cdot\nabla\mathbf{u}, \tag{2.62}$$

where $I$ is now simply a potential which maintains $\nabla\cdot\mathbf{u} = 0$. On the other hand, $\nabla\cdot\mathbf{b} = 0$ is maintained by the antisymmetry of $\mathbf{ub} - \mathbf{bu}$ as a tensor. We may define the 'Elsässer variables'

$$\mathbf{u}_\pm = \mathbf{u} \pm \mathbf{b}, \tag{2.63}$$

which satisfy

$$\frac{1}{v_{A0}}\frac{\partial \mathbf{u}_\pm}{\partial t} + \mathbf{u}_\mp\cdot\nabla\mathbf{u}_\pm = -\nabla I. \tag{2.64}$$

Computations of MHD turbulence often use this model, which allows for very simple numerical schemes such as are used for the hydrodynamic model of Eq. (2.19). The force potential gradient $\nabla I$ need not be computed specifically, though it is often used to subtract off the part of the nonlinearity which has a finite divergence (in computational fluid dynamics this is called the 'Laplacian pressure method'). We note that there is only one force potential since

$$\nabla\cdot(\mathbf{u}_-\cdot\nabla\mathbf{u}_+) = \nabla\cdot(\mathbf{u}_+\cdot\nabla\mathbf{u}_-), \tag{2.65}$$

although some numerical schemes apply two of them independently.

The main difference between MHD and fluid turbulence is that there are two equations each describing advection of one of the Elsässer variables by the other. The three-wave coupling condition is no longer among different wavenumber components of the same variable but is now also among the variables. Several consequences regarding the relationship between the velocity and the magnetic field dynamics follow.

There are some additional conserved quantities. The first is 'magnetic helicity', whose spatial density is given by $\mathbf{A}\cdot\mathbf{B}$, where $\mathbf{A}$ is the magnetic vector potential defined by $\mathbf{B} = \nabla\times\mathbf{A}$. Magnetic helicity can be shown to satisfy a transport equation. The definition of **E** in terms of the vector and scalar potentials, and Faraday's law are

$$\mathbf{E} = -\frac{\partial \mathbf{A}}{\partial t} - \nabla \phi \qquad \frac{\partial \mathbf{B}}{\partial t} = -\nabla \times \mathbf{E}. \tag{2.66}$$

The Lorentz force law says that **E** parallel to **B** vanishes,

$$\mathbf{E} + \mathbf{v} \times \mathbf{B} = 0 \implies \mathbf{E} \cdot \mathbf{B} = 0. \tag{2.67}$$

Using these we can find the result

$$\frac{\partial}{\partial t}(\mathbf{A} \cdot \mathbf{B}) + \nabla \cdot (\mathbf{E} \times \mathbf{A} + \phi \mathbf{B}) = 0, \tag{2.68}$$

and hence the integral conservation law

$$\frac{\partial}{\partial t} \int d\mathcal{V} \; \mathbf{A} \cdot \mathbf{B} = 0, \tag{2.69}$$

for total magnetic helicity. The second one is 'cross helicity', whose spatial density is given by $\mathbf{v} \cdot \mathbf{B}$. Its equation is found by combining Eqs (2.52) and (2.59) to form

$$\frac{\partial}{\partial t}(\mathbf{v} \cdot \mathbf{B}) + \nabla \cdot [\mathbf{v} \times (\mathbf{v} \times \mathbf{B})] + \frac{1}{\rho}\mathbf{B} \cdot \nabla p = 0. \tag{2.70}$$

If the dynamics including the background is barotropic ($p$ is a function of $\rho$ in all respects), or if the density is taken to be constant either by approximation or flow incompressibility, then this may be expressed as

$$\frac{\partial}{\partial t}(\mathbf{v} \cdot \mathbf{B}) + \nabla \cdot [\mathbf{v} \times (\mathbf{v} \times \mathbf{B}) + \Phi \mathbf{B}] = 0, \tag{2.71}$$

where $\nabla \Phi = (1/\rho)\nabla p$ defines the enthalpy. The integral conservation law

$$\frac{\partial}{\partial t} \int d\mathcal{V} \; \mathbf{v} \cdot \mathbf{B} = 0, \tag{2.72}$$

for total flow cross helicity then follows.

Both of these additional conservation laws lead to a more complicated system than simply an energy cascade in MHD, especially in three dimensions. Nonlinear interactions which would tend towards equipartition of energy can become inhibited by the requirement that extra quantities be conserved along the way. Under the small-scale incompressible flow assumptions, the form of the Elsässer variable nonlinearities results in the squared amplitude of each of them being separately conserved:

$$\frac{1}{v_{A0}} \frac{\partial}{\partial t} \frac{u_\pm^2}{2} + \nabla \cdot \left( \frac{u_\pm^2}{2} \mathbf{u}_\mp + I \mathbf{u}_\pm \right) = 0, \tag{2.73}$$

due to the linearity of the $\nabla$ operator and the incompressibility of $\mathbf{u}_\pm$. There is, therefore, some memory of the initial conditions in an ideal sense. With long term evolution under consistent drive and dissipation, the drive mechanism can stamp some form of cross correlation onto the system which persists.

The spectral transfer of both the energy and the Elsässer amplitudes to high-$k$ dissipation results in an aligning effect: the nonlinear transfer rate of $\mathbf{u}_+$ is given by the vigour of $\mathbf{u}_-$, and vice versa. So if there is an initial inequality in the Elsässer amplitudes, the one which is smaller is transferred to dissipation faster, leaving all the energy in the one which was larger. As a result, $\mathbf{b}$ becomes either aligned or anti-aligned to $\mathbf{u}$. This tendency has been actually observed in the solar wind. Finally, we observe that if one of the Elsässer amplitudes is zero the 'mode structure' is that of a shear Alfvén wave, with $\omega = kv_{A0}$, and the nonlinearity vanishes. Therefore, pure shear Alfvén waves have no nonlinear interaction. This can be described as a cancellation between the 'Reynolds stress', from $\mathbf{u} \cdot \nabla\mathbf{u}$, and the 'Maxwell stress', from $\mathbf{b} \cdot \nabla\mathbf{b}$. We will not go further into the details of these points, except to note that their basis is a possibility for a small-scale vortex and a small-scale magnetic disturbance to drive a rather larger-scale magnetic disturbance, or for two similarly small-scale magnetic disturbances to drive a rather larger-scale flow, and for the flows to be particular sensitive to the direction of the local magnetic field. This affords the possibility of a long-range spectral interaction, over about one decade corresponding to $k_1 \sim k_2 \sim 10k_3$. Interactions like these in three dimensions are thought to be responsible for the generation of large-scale magnetic field energy (the dynamo process) or large-scale flows (magnetic reconnection) starting with small-scale turbulence.

## 2.7 Selective decay

If one of several of the conserved quantities in the turbulence tends to collect at large scales, it decays much more slowly due to dissipation than the other quantities. If the dynamics were laminar the dissipation would be the same for these quantities because there is only one scale of interest: the system scale. In high Reynolds number turbulence on the other hand, the dissipation occurs at small scales, so that quantities with a more direct cascade tendency, or at least a weaker inverse cascade tendency, decay all the more rapidly. For two-dimensional turbulence we have the result in Eq. (2.34), which indicates that for small viscous (diffusive) dissipation the energy decay can go to zero while enstrophy decay remains finite. Selective decay is more general than this, though, since this result is also found for other forms of the high-$k$ dissipation operator. Examples of this will be given in the next Chapter.

## 2.8 How the turbulence becomes two-dimensional

Early work on turbulence was concerned with rotating systems as these provided a useful way to study turbulence in the laboratory. If we consider fluid dynamics in a rotating background with constant vorticity $\mathbf{W}_0$ we can write

$$\frac{\partial \mathbf{v}}{\partial t} + \mathbf{v} \cdot \nabla\mathbf{v} + 2\mathbf{W}_0 \times \mathbf{v} = -\frac{1}{\rho}\nabla p - \mathbf{W}_0 \times \mathbf{W}_0 \times \mathbf{r}. \tag{2.74}$$

Assuming the two forces on the right can be written in terms of a potential whose curl vanishes, the vorticity equation is given by

$$\frac{\partial \mathbf{W}}{\partial t} + \mathbf{v} \cdot \nabla \mathbf{W} + 2\nabla \times (\mathbf{W}_0 \times \mathbf{v}) = \mathbf{W} \cdot \nabla \mathbf{v}, \tag{2.75}$$

assuming $\nabla \cdot \mathbf{v} = 0$ as before (Eq. (2.22), with the new term on the left-hand side). If the background vorticity represents a frequency faster than anything in the dynamical system, the result is

$$(\mathbf{W}_0 \cdot \nabla)\mathbf{v} = 0, \tag{2.76}$$

namely, that all components of the dynamical velocity should be constant on lines parallel to $\mathbf{W}_0$, which now serves as a guide for two-dimensional motion. This is known as the Proudman–Taylor (or Taylor–Proudman) theorem for flows in rotating systems. It is the starting point for the study of geophysical fluid dynamics and applications to astrophysics and planetary sciences, all of which are concerned with flows in strongly rotating backgrounds.

In the case of a plasma, the background magnetic field has a similar effect, though the differences are significant. The Lorentz force substitutes for the Coriolis force. Assuming that the Lorentz force is dominant, namely that any dynamical frequency be slow compared to the slowest gyrofrequency (i.e. the ion one), the resulting flow is mostly two-dimensional. However, the curl operation on the equation of motion is now on the MHD force equation (Eq. 2.55), whose Lorentz force is charge-averaged, namely with the current $(\mathbf{J} \times \mathbf{B})$ than with the flow. Taking the curl produces

$$\frac{\partial \mathbf{W}}{\partial t} + \mathbf{v} \cdot \nabla \mathbf{W} = \mathbf{W} \cdot \nabla \mathbf{v} + \frac{1}{\rho}(\mathbf{B} \cdot \nabla \mathbf{J} - \mathbf{J} \cdot \nabla \mathbf{B}). \tag{2.77}$$

If the background magnetic field is constant or slowly varying, the last term in parentheses adds to the forcing terms, to which we arrange all effects from $\mathbf{v}$ and $\mathbf{W}$, that is, turbulence which tries to become three-dimensional in the face of the magnetic field. Then, the limit of the strong guide field $(\mathbf{B}_0)$ would say

$$(\mathbf{B}_0 \cdot \nabla)\mathbf{J} \to 0, \tag{2.78}$$

which is different from the statement of the Proudman–Taylor theorem, Eq. (2.76). However, given that in a coupled model the dynamical currents result from all the dependent variables, the strong guide field limit would tend to result in all the dynamics becoming restricted to the perpendicular plane. In the component of the vector Eq. (2.78) parallel to $\mathbf{B}_0$, the corrections are all which lead to a finite $\mathbf{B} \cdot \nabla J_\parallel$, and the action of the guide field is to limit these responses such that $k_\parallel \ll k_\perp$ and eventually that the dynamics in the parallel direction, or parallel dynamics (as a concept), will be slowly varying spatially but with frequencies comparable to those of the turbulence. Hence we have a situation of 2D turbulence in the perpendicular plane, but in which the dependent variables are all coupled through wave-like or dissipative relaxation in the parallel direction. Or, 'three-dimensional' to us does not mean the same thing as it does in the usual discussions of 2D versus 3D turbulence. We will introduce the concept of constraining by parallel dynamics in Chapter 4 and then explore the dynamical tendency explicitly in Chapter 11.

## 2.9 Plan

Subsequent Chapters will treat the two-dimensional model systems relevant to each limit of all these cases, to give some experience with the morphology and energetics, hence basic physical character. The methods of energetic and mode structure diagnosis are introduced as needed, with the two-dimensional models first. Most of the examples given can be run nowadays on simple computational platforms, such as a reasonable laptop, so they have become very accessible.

With this in hand we will turn to systematic derivation of the equations, including what limits of 'low frequency' and 'quasineutrality' really mean and how they are maintained. After going through the wave-like responses in the (third) direction parallel to the background magnetic field, we will bring in the three-dimensionality of turbulence in a properly magnetised plasma. In this context 'magnetised plasma' has the following consequence: 'three-dimensional turbulence' does not mean homogeneous three-dimensional turbulence as in many MHD studies, but mostly two-dimensional turbulence in the perpendicular plane constrained by mostly wave-like parallel responses in a mutually self-consistent system. Since there is usually more than one energetic coupling and transfer mechanism active, we will study them one by one in simplified limits, learning each form of mode structure and dynamical character. Then we will generalise progressively while diagnosing the physics to distinguish the various processes. In Volume 2, we will explore some of the effects of temperature fluctuation dynamics and then will have to return to some more sophisticated theory to handle the weakly collisional and finite-gyroradius nature of the dynamics encountered. This will allow a summary of the current state of the art.

The following Chapter starts this by introducing and comparing hydrodynamic and MHD turbulence in simplified two-dimensional models.

## Further Reading

Turbulence as a subject goes back to works such as J Boussinesq 1897 *Théorie de l'écoulement tourbillonnant et tumultueux des liquides dans les lits rectilignes a grande Section (Theory of the Swirling and Agitated Flow of Liquids)* (Paris: Gauthier-Villars); G I Taylor 'Eddy motion in the atmosphere' *Philos. Trans. R. Soc. A* **215** (1915) 1; O Reynolds 'An experimental investigation of the circumstances which determine whether the motion of water shall be direct or sinuous, and of the law of resistance in parallel channels' *Philos. Trans. R. Soc. A* **174** (1883) 935; including such concepts as the Reynolds number, scale separation, and fluctuations on a background whose slowly varying inhomogeneity can be represented by external drive.

Statistical treatments were introduced by J M Burgers 1995 On the application of statistical mechanics to the theory of turbulent fluid motion (orig. 1929), *Selected Papers of J. M. Burgers* ed F Nieuwstadt and J Steketee (Berlin: Springer) p 91; L Onsager 'Statistical hydrodynamics' *Nuovo Cimento Suppl.* **6** (1949) 279; T D Lee 'On some statistical properties of hydrodynamical and magneto-hydrodynamical fields' *Q. Appl. Math.* **10** (1952) 69.

The turbulent cascade model was introduced independently by: A N Kolmogorov 'The local structure of turbulence in incompressible viscous fluid for very large Reynolds numbers' *Dokl. Akad. Nauk SSSR* **30** (1941) 301 [*J. Fluid Mech.* **13** (1962) 82] (transl. V Levin *Proc. R. Soc. Lond. A* **434** (1991) 9); L Onsager The distribution of energy in turbulence *Minutes of the Meeting of the Metropolitan section held at Columbia University, New York, November 9 and 10, 1945 Phys. Rev.* **68** (1945) 281; W Heisenberg 'On the theory of statistical and isotropic turbulence' *Proc. R. Soc. Lond. A* **195** (1948) 402; C F von Weisäcker 'Das Spektrum der Turbulenz bei großen Reynoldsschen Zahlen' *Z. Physik* **124** (1948) 614.

Vortex tube stretching was treated by H Tennekes 'Simple model for the small-scale structure of turbulence' *Phys. Fluids* **11** (1968) 669. A related topic in 3D turbulence, which we will not treat here, is helicity in flows and magnetic fields. Conservation of magnetic helicity and magnetic flow cross helicity were proved by L Woltjer 'A theorem on force-free magnetic fields' *PNAS* **44** (1958) 489, and 'On hydromagnetic equilibrium' *PNAS* **44** (1958) 833, respectively. Conservation of flow helicity was proved by J J Moreau 'Constantes d'un îlot tourbillonnaire en fluide parfait barotrope' *C. R. Hebd. Séances Acad. Sci.* **252** (1961) 2810. See also R Betchov 'Semi-isotropic turbulence and helicoidal flows' *Phys. Fluids* **4** (1961) 925; and for MHD see H K Moffat 1978 *Magnetic Field Generation in Electrically Conducting Fluids* (Cambridge: Cambridge University Press). A seminal reference connecting helicity to the knottedness of flows is given by H K Moffat 'The degree of knottedness of tangled vortex lines' *J. Fluid Mech.* **35** (1969) 117. Simultaneous consideration of the three invariants given by energy, helicity, and cross helicity leads to very different cascade dynamic in MHD flows as compared to fluid flows. Treatment of this with specific application to turbulence in general and the turbulent dynamo in particular was given by U Frisch, A Pouquet, J Léorat and A Mazure 'Possibility of an inverse cascade of magnetic helicity in magnetohydrodynamic turbulence' *J. Fluid Mech.* **68** (1975) 769.

Intermittence in turbulence was treated by B Mandelbrot 'Intermittent turbulence in self-similar cascades: divergence of high moments and dimension of the carrier' *J. Fluid Mech.* **62** (1974) 331; U Frisch, P-L Sulem and M Nelkin 'A simple dynamical model of intermittent fully developed turbulence' *J. Fluid Mech.* **87** (1978) 719; Monin and Yaglom *op cit* (below). The special features of the statistics of a passive scalar advected by divergence free turbulence were treated by Ya G Sinai and V Yakhot 'Limiting probability distributions of a passive scalar in a random velocity field' *Phys. Rev. Lett.* **63** (1989) 1962.

The disparate dissipation rates of energy and enstrophy in weakly dissipative 2D turbulence was noted by G K Batchelor 'Computation of the energy spectrum in homogeneous two-dimensional turbulence. High-speed computing in fluid dynamics' *Phys. Fluids Suppl.* **11** (1969) 223. The three-wave analysis of the dual cascade was done by R Kraichnan 'Inertial ranges in two-dimensional turbulence' *Phys. Fluids* **10** (1967) 1417.

Monographs and texts include: G Batchelor 1963, repr. 1982 *Homogeneous Turbulence* (Cambridge: Cambridge University Press); H Tennekes and J Lumley 1972 *A First Course in Turbulence* (Cambridge, MA: MIT Press); A S Monin and A M Yaglom 1975 *Statistical Fluid Mechanics* vol 2, ed J Lumley (Cambridge, MA: MIT Press); U Frisch 1979 *Turbulence* (New York: Dover); U Frisch 1995 *Turbulence: The legacy of A.N. Kolmogorov* (Cambridge: Cambridge University Press); M Lesieur 1997 *Turbulence in Fluids* 3rd edn (Dordrecht: Kluwer). Most of these treat the three-dimensional homogeneous case as that has the most general interest.

**IOP** Publishing

Turbulence and Instabilities in Magnetised Plasmas, Volume 1
Fluid drift turbulence
**Bruce Scott**

# Chapter 3

# Turbulence in two-dimensional systems

## 3.1 Various model systems

We start by introducing the phenomenology of native turbulence in several idealised computational systems. The work was pioneered in the 1970s and 1980s, with newer variants in the 1990s. The pseudospectral method was the principal one used then. Concurrently, in geophysical fluid dynamics some finite difference schemes were formulated which had the same symmetry properties, and during the end of this time frame some new time stepping schemes were found. We will illustrate two-dimensional turbulence and some of its properties with a combination of the latter, which was developed at the end of the 1990s and became mature after that. We benefit as well from the fact that leading-edge, high performance computing of those days is simple cluster work nowadays, so that it is much easier to reproduce the work than it was to do it in the first place. We also have the hindsight of over forty years of experience in computational turbulence to work with. Native turbulence refers to a pure representation of the turbulent nonlinear dynamical interactions in the absence of mid-spectrum forcing mechanisms, not only drive and dissipation but also coupling. These effects will be introduced one at a time in subsequent Chapters.

We will refer to the cases of Eulerian hydrodynamics, magnetohydrodynamics, electron magnetohydrodynamics, and four-field Hall magnetohydrodynamics or low-frequency fluid drift cases as the fluid case, the MHD case, the EMHD case, and the Hall MHD or two-fluid cases, respectively. They will be qualified as two-dimensional (2D) since we are assuming the presence of a homogeneous background magnetic field which defines, and suppresses turbulent motion in, the third direction. In these 2D cases variation in the direction along the guide field is neglected so as to study the native turbulence in isolation before bringing in the wave-like dissipative responses along the guide field in later Chapters.

doi:10.1088/978-0-7503-2504-2ch3

## 3.2 2D hydrodynamic turbulence

This is the model readers are most likely to be familiar with. The simplest version of it is a Euler fluid equation, in which the velocity is expected to be incompressible and the density taken to be constant, so that the physics of mixing by shearing eddies is at the forefront. The fluid equation of motion is simple self-advection by the flow with the possibility of a pressure gradient. The equation is

$$\rho\left(\frac{\partial \mathbf{u}}{\partial t} + \mathbf{u} \cdot \nabla\mathbf{u}\right) = -\nabla p. \tag{3.1}$$

Taking the curl to find the vorticity equation, we have

$$\frac{\partial \mathbf{W}}{\partial t} + \mathbf{u} \cdot \nabla\mathbf{W} = 0 \qquad \mathbf{W} = \nabla \times \mathbf{u}. \tag{3.2}$$

with the right-hand side zero because of the two-dimensionality. By the Helmholtz decomposition theorem, the velocity vector may be written as a sum of solenoidal and potential parts,

$$\mathbf{u} = \nabla\zeta \times \nabla\phi - \nabla\chi, \tag{3.3}$$

where the first term is written in what is called a Clebsch representation, and $\zeta$, $\phi$, and $\chi$ are scalar fields. Taking an incompressible flow in two dimensions implies that $\chi$ vanishes and $\zeta$ may be taken as the ignorable coordinate, so that $\phi$ becomes the flow stream function. Letting the motion occur in the $xy$-plane perpendicular to $\hat{\mathbf{z}} \equiv \nabla z$ in Cartesian geometry, we rewrite the velocity as

$$\mathbf{u} = \hat{\mathbf{z}} \times \nabla\phi \qquad \mathbf{W} = \varpi\hat{\mathbf{z}} \qquad \varpi = \nabla_\perp^2\phi, \tag{3.4}$$

and therefore the vorticity equation as

$$\frac{\partial \varpi}{\partial t} + \mathbf{u} \cdot \nabla\varpi = 0, \tag{3.5}$$

where the perpendicular symbol refers to the plane perpendicular to $\hat{\mathbf{z}}$. The operator denoted $\nabla_\perp^2$ is the perpendicular Laplacian defined by

$$\nabla_\perp^2 = \nabla \cdot (\nabla_\perp) \qquad \text{where} \quad \nabla_\perp = -\hat{\mathbf{z}} \times \hat{\mathbf{z}} \times \nabla, \tag{3.6}$$

and $\nabla_\perp$ is the perpendicular gradient. In this case, perpendicular means perpendicular to $\hat{\mathbf{z}}$, while in general it is perpendicular to the background magnetic field in a magnetised plasma. The methods to do this for a spatially varying background magnetic field will be developed as they are needed in later Chapters treating the three-dimensional models.

### 3.2.1 Bracket notation for the advection nonlinearity

At this point we introduce bracket notation for the nonlinear term, which will be common throughout. We define

$$\nabla z \times \nabla\phi \cdot \nabla\varpi \equiv [\phi, \varpi] = \frac{\partial\phi}{\partial x}\frac{\partial\varpi}{\partial y} - \frac{\partial\varpi}{\partial x}\frac{\partial\phi}{\partial y}, \tag{3.7}$$

which defines the bracket structure. We will be using this for essentially any quadratic nonlinearity which can be written as combinations of derivatives in this form. Ultimately, since these models descend from a Hamiltonian system, almost any nondissipative process acting in the perpendicular plane may be represented by a term written in this form. The properties satisfied are antisymmetry, linearity, associativity (also called anticommutativity, bilinearity, and Leibnitz's rule, respectively),

$$[a, b] = -[b, a] \quad [a, (b + c)] = [a, b] + [a, c] \quad [a, bc] = c[a, b] + b[a, c], \tag{3.8}$$

and the Jacobi identity,

$$[a, [b, c]] + [b, [c, a]] + [c, [a, b]] = 0, \tag{3.9}$$

where $a$, $b$, and $c$ are three scalar fields (or constants) on the same space (technically, functions of the coordinates, in this case the spatial coordinates and time). This set of properties identifies the structure as a special case of Poisson bracket (in general they can be defined on any space, not just a spatial one or a space–time one).

For spatial Poisson brackets of this type, we also require the pure divergence property,

$$[a, b] = \nabla \cdot (b\nabla z \times \nabla a) \qquad \int d\mathcal{V}\, [a, b] = 0. \tag{3.10}$$

The integration is over the entire spatial domain, $d\mathcal{V}$, which in this case is $dx\, dy$. In general $d\mathcal{V}$ is given by the product of the differentials in all the coordinates multiplied by $\sqrt{g}$, where $g$ is the determinant of the covariant metric coefficients. In simple terms in three dimensions, $g$ is given by

$$\frac{1}{\sqrt{g}} = \nabla z \times \nabla x \cdot \nabla y, \tag{3.11}$$

which is often called the coordinate Jacobian. Note that the orientation is positive for positive permutations of the three coordinates $xyz$ in a right-handed coordinate system.

In all of the turbulence computations in this work, any evaluation of a volume integral (over the entire spatial domain in however many dimensions one considers) will be given as a domain average, so that any quantitative result does not depend on the domain size. This means that

$$\int d\mathcal{V} \quad \rightarrow \quad \mathcal{V}^{-1} \int d\mathcal{V}, \tag{3.12}$$

will be the way we express these averages in computational results. Sometimes, the phrase 'in volume average' is used as shorthand. Keeping this distinction, we use $E_T$ instead of $\mathcal{E}$ and $W_T$ instead of $\mathcal{W}$, for the averages rather than the global totals.

At minimum, the derivative in one of the directions is assumed to vanish on any non-periodic boundary. Use of this will greatly facilitate many of the derivations we have to do. All of the conservation laws we will discuss depend upon these properties. This is also true of any numerical computations—we wish to satisfy these properties as closely as possible.

### 3.2.2 Dissipation

Physical dissipation is assumed to be arbitrarily small but not zero, as ultimately it works to absorb energy and enstrophy at small scale. In this context, scale separation means that there is a wide region in wavenumber space between the energy-containing scales and this dissipation. So we do not try to model physical viscosity or other dissipation mechanisms, but we do need something which represents positive definite dissipation of free energy (equivalently, increase of entropy, i.e. the Boltzmann H-theorem), and act primarily at small scale. This is called 'subgrid dissipation' and it is implemented through the 'subgrid dissipation operator' acting upon the dependent variables.

Work in the 1980s introduced hyperviscosity, more generally hyperdiffusion, to achieve this:

$$D(\varpi) = -(-1)^n \nu_\perp (\nabla^2)^n, \tag{3.13}$$

where $n = 1$ corresponds to diffusion and $n > 1$ to various extremities of hyperdiffusion. The purpose is to narrow the spectral range over which dissipation occurs, hence widening the range for turbulence, attempting to establish an inertial range. Various experiments showed that making the dissipation range too narrow became equivalent to having no dissipation but a hard resolution boundary, from which free energy would reflect and especially through the inverse energy cascade, pollutes the energy-containing scales. For turbulence, the $n = 2$ case turned out to be the most effective, and the work presented here will simply follow that. Readers who would like to experiment are encouraged to do that.

### 3.2.3 The numerical scheme

These equations and all the others we will explore follow the same basic structure. A set of dependent variables is evolved forward in time for a given step, one equation per variable, and then at the end of each time step the field variables are solved for using the new values of the dependent variables. The equations are mostly hyperbolic (property of $[a, b]$), with dissipation effects serving as corrections. The time step method of choice is a 'stiffly stable' method worked out around 1970, filed away, and then resurrected by Karniadakis (1991). The spatial discretisation method for the Poisson bracket structures $[a, b]$, satisfying the above properties, especially the antisymmetry and divergence properties, was constructed explicitly to do this by Arakawa (1966). This combination of space and time discretisation is due to Naulin (2002). Any first-order derivative terms, such as we will see later, are done via fourth-order centred differences. Any other effects such as dissipation and sources, acting as scalars, are simply added.

For the time step, a set of dependent variables $u$ are advanced in time with the right-hand sides labelled $S$. The equations are labelled

$$\frac{\partial u}{\partial t} = S(u), \tag{3.14}$$

where $u$ means the entire set of variables, and $S$ may be arbitrarily complicated in terms of spatial derivatives and even integrals. Both sides are Taylor expanded three orders deep,

$$\sum_{n=1}^{3} a_n \frac{u_0 - u_n}{\tau} = \sum_{n=1}^{3} b_n\, S_n, \tag{3.15}$$

where the subscript refers to the time step at which each piece is evaluated, and the $a_n$ and $b_n$ are numerical coefficients, and $\tau$ is the time step, assumed to be equidistant. Subscript zero is the current time step, at which the newest $u$ is to be found, and subscripts $n = 1, 2, 3$ refer to one, two, or three time steps backward, that is,

$$u_n = u(t - n\tau) \qquad S_n = S(u_n). \tag{3.16}$$

For order three, the coefficients are

$$a_{1,2,3} = 3 \quad -\frac{3}{2} \quad \frac{1}{3} \qquad b_{1,2,3} = 3 \quad -3 \quad 1. \tag{3.17}$$

If there is an implicit set of terms, labelled $L(u)$, these are evaluated with $u_0$ and set up as

$$\sum_{n=1}^{3} a_n \left( \frac{u_0 - u_n}{\tau} + L(u_0) \right) = \sum_{n=1}^{3} b_n\, S_n, \tag{3.18}$$

and then the whole solved for $u_0$ knowing the other information (dependent variables are usually coupled by doing this). The above gives the third-order Karniadakis time stepping scheme.

For the spatial bracket structure, a set of discretisations is set up

$$\begin{aligned} J^{++} = {} & \left(\phi_{i+1,j} - \phi_{i-1,j}\right)\left(\varpi_{i,j+1} - \varpi_{i,j-1}\right) \\ & - \left(\varpi_{i+1,j} - \varpi_{i-1,j}\right)\left(\phi_{i,j+1} - \phi_{i,j-1}\right), \end{aligned} \tag{3.19}$$

$$\begin{aligned} J^{+\times} = {} & \phi_{i+1,j}\left(\varpi_{i+1,j+1} - \varpi_{i+1,j-1}\right) - \phi_{i-1,j}\left(\varpi_{i-1,j+1} - \varpi_{i-1,j-1}\right) \\ & - \phi_{i,j+1}\left(\varpi_{i+1,j+1} - \varpi_{i-1,j+1}\right) + \phi_{i,j-1}\left(\varpi_{i+1,j-1} - \varpi_{i-1,j-1}\right), \end{aligned} \tag{3.20}$$

$$\begin{aligned} J^{\times +} = {} & \phi_{i+1,j+1}\left(\varpi_{i,j+1} - \varpi_{i+1,j}\right) - \phi_{i-1,j-1}\left(\varpi_{i-1,j} - \varpi_{i,j-1}\right) \\ & - \phi_{i-1,j+1}\left(\varpi_{i,j+1} - \varpi_{i-1,j}\right) + \phi_{i+1,j-1}\left(\varpi_{i+1,j} - \varpi_{i,j-1}\right), \end{aligned} \tag{3.21}$$

$$J^{\times\times} = \left(\phi_{i+1,j+1} - \phi_{i-1,j-1}\right)\left(\varpi_{i-1,j+1} - \varpi_{i+1,j-1}\right) - \left(\phi_{i-1,j+1} - \phi_{i+1,j-1}\right)\left(\varpi_{i+1,j+1} - \varpi_{i-1,j-1}\right), \tag{3.22}$$

$$J_2^{+\times} = \phi_{i+2,j}\left(\varpi_{i+1,j+1} - \varpi_{i+1,j-1}\right) - \phi_{i-2,j}\left(\varpi_{i-1,j+1} - \varpi_{i-1,j-1}\right) - \phi_{i,j+2}\left(\varpi_{i+1,j+1} - \varpi_{i-1,j+1}\right) + \phi_{i,j-2}\left(\varpi_{i+1,j-1} - \varpi_{i-1,j-1}\right), \tag{3.23}$$

$$J_2^{\times+} = \phi_{i+1,j+1}\left(\varpi_{i,j+2} - \varpi_{i+2,j}\right) - \phi_{i-1,j-1}\left(\varpi_{i-2,j} - \varpi_{i,j-2}\right) - \phi_{i-1,j+1}\left(\varpi_{i,j+2} - \varpi_{i-2,j}\right) + \phi_{i+1,j-1}\left(\varpi_{i+2,j} - \varpi_{i,j-2}\right), \tag{3.24}$$

and then the combination

$$[\phi, \varpi] = \frac{1}{12h_x h_y}\left[2(J^{++} + J^{+\times} + J^{\times+}) - \frac{1}{2}(J^{\times\times} + J_2^{+\times} + J_2^{\times+})\right], \tag{3.25}$$

is found to be a fourth-order approximation which satisfies the properties listed above to machine accuracy. Subscripts $ij$ refer to the collocation points, or grid points, given by

$$\phi(x, y) \rightarrow \phi(x_i, y_j) \quad \text{at} \quad x = x_i \quad y = y_j, \tag{3.26}$$

and the $i$ and $j$ refer to the grid point in the $x$- and $y$-direction, respectively,

$$\phi = \phi(x, y) \qquad \phi_{ij} = \phi(x_i, y_j) \qquad x_{i+1} - x_i = h_x \qquad y_{j+1} - y_j = h_y. \tag{3.27}$$

The step sizes on the grid, $h_x$ and $h_y$, are assumed to be equidistant. The structure in Eq. (3.25), the fourth-order Arakawa scheme for discretising the bracket structure, can be used in any geometry if it is divided by $\sqrt{g}$, which then cancels under the integration over the spatial domain. The Jacobi identity Eq. (3.9) is not satisfied exactly for the Arakawa scheme, but is to high accuracy for the fourth-order version with reasonable grids (64 × 64 and higher).

The hyperdiffusion is put in through perpendicular Laplacian operators, applied twice. The standard 5-point stencil is used,

$$\nabla_\perp^2 \varpi = \frac{1}{h_x^2}\left(\varpi_{i+1,j} - 2\varpi_{i,j} + \varpi_{i-1,j}\right) + \frac{1}{h_y^2}\left(\varpi_{i,j+1} - 2\varpi_{i,j} + \varpi_{i,j-1}\right), \tag{3.28}$$

and then applied again to form the squared perpendicular Laplacian with the coefficient $\nu_\perp$.

### 3.2.4 Energetics

The dissipation is put onto the dependent variables but if a potential appears by itself under $\partial/\partial t$ it will be important not to have $D$ act directly upon it. The reason for this is energy conservation, so we will address that model by model.

In the fluid case the energy theorem is simple. Multiply Eq. (3.5) by $-\phi$ to find

$$\frac{\partial}{\partial t}\frac{1}{2}\mid \nabla_\perp \phi \mid^2 + \nabla \cdot \left(-\phi \nabla_\perp \frac{\partial \phi}{\partial t}\right) + [\phi, \phi\varpi] = 0, \tag{3.29}$$

which can serve as a local energy theorem. The quantity $-\phi(\nabla_\perp^2 \phi)$ is effectively (locally) integrated by parts to form the first two terms as written: the goal is to find the operation which turns this term into the time derivative of a positive definite quantity and a total divergence of a quantity which becomes one of the fluxes. The combination of the two bracket structures into $[\phi, (\cdots)]$ proceeds according to the Leibnitz rule in Eq. (3.8). This gives the total divergence of the main kinetic energy flux, which is usually found to be the largest. But it is important to cast equations of this type as exact conservation laws and not neglect the terms whose role is to preserve energetic consistency.

Integrating over the entire spatial domain, we find the global energy theorem,

$$\frac{\partial E_T}{\partial t} = 0 \qquad \text{where} \quad E_T = \mathscr{V}^{-1}\int d\mathscr{V}\, \frac{1}{2}\mid \nabla_\perp \phi \mid^2, \tag{3.30}$$

neglecting surface terms, which for these models are always assumed to vanish due to boundary conditions. Similarly, we find an enstrophy theorem. Multiply Eq. (3.5) by $\varpi$ to find

$$\frac{\partial}{\partial t}\frac{1}{2}\varpi^2 + \frac{1}{2}[\phi, \varpi^2] = 0, \tag{3.31}$$

so that the global enstrophy theorem is

$$\frac{\partial W_T}{\partial t} = 0 \qquad \text{where} \quad W_T = \mathscr{V}^{-1}\int d\mathscr{V}\, \frac{1}{2}\varpi^2. \tag{3.32}$$

These are quadratic quantities, and our domain is 2D periodic, so we can express them in terms of the Fourier components. Every scalar quantity satisfies

$$\phi(x + L_x, y) = \phi(x, y + L_y) = \phi(x, y), \tag{3.33}$$

where $L_x$ and $L_y$ are the periodicity lengths in each direction. We can decompose each of them according to

$$\phi(x, y) = \sum_{\mathbf{k}} \phi_{\mathbf{k}} \exp\left[i(k_x x + k_y y)\right], \tag{3.34}$$

where $\mathbf{k} = k_x \nabla x + k_y \nabla y$ is the vector wavenumber. The inverse transform is

$$\phi_{\mathbf{k}} = \mathscr{V}^{-1}\int d\mathscr{V}\, \exp\left[-i(k_x x + k_y y)\right] \phi(x, y), \tag{3.35}$$

averaged over the domain volume $V$ for normalisation. The wavenumbers $k_x$ and $k_y$ are integral multipliers of $2\pi/L_x$ and $2\pi/L_y$, respectively. The wavenumbers $\mathbf{k}$ are binned within shells $\mathscr{K}$, defined as the set of $\mathbf{k}$ with magnitude closest to $k$, which is an integer multiple of $k_0$, the fundamental. This is defined as

$$\mathbf{k} \in \mathscr{K} \quad \text{if} \quad k - \frac{1}{2}k_0 < \left(k_x^2 + k_y^2\right)^{1/2} < k + \frac{1}{2}k_0. \tag{3.36}$$

The fundamental, $k_0$, is usually defined as $2\pi$ divided by the shorter of $L_x$ or $L_y$ since this is what constrains the largest isotropic eddy size. For these examples we will use a square domain with $L_y = L_x = L$ so that $k_0 = 2\pi/L$. The energy spectrum is therefore expressed as

$$E_k = \sum_{\mathbf{k}\in\mathscr{K}} \text{Re}[(-\phi_\mathbf{k})^* \varpi_\mathbf{k}], \tag{3.37}$$

referred to as the isotropic energy spectrum. Similarly, the enstrophy spectrum is expressed as

$$W_k = \sum_{\mathbf{k}\in\mathscr{K}} \text{Re}[(\varpi_\mathbf{k})^* \varpi_\mathbf{k}]. \tag{3.38}$$

In addition to $E_T$ and $W_T$ we measure the volume average and spectrum of the half-squared amplitude of $\phi$,

$$A_p = \mathscr{V}^{-1} \int d\mathscr{V}\, \frac{1}{2}\phi^2 \qquad A_p(k) = \sum_{\mathbf{k}\in\mathscr{K}} \text{Re}[(\phi_\mathbf{k})^* \phi_\mathbf{k}], \tag{3.39}$$

for comparison. The character of the turbulence will largely be shown by the relative run of these quantities.

### 3.2.5 Dynamical range

The number of grid points in each direction defines the dynamical range $k_{\max}/k_{\min}$ of the spectrum. Always think of a spectrum in logarithmic space, and the dynamical range in terms of powers of two or in decades (powers of ten). Turbulent activity starting at some $k > k_{\max}$ needs finite time to develop structure and correlation as the spectrum spreads out and dissipation acts upon it. Typical cases require at least one decade to even begin to represent this, and the minimum is closer to two. It is always subject to testing, whether anything in the results, especially the spectrum, is convergent against finite resolution. The dynamical range is defined by

$$k_{\min} = k_0 \quad k_{\max} = k_0 \frac{N}{2} \qquad \text{hence} \quad k_{\max}/k_{\min} = N/2, \tag{3.40}$$

where $N$ is the number of points in the *overlap range*. If $L$ is the *shortest* of the domain sizes and $h$ is the *coarsest* of the step sizes, then the overlap range is defined as

$$N = L/h \qquad \text{where} \qquad N = \text{Min}(N_{i,j}) \qquad \text{and} \qquad h = \text{Max}(h_{i,j}), \tag{3.41}$$

and the dynamical range is then given by $k_{\max}/k_{\min} = N/2$. When quoting spatial resolution in a computation, this overlap range and hence the dynamical range should always be given. We describe these as such ($xy$ not $xyz$) since all the cases we treat in turbulence in a magnetised plasma will involve turbulence in only two of the dimensions, with the third one along the guiding magnetic field serving to apply constraints.

### 3.2.6 The base case for fluid decaying turbulence

As a starting point we normalise length in terms of a scale $d$ and velocity in terms of an amplitude $V_0$. The unit of time is $d/V_0$ hence transit frequency $V_0/d$. One approach is to set the step sizes $h_x = h_y = d$, so that if we have $N$ collocation points in each direction the domain length in each direction is $L_x = L_y = L = Nd$, equal to a single length $L$. In our case, we set the domain and grid such that $d$ is roughly in the middle of the spectrum. The fundamental wavenumber in terms of $d$ is set to $K = k_0 d$. If we choose $K = 0.05$ then $L/d = 40\pi$, which is about 128 in terms of powers of two. Then, with $N$ points in each direction we have $h_x = h_y = h = 2\pi/KN$ in terms of $d$. The scale at which $kd = 1$ is then mode 20, and with 256 points we would have $kd < 6$ approximately.

The equation with the dissipation operator is

$$\frac{\partial \varpi}{\partial t} + [\phi, \varpi] = -\nu_\perp (\nabla_\perp^2)^2 \varpi \qquad \varpi = \nabla_\perp^2 \phi. \tag{3.42}$$

All quantities are dimensionless. The only parameters are $K$ and $\nu_\perp$.

$$K = 0.05 \qquad \nu_\perp = \nu_0 \frac{h^4}{16}, \tag{3.43}$$

that is, the dissipation is tied to the spatial resolution within a parameter $\nu_0$ which will be set nominally to 1.0 for driven cases, but the decaying ones require lower values, and the fluid case especially does so, to let the vorticity spectrum develop before its total damps away. The trial and error in this is discussed below.

The initial state is a random-phase distribution with a prescribed initial spectrum, according to Eq. (3.34) with

$$\phi_{\mathbf{k}} = A_0\, k \left(1 + (k/k_1)^6\right)^{-1} \exp(i\Theta), \tag{3.44}$$

with $\Theta$ a random number on $[0, 2\pi]$ giving the phases and $k_1 = 5k_0$ defining where the initial spectrum should peak. The steepness gives reasonable time for evolution of the high-$k$ side (recall $k^2 = k_x^2 + k_y^2$). The number of grid points in each direction is set to $N = N_x = N_y = 256$. This is sufficient to display the properties of the turbulence, but the tendency of the energy to remain at the low-$k$ end of the spectral range makes proving the essential statements difficult unless more spatial resolution is used. Nevertheless, the aim is introduction: learn the look and feel first, and then do the necessary diagnosis thoroughly. There is no substitute for experience in understanding.

The initial amplitude $A_0$ is set such that the RMS $\phi$ defined as $\phi_0$ is set to unity,

$$\phi_0^2 = (N_x N_y)^{-1} \sum_{ij} \left| \phi_{ij} \right|^2 = 1, \tag{3.45}$$

namely, the RMS of any quantity is the square root of the volume average of the square. This assures a nonlinear start with all degrees of freedom excited such that vigorously decaying turbulence results. The time step is checked such that any velocity times $\tau/h$ are less than 0.2, which is the limit of stability and accuracy for this stepping scheme. If the dynamics is too strong, the value of $\tau$ is reduced and the run is repeated. Corresponding effort by trial and error is undergone. For these cases $\tau = 10^{-2}$ was found to be sufficient.

We start with the case $k_1 = 5k_0$ and $\nu_0 = 10^{-2}$. The time traces are shown in figure 3.1. These are sampled every $\Delta t = 0.2$ in normalised units. The fluid energy $E_T$ remains nearly constant while the enstrophy $W_T$ decays more visibly. The amplitude of $\phi$ tracked by $A_p$ increases—note that $A_p$ is a diagnostic not a conserved quantity. The next line shows the energy and enstrophy decay rates and the energetic error rate. The decay rates result from

$$\Gamma_T = \frac{1}{2E_T}\frac{\partial E_T}{\partial t} \qquad \Gamma_W = \frac{1}{2W_T}\frac{\partial W_T}{\partial t}, \tag{3.46}$$

that is, growth rates which are found to be negative (the factor of two arises from the fact that $E_T$ and $W_T$ are quadratic in the dependent variables which contribute). These show the disparity in the decay rates of $E_T$ and $W_T$, in this case a factor of

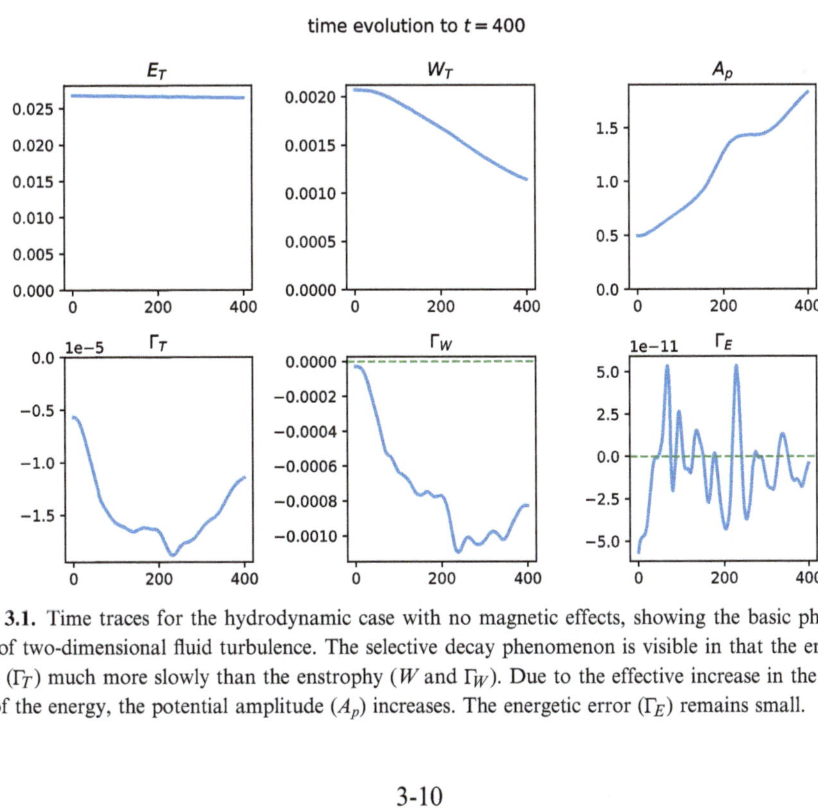


**Figure 3.1.** Time traces for the hydrodynamic case with no magnetic effects, showing the basic phenomenology of two-dimensional fluid turbulence. The selective decay phenomenon is visible in that the energy ($E$) decays ($\Gamma_T$) much more slowly than the enstrophy ($W$ and $\Gamma_W$). Due to the effective increase in the effective scale of the energy, the potential amplitude ($A_p$) increases. The energetic error ($\Gamma_E$) remains small.

60 between them. The energetic error is defined as the measured energy growth rate compared with the forcing rates given by drive and sink mechanisms. It is given by

$$\Gamma_E = \Gamma_T - (\Gamma_+ - \Gamma_-), \tag{3.47}$$

where $\Gamma_\pm$ give the total drive and dissipation rates determined by the equations (for this case, $\Gamma_+ = 0$ and $\Gamma_-$ is the subgrid dissipation due to $\nu_\perp$) and $\Gamma_T$ gives the total growth rate measured from the time trace of $E_T(t)$. The difference is the error. We see that this is not zero, since the numerical scheme is not automatically conservative. But its smallness indicates a well-performing numerical scheme.

The isotropic spectra are given in figure 3.2. The initial state at $t = 0$ is shown, then an early phase at $t = 50$, and then a later phase at $t = 300$, when the turbulence is fully developed. The enstrophy spectrum is seen to spread out into a power-law form. The computation has barely enough resolution to see this. Basically, the initial peaked spectrum spreads out up to the dissipation range, and in this case there is just enough time for this to find its form before decay sets in enough to change the spectral form. The power-law form develops within about $0 < t < 150$ and persists somewhat beyond $t = 300$. A case with $\nu_0 = 1$ enters the strong decay phase too soon, and a case with $\nu_0 = 10^{-3}$ runs out of resolution on the high-$k$ side such that the enstrophy spectrum does not turn down properly. A case with $\nu_0 = 10^{-1}$ is more

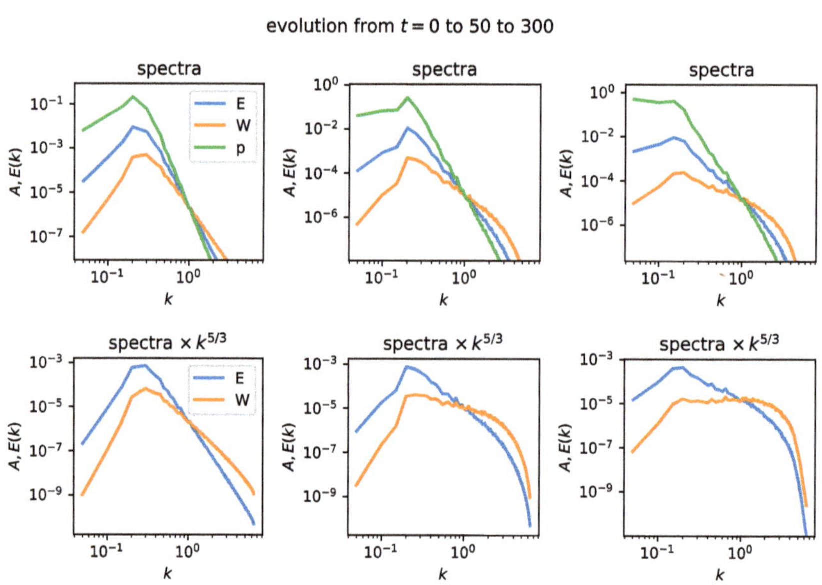


**Figure 3.2.** Isotropic spectra for the hydrodynamic case with no magnetic effects, showing the basic phenomenology of two-dimensional fluid turbulence as it develops out of the initial state. The curves are for the energy $E$ ('E'), enstrophy $W$ ('W'), and the potential amplitude $A_p$ ('p'). The initial state (left frames) indicates how the short-term averaged state in the developed phase (right) has evolved. The enstrophy spectrum, not the energy, is the one showing a self-similar cascade towards smaller scales, lining up with a Kolmogorov $k^{-5/3}$ curve.

marginal but the power-law form is still visible. The diagnostic for this is the enstrophy spectrum multiplied by $(k/k_0)^{5/3}$. The flat line shows the Kolmogorov $(k/k_0)^{-5/3}$ power law and the turn-down is strong enough that the activity does not pile up at the highest $k$. This spectrum shows the Kolmogorov power law indicative of a cascade process towards smaller scale for enstrophy, rather than energy, as expected for a process which conserves both quantities.

The spatial morphology is shown in figure 3.3, at the same three times, $t = 0$, 50, and 300. The potential shifts to large scale, which is why its amplitude time trace rises. The vorticity shows first sheets and then an isolated, intermittent structure. Note that the potential and the vorticity, despite being tied through their relation in Eq. (3.42), form a structure at quite different scales. This is why the spread in $k$-space in the spectrum is essential, without which the power-law range and the vorticity structure would not be able to form. The visual smoothness or choppiness of the vorticity contours is an important diagnostic along with the turn-down of its spectrum. Ultimately, it is the vorticity and not the field potential that indicates whether a turbulence run is well-behaved or not. In MHD cases, the current also serves this purpose (and not the magnetic field or its potential).

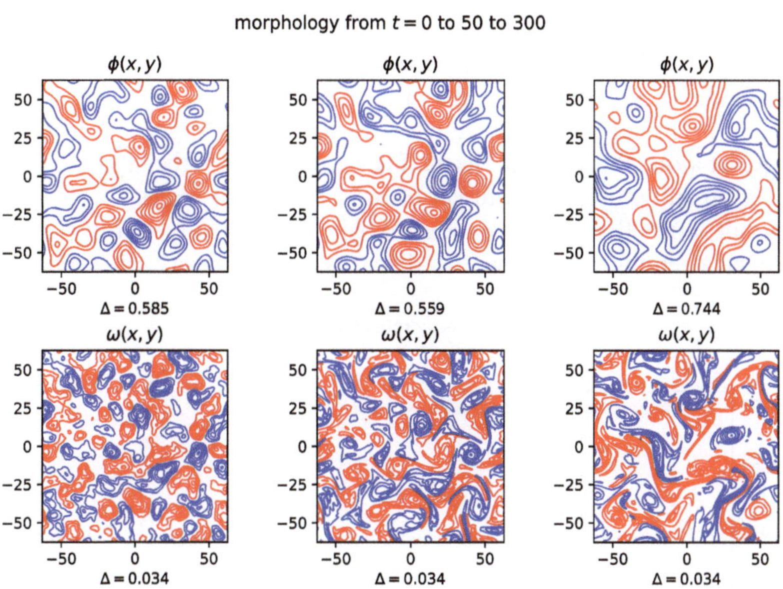


**Figure 3.3.** The spatial morphology for the hydrodynamic case with no magnetic effects, showing the basic phenomenology of two-dimensional fluid turbulence. The initial state at $t = 0$ and an early phase at $t = 50$ are shown in comparison to the developed turbulence at $t = 300$. In this and every other contour plot in this work, the $x$ and $y$ labels are understood by the title (e.g. $\phi(x, y)$), and the interval, which is always 0.16 of the maximum absolute value, is given at the bottom of every frame. Also, positive contours are in drawn in red; negative, blue. The vorticity ($\varpi$) has formed sheet-like structures, where most of the dissipation occurs, and begins to show an isolated, intermittent structure. The scale of the flow eddies, tracked by the potential ($\phi$), which serves as an instantaneous stream function for the flow, has visibly increased as part of the dual cascade which sends enstrophy towards small scales and energy towards large.

### 3.2.7 Self similarity

The properties of the equations in the fluid case is that there is no preferred scale—no small parameter upon which to anchor the grid and resolution. Hence for this case $d$ and $V_0$ are chosen arbitrarily. If $d$ and $V_0$ are re-defined then the time and space units may be re-set such that the same equation in normalised units results. If $k_0 d$ is kept constant then the $K$ doesn't change and the same domain size and grid are used. This self similarity is broken in cases containing an intrinsic physical scale, which occurs in all of the plasma models except for the pure-fluid or MHD ones.

Under self similarity, the spectrum has the same shape at all scales except for lowest (too few degrees of freedom) and highest (dissipation range). The manifestation of self similarity in the energetics is that the specific decay rate approaches a finite value which is held for finite time. The specific decay rates for energy and enstrophy are given by

$$Q_E = \frac{1}{2E_T^2}\frac{\partial E_T}{\partial t}, \qquad Q_W = \frac{1}{2W_T^2}\frac{\partial W_T}{\partial t}. \tag{3.48}$$

We see this in the results, especially $Q_W$ since the energy decay is so slow, if the dynamical range in $k$ is large enough. The time traces of $Q_T$ and $Q_W$ for various $\nu_0$ are shown in figure 3.4. The middle two values $10^{-1}$ and $10^{-2}$ show similar specific enstrophy decay, while $10^0$ is confirmed as being too large and $10^{-3}$ too small. This fits with the same impression from the turn-down at high $k$ in their vorticity spectra. Note also how enstrophy puts more pressure on the convergence testing than does the energy, let alone the potential itself. The cause of this is the higher $k$-dependence and the resultingly more robust direct transfer to smaller scales.

This result shows that the resolution we have used is really marginal for the fluid case. We have enough to show the disparity in the decay rates and the scale

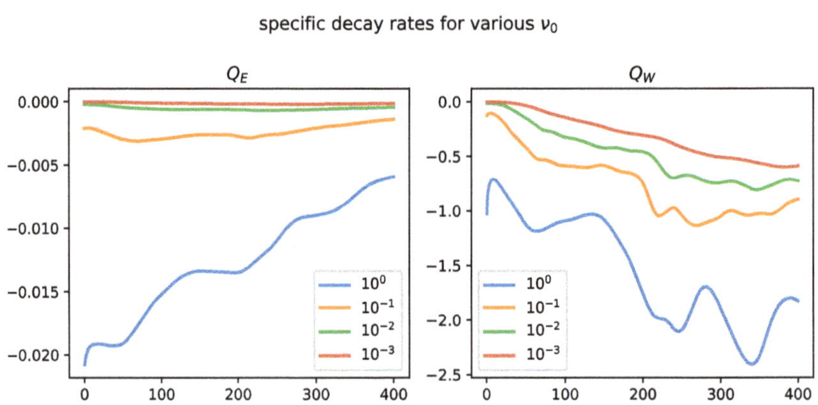


**Figure 3.4.** Specific decay rates of the energy (left) and enstrophy (right) for the hydrodynamic case, for different values of the subgrid dissipation (coefficient $\nu_0$). The case with $\nu_0 = 10^0$ has the dissipation set too high, while $\nu = 10^{-3}$ has it too small. The curves of the middle two cases have enstrophy decay levels a factor of ten apart but specific decay curves with similar form and rates. This also confirms this grid size (256 × 256) as marginal for this problem.

separation between the energy and enstrophy but it takes a larger grid to confirm self similarity to a more reasonable degree than this. If $\nu_0$ is always set to $10^{-2}$ then a variation of $k_{max}/k_{min}$ is necessary. For different $k_{max}$ for the same $k_{min}$ we would tie the dissipation to the resolution and find the same value of $Q$ and the same spectral shapes for $E_k$ and $W_k$. With the model, though, the choices of $d$ and $V_0$ are arbitrary: renormalising with $d \rightarrow 2d$, for example, re-produces the same equation and therefore the only essential parameter is the dynamical range. Nevertheless, in our case the difference in the specific decay rate is much less than the factor of ten difference in the decay coefficient.

### 3.2.8 Selective decay

More than one quantity is conserved, but those with higher $k$-dependence decay faster. In this case we have the energy and enstrophy $E_T$ and $W_T$, with $W_k$ having the higher $k$-dependence. So we compare the decay rates of these two quantities in a time trace. We already see this in the time traces of figure 3.1 for the fluid case. As long as there is reasonable dynamical range $W_T$ always decays faster than $E_T$. It builds the self-similar $k^{-5/3}$ spectrum, transferring to smaller scale more robustly than $E_T$ even under the general decay. The limit of this is the result in the previous Chapter, where in the case of fluid viscosity as the decay mechanism, the energy decay rate goes to zero in the limit where viscosity goes to zero, whereas the vorticity decay rate approaches a finite limit with small viscosity multiplying large vorticity gradient. Even in the case of hyperdiffusion, the faster vorticity dynamics in the dual cascade leads to faster enstrophy decay with any dissipation model, leaving energy to decay slowly. Only after the decay has proceeded long enough for the amplitude relative to the dissipation parameters to decrease, therefore for the effective Reynolds number to drop, does the decay process encroach upon the energy-containing range of the turbulence. The dynamical range reduces enough that the situation is then no longer turbulent.

### 3.2.9 Structure formation

As the decay of the turbulence progresses, the nonlinear correlations become ever more complicated. The initial development of the vorticity changes from sheet structures to isolated vortices whose size decreases with time but whose magnitude increases. Overall the enstrophy decreases. The physical nature of the nonlinear interactions is vortex merger for the flow eddies and shearing apart of the structures for the quantities directly advected by the eddies. As this includes the vorticity, the stream function field is involved in both the flows and the advected quantities. The Euler equation's dynamics is essentially this merger/shearing process, until at late times (if the Reynolds number is high enough) one is left with a shear-free flow field. This shear-free flow field manifests itself as a collection of isolated, long-lived coherent vortices.

This development is only marginally observed in the case above with $k_1/k_0 = 5$, but another with $k_1/k_0 = 10$ shows it more clearly. More of the initial signal is at higher $k$ so that the ratio $W_T/E_T$ is larger. The development into isolated vortices is

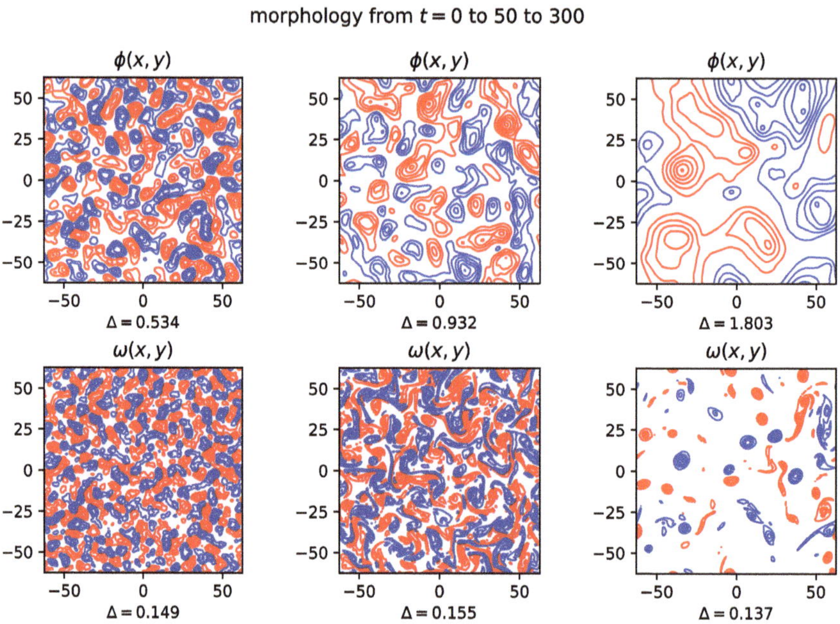


**Figure 3.5.** Spatial morphology for the hydrodynamic case, started with the spectrum peaked at smaller scale ($k_1/k_0 = 10$ rather than 5), showing development into isolated vortices more visibly. Obtaining convergent decay rates, as for the previous cases, and also this feature of isolated vortices, simultaneously would require at least doubled resolution, or a more highly tailored initial state.

faster and more visible. Figure 3.5 shows the spatial morphology of this case for comparison to figure 3.3. In both cases $\nu_0 = 10^{-2}$ and $K = 0.05$ but the start with $k_1/k_0 = 5$ gave enough free play in $\nu_0$ to show the independence of the result on it, while $k_1/k_0 = 10$ gave enough mid-range vorticity to set up within finite time the nonlinear correlations necessary to describe the structure inherent in the long-lived coherent vortices in the freely decaying turbulence. Showing both features with the same initialisation would require larger $N$, say, $N = 1024$. Then, there is enough time to form the vortices and there is a wide enough window to show independence against $\nu_0$ of the spectral form in both initialisations.

Later, when we claim that the coherent vortices are eliminated by dissipative forcing, we will add the forcing to a fluid case which showed these vortices. The confirmation of coherent vortices in the fluid case serves as the necessary control case for the eventual conclusion.

### 3.2.10 Intermittence and structure functions

The turbulent cascade, as part of building structures, does not fill all space as activity at scale $k$ descends to scale $2k$. For the fluid case this is most apparent in the structure the vorticity builds as it cascades towards higher $k$ and the dissipation range. Intermittence in general was considered in a three-dimensional fluid model but one which produced diagnostics that can be used in any model. We will not develop this herein, since as soon as we add magnetic dynamics and more coupling mechanisms between the variables, the nonlinearities will not be acting in isolation and these coherent vortices will disappear.

## 3.3 2D MHD turbulence

The magnetohydrodynamics (MHD) model is also expected to be somewhat familiar. Again we assume the velocity to be incompressible and the density taken to be constant. The equations are then just those for the velocity, with fluid inertia, the Lorentz force, and the fluid pressure,

$$\rho\left(\frac{\partial \mathbf{u}}{\partial t} + \mathbf{u} \cdot \nabla \mathbf{u}\right) = \mathbf{J} \times \mathbf{B} - \nabla p. \tag{3.49}$$

Faraday's law for the magnetic field,

$$\frac{\partial \mathbf{B}}{\partial t} = -\nabla \times \mathbf{E}. \tag{3.50}$$

Ampère's law for the current,

$$\mu_0 \mathbf{J} = \nabla \times \mathbf{B}, \tag{3.51}$$

and the Ohm's law for each charged species,

$$\mathbf{E} + \mathbf{u} \times \mathbf{B} = 0, \tag{3.52}$$

with the main MHD assumption being that of a single fluid with a single mass density, pressure, and velocity. Decomposing $\mathbf{B}$ the same way as previously for $\mathbf{u}$ we have, along with Eq. (3.4),

$$\mathbf{B} = \nabla\psi \times \hat{\mathbf{z}} \qquad \mathbf{J} = J\hat{\mathbf{z}} \qquad \mu_0 J = -\nabla_\perp^2 \psi, \tag{3.53}$$

where in this case we neglect compressibility and therefore the third component (along $\hat{\mathbf{z}}$) of both $\mathbf{u}$ and $\mathbf{B}$. We therefore have

$$\mathbf{B} \cdot \nabla \equiv B\nabla_\parallel = [-\psi, ], \tag{3.54}$$

for the parallel gradient representing variation along the nonlinear magnetic field. In two dimensions, therefore, the parallel gradient is written in terms of the same bracket structure as in advection, with $-\psi$ as the advecting field instead of the stream function $\phi$ in the case of the flow. The only difference is in terms of normalising factors, which in the 2D models are constants. Parallel gradient and divergence are taken interchangeably because $\nabla \cdot \mathbf{B} = 0$ in all cases. The differences incurred with spatially variable background magnetic field will be treated later on, with the three-dimensional models.

We get this parallel gradient first as a divergence when computing $-\nabla \cdot \hat{\mathbf{z}}\times$ on the original velocity equation Eq. (3.49). Similarly, we apply $\nabla \cdot \hat{\mathbf{z}}\times$ on Eq. (3.50) and then drop the common operation by $\nabla_\perp^2$ to get the equation for $\psi$, which depends on parallel gradients of scalar quantities which serve as force potentials. The normalisation scheme is to choose a length scale $d$, and scale the velocity like $v_A$, the magnetic field and current like $B$ and $B/\mu_0 d$, and the time like $d/v_A$, where $v_A$ is the Alfvén velocity given by

$$v_A^2 = \frac{B^2}{\mu_0 \rho}. \tag{3.55}$$

Having done this we find that all the units cancel out and $d$ can be chosen arbitrarily. In other words, the MHD model is also self similar as in the fluid case.

The equations in normalised units are

$$\frac{\partial \varpi}{\partial t} + [\phi, \varpi] = [-\psi, J], \tag{3.56}$$

(compare with Eq. (3.4)) for $\phi$ and

$$\frac{\partial \psi}{\partial t} + [\phi, \psi] = 0, \tag{3.57}$$

for $\psi$, where

$$\varpi = \nabla_\perp^2 \phi \qquad J = -\nabla_\perp^2 \psi, \tag{3.58}$$

are auxiliary quantities. It is useful to note that

$$\mathbf{u} \cdot \nabla \psi = \mathbf{B} \cdot \nabla \phi = [\phi, \psi] = [-\psi, \phi], \tag{3.59}$$

in two dimensions. When we apply dissipation it will be to $\varpi$ and $J$, not to $\psi$ directly. The time step advances $\varpi$ and $\psi$ according to Eqs (3.56) and (3.57), after which $\phi$ and $J$ are found by solving and applying $\nabla_\perp^2$ according to Eq. (3.58).

### 3.3.1 MHD energetics

As in the fluid case, the energy theorem is built by constructing positive definite quantities under the partial time derivative and showing that all other terms combine into a total divergence. Hence we have a density of a quantity which can be called free energy and a flux of that quantity. In the fluid case the energy density is easily identified as the fluid kinetic energy. In the MHD case we have additionally the magnetic energy. The usefulness of this is that the method we use to prove these can be used on other systems whose web of conservative energetic coupling is more complicated. In this case we multiply the vorticity equation Eq. (3.56) by $-\phi$ and the magnetic flux Eq. (3.57) by $J$ to find

$$-\phi \frac{\partial \varpi}{\partial t} + [\phi, \phi\varpi] + [-\psi, \phi J] = J[-\psi, \phi] \tag{3.60}$$

$$J \frac{\partial \psi}{\partial t} = -J[-\psi, \phi], \tag{3.61}$$

where the Leibnitz rule and antisymmetry in Eq. (3.8) are used to rewrite the bracket terms in terms of transport terms and the conservative transfer due to

$$J\nabla_{\|}\phi = J[-\psi,\, \phi]. \tag{3.62}$$

This clarifies the energetic symmetry. The time derivative terms are locally integrated by parts as in Eq. (3.29) and the two parallel gradient terms are combined into one parallel divergence, such that

$$\begin{aligned}&\frac{\partial}{\partial t}\frac{1}{2}(|\nabla_{\perp}\phi|^2 + |\nabla_{\perp}\psi|^2) + \nabla\cdot\left(-\phi\nabla_{\perp}\frac{\partial\phi}{\partial t} - \frac{\partial\psi}{\partial t}\nabla_{\perp}\psi\right)\\ &\quad + ([\phi,\, \phi\varpi] + [-\psi,\, \phi J]) = 0,\end{aligned} \tag{3.63}$$

where the parentheses denote grouping into three sets of terms: the energy density, the wave fluxes, and the main fluxes, which are called ExB advection (carrying of quantity densities by the ExB velocity) and magnetic flutter (from fluttering field lines) fluxes. Which of the latter is the larger depends on the model and sometimes the choice of parameters within the model.

The 2D MHD model has another conserved quantity which becomes relevant in nonlinear interactions. In ideal MHD the helicity, whose density is $\mathbf{A}\cdot\mathbf{B}$, is conserved generally, as we saw in the previous Chapter. In the purely 2D case, however, $\mathbf{A}$ is perpendicular to the plane (it is $\psi\hat{\mathbf{z}}$) while $\mathbf{B}$ is within the plane, so $\mathbf{A}\cdot\mathbf{B}$ vanishes. The 2D analogue is related to the flux conservation, in this case the volume integral over $\psi$, which is conserved in the dissipation-free limit (see Eq. (3.57)). In fact, the integral over any function $G(\psi)$ is also conserved since $G$ satisfies a simple transport equation (this follows from the Leibnitz rule in Eq. (3.8)). The quadratic form relevant to this is

$$A_a = \mathscr{V}^{-1}\int d\mathscr{V}\,\frac{1}{2}\psi^2, \tag{3.64}$$

which is invariant in the absence of dissipation.

The cross helicity for this system is given by

$$\mathbf{u}\cdot\mathbf{B} = (\hat{\mathbf{z}}\times\nabla\phi)\cdot(\nabla\psi\times\hat{\mathbf{z}}). = -\nabla\psi\cdot\nabla\phi. \tag{3.65}$$

The manifestation of conservation is found by multiplying Eqs (3.56) and (3.57) by $\psi$ and $\varpi$, respectively, to find

$$\frac{\partial}{\partial t}(\varpi\psi) + [\phi,\, \varpi\psi] + [\psi,\, \psi J] = 0, \tag{3.66}$$

using the Leibnitz rule in Eq. (3.8), and noting that

$$\varpi\psi = \mathbf{B}\cdot\mathbf{u} + \nabla\cdot(\psi\nabla_{\perp}\phi). \tag{3.67}$$

We will measure the volume averaged quantity

$$C_H = \mathscr{V}^{-1}\int d\mathscr{V}\,\varpi\psi, \tag{3.68}$$

along with $A_a$ given above. These quantities are conserved by the nonlinear interactions but are affected by the dissipation. Their consideration is most interesting when issues of equipartition are encountered, as we will see in the energetics.

### 3.3.2 The base case for MHD decaying turbulence

The grid and scales are chosen as in the fluid case for close comparison. For $V_0$ the Alfvén velocity $v_A$ is substituted. The same grid and use of $K$ also apply here. There are two choices for initialisation. In a magnetic case we set $\psi$ the same as for $\phi$ in the fluid case, with amplitude $\psi_0 = 1$, and set $\phi = 0$, therefore with no flow. The flow emerges naturally through the magnetic stress nonlinearity (field line bending: $\nabla_{\|}J_{\|} = [-\psi, J]$), and then produces turbulence. In a kinematic case we set $\phi$ the same as in the fluid case, and $\psi$ similarly. Each has the same spectrum, and the amplitudes are $\phi_0 = 1$ and $\psi_0 = 0.1$, hence with finite magnetic energy. The latter is set to be nonzero since $\psi$ will remain zero for all time if $\psi_0$ is zero. In this case each field $\psi$ or $\phi$ is set up with independent random phases $\Theta$, i.e. the random-phase initial states are set up anew for each field sequentially. The flow proceeds as in the fluid case and the magnetic turbulence proceeds with conserved flux (total $\psi$) but with magnetic energy evolving according to how $\psi$ is arranged by the flow. This is the kinematic regime, since the back reaction of $\psi$ and $J$ back to $\varpi$ hence $\phi$ is limited. The spectrum is chosen as in the fluid case with $k_1/k_0 = 5$ (Eq. 3.44), once more to try to get a true inertial range. Again, trial and error experience is necessary to understand the setup before one gets a chance to understand the energetics in turbulence.

The equations with the subgrid dissipation operator are

$$\frac{\partial \varpi}{\partial t} + [\phi, \varpi] = [-\psi, J] - \nu_{\perp}(\nabla_{\perp}^2)^2 \varpi, \tag{3.69}$$

(compare with Eq. (3.56)) and

$$\frac{\partial}{\partial t}(\psi + d_e^2 J) + [\phi, (d_e^2 J)] = -[-\psi, \phi] - \nu_{\perp}(\nabla_{\perp}^2)^2 (d_e^2 J), \tag{3.70}$$

(compare with Eq. (3.57)), with the field Eqs (3.58). Anticipating the finite electron mass models introduced below, a finite parameter $d_e = 0.025$ is set for the normalised skin depth. In this case it is to facilitate the numerical scheme: the quantity $\psi + d_e^2 J$ is advanced, then $\psi$ is solved for according to

$$(1 - d_e^2 \nabla_{\perp}^2)\psi = \psi + d_e^2 J \equiv F_{\psi} \qquad J = d_e^{-2}(F_{\psi} - \psi). \tag{3.71}$$

The dissipation operator itself acts on $\varpi$ and $J$ only, not on the fields directly.

Among quadratic quantities, only the total energy defined by

$$E_T = E_e + E_b \quad E_e = \mathcal{V}^{-1} \int d\mathcal{V} \, \frac{1}{2} \mid \nabla_{\perp}\phi \mid^2 \quad E_b = \frac{1}{2} \mathcal{V}^{-1} \int d\mathcal{V} \, J F_{\psi} \tag{3.72}$$

is conserved except for dissipation. We also measure

$$E_j = \mathscr{V}^{-1} \int d\mathscr{V} \; \frac{1}{2} d_e^2 J^2, \tag{3.73}$$

to check the spectrum of the current and to ensure that the contribution due to $J^2$ remains small. The energetic error is given as in Eq. (3.47), but now with this definition of $E_T$ and with

$$\Gamma_- = \frac{1}{2E_T} \mathscr{V}^{-1} \int d\mathscr{V} \; \nu_\perp\left( | \nabla_\perp \varpi |^2 + | d_e \nabla_\perp^2 J |^2 \right), \tag{3.74}$$

from the subgrid dissipation. The specific decay rate is given as in Eq. (3.48) but with this definition of $E_T$.

Larger values of $\nu_0$ are needed than in the fluid case due to the general direct transfer to high $k$. Since the magnetic stresses break enstrophy conservation the dual conservation constraint no longer exists on the fluid energy. We start with the magnetic case with $\nu_0 = 10^0$. The time traces are shown in figure 3.6. Sampling is as in the fluid case. The fluid energy $E_e$ rises quickly until about $t = 30$ as magnetic energy drives the fluid response, but not all the way to the $E_b$ level. The initial reaction tends toward equipartition but once the turbulence has started both energy components decay together. Subsequently, the rise of $A_p$ is weak compared with the fluid case, and both pieces of the energy and the enstrophy $A_w$ all decay at a similar rate. The specific decay rate of the total energy ($Q_T$, defined as for $Q_E$ in the fluid

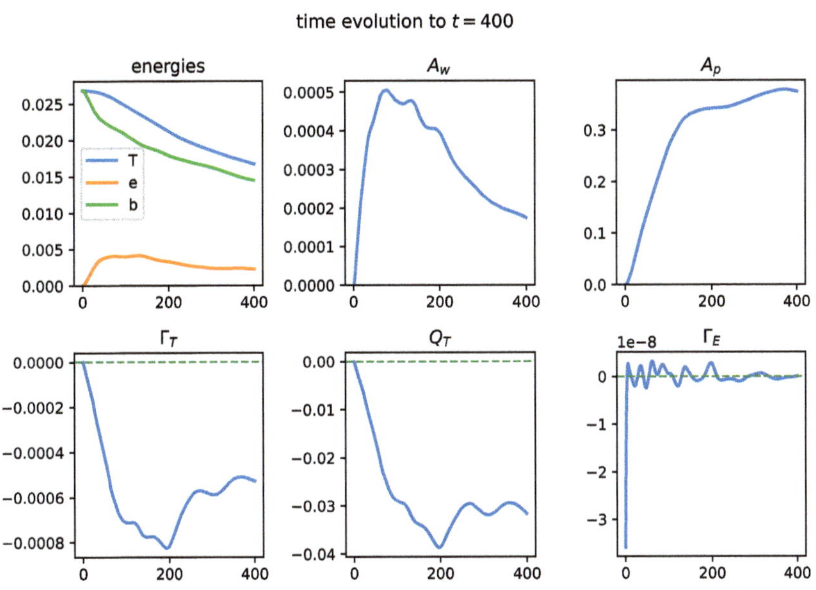


**Figure 3.6.** Time traces for the two-dimensional MHD magnetic case (initialising $\psi$ only, with $\phi = 0$ at $t = 0$). The initial tendency towards equipartition between fluid ('e') and magnetic ('b') energy has led to both decaying together. Now, with coupling to the magnetic energy, the energy and enstrophy decay together. Over this short time the decay $\Gamma_T$ and specific decay $Q_T$ of the energy are similar, although $Q_T$ is flatter after the maximum decay point near $t = 200$. More of this equipartition would be seen if the dissipation were out of the way (equipartition and statistical equilibration are the subject of Chapter 5).

case, Eq. (3.48) is in the range $-0.04 < Q_T < -0.03$, somewhat slower than in the fluid case. The kinematic case in figure 3.7 is similar but inverted. The magnetic energy comes up, if more slowly, but short of equipartition. Both the energy and enstrophy decay somewhat faster in the kinematic case, since there is more vorticity relative to total energy in this case. Companion runs for both cases with $\nu_0 = 0.1$ showed closer convergence in the specific decay than in the fluid case. The time trace for the energetic error shows values in the range $|\Gamma_E| < 10 - 8$ for all time except $10^{-7}$ in some initial transients.

The isotropic spectra are given in figure 3.8 for the magnetic case, for the time points $t = 0$, 50, and 300. The evolution out of the initial state through the early phase and into the developed phase shows the magnetic energy $E_b$ spectrum spreading out into a power-law form which is still close to the $k^{-5/3}$ found for enstrophy in the fluid case. This indicates a robust direct cascade for energy, together with the decay findings above, constituting significant contrast to the fluid case. The initial peaked spectrum spreads out up to the dissipation range with about the same speed as for the enstrophy in the fluid case. Comparison to figure 3.9 for the kinematic case shows some interesting differences. The magnetic energy spreads out even more in this case, since in the kinematic regime the vorticity is stronger in relative terms, and the back reaction from $E_b$ back to $E_e$ is weaker. But it is $E_b$ which puts the strongest pressure on the resolution of the grid, since the kinematic regime yields a rising spectrum for $E_b$ versus $k$ well to the high-$k$ side of the peak in the enstrophy. The spectra of $E_j$, the contribution $d_e^2 J^2/2$ to the total of $E_b$, confirms that

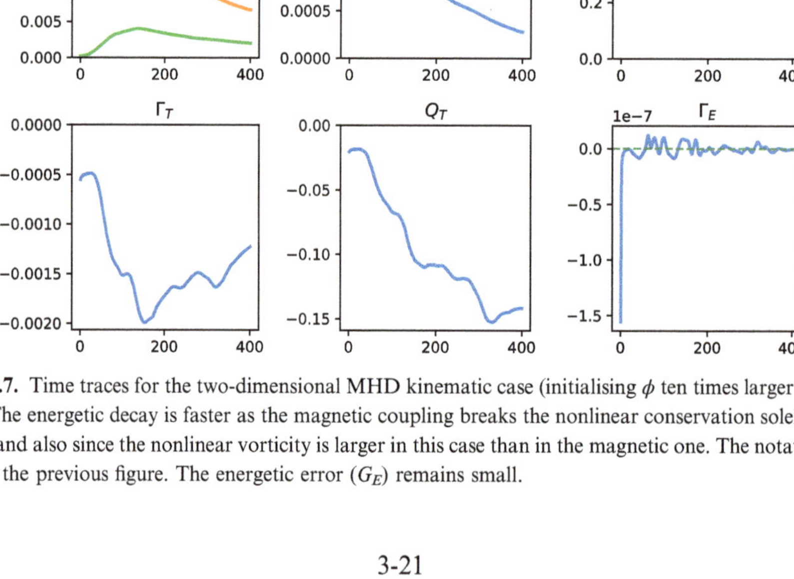


**Figure 3.7.** Time traces for the two-dimensional MHD kinematic case (initialising $\phi$ ten times larger than $\psi$ at $t = 0$). The energetic decay is faster as the magnetic coupling breaks the nonlinear conservation solely of fluid energy, and also since the nonlinear vorticity is larger in this case than in the magnetic one. The notation is the same as the previous figure. The energetic error ($G_E$) remains small.

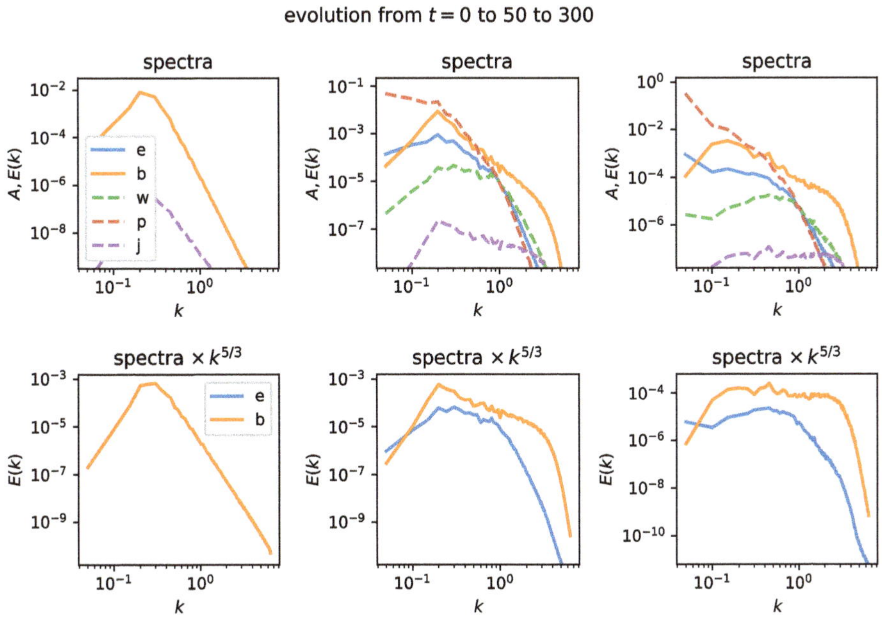


**Figure 3.8.** Isotropic spectra for the kinematic MHD case, showing the development out of the initial state, with the same format as in figure 3.2, with which it is compared. The curves give the ExB ('e') and magnetic energies ('b'), and the vorticity ('w'), potential ('p'), and current ('j') amplitudes. The magnetic energy, as the dominant one, follows the Kolmogorov line, but with the vorticity amplitude at low level the overall decay is still slow.

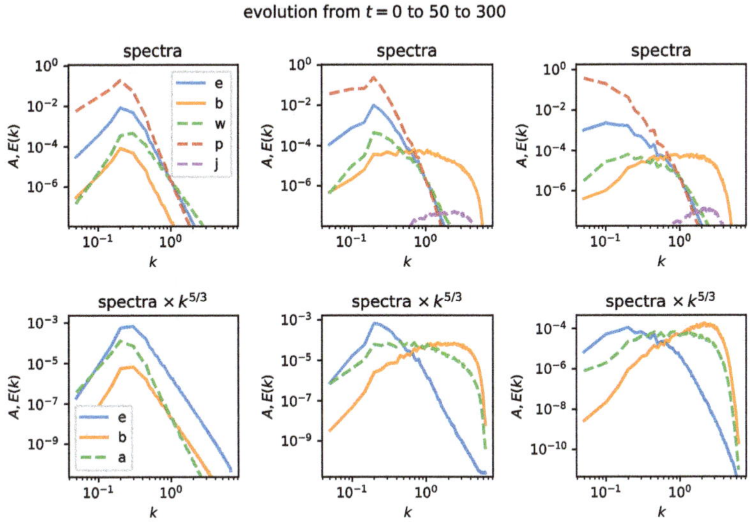


**Figure 3.9.** Isotropic spectra for the kinematic MHD case, showing the development out of the initial state, with the same format as in figures 3.2 and 3.8, with which it is compared. The curves give the ExB ('e') and magnetic energies ('b'), and the vorticity ('w'), potential ('p'), and current ('j') amplitudes. Now the magnetic energy built up by the dynamics is concentrated at small scale, and with the fluid component initialised at the same amplitude as in the fluid case, the vorticity is larger and cascades the energy more effectively. The roundedness of the spectra is due to the decay of the largest scales in this case having been more effective than in the hydrodynamic case.

$E_j \ll E_b$ for all $k < k_{\max}$ so that the finite $d_e$ is not physically significant. The turn-down into the dissipation range is strong in both cases, even for $E_j$, but for $\nu_0 = 0.1$ it was found to be insufficient especially for $E_j$. For all other quantities both values of $\nu_0$ produced the same relative spectra, and as in the fluid case the specific decay rate was found to differ much less than the factor of ten in the coefficient. The real reason for this high-$k$ development of the magnetic energy in a kinematic case is that the magnetic flux is conserved by itself and remains in statistical equilibration, preventing magnetic energy from collecting at larger scale.

The spatial morphology for the magnetic case is shown in figure 3.10 and for the kinematic case in figure 3.11, in the late phase at $t = 400$. The difference in the vorticity in the magnetic case to the fluid case in figure 3.3 and to the kinematic case is essential. In the magnetic case the flow is directed by the magnetic disturbance field with the vorticity collected into sheets and the current into thin layers in the interstitial regions between the magnetic structures. The latter are mostly isotropic pairs which squeeze thin current layers between them. The pairs of flux bundles are a double-wire structure, reflecting the tendency of like currents to attract. The current layers between them are magnetic reconnection sites—as magnetic flux is destroyed there by the dissipation, the flux bundles try to merge over time (on this reconnection, see the Section Further Reading). The difference on these forms between cases with $\nu_0 = 0.1$ and 1 is the thickness of the vortex sheets and current layers; otherwise the evolution is the same. In the kinematic case the morphology is different. The magnetic flux is conserved, so the energy, which does transfer from the

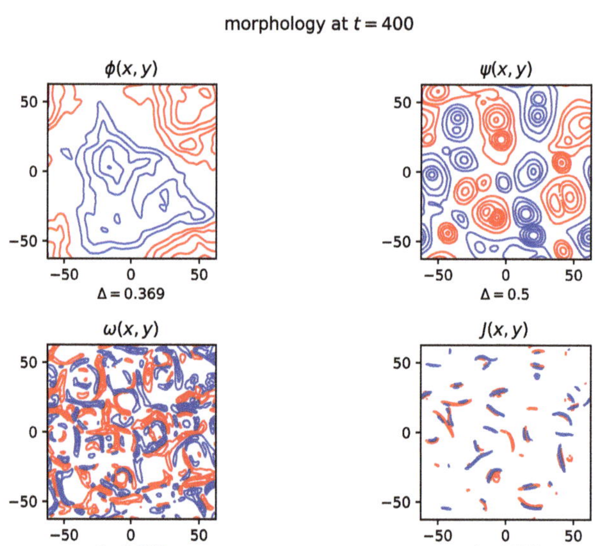


**Figure 3.10.** Spatial morphology for the magnetic MHD case, showing the developed phase at $t = 400$. With the magnetic disturbances controlling the dynamics, the flux $\psi$ (magnetic stream function) forms individual 'wires' with the vorticity $\varpi$ and current $J$ expelled into interstitial sheet structures. The potential $\phi$ indicates eddies at the largest available scale.

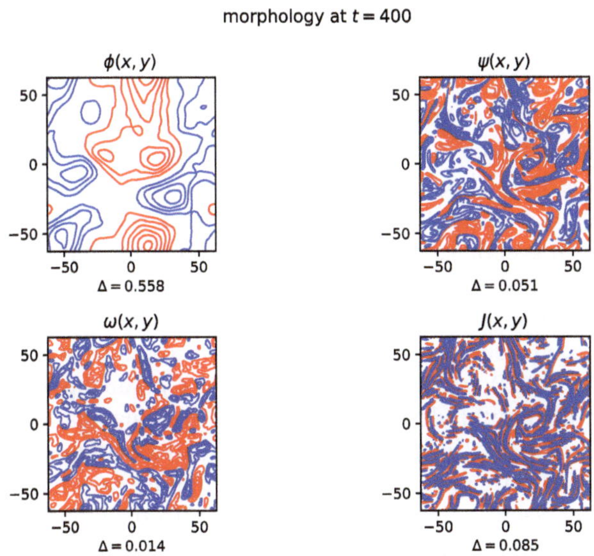


**Figure 3.11.** Spatial morphology for the kinematic MHD case, showing the developed phase at $t = 400$, with the same format as in the magnetic case in the previous figure. The flow eddies are in control of the dynamics, but the coupling to the magnetic disturbances prevents the vorticity from forming individual coherent structures. Compare with the right frames of figure 3.3. The magnetic field and current form sheet structures but not individual wires.

flows into the magnetic field, is constrained to be more at small scale. The manifestation of this is the sheet-like structures in both $\psi$ and $J$. The energy cannot go into the double-wire structures seen in the magnetic case, so the current does not form layers between wires. The vorticity no longer forms coherent vortices in either of these cases. In the magnetic case vorticity is excluded from the flux bundles. In the kinematic case there is still enough nonlinear exchange between the fluid and magnetic energy that the relaxation of enstrophy conservation makes the magnetic stresses function as a dissipation channel at late times, accelerating decay of fluid energy. The tendency to collect into larger structures is still present in the spatial morphology, but it does not result in an inverse cascade and lack of decay in $E_e$.

### 3.3.3 Equipartition in the MHD model

The MHD model no longer conserves fluid (ExB) energy separately, due to the presence of the field line bending effect (right-hand side of Eq. (3.69)), which exchanges energy with the magnetic field. Conversely, the static force term in the Ohm's law conserves flux (and $A_a$) but not energy, since to get that it is multiplied by $J$ (right-hand side of Eq. (3.70)). These two act as a conservative exchange, as can be seen from Eq. (3.61). This is why only the total energy $E_T$ is conserved by the nonlinear interactions. In the absence of $d_e$, the other quantities $A_a$ and $C_H$ are also conserved except for dissipation. These are all involved when, if as usual in natural cases, the fluid and magnetic energies are started or driven at disparate amplitudes.

Since we run a lot of different cases to study the effect of not only dissipation and the disparate start but also the finiteness of $d_e$, we can run them with small domains and lower resolution, which may not be fully converged but is well enough to see the effects of changing both $d_e$ and $\psi_0$ in the kinematic cases. These sub-cases are run with $N = N_x = N_y = 64$ and $K = 0.1$, with the other quantities set as before. To make these points it is sufficient to run up to $t = 200$ as the initial decay phase is the main one.

In the magnetic case ($\psi_0 = 1$ and $\phi_0 = 0$) we do find some evolution of both $A_a$ and $C_H$ with time, although in $A_a$ it is minimal. Averaging over $100 < t < 200$ we find $A_a = 0.494 \pm 0.0021$ from a start at 0.5, and $C_H$ starting at 0 and oscillating about zero with an amplitude of about $10^{-3}$. The equal-start case ($\psi_0 = \phi_0 = 1$) shows $A_a = 0.494 \pm 0.0021$ and $C_H$ starting at –0.015 evolving through zero. The kinematic case ($\psi_0 = 0.1$ and $\phi_0 = 1$) shows $A_a$ dropping by a factor of about 2 (see the energy transfer Section, below). Varying $d_e$ up to 0.25 and down to 0.0025 we find the strongest effect on $A_a$, indicating that it is conserved in the limit that $\psi \gg d_e^2 J$, as expected. The time traces for the equal-start case together with two variants setting $\nu_0 = 0$ and $d_e = 0.25$ and 0.0025 are shown in figures 3.12, 3.13, and 3.14, respectively. Here we see clearly that $A_a$ and $C_H$ are conserved in the limit of no dissipation and small $d_e$ but are affected slightly by larger $d_e$ and more strongly by dissipation.

The sharp drop in the fluid energy seen in the equal-start case with dissipation is explained by the energy transfer spectra, to which we now turn. For these we restore the $256 \times 256$ grid with $K = 0.05$ as in the other cases shown in this Chapter.

The normalised transfer from fluid to magnetic energy is given by

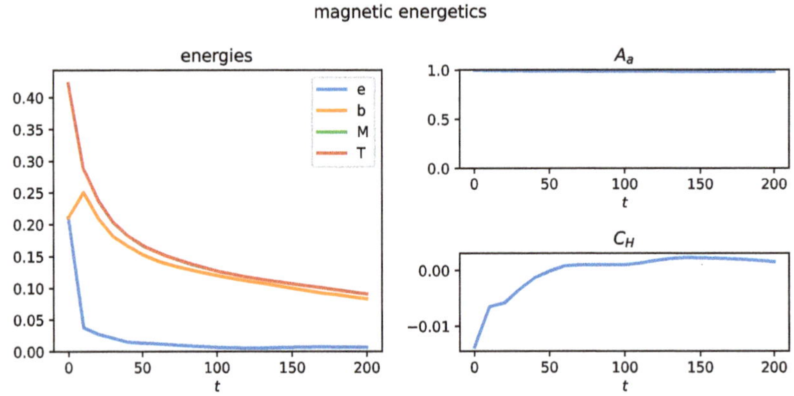


**Figure 3.12.** Time traces of fluid energy ($E_e$), magnetic energy ($E_b$), and the total ($E_T$), and the total without measuring the $d_e^2 J$ term ($E_M$), for the $64 \times 64$ equal-start case with $d_e = 0.025$. The fluid energy drops sharply in the beginning but then the two components decay together (see the following figures with energy transfer spectra). Also shown are the half-squared flux ($A_a$), which decays minimally, and the cross helicity ($C_H$), which is affected by the dissipation.

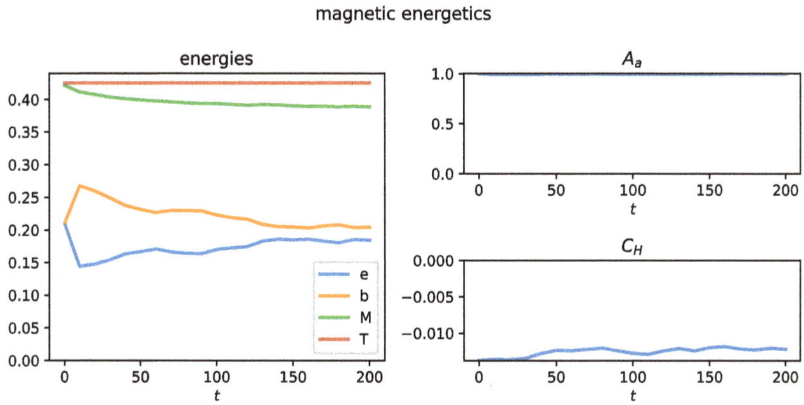


**Figure 3.13.** Time traces of fluid energy ($E_e$), magnetic energy ($E_b$), and the total ($E_T$), and the total without measuring the $d_e^2 J$ term ($E_M$), for the 64 × 64 equal-start case without dissipation and with $d_e = 0.25$. The half-squared flux ($A_a$) and cross helicity ($C_H$) are also shown. In this case the latter is conserved approximately, with some finite evolution. Note that especially $C_H$ is not conserved by the electron kinetic energy nonlinearity.

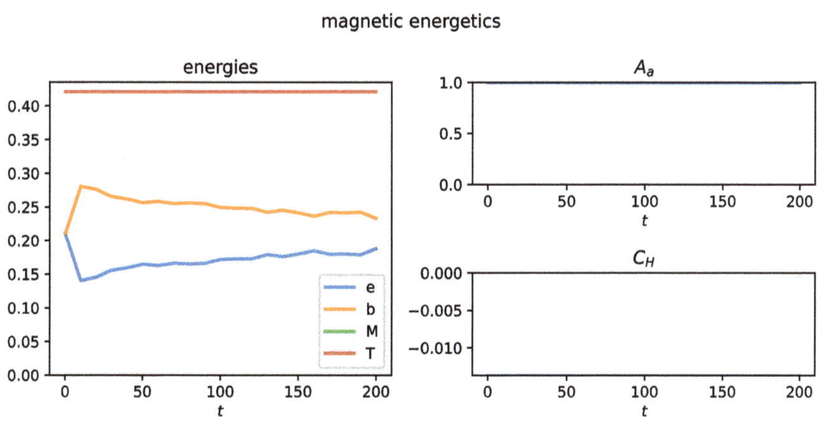


**Figure 3.14.** Time traces of fluid energy ($E_e$), magnetic energy ($E_b$), and the total ($E_T$), and the total without measuring the $d_e^2 J$ term ($E_M$), for the 64 × 64 equal-start case without dissipation and with $d_e = 0.0025$. The half-squared flux ($A_a$) and cross helicity ($C_H$) are also shown. In this case, with the electron kinetic energy artificially small, the latter is conserved to high accuracy.

$$T_J = \frac{1}{2E_T}\mathscr{V}^{-1}\int d\mathscr{V}\ \phi[-\psi, J] = \frac{1}{2E_T}\mathscr{V}^{-1}\int d\mathscr{V}(-J)[-\psi, \phi], \tag{3.75}$$

from Eq. (3.61). Similarly, the normalised transfer from magnetic to fluid energy is given by

$$T_A = \frac{1}{2E_T}\mathscr{V}^{-1}\int d\mathscr{V}(-\phi)[-\psi, J], \tag{3.76}$$

(here, '$J$' for transfer exchange over the current and '$A$' for Alfvén energetics including the transport term $[-\psi, \phi J]$). It becomes interesting to note that while these two expressions are equal under the integral, they are not necessarily equal pointwise and can even have different spectra. The difference is the transport term $[-\psi, \phi J]$ given in Eq. (3.60). So we measure them both. We also need the fluid energy transfer within the vorticity dynamics,

$$T_E = \frac{1}{2E_T}\mathscr{V}^{-1}\int d\mathscr{V}\ \phi[\phi, \varpi], \tag{3.77}$$

to compare with the Alfvénic transfer (here, '$E$' for ExB energy transfer, which vanishes in total but has a finite spectrum for transfer between scales). Their isotropic spectra, done as in Eqs (3.37)–(3.39), are

$$T_J(k) = \frac{1}{2E_T}\sum_{\mathbf{k}\in\mathscr{K}} \mathrm{Re}\big[(-J_\mathbf{k})^*([-\psi, \phi])_\mathbf{k}\big], \tag{3.78}$$

for $T_J(k)$ and similarly for $T_A(k)$ and $T_E(k)$. These are to be compared with the normalised dissipation in each equation, developed similarly from

$$\Gamma_w = \frac{1}{2E_T}\mathscr{V}^{-1}\int d\mathscr{V}\ \nu_\perp \mid \nabla_\perp\varpi \mid^2 \quad \Gamma_j = \frac{1}{2E_T}\mathscr{V}^{-1}\int d\mathscr{V}\ \nu_\perp \mid d_e\nabla_\perp^2 J \mid^2. \tag{3.79}$$

These quantities are shown in figures 3.15, 3.16, and 3.17, for the magnetic, kinematic, and equal-start cases. In each figure, the left frame shows the fluid energetics and the right frame shows the magnetic energetics. The dissipation channels in each equation are shown to be negative making the balance visible. The format is called symmetric-log, in which a small region near zero is done with a linear scale, which then switches to the logarithmic one at $10^{-7}$ of the maximum. This way, the transition from positive to negative values is very visible, and the details of the zero crossing are not significant.

Comparing these shows that the magnetic energy transfers to the flow at small scale but takes energy out of the flow at large scale. It is important to note that the sums over all modes for $T_A$ and $T_J$ are equal to machine round-off error, though looking at the curves can render that a little deceptive. For both fluid and magnetic energy the balance between dissipation and transfer-in through the MHD coupling is very close (the temporal decay is slow after initial adjustment over the first $0 < t < 50$ interval). The intra-fluid energy transfer ($T_E$) is subdominant at all but the largest scales. It plays the largest role for the kinematic case, where its transfer to large scales balances the Alfvénic losses there. The transport term $[-\psi, \phi J]$ represents an interscale transfer so that most of the exchange through $T_J$ with the magnetic energy is at small scale. In all three cases the balance between magnetic nonlinearity and dissipation is close. Note how the presence of $T_A$ drives $T_E$ positive at high $k$ even though $T_E$ remains subdominant. Even in the kinematic case the magnetic nonlinearity is stronger. Changing that would require much smaller initial ratios of $E_b/E_e$. The main difference in the magnetic case is that the positive spectra range for $T_A$ is wider, reflecting its overall positive sign for $E_b > E_e$.

This is why the decay dynamics of the hydrodynamic and MHD cases is so different. In the kinematic case the fluid transfer has its inverse effect, transferring

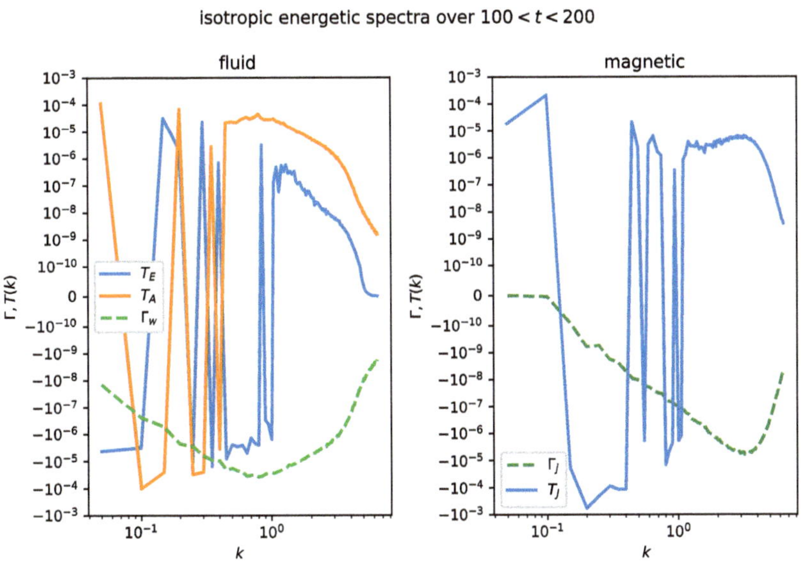


**Figure 3.15.** Fluid (left) and magnetic (right) energetics for the magnetic case, with $A_0 = 1.0$ for $\psi$ and 0 for $\phi$. The curves are in symmetric-log form, as some of them are of indeterminate sign. The dashed curves are subgrid dissipation in each equation ($\Gamma_w$ and $\Gamma_j$, respectively). The solid curves are the transfer terms ('$A$' for Alfvén, '$E$' for ExB fluid, and '$J$' for the current). These spectra in this and the next two figures are compared with a picture of the energetics in an MHD system, as explained in the text. Note the width of the Alfvén transfer $T_A$ as an effective drive for fluid energy, as its positive region is widest for this case.

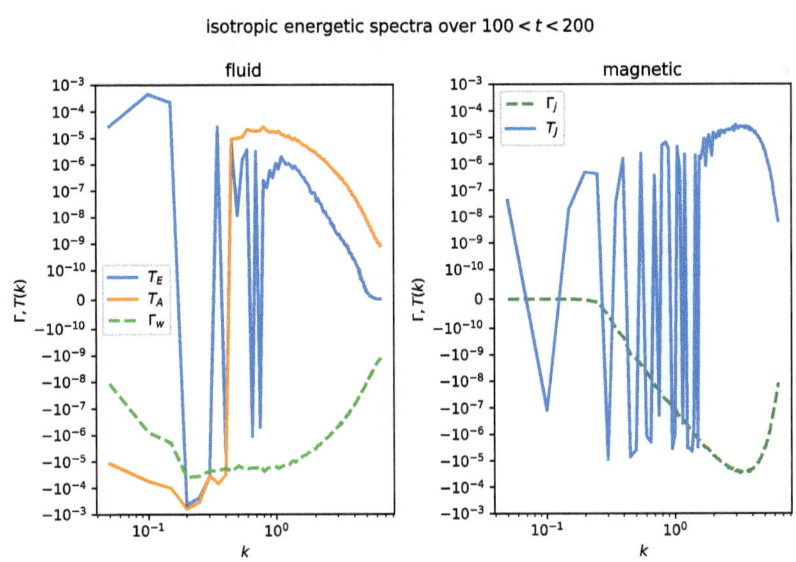


**Figure 3.16.** Fluid (left) and magnetic (right) energetics for the kinematic case, with $A_0 = 0.1$ for $\psi$ and 1.0 for $\phi$. The curves are in symmetric-log form, as some of them are of indeterminate sign. The dashed curves are subgrid dissipation in each equation ($\Gamma_w$ and $\Gamma_j$, respectively). The solid curves are the transfer terms ('$A$' for Alfvén, '$E$' for ExB fluid, and '$J$' for the current). Note the stronger fluid energy transfer $T_E$ to large scale, but even here the magnetic nonlinearity controls dissipation of fluid energy.

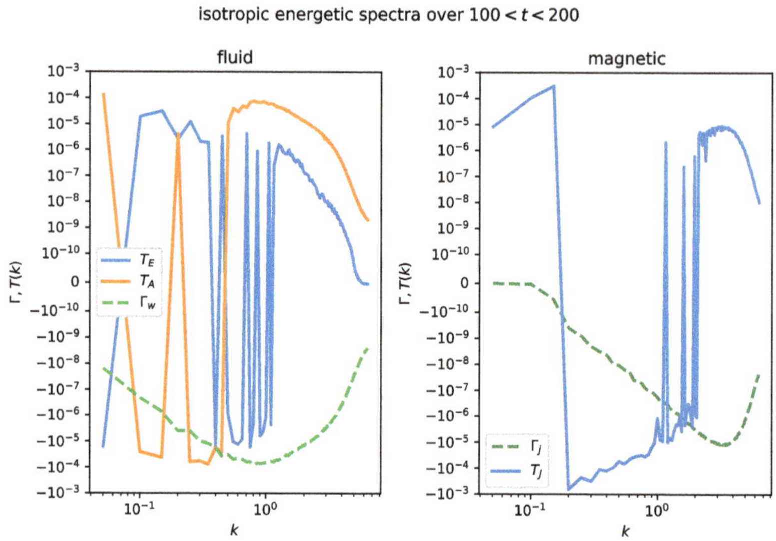


**Figure 3.17.** Fluid (left) and magnetic (right) energetics for the equal-start case, with $A_0 = 1.0$ for both $\psi$ and $\phi$. The curves are in symmetric-log form, as some of them are of indeterminate sign. The dashed curves are subgrid dissipation in each equation ($\Gamma_w$ and $\Gamma_j$, respectively). The solid curves are the transfer terms ('*A*' for Alfvén, '*E*' for ExB fluid, and '*J*' for the current). Note the strength of $T_J$ as an effective mid-range sink for magnetic energy.

from small to large scale, and the magnetic transfer is in balance with dissipation at all scales. In the magnetic case, the transfer out to the fluid energy is larger for the intermediate scales, indicating that this is how the fluid energy is maintained. In the equal-start case this feature is even more pronounced since the energies are maintained closer together during overall decay. The dominance of the MHD exchange transfer over the pure-fluid processes is the reason for the sharp drop in fluid energy at early times in all these cases. Equilibration stops when the fluid and magnetic energies are in mode structure balance, decaying at a similar rate. The decaying state is then characterised by a few large-scale fluid eddies, with small-scale magnetic fluctuations with the current concentrated into thin sheets.

Several of these issues, how the magnetic helicity conservation constrains both cascade and equilibration of the magnetic energy, how the magnetic nonlinearity changes even the fluid energy cascade tendency from inverse to direct, how the synergy between the nonlinearities works when the content of the fluid and magnetic energy are comparable, and why the presence of dissipation at high $k$ is required for turbulence to produce a physically realistic spectrum, will be addressed again through the study of absolute equilibrium ensembles in Chapter 5.

## 3.4 2D electron MHD turbulence

The usual expression for the MHD electric field—a statement that the Lorentz force on every species vanishes since other terms are considered to be small—holds only if the drift velocity represented by the current is small compared with the fluid velocity. If this is not the case, then the magnetic field lines are frozen into the fluid of the electrons, not the ions. As pointed out by Lighthill (1960) this makes more sense because the electrons are the charge carriers. The extreme case is the one with static ions and mobile electrons, such as the classical model for metals. This defines the electron MHD (EMHD) case.

Following Lighthill we consider a resistive Ohm's law

$$\mathbf{E} = \eta \mathbf{J} + \frac{1}{n_e e}\mathbf{J} \times \mathbf{B} \qquad \eta = \frac{m_e \nu_e}{n_e e^2}, \tag{3.80}$$

where $\eta$ is the resistivity due to Coulomb collisions between electrons and ions. Solving this equation for $\mathbf{J}$ gives

$$\mathbf{J} = \frac{1}{\eta_*^2}(\eta \mathbf{E} - n_e e \mathbf{E} \times \mathbf{B}), \tag{3.81}$$

where

$$\eta_*^2 = \eta^2 + \frac{B^2}{n_e^2 e^2}. \tag{3.82}$$

If the first term is dominant, this gives the simple form corrected by the Hall effect current. However, in a plasma the two terms track the Coulomb collision frequency $\nu_e$ and the electron cyclotron frequency $\Omega_e$, respectively, and the latter is usually far larger. In that case the current is simply the electron charge density $n_e e$ times the ExB velocity $\mathbf{v}_E$. The vision of the dynamics when the electrons are so fast the ions cannot keep up is defined as electron MHD and was introduced by the review by Kingsep *et al* (1990). The basic interest in it followed from the prospect that a magnetic field could be transported by a current. The motion given by the ExB velocity of the electrons makes sense in a plasma only if the ions are unmagnetised, with time scales faster than the ion gyrofrequency. Below, when considering Hall MHD with both species active we will find additional limitations.

Turbulence in this model was considered by Biskamp *et al* (1996). The key step is to treat the current as a velocity in the same form used above for two-dimensional MHD, namely, to keep the variability of the magnetic field component parallel to $\hat{\mathbf{z}}$,

$$\mathbf{B} = I\hat{\mathbf{z}} + \nabla\psi \times \hat{\mathbf{z}} \qquad \mu_0 \mathbf{J} = \mu_0 J \hat{\mathbf{z}} + \nabla I \times \hat{\mathbf{z}} \qquad \mu_0 J = -\nabla_\perp^2 \psi. \tag{3.83}$$

We envisage a background magnetic field $I_0 \gg I$ such that $I$ and $\psi$ represent small fluctuations, so this in principle represents a compressible magnetic field, but the electron flow is incompressible since both $\mathbf{B}$ and $\mathbf{J}$ are divergence free. The equations of electron MHD are the same Faraday's and Ampère's laws as in Eqs (3.50) and (3.51), but now with an Ohm's law for electrons,

$$\mathbf{E} = \frac{1}{n_e e}\left(\mathbf{J} \times \mathbf{B} - \nabla p_e\right) + \frac{m_e}{n_e e^2}\frac{d_e \mathbf{J}}{dt}, \tag{3.84}$$

in which the velocity is given by the current. Herein, the resistivity is neglected as in the previous cases.

The normalisation scheme is similar to the MHD case, except we now have a non-arbitrary coefficient due to the electron flow and inertia. The quantity $\psi + (m_e/n_e e^2)J$ due to the inertial correction to $\mathbf{E}$ is evaluated using Eq. (3.83), yielding

$$\psi + \frac{m_e}{n_e e^2}J = (1 - d_e^2)\nabla_\perp^2\psi, \tag{3.85}$$

where $d_e$ is the electron skin depth given by

$$d_e^2 = \frac{m_e}{\mu_0 n_e e^2}. \tag{3.86}$$

This makes $d_e$ the natural space scale, and this time it is not an artificial construct for numerics as we used it in the MHD case. The advective time derivative for electrons is given by

$$\frac{d_e}{dt} = \frac{\partial}{\partial t} - \frac{1}{n_e e}\mathbf{J} \cdot \nabla. \tag{3.87}$$

This sets the time normalisation

$$\frac{\partial}{\partial t} \leftrightarrow \frac{I_0}{n_e e d^2} = \frac{e I_0}{m_e}\frac{m_e}{n_e e^2 d^2} = \Omega_e \frac{d_e^2}{d^2}. \tag{3.88}$$

The inverse of this was referred to as the whistler time by Biskamp (1996), noting that we are still free to set $d$. In our case, however, we set $d = d_e$ to eliminate the coefficients, which also sets $\Omega_e^{-1}$ as the time scale.

The ratio $\Omega_e/\Omega_i = m_i/m_e$ is simply the ion/electron mass ratio. So for time scales longer than about $10^3$ the dynamics is not really safely faster than the ion gyrofrequency, and we will see in Chapter 6 what difficulties that entails. Nevertheless, it is instructive to examine the differences in results produced by this model and the native MHD one. The effective vorticity is what matters the most, and we will check to see whether this is larger than $\Omega_i$ in these units.

Putting the representation in Eqs (3.83) into Faraday's law (Eq. 3.50), this gives the normalised equations

$$\frac{\partial F_I}{\partial t} + [I, F_I] - [\psi, J] = 0 \qquad \frac{\partial F_\psi}{\partial t} + [I, F_\psi] = 0, \tag{3.89}$$

in which

$$F_I = \left(1 - \nabla_\perp^2\right)I \equiv I + U \qquad F_\psi = \left(1 - \nabla_\perp^2\right)\psi \equiv \psi + J, \tag{3.90}$$

with two components for Ampère's law given by

$$U = -\nabla_\perp^2 I \qquad J = -\nabla_\perp^2\psi, \tag{3.91}$$

serving as field equations.

Solving for $\psi$ proceeds according to

$$\psi = \left(1 - \nabla_\perp^2\right)^{-1} F_\psi \qquad \text{then} \quad J = F_\psi - \psi, \tag{3.92}$$

that is, we obtain $J$ without having to apply $\nabla_\perp^2$ directly. The same method is used to get $I$ and $U$.

### 3.4.1 Electron MHD energetics

The energy theorem for Eq. (3.89) is constructed by multiplying, respectively, by $I$ and $J$, to form

$$\begin{aligned} &\frac{1}{2}\frac{\partial}{\partial t}(I^2 + |\nabla_\perp I|^2 + |\nabla_\perp \psi|^2 + J^2) + \nabla \cdot \left(-I\nabla_\perp \frac{\partial I}{\partial t}\right) \\ &\quad + \left([-\psi, IJ] + [I, IU] + \frac{1}{2}[I, J^2]\right) = 0, \end{aligned} \tag{3.93}$$

following the steps leading to Eq. (3.63). As in the MHD case there are more pieces but only two are independent, $E_i$ and $E_b$, determined by the field variables $I$ and $\psi$ (here $I$ instead of $\phi$). These are given by

$$E_i = \mathcal{V}^{-1} \int d\mathcal{V} \, \frac{1}{2} I F_I \qquad E_b = \mathcal{V}^{-1} \int d\mathcal{V} \, \frac{1}{2} J F_\psi. \tag{3.94}$$

The total energy

$$E_T = E_b + E_i, \tag{3.95}$$

(alone among quadratic quantities, as in the MHD case) is conserved except for dissipation, with

$$\Gamma_- = \frac{1}{2E_T} \mathcal{V}^{-1} \int d\mathcal{V} \, \nu_\perp\left(|\nabla_\perp^2 I|^2 + |\nabla_\perp^2 J|^2\right), \tag{3.96}$$

in the EMHD version of Eq. (3.47). We also measure the half-squared amplitudes $A_J$, $A_U$ and $A_I$,

$$A_J = \mathcal{V}^{-1} \int d\mathcal{V} \, \frac{1}{2} J^2 \quad A_I = \mathcal{V}^{-1} \int d\mathcal{V} \, \frac{1}{2} I^2 \quad A_U = \mathcal{V}^{-1} \int d\mathcal{V} \, \frac{1}{2} U^2, \tag{3.97}$$

in the diagnostics.

### 3.4.2 The base case for EMHD decaying turbulence

As in the MHD case we run with the grid and scales as in the fluid case for close comparison. Note that the equations for EMHD are nearly the same as for MHD except for the $(1 - \nabla_\perp^2)$ on $I$ instead of $\nabla_\perp^2$ on $\phi$, while in this case $I$ serves the same role as stream function for the electron flow as $\phi$ does for the MHD ExB mass flow.

The responses at low $k$ are correspondingly slower, since $(1 + k^2)$ is much larger than $k^2$ there. The structures in $I$ remain mostly at small scale, where the strength of the nonlinearity $[I, U]$ is larger. This slows the energetic cascade for the low $k < 1$ side. We find running with $\nu_0 = 10^{-2}$ to be sufficient judged by the turn-down in the spectra for $U$ and $J$.

The time traces are shown in figure 3.18 with the same sampling scheme as in the fluid and MHD cases. The initialisation is purely within $\psi$ and $J$, giving the $E_b$ energy component, with $I = U = 0$. The component $E_i$ rises from zero to find a level somewhat below $E_b$, after which both decay together, even though the initial equilibration transient is much weaker. The energetic error is below $10^{-9}$ throughout. The specific decay rate is slower than for the MHD case, with the nonlinear magnetic stresses less effective in generating electron flow (stream function $I$) and vorticity ($\varpi = -U$ for this case) due to the governance of the response by $(1 + k^2)$ rather than $k^2$ as the effective coefficient.

The isotropic spectra are shown in figure 3.19 for the EMHD case, for the time points $t = 0$, 100, and 400. The strong turn-down at high $k$ is apparent in all quantities, especially $U$ and $J$. The spectrum of $J$ (the parallel current) is the flattest and hence puts the most pressure on resolution, but $\nu_0 = 10^{-2}$ is sufficient to contain the cascade. The spectra of $E_i$ and $E_b$ multiplied by $k^{7/3}$ show a power-law form of $k^{-7/3}$ for $E_T$, which was also found by Biskamp (1996). The energy components are equipartitioned in the $k \sim 1$ range which translates to lower $E_i$ at low $k$ due to its shallower cascade spectrum.

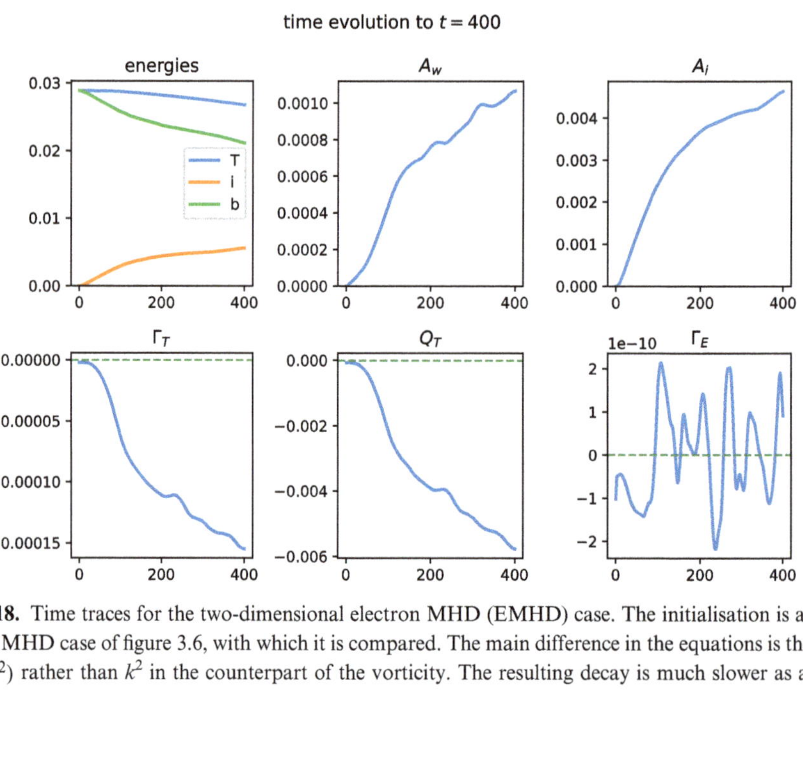


**Figure 3.18.** Time traces for the two-dimensional electron MHD (EMHD) case. The initialisation is as in the magnetic MHD case of figure 3.6, with which it is compared. The main difference in the equations is the factor of $(1 + k^2)$ rather than $k^2$ in the counterpart of the vorticity. The resulting decay is much slower as a result.

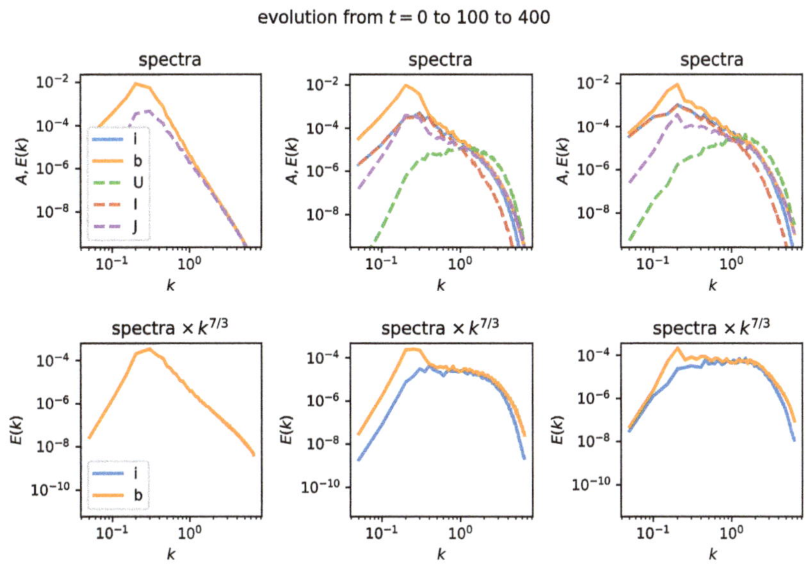


**Figure 3.19.** Isotropic spectra for the electron EMHD case, showing the initial state and early phases (left frames) developing into a more self-similar one (right) The curves give the energies and amplitudes (shown dashed) with the same format as in figure 3.8, with which it is compared. With the overall decay weaker, the energy in both channels is spread out more effectively. But the Kolmogorov curve for this case is $k^{-7/3}$ not the $k^{-5/3}$ one from before. This is the effect of the $(1 + k^2)$ rather than $k^2$ in the counterpart of the vorticity.

The spatial morphology in figure 3.20 at $t = 400$ shows the main difference to the MHD case, that is the lack of large-scale structure in $I$ here as opposed to $\phi$ there (the stream functions for the respective flows). Here, the vorticity shows a more intermittent structure than $J$, which is opposite to the MHD case. The parallel magnetic flux $\psi$ shows less of a tendency to form the double-wire pairs with current layers between them. Looking at the equation sets at low $k$, the equations are isomorphic except for the $(1 + k^2)$ governing $\partial I/\partial t$ here in contrast to the $k^2$ for $\partial\phi/\partial t$ there. We will find the same form as the EMHD model in some of the fluid models to be encountered later, which contain coupling processes between the vorticity $\varpi$ and another variable which depends on the stream function $\phi$ of the flow. This does not do justice to EMHD as a subject, which is very broad but much of which is orthogonal to our interest in understanding the basic difference among models that can be caused by a single term. Interested readers can consult the references given in the Section Further Reading.

## 3.5 2D Hall MHD turbulence

Having explored the MHD and EMHD models, we notice that the pairs of variables in those may be combined naturally into another model with four dependent variables, often called a four-field model. Both the magnetic field and the ion mass flow depend on two scalar variables, making a total of four. The extension is to keep the parallel flow for the ions as well as the parallel current for the electrons.

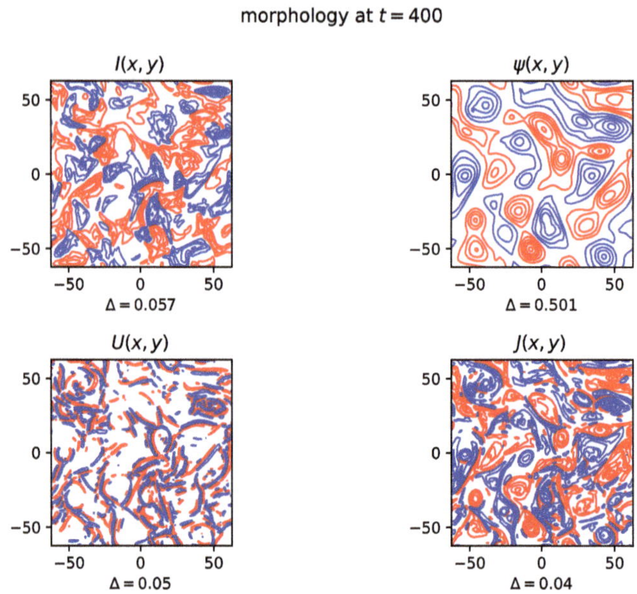


**Figure 3.20.** Spatial morphology for the electron EMHD case, showing the developed phase at $t = 400$, with the same format as in the MHD case in figure 3.10. The quantities $I$ and $U$ here correspond to $\phi$ and $\varpi$ there. The sheet structure is now more pronounced in $U$ than in $J$, with $\psi$ not forming wire structures as effectively as in the MHD case.

In the 2D Hall MHD case we keep the motion of both the electrons and ions, doing them separately. We have two velocities, but one is given by the current as in EMHD. In terms of the vector quantities, we have $\mathbf{E}$ and $\mathbf{B}$ for the fields, $\mathbf{u}$ and $\mathbf{v}$ for the ion and electron velocities, respectively, and we also have the current, given simultaneously by $\mathbf{B}$ through Ampère's law and the relative drift velocity $\mathbf{u} - \mathbf{v}$. The same treatment for $\mathbf{B}$ as in EMHD is used, with $I$ and $\psi$ the dependent variables, and similarly for the ion velocity $V$ and $\phi$. The vector equations are (3.49)–(3.51) for the ion velocity, magnetic field, and current, with Eq. (3.52) replaced by an electron Ohm's law with their velocity and inertia

$$\mathbf{E} + \mathbf{v} \times \mathbf{B} = -\frac{1}{n_e e}\nabla p_e + \frac{m_e}{n_e e^2}\frac{d_e \mathbf{J}}{dt}. \tag{3.98}$$

For energetic consistency we still keep only $\mathbf{J} \cdot \nabla$ in the electron inertia term, whose advective time derivative is given by Eq. (3.87). As previously, we are neglecting changes to the density and pressure for both species. This is essentially an ordering on the dynamical plasma beta as well as the time scales:

$$\beta_e = \frac{\mu_0 p_e}{B^2}. \tag{3.99}$$

The dynamical beta is the square of the ratio between the isothermal sound speed and the Alfvén velocity.

The decomposition for the vectors is given by

$$\mathbf{B} = I\hat{\mathbf{z}} + \nabla\psi \times \hat{\mathbf{z}} \qquad \mathbf{u} = V\hat{\mathbf{z}} + \hat{\mathbf{z}} \times \nabla\phi, \tag{3.100}$$

$$\mu_0\mathbf{J} = \mu_0 J\hat{\mathbf{z}} + \nabla I \times \hat{\mathbf{z}} \qquad \mu_0 J = -\nabla_\perp^2\psi, \tag{3.101}$$

in which the forms for **B** and **J** are the same as in EMHD, namely Eq. (3.83). Now, there are four independent quantities, $I$ and $\psi$ for the magnetic field, and $V$ and $\phi$ for the ion velocity, with the electron velocity given by the ion velocity and the current. The EMHD model used just the first two, with the ion velocity neglected. Now, we have the two velocities on an equal footing in a coupled system.

The normalisation runs differently this time. The current is simultaneously a magnetic field strength divided by $\mu_0$ and length $d$, and also velocity times $n_e e$, namely the electron charge density. If we set these to be equal using the Alfvén velocity we get the native form for $d$:

$$\frac{B}{\mu_0 d} = n_e e v_A \implies d = \frac{v_A}{\Omega_i} \equiv d_i \qquad \text{noting that} \quad d_i^2 = \frac{m_i}{\mu_0 n_e e^2}, \tag{3.102}$$

using the definitions of $v_A$ and $\Omega_i$ given above (solve the definition of $v_A$ for $\mu_0$ and insert the result into this equation). This has the same form as a skin depth but is better thought of as an Alfvénic gyroradius, although it is called the ion skin depth in the literature. The natural time scale is the inverse ion gyrofrequency. If we normalise in terms of $d_i$ and $v_A$ and neglect the electron inertia corrections we would have the equations without any coefficients. As in the EMHD model with the electron skin depth $d_e$ used for $d$, this preferred scale $d_i$ appears only in setting the domain size. However, with the electron inertia corrections included in the model, $d_e^2$ appears as an additional coefficient. The ratio $d_i/d_e$ for proton ions is 42 and with deuterium 60, so it is more involved to keep both scales within the spectrum in a computation. For our purposes, however, we will run as in the MHD case with small $d_e$ placing $d = d_i$ roughly in the middle of the spectrum. The sole normalised parameter becomes

$$\frac{d_e^2}{d_i^2} = \frac{m_e}{m_i} \equiv \mu_e, \tag{3.103}$$

namely, the electron/ion mass ratio.

The normalised equations are

$$\frac{\partial F_I}{\partial t} + [\phi, I] + \mu_e[I, U] + [\psi, V - J] = 0, \tag{3.104}$$

$$\frac{\partial F_\psi}{\partial t} + \mu_e[\phi + I, J] - [\psi, \phi + I] = 0, \tag{3.105}$$

$$\frac{\partial \varpi}{\partial t} + [\phi, \varpi] + [\psi, J] = 0, \tag{3.106}$$

$$\frac{\partial V}{\partial t} + [\phi, V] + [\psi, I] = 0, \tag{3.107}$$

where

$$F_I = I + \mu_e U \qquad F_\psi = \psi + \mu_e J, \tag{3.108}$$

$$U = -\nabla_\perp^2 I \qquad J = -\nabla_\perp^2 \psi \qquad \varpi = \nabla_\perp^2 \phi, \tag{3.109}$$

are auxiliaries. Seen another way, the dependent variables are those under the $\partial/\partial t$, namely $F_I$, $F_\psi$, $\varpi$, and $V$, with $I$, $\psi$, and $\phi$ serving as field variables. The advance of $F_I$ and $F_\psi$, solution for $I$ and $\psi$, and recovery of $U$ and $J$ proceed as in the EMHD case.

### 3.5.1 Hall MHD energetics

The energy theorem for Eqs (3.104)–(3.107) is constructed by multiplying, respectively, by $I$, $J$, $-\phi$, and $V$ to form

$$\begin{aligned} &\frac{1}{2}\frac{\partial}{\partial t}\left(I^2 + |\nabla_\perp \psi|^2 + |\nabla_\perp \phi|^2 + V^2 + \mu_e J^2 + \mu_e |\nabla_\perp I|^2\right), \\ &\qquad -\nabla\cdot\left(\mu_e I \nabla_\perp \frac{\partial I}{\partial t} + \frac{\partial \psi}{\partial t}\nabla_\perp \psi + \phi \nabla_\perp \frac{\partial \phi}{\partial t}\right), \\ &+\frac{1}{2}[\phi, I^2 + V^2 - 2\phi\varpi] + [\psi, I(V-J) - \phi J] + \mu_e[I, IU] \\ &+\frac{1}{2}\mu_e[I, J^2] = 0, \end{aligned} \tag{3.110}$$

following the steps leading to Eq. (3.63).

### 3.5.2 Contrast to the MHD cases

We find that the Hall MHD case is almost identical in result to the MHD case, since $d_e$ is almost out of the spectrum. But the spectra show finite energetic content for $I$ and $V$ in the scales $k \sim 1$ (that is, $d_i$ scales) becoming larger for smaller scales. The vorticity spectrum peaks in this range as well, but for these cases the decay rate for Hall MHD is about the same as in the MHD case.

We highlight the case with initial state with only $\psi$ and $J$ nonzero, with $\psi_0 = 1$ as in the corresponding MHD case. The subgrid dissipation is kept at $\nu_0 = 1$ due to the stronger high-$k$ dynamics involving especially $I$ due to the coupling between $I$ and $J$ (see $[\psi, J]$ in Eq. (3.104) and $\psi$, $I$ in Eq. (3.105)). In all other respects the initial state is the same as in the MHD case, but the evolution now involves $I$ and $V$ as well as $\phi$, and we give special attention to whether these components acquire significant energy. The energetic quantities tracked are the four energy components

$$E_e = \mathcal{V}^{-1}\int d\mathcal{V}\,\frac{1}{2}(-\phi)\varpi \quad E_b = \mathcal{V}^{-1}\int d\mathcal{V}\,\frac{1}{2}JF_\psi, \tag{3.111}$$

$$E_i = \mathscr{V}^{-1}\int d\mathscr{V}\ \frac{1}{2}IF_I \quad E_v = \mathscr{V}^{-1}\int d\mathscr{V}\ \frac{1}{2}V^2, \tag{3.112}$$

and the total energy

$$E_T = E_e + E_b + E_i + E_v, \tag{3.113}$$

which is conserved except for dissipation, and the amplitudes tracked by

$$A_p = \mathscr{V}^{-1}\int d\mathscr{V}\ \frac{1}{2}\phi^2 \quad A_w = \mathscr{V}^{-1}\int d\mathscr{V}\ \frac{1}{2}\varpi^2 \quad A_J = \mathscr{V}^{-1}\int d\mathscr{V}\ \frac{1}{2}\mu_e J^2, \tag{3.114}$$

in the diagnostics. Note that this expression for $E_e$ is the same as in Eq. (3.72) as they differ by a periodic integral over a divergence, which vanishes. The dissipation of $E_T$ in the Hall MHD case is

$$\begin{aligned}\Gamma_- = \frac{1}{2E}\mathscr{V}^{-1}\int d\mathscr{V}\ \nu_\perp\Big[\Big(|\nabla_\perp\varpi|^2 + |\nabla_\perp^2 V|^2 + U^2\Big)\\ + \mu_e\Big(|\nabla_\perp U|^2 + |\nabla_\perp^2 J|^2\Big)\Big].\end{aligned} \tag{3.115}$$

The specific decay rate is given as in Eq. (3.48) but with these definitions of $E_T$ and $\Gamma_-$ in the energetics.

All run considerations are as in the MHD case, with $\nu_0 = 10^0$ for the subgrid dissipation. The time traces are shown in figure 3.21. Only $E_b$ is nonzero initially, but the other three rise into balance and then all four decay together. This takes longer for the two subdominant channels $E_i$ and $A_v$. The overall decay rate is similar to the MHD case, and the energetic error is small. The most obvious difference is that the increase in amplitude for $\phi$, here tracked by $A_p$, is less than in the corresponding MHD case in figure 3.6.

The isotropic spectra are shown in figure 3.22, for the Hall MHD case, for the time points $t = 0$, 50, and 400. The strong turn-down at high $k$ is apparent in all quantities. The strongest pressure on resolution is the magnetic field and current fluctuations, as in the MHD case. The spectrum of $I$ is small at low $k$ but becomes the second energy channel for $k > 1$. At large scale Hall MHD is like MHD but at small scale it is more like EMHD. The sharpest turn-down at high $k$ is in $V$, so that parallel flow energy does not really enter the dynamics.

The spatial morphology is shown in figure 3.23 at $t = 400$, for comparison to the MHD case. The visible evolution of $\psi$ and $\phi$ is as in the MHD case, but there are subtle form differences in the vorticity since much of it is in the $k > 1$ scale range. The form of $I$ and also of $V$ follows the vorticity as indicated by the spectrum: thin sheets around voids where the main magnetic structures are. The parallel current $J$ forms the same isolated sheets between those structures seen in the MHD case.

## 3.6 Compressibility in MHD

The assumption that flow is incompressible in MHD or Hall MHD is something we should check, especially in the latter case where we treat the changes $I \neq I_0$ as a finite

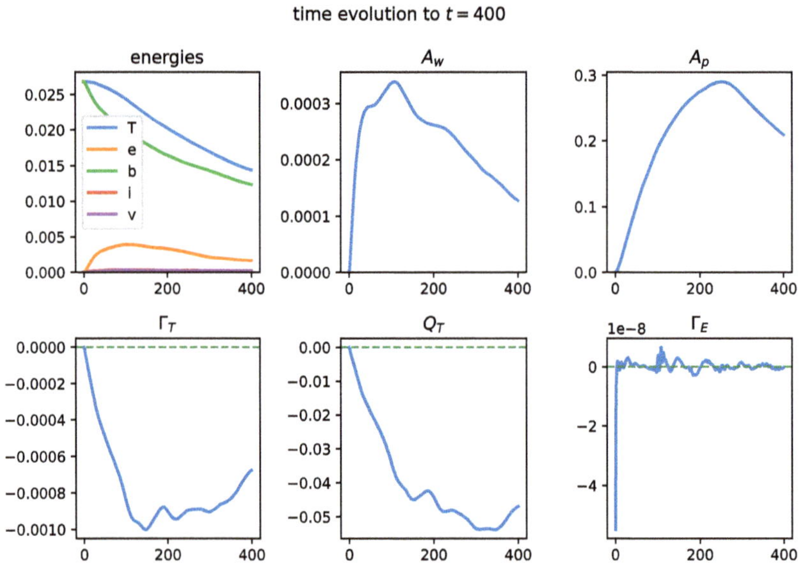


**Figure 3.21.** Time traces for the four-field Hall MHD case, with the same format as in the corresponding MHD case in figure 3.21. The fluid component ($E_e$) and the enstrophy $A_w$ rise slightly less and decay faster than in the MHD case, and the increase in amplitude of the potential ($A_p$) is less, with $A_p$ turning around to decay with the rest of the system. The energetic content in $I$ and $V$ remains rather small.

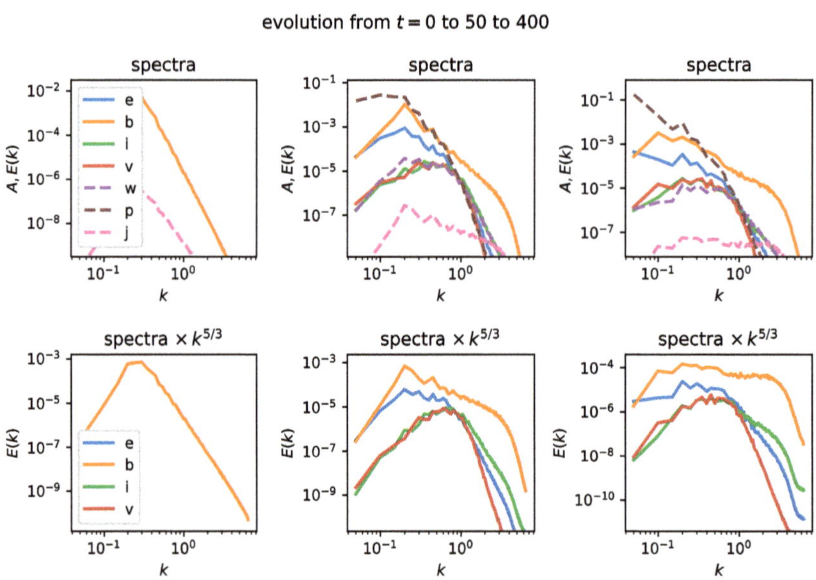


**Figure 3.22.** Isotropic spectra for the four-field Hall MHD case, showing the initial state and early phases (left frames) developing into a more self-similar one (right) The curves give the energies and amplitudes (shown dashed) with the same format as in figure 3.6, with which they are compared. The shapes of the spectra are more or less the same.

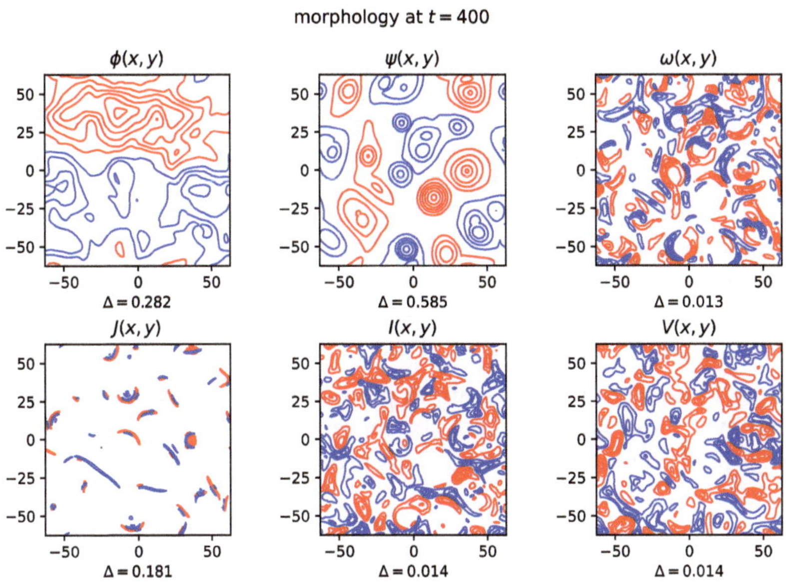


**Figure 3.23.** Spatial morphology for the four-field Hall MHD case, with the same format as in the corresponding MHD case in figure 3.10. The magnetic disturbances and current ($\psi$ and $J$) corresponding to MHD form a similar structure, while the parallel magnetic fluctuation $I$ and the parallel ion flow $V$ both have a structure closer to the ion vorticity $\varpi$ than to $\psi$ and $J$.

quantity. Assume *a priori* that the mass density can be treated as constant, i.e. flow compression should be small, we start with ideal MHD and find equations for the flow divergence and the magnetic field strength by taking the divergence of Eq. (3.49) and contracting Eq. (3.50) with $\mathbf{B}/\mu_0$, using Eq. (3.52), to find

$$\rho\frac{\partial}{\partial t}(\nabla\cdot\mathbf{u}) + \nabla\cdot\left(\rho\mathbf{u}\cdot\nabla\mathbf{u} - \frac{1}{\mu_0}\mathbf{B}\cdot\nabla\mathbf{B}\right) = -\nabla^2\left(p + \frac{B^2}{2\mu_0}\right), \tag{3.116}$$

$$\left(\frac{\partial}{\partial t} + \mathbf{u}\cdot\nabla\right)\frac{B^2}{2\mu_0} + \frac{B^2}{\mu_0}(\nabla\cdot\mathbf{u}) = \frac{1}{\mu_0}\mathbf{BB}\colon\nabla\mathbf{u}. \tag{3.117}$$

If the background scale length is large compared with the dynamical scales and the flow is compressional ($\nabla\cdot\mathbf{u}\sim\mathbf{u}\cdot\nabla$) then the terms with $B^2/2\mu_0$ and $(\nabla\cdot\mathbf{u})$ are the main ones, it follows that

$$\frac{\partial^2}{\partial t^2}\sim k_\perp^2 v_A^2, \tag{3.118}$$

where $k_\perp$ is the scale of $\nabla$ in the dynamics. Therefore, if the dynamics is forced by other phenomena, and $\partial/\partial t \ll k_\perp v_A$, it follows that the $\partial/\partial t$ terms are small and $B^2/2\mu_0$ and $(\nabla\cdot\mathbf{u})$ are determined by balances. If the parallel gradient against the background magnetic field is small compared with $k_\perp$ then

$$\nabla \cdot \mathbf{u} \approx 0 \qquad \nabla \frac{B^2}{2\mu_0} \approx -\nabla p. \tag{3.119}$$

This condition is called perpendicular force balance. Practically all the dynamics involved in turbulence in magnetised plasmas is well within this regime. We will see this more when systematically exploring fluid drift dynamics in Chapter 7. For now, we can take the Hall MHD and replace $I$ with $p$ appropriately everywhere it occurs. Then, using the factor of $\beta_e$ which appears when the gradient balance is normalised, we re-normalise the equations around $c_s$ instead of $v_A$, where

$$c_s^2 = \beta_e v_A^2, \tag{3.120}$$

with $\beta_e$ given in Eq. (3.99). The quantities $\phi$, $\varpi$, $V$, $\psi$, and $J$ get re-normalised while $I$ is replaced by $-p$ and the skin depth correction to $I$ is dropped because it does not appear in a fluid continuity equation. This produces the normalised equation set

$$\frac{\partial \varpi}{\partial t} + [\phi, \varpi] + [\psi, J] = 0, \tag{3.121}$$

$$\frac{\partial F_\psi}{\partial t} + \mu_e[\phi, J] - [\psi, \phi - p] = 0, \tag{3.122}$$

$$\frac{\partial p}{\partial t} + [\phi, p] - [\psi, V - J] = 0, \tag{3.123}$$

$$\frac{\partial V}{\partial t} + [\phi, V] - [\psi, p] = 0, \tag{3.124}$$

with $F_\psi$, and $\varpi$ and $J$, defined in Eqs (3.108) and (3.109), respectively. This system is referred to as the normalised four-field fluid drift model. We have defined it heuristically here while side-stepping the role of finite ion temperature. The systematic treatments will occur in Chapters 6 and 7. We include it here because the properties of this type of dynamics form the centrepiece of turbulence in magnetised plasmas. If the turbulence is driven by thermal/pressure gradients the demand is $\beta_e \ll 1$, and if it is driven by external flows the demand is $V_0 \ll v_A$ for the flow strength $V_0$ with an Alfvén velocity $v_A$ given by a background magnetic guide field. Furthermore, to use drifts we will require $\partial/\partial t \ll \Omega_i$ so that the ion fluid mass flow is solved for via the Lorentz force and not the inertia. In this case the inertia is a correction describing the polarisation drift introduced in the first Chapter. This means that Hall MHD *per se* loses relevance and is replaced by the four-field fluid drift model under perpendicular force balance in anything that will concern us herein.

## Further Reading

Early papers on computational 2D fluid turbulence include G K Batchelor 'Computation of the energy spectrum in homogeneous two-dimensional turbulence. High-speed computing in fluid dynamics' *Phys. Fluids Suppl.* **11** (1969) 223;

S A Orszag 'Numerical methods for the simulation of turbulence' *Phys. Fluids Suppl.* **11** (1969) 250; D K Lilly 'Numerical simulation of two-dimensional turbulence' *Phys. Fluids Suppl.* **11** (1969) 240, and 'Numerical simulation of developing and decaying two-dimensional turbulence' *J. Fluid Mech.* **45** (1971) 395; J Herring, S Orszag, R Kraichnan and D Fox 'Decay of two-dimensional homogeneous turbulence' *J. Fluid Mech.* **66** (1974) 417.

Stronger computational capability led to the higher-resolution studies in the 1980s including demonstrations of the cascade process, intermittency, and long-lived vortex formation, such as: U Frisch and P-L Sulem 'Numerical simulation of the inverse cascade in two-dimensional turbulence' *Phys. Fluids* **27** (1984) 1921; J C McWilliams 'The emergence of isolated coherent vortices in turbulent flow' *J. Fluid Mech.* **146** (1984) 21.

The time step scheme used herein was given by G E Karniadakis, M Israeli and S A Orszag 'High-order splitting methods for the incompressible Navier–Stokes equations' *J. Comput. Phys.* **97** (1991) 414. They cite the 1970s precedent. The spatial bracket scheme is due to A Arakawa 'Computational design for long-term numerical integration of the equations of fluid motion: two-dimensional incompressible flow. Part I' *J. Comput. Phys.* **1** (1966) 119. The combination by V Naulin 'Aspects of flow generation and saturation in drift-wave turbulence' *New J. Phys.* **4** (2002) 28.

The 2D decomposition leading to 2D MHD turbulence was done for reduced MHD by H R Strauss 'Nonlinear, three-dimensional magnetohydrodynamics of noncircular tokamaks' *Phys. Fluids* **19** (1976) 134. The cascade dynamics was studied by A Pouquet 'On two-dimensional magnetohydrodynamic turbulence' *J. Fluid Mech.* **88** (1978) 1; S Orszag and C-M Tang 'Small-scale structure of two-dimensional magnetohydrodynamic turbulence' *J. Fluid Mech.* **90** (1979) 129. The methods used here are from D Biskamp and H Welter 'Dynamics of decaying two-dimensional magnetohydrodynamic turbulence' *Phys. Fluids B* **1** (1989) 1964.

References pertinent to magnetic helicity and cross helicity were given at the end of Chapter 2.

Electron MHD was reviewed by A S Kingsep, K V Chukbar and V V Yan'kov 'Electron magnetohydrodynamics' *Rev. Plasma Phys.* **19** (1990) 243. In 2D turbulence the essential step was to allow fluctuation of the guide field to get the electron flow. The methods used here are from D Biskamp, E Schwarz and J F Drake 'Two-dimensional electron magnetohydrodynamic turbulence' *Phys. Rev. Lett.* **76** (1996) 1264.

Hall MHD was coined and introduced by M J Lighthill 'Studies on magneto-hydrodynamic waves and other anisotropic wave motions' *Proc. R. Soc. Lond. A* **252** (1960) 397. A systematic study of the conservation laws and consistency principles, including the two-fluid fundamentals, electromagnetic momentum transfer, and the consequences of quasineutrality in Hall MHD was given by E A Witalis 'Hall magnetohydrodynamics and its applications to laboratory and cosmic plasma' *IEEE Trans. Plasma Sci.* **PS-14** (1986) 842. Two-dimensional Hall MHD was studied by R Fitzpatrick 'Scaling of forced magnetic reconnection in the Hall-magnetohydrodynamic Taylor problem' *Phys. Plasmas* **11** (2004) 937; and by D O Gómez, S M Mahajan and P Dmitruk 'Hall magnetohydrodynamics in a strong magnetic field' *Phys. Plasmas* **15** (2008) 102303. (These equation sets are like ours but without the finite-$d_e$ corrections.) The connection between Hall MHD and low frequency acoustic turbulence with adiabatic responses was made by the 2004 work by R Fitzpatrick cited above, and by N H Bian and D Tsiklauri 'Compressible Hall magnetohydrodynamics in a strong magnetic field' *Phys. Plasmas* **16** (2009) 064503.

**IOP** Publishing

Turbulence and Instabilities in Magnetised Plasmas, Volume 1
Fluid drift turbulence
**Bruce Scott**

# Chapter 4

# Driven/dissipative turbulence

## 4.1 Parallel dynamics along the guide field

So far we have considered basic two-dimensional systems. Parallel dynamics in these refers to the fluctuating magnetic field component which is self consistent to the rest of the turbulence. Flows and Alfvén responses move along the magnetic field, in general along either the assumed background field or the fluctuations. Parallel dynamics in a magnetised plasma most commonly refers to the third dimension, which in our context means parallel to the background magnetic field, often referred to as a guide field. This field is what makes the dynamics mostly two-dimensional in the first place. The reason is that the responses parallel to the guide field are so much faster than the turbulence itself that the wavelengths of fluctuations parallel to the field are correspondingly longer than those in the turbulence. In fact, the limiting case is that the parallel variation should go to zero if the Alfvén speed is arbitrarily large. Forcing of finite parallel wavelengths depends on the turbulence itself, which opens up the parallel wavenumber spectrum only to the extent that

$$k_{\|}v_A \sim \varpi, \tag{4.1}$$

that is, the finite turbulent vorticity is what maintains the finite width of the parallel spectrum. This is generally due to the drive mechanisms acting across the magnetic field and the dissipation/sink mechanisms acting through the parallel dynamics, i.e., responses. In vector terms we have

$$v_A\mathbf{b}\cdot\nabla \sim \mathbf{v}_E\cdot\nabla, \tag{4.2}$$

with the ExB velocity the main actor across the field and the Alfvén dynamics the main response along the field. Both of these are in general nonlinear. In fact, the two-dimensional systems of the previous Chapter are set up with the assumption that only the nonlinear parts of these dynamics are present. However, a magnetised plasma usually refers to a system with a strong guide field. In general, the nonlinear Alfvén response will have all three components of the parallel gradient comparable to each other. But for some parameter combinations the strongest piece is the linear guide field response to the nonlinearities in the turbulence. The parallel response can be modelled in a simple way to produce a two-dimensional model, in which a turbulent velocity field is coupled to a scalar field representing the electron density (synonymous with pressure in an isothermal model). The drive is due to the cross-field gradient and the dissipation is due to the coupling. This model provides a useful introduction to the phenomenology of turbulence in magnetised plasmas. Quite the opposite to the MHD systems of the previous Chapter, it represents a minimal version of the electrostatic limit of the dynamical system, which we will study with more generality later on.

## 4.2 The model system for dissipative ExB turbulence

We have already seen the Hall MHD system and its force balance counterpart, which is the one of relevance here. However, it might be helpful to give a purely heuristic description of coupled turbulence for the purposes of understanding the nonlinear dynamics. In later Chapters these equations will be seen to emerge from more general ones that we will derive systematically.

We start with two-dimensional Euler hydrodynamics, the decaying fluid case from the previous Chapter, and then add a passive scalar advection equation. The flow represents the ExB velocity, taken to be incompressible, with the electrostatic potential acting as its stream function. The definition and simplifications are

$$\mathbf{v}_E = \frac{1}{B^2}\mathbf{E} \times \mathbf{B} = \frac{1}{B}\mathbf{b} \times \nabla\phi, \tag{4.3}$$

with $\mathbf{B}$ and its unit vector $\mathbf{b}$ constant. The scalar is taken to represent fluctuations of the electron density evolving kinematically in the ExB flow field as in

$$\frac{\partial n}{\partial t} + \mathbf{v}_E \cdot \nabla n = 0. \tag{4.4}$$

This is the 2D fluid model with a passively advected scalar. Then, a simple dissipative coupling mechanism is added, which in our case represents resistive friction of currents flowing parallel to the magnetic field. The current is in the third dimension, but in a two-dimensional model its action is represented by a coupling coefficient between the electrostatic potential and the density, whose difference represents the force potential on the electrons parallel to the magnetic field, balanced by resistivity. This gives a closed system between two scalar quantities, each with its own evolution equation.

The dynamics is in the $xy$-plane perpendicular to a homogeneous guide field which is assumed to be in the $z$-direction in a right-handed Cartesian coordinate system. The constant background density/temperature parameters are written as $n_e$ and $T_e$, while $n$ represents any departure from homogeneity, including a slowly varying gradient. The electrostatic potential is denoted $\phi$ and the flow stream function is $\phi/B$. The equation for the ExB flow is

$$\frac{\partial \varpi}{\partial t} + [\phi, \varpi] = \nabla_{\parallel} J_{\parallel}, \tag{4.5}$$

as in the MHD case. In this case we write the parallel gradient in terms of the third coordinate rather than in terms of the magnetic stress nonlinearity,

$$\nabla_{\parallel} = \frac{\partial}{\partial z}. \tag{4.6}$$

The assumption is of electrostatic responses along the field and negligible magnetic induction, therefore $\psi$ is neglected. The electron density evolves according to its flow field, ExB across the field and the parallel current along it, given by

$$\frac{\partial n}{\partial t} + [\phi, n] = \frac{1}{e} \nabla_{\parallel} J_{\parallel}. \tag{4.7}$$

The field equation to obtain the potential from the vorticity variable is

$$\frac{\rho}{B^2} \nabla_{\perp}^2 \phi = \varpi. \tag{4.8}$$

The current under resistive evolution is given constitutively, according to

$$\eta_{\parallel} J_{\parallel} = \frac{T_e}{n_e e} \nabla_{\parallel} n - \nabla_{\parallel} \phi. \tag{4.9}$$

In other words, the current flows in response to any force imbalances between $n$ (pressure gradient) and $\phi$ (electric field) along the field, caused by turbulent dynamics across it. The current is quasistatically evolving, controlled by the resistivity, given by

$$\eta_{\parallel} = 0.51 \frac{m_e \nu_e}{n_e e^2}, \tag{4.10}$$

representing electron–ion friction due to Coulomb collisions. Eq. (4.9) is referred to as the 'Ohm's law' qualified as 'the' since it is different in different models. For this simplified model the inductive part of the electric field and the electron inertia are negligible (to be re-introduced later). Together, Eqs (4.5), (4.7)–(4.9) would be the desired equations in three dimensions.

### 4.2.1 The adiabatic response and the drift scale

Before going to two dimensions it is interesting to note the nature of this response. First, it acts as a restoring force, since all eigenvalues of $\nabla_{\perp}^2$ are negative. A force imbalance causes a current which acts to reduce the imbalance. Second, if $\eta_{\parallel}$ is small then the electrons remain in quasistatically evolving force balance. This is called 'adiabatic electrons' and the tendency to dissipate towards it is called the 'adiabatic response'. It is fundamentally outside of conventional MHD because in that model the pressure forces on the electrons are neglected and only the ion mass flow (sound waves) is considered along the guide field. Second, it is this force which defines the physical scale. If the adiabatic response is at least concurrent, we always have similar magnitudes among

$$h_e \equiv n - n_e\frac{e\phi}{T_e} \sim n \sim n_e\frac{e\phi}{T_e}. \tag{4.11}$$

Multiplying the vorticity equation by $T_e/e$ and the density equation by $e$ and then dividing both by $n_e e$, we have the two right-hand sides in the same units and hence the two opposite responses on the two left-hand sides,

$$\rho_s^2\frac{e}{T_e}\frac{d}{dt}\nabla_\perp^2\phi = \frac{1}{n_e e}\nabla_\parallel J_\parallel = \frac{1}{n_e}\frac{dn}{dt} \qquad \text{where} \qquad \frac{d}{dt} = \frac{\partial}{\partial t} + [\phi, \,], \tag{4.12}$$

which holds under the ExB advective derivative $d/dt$, that is, any level of non-linearity. With $n$ and $\phi$ coupled by the adiabatic response and evaluating the parameters, we have found the natural scale $\rho_s$ given by

$$\rho_s^2 = \frac{T_e}{n_e e^2}\frac{\rho}{B^2} \qquad \text{hence} \qquad \rho_s = \frac{c_s}{\Omega_i}, \tag{4.13}$$

according to the parameter definitions

$$c_s^2 = \frac{T_e}{m_i} \qquad \Omega_i = \frac{eB}{m_i} \qquad \rho = n_i m_i \qquad n_e = n_i, \tag{4.14}$$

with the sound speed $c_s$ defined in terms of the electron temperature and ion mass, reflecting electron mobility and ion inertia.

### 4.2.2 The causal direction of the adiabatic response

From Eq. (4.12), we also find that for scales of motion larger than $\rho_s$, such that $k_\perp\rho_s < 1$, the response of $\phi$ to any non-adiabatic imbalance with $n$ is faster than the reverse (here recall that $\nabla_\perp^2$ has all negative eigenvalues $-k_\perp^2$ in the perpendicular wavenumber spectrum). This force balance adiabatic response works opposite to the usual test particle assumptions that the electrons moving find a statistical equilibrium in a prescribed field. In magnetised plasma turbulence the adiabatic response affects the field variable first and the plasma second, because the coefficient ratio satisfies $k_\perp^2\rho_s^2 < 1$ for most of the spectrum. Hence the field follows the plasma rather than the plasma following the field. In the extreme case—the low-beta MHD limit given by $k_\perp^2\rho_s^2 \ll 1$ together with $v_A \gg c_s$—the plasma doesn't respond at all. From a non-adiabatic initial state, the field self-consistently adjusts toward an end state of adiabatic electron force balance. The plasma then evolves quasistatically according to sound wave and thermal responses. We will see more of this later when considering three-dimensional electromagnetic models of this same type of turbulence.

### 4.2.3 Minimal model of adiabatic coupling

Having defined $\rho_s$ and explored the adiabatic response, we can insert the Ohm's law (Eq. 4.9) into the vorticity and density Eqs (4.5) and (4.7), and normalise the units such that

$$\frac{\partial \varpi}{\partial t} + [\phi, \varpi] = D(\phi - n), \tag{4.15}$$

$$\frac{\partial n}{\partial t} + [\phi, n + n_0] = D(\phi - n), \tag{4.16}$$

where the coupling constant is given by

$$D = -D_{\|}k_{\|}^2, \tag{4.17}$$

following a parallel diffusivity given by $D_{\|} = T_e/0.51 m_e \nu_e$ and $k_{\|}^2$ given in terms of a parallel scale. The field equation and gradient are given in normalised units by

$$\nabla_\perp^2 \phi = \varpi \qquad \frac{\partial n_0}{\partial x} = -\omega_n. \tag{4.18}$$

The length is in units of $\rho_s$ and time in $L_\perp/c_s$, the fields in terms of $\delta_0 T_e/e$ for $\phi$ and $\delta_0 n_e$ for $n$. The parameters are $D$, $\omega_n$, and $\delta_0 = \rho_s/L_\perp$, the latter of which is normalised away.

The coupling constant $D$ denotes the resistive coupling strength. We will sweep this parameter from large to small values to explore the relevant limits:

$$\text{'hydrodynamic:'} \qquad D \to 0 \quad \text{hence arbitrary} \quad h_e, \tag{4.19}$$

$$\text{'adiabatic:'} \qquad D \gg 1 \quad \text{hence} \quad h_e \to 0. \tag{4.20}$$

The hydrodynamic limit reduces to the 2D fluid model with a passively advected scalar. In the adiabatic limit, $n$ and $\phi$ become synonymous, and the equations reduce to the one-field model

$$\frac{\partial N}{\partial t} + [\phi, N + n_0] = 0 \qquad \left(1 - \nabla_\perp^2\right)\phi = N. \tag{4.21}$$

For this case in the absence of subgrid dissipation the presence of $n_0$ serves only to give a drifting frame, as the $n_0$ term in the bracket may be transformed away, using

$$dt' = dt - \omega_n \, dx. \tag{4.22}$$

This velocity $\omega_n$ in normalised units translates in physical units back to

$$v_D = \frac{\rho_s c_s}{L_n} \qquad \text{where} \quad L_n = |\nabla \log n_e|^{-1}, \tag{4.23}$$

which are the diamagnetic drift velocity and the density gradient scale length, respectively. We will see more detail on this in Chapter 9. This drift is why this dynamics was called 'drift wave turbulence' in earlier papers.

These models form the prototypical cases for 2D turbulence in magnetised plasma. The set of Eqs (4.15)–(4.18) is called the Hasegawa–Wakatani model (1983/4) and the adiabatic case of Eq. (4.21) is called the Hasegawa–Mima equation (1977/8) in the plasma physics literature and the Charney equation (1948/50) in the geophysical literature. These systems have attracted detailed attention in the scientific literature and for our purposes they serve as useful control cases against which to compare all subsequent models.

## 4.3 Turbulence in the adiabatic and hydrodynamic limits

We first take each of the two extremes, $D \to \infty$ and $D = 0$, using Eqs (4.21) and (4.15)–(4.18), respectively. This serves to set the control cases. We will later examine various $D > 0$ and allow evolution out of a state with finite $n$ but zero $\phi$. Particular attention should be given to how the spectra of the various quantities, especially $\varpi$, spread in time, and whether they find statistical equilibrium, i.e., saturate. For these studies we eliminate flows by setting the $k_y = 0$ components to zero in the Fourier transform operations[1]. We will address flows in Chapter 16 after the three-dimensional models have been introduced.

The initial state is adiabatic, with $\phi = n$ chosen as a random-phase field with prescribed spectrum in normalised units, with the constant $A_0$ chosen such that the RMS amplitude is set to unity. The domain is chosen such that the minimum wavenumber in each direction is $k_0 = 0.05$, with $256 \times 256$ equidistant grid points over the plane, with periodic boundary conditions on the domain size in each direction. This means

$$-40\pi < \{x, y\} < 40\pi, \tag{4.24}$$

and the maximum value of $k$ in each direction is $k_0 \times 128 = 6.4$ with the dissipation range covering about the last factor of two. The subgrid dissipation is given by

$$\frac{\partial f}{\partial t} = \cdots - \nu_0 \frac{h^4}{16}\left(\nabla_\perp^2\right)^2 f \qquad \nu_0 = 10^{-2}, \tag{4.25}$$

where $h$ is the spatial stepsize and the scalar $f$ is $\varpi$, or $n$, or $N$ in each of their equations. The value of $\nu_0$ is the same as in the decaying fluid case. The time step for these cases is 0.01 in normalised units. The prescription for the initial state given by

$$\phi_{\mathbf{k}} = n_{\mathbf{k}} = A_0\, k \left(1 + (k/k_1)^6\right)^{-1} \exp(i\Theta), \tag{4.26}$$

with the expression involving $k$ giving the spectrum and $\Theta$ a random number on $[0, 2\pi]$ giving the phases. We choose $k_1/k_0 = 10$ to get a reasonable peak with space

[1] The special nature of $k_y = 0$ rather than $k_x = 0$ is provided by the gradient in the $x$-direction, and hence drifts in the $y$-direction.

on both sides for the activity to spread out. Unless otherwise noted, the drive and dissipation are set to $\omega_n = 1$ and $\nu_0 = 10^{-1}$.

Energetically we have, with the subgrid dissipation,

$$\frac{1}{2}\frac{\partial}{\partial t}(n^2 + |\nabla_\perp \phi|^2) - \nabla \cdot \left(\phi \nabla_\perp \frac{\partial \phi}{\partial t}\right) = \omega_n(n v_E^x) - D(n - \phi)^2 \\ - \nu_0(|\nabla_\perp \varpi|^2 + |\nabla_\perp^2 n|^2), \tag{4.27}$$

where $v_E^x = -\partial\phi/\partial y$ is the component of the ExB velocity down the gradient. In this case $D = 0$ and the only sink is the subgrid dissipation. The free energy pieces and the total are

$$E_e = \mathcal{V}^{-1} \int d\mathcal{V} \, \frac{1}{2} |\nabla_\perp \phi|^2 \quad E_n = \mathcal{V}^{-1} \int d\mathcal{V} \, \frac{1}{2} n^2 \quad \text{with} \quad E_T = E_e + E_n. \tag{4.28}$$

The amplitude functionals for $\phi$ and $\varpi$ are

$$A_p = \mathcal{V}^{-1} \int d\mathcal{V} \, \frac{1}{2} \phi^2 \qquad A_w = \mathcal{V}^{-1} \int d\mathcal{V} \, \frac{1}{2} \varpi^2. \tag{4.29}$$

The drive and dissipation terms in the physics are

$$\Gamma_n = \omega_n \frac{1}{2E} \mathcal{V}^{-1} \int d\mathcal{V} \, n v_E^x \qquad \Gamma_c = D \frac{1}{2E} \mathcal{V}^{-1} \int d\mathcal{V} \, (n - \phi)^2. \tag{4.30}$$

The subgrid dissipation rate is given by

$$\Gamma_v = \nu_0 \frac{1}{2E} \mathcal{V}^{-1} \int d\mathcal{V} \left(|\nabla_\perp \varpi|^2 + |\nabla_\perp^2 n|^2\right). \tag{4.31}$$

The total drive rate and energetic error are given by

$$\Gamma_T = \frac{1}{2E_T} \frac{\partial E_T}{\partial t} \qquad \Gamma_E = \Gamma_T - \Gamma_n + \Gamma_c + \Gamma_v. \tag{4.32}$$

For saturated cases we also measure the transport flux caused by the turbulent ExB advection, given by

$$F_e = \mathcal{V}^{-1} \int d\mathcal{V} \, n v_E^x. \tag{4.33}$$

These are used as diagnostics of the physics as well as the numerics.

### 4.3.1 Decaying turbulence in the hydrodynamic limit

We start with decaying cases without the gradient ($\omega_n = 0$) to show how $n$ and $\phi$ evolve apart naturally in the hydrodynamic limit ($D = 0$) but are tied together in the adiabatic limit ($D \to \infty$). For adiabatic cases it is sufficient to run with $D = 10^3$, which is able to tie $\phi \leftrightarrow n$ even at the highest $k$ being kept.

The decaying hydrodynamic case is essentially the same as the fluid case with $k_1/k_0 = 10$ used for the coherent vortices in the last Chapter. The only difference is the suppression of $k_y = 0$ done here as well to compare with the later cases. Due to the fact that there is no coupling from $n$ back to $\phi$ in the hydrodynamic model, the two dynamical fields evolve separately. The ExB eddies tracked by $\phi$ evolve according to the Euler equation while $n$ is passively advected. The time traces are shown in figure 4.1, with the energetic quantities listed above sampled every $\Delta t = 0.2$ in normalised units, as in the cases of the previous Chapter. The selective decay of the energy components and the enstrophy is as in the fluid case, since without the gradient forcing $n$ follows the same equation as $\varpi$. The rise in $A_p$ is indicative of the shift of ExB energy to larger scale. The decay rate is robust earlier on as the isotropic spectra (below) initially spread out towards high $k$, the amplitudes there increasing in response before being brought down by the subgrid dissipation. Then, the decay rates decrease with the fall in the overall amplitudes of $n$ and $\varpi$ (in $E_n$ and $A_w$, respectively). The decay of $E_e$ itself is minimal, so at later times it is the larger channel, and thereafter the overall energy decays more slowly than the enstrophy. The energetic error $\Gamma_E$ remains small.

The isotropic spectra, given in figure 4.2, for the initial state, the state at $t = 30$ and at $t = 300$, show $n$ and $\phi$ evolving apart as the nonlinear transfer dynamics destroys their alignment, with broadening, respectively, to the high-$k$ and low-$k$ side. The high-$k$ sides of $n$ and $\varpi$ spread out together following their identical dynamics. Most of the initial redistribution has already occurred by $t = 30$ with the $n$ and $\varpi$ spectra becoming much flatter while $\phi$ steepens.

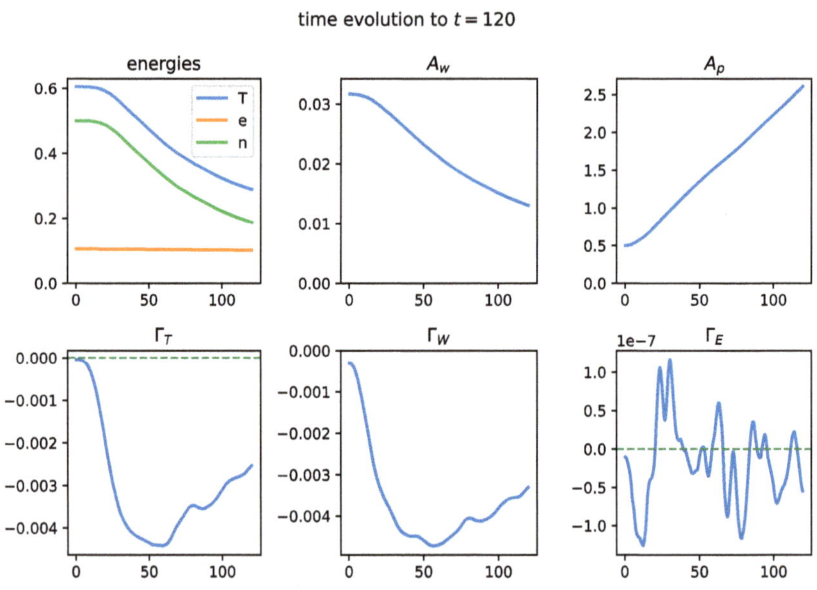


**Figure 4.1.** Time traces for the hydrodynamic decaying case. This is the same as the 2D fluid one from the previous Chapter, here also with the density $n$ as a passive scalar field. The evolution of the ExB energy ($E_e$, here, 'e') is minimal, but now the thermal free energy ($E_n$, 'n') is the largest component of the total ($E_T$, 'T') and decays more like the enstrophy (twice $A_w$)—the decay rates are in $\Gamma_T$ and $\Gamma_W$, respectively. Note how both decay rates reach their maxima only after the turbulence correlations (i.e. differences from a random-phase state) develop. The energetic error ($\Gamma_E$) is several orders of magnitude smaller.

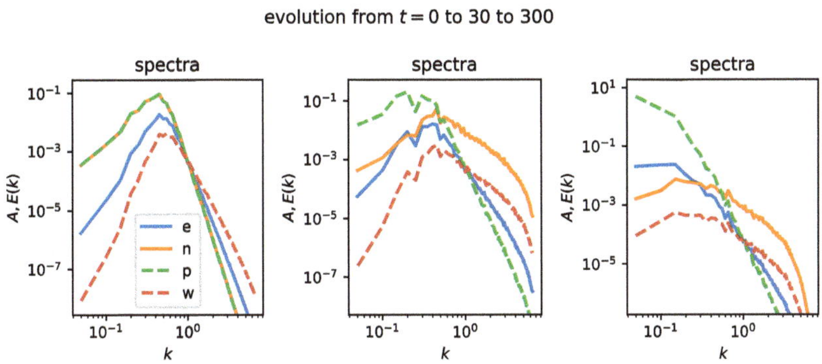


**Figure 4.2.** Isotropic perpendicular spectra for the hydrodynamic decaying case, with the initial state (left frame), early phase at $t = 30$ (centre), and the developed phase at $t = 300$ (right). The curves are for the energy channels $E_e$ ('e') and $E_n$ ('n'), drawn solid, and for the amplitudes $A_p$ ('p') and $A_w$ ('w'), drawn dashed. The curve legends are suppressed in the latter frames so as not to obscure the details of the evolution. The density and potential start equal, but evolve apart according to their differing cascade tendencies.

The spatial morphology, given in figure 4.3, for the $t = 30$ and $t = 300$ time points, show rapid spatial de-alignment between $n$ and $\phi$ as the two fields build large cells for the ExB flow and the same structure as the isolated vortices in $\varpi$ (the same as in the decaying fluid case). The vortices replace the initial vortex sheets on a slower time scale but by $t = 300$ the qualitative changes in the evolution are mostly over. These interactions represent the physical manifestations of the cascade dynamics in the decaying hydrodynamic case, which is the same as the decaying fluid case with a passive scalar.

### 4.3.2 Decaying turbulence in the adiabatic limit

In the adiabatic model the two fields do not evolve apart because they are constrained to be equal. This eliminates the direct cascade in the free energy associated with $n$ since the advective nonlinearity on that quantity ($[\phi, n]$ with $\phi = n$) vanishes. There remains the vorticity nonlinearity $[\phi, \varpi]$ to represent the turbulence. For the evolution of the flow field, the difference to the decaying fluid case is the effective coefficient of $\partial\phi_{\mathbf{k}}/\partial t$ is $1 + k^2$ here rather than $k^2$ there, so nonlinear action, hence also the dual cascade, at low $k$ is effectively suppressed. This is the new twist in the dynamics under the adiabatic response.

The time traces for this case are shown in figure 4.4, sampled as above. The rise in $\phi$ is minimal since now $A_p$ is the same as $E_n$, which is part of the energy. The suppression of even the vorticity nonlinear advection at low $k$ prevents $\phi$ from reaching large amplitudes in the generally decaying system. The total energy $E_T$ still drops since $E_e$ now drops faster—there is no longer the strong shift to low $k$ to compensate the energy decay due to the vorticity cascade. The decay in $E_T$ is much slower in this case since the direct cascade has to go through $\varpi$ as the direct advection of $n = \phi$ now vanishes. The enstrophy decay is still robust but somewhat slower than in the hydrodynamic case. The energetic error remains small.

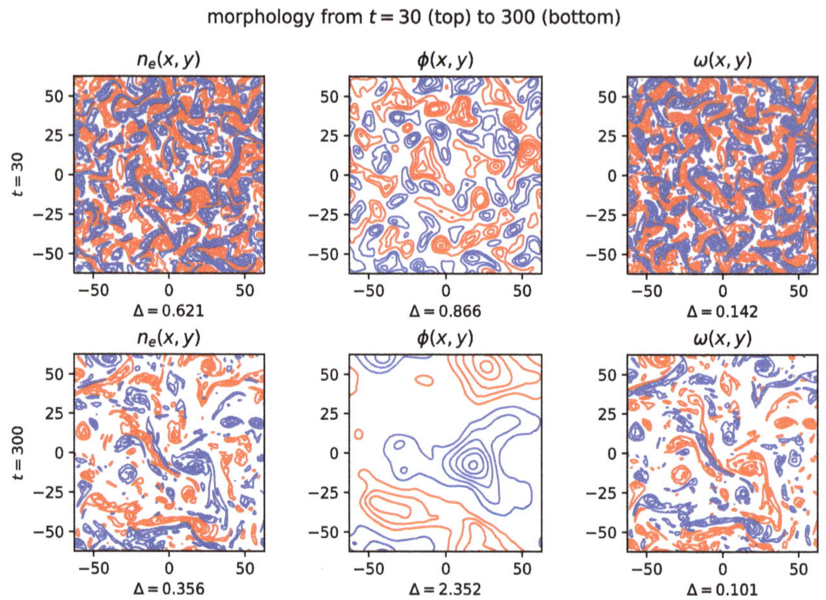


**Figure 4.3.** Spatial morphology of the dependent variables for the hydrodynamic decaying case, in the early phase at $t = 30$ (top row) and the developed late phase $t = 300$ (bottom). In each frame the axes are $x$ and $y$ as implied by the title, and the contour interval, always 0.16 times the maximum absolute value, is given at the bottom. Positive values are coded red, and negative are blue, and the contour interval is given at the bottom of every frame. The time is given at the left. While the potential (hence the turbulent eddy flows) build ever-larger scales up to the largest available, the density and vorticity evolve together, following the same equation, exhibiting intermittency and smaller scale as the nonlinear correlations between the degrees of freedom intensify, unimpeded by any linear forcing or coupling.

The evolution of the isotropic spectra is shown in figure 4.5, at the same three time points as before. Most of the evolution is on the high-$k$ side of the initial peak in this case due to the suppression of the low-$k$ nonlinear dynamics by the adiabatic constraint and its effect on the vorticity response. The vorticity spectrum spreads out towards high $k$ but with the suppressed low-$k$ nonlinearity the effect is not as robust. Nevertheless the high-$k$ turn-down in $A_w(k)$ is no sharper than before, so there is still effective nonlinearity to oppose the dissipation there.

The spatial morphology, in figure 4.6, for the $t = 30$ and $t = 300$ time points, shows very slow evolution. The vortex pair near the top right corner (roughly 45, 40) is seen to turn only part-way over by $t = 300$ though much of the vorticity has been drawn out into sheets by this time. What stands out is the lack of large-scale structure building, as compared with the hydrodynamic case. One also does not find isolated vortices, suggesting that the entire range of the spectrum is involved in setting up the nonlinear correlations that represent them.

A computation must involve scales in the range of $\rho_s$ and smaller, or not much happens in the adiabatic model. Cases with smaller domain sizes ($L < 30$ or so) show much stronger evolution before this slowly evolving end state is reached (i.e. much of $\varpi$ is ejected to dissipation), and those with weaker resolution ($h > 3$ or so)

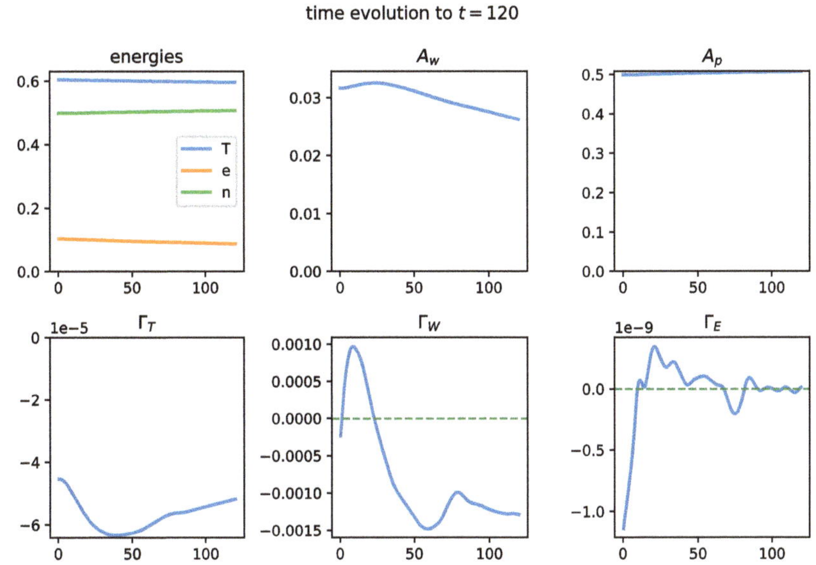


**Figure 4.4.** Time traces for the adiabatic decaying case. Here, $n$ is not a passive scalar but is tightly coupled through the dissipation to $\phi$. Now, $\phi$ does not rise in amplitude because the dual cascade is suppressed when the $k^2$ dependence of vorticity is effectively replaced by $1 + k^2$. Even the enstrophy decay is slower. All differences to the hydrodynamic case are due to the adiabatic coupling.

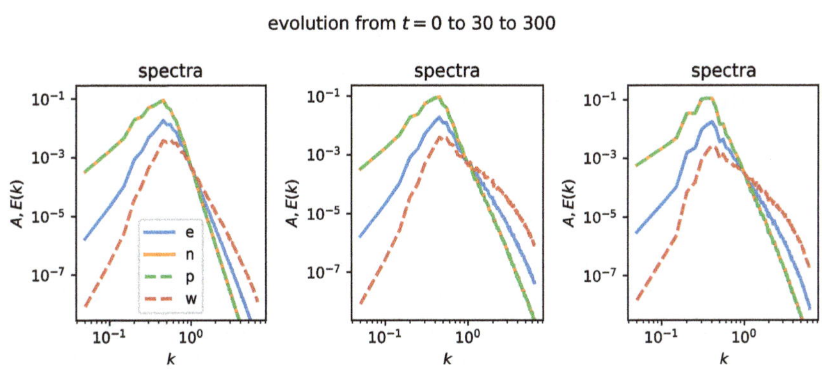


**Figure 4.5.** Isotropic perpendicular spectra for the adiabatic decaying case, with the initial state (left frame), early phase at $t = 30$ (centre), and the developed phase at $t = 300$ (right), with the same format as in figure 4.2, with which it is compared. With the adiabatic coupling dominant, $\phi$ and $n$ are tied together by the dissipative coupling and the evolution is due to the moderate decay of enstrophy. The spread to high $k$, small scale, in the vorticity is much less in this case.

show even less activity than is seen here. In any of the cases there is little spectrum evolution for $k_y < 0.3$ or so. The inverse cascade via vortex merger stops when the scales get too large, that is, when the nonlinearity is no longer sizeable compared with the energy content. The enstrophy content at high $k$ drops faster with time than the energy content, which in turn decays faster than the $\phi$-field amplitude. The hydrodynamic cases by contrast always see the inverse $\tilde{\phi}$ cascade continue until the

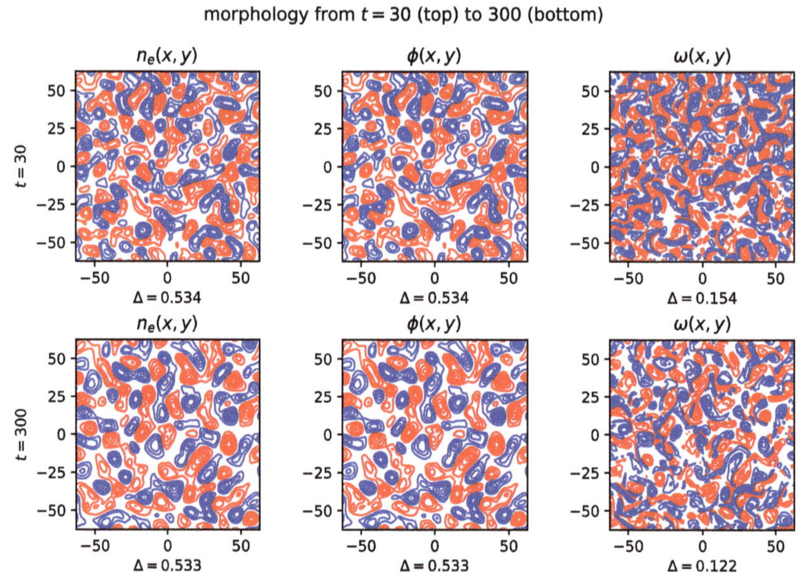


**Figure 4.6.** Spatial morphology of the dependent variables for the adiabatic decaying case, with the same format as in figure 4.3, with which it is compared. The large-scale dynamics is largely suppressed by the coupling, and as a result there is much less evolution visible at small scale.

last vortex merger allowed by the boundary conditions takes place, if the computations are run long enough. If one is to do computations in both regimes and show the contrasts, these control cases are necessary to determine that the observed differences are actually due to the physics and not the limitations of the computations.

### 4.3.3 Driven turbulence in the hydrodynamic limit

Before putting in the dissipative coupling at moderate levels we look at the two limits in cases driven by the background gradient, so $\omega_n = 1$ is set. With no coupling there is no guarantee the dynamics will saturate into a statistically equilibrated state, but the structure found in the density field can show interesting differences to the decaying cases.

Driven turbulence in the hydrodynamic limit is run with $D = 0$. In this case the passive scalar also has a background gradient. The energetic drive term $\Gamma_n$ is determined by the flux down the gradient, as $\omega_n v_E^x$ is a direct drive of $n$ with the same phase in the $y$-direction as $v_E^x$, the flow in the $x$-direction. In all other respects the driven and decaying hydrodynamic cases are much the same.

The time traces for this case are shown in figure 4.7, sampled as above. Since there is no coupling from $n$ back to $\phi$, the fluid part with $\phi$ and $\varpi$ evolves as in the decaying fluid and hydrodynamic cases. The quantities $A_p$, $E_e$, and $A_w$ rise, fall slightly, and fall robustly, respectively. The only difference between this case and that in figure 4.1 is the drive, due to which $E_n$ dominates the total energy, and $n$ no longer follows $\varpi$ but increases continuously. The second line shows the energetic

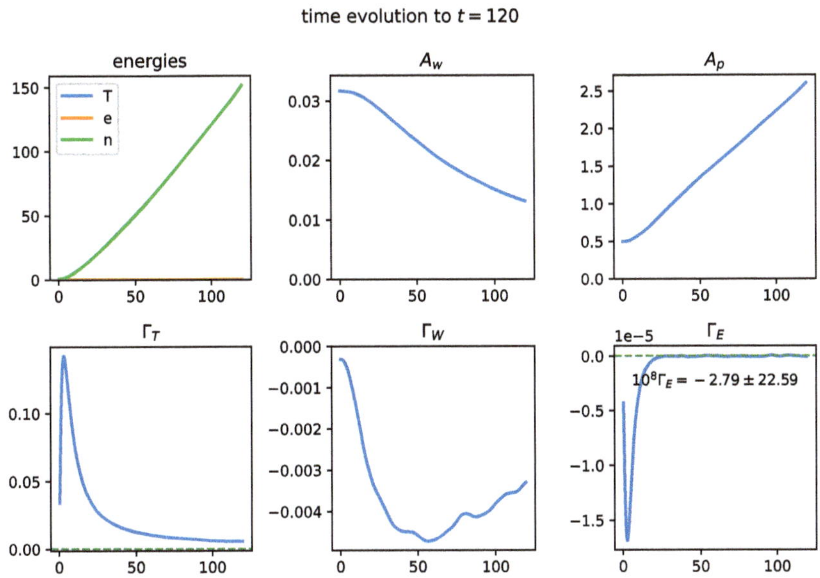


**Figure 4.7.** Time traces for the hydrodynamic driven case. The total energy ($E_T$) is mostly all density ($E_n$), which is driven without any coupling channel back to the ExB energy ($E_e$ at upper left). The decay of the fluid component is the same as before. For driven non-adiabatic cases the error spikes early to values which are still comparatively small. Averaged over the last half of the time range, the value given, $\Gamma_E = (-1.13 \pm 8.85) \times 10^{-8}$, is very small.

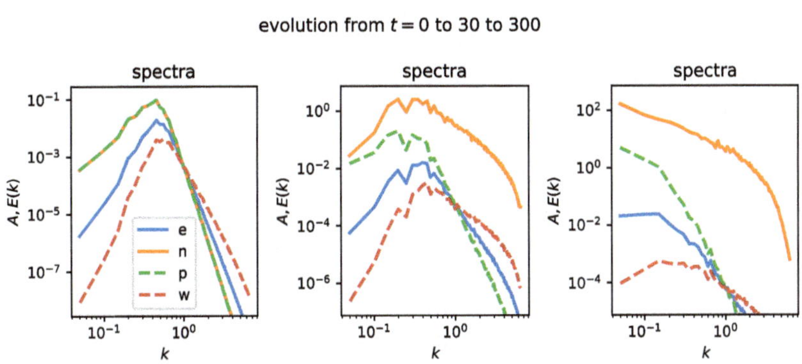


**Figure 4.8.** Isotropic perpendicular spectra for the hydrodynamic driven case, with the same format as in figure 4.2, with which it is compared. The fluid component is the same but the drive increases the density ('n') to ever-larger amplitude. The concurrent decay of the vorticity ('w') weakens the nonlinear decorrelation effect at late times, causing the low-$k$ peak in $E_n$ to grow the fastest.

quantities, for which $\Gamma_T \approx \Gamma_n \gg \Gamma_v$ and $\Gamma_c = 0$ in this case. The drive rate falls because $E_n$ (hence $n$) rises while $E_e$ and especially $A_w$ (hence dynamics) falls. The energetic error is very small compared with $\Gamma_T$ at all times.

The isotropic spectra are shown in figure 4.8, for the initial state, the state at $t = 30$ and at $t = 300$. The evolution of $E_n$ (hence $n$) is initially similar to $A_w$ (hence $\varpi$)

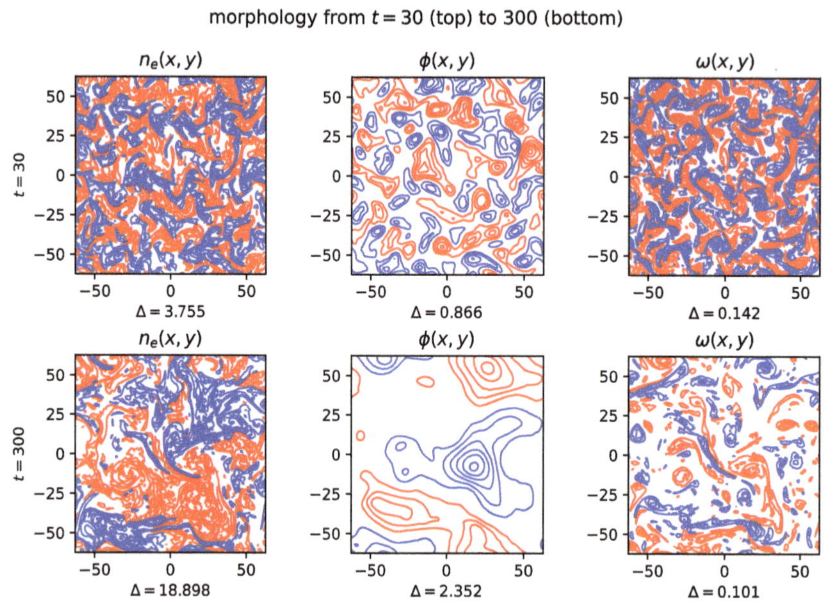


**Figure 4.9.** Spatial morphology of the dependent variables for the hydrodynamic driven case, with the same format as in figure 4.3, with which it is compared. The density (driven) now builds both large scales, while the fluid component evolves as before due to the lack of coupling.

but then, due to the drive, it eventually follows $\phi$ more at long wavelength and moreover rises above all the other quantities. The change in the shapes of the spectra with time make display of snapshots at individual times more useful than time averages. This will change again when drive and dissipation can balance and the energetics is statistically stationary. The turn-down at high $k$ in all quantities remains adequate.

The spatial morphology is shown in figure 4.9, for same time points as before. The evolution of the spectra is reflected well in the structure one sees here. The potential builds ever-larger structures while the vorticity develops first sheets and then isolated vortices as in the decaying fluid case. This is again the manifestation of the dual cascade process. Due to the gradient drive, the density builds large-scale sweeps but there is structure at all scales. The drive is mostly at large scale but cascades robustly down towards high $k$ as indicated by the spectra. All scales are in fact driven.

### 4.3.4 Driven turbulence in the adiabatic limit

The adiabatic case represents $D \to \infty$, which results in $N = n - \varpi$ and $n = \phi$, so that both drive and dissipation vanish, leaving only the subgrid dissipation as a sink. In the driven adiabatic case the finite $\omega_n$ leads to the drift discussed above, corrected only by the minimal action of the subgrid dissipation. Hence adiabatic drift wave turbulence is essentially a decaying case.

The time traces for this case are shown in figure 4.10. The rate of evolution is much slower than in the hydrodynamic cases, since the nonlinearity $[\phi, n]$ with

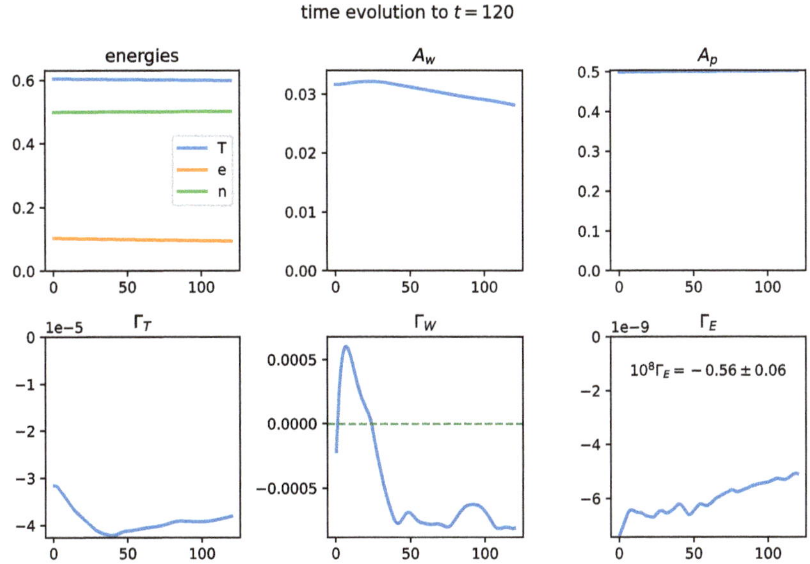


**Figure 4.10.** Time traces for the adiabatic driven case, with the same format as in figure 4.4, with which it is compared. With $n$ and $\phi$ coupled adiabatically, their lack of a phase shift means that there is little drive, and the enstrophy decays moderately. Averaged over the last half of the time range, the error is in the same range of the previous case, but this time there is no early spike.

$\phi = n$ vanishes. There remains the $[\phi, \varpi]$ one, but while it has a strong effect in the pure fluid case, in the adiabatic drift wave case it has to compete with the factor of 1 in $1 + k^2$, which at low $k$ is a weak effect, as seen already in the decaying case. The disparate decay rates for $E_e$, and $A_w$ are as in the fluid case but the rise in $A_p$ is slower, as in the decaying adiabatic case. The energetic error $\Gamma_E$ is small at all times. The main takeaway for this is that there is no real energetics in the adiabatic case since both $\Gamma_n$ and $\Gamma_c$ vanish. However, the enstrophy itself reaches saturation by about $t = 200$.

The isotropic spectra are shown in figure 4.11, for the initial state, the state at $t = 30$ and at $t = 300$. The evolution out of the initial state is very slow at low $k$ for the same reasons as given for the time traces. The vorticity spreads out towards high $k$, but not so strongly as even in the decaying fluid case, which has more large-scale flow to drive the dynamics. The steepness of the spectra is deceptive, as the turn-down in the vorticity is barely sufficient. Indeed, a trial case with $\nu_0 = 10^{-3}$ gave a spectrum that showed insufficient subgrid dissipation—too flat, especially in $A_w$. This is the same experience as in the decaying fluid case—even though the cascade dynamics is globally weaker, it is about the same at high $k$ here as there.

The spatial morphology is shown in figure 4.12, for the $t = 30$ and $t = 300$ time points, as before. With the $1 + k^2$ response instead of $k^2$, the large-scale structure seen in the hydrodynamic case does not form. The vorticity no longer forms isolated structures, indicating that those are a feature of strong nonlinearity everywhere in the spectrum. The density, by construct, is the same as the potential in the adiabatic case. The driven and decaying adiabatic cases are similar, except for the drift in the

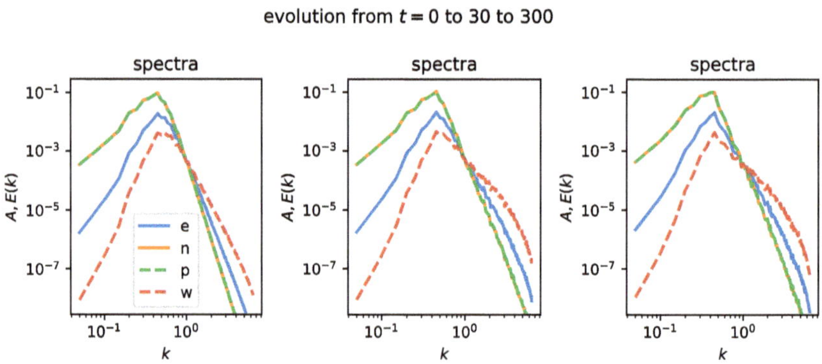


**Figure 4.11.** Isotropic perpendicular spectra for the adiabatic driven case, with the same format as in figure 4.5, with which it is compared. Apart from a very slight rise and fall in the spectrum peak (the amplitudes satisfy $\phi \sim n$), and a broadening of the enstrophy spectrum ('w') there are very few changes.

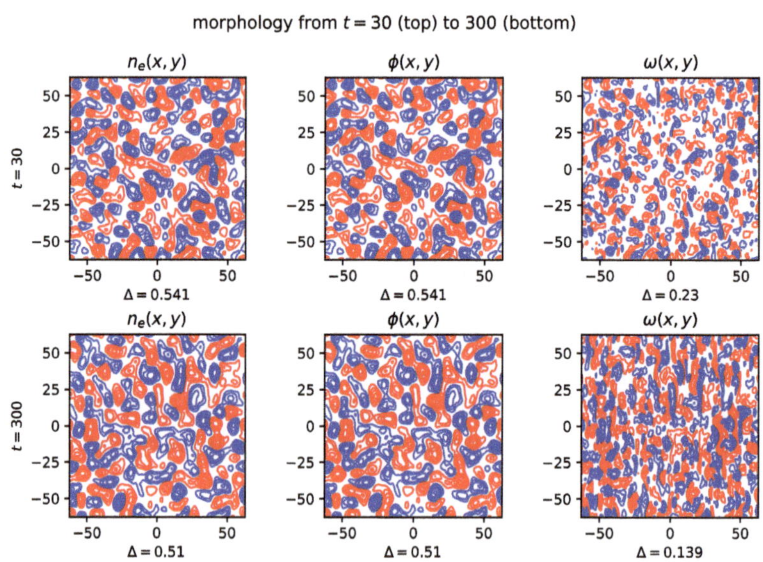


**Figure 4.12.** Spatial morphology of the dependent variables for the adiabatic driven case, with the same format as in figure 4.6, with which it is compared. Again, evolution is very slow and there is very little difference to the adiabatic decaying case.

$y$-direction in the case with the gradient but also the increased enstrophy in this case since it reaches saturation. Therefore, the individual structures undergo shearing more rapidly.

The contrast between the two driven cases is essentially in the difference in the vorticity response, namely the factor of $1 + k^2$ in the adiabatic case versus $k^2$ in the hydrodynamic case. In the latter, the large-scale structure which forms for $\phi$ assists the nonlinearity at all scales and produces the isolated vortices seen in the pure fluid case. In the hydrodynamic case $n$ and $\phi$ evolve apart naturally if they start together. This indicates that if the adiabatic response is to hold it must be forced.

The nonlinear responses of $n$ and $\varpi$ are similar in decaying turbulence but $\phi$ is different, due to the dual cascade. When the system is coupled together in a finite manner, that is, the adiabatic response acts but instead of forcing an adiabatic state finds a balance with the turbulence, we obtain a different structure yet. And it is this one that will have the most relevance to the general situation of turbulence forced in some manner via an instability mechanism, be it 2D or 3D, electrostatic or electromagnetic.

### 4.3.5 Dissipative drift wave turbulence

As noted above, this dynamics is what we have known as drift wave turbulence over the years since around 1980. Further work has shown that 3D cases, especially in toroidal configurations, are fundamentally different, but this system is still useful as a learning tool to understand the basic processes of driven/dissipative turbulence in a magnetised plasma, especially the cascade dynamics and the energetics.

In these cases, with both drive and dissipation mechanisms present in the intrinsic physics (i.e. not just subgrid dissipation) and potentially active at all scales, the dynamics is expected to saturate at reasonably nonlinear levels. Since $n$ and $\phi$ are coupled to each other at all scales, the direct cascade in $n$ also pumps up the vorticity spectrum to higher levels at high $k$ than otherwise seen. For this reason it is necessary to lower the time step to 0.005 in normalised units.

The role of the coupling parameter $D$ is to effect a continuous transition between the deep hydrodynamic ($D = 0$) and adiabatic ($D = \infty$) limits. We examine the cases with $D = 0.1$ and $D = 0.01$ to explore the role of coupling at different strengths. These are both robustly nonlinear cases, and their differences will elucidate the role of the adiabatic response in the turbulence. As before, the initial state is with $n = \phi$ so that we can find the degree to which the dynamics departs from the adiabatic state.

These cases evolve initially as they would do in the hydrodynamic limit, but then the dissipative coupling prevents $n$ and $\phi$ from separating completely. Most of the activity is in a driven range of the spectrum, and the action of the dissipation involving $D$ prevents the 'long term' forms of the decaying systems from establishing themselves. Larger values of $D$ lead to more constraint at low $k$, approaching the adiabatic limit, but any reasonable value[2] of $D$ establishes a floor on the energetically active range of $k$, so evolution towards ever-larger scales as in the driven hydrodynamic case is stopped. The turbulence then relaxes statistically into its saturated state. The turbulence approaches saturation after about $t = 100$, and averaged quantities are measured over the interval $200 < t < 400$, with sampling every $\Delta t = 0.2$ as before. The time traces for both the $D = 0.1$ and $D = 0.01$ cases are shown in figures 4.13 and 4.14. Subsequent to a driven growth phase over $0 < t < 40$, by the averaging interval all quantities are well saturated. In all cases the standard deviation of the sample (error bar) is in the range of 5 to 10 percent of the mean. The exception is $\Gamma_T = (0.252 \pm 3.37) \times 10^{-3}$, i.e. the standard deviation is

[2] We can consider 0.01 to be a lower limit, as the spectra will suggest.

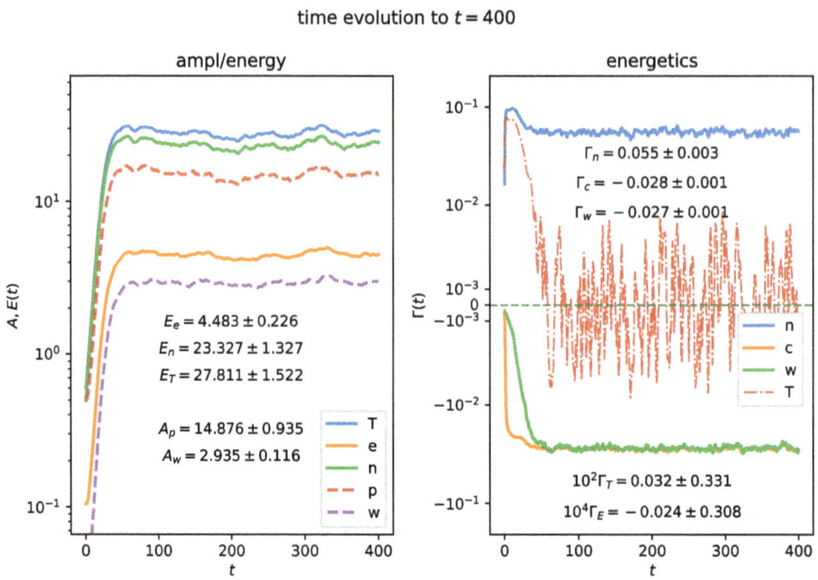


**Figure 4.13.** Time traces of the energy channels and amplitudes (left) and energetics (right), for the dissipative drift wave case with $D = 0.1$, a bit less hydrodynamic than $D = 0.01$, which is shown in figure 4.14. With $\phi$ and $n$ tied together by the coupling, the nonlinearities work together in a complementary manner, and the turbulence saturates. Neither does the density experience runaway growth nor does the enstrophy decay away. The values given are the average and standard deviation over the last half of the time range. All channels are flat, and saturation is indicated in the mean of the total growth rate ($\Gamma_T$) being much smaller than its standard deviation. The energetic error remains very small, even though, due to more robust nonlinearity reaching the resolution limit, it is larger than it was before.

more than ten times the mean, indicating saturation. In the energy pieces $E_n$ is the larger, 23.2 versus 4.47. In the amplitudes of $\phi$ and $n$, $A_p$ is smaller, 14.8 versus the value for $E_n$. The drive $\Gamma_n$ is balanced by $\Gamma_c$ and $\Gamma_v$ in roughly equal measure. Approaching saturation, $\Gamma_c$ rises first into the turbulence as the difference $h_e = n - \phi$ is excited by the vorticity nonlinearity $[\phi, \varpi]$ acting against the dissipation. Then, when the spectrum broadens out, the subgrid dissipation rises into saturation. The energetic error $\Gamma_E$ is small compared with the standard deviation in $\Gamma_T$, which is the main benchmark for driven/dissipative cases. The transport flux is $F_e = 3.06$, which is this times $v_D = \rho_s c_s / L$ in physical units. Moreover, $A_w$ is 2.92, hence the ExB vorticity $\Omega_E$ is about $2.4 \times c_s/L$ in physical units, and is well over ten times the gradient drive (always compare $A_w$ with $\Gamma_+$ in saturation). This is robust nonlinearity.

The isotropic spectra are shown in figure 4.15, time averaged with 20 samples over $200 < t < 400$ corresponding to the averaging interval in the time traces. These reflect the varying effectiveness of the dissipative coupling at larger and smaller scales. The spectra look more like the adiabatic limit at low $k$ but then separate at smaller scales at which the nonlinear interactions are more able to compete with the coupling. At smaller scales $E_n$ and $A_w$ have the same shape, but at larger it is $E_n$ and $A_p$ which track each other. At larger scales the vorticity ($A_w$) is very weak but enters

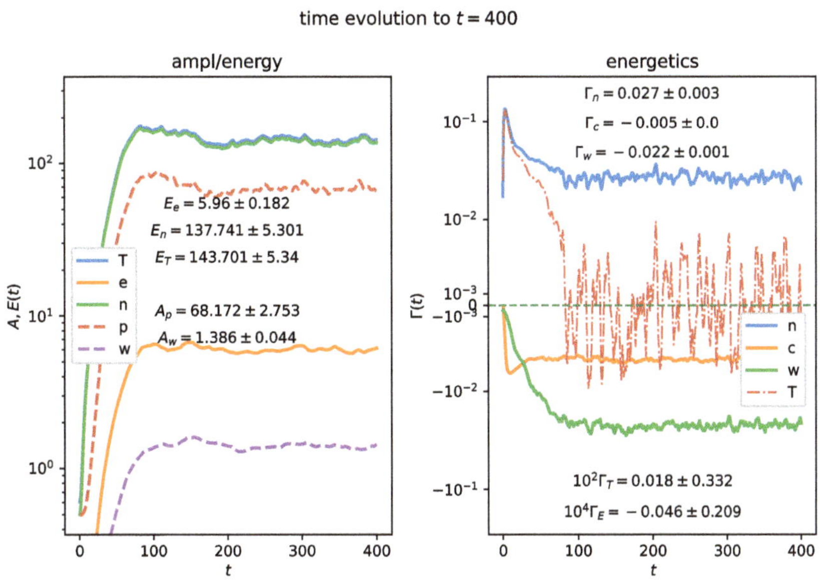


**Figure 4.14.** Time traces of the energy channels and amplitudes (left) and energetics (right), for the dissipative drift wave case with $D = 0.01$, a bit more hydrodynamic than $D = 0.1$, which is shown in figure 4.13. In both cases the total energy ($E_T$) is mostly thermal free energy ($E_n$) with the ExB energy ($E_e$) smaller, due to the fact that $k_\perp\rho_s$ is below unity and $n$ and $\phi$ are comparable, more closely so in the more adiabatic case with stronger coupling. The enstrophy saturates at values greater than unity, corresponding to $c_s/L_\perp$, in both cases. Even for the large amplitudes in the more hydrodynamic case, the amplitudes of $\phi$ and $n$ are within a factor of about 1.4 ($E_n$ and $A_p$ disparate by only two).

more at and past the peak. For $D = 0.01$ the peak is at larger scale, and the separation of $n$ away from $\phi$ on the high-$k$ side of it is much larger. The stronger dynamics makes $\Gamma_c$ higher at low $k$ in the $D = 0.01$ case despite the lower coefficient. The overall vorticity is actually lower, indicating that the coupling for moderate values of $D$ is what produces the most vorticity relative to energy. On the other hand, if $D$ is higher the nonlinearity in $n$ is shut off and the strength of the dynamics decreases again. This is why the transport flux $F_e$ for $D = 0.01$ is 7.21 versus 3.06 for $D = 0.1$, namely only twice as large even though the energy content ($E_n = 125$ versus 23.2) is much larger. But the flow energy itself ($E_e = 5.63$ versus 4.47) is only marginally larger. This reflects less efficient coupling despite more content in the $D = 0.01$ case.

The spatial morphology is shown for the $D = 0.1$ case in figure 4.16 and for the $D = 0.01$ case in figure 4.17, both at the $t = 400$ time point. The structure in the two cases looks similar but for $D = 0.01$ is at a larger scale, reflecting the respective peaks in the spectra. One can still see the correlation in the larger scales between $n$ and $\phi$ for both cases. The vortices are not so isolated since they are being continuously generated by the forcing. Some of them are connected by sheet structures which represent vortices having been destroyed by the shearing motion (watch for this again in the three dimensional cases in Chapters 11 and 12). The correlation between

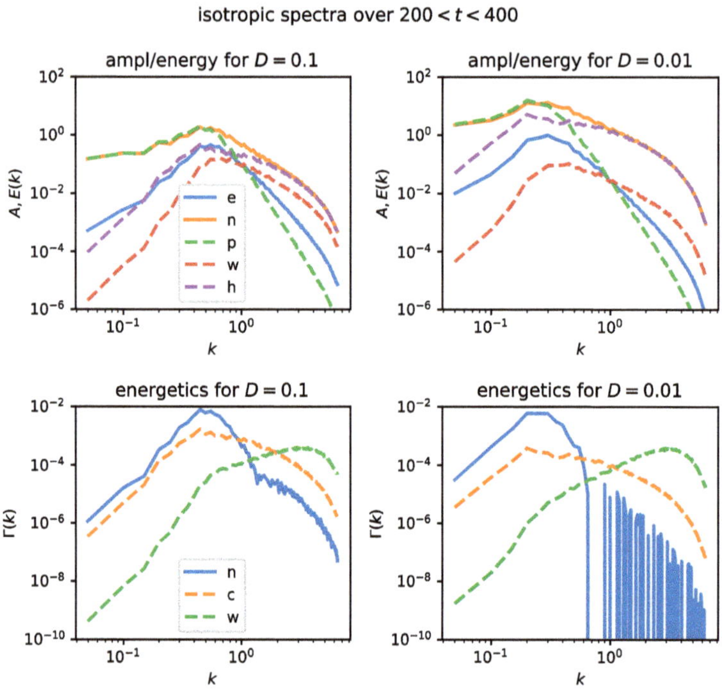


**Figure 4.15.** Perpendicular isotropic spectra for dissipative drift wave cases with $D = 0.1$ and 0.01, showing the relative effect of the coupling. The format is similar to the spectra in figures 4.2 or 4.11, with the addition of $A_h$ tracking $h = n - \phi$ and the energetics $\Gamma_n$ giving the gradient drive, and $\Gamma_c$, $w$ giving the resistive and subgrid dissipation, respectively. One sees that the adiabatic coupling is closer at larger scale than at small, due to the $k^2$ factors in the vorticity nonlinearity. Less coupling ($D = 0.1$) opens the spectral range of non-adiabaticity towards larger scale, and correspondingly more drive, as the time traces of the previous figures show.

the tendency towards isolated vortices for the lowest value of $D$ is clear though, since their elimination by the coupling is more obvious in the $D = 0.1$ case. In both cases, the adiabatic coupling reduces the basic tendency for all the ExB energy to go to larger scales.

### 4.3.6 Energy transfer dynamics

In setting up the isotropic spectra, the net energy transfer in and out of each scale is measured. Net spectral transfer in three quantities $E_e$, $E_n$, and $A_w$ (flow energy, thermal free energy, and enstrophy) is given by

$$T_E(k) = \mathscr{V}^{-1} \int d\mathscr{V}\ \phi^*_{\mathbf{k}}([\phi, \varpi])_{\mathbf{k}}, \tag{4.34}$$

$$T_n(k) = \mathscr{V}^{-1} \int d\mathscr{V}\ (-n)^*_{\mathbf{k}}([\phi, n])_{\mathbf{k}}, \tag{4.35}$$

$$T_w(k) = \mathscr{V}^{-1} \int d\mathscr{V}\ (-\varpi)^*_{\mathbf{k}}([\phi, \varpi])_{\mathbf{k}}, \tag{4.36}$$

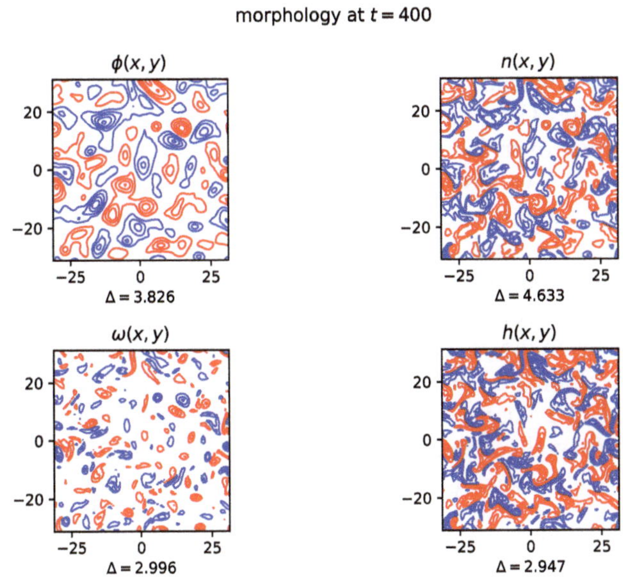


**Figure 4.16.** Spatial morphology of the dependent variables in the dissipative case with $D = 0.1$, relatively tight coupling, especially at long wavelength. The flow eddies and intermittent vortices are confined to very small scale, as the spectra in figure 4.15 indicate. The density remains at smaller scale, due to the coupling.

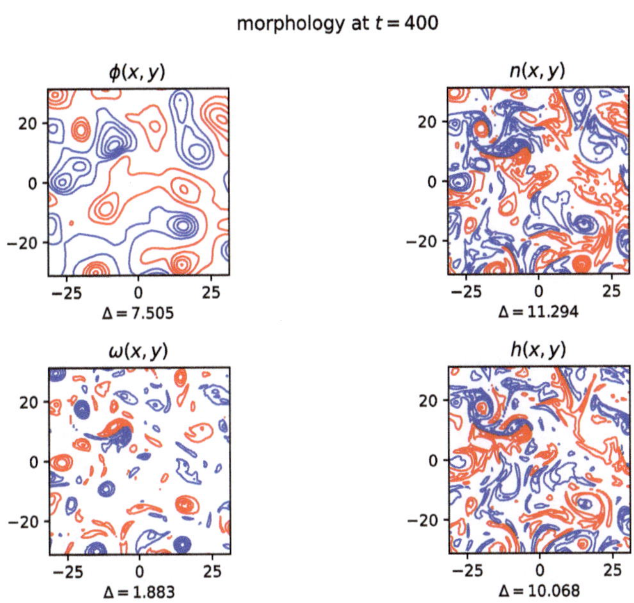


**Figure 4.17.** Spatial morphology of the dependent variables in the dissipative case with $D = 0.01$, weaker coupling, especially at long wavelength. Compare to the previous figure. Note the larger scale of the flow eddies ($\phi$) and the more sheet-like structure of the vorticity.

respectively, where in the second factor the bracket is first calculated and then its Fourier transform is taken. If $T_E(k)$ is positive, there is net transfer into that spectral range through the flow energy, similarly with $T_n(k)$ for the thermal free energy, and $T_w$ for the enstrophy. These are shown in figure 4.18, for the $D = 0.1$ and 0.01 cases, averaged as before over $200 < t < 400$. The symmetric-log format allows both positive and negative values to be shown together. There is continual transfer of energy out of low $k$ and into high $k$ through $T_n$ for $E_n$, while it is out of high $k$ and into low $k$ through $T_E$ for $E_e$. These are balanced by the coupling, the direction of which is $n \to \phi$ due to the drive but which is controlled by the nonlinear processes in the saturated turbulence. The enstrophy $A_w$ is transferred through $T_w$, and we see that some of it goes to low $k$. This only occurs due to the dissipative coupling, which sends excess energy from the thermal free energy, which has the strongest nonlinearity, into the ExB channel at high $k$, forcing more of it to transfer to low $k$, and the enstrophy is brought along with it.

Some of this is elucidated if we look at the details of the transfer. Each three-wave triplet $\mathbf{k} + \mathbf{k}' + \mathbf{k}'' = 0$ in the set contributes to the total. The net transfer from a range $k'$ to $k$ is found by finding the transfer $\mathbf{k}' \to \mathbf{k}$ for each triplet and then finding the transfer between shells $\mathbf{k} \in \mathcal{K}$ for nearest-integer values of $k/k_0$ and $k'/k_0$ by binning the contributions according to $\mathbf{k} \to k$ according to the method used to get the isotropic spectra in the first place (Chapter 3, Eq. (3.36)). The sense is defined as positive for $k' \to k$.

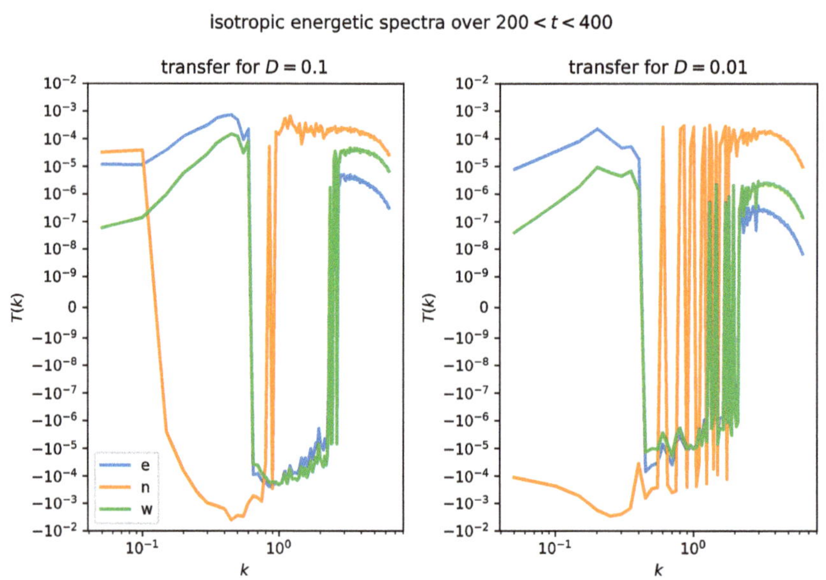


**Figure 4.18.** Isotropic perpendicular energy transfer spectra for the ExB nonlinearities, drawn separately where positive (top row) or negative (bottom). The averaging is over wavenumber shells of a given magnitude and over $200 < t < 400$ in time. The left side is for $D = 0.1$ and the result for $D = 0.01$ is given on the right side. The dual cascade tendency is reflected in the transfer out of $k_\perp \rho_s \sim 1$ to mostly larger scale for ExB energy ($T_e$, 'e'), with simpler transfer out of large scales into small for thermal free energy ($T_n$, 'n'). The enstrophy transfer ($T_w$, 'w') is similar to $T_n$ due to the structures of their respective nonlinearities (cf. Chapter 2).

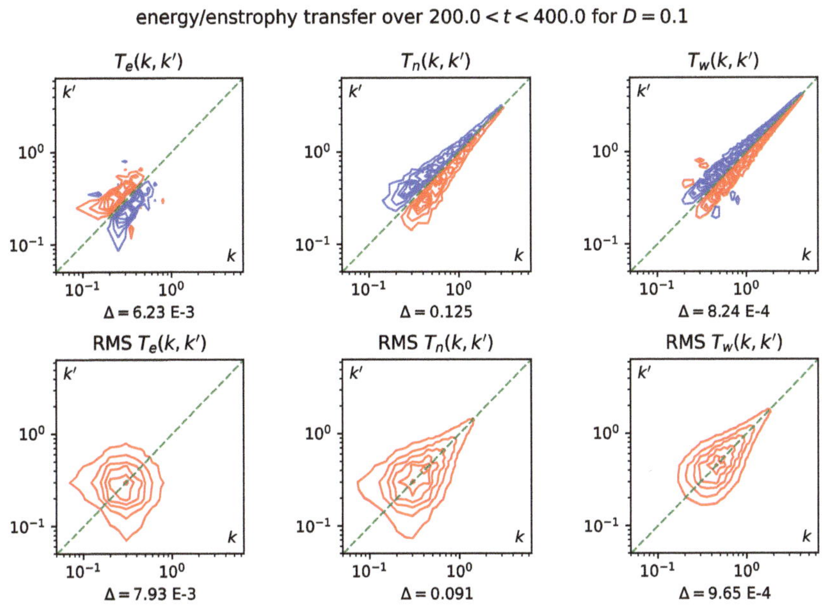


**Figure 4.19.** Isotropic energy transfer spectra, sensed positive for energy transfer from modes $k'$ to modes $k$ within the same spectrum, for the $D = 0.1$ case. The averaging is over wavenumber shells of a given magnitude and over $200 < t < 400$ in time. The ExB energy (left), thermal free energy (centre), and enstrophy (right) transfer is given separately (top row). Positive (red), (negative in blue), above the $k' = k$ line indicates inverse transfer; if below, direct. The RMS transfer is given by contrast in the bottom row. The mean is small compared to the actual activity. Compare with the $D = 0.01$ case in figure 4.20.

The energy transfer dynamics is shown shell by shell in figure 4.19, for $D = 0.1$ and figure 4.20, for $D = 0.01$. In each channel the transfer is defined as positive if the free energy leaves a spectral range $k'$ and enters the spectral range $k$. For the thermal free energy $E_n$ and the enstrophy $A_w$ the passive scalar form is used,

$$T_n(\mathbf{k} \leftarrow \mathbf{k}') = 2C_{\mathbf{kk}'}\mathrm{Re}\left[-n_{\mathbf{k}}n_{\mathbf{k}'}\phi_{\mathbf{k}''}\right], \tag{4.37}$$

$$T_w(\mathbf{k} \leftarrow \mathbf{k}') = 2C_{\mathbf{kk}'}\mathrm{Re}\left[-\varpi_{\mathbf{k}}\varpi_{\mathbf{k}'}\phi_{\mathbf{k}''}\right]. \tag{4.38}$$

For the flow energy we have

$$T_E(\mathbf{k} \leftarrow \mathbf{k}') = 2C_{\mathbf{kk}'}\,\mathrm{Re}\left[-\phi_{\mathbf{k}}\phi_{\mathbf{k}'}\varpi_{\mathbf{k}''}\right], \tag{4.39}$$

and then, for example for $E_e$, the isotropic transfer is given by

$$T_E(k \leftarrow k') = \sum_{\mathbf{k}\in k}\sum_{\mathbf{k}'\in k'} T_E(\mathbf{k} \leftarrow \mathbf{k}'), \tag{4.40}$$

(these are as derived in Chapter 2). The symmetry about $k' = k$ is obvious, and clarified by showing $k = k'$ as a dashed line. Positive activity above the $k' = k$ line indicates an inverse transfer tendency, while a direct transfer is indicated by $T$ positive below the line. The quantitative level of each transfer is given by the contour

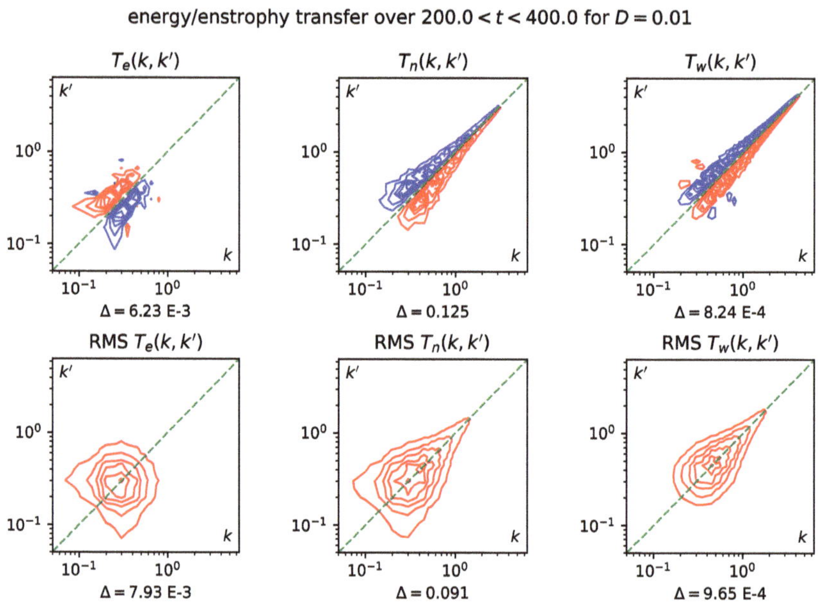


**Figure 4.20.** Isotropic energy transfer spectra, sensed positive for energy transfer from modes $k'$ to modes $k$ within the same spectrum, for the $D = 0.01$ case. The format is the same as in figure 4.19, which shows the $D = 0.1$ case. The basic character is the same, with only the magnitudes sensitive to the strength of the coupling.

interval, which as in other cases is 0.16 of the maximum absolute value. A local cascade is a successive transfer between spectral ranges corresponding to scales within a factor of two of each other, and indeed this is where we find the activity. The RMS values are also shown. Together we find that the standard deviation of any of these quantities is significantly larger than the mean, especially away from the $k = k'$ line. The enstrophy shows small inverse signals at low $k$. The lower the value of $D$, weaker coupling, the larger the transfer dynamics in $T_n$ relative to the others. At larger $D$, the vorticity relative to the overall energy (mostly $E_n$) is much larger, and indeed the absolute value of the vorticity is larger for $D = 0.1$ than it is for $D = 0.01$, which is also driving some of the differences between the coupled case and the hydrodynamic case. The basic form of the isotropic transfer spectra is the same for different values of $D$ with mostly the amplitudes of it changing.

The statistics of the transfer can discern the nature of the process—whether more coherent or more random. The standard deviation of the transfer reflects its magnitude rather than the average, tracking the level of activity, including back-and-forth transfer that can be referred to as scattering. In figure 4.21, the statistics of a representative pair of $k$-shells, transfer between $8k_0$ and $16k_0$, defined positive for inverse transfer towards lower $k$, are given in terms of the PDF of the same three quantities, for both cases. The mean and standard deviation of the PDF are measured and used to normalise them. The blue dashed curves are the Gaussians for the same mean and standard deviation in each case. The direct transfer in $T_n$ is the most statistically significant (the standard deviation is only about three times as

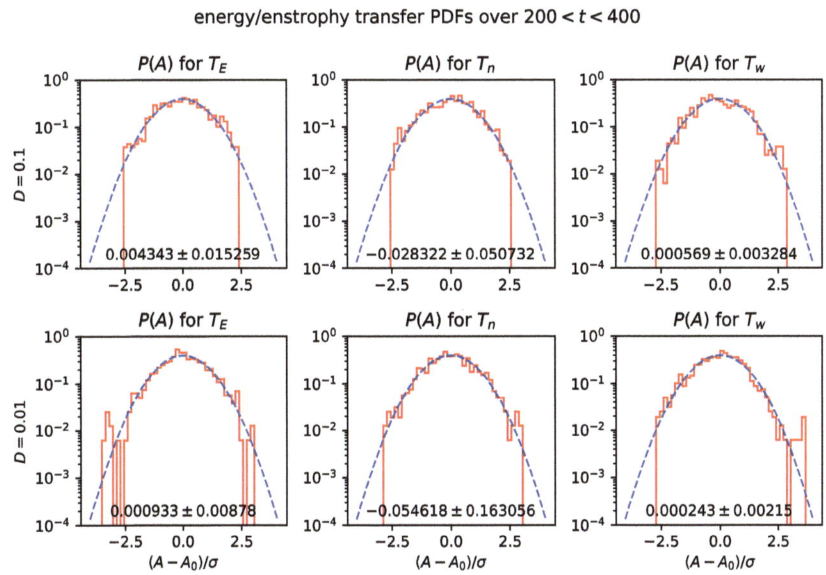


**Figure 4.21.** Statistics of the nonlinear ExB transfer for ExB energy (left frame), thermal free energy (centre), and enstrophy (right), for the case with $D = 0.1$ (top row), and for $D = 0.01$ (bottom). The direct transfer in thermal free energy is the most statistically significant, but the mean is less than the standard deviation in each case.

large as the mean for $D = 0.01$ and twice for $D = 0.1$), but it still has the shape of a Gaussian curve shifted by the mean value. One also sees in this diagnostic the slightly larger values for $T_w$ in the $D = 0.1$ case than for $D = 0.01$, showing directly that the coupling can act to enhance vorticity and make the dynamics more nonlinear compared to linear gradient forcing. The standard deviation is much stronger than the mean in every transfer channel, indicating that the net transfer is a correction on a mostly randomised process.

We learn from all this information that (1) the nonlinear process is turbulence, given by the standard deviation being greater than the mean in the PDFs of all the energy and enstrophy transfer channels, and the near-Gaussian shape in the transfer PDFs, and (2) it is a cascade process, given by the net transfer $k' \leftarrow k$ being mostly near $k' = k$, namely, transfer between similar scales indicative of open triangles $k \sim k' \sim k''$ is dominant. Moreover, (3) it is fundamentally nonlinear, as linear forcing at $k \sim k_0$ would demand that the dissipative coupling give transfer from $n$ to $\phi$. But the nonlinear transfer dynamics is stronger and determines the flow of free energy around the dynamical system.

## 4.4 Implication of the ion gyroradius

This type of dynamics is effectively tied to the scale $\rho_s$ given in Eq. (4.13). It arises because the coupling enforces a state comparable to the adiabatic one as expressed in Eq. (4.11). When this scaling is applied in Eqs (4.5) and (4.8), the perpendicular scale of relevance becomes $\rho_s$. For strongly driven turbulence the energy shifts to larger scales, due to both the inverse cascade tendency in the ExB energy and the larger power opposing the coupling at longer wavelength. But the vorticity remains in the vicinity of $k_\perp\rho_s = 1$ or somewhat less. For cases of interest we will almost never see the vorticity peak as low as $k_\perp\rho_s = 0.3$ let alone low enough that $\rho_s$ can be considered small. The enhanced vorticity due to both the direct cascade tendency in the thermal free energy and then the coupling back to the ExB dynamics at higher $k_\perp$ is usually powerful enough that nonlinear dynamics in the vicinity of $\rho_s$ scales is always involved. Now, $\rho_s$ is not a gyroradius, but $\rho_i$ is, and if $T_i \sim T_e$ then $\rho_i \sim \rho_s$, and we start having to face the finite gyro-(Larmor) radius, what we call FLR effects. It will be helpful to develop the physics of magnetised plasma turbulence before we face this, however, so until later we will be leaving $\rho_i$ out of the computations. Reduced MHD does this by having $e\phi/T_0 \gg n/n_e$ in the above variables, thereby ordering $\nabla_\| n$ to be small in the Ohm's law. But for turbulence driven by thermodynamic gradients the adiabatic response is never negligible. So we will have to consider 'cold-ion' cases with $T_i \ll T_e$ until the time comes to treat FLR effects.

The next Chapter will cover statistical equilibrium and the role of high-$k$ dissipation, with which we complete our study of the purely two-dimensional systems. Then we will go back to the fundamentals of the equations to treat things like quasineutrality and the meaning of low-frequency dynamics in a magnetised plasma, the drift approximation for velocities and heat fluxes, and then the interplay between perpendicular and parallel dynamics in three-dimensional magnetised plasma turbulence.

## Further Reading

The adiabatic and hydrodynamic limits for drift wave turbulence arise from the earliest standard models, developed by A Hasegawa and K Mima 'Stationary spectrum of strong turbulence in magnetized nonuniform plasma' *Phys. Rev. Lett.* **39** (1977) 205; 'Pseudo-three-dimensional turbulence in magnetized nonuniform plasma' *Phys. Fluids* **21** (1978) 87; and by A Hasegawa and M Wakatani 'Plasma edge turbulence' *Phys. Rev. Lett.* **50** (1983) 682; M Wakatani and A Hasegawa 'A collisional drift wave description of plasma edge turbulence' *Phys. Fluids* **27** (1984) 611; respectively.

In a different context, geophysical fluid dynamics, the same equation was found earlier by J G Charney 'On the scale of atmospheric motion' *Geofys. Publ. Norske Vid.-Akad. Oslo* **17** (1948) 1. The earliest attempt at computation was J G Charney, R Fjørtoft and J von Neumann 'Numerical integration of the barotropic vorticity equation' *Tellus* **2** (1950) 237.

# Chapter 5

# Absolute equilibrium ensembles

## 5.1 AEQ and the role of dissipation in turbulence

We have been discussing the spectrum of fully developed turbulence in several different models, including under drive in the dissipatively coupled ExB fluid case. However, we might ask the question of whether turbulence in cases of very small dissipation can function under a complete lack of dissipation. Indeed, there is a line of thought within research on high-temperature plasmas that the physical situation is a collisionless plasma with no dissipation at all. Moreover, in the computational side of that there is another line of thought that any dissipation in the numerical scheme is harmful to the integrity of the result. As it happens, the fate of a finite ensemble of degrees of freedom in turbulent nonlinear interaction has been studied. Such a system is found to reach a statistical equilibrated state, in the sense that the spectrum of each of the conserved quantities fluctuates about a mean which, once established, does not further evolve in time. These are called absolute equilibrium ensembles (AEQ). Initially it was supposed that the energy should be equipartitioned among degrees of freedom. Later the equipartition was taken to be a linear combination of the conserved quantities, should there be more than one, with the ratio among coefficients given by initial conditions. With some invariants being of indefinite sign, such as magnetic helicity in MHD, the equipartition was taken among robust invariants, namely, the positive-definite ones. The set of degrees of freedom as a phase space was taken seriously, in the sense that phase-space conservation should hold for the AEQ concept to be valid. The resulting spectra of the conserved quantities in the AEQ of any system was held to be related to whether each quantity should exhibit direct or inverse cascade tendencies. One also finds interesting consequences in the study of equipartition among each energy channel within a set. Re-introducing dissipation progressively away from zero changes the spectrum from the AEQ result to the more physical one, ultimately showing the role that dissipation takes in physical turbulence. Ultimately, we find that very weak dissipation coefficients, or collisionality, is quite different from zero

doi:10.1088/978-0-7503-2504-2ch5

collisionality. High Reynolds number turbulence is not a situation of zero collisionality, but one in the limit as collisionality goes to zero.

## 5.2 The conserved quantities and equipartition

Equipartition in turbulence was initially conceptualised for the three-dimensional homogeneous case. If the energy is equipartitioned among all degrees of freedom, the meaning is that

$$U_{\mathbf{k}} = \alpha \qquad \text{where} \quad U_{\mathbf{k}} = \frac{1}{2}\,|\mathbf{u}_{\mathbf{k}}|^2. \tag{5.1}$$

These are binned into three-dimensional shells defined by

$$\mathbf{k} \in \mathscr{K} \quad \text{if} \quad k - \frac{1}{2}k_0 < \left(k_x^2 + k_y^2 + k_z^2\right)^{1/2} < k + \frac{1}{2}k_0, \tag{5.2}$$

with Cartesian coordinates $xyz$, where $k_0$ is the fundamental wavenumber (lowest $k$ of the shortest domain size side of the grid). The isotropic spectrum is then

$$U_k = \sum_{\mathbf{k}\in\mathscr{K}} U_{\mathbf{k}}. \tag{5.3}$$

The equipartition in Eq. (5.1) would then exhibit a spectrum of

$$U_k = \alpha' k^2, \tag{5.4}$$

since the factor of $k^2$ gives the relative density of states (number of degrees of freedom within $\mathscr{K}$ for each $k$), and the prime symbol covers all considerations of normalisation. Since this rises with $k$ it cannot result form a cascade process as discussed before. The energy would only find this spectrum after complete redistribution among all available $\mathbf{k}$, i.e. approaching an infinite number of degrees of freedom. The cascade process is assumed to have a sink at higher $k$ even if that is infinitely far off. On the other hand, the AEQ can only be realised in a finite system with a high-$k$ cutoff that in the real world is artificial. Nevertheless, the tendency to build the AEQ spectrum was taken as a manifestation of the cascade process. That is, the dynamics continually tries to fill out the $k^2$ spectrum but is never able to.

In the two-dimensional (2D) case both energy and enstrophy are conserved, with each degree of freedom given by

$$U_{\mathbf{k}} = \frac{1}{2}\,|\mathbf{u}_{\mathbf{k}}|^2 \qquad W_{\mathbf{k}} = \frac{1}{2}k^2\,|\mathbf{u}_{\mathbf{k}}|^2, \tag{5.5}$$

hence the latter is given as $k^2$ times the former at each vector $\mathbf{k}$. The turbulence is only homogeneous in the two dimensions, with the third controlled in our case by the guide field, so the shells are given by

$$k \in \mathscr{K} \quad \text{if} \quad k - \frac{1}{2}k_0 < \left(k_x^2 + k_y^2\right)^{1/2} < k + \frac{1}{2}k_0, \tag{5.6}$$

with Cartesian coordinates $xy$, where $k_0$ is the fundamental wavenumber. The isotropic spectra for energy and enstrophy are

$$U_k = \sum_{\mathbf{k}\in\mathscr{K}} U_{\mathbf{k}} \qquad W_k = \sum_{\mathbf{k}\in\mathscr{K}} W_{\mathbf{k}}. \tag{5.7}$$

The general combination of both invariants is given by

$$\sum_{\mathbf{k}\in\mathscr{K}} (\alpha U_{\mathbf{k}} + \beta W_{\mathbf{k}}) = \sum_{\mathbf{k}\in\mathscr{K}} (\alpha + \beta k^2) U_{\mathbf{k}}. \tag{5.8}$$

Given the equipartition of this quantity among degrees of freedom, we have

$$U_{\mathbf{k}} = \frac{1}{\alpha + \beta k^2} \qquad U_k = \frac{k}{\alpha' + \beta' k^2}, \tag{5.9}$$

where the factor of $k$ gives the relative density of states in the 2D case, and the prime symbol covers all considerations of normalisation. The $\alpha'$ and $\beta'$ can be determined from the value of $E_T$ and $W_T$ in the initial state, given that these are invariant. If $\alpha' \gg \beta'$ then the energy is equipartitioned. If $\beta' \gg \alpha'$ then the enstrophy is equipartitioned. In general, the equipartitioning is mixed, according to the ratio of $\alpha'$ to $\beta'$. At ever larger $k$ in a high-resolution system, the $\beta' k^2$ term should usually become the largest, with enstrophy being equipartitioned. Then, the AEQ energy and enstrophy spectra should be

$$U_k \sim k^{-1} \qquad W_k \sim k^{+1}. \tag{5.10}$$

This was an argument given to support the inverse cascade model for 2D homogeneous turbulence, and afterwards it was used to try to identify inverse or direct cascade tendencies in more complicated cases.

## 5.3 The phase space of degrees of freedom

The ideas of statistical physics supporting the AEQ concept ultimately depend on phase-space conservation. We need to show that the flow in phase space is incompressible, that it, that phase-space volume elements are conserved under the evolution. For the fluid case (2D hydrodynamics) the evolution is given by

$$\frac{\partial \varpi}{\partial t} + [\phi, \varpi] = 0 \qquad \nabla_\perp^2 \phi = \varpi, \tag{5.11}$$

which in terms of the wavenumbers is

$$\frac{\partial \varpi_{\mathbf{k}}}{\partial t} = C_{\mathbf{k}\mathbf{k}'}\left(\varpi_{-\mathbf{k}''} \phi_{-\mathbf{k}'} - \varpi_{-\mathbf{k}'} \phi_{-\mathbf{k}''}\right), \tag{5.12}$$

where $\mathbf{k} + \mathbf{k}' + \mathbf{k}'' = 0$ and $C_{\mathbf{k}\mathbf{k}'} = \frac{1}{2}\,\hat{\mathbf{s}} \cdot (\mathbf{k} \times \mathbf{k}')$ serves as a coupling coefficient. The antisymmetry of the spatial Poisson bracket translates to the pair of terms, which is anti-symmetric in the $\mathbf{k}' \leftrightarrow \mathbf{k}''$ exchange. The divergence of the phase-space flow is given by

$$\sum_{\mathbf{k}} \frac{\partial}{\partial \varpi_{\mathbf{k}}} \frac{\partial \varpi_{\mathbf{k}}}{\partial t}, \tag{5.13}$$

with $\phi_{\mathbf{k}} = -k^2 \varpi_{\mathbf{k}}$. This is zero unless one of the $\mathbf{k}'$ or $\mathbf{k}''$ is either $\mathbf{k}$ or $-\mathbf{k}$. The only possibilities for that are in either the triplet $\{\mathbf{k}, -\mathbf{k}, 0\}$ or $\{\mathbf{k}, \mathbf{k}, -2\mathbf{k}\}$, in which case the $C_{\mathbf{k}\mathbf{k}'}$ vanishes. The phase-space incompressibility condition is therefore satisfied.

## 5.4 Computational verification

The AEQ concept was always regarded as difficult to verify via direct numerical simulation, since the rising spectra in $k$ cause difficulty at the high-$k$ end. The nonlinearities are more powerful there in an AEQ than in evolved turbulence, which is given a dissipation range. We will see, however, that the numerical scheme we have been using is robust under these conditions. It can be regarded to be a shame that the early studies in 2D computational turbulence were done with time step schemes that were suboptimal, either with respect to stability or to dispersion or, even worse, to a scheme-dependent dissipation which manifested itself in some form of artificial structure. The time of peak interest in AEQ systems was prior to the establishment of the stiffly stable time step scheme in the years after 1990. The combination of this time step with the Arakawa bracket discretisation allows one to run for an essentially indefinite time. It turns out that the AEQ results are readily accessible at least in 2D model systems, which can be evolved without additional artificial subgrid dissipation under a high degree of maintenance of all conserved quantities in the model. We will do this with the fluid case first.

### 5.4.1 The fluid case with passive scalar

This case uses the limit of the dissipatively coupled ExB fluid case in which the coefficient is zero, so that the fluid is hydrodynamic with conserved enstrophy and the density variable is kinematically (passively) advected. The equations are

$$\frac{\partial \varpi}{\partial t} + [\phi, \varpi] = 0, \tag{5.14}$$

$$\frac{\partial n}{\partial t} + [\phi, n] = 0, \tag{5.15}$$

$$\nabla_\perp^2 \phi = \varpi, \tag{5.16}$$

with all forms of drive and dissipation set to zero.

The three conserved quantities are ExB energy, enstrophy, and thermal free energy given by

$$E_e = \mathcal{V}^{-1} \int d\mathcal{V}\, \frac{1}{2}(-\phi)\varpi \quad A_w = \mathcal{V}^{-1} \int d\mathcal{V}\, \frac{1}{2}\varpi^2 \quad E_n = \mathcal{V}^{-1} \int d\mathcal{V}\, \frac{1}{2} n^2, \tag{5.17}$$

respectively. Their time derivatives are found by multiplying Eq. (5.14) by $-\phi$, Eq. (5.15) by $n$, and Eq. (5.14) by $\varpi$, respectively, and averaging over the spatial domain in the same way as obtaining $E_e$, $A_w$, and $E_n$, through their densities given by their respective integrands. In addition to $E_e$, $A_w$, and $E_n$ we measure

$$A_p = \mathcal{V}^{-1} \int d\mathcal{V} \, \frac{1}{2}\phi^2, \tag{5.18}$$

for comparison. Each of these have spectra measured in the way given above: with $k_0$ chosen with the grid, the spectrum amplitudes of individual degrees of freedom is calculated first, and then binned according to integer multiples of $k_0$ as

$$A_p(k) = \sum_{\mathbf{k}\in\mathcal{K}} \frac{1}{2} \, |\phi_{\mathbf{k}}|^2. \tag{5.19}$$

All quantities are done accordingly.

Since there is no energy transfer between the equations (i.e. between $E_e$ and $E_n$) it matters not what the initial relative levels of $\phi$ and $n$ are. It will be instructive to launch them out of an adiabatic initial state with a spectrum peaked in the middle of the range. A random-phase initial state is set up for $n = \phi$ with $\varpi$ constructed from Eq. (5.16). For illustrative purposes we can start small, with a domain of $64 \times 64$ grid points in $x$, $y$ and verify results on larger systems afterwards. The spectral range is

$$0.1 < k_x, \, k_y < \pi, \tag{5.20}$$

and $n$ and $\phi$ are set up with an RMS amplitude of 1.0 with spectrum and random phase ($\Theta \in [-\pi, \pi]$) given by

$$\phi_{\mathbf{k}} = A_0 k (1 + k^6)^{-1} \exp(i\Theta), \tag{5.21}$$

with the steepness set to give reasonable time for evolution of the high-$k$ side (recall $k^2 = k_x^2 + k_y^2$). The system is evolved from $t = 0$ with a time step of $10^{-2}$ with short runs going only to 20 or 50 and long ones to 500 or 5000, always recalling the experience that the saturation time for reasonable driven/dissipative cases is in the 200 range. In other words, an equilibration that takes well over 1000 is of less interest than the signatures we will find for the shorter cases. Readers of computational bent will be able to reproduce most of these in a few minutes each at the slowest, on a reasonable laptop.

We examine a short run to $t = 120$ with time traces sampled every $\Delta t = 0.2$ and snapshots every 2.0 to see the significant early evolution out of the steeply peaked initial state. Time traces of the four quantities given above are shown in figure 5.1. The squared amplitude of $\phi$, given by $A_p$, rises with time due to the shift to larger scale of the ExB activity. The other three quantities, $E_e$, $A_w$, and $E_n$, are conserved to high accuracy. The evolution of the isotropic spectra is given in figure 5.2, for the initial state $t = 0$, and the time points $t = 2$, 4, and 6. The dotted lines show the $k^{\pm 1}$ signatures for comparison. The spectra of $E_e$, $E_n$, and $A_w$ (labelled 'e', 'n', and 'w', respectively) evolve separately, with the high-$k$ side changing rapidly. $E_n(k)$ quickly follows $A_w(k)$ while $E_e(k)$ rises more slowly at high $k$. On a more moderate time

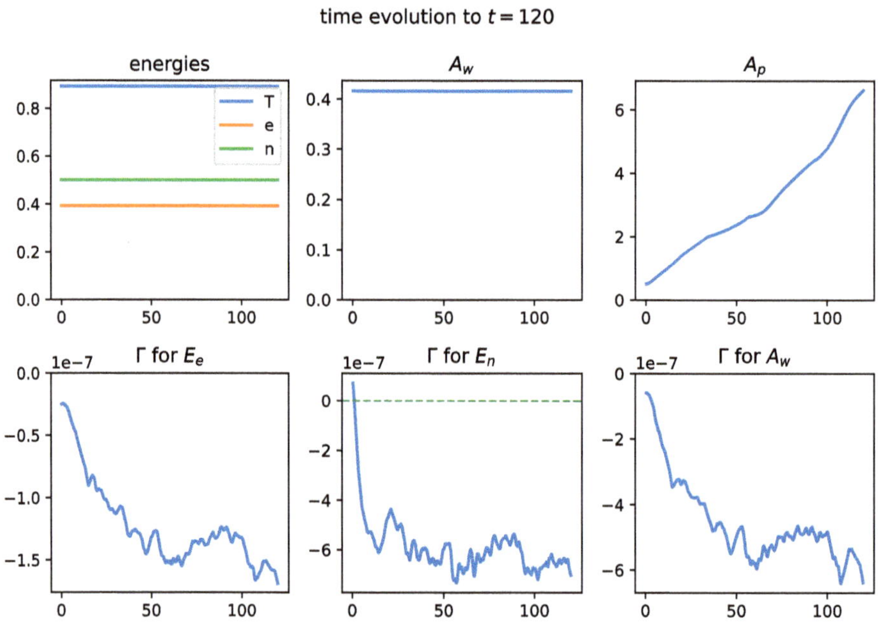


**Figure 5.1.** Time traces of the evolution of the hydrodynamic case out of a peaked spectrum initial state. The statistics shown are measured over the last half of the run (also true in all later cases). The three conserved quantities, ExB energy ($E_e$, 'e'), thermal free energy ($E_n$, 'n'), and enstrophy ($A_w$, 'w'), are static to high accuracy. The potential amplitude ($A_p$, 'p') rises following the dual cascade of energy to large scale and enstrophy to small.

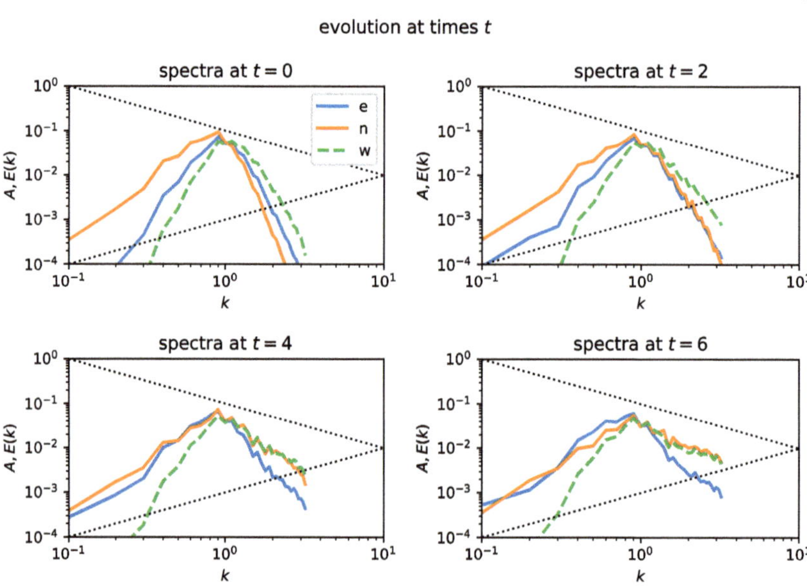


**Figure 5.2.** Isotropic perpendicular spectra for the hydrodynamic case as it evolves out of the peaked initial state (left to right, top to bottom). The $k^{\pm 1}$ lines are shown for comparison. The enstrophy ('w') and density ('n') equilibrate at high $k$ the fastest, with the energy ('e') shifting to lower $k$ more slowly.

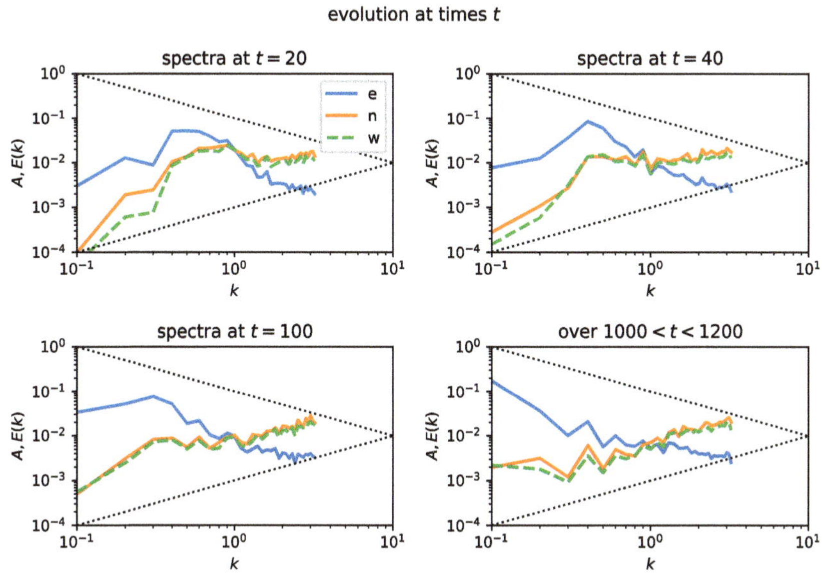


**Figure 5.3.** Isotropic perpendicular spectra for the hydrodynamic case as it continues to evolve out of the peaked initial state (format as in figure 5.2). The enstrophy and density are rigidly equilibrated at high $k$ while the low-$k$ portion settles in. Due to the multiple factors of $k$ in the nonlinear advection, the relaxation time scales are wildly disparate, in a way similar to diffusion.

scale, shown in figure 5.3, the central peak spreads out as the equilibrated spectra start to form, passing through the time points $t = 20$, 40, and 100, by which time the range $k > 1$ has already found the equilibrated form. The last frame shows the eventual equilibrated form in the spectra averaged over $1000 < t < 1200$ with 11 snapshots, and these are in fact the expected AEQ signatures of $k^{+1}$ for $E_n$ and $A_w$, and $k^{-1}$ for $E_e$. Most of the evolution has already occurred up to the $100 < t < 200$ range, but it takes longer to equilibrate the entire spectrum. This is a little like watching diffusion: one knows the result and understands the process early within the evolution, but putting the last part of the low-$k$ range into place takes longer than is interesting. As with diffusion, the time scale of changing evolution scales with $k^{-2}$, progressively slower for the larger scales. The main point of this is that the AEQ indeed do form as expected, very quickly for the short wavelengths but eventually so for the entire range.

## 5.5 Equipartition among the energies

We have seen that the absolute equilibrium ensemble produces a sort of equipartition among the degrees of freedom, in that each vector wavenumber acquires the same enstrophy in the statistical sense. The enstrophy is the one that equipartitions because its redistribution rate is faster while both energy and enstrophy are conserved. We are also interested in other systems, the next level of complication being the MHD one. In this case enstrophy is no longer conserved because of the magnetic stress term represented by $\mathbf{B} \cdot \nabla J = [-\psi, J]$, so we might expect the MHD energy spectrum to exhibit the $k^{+1}$ dependence of the hydrodynamic enstrophy rather than the $k^{-1}$ distribution taken on by the energy in the fluid case. In addition to the issue of the overall spectrum, we are interested in the various fates of the kinetic (ExB fluid) and magnetic energies. Should these evolve into a final state in which these two channels have equal energy or not? We will find an interesting result of directionality in this question. The result from an initial state of pure magnetic energy is different from the opposite case, even if the initial magnetic energy is small but finite. That is, zero magnetic energy always stays zero, because the evolution of the magnetic flux $\psi$ depends on magnetic stress nonlinearities which contain $\psi$ as one of the pieces. This will point to some of the properties of the three-dimensional guide field case which we will look at later.

### 5.5.1 The MHD case

In the MHD case the equations are

$$\frac{\partial \varpi}{\partial t} + [\phi, \varpi] = [-\psi, J], \tag{5.22}$$

$$\frac{\partial F_\psi}{\partial t} + [\phi, F_\psi] = 0, \tag{5.23}$$

$$\nabla_\perp^2 \phi = \varpi \qquad \psi + d_e^2 J \equiv \left(1 - d_e^2 \nabla_\perp^2\right)\psi = F_\psi, \tag{5.24}$$

with all forms of drive and dissipation set to zero. The same grid of $64 \times 64$ and $k_0 = 0.1$ is used. The parameter $d_e$ is set to 0.025, placing the skin depth point outside the spectrum to the high-$k$ side.

The enstrophy is no longer conserved due to the $[-\psi, J]$ term, so we concentrate on the energy components given by

$$E_e = \mathscr{V}^{-1} \int d\mathscr{V}\, \frac{1}{2}(-\phi)\varpi \qquad E_b = \mathscr{V}^{-1} \int d\mathscr{V}\, \frac{1}{2} J F_\psi. \tag{5.25}$$

These are the ExB and magnetic energies, respectively, with the latter containing the electron flow kinetic energy density $d_e^2 J^2/2$. Their time derivatives are found by multiplying Eq. (5.22) by $-\phi$, Eq. (5.23) by $J$, respectively, and averaging over the spatial domain in the same way as obtaining $E_e$ and $E_b$ through their densities given by their respective integrands. The energy pieces $E_e$ and $E_b$ exchange through the $[-\psi, J]$ term, shown by constructing

$$\frac{\partial E_e}{\partial t} = \mathscr{V}^{-1} \int d\mathscr{V}\, (-\phi)[-\psi, J] = \mathscr{V}^{-1} \int d\mathscr{V}\, J[-\psi, \phi], \tag{5.26}$$

$$\frac{\partial E_b}{\partial t} = \mathscr{V}^{-1} \int d\mathscr{V}\, (-J)[\phi, F_\psi] = \mathscr{V}^{-1} \int d\mathscr{V}\, (-J)[-\psi, \phi], \tag{5.27}$$

as studied in Section 3.3.3 of Chapter 3, in which the divergence property of the spatial bracket is used. Due to this magnetic stress coupling, only the total energy given by

$$E_T = E_e + E_b, \tag{5.28}$$

is conserved, so finite change in $E_T$ is used as a diagnostic.

The issue of equipartition therefore arises, i.e. whether we should expect that the AEQ state should have equal $E_e$ and $E_b$. The limiting case of zero initial $\psi$ clearly exhibits $E_b = 0$ for all time, but the issue remains for any finite initial $\psi$. We can test this by setting $\phi$ with the same initial state as for the fluid case, with RMS level 1.0, and $\psi$ with the same spectrum but with different RMS levels. These will be the kinematic cases (especially so if $E_b \ll E_e$). Alternatively, we set up $\psi$ with the RMS level 1.0, and then $\phi$ with the same spectrum but with different RMS levels. These will be the magnetic cases. The equal-start case is with $\psi_0 = \phi_0 = 1$, forming the transitional case.

We start with the magnetic cases. The equal-start case with $\psi_0 = \phi_0 = 1$ is showcased. While the energies start equal in total, the details of the distribution is not equilibrated. Time traces of the energy components are shown in figure 5.4, sampled every $\Delta t = 0.2$, with averages over $1000 < t < 1200$ given. Initially they separate before evolving back together, with statistical equilibration setting in later on. For $t > 10$ the convergence starts, and after $t > 100$ the energy components remain close. (One can compare the early stages of this with the cases in Section 3.3.3 of Chapter 3.) The ratio $E_b/E_e$ is 0.409/0.375 for $100 < t < 110$ and 0.400/0.384 for $190 < t < 200$. Afterwards the relaxation is much slower. The evolution of the

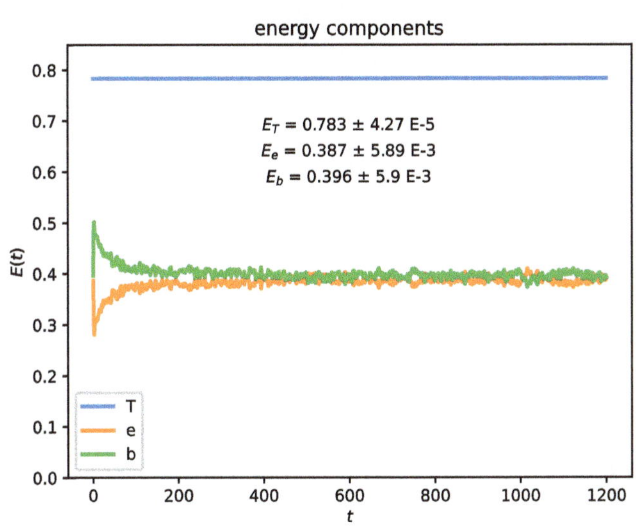


**Figure 5.4.** Time traces of the evolution of the equal-start ($\phi_0 = \psi_0$, in RMS level) MHD case out of a peaked spectrum initial state. Evolution of the ExB ($E_e$, 'e') and magnetic ($E_b$, 'b') jumps in opposite directions before returning to a level close to statistical equilibration. We will see later that this only occurs when $E_b > E_e$ in the beginning.

isotropic spectra shows how this works. These are given in figure 5.5, for the $t = 0$, 20, and 60 time points and the average over $1000 < t < 1200$ after equilibration (11 snapshots). The initial state shows the spectra of $E_e = E_b$ with those of $A_p$ and $A_w$ as in the fluid case. Already by $t = 20$ the range $k > 2$ is already equipartitioned. While the signature of $k^{+1}$ is not yet reached, it is significant that the high-$k$ part of $E_e$ is already well away from the fluid case's $k^{-1}$ and following $E_b$. By $t = 60$ the range $k > 1$ is equilibrated and also follows the $k^{+1}$ signature. These spectra averaged over $1000 < t < 1200$ (11 snapshots) are also shown for comparison, with the spectra in the $k^{+1}$ signature and well equilibrated over most of the range. This shows that the AEQ for the MHD case with approximately equal energies is very different from the fluid case. The ability to exchange energy through the magnetic stress term allows the ExB (fluid kinetic) energy to accumulate at the high-$k$ side representing shorter wavelengths.

The additional magnetic cases are with

$$\phi_0 = \{10^{-2}, 10^{-6}\} \qquad \psi_0 = 1. \tag{5.29}$$

The time traces for the $10^{-6}$ case are shown in figure 5.6, sampled and averaged as before. With $E_e$ initially the smaller, these cases equipartition quickly, on the same time scale as equilibration. The time traces are much the same after the time point

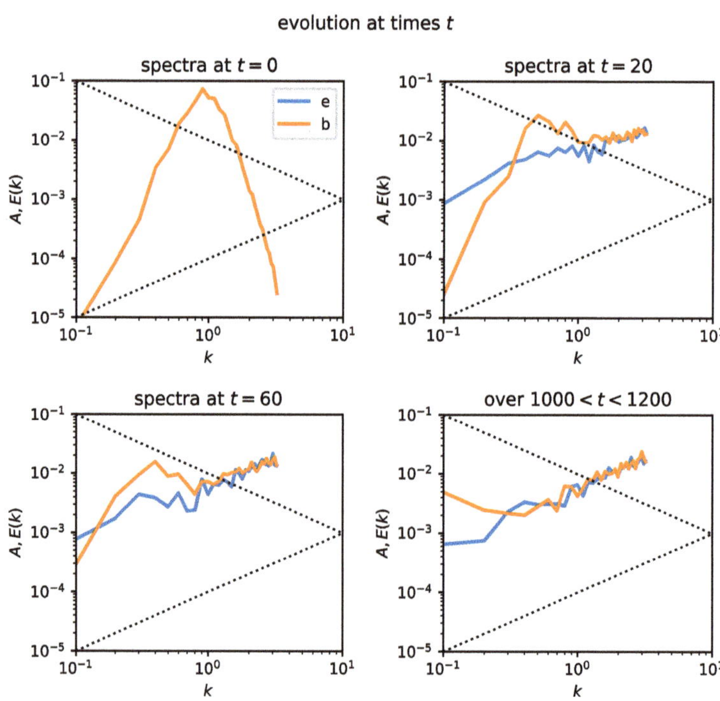


**Figure 5.5.** Isotropic perpendicular spectra for the equal-start MHD case as it relaxes toward equipartition of the ExB and magnetic energy components. The high-$k$ portion of the magnetic energy statistically equilibrates to the $k^{+1}$ line, taking the fluid energy with it, as otherwise the latter would go to a $k^{-1}$ line, as seen for the hydrodynamic case in the previous figures.

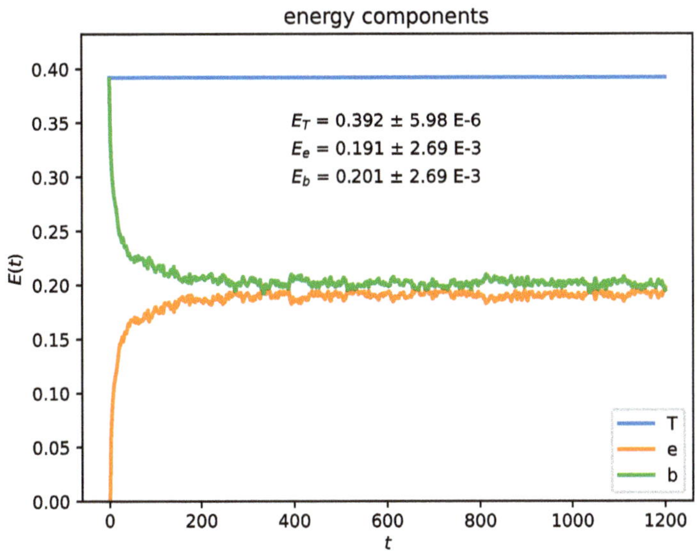


**Figure 5.6.** Time traces of the evolution of the MHD magnetic case with $\phi_0 = 10^{-6}\psi_0$, in RMS. Compare with figure 5.4. The magnetic energy drops and the ExB energy rises, with the total well conserved. The two channels evolve toward and then remain in near equipartition.

$t = 200$, with an initial reaction phase over by $t = 40$ and most of the evolution having passed by $t = 100$. The process is shown by the isotropic spectra, given in figure 5.7, for the $t = 0$, 20, and 60 time points and the average over $1000 < t < 1200$ after equilibration (11 snapshots). The spectra are in the same equipartitioned state with both $E_e$ and $E_b$ showing the $k^{+1}$ signature. All cases with $\psi_0 > \phi_0$ with the same initial spectrum end in this equipartitioned AEQ state.

The kinematic cases now receive the showcase, because they will decide this issue. We have seen that initial magnetic energy transfers into fluid kinetic energy through the field line bending nonlinearity. On the other hand, initial fluid energy has to transfer through the parallel gradient nonlinearity $[-\psi, \phi]$ involving both stream potentials (for **B** and **u**, respectively). And in the Ohm's law, there is an additional conserved quantity to address, namely

$$\frac{\partial A_a}{\partial t} = 0 \qquad \text{where} \quad A_a = \mathcal{V}^{-1} \int d\mathcal{V} \, \frac{1}{2}\psi^2, \tag{5.30}$$

the amplitude of the flux variable, whose square is conserved in two-dimensional MHD, as noted in Chapter 3. The equation for $\psi$ involves only parallel gradients, or brackets of variables with $\psi$, and so $\psi$ and also $\psi^2$ are passively advected by the nonlinearities. Conservation of $A_a$ is broken only by finite electron inertia or resistive dissipation.

Comparing with the magnetic cases, and the equal-start case, we assign

$$\phi_0 = 1 \qquad \psi_0 = \{1, 10^{-2}, 10^{-6}\}, \tag{5.31}$$

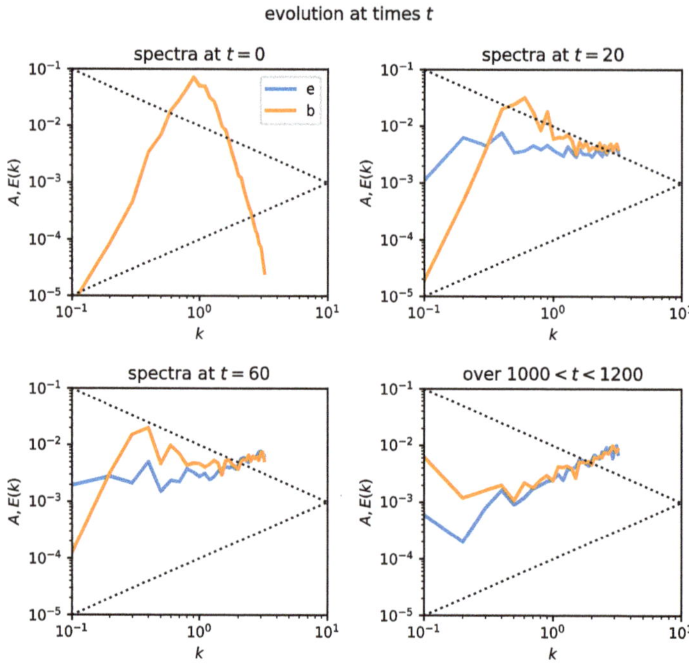


**Figure 5.7.** Isotropic perpendicular spectra for the MHD magnetic case with $\phi_0 = 10^{-6}\psi_0$, in RMS. Compare with figure 5.5. The ExB and magnetic energy components relax to equilibration and equipartition in the same way as in the equal-start case.

with $\varpi = \nabla_\perp^2\phi$ and $J = -\nabla_\perp^2\psi$ set accordingly. The time traces for the $10^{-2}$ and $10^{-6}$ cases are shown in figures 5.8 and 5.9, respectively, sampled and averaged as before. For these cases we also follow $A_a$ to see how it evolves when $E_b$ would tend to rise to equipartition with $E_e$. Statistical equilibration of the spectrum proceeds on similar time scales to the $\psi_0 = 1$ case, but the equipartition between $E_e$ and $E_b$ is now seen not to be universal. The cases with lower values of initial $E_b$ show it coming up, but only to an equilibrated value well below $E_e$, with $E_e$ becoming more and more nearly conserved when $E_b$ stays small.

The equilibrated isotropic spectra averaged over $1000 < t < 1200$ (11 snapshots) are shown in figure 5.10, in which for visibility $E_b$ is multiplied by a scale factor equal to $\psi_0/\phi_0$ squared. It is interesting to note that the shapes of these spectra evolve differently now, especially in the case with very low $E_b$. Instead of the $k^{+1}$ signature, the $E_e$ spectrum in the $\psi_0 = 10^{-2}$ and $10^{-6}$ cases equilibrates much as in the fluid case towards the $k^{-1}$ signature given by the limiting fluid AEQ. The resolution of these findings concerning the evolution of the energy components is given by the evolution of the flux amplitude. In the time traces we saw that $A_a$ remains constant while $E_b$ is in evolution. The spectra show that this evolution proceeds mostly by shifting as much of the flux as possible to smaller scale. This increases the magnetic energy, but only so far as the flux amplitude itself reaches the equipartition spectrum with the $k^{+1}$ signature. Once the flux amplitude has equilibrated, the magnetic energy ceases to evolve.

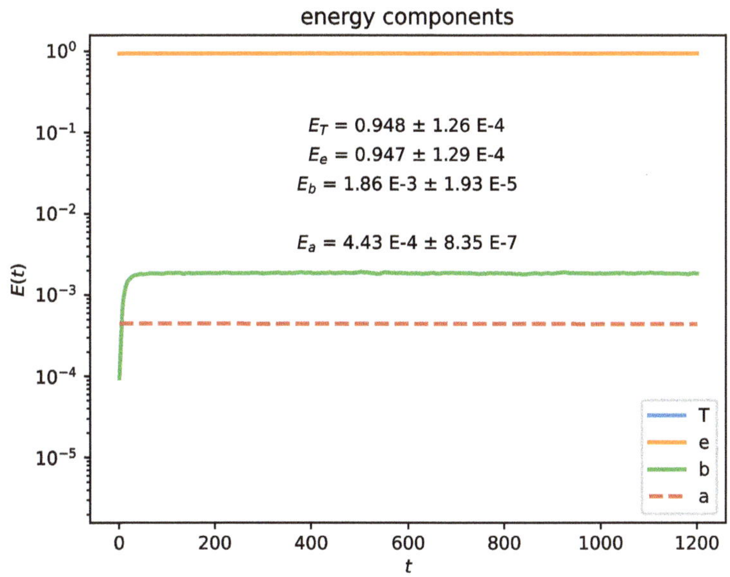


**Figure 5.8.** Time traces of the evolution of the MHD kinematic case with $\psi_0 = 10^{-2}\phi_0$, in RMS. Compare with figure 5.6. The magnetic energy ('b') tries to come up but it is stopped by the constraint that the flux amplitude ('a') has to be conserved. We investigate the details of this action as a constraint next.

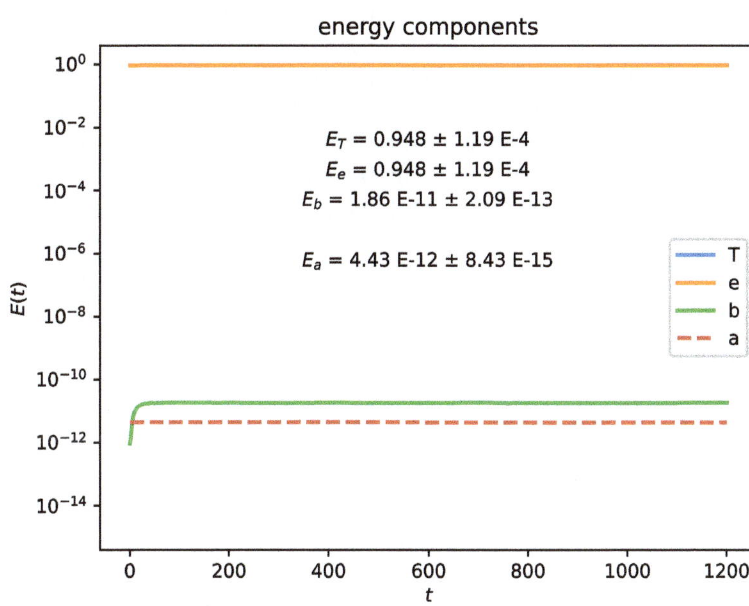


**Figure 5.9.** Time traces of the evolution of the MHD kinematic case with $\psi_0 = 10^{-6}\phi_0$, in RMS. Compare with figures 5.6 and 5.8. The magnetic energy ('b') tries to come up but it is stopped even more effectively than for the case with more magnetic energy in the beginning. The constraint due to conservation of the flux amplitude ('a') is tighter here.

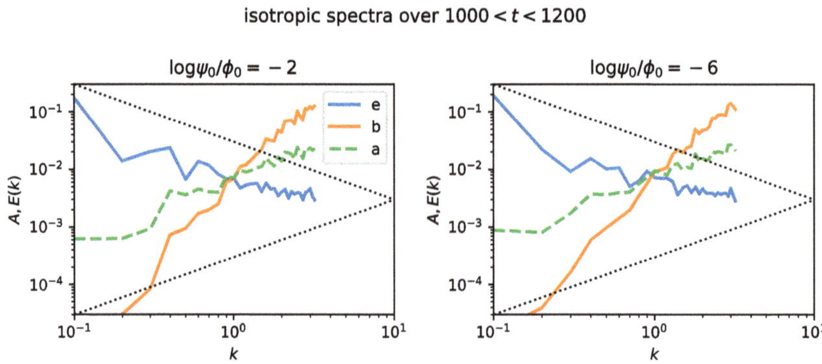


**Figure 5.10.** Isotropic perpendicular spectra for the MHD kinematic case with $\psi_0 = 10^{-2}\phi_0$, in RMS. Compare with figures 5.2 and 5.7. Now the fluid energy ('e') equilibrates as if it were by itself, and the magnetic energy ('b') assumes a steep positive-power $k$-spectrum. The latter is caused by rigid conservation of the magnetic flux amplitude ('a'), while the energy can evolve. Hence, it is $A_a$, not $E_b$, which sets into the equilibrated spectrum.

The most interesting intermediate cases are those with $\psi_0 = 0.1$, 0.15, 0.3, and 0.5. The case with $\psi_0 = 0.3$ has $E_b$ rising to about 0.19 and $E_e$ falling to 0.28 from the initial 0.4, while the case with $\psi_0 = 0.5$ shows $E_b$ rising to about 0.24 and $E_e$ falling to about 0.25 (numbers given are averages over the range $100 < t < 120$). By overall amplitude, this 0.5 case looks equipartitioned. The 0.1 case shows very small $E_b$, less than 0.02. There is apparently a transition point for the magnetic energy below which the constraint of the flux amplitude conservation effectively stops the equipartition.

The role of the flux amplitude conservation appears next, in the equilibrated isotropic spectra for these cases, in figure 5.11. The tendency to equipartition starts at the shortest wavelength, and when the amplitude of $E_b$ is large enough to compete with $E_e$ somewhere in the spectrum. The 0.5 case shows close equipartition hence also the same $k^{+1}$ signature in both components. The 0.3 case shows $E_b(k) < E_e(k)$ throughout the range, but the influence of the magnetic stresses is enough for $E_e(k)$ to be still close to the $k^{+1}$ signature. The 0.1 case has $E_e(k)$ mostly close to $k^{-1}$ but flattening slightly at the high-$k$ end. The 0.15 case is really the only one that is transitional between the two signatures. The key to this is the spectrum of the flux amplitude, $A_a(k)$. For small magnetic energy, $E_b \ll E_e$, the magnetic flux equilibrates by itself, unlike the magnetic energy being conserved, and so does the fluid energy. And the reason for that is that $A_a$ cannot change. As $E_b$ rises by shifting as much as possible to higher $k$, it is constrained by the constancy of $A_a$ that it cannot rise uniformly. We now see why the fluid energy does not transfer effectively into magnetic energy in these two-dimensional cases.

For cases with dominant or concurrent magnetic energy, that quantity controls statistical equilibration, and because neither fluid energy nor enstrophy are conserved in the face of the magnetic nonlinearity, the fluid energy is able to rise to meet the magnetic component. The magnetic component drops by half in turn, which now sends more of the flux to lower $k$, larger scale. Hence the energy, not the flux, is the

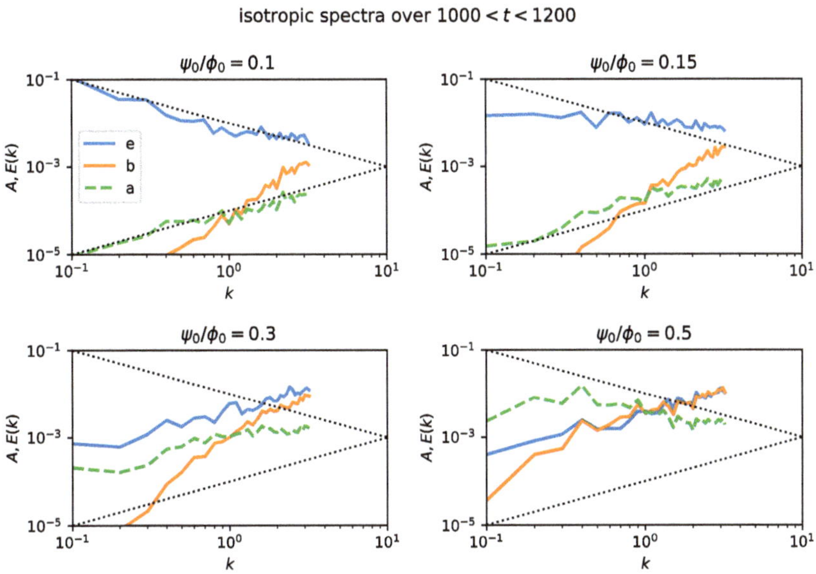


**Figure 5.11.** Isotropic perpendicular spectra for several MHD kinematic cases with varying relative levels $\psi_0/\phi_0$, in RMS. Compare with figure 5.7 and 5.10. At levels below $10^{-1}$ the ExB fluid energy evolves as if by itself, but with more magnetic energy present the field line bending effect opens a transfer with the magnetic energy, and the equipartition process starts, at least at high $k$. Closer equipartition requires more initial magnetic energy, so that it can occur while still conserving the flux amplitude.

quantity reaching statistical equilibration, and the fluid energy equipartitions with it at all scales. When the energy components are comparable, some mix of these states occurs, according to the competition between the pieces which have the strongest tendency to equilibrate.

We find two things in the end: magnetic stresses make a difference between the fluid and MHD AEQ states, with a sharp transition occurring in the vicinity $E_b/E_e \sim 0.1$. Then, cases with $\psi_0/\phi_0 < 0.5$ do not find energy equipartition, and cases with $E_b/E_e < 0.1$ show the AEQ state of the fluid case for fluid energy and enstrophy, and a steep signature of $E_b(k) \sim k^{+2}$ for the magnetic energy. In both the latter two sets of cases, the outcome is caused by realisation of the AEQ state for the magnetic flux amplitude, not the energy. The relevance of $A_a$ as a conserved quantity is to constrain the evolution of $E_b$ due to other factors. This provides the reason why the magnetic energy cannot equilibrate if it starts out much smaller than the fluid energy. It then stays there. We can call this the kinematic MHD regime.

### 5.5.2 The four-field two-fluid MHD case

We briefly repeat this exercise for the four-field MHD case. The equations are

$$\frac{\partial \varpi}{\partial t} + [\phi, \varpi] = [-\psi, J], \tag{5.32}$$

$$\frac{\partial F_\psi}{\partial t} + [\phi, d_e^2 J] = [\psi, \phi - n], \tag{5.33}$$

$$\frac{\partial n}{\partial t} + [\phi, n] = [\psi, u - J], \tag{5.34}$$

$$\frac{\partial u}{\partial t} + [\phi, u] = [\psi, n], \tag{5.35}$$

$$\nabla_\perp^2 \phi = \varpi \qquad \psi + d_e^2 J \equiv \left(1 - d_e^2 \nabla_\perp^2\right)\psi = F_\psi, \tag{5.36}$$

with all forms of drive and dissipation set to zero. The same grid of 64 × 64 and $k_0 = 0.1$ is used. The parameter $d_e$ is set to 0.025, placing the skin depth point outside the spectrum to the high-$k$ side.

This time there are four energy components given by

$$E_e = \mathscr{V}^{-1} \int d\mathscr{V}\, \frac{1}{2}(-\phi)\varpi \qquad E_b = \mathscr{V}^{-1} \int d\mathscr{V}\, \frac{1}{2} J F_\psi, \tag{5.37}$$

$$E_n = \mathscr{V}^{-1} \int d\mathscr{V}\, \frac{1}{2} n^2 \qquad E_u = \mathscr{V}^{-1} \int d\mathscr{V}\, \frac{1}{2} u^2, \tag{5.38}$$

respectively. Their time derivatives are found by multiplying Eqs (5.32)–(5.35) by $-\phi$, $J$, $n$, and $u$, respectively, and averaging over the spatial domain as before. The spectrum of each is done as in the fluid and MHD cases. Since the magnetic stresses couple all the equations, only the total energy given by

$$E_T = E_e + E_b + E_n + E_u, \tag{5.39}$$

is conserved, with $E_T$ used as a diagnostic. The grid size, spectral range, time step, and sampling are set the same as before.

As in the MHD case the magnetic stresses suffice to excite all the other degrees of freedom, while if $\psi_0 = 0$ then no magnetic disturbances will evolve. The case with $\psi_0 = 1$ (setting $J = -\nabla_\perp^2 \psi$) and all others zero then serves as the nominal one. Only the magnetic energy $E_b$ starts nonzero but the time traces in figure 5.12 show that the others rise and equilibrate on the same time scales as in the MHD case with only $\psi_0$ finite. No new time scales are introduced. The equilibration is on the 100 or 200 time scale, with approximate equipartition such that $E_u$ is about 10% lower than the others. The process is shown by the isotropic spectra, given in figure 5.13, for the $t = 0$, 20, and 60 time points and the average over $1000 < t < 1200$ after equilibration (11 snapshots). Equipartition is evident in all but the first few $k$-shells. The low-$k$ pieces are deceptive, as most of the energy is at high $k$ due to the density of states there. The MHD component of this case is very similar to the MHD case with the same initial state. Without drive terms in the thermal free energy component there is no new effect caused. We will see more energetic detail in the three-dimensional cases in later Chapters. The main lesson for now is that free energy components tend to take the $k^{+1}$ AEQ signature except for the dual conservation of

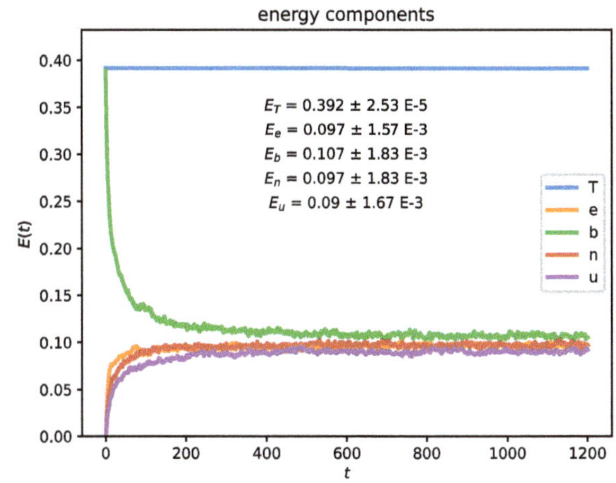


**Figure 5.12.** Time traces of the evolution of the four-field two-fluid MHD magnetic case with $\psi_0 = 1$ and all others zero, in RMS. Compare with figure 5.4. All channels try to reach equipartition, even the parallel flow energy $E_u$ as well as the thermal free energy $E_n$, the latter comparable to the ExB energy $E_e$, but the magnetic energy $E_b$ remains a little higher.

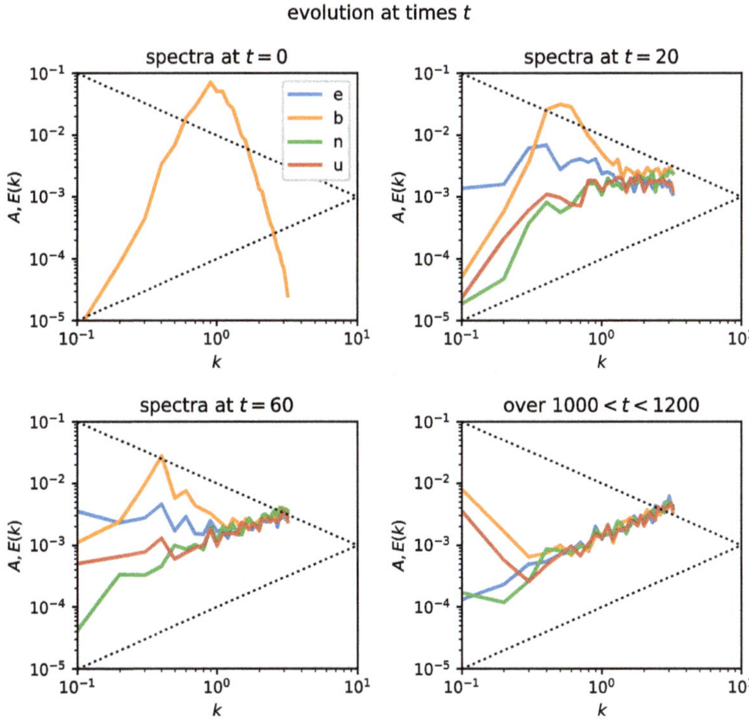


**Figure 5.13.** Isotropic perpendicular spectra for the four-field two-fluid MHD magnetic case with $\psi_0 = 1$ and all others zero, in RMS. Compare with figure 5.5. As in the MHD case, all channels equilibrate and equipartition fastest at high $k$, with the low-$k$ part coming into line much more slowly. Equipartition requires reasonable statistical decorrelation, for which at $k/k_0 = 1$ or 2 the density of states is too low.

energy and enstrophy in the fluid case. This is of interest since the tendency to collect at high rather than low $k$ leads to faster nonlinear processing in general cases, where it will be seen to compete effectively against linear drive mechanisms.

## 5.6 Reintroduction of dissipation

The results from these models with subgrid dissipation were shown in Chapter 3 with dissipation and in this Chapter so far as AEQ ensembles. The former results are the physically realistic ones. What turns the AEQ state into the physical result from any of these models is the presence of high-$k$ dissipation (viscosity, diffusion). The high Reynolds number limit is not a state of zero dissipation, but one in which the coefficient of dissipation is arbitrarily small. The energetic dissipation does not tend towards zero but converges asymptotically to a finite level, no matter how small. Recall that even in the 2D fluid case the enstrophy cascades to small scales, hence the relevance of finiteness of dissipation there as well. It is especially relevant when density fluctuations are present in a magnetised plasma—even more so since in realistic cases there is always a coupling mechanism from the electron density back to the ExB flow through currents. These currents effectively break the conservation of enstrophy, which is what sets up the 2D fluid dual cascade in the first place. The addition of successive detail that we will do makes the situation more complicated, but it serves to add coupling mechanisms, not obviate them.

The role of dissipation may be studied by gradually re-introducing it. Just how much or how little is required to change the AEQ state, for the fluid case or for the four-field two-fluid MHD case, is the question. For both the fluid and four-field cases the values of the dissipation considered are

$$\nu_0 = \{10^0, 10^{-1}, 10^{-2}, 10^{-4}, 10^{-6}, 10^{-8}\}, \tag{5.40}$$

which is enough to display the transition from too little dissipation to too much. For these purposes it is sufficient to give the isotropic spectra at $t = 20$ to see if the high-$k$ side settles down into a dissipation-resolved state or if it tends towards the AEQ state. The situation in the fluid case is given in figure 5.14. These show clearly what the effect of increasing subgrid dissipation is. The lowest-$\nu_0$ values show a tendency towards the AEQ state, which is why they were judged in Chapters 3 and 4 to be insufficient. The larger values $\nu_0 = 10^0$ and $10^{-1}$ show a turn-down at high $k$ which is too steep. This is how we find that the $10^{-2}$ value is optimal for the decaying fluid case. For the four-field two-fluid MHD case we take $\psi_0 = 1$ and others zero, which guarantees a situation with all four energy channels excited, as the equipartition study showed. This case with the six values of $\nu_0$ as above is given in figure 5.15. Again, the lower values show a tendency towards the AEQ state with the high-$k$ equipartition even at this early time. Due to the high-$k$ behaviour of the magnetic energy the $10^{-1}$ and $10^{-2}$ values are marginal and for driven cases would not be enough. Hence the value $\nu_0 = 10^0$ is the one needed. Larger values would tend to erase the fluid energy at high $k$, and the only cure in robustly driven cases would be to increase the grid resolution such that this value gives a good turn-down for all the channels. As ever, this is a matter of trial and error for any new case and one must

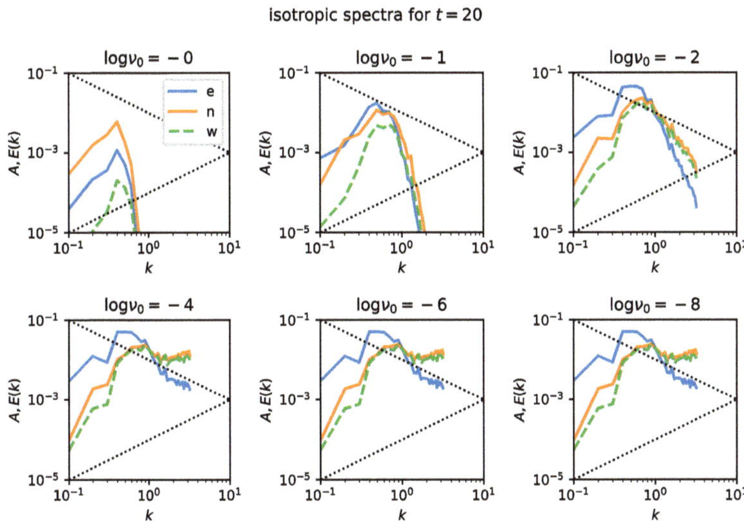


**Figure 5.14.** Initial evolution of the isotropic perpendicular spectra for the hydrodynamic case with varying degrees of subgrid dissipation controlled by $\nu_0$. Compare with figure 5.2, which gives the case with $\nu_0 = 0$. The statistical equilibration process results in the direct cascade tendencies of the quantities tending towards $k^{+1}$ setting the spatial resolution demands. Reasonable levels of $\nu_0$ are found between the extremes.

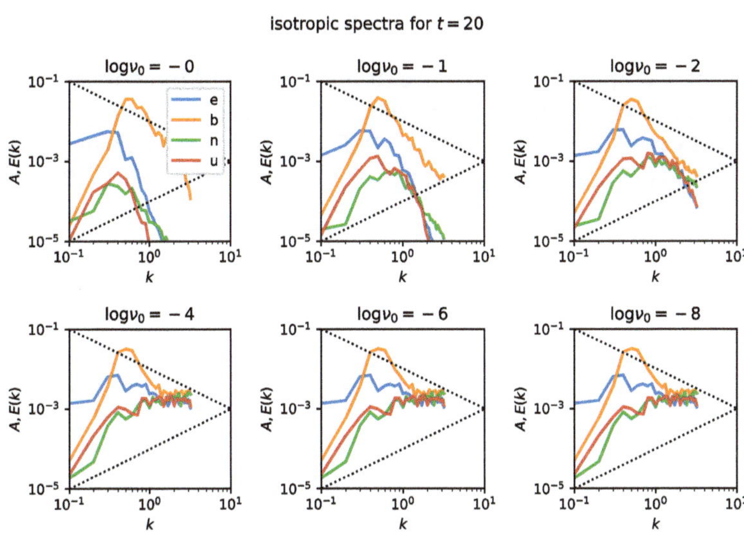


**Figure 5.15.** Initial evolution of the isotropic perpendicular spectra for the four-field two-fluid MHD magnetic case with varying degrees of subgrid dissipation controlled by $\nu_0$. Compare with figure 5.13, which gives the case with $\nu_0 = 0$. The statistical equilibration process decides which channels pressure the spatial resolution, with those going to $k^{+1}$ or higher doing so the most. Reasonable levels are found between the extremes, where the spectrum turns down properly at high $k$, but not so much that the nonlinear activity is shut off. The resolution demand is to have a wide enough spectrum to contain the activity while converging against the value of $\nu_0$, for any particular case, the details of which making case by case testing imperative.

guard against complacency based on experience which is too different for any new case.

The role of dissipation in turbulence is to make the difference between low viscosity and no viscosity, and it is important to have it act on all channels. Ideally, the dissipation is diagonal in the sense that the same operator is placed on any quantity mixed by the advective derivative (i.e. $\varpi$ not $\phi$, and $J$ not $\psi$). In the physical world, no turbulent system is dissipation-free, and the high Reynolds number limit is one which the high-$k$ dissipation is far enough down the spectrum that the value of the coefficient does not affect the result. We will be interested in the saturated state in driven/dissipative cases, which is a statistical equilibration rather than a relaxed one, and for this reason no computation without dissipation can give anything other than an initial transient response—it cannot represent a saturated state and usually does not give good guidance towards one.

## Further Reading

For statistical physics and energetic equipartition, see R C Tolman1938 *Principles of Statistical Physics* (Oxford: Clarendon Press), (repr. New York: Dover, 1979).

Absolute equilibrium ensembles were introduced and developed by L Onsager 'Statistical hydrodynamics' *Nuovo Cimento Suppl.* **6** (1949) 279; T D Lee 'On some statistical properties of hydrodynamical and magneto-hydrodynamical fields' *Q. Appl. Math* **10** (1952) 69; R Kraichnan 'Inertial ranges in two-dimensional turbulence' *Phys. Fluids* **10** (1967) 1417; U Frisch, A Pouquet, J Léorat and A Mazure 'Possibility of an inverse cascade of magnetic helicity in magnetohydrodynamic turbulence' *J. Fluid Mech.* **68** (1975) 769.

The concept was applied to drift wave turbulence by D Montgomery and L Turner 'Two-dimensional electrostatic turbulence with variable density and pressure' *Phys. Fluids* **23** (1980) 264; F Y Gang, B D Scott and P H Diamond 'Statistical mechanics of a two-field model of drift wave turbulence' *Phys. Fluids B* **1** (1989) 1331.

**IOP** Publishing

Turbulence and Instabilities in Magnetised Plasmas, Volume 1
Fluid drift turbulence
**Bruce Scott**

# Chapter 6

# Fluid electrodynamics in a plasma

## 6.1 Introduction

The classical picture of natural phenomena in physics is one of particles and fields. Plasma physics, however, deals with collective phenomena. The number of particles in any situation of interest is arbitrarily large, and this renders the solution of a Sturm–Liouville equation for each particle to be impractical. We therefore use a statistical picture, which replaces the discrete particle model with a continuous model of a distribution function in position/velocity phase space for each species of particle. We understand the distribution function as a $C^\infty$ field much like the electromagnetic field itself. A $C^\infty$ field is one which is assumed to be differentiable as many times as deemed necessary; it is a function of several independent variables, position and velocity space for a distribution function or just position space for the dynamical fields, or for a fluid.

The situations which we will describe are strongly ruled by self consistency, especially for the electrons, which are the principal carriers of electric current. This means that the thermodynamics of electrons cannot be separated from the electromagnetic phenomena for which they act as sources. In order to best understand this with a minimum of extra complication we will treat a plasma as a collection of charged fluids. We therefore replace the distribution function with a small number of field quantities which can reflect the overall nature of the dynamics without having to treat what can become enormously complicated details of kinetic phenomena (specialised reactions of particles with particular velocities). In many cases it is possible to work with ideal fluids, in which only a density, velocity, and temperature are retained. At most, we will require a few extra quantities such as conductive heat fluxes. The emphasis will be on the interplay between the thermodynamics of each fluid and the electromagnetic field.

We introduce a hierarchy of three dynamical systems: The first is fluid electrodynamics, in which the fluid motion is considered as ideal and nonrelativistic but the entire electrodynamics is retained. This can treat arbitrary frequencies since the wave

doi:10.1088/978-0-7503-2504-2ch6

propagation part does not require approximation, and allows finite charge densities to appear even if the equilibrium about which one expands is net charge neutral. It is useful in setting the parameter limits on the more restricted cases. The second can be called fluid plasma dynamics. Here, the nonrelativistic approximation is extended to the fields; we will find that this introduces several simplifications that enter together: the fluid system becomes quasineutral and the fields evolve quasistatically in response to the varying source distributions. By quasineutral we refer to the important property that the total charge density can be neglected but the electric field still has a primary role in the dynamics. The third system is magnetohydrodynamics, also called MHD. This is a much simplified picture which replaces several charged fluids with a single fluid that can carry current; that is, we retain only one mass density, one velocity field, and one temperature (equivalently, pressure). We discuss MHD at length for two reasons: (1) to understand both the incompressible and compressible classes of its dynamics, which will be our route to the low frequency approximations in Chapter 7; and (2) to note explicitly upon which assumptions this important and widely used model depends. Although it is applicable to many problems of general interest, the fact that it is limited—especially at small scales and in dynamics for which the thermodynamic gradient is the main drive mechanism—is unfortunately overlooked, to the point that untenable treatments still receive attention in the scientific literature. The aim is to give a presentation of the broad scope of fluid plasma dynamical models, so that one understands the conditions under which each is applicable.

## 6.2 Ideal fluid equations and electrodynamics

Ideal fluid equations can be derived heuristically according to simple considerations. If all we need to consider is a closed system with the three field quantities—a scalar density, a vector velocity field, and a scalar pressure—the matter can be reduced to bookkeeping via the conservation of material, momentum, and thermal energy, respectively. Along the way we encounter an important part of fluid dynamics: the velocity can advect ('carry') itself. This property gives the dynamics a nonlinear character which leads to the richness of problems such as shock waves and turbulence, making it quite unusual if one has spent one's time principally with single particle motion.

The density gets its equation from the consideration of material conservation. We specify $n$ as the material density, which is equivalent to a mass density $\rho = nm$ or charge density $\rho_{ch} = ne$ if $m$ and $e$ are the unit mass and charge for the fluid, respectively. At this point we assume that the velocity field at a given time is known, and ask simply how the resulting material flux affects the evolution of the density within a small interval of time. We can look at this in two ways: consider a volume of fluid which is fixed in space and sum up all the material which enters and leaves, or a volume whose bounding surface moves continuously with the fluid. In the first case the volume is fixed but the amount of material changes; in the second the amount of material is fixed but the volume can change. Either situation tells us how the density changes at a given point, and we can learn about the advection property by showing how they are equivalent.

For the fixed volume, we assume that a fluid in the vicinity of a point $\mathscr{P}$ and time $\mathscr{T}$ has density $n(\mathbf{x}, t)$ and flows with velocity $\mathbf{v}(\mathbf{x}, t)$. Let $\mathscr{V}$ be a fixed volume which contains $\mathscr{P}$ and is bounded by the surface $\mathscr{S}$ with outwardly directed element $\mathbf{dS}$. Let $N$ be the total amount of material within $\mathscr{V}$ at time $\mathscr{T}$. The total change in the material within the interval $\delta t$ is then given by the total amount exiting $\mathscr{V}$ through $\mathscr{S}$:

$$\delta N = \oint d\mathscr{V}\ \delta n = -\oint \mathbf{dS} \cdot n\mathbf{v}\delta t. \tag{6.1}$$

We can let $\delta t$ go to zero and then rewrite the change in $N$ as a time derivative, which may be exchanged with the integration since $\mathscr{V}$ is fixed. The right side can be reformed with the use of Gauss's theorem as a volume integral over the divergence of the material flux represented by $n\mathbf{v}$,

$$\delta N = -\oint \mathbf{dS} \cdot n\mathbf{v}\delta t = -\oint d\mathscr{V}\ \nabla \cdot n\mathbf{v}\delta t. \tag{6.2}$$

Finally, we note that $\mathscr{V}$ is arbitrary and so we can shrink its extent to zero and thus neglect second-order differentials. Since $\mathscr{P}$ and $\mathscr{T}$ themselves are arbitrary, the integrand then vanishes for all $\mathbf{x}$ and $t$:

$$\frac{\partial n}{\partial t} + \nabla \cdot n\mathbf{v} = 0. \tag{6.3}$$

The divergence of the material flux vector then gives the time evolution of the material density. This is the general form of a conservation law for field quantities.

Using the comoving volume, the one whose surface elements move with the fluid, we can find the same result but in a different language. We now specify $\mathscr{V}$ as a volume in the neighbourhood of $\mathscr{P}$ and $\mathscr{T}$, but since it is comoving, no material enters or leaves. Instead, $\mathscr{V}$ deforms. The demand that $N$ does not change is expressed as

$$\delta N = 0 = \delta n \mathscr{V} + n\, \delta\mathscr{V}. \tag{6.4}$$

From a theorem of differential geometry we can find the change in the volume in terms of the divergence of the surface displacements,

$$\delta\mathscr{V} = \mathscr{V}\nabla \cdot \delta\mathbf{x}. \tag{6.5}$$

Using these two equations we can find the change in the density,

$$\delta n = -n\frac{\delta\mathscr{V}}{\mathscr{V}} = -n\nabla \cdot \delta\mathbf{x} = -n\ \delta t\ \nabla \cdot \mathbf{v}, \tag{6.6}$$

with the help of the relation $\delta\mathbf{x} = \mathbf{v}\delta t$. As before we let $\delta t \to 0$ and then we have the change in the density:

$$\frac{dn}{dt} + n\nabla \cdot \mathbf{v} = 0, \tag{6.7}$$

where the symbol $d/dt$ is used to distinguish this case from the one in which $\mathcal{V}$ is fixed. This relation says that as one follows the fluid, the density changes according to how the fluid is compressed.

These two forms of material conservation form the two reference frames used in fluid dynamics. Equations in the fixed volume system are called the Eulerian form and in the comoving system they are called the Lagrangian form. They are related simply through a local coordinate transformation, which is a one-to-one and onto map between the infinitesimal space about a point in space and time, and itself. Since the fluid moves arbitrarily we can only specify such a transform locally, i.e. through the coordinate differentials, not the coordinates themselves. But since we are transforming derivatives, this is sufficient. The transform between the fixed and comoving reference frames is

$$dt' = dt \qquad d\mathbf{x}' = d\mathbf{x} - \mathbf{v}dt \tag{6.8}$$

$$dt = dt' \qquad d\mathbf{x} = d\mathbf{x}' + \mathbf{v}dt', \tag{6.9}$$

where the prime denotes the comoving one. The corresponding transform of the derivatives is

$$\left.\frac{d}{dt'}\right|_{\mathbf{x}'} = \left.\frac{dt}{dt'}\right|_{\mathbf{x}'}\left.\frac{\partial}{\partial t}\right|_{\mathbf{x}} + \left.\frac{d\mathbf{x}}{dt'}\right|_{\mathbf{x}'}\cdot\left.\frac{\partial}{\partial \mathbf{x}}\right|_{t} \qquad \left.\frac{\partial}{\partial \mathbf{x}'}\right|_{t'} = \left.\frac{\partial}{\partial \mathbf{x}}\right|_{t}. \tag{6.10}$$

The spatial gradient remains unchanged, but the time derivative is affected:

$$\frac{d}{dt} = \frac{\partial}{\partial t} + \mathbf{v}\cdot\nabla, \tag{6.11}$$

where in general we will write these in different form if there are more than one of them (i.e. two distinct advection velocities for two sets of dependent variable quantities).

This form of the comoving time derivative is called the advective derivative, since it refers to how a field changes when it is carried by a 'wind.' One can think, for example, of the air at a given place cooling in time when the wind blows from a place which is colder. An advantage of the ability to move back and forth between the fixed and comoving reference systems is that the equations are easier to construct in the comoving frame. We can then use the transformations to convert to the fixed frame, which we need in order to do a computation involving places at which the velocity field is different.

With this to hand we can simply write the next two equations, but first we come to the other property of a fluid: it moves not only according to external forces but also in response to its own pressure. To get the change of the fluid velocity in the comoving reference frame we simply find the acceleration due to the electric force and the pressure—since the volume is comoving, the local mean velocity is zero and so there is no magnetic force until we transform back to the fixed frame. The acceleration of the fluid element in its local rest frame is

$$Nm\frac{d\mathbf{v}}{dt} = Ne\mathbf{E} - \oint \mathbf{dS}p, \tag{6.12}$$

where the pressure acts on the bounding surface of the comoving volume. Explicitly writing the volume integration and using Stokes's theorem to rewrite the surface integral, we find

$$\oint d\mathcal{V}\ nm\frac{d\mathbf{v}}{dt} = \oint d\mathcal{V}\ ne\mathbf{E} - \oint d\mathcal{V}\ \nabla p. \tag{6.13}$$

We once again take the extent of the volume to zero, finding the acceleration in the local rest frame,

$$nm\frac{d\mathbf{v}}{dt} = ne\mathbf{E} - \nabla p. \tag{6.14}$$

At this point we note that in transforming frames with the electromagnetic field we always have to use the Lorentz transform, if only to first order in $v/c$; this is how we pick up the magnetic force. In the fixed frame we then find

$$nm\left(\frac{\partial}{\partial t} + \mathbf{v} \cdot \nabla\right)\mathbf{v} = ne(\mathbf{E} + \mathbf{v} \times \mathbf{B}) - \nabla p, \tag{6.15}$$

where we also write in the advective derivative explicitly.

We see now the two important differences between the dynamics of a fluid and that of a particle. First, the fluid has an internal pressure, whose gradient acts as an additional force. Second, the velocity advects itself. Both of these are consequences of the $C^\infty$ nature of the fluid as a set of field quantities rather than as a set of particles.

The final equation is conservation of thermal energy, simply the first law of thermodynamics for a fluid. We can use either temperature $T$ or pressure $p$ for an ideal fluid, but due to the later necessity of handling dissipation we will use $T$. We now restrict to an ideal gas, and note that the internal energy per unit of material is $\frac{3}{2}T$ and the pressure is given by $p = nT$, both with temperature written in units of energy. The first law of thermodynamics is in general

$$\delta U + p\delta\mathcal{V} = \delta Q, \tag{6.16}$$

which relates changes to the internal energy $\mathcal{U}$ resulting from the pressure doing work on changes in the volume, or from the addition of heat given by $\delta Q$. In fluid dynamics it is more usual to work with exit fluxes, so we will rewrite $\delta Q$ in terms of the heat flux $\mathbf{q}$. For a comoving volume the material quantity $N$ does not change, so the internal energy simply changes according to the temperature,

$$\delta U = \frac{3}{2}N\delta T = \oint d\mathcal{V}\ \frac{3}{2}n\delta T. \tag{6.17}$$

The change in the volume is given as before in terms of the divergence of the displacements, and the heat change is given by

$$-\delta Q = \delta t \oint \mathbf{dS} \cdot \mathbf{q} = \delta t \oint d\mathscr{V}\, \nabla \cdot \mathbf{q}. \tag{6.18}$$

Allowing the extent of the volume and the time interval to go to zero, we have

$$\frac{3}{2}n\frac{dT}{dt} + p\nabla \cdot \mathbf{v} = -\nabla \cdot \mathbf{q}. \tag{6.19}$$

With $d/dt$ as shorthand for the advective derivative, this same equation then holds for the fixed frame:

$$\frac{3}{2}n\left(\frac{\partial}{\partial t} + \mathbf{v} \cdot \nabla\right)T + p\nabla \cdot \mathbf{v} = -\nabla \cdot \mathbf{q}. \tag{6.20}$$

But the role of $\mathbf{q}$ is clarified by these considerations: in the comoving frame no fluid material flows past the bounding surface of the volume, so $\mathbf{q}$ is understood as the conductive heat flux, not the total. This is the part of the energy flux that is not carried by advection: if two bodies at rest are brought together at different temperature, the heat will flow from hot to cold by conduction. By writing the equation for the pressure, using the equations for $n$ and $T$,

$$\frac{\partial}{\partial t}\left(\frac{3}{2}nT\right) + \nabla \cdot \left(\frac{5}{2}p\mathbf{v} + \mathbf{q}\right) = \mathbf{v} \cdot \nabla p, \tag{6.21}$$

we find the total flux as the quantity whose divergence appears, and the work done by pressure. In the total heat flux the pressure appears with a factor of 5/2 instead of 3/2 due to its ability to do work by compressing the fluid. The alternative form is

$$\frac{\partial}{\partial t}\left(\frac{3}{2}nT\right) + \nabla \cdot \left(\frac{3}{2}p\mathbf{v} + \mathbf{q}\right) = -p\nabla \cdot \mathbf{v}. \tag{6.22}$$

If the velocity is divergence free, the total heat flux appears with a factor of 3/2, since in the absence of compressions the pressure does no work. For an ideal fluid we have $\mathbf{q} = 0$ and heat can only be carried by advection.

The foregoing can also be derived using the Principle of Least Action, using a suitable Lagrangian density. As this is the way one best derives the Maxwell equations of electrodynamics, we present an outline of the method. The usual application is to specify a Lagrangian $L(x^i, \dot{x}^i)$, dependent on path coordinates $x^i(t)$ and their total time derivatives $\dot{x}^i$, such that the action is given by

$$S = \int dt\, L, \tag{6.23}$$

where the integral is over the path $x^i(t)$ the system takes between two time points. The action is varied with respect to changes in the $x^i \rightarrow x^i + \delta x^i$, keeping the path endpoints fixed. The resulting variation in the action is set to zero in order to find the extremal action:

$$\delta S = \int dt\, \delta x^i \left(\frac{\partial L}{\partial x^i} - \frac{d}{dt}\frac{\partial L}{\partial \dot{x}^i}\right) = 0, \tag{6.24}$$

where $d/dt$ is the total time derivative (i.e. the 'dot'). As the variations $\delta x^i$ are arbitrary, the quantity in parentheses must vanish. Setting it to zero for each $i$ yields the Euler–Lagrange equations for the system. This is the textbook version which is most familiar.

With fields one expresses the Lagrangian as an integral over all space of a Lagrangian density, $\mathscr{L}$, which now depends on a set of field quantities. For the Maxwell equations these are the potentials ($\phi$, $\mathbf{A}$) which together form a four-potential in the framework of a relativistic treatment. For a nonrelativistic treatment we can simply write the Lagrangian in terms of the electric and magnetic fields, using the defining relations

$$\mathbf{E} = -\frac{\partial \mathbf{A}}{\partial t} - \nabla\phi \qquad \mathbf{B} = \nabla \times \mathbf{A}. \tag{6.25}$$

For the fluid we treat the thermal energy as a sort of potential energy, and write the Lagrangian density in the usual form of kinetic minus potential energy. The action becomes

$$S = \int dt\, L = \int dt \int d^3x\, \mathscr{L}, \tag{6.26}$$

with $\mathscr{L}$ given by

$$\mathscr{L} = nm\frac{v^2}{2} - \frac{3}{2}nT + \varepsilon_0\frac{E^2}{2} - \frac{B^2}{2\mu_0} + ne\mathbf{v}\cdot\mathbf{A} - ne\phi, \tag{6.27}$$

for a single charged fluid whose field variables are the density, velocity, and pressure, with mass and charge as given above. The first two terms are for the free fluid, the next two are for the free field, and the last two are for the interaction. The fluid has pressure, and charge density and current given by

$$p = nT \qquad \rho_{ch} = ne \qquad \mathbf{J} = ne\mathbf{v}. \tag{6.28}$$

The field variables for the Lagrangian density are the fluid element displacements $\boldsymbol{\xi}$ and the potentials $\phi$ and $\mathbf{A}$. When the $\boldsymbol{\xi}$ are given, the fluid field quantities vary as

$$\delta n = -\boldsymbol{\xi}\cdot\nabla n - n\nabla\cdot\boldsymbol{\xi}, \tag{6.29}$$

$$\delta T = -\boldsymbol{\xi}\cdot\nabla T - \frac{2}{3}T\nabla\cdot\boldsymbol{\xi}, \tag{6.30}$$

$$\delta\mathbf{v} = -\boldsymbol{\xi}\cdot\nabla\mathbf{v} + \mathbf{v}\cdot\nabla\boldsymbol{\xi} + \frac{\partial\boldsymbol{\xi}}{\partial t}. \tag{6.31}$$

Variations involving $\boldsymbol{\xi}$ gives the fluid equations, while varying $\phi$ and $\mathbf{A}$ gives the Maxwell equations, as shown by the variation of the action:

$$\delta S = \int dt \int d^3x \Bigg\{ \delta\phi \left[ \varepsilon_0 \left( \frac{1}{c^2} \frac{\partial^2 \phi}{\partial t^2} - \nabla^2 \phi \right) - ne \right]$$
$$+ \mu_0^{-1} \delta \mathbf{A} \cdot \left[ \frac{1}{c^2} \frac{\partial^2 \mathbf{A}}{\partial t^2} - \nabla^2 \mathbf{A} - \mu_0 ne\mathbf{v} \right] \tag{6.32}$$
$$+ \boldsymbol{\zeta} \cdot \left[ -nm \frac{d\mathbf{v}}{dt} - \nabla p - ne \left( \frac{\partial \mathbf{A}}{\partial t} + \nabla \phi \right) + ne(\nabla \mathbf{A} \cdot \mathbf{v} - \mathbf{v} \cdot \nabla \mathbf{A}) \right] \Bigg\},$$

where $1/c^2 = \mu_0 \varepsilon_0$ and $d/dt$ is the advective derivative for each fluid. These yield, respectively, the fluid equation for $\mathbf{v}$,

$$nm\left( \frac{\partial}{\partial t} + \mathbf{v} \cdot \nabla \right)\mathbf{v} = -\nabla p - ne\left( \frac{\partial \mathbf{A}}{\partial t} + \nabla \phi \right) + ne\mathbf{v} \times \nabla \times \mathbf{A}, \tag{6.33}$$

which is the same as that already given, the scalar field equation for $\phi$,

$$\frac{1}{c^2} \frac{\partial^2 \phi}{\partial t^2} - \nabla^2 \phi = \frac{1}{\varepsilon_0} \rho_{ch}, \tag{6.34}$$

and the vector field equation for $\mathbf{A}$,

$$\frac{1}{c^2} \frac{\partial^2 \mathbf{A}}{\partial t^2} - \nabla^2 \mathbf{A} = \mu_0 \mathbf{J}. \tag{6.35}$$

Using the vector identity

$$\nabla \times \nabla \times \mathbf{A} = \nabla(\nabla \cdot \mathbf{A}) - \nabla^2 \mathbf{A}, \tag{6.36}$$

this is subject to the condition

$$\frac{1}{c^2} \frac{\partial \phi}{\partial t} + \nabla \cdot \mathbf{A} = 0, \tag{6.37}$$

called the Lorentz gauge. These equations for $\phi$ and $\mathbf{A}$ form two of the Maxwell equations:

$$\mu_0 \varepsilon_0 \frac{\partial \mathbf{E}}{\partial t} = \nabla \times \mathbf{B} - \mu_0 \mathbf{J} \qquad \varepsilon_0 \nabla \cdot \mathbf{E} = \rho_{ch}. \tag{6.38}$$

The other two follow from the way in which we wrote $\mathbf{E}$ and $\mathbf{B}$ in terms of the potentials:

$$\frac{\partial \mathbf{B}}{\partial t} = -\nabla \times \mathbf{E} \qquad \nabla \cdot \mathbf{B} = 0. \tag{6.39}$$

We write again the equations for each charged fluid of species $j$,

$$\frac{\partial n_j}{\partial t} + \nabla \cdot n_j \mathbf{v}_j = 0, \tag{6.40}$$

$$n_j m_j \left( \frac{\partial}{\partial t} + \mathbf{v}_j \cdot \nabla \right) \mathbf{v}_j = n_j e_j \left( \mathbf{E} + \mathbf{v}_j \times \mathbf{B} \right) - \nabla p_j, \tag{6.41}$$

$$\frac{3}{2} \left( \frac{\partial p_j}{\partial t} + \mathbf{v}_j \cdot \nabla p_j \right) + \frac{5}{2} p_j \nabla \cdot \mathbf{v}_j = 0, \tag{6.42}$$

and note that the charge density and current are summed up over all fluid species; explicitly, we have

$$\mathbf{J} = \sum_j n_j e_j \mathbf{v}_j \qquad \rho_{ch} = \sum_j n_j e_j, \tag{6.43}$$

and for each species $T_j$ is given by $p_j / n_j$.

The Maxwell Eqs (6.38) and (6.39) and the fluid Eqs (6.40)–(6.42) form a closed system, with obvious generalisation to an arbitrary number of charged fluids. We have written these for ideal fluids, and will delay discussion of dissipation until after making the low frequency approximations.

## 6.3 High frequency motion under fluid electrodynamics

The dynamical system just presented has a rich character in its own right, but since our purpose is to proceed to the low frequency regime we will single out the ingredients needed to motivate the necessary approximations. These are the plasma waves, for motion along the field lines, and the compressional magnetosonic waves, for motion across them. Since the associated frequencies are very fast, we can neglect thermal effects and treat the cold fluid system.

We assume a homogeneous background with finite **B**, zero **E**, and a neutral two-component plasma at rest. For simplicity let this be made of one singly charged ion species, and an equal number density of electrons. We investigate the resulting wave-like motion of small disturbances on this background. The results are summarised in figure 6.1.

To obtain the first set of waves, we assume that the mass ratio $\mu_e = m_e / m_i$ is very small, and hence that the ions are stationary. We linearise the dynamical system as follows. Let each quantity be given by an equilibrium value and a perturbation thereof; for example, the electron density $n_e$ becomes

$$n_e \to n_e + \tilde{n}_e \tag{6.44}$$

with the tilde used to denote the disturbances. In addition, we note that **v** and **E** belong purely to the disturbances. As the equilibrium is trivial, we need only write the equations for the disturbances, dropping all terms of order higher than one in the disturbance amplitudes, e.g. neglecting terms such as $\mathbf{v} \cdot \nabla \mathbf{v}$ or $\tilde{n}_e \mathbf{E}$. Furthermore, we choose a single degree of freedom of oscillation. Since the equilibrium is homogeneous we may assume periodicity for small-scale motion in time and all three dimensions, and then choose a single Fourier component. Again, for the density we have

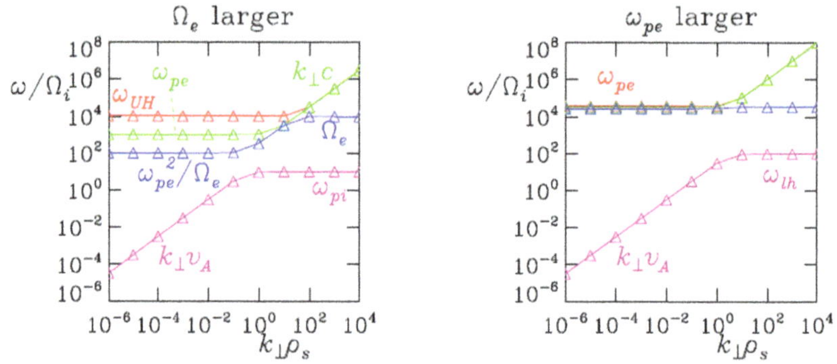


**Figure 6.1.** Fast perpendicular waves in a magnetised plasma, for the two cases in which either $\omega_{pe}$ or $\Omega_e$ is the larger. The latter case is more prevalent, such that the magnetosonic waves turn over into non-quasineutral ion plasma waves. For $\omega_{pe} > \Omega_e$ (less magnetised) one finds the more conventional lower hybrid waves at $\omega_{lh}$. Below the lowest of the four curves in each graph, motion is dynamically incompressible perpendicular to the magnetic field and is quasineutral.

$$\tilde{n}_e = \tilde{n}_{e\mathbf{k}} \exp[i(\mathbf{k} \cdot \mathbf{x} - \omega t)], \tag{6.45}$$

with the condition $\tilde{n}_{e-\mathbf{k}} = \tilde{n}^*_{e\mathbf{k}}$ such that the disturbance is real valued. We may drop the subscripts $\mathbf{k}$ for clarity.

The Eqs (6.38)–(6.41) become

$$\frac{i\omega}{c^2}\mathbf{E} = i\mathbf{k} \times \tilde{\mathbf{B}} + \mu_0 n_e e \mathbf{v}, \tag{6.46}$$

$$-i\varepsilon_0 \mathbf{k} \cdot \mathbf{E} = e\tilde{n}_e, \tag{6.47}$$

$$i\mathbf{k} \cdot \tilde{\mathbf{B}} = 0, \tag{6.48}$$

$$\omega\tilde{\mathbf{B}} = \mathbf{k} \times \mathbf{E}, \tag{6.49}$$

$$\omega\tilde{n}_e = n_e \mathbf{k} \cdot \mathbf{v}, \tag{6.50}$$

$$\omega\mathbf{v} = -i\frac{e}{m_e}\mathbf{E} + i\frac{e\mathbf{B}}{m_e} \times \mathbf{v}, \tag{6.51}$$

where we note the fundamental role of the electron gyrofrequency,

$$\Omega_e = \frac{eB}{m_e}, \tag{6.52}$$

in the equation for $\mathbf{v}$. There is more generality here than we need; besides choosing a direction for $\mathbf{k}$ we can choose a polarisation, and this reduces the number of independent quantities to manageable levels.

For wave propagation parallel to $\mathbf{B} = B\mathbf{b}$ we let $\mathbf{k} = k_\parallel \mathbf{b}$. The system splits into three sets of modes: for motion along the magnetic field $(\mathbf{v}\|\mathbf{B})$, we have $\mathbf{E}\|\mathbf{B}$ and hence the magnetic perturbation vanishes. This gives longitudinal propagation $(\mathbf{k}\|\mathbf{v})$

involving the charge density and current, with the electric field determined constitutively by the density. There are two waves, each with the electron plasma frequency $\omega_{pe}$ regardless of the size of $k_\parallel$,

$$\omega^2 = \omega_{pe}^2 = \frac{n_e e^2}{\varepsilon_0 m_e}. \tag{6.53}$$

For motion across the field ($\mathbf{v} \perp \mathbf{B}$) we have $\mathbf{k} \cdot \mathbf{v} = 0$ and therefore two sets of three incompressible waves (electric field, current, and magnetic field but no density disturbance), one for each polarisation. Denoting $x$ and $y$ as the two perpendicular coordinates, with $\nabla x \times \nabla y \cdot \mathbf{b} = 1$, we may express the electric field disturbance, for example, as

$$E_\pm = E_x \pm iE_y. \tag{6.54}$$

For the plus and minus polarisations we find in turn,

$$(\omega \pm \Omega_e)(\omega^2 - k_\parallel^2 c^2) = \omega\omega_{pe}^2. \tag{6.55}$$

In each case there are two limits: $\Omega_e \gg \omega_{pe}$ or vice versa. The interesting case is the only one in which we obtain frequencies slower than light waves ($\omega \sim k_\parallel c$) or plasma waves; this is for $\Omega_e \gg \omega_{pe}$ and $k_\parallel \to 0$, where we find motion at another frequency,

$$\omega = \pm\omega_{pe}^2/\Omega_e, \tag{6.56}$$

for the plus and minus polarisations, respectively.

For wave propagation perpendicular to $\mathbf{B}$ we can choose the plane of propagation. We choose coordinates such that $\mathbf{k} = k_\perp \nabla x$. For motion along the field ($\mathbf{v} \| \mathbf{B}$) we find the same result as for electromagnetic waves in the unmagnetised case,

$$\omega^2 = k_\perp^2 c^2 + \omega_{pe}^2, \tag{6.57}$$

with only two roots since with $\mathbf{k} \perp \mathbf{v}$ the charge density remains zero. For motion across the field ($\mathbf{v} \perp \mathbf{B}$) the system couples compressional motion along $\nabla x$ with transverse motion along $\nabla y$, through the finite $\Omega_e$, and so all four remaining roots are coupled. The slowest mode of interest is for $\Omega_e \gg \omega_{pe}$ and $k_\perp \to 0$, where we find two modes each at

$$\omega = \pm\omega_{pe}^2/\Omega_e \qquad \text{and} \qquad \omega = \pm\Omega_e. \tag{6.58}$$

The fastest mode is at the upper hybrid frequency $\omega_{UH}$ given by $\omega_{UH}^2 = \Omega_e^2 + \omega_{pe}^2$.

These frequencies are very large, involving only two frequencies whose dependence is inverse to the (small) electron mass. If we consider ion motion but still at the stage of high frequencies, we obtain the dynamics which puts limits on what we are to do next. Considering only propagation and motion perpendicular to $\mathbf{B}$, we can reduce the set of equations to three, involving only $\mathbf{E}$ and the two velocities, $\mathbf{v}$ for electrons and $\mathbf{u}$ for ions:

$$\left(k_\perp^2 - \frac{\omega^2}{c^2}\right)\frac{e\mathbf{E}}{m_i} - \mathbf{k}\mathbf{k} \cdot \frac{e\mathbf{E}}{m_i} = i\omega\frac{\omega_{pi}^2}{c^2}(\mathbf{u} - \mathbf{v}), \tag{6.59}$$

$$-i\mu_e\omega\mathbf{v} = -\frac{e\mathbf{E}}{m_i} + \Omega_i\mathbf{b}\times\mathbf{v}, \tag{6.60}$$

$$-i\omega\mathbf{u} = \frac{e\mathbf{E}}{m_i} - \Omega_i\mathbf{b}\times\mathbf{u}, \tag{6.61}$$

valid for all of **k**, **E**, **u**, and **v** perpendicular to **b**, where $\mu_e$ is the mass ratio, $\Omega_i$ is the ion gyrofrequency, and $\omega_{pi}$ is the ion plasma frequency, respectively, given by

$$\mu_e = \frac{m_e}{m_i} \qquad \Omega_i = \frac{eB}{m_i} \qquad \omega_{pi}^2 = \frac{n_i e^2}{\varepsilon_0 m_i}. \tag{6.62}$$

The two lowest frequency modes of interest are those found by setting

$$\mu_e \to 0 \qquad \text{and} \qquad \omega^2 \ll k_\perp^2 c^2. \tag{6.63}$$

That is, we neglect the four roots found above. Then we replace $\mathbf{v} = \mathbf{u} - \mathbf{J}/n_e e$ and add the equations of motion neglecting finite $\mu_e$ to obtain

$$\frac{e\mathbf{E}}{m_i} + \Omega_i\mathbf{u}\times\mathbf{b} = \frac{\Omega_i}{n_e e}\mathbf{J}\times\mathbf{b}, \tag{6.64}$$

$$-i\omega\mathbf{u} = \frac{\Omega_i}{n_e e}\mathbf{J}\times\mathbf{b}. \tag{6.65}$$

What we will want to know is whether we can neglect **J** in the direction along **k**; that is, whether we can take the current as divergence free. Choosing **k** in the *x*-direction, we can find a relation between the components of the current,

$$J^x = \frac{i\Omega_i\,\omega}{\omega_{pi}^2 + \Omega_i^2}J^y, \tag{6.66}$$

noting the cancellation between $k^2$ and **kk** for $J^x$ and the use of $\omega^2 \ll k^2c^2$ in Eq. (6.59), and then write the rest of the system for the velocity in the direction along **k**, and the current in the direction along $\mathbf{b}\times\mathbf{k}$,

$$-i\omega u^x = \Omega_i\,\frac{J^y}{n_e e}, \tag{6.67}$$

$$\omega\frac{J^y}{n_e e} = -i\Omega_i\,\frac{k_\perp^2 c^2}{\omega_{pi}^2}\left(u^x - \frac{J^x}{n_e e}\right). \tag{6.68}$$

If $J^x$ can be ignored then the frequency we get is consistent with propagation at the Alfvén velocity along **k**,

$$\omega^2 = \frac{k_\perp^2 c^2\Omega_i^2}{\omega_{pi}^2} = k_\perp^2 v_A^2 \qquad \text{where} \quad v_A^2 = \frac{B^2}{\mu_0 n_i m_i}. \tag{6.69}$$

But for progressively smaller scales this frequency rises to the ion gyro and plasma frequencies, $J^x$ becomes significant, and the resulting frequency is modified to

$$\omega^2 = \frac{\left(\omega_{pi}^2 + \Omega_i^2\right)k_\perp^2 v_A^2}{\omega_{pi}^2 + \Omega_i^2 + k_\perp^2 v_A^2}. \tag{6.70}$$

For sufficiently small values of $k$ we get waves propagating at $v_A$ but the frequency then turns over at what can be called a low-density generalisation of the lower hybrid frequency:

$$\omega_{LH} = \left(\omega_{pi}^2 + \Omega_i^2\right)^{1/2}. \tag{6.71}$$

The textbook lower hybrid wave is something different, entering at $\omega_{LH} = (\Omega_i\Omega_e)^{1/2}$, given as the textbook lower hybrid frequency. It is generalised to

$$\omega_{LH} = \left[(\Omega_i\Omega_e)^{-1} + \omega_{pi}^{-2}\right]^{-1/2}. \tag{6.72}$$

On the other hand, for these slower modes we have neglected the electron motion one needs to derive $\omega_{LH}$. Note that the boundary $\omega_{LH} < \omega_{pi}$ is the same as $\Omega_e < \omega_{pe}$, while it is more common in magnetised plasmas that $\Omega_e > \omega_{pe}$. In the latter limit, coupled with $v_A < c$, the turnover is at the ion plasma frequency $\omega_{pi}$.

The ion plasma frequency is the limit at which space charge effects become important, since it is a reflection that $\nabla \cdot \mathbf{J} = 0$ no longer holds. We finally note that this simplified analysis of compressional magnetosonic waves is only as valid as our two assumptions (Eq. 6.63). For these we require

$$\frac{m_e}{m_i} \ll \frac{v_A^2}{c^2} \ll 1. \tag{6.73}$$

The first decouples the magnetosonic waves from the faster electrons-only motion, thereby also ensuring that $\omega_{pi} < \omega_{LH}$, and the second allows the neglect of light speed effects on the Alfvén waves at low $k_\perp^2$, and furthermore ensures that $\omega_{pi} > \Omega_i$.

## 6.4 Quasineutral motion in a neutral plasma

The ion plasma frequency is important because it tells us the slowest time scale on which motion in a plasma involves the charge density as such, which in the cases we will address should be negligible even though we will have fluctuating electric fields. This latter situation is called quasineutrality, so named because the plasma is neutral enough so that the total current is divergence free, but not so much that the electric field has to be divergence free. Adding the fluid continuity equations we have

$$\frac{\partial\rho_{ch}}{\partial t} = -\nabla \cdot (n_i e\mathbf{u} - n_e e\mathbf{v}) = -\nabla \cdot \mathbf{J}, \tag{6.74}$$

so the ability to neglect the charge density in wave-like motion requires

$$\frac{\partial \rho_{ch}}{\partial t} \to 0 \implies \nabla \cdot \mathbf{J} = 0. \tag{6.75}$$

It is interesting to note the consequences to the electromagnetic field of making this approximation on the motion of the plasma. The electromagnetic field Eqs (6.38) and (6.39) admit a gauge freedom, in the sense that for any scalar field $\chi$ we can transform the potentials according to

$$\phi \to \phi - \frac{1}{c^2}\frac{\partial \chi}{\partial t} \qquad\qquad \mathbf{A} \to \mathbf{A} + \nabla\chi, \tag{6.76}$$

while leaving the electric and magnetic fields unchanged, and hence the Maxwell equations unchanged. If $\mathbf{A}$ has a divergence, then, we can easily find a $\chi$ for which $\nabla^2\chi$ will cancel that divergence. On the other hand, working strictly within the Lorentz gauge, we find by examining Eq. (6.35) that

$$\nabla \cdot \mathbf{J} = 0 \implies \nabla \cdot \mathbf{A} = 0. \tag{6.77}$$

In some sense, then, $(1/c^2)\partial\phi/\partial t$ is 'small' compared to $\nabla \cdot \mathbf{A}$. This is actually the same sense in which the displacement current is small compared to $\mu_0^{-1}\nabla \times \mathbf{B}$ and the plasma current $\mathbf{J}$. All together, we then have

$$\nabla \cdot \mathbf{J} \to 0 \implies \begin{cases} \dfrac{1}{c^2}\dfrac{\partial \mathbf{E}}{\partial t} & \ll \mu_0 \mathbf{J} \\ \dfrac{1}{c^2}\dfrac{\partial^2 \mathbf{A}}{\partial t^2} & \ll \nabla^2 \mathbf{A} \\ \dfrac{1}{c^2}\dfrac{\partial^2 \phi}{\partial t^2} & \ll \nabla^2 \phi \\ \dfrac{1}{c^2}\dfrac{\partial \phi}{\partial t} & \ll \nabla \cdot \mathbf{A} \\ \rho_{ch} & \ll n_e e \end{cases}, \tag{6.78}$$

all of which follow if both $\omega \ll \omega_{pi}$ and $v_A \ll c$, which in turn hold if $\epsilon_0$ is taken to zero. Quasineutrality is clearly tied together with the requirement that the electromagnetic field itself remains nonrelativistic, since it implies neglect of the displacement current. Ultimately, the meaning of it is

$$\text{quasineutrality} \equiv \lim \varepsilon_0 \to 0, \tag{6.79}$$

on all of plasma electrodynamics, together with all the associated consequences. The specific meaning of this for ExB motion, including both MHD and fluid drift dynamics, will become more clear after we examine the polarisation current in detail in Section 7.12 of the next Chapter.

## 6.5 Fluid plasma dynamics under quasineutrality

Fluid plasma dynamics differs from fluid electrodynamics principally in the sense that the electromagnetic field potentials are statically given by their sources: the small but finite $\rho_{ch}$ gives $\phi$ while **J** gives **A**. In this sense we can regard the electromagnetic field as nonrelativistic. We also have the dynamics of a plasma in the sense that for a neutral background, space charge effects are not important to the dynamics of disturbances. Although electric fields do arise, the dynamics of the current and magnetic field is what is most important. This situation still differs from the more familiar MHD, however, since it allows qualitatively different dynamics in the electron and ion pressures, and in their velocities.

With the restriction that all the dynamics under study satisfies quasineutrality, we can simplify the governing equations. We eliminate the density equation for the electrons or one of the ion species, since the requirement that $\nabla \cdot \mathbf{J} = 0$ reduces the number of necessary continuity equations by one. We can find the density whose equation is left out, by solving for its contribution to $\rho_{ch} = 0$. Without the displacement current, we find **J** from **B**, so by using the definition of **J** in terms of the fluid velocities we may dispense with the electron equation of motion. The latter is used simply to eliminate **E** in the equation for **B** and in the equation of motion for each ion species. For an ideal plasma with one ion species of unit charge $e$ we have $n_i = n_e$, and then

$$\frac{\partial n_i}{\partial t} + \nabla \cdot n_i \mathbf{u} = 0, \tag{6.80}$$

$$n_i m_i \frac{d\mathbf{u}}{dt} = -\nabla(p_i + p_e) + \frac{\mathbf{J} \times \mathbf{B}}{c} - n_e m_e \frac{d_e \mathbf{v}}{dt}, \tag{6.81}$$

$$\frac{dp_i}{dt} + \frac{5}{3} p_i \nabla \cdot \mathbf{u} = 0, \tag{6.82}$$

$$\frac{d_e p_e}{dt} + \frac{5}{3} p_e \nabla \cdot \mathbf{v} = 0, \tag{6.83}$$

$$\frac{\partial \mathbf{B}}{\partial t} = \nabla \times (\mathbf{v} \times \mathbf{B}) + \nabla\left(\frac{1}{n_e e}\right) \times \nabla p_e + \nabla \times \left(\frac{m_e}{e} \frac{d_e \mathbf{v}}{dt}\right), \tag{6.84}$$

with the plasma current and electron velocity both determined by the magnetic field and ion velocity via

$$\mu_0 \mathbf{J} = \nabla \times \mathbf{B} \qquad \mathbf{v} = \mathbf{u} - \frac{\mathbf{J}}{n_e e}, \tag{6.85}$$

and the ion and electron advective derivatives given by

$$\frac{d}{dt} = \frac{\partial}{\partial t} + \mathbf{u} \cdot \nabla \qquad \frac{d_e}{dt} = \frac{\partial}{\partial t} + \mathbf{v} \cdot \nabla. \tag{6.86}$$

The dependent variables are the set $\{n_i, \mathbf{u}, p_e, p_i, \mathbf{B}\}$, with $n_e$, $\mathbf{J}$, and $\mathbf{v}$ as auxiliaries. Note that we could just as easily use the electron density equation instead,

$$\frac{\partial n_e}{\partial t} + \nabla \cdot n_e \mathbf{v} = 0, \tag{6.87}$$

noting that for the single ion species with unit charge we have $\nabla \cdot n_i \mathbf{u} = \nabla \cdot n_e \mathbf{v}$. In some applications it is more expedient to use the electron density, as we will see in the next Chapter.

In this form of the equations the electron dynamics is largely submerged, appearing explicitly only in the equation for $p_e$. This is because it is controlled by the dynamics of the magnetic field, a limit which becomes close to complete if we can neglect the electron mass corrections represented by the two appearances of the electron inertia. In that case we still have essentially two fluid dynamics, however, because the difference between $\mathbf{u}$ and $\mathbf{v}$ is still present. If that is important, the electron and ion pressures evolve differently, and most importantly the magnetic field responds to $\mathbf{v}$ rather than $\mathbf{u}$. This is the origin of the drift dynamical effects presented in the rest of this work. Under certain circumstances $\mathbf{J}$ can actually cancel $\mathbf{u}$ in the equation for $\mathbf{B}$, removing the field line dragging term (the first one on the right side). Ultimately, this means the same thing as $\mathbf{E}$ cancelling $\nabla p_e$, leaving $\mathbf{v}$ at or close to zero. So even in a situation in which $\nabla n_e$ and $\nabla p_e$ are parallel (and hence the curl of the $\nabla p_e$ term vanishes), we can still have the adiabatic response covered in Chapter 4 and derived explicitly in Eq. (7.81) of the next Chapter.

Fluid plasma dynamics is the starting point for the study of not only low frequency but also dissipative effects treatable by such macroscopic considerations. The step from here to fluid drift dynamics effects is straightforward, but before doing that it is useful to go straight to single-fluid MHD and its principal properties to make the relevant time scales most transparent. This will tell us what we have to satisfy in order to take the drift approximation.

## 6.6 E Pluribus Unum—the steps to MHD

'Out of many, one' is a good description of the derivation of MHD from the dynamics of several charged fluids we have just seen. The most familiar picture of plasma dynamics is and remains magnetohydrodynamics, or MHD. Even for those whose work never goes beyond that, however, it is important to know through which assumptions and their associated parameter restrictions the validity of MHD is limited. It is a question of what we have to do to take a plasma made up of several fields and end with a model of a single fluid with a single density, velocity, and temperature, but which due to its make-up of charged particles can carry an electric current. The effect of the current is to force the motion of the fluid by means of the total Lorentz force ($\mathbf{J} \times \mathbf{B}$), while the effect of the fluid back on the current is to advect the lines of force of the magnetic field. Under many applications, the pressure force can be negligible compared to the $\mathbf{J} \times \mathbf{B}$ force and the dynamics is very different from that of a neutral fluid, principally in the ability of the plasma to support transverse waves (the shear Alfvén waves). The opposite situation is kinematic MHD. If the flow is prescribed, or results from fluid motion for which the $\mathbf{J} \times \mathbf{B}$ force is negligible, the magnetic field can evolve without self-consistent back reaction on the fluid. This is the typical case of a dynamo plasma.

A cursory list of the assumptions needed to realise this picture is the following: the plasma should be neutral so that the electric field and space charge effects do not dominate the dynamics. The electron mass should be strictly negligible, so as not to affect the inertia or the dynamics of the magnetic field (the two appearances of $m_e$ in Eqs (6.81) and (6.84)). The normalised current should be small, such that we can treat each constituent of the plasma as flowing with the same velocity—the only exception is in the Lorentz force, which due to the neutrality of the plasma would otherwise vanish. This allows the thermodynamics of the electron and ion pressures to react equally to the motion of the plasma. Lastly, the pressure gradient should be small compared to the electric field in the forces on any of the fluids, so that each fluid reacts the same way to the Lorentz force. Quantitatively speaking, we require

$$\rho_{ch}\mathbf{E} \ll \mathbf{J} \times \mathbf{B} \qquad n_e m_e \ll n_i m_i, \tag{6.88}$$

$$\frac{\mathbf{J}}{n_e e} \ll \mathbf{u} \approx \mathbf{v} \qquad \frac{\nabla p_e}{n_e e} \ll \mathbf{E} \approx -\mathbf{v} \times \mathbf{B}, \tag{6.89}$$

in the case of a single component plasma whose ions have unit charge $e$.

The first two of these points are easily satisfied for most applications, and so they are the most convincing. The last two points are clearly related: if the pressure forces are substantially different then the fluids acquire differing velocities, or if the velocities have different properties then the pressures will evolve differently and do different thermodynamic work on each fluid in the plasma. They also present the greatest difficulty, since the changes they bring to an MHD situation cannot be judged from within MHD, which although parametrically limited is a perfectly self-consistent model. The procedure we will use here is to outline the basic properties of the single-fluid dynamics, most of which are familiar, and then qualify the situation within the framework of drift dynamics, which is where the limitations of MHD are most important in the laboratory or in nature.

Under the above assumptions the steps to MHD from the plasma dynamics of several fluids are straightforward, whether there is one ion species or several. We add the density equations, incorporating the unit masses for each species to get a mass density equation,

$$\frac{\partial \rho}{\partial t} + \nabla \cdot \rho \mathbf{u} = 0, \tag{6.90}$$

where $\mathbf{u}$ is simply the velocity of the entire fluid. We add the equations of motion of all species, including the electrons which cancel the electric field, to obtain

$$\rho \frac{d\mathbf{u}}{dt} = -\nabla p + \mathbf{J} \times \mathbf{B}, \tag{6.91}$$

where

$$\frac{d}{dt} = \frac{\partial}{\partial t} + \mathbf{u} \cdot \nabla, \tag{6.92}$$

is the advective derivative for the single fluid. This is called the MHD force equation. It is the same as the equation of motion under fluid plasma dynamics except that we neglect the electron inertia, and we write the total pressure as $p$, simply summing all the pressures including that of the electrons. Finally, we sum the pressure equations after the same fashion,

$$\frac{\partial p}{\partial t} + \frac{5}{3} p \nabla \cdot \mathbf{u} = 0, \tag{6.93}$$

noting especially the neglect of the extra terms involving $\mathbf{J}$. This is the complete set for the plasma itself, and it is closed except for the presence of the $\mathbf{J} \times \mathbf{B}$ force.

The most important statement of the electrodynamics part of MHD is that the Lorentz force on each fluid should self-cancel up to negligible corrections, so that each fluid moves with the ExB and parallel velocity:

$$\mathbf{E} + \mathbf{u} \times \mathbf{B} = 0 \implies \mathbf{u} = \frac{1}{B^2} \mathbf{E} \times \mathbf{B} + u_{\parallel} \mathbf{b}, \tag{6.94}$$

which we derive by assuming that the electron inertia and $\nabla p_e$ are small compared to the magnetic and electric parts of the Lorentz force, respectively, and that $\mathbf{J} \ll n_e e \mathbf{u}$. This fixes $\mathbf{E}$ in terms of $\mathbf{u}$ and $\mathbf{B}$, which immediately gives the equation for the magnetic field:

$$\frac{\partial \mathbf{B}}{\partial t} = \nabla \times (\mathbf{u} \times \mathbf{B}). \tag{6.95}$$

This is called the MHD kinematic equation. The current is again given by

$$\mu_0 \mathbf{J} = \nabla \times \mathbf{B}. \tag{6.96}$$

In terms of the density, temperature, and unit mass of each fluid, we have as equations of state

$$\rho = \sum_j n_j m_j \qquad p = \sum_j p_j \qquad p_j = n_j T_j, \tag{6.97}$$

where $j$ is the species label (including the electrons). The temperatures are allowed to differ, but only if their relaxation dynamics does not affect the rest of the system. The set of dependent variables reduces to $\{\rho, \mathbf{u}, p, \mathbf{B}\}$, whose equations form a closed system with the help of the constitutive relation for $\mathbf{J}$.

## 6.7 MHD waves—Alfvén waves

There are three basic types of wave motion in MHD. The first is sound waves familiar from the neutral fluid. The Lorentz force gives rise to two further types of waves, which act differently depending on whether they propagate along or across the magnetic field. These are the Alfvén waves. Together, sound waves and Alfvén waves determine the basic time scales of MHD.

The two forms of Alfvén waves involve quite different types of dynamics. To see this most clearly we can eliminate **J** in the MHD force equation, writing

$$\rho\left(\frac{\partial}{\partial t} + \mathbf{u}\cdot\nabla\right)\mathbf{u} = -\nabla\left(p + \frac{B^2}{2\mu_0}\right) + \frac{\mathbf{B}\cdot\nabla\mathbf{B}}{\mu_0}, \tag{6.98}$$

having expanded the advective derivative. And we can expand the vector operations in the MHD kinematic equation, writing

$$\frac{\partial\mathbf{B}}{\partial t} = -\mathbf{u}\cdot\nabla\mathbf{B} + \mathbf{B}\cdot\nabla\mathbf{u} - \mathbf{B}\nabla\cdot\mathbf{u}. \tag{6.99}$$

Plasma flow parallel to the magnetic field does not affect the kinematic equation at all, since $\mathbf{B}\cdot\nabla\mathbf{u}$ cancels the divergence in that case. This yields the sound waves. Motion perpendicular to the magnetic field is of the two types depending on whether it is compressional or not. A squeezing disturbance (both $\nabla$ and $\mathbf{u}\perp\mathbf{B}$) affects the magnetic field strength ($\mathbf{B}\nabla\cdot\mathbf{u}$) and causes it to act as part of a total pressure ($\nabla B^2$). This gives rise to the compressional Alfvén waves. A shearing disturbance ($\mathbf{u}\perp\mathbf{B}$ but $\nabla\|\mathbf{B}$) does not affect the magnetic field strength, but bends the field lines in the direction of the flow ($\mathbf{B}\cdot\nabla\mathbf{u}$). The restoring force is provided by magnetic field line tension ($\mathbf{B}\cdot\nabla\mathbf{B}$), through which the magnetic field lines act as wires threading the plasma. This yields the shear Alfvén waves which provide the most characteristic difference between MHD and fluid dynamics, the ability of a magnetised plasma to support transverse waves.

To do this quantitatively we take the same arrangement as in Section 6.3, namely a homogeneous plasma at rest in the presence of a magnetic field, and linearised disturbances for a single Fourier component. The equations for the disturbances are

$$\omega\tilde{\rho} = \rho\mathbf{k}\cdot\mathbf{u}, \tag{6.100}$$

$$\omega\rho\mathbf{u} = \mathbf{k}\left(\tilde{p} + \frac{\mathbf{B}\cdot\tilde{\mathbf{B}}}{\mu_0}\right) - \frac{\mathbf{B}\cdot\mathbf{k}}{\mu_0}\tilde{\mathbf{B}}, \tag{6.101}$$

$$\omega\tilde{p} = \frac{5}{3}p\mathbf{k}\cdot\mathbf{u}, \tag{6.102}$$

$$\omega\tilde{\mathbf{B}} = -(\mathbf{B}\cdot\mathbf{k})\mathbf{u} + \mathbf{B}\mathbf{k}\cdot\mathbf{u}. \tag{6.103}$$

Since the pressure dynamics is adiabatic the density equation is superfluous. For motion along the magnetic field ($\mathbf{u}\|\mathbf{B}$), there is no reaction by $\tilde{\mathbf{B}}$. If the motion is shearing ($\mathbf{k}\perp\mathbf{u}$) then there is no reaction at all, and if it is compressional ($\mathbf{k}\|\mathbf{u}$) then sound waves result,

$$\omega^2 = k_\|^2\frac{5}{3}\frac{p}{\rho} = k_\|^2 v_s^2 \qquad v_s^2 = \frac{5}{3}\frac{p}{\rho}, \tag{6.104}$$

just like in a neutral fluid.

For motion across the field ($\mathbf{u} \perp \mathbf{B}$) we get the Alfvén waves. If the motion is compressional then we have longitudinal waves with

$$\mathbf{u} \perp \mathbf{B} \qquad \mathbf{k} \perp \mathbf{B} \qquad \mathbf{k} \| \mathbf{u} \qquad \tilde{\mathbf{B}} \| \mathbf{B}, \tag{6.105}$$

and the pressure and MHD kinematic equations may be combined as

$$\omega\left(\tilde{p} + \frac{\mathbf{B} \cdot \tilde{\mathbf{B}}}{\mu_0}\right) = \left(\frac{5}{3}p + \frac{B^2}{\mu_0}\right)\mathbf{k} \cdot \mathbf{u}, \tag{6.106}$$

and the MHD force equation may be rewritten as

$$\omega \rho \mathbf{k} \cdot \mathbf{u} = k_\perp^2\left(\tilde{p} + \frac{\mathbf{B} \cdot \tilde{\mathbf{B}}}{\mu_0}\right), \tag{6.107}$$

with $\mathbf{k}$ explicitly written as perpendicular to $\mathbf{B}$. This yields the compressional Alfvén waves with both plasma pressure and magnetic pressure contributing to the phase velocity,

$$\omega^2 = k_\perp^2(v_s^2 + v_A^2). \tag{6.108}$$

This is the low frequency, finite pressure limit of the compressional magnetosonic waves discussed above (cf. Eq. 6.70), and it tells us that the cutoff at $\omega = \omega_{pi}$ limits MHD in the same way as it does the plasma dynamics of several fluids—ultimately this is a space scale limitation, as we require

$$\omega \ll \omega_{pi} \implies k_\perp \ll \frac{\omega_{pi}}{v_A}, \tag{6.109}$$

for quasineutrality in compressional propagation perpendicular to $\mathbf{B}$.

If the motion across the field is shearing then we have transverse waves with

$$\mathbf{u} \perp \mathbf{B} \qquad \mathbf{k} \| \mathbf{B} \qquad \mathbf{k} \perp \mathbf{u} \qquad \tilde{\mathbf{B}} \perp \mathbf{B}. \tag{6.110}$$

Since the flow is divergence free the pressure does not react, and the existence of waves depends entirely on the finite $\mathbf{B}$. Since the magnetic pressure also does not react, we may rewrite the equations as

$$\omega \rho \mathbf{u} = -\frac{Bk_\|}{\mu_0}\tilde{\mathbf{B}}, \tag{6.111}$$

$$\omega \tilde{\mathbf{B}} = -(Bk_\|)\mathbf{u}, \tag{6.112}$$

with $\mathbf{k}$ explicitly written as parallel to $\mathbf{B}$. This yields the shear Alfvén waves with propagation at the Alfvén velocity,

$$\omega^2 = k_\|^2 v_A^2. \tag{6.113}$$

Besides the differing qualitative character of compressional and shearing Alfvén dynamics, the high aspect ratio character of most magnetised plasmas (in nature or in the laboratory) leads to a significant difference in their time scales. The distance across the magnetic field is significantly less than that along the field lines, so the compressional waves are faster. For this reason they are often called 'fast Alfvén waves' while the shear waves are called 'slow Alfvén waves.'

The other important difference between compressional and shear Alfvén waves is that the former go over into sound waves if $B \to 0$. It is interesting to note that the high frequency limit of these sound waves is also the ion plasma frequency, $\omega_{pi}$, limiting $\omega$ for $k_\perp \lambda_D > 1$ (take $\Omega_i \to 0$ and $v_A \to c_s$ in Eq. (6.70)). Whether the $B \to 0$ limits are reached is a function of the plasma beta, defined as the ratio of plasma to magnetic pressures,

$$\beta = \frac{2\mu_0 p}{B^2} \sim \frac{v_s^2}{v_A^2}, \tag{6.114}$$

also related to the ratio of the sound and Alfvén velocities (take $\mu_0 p_e$ instead, and $\beta$ becomes $\beta_e$). Except for the kinematic regime of natural dynamo phenomena, this parameter is usually smaller than unity, and it is this fact which makes MHD Alfvén waves as important as they are.

For the low frequency drift motion involved in magnetised plasma, the perpendicular dynamics takes place on scales rather smaller than those of the global system size. Disparity ratios of parallel to perpendicular scale of motion in the range of $10^3$ are common. Motion across the magnetic field with sound velocities or slower is too slow to effectively excite compressional motion, since $v_s \ll v_A$. On the other hand, the shear Alfvén motion can be excited since the aspect ratio can be comparable to the ratio between $v_A$ and $c_s$. Expressed as frequencies, we have

$$k_\perp v_A \gg k_\parallel v_A \sim \omega. \tag{6.115}$$

If the dynamical frequencies of interest are always much smaller than the minimum of $k_\perp v_A$ and $\omega_{pi}$, then the motion across the field can be treated as dynamically incompressible, even if quasistatic compression (due to an inhomogeneous **B**) can still enter the dynamics. This leads directly to the fluid drift picture to be introduced in Section 7.2 of the next Chapter. But before that we treat the energy and momentum conservation properties of the fluid electrodynamic and fluid plasma models, and then dissipation.

## 6.8 Energetics of the ideal fluid dynamical systems

In the fluid electrodynamic model, energy in total is conserved, but it can be exchanged between fluid and electromagnetic field energy, in addition to its exchange between thermal (random) and kinetic (directed) energy through work done by pressure. Each piece of the total can also be transported from place to place through its own energy flux. We have already seen this type of conservation equation for material and thermal energy densities, during the ideal fluid derivation (Eqs 6.21 and 6.22). Electromagnetic field energy is carried at the speed of light, while thermal and kinetic energy is carried at the speed of the fluid. In fluid plasma dynamics, the energy content of the electric field becomes negligible along with the charge density, leaving magnetic energy alone, which is now carried at the Alfvén speed since Alfvén waves are the fastest things that model retains. The same is true of single-fluid MHD. Otherwise, the form of energy conservation is the same for all three models. Momentum conservation, on the other hand, is progressively more trivial, especially for the electromagnetic field. In MHD the step between Eqs (6.91) and (6.98) expresses the entire content of electromagnetic momentum conservation.

In electrodynamics the electromagnetic field energy density is given by

$$U_{em} = U_E + U_B \qquad U_E = \varepsilon_0 \frac{E^2}{2} \qquad U_B = \frac{B^2}{2\mu_0}. \tag{6.116}$$

Its evolution according to the Maxwell Eqs (6.38) and (6.39) is given by

$$\frac{\partial U_{em}}{\partial t} = \varepsilon_0 \mathbf{E} \cdot \frac{\partial \mathbf{E}}{\partial t} + \frac{\mathbf{B}}{\mu_0} \cdot \frac{\partial \mathbf{B}}{\partial t}, \tag{6.117}$$

in which the gradient operations are manipulated into a total divergence,

$$\frac{\partial U_{em}}{\partial t} + \nabla \cdot \left( \frac{\mathbf{E} \times \mathbf{B}}{\mu_0} \right) = -\mathbf{E} \cdot \mathbf{J}. \tag{6.118}$$

The field energy is transported by the Poynting energy flux, given by $c^2$ times the Poynting momentum, reflecting transport at the speed of light. The term on the right side gives the transfer to the fluid; it is called the Joule work. The kinetic energy density in total and for each fluid of species $j$ is

$$V_T = \sum_j V_j \qquad V_j = n_j m_j \frac{v_j^2}{2}, \tag{6.119}$$

respectively. The evolution of $V_j$ follows by contracting each Eq. (6.41) with $\mathbf{v}_j$, yielding

$$\frac{\partial V_j}{\partial t} + \nabla \cdot \left( V_j \mathbf{v}_j \right) = n_j e_j \mathbf{v}_j \cdot \mathbf{E} - \mathbf{v}_j \cdot \nabla p_j, \tag{6.120}$$

where we note here also that the magnetic force is not involved in the transfer between the fluid and the field. The thermal energy densities for each fluid of species $j$ and for the total fluid are

$$U_j = \frac{3}{2} p_j \qquad U_T = \sum_j U_j, \tag{6.121}$$

and its evolution follows from Eq. (6.42), to which we can add an arbitrary conductive heat flux $\mathbf{q}_j$

$$\frac{\partial U_j}{\partial t} + \nabla \cdot \left[ (U_j + p_j)\mathbf{v}_j + \mathbf{q}_j \right] = \mathbf{v}_j \cdot \nabla p_j. \tag{6.122}$$

Here, the total heat flux is built up from advective and conductive pieces and the transfer with kinetic energy is through the work done by the pressure, just as for a neutral fluid. In the fluid electrodynamics model the energy transfer between fluids and field is entirely through the electric Lorentz force, with the transfer for each fluid given by its contribution to the total current along **E**. For the total system of fluids and field the conservation law for energy reads

$$\frac{\partial}{\partial t}(U_{em} + V_T + U_T) + \nabla \cdot \left[\frac{\mathbf{E} \times \mathbf{B}}{\mu_0} + \sum_j \left(V_j \mathbf{v}_j + [U_j + p_j]\mathbf{v}_j + \mathbf{q}_j\right)\right] = 0, \quad (6.123)$$

which is in the local form of a conservation law (partial time derivative of a density quantity plus the divergence of a flux quantity equals zero). Then, the global form is just the spatial domain (volume) integral or average over the density quantity, where the divergence quantity is expected to vanish under the integral.

Momentum conservation is derived in a similar fashion. The field momentum density is given by the Poynting flux, whose conservation law reads

$$\frac{\partial}{\partial t}(\varepsilon_0 \mathbf{E} \times \mathbf{B}) + \nabla \cdot \left(U_E \mathbf{g} - \varepsilon_0 \mathbf{E}\mathbf{E} - \frac{\mathbf{B}\mathbf{B}}{\mu_0}\right) = -\left(\rho_{ch}\mathbf{E} + \mathbf{J} \times \mathbf{B}\right), \quad (6.124)$$

where **g** is the metric tensor and the quantity under the divergence is called the Maxwell stress tensor. All four Maxwell equations and the identity

$$\mathbf{E} \times (\nabla \times \mathbf{E}) = \nabla \frac{E^2}{2} - \nabla \cdot (\mathbf{E}\mathbf{E}) + \mathbf{E}(\nabla \cdot \mathbf{E}), \quad (6.125)$$

for both **E** and **B** have been used. The exchange with the fluid over the Lorentz force is clear from Eq. (6.41). For the total system of fluids and field the conservation law for momentum reads

$$\frac{\partial}{\partial t}\left[\varepsilon_0 \mathbf{E} \times \mathbf{B} + \sum_j (n_j m_j \mathbf{v}_j)\right] + \nabla \cdot \left[(p + U_E)\mathbf{g} - \varepsilon_0 \mathbf{E}\mathbf{E} - \frac{\mathbf{B}\mathbf{B}}{\mu_0} + \sum_j (n_j m_j \mathbf{v}_j \mathbf{v}_j)\right] = 0, \quad (6.126)$$

which is also in the local form.

Under quasineutrality ($\varepsilon_0 \to 0$), the neglect of the displacement current in the dynamics is equivalent to neglecting **E** compared to **B**, so that **E** no longer has a role in the energy, except for the Poynting flux and the Joule work. Otherwise, the energy conservation dynamics is largely the same, with the contribution by $E^2$ to $U_E$ simply dropped. Momentum conservation, however, is greatly simplified. In the case that $v_A \ll c$ we note that since

$$\frac{\mathbf{E} \times \mathbf{B}}{\mu_0} \sim n_i m_i \frac{\mathbf{E} \times \mathbf{B}}{B^2} \frac{v_A^2}{c^2} \ll n_i m_i \frac{\mathbf{E} \times \mathbf{B}}{B^2} \sim n_i m_i \mathbf{u}, \quad (6.127)$$

the field momentum is rendered negligible. We then have a quasistatic equilibrium between the transfer and the transport of field momentum,

$$\nabla \cdot \left( \frac{B^2}{2\mu_0} \mathbf{g} - \frac{\mathbf{BB}}{\mu_0} \right) = \mathbf{B} \times \frac{\nabla \times \mathbf{B}}{\mu_0} = -\mathbf{J} \times \mathbf{B}, \tag{6.128}$$

which is a simple restatement of Eq. (6.96). In either plasma dynamics or MHD, then, momentum conservation becomes relatively trivial, simply built in by Eq. (6.96). In terms of the energy transport, we compare the Poynting energy flux to the thermal transport,

$$\frac{\mathbf{E} \times \mathbf{B}}{\mu_0} \sim \frac{B^2}{\mu_0} \frac{\mathbf{E} \times \mathbf{B}}{B^2} \sim \frac{B^2}{\mu_0} \mathbf{u} \sim \beta^{-1} p \mathbf{u}, \tag{6.129}$$

assuming the fluid velocity is comparable to the ExB velocity. The Poynting flux is nominally the larger by a factor of $\beta^{-1}$. In the case of motion being sufficiently slow that $\nabla \cdot \mathbf{u} \approx 0$, i.e. divergence free perpendicular flows, however, the total force potential of longitudinal motion remains undisturbed, so that we have

$$\tilde{p} + \frac{B\tilde{B}}{\mu_0} \approx 0, \tag{6.130}$$

and hence rough equality in the divergences, i.e. the actual transport, between the thermal and Poynting energy fluxes, but of opposite sign. For fluid drift motion the role of the Poynting flux is simply to cancel one factor out of the total 5/2 in the thermal flux, leaving a total of 3/2. This result,

$$\frac{\mathbf{E} \times \mathbf{B}}{\mu_0} + p\mathbf{v}_E \approx 0 \qquad \mathbf{v}_E = \frac{\mathbf{E} \times \mathbf{B}}{B^2} \tag{6.131}$$

valid for low frequency force balance, that is, drift motion, is called the 'Poynting cancellation'.

The only real difference in the conservation properties between fluid electrodynamics and fluid plasma dynamics (or MHD), then, is that there are no space charge or light speed effects and hence no propagation of energy or momentum through high frequencies at the speed of light. As long as $v_A \ll c$ the important transfer is through the fluid motions, to which only the Poynting energy flux (momentum times $c^2$) contributes at a comparable level. Joule work remains as well, as the sole energy transfer mechanism between the fluids and the field.

## 6.9 Dissipation—corrections to the ideal plasma

The above model systems are all useful as a framework for understanding how ideal fluid dynamics and related systems for plasmas are derived and how their fundamental properties compare. For lower frequency motion, however (Eq. 6.115), we need to introduce dissipation, and the methods used to derive the ideal systems are not sufficient for that. Although simple models can be heuristically constructed, we will find that there are several important properties of the corrections to ideal dynamics that are not easily understood without a formal derivation, which starts with collections of particles that are described by distribution functions in position/velocity phase space. A simple example is the fact that an electric field can cause a conductive heat flux, or that a temperature gradient can cause a current. A not so simple example is the existence of special dissipation-free momentum and thermal fluxes directed perpendicular both to the magnetic field and to the gradients of the quantities giving rise to them. Besides these, it is desirable to know how the fluid model is related to the underlying kinetic physics.

There are two main procedures by which fluid equations can be rigorously derived from the kinetic model. Each is most useful for a certain purpose. For actual dissipation such as friction and conduction, we start with a Maxwellian distribution function which, due to its maximal entropy, is not changed by a statistical collision process. We then find corrections that are assumed to be small in an ordering parameter, which gives the ratio of the time scales between the dynamics of interest and the relaxation time on which the collision process forces the distribution function back towards the Maxwellian. Afterwards, the parameter regime is found from the requirements that the corrections be small. This is called the Chapman–Enskog procedure, after the inventors who applied it to the viscosity of neutral fluids.

For the non-dissipative fluxes that arise from the corrections due a finite particle gyroradius in the presence of an inhomogeneity, the moment approach is most transparent. The details of the distribution function are not followed, but the lower-order moments are calculated in a hierarchy which is then closed at some level. A moment of the distribution function is an integral of it times powers of the velocity coordinate over the velocity part of the phase space. The density is the zeroth moment, simply an average. The velocity is the first moment, and the pressure is the isotropic second moment. The hierarchy arises due to the presence of factors of the velocity coordinate in the kinetic equation, so that the equation for the $n$th moment is determined at least partly by the $(n + 1)$ th or higher moments. A closure approximation enters, by which one expresses the moments at some level and higher as the result which would be obtained if the distribution function were Maxwellian. An example here is the ideal fluid, in which off-diagonal second and all third moments are taken to vanish, leaving a closed set of equations for the density, velocity, and pressure (or temperature). For collisional processes the moment approach is equivalent to but more difficult to carry out than the Chapman–Enskog procedure. For gyroradius corrections, especially in nontrivial magnetic geometries, the Chapman–Enskog procedure is in turn more difficult.

## 6.10 Chapman–Enskog procedure—dissipation

For a collisional process between electrons and ions we apply the Chapman–Enskog procedure to find the conductive heat flux and the resistive friction which we need to put into the fluid equations. We start with a collision process, which is assumed to take place faster than any dynamical scale of interest—the usefulness of this principally for low frequency motion is clear. The time scale comparison applies to the rest frame of the particles, so we require both a fast collision frequency and a short mean free path:

$$\frac{\partial}{\partial t} \sim v_t \nabla_{\|} \ll \nu \implies \lambda_{mfp} \nabla_{\|} \ll 1 \qquad \lambda_{mfp} = \frac{v_t}{\nu}, \tag{6.132}$$

where $v_t$ is the particle thermal velocity, giving the mean free path $\lambda_{mfp}$. Here, we take the maximal ordering on the dissipation-free part of the dynamics, implicitly assuming all 'ideal' terms in the fluid equations to be equally important. Since what we expect to obtain are parallel fluxes, a spatially one-dimensional treatment is sufficient to outline the method. For most applications the perpendicular fluxes arising from collisional dissipation are so limited by the gyromotion as to be negligibly small.

This is an ordering procedure in the small parameter $\omega/\nu$. An additional ingredient to the ordering hierarchy is the application of consistency conditions at each level of the expansion. At each level the solution is not a single distribution function but a class of them whose parameters are constrained by the consistency conditions—application of operators which 'annihilate' the largest term, the collision operator. An example is simple averaging over velocity space, since the collision process conserves particles.

The form of the kinetic equation is

$$\mathscr{L}[f^{(n-1)}] = \mathscr{C}[f^{(n)}], \tag{6.133}$$

where $f = f(\mathbf{x}, \mathbf{w})$ is a distribution function of the six phase space coordinates $\{\mathbf{x}, \mathbf{w}\}$ and time, $(n)$ is the current order of the expansion, $\mathscr{L}$ is the operator containing all terms except $\mathscr{C}$, and $\mathscr{C}$ is the collision operator which is assumed to conserve particles and energy, and at most transfer momentum between the particle species.

The logic of the procedure is the establishment of *a priori* statements about $f$ at each order in the expansion, which are then further qualified by the consistency conditions. This is done by placing the $f$ at the current order into $\mathscr{L}$ and the $f$ at the next order into $\mathscr{C}$ and then applying operations whose action on $\mathscr{C}$ vanishes for any $f$; this further constrains what the $f$ at the current order can be. At the end of the calculation, all outstanding assumptions are checked *a posteriori*.

The steps up to second order take place according to the following sequence ⋯

- To zeroth order $\mathscr{C}[f]$ must vanish; this tells us that the form of $f^{(0)}$ to zeroth order is that of a Maxwellian in the particle velocity $\mathbf{w}$, with parameters $n_e$, $\mathbf{v}$, and $T_e$. At zeroth order there is no current since for a relative friction $\mathscr{C}$ vanishes only if $\mathbf{v} - \mathbf{u} = 0$, where as before $\mathbf{v}$ and $\mathbf{u}$ are the velocities of the electron and ion fluids and hence the mean velocities of their distributions. The existence of a current at zeroth order would imply interesting dynamics on the scale of $\nu$, which is excluded. With $\mathbf{v} = \mathbf{u}$, we can do the calculation in the local rest frame of the ion fluid.
- Write the equation at first order, inserting $f^{(0)}$ into $\mathscr{L}[f]$ and the still unknown $f^{(1)}$ into $\mathscr{C}[f]$.
- With momentum conservation already satisfied, the consistency conditions which constrain $n_e$ and $T_e$ are conservation of particles and energy. Through application of the annihilation operators represented by integrating $\mathscr{C}[f]$ times 1 and $m_e w^2/2$, respectively, we find the equations for $n_e$ and $T_e$ at order zero.
- Solve the equation for $f^{(1)}$.

- Steps (2) and (3) at first order give the dissipative fluxes represented by the current and conduction and hence the equations for $n_e$ and $T_e$ at order one. At this point the procedure stops if we are satisfied with the equations at first order.

To illustrate this procedure it is sufficient to specify a particle and energy conserving collision operator describing frictional drag of the electrons against the ions. To lowest order in $\mu_e = m_e/m_i$, the corresponding operator is the simple Lorentz one describing elastic pitch angle scattering. With $\mu_e \ll 1$ the collision process is one of electrons scattering against a fixed ion background. Momentum is transferred conservatively from the electrons to the ions, tending to relax the relative drift towards zero. With the ion temperature satisfying $m_e T_i \ll m_i T_e$, the velocities of the ions can be neglected. One can show that the energy is largely conserved in the electrons by considering a head-on collision between a fixed ion and an electron moving with velocity $v_1$. With the change in the electron momentum given by $2m_e v_1$ the ion acquires velocity $u_1 = 2(m_e/m_i)v_1$. The corresponding kinetic energy is $m_i u_1^2/2 = 2m_e v_1^2 \mu_e$, that is, one order down in $\mu_e$ from the electron energy. To this order we can neglect the energy transfer (which, however, is necessary if we want to consider collisional relaxation of a temperature difference). The collision process is described as pitch angle scattering because the degree of freedom which changes for the electron is the angle between its velocity vector and the magnetic field. For a distribution of electrons this represents a diffusion in velocity space, which for $\nu \ll \Omega_e$ is described by

$$\mathscr{C}[f_e] = \frac{\nu_p v_t^3}{w^3}\frac{\partial}{\partial \zeta}(1 - \zeta^2)\frac{\partial f_e}{\partial \zeta}, \tag{6.134}$$

where

$$\nu_p = \frac{3}{\sqrt{8}}\sqrt{\pi}\,\nu_e, \tag{6.135}$$

is the scattering rate for thermal electrons and $\zeta = w_\| / w$ is the pitch angle. The $w^{-3}$ dependence arises from the inverse square nature of the Coulomb force. Since we are taking $\nu \ll \Omega_e$, the distribution function has dependence on two degrees of freedom in velocity space; we denote these as the total and parallel velocities, $w$ and $w_\|$. Since only $w_\|$ is directed, we can illustrate the procedure by doing a one-dimensional calculation considering only parallel dynamics.

To zeroth order we have

$$\mathscr{C}[f_e^{(0)}] = 0, \tag{6.136}$$

and $f_e^{(0)}$ is a Maxwellian with zero drift relative to the ions

$$f_e^{(0)} = \left(\frac{n_e}{(2\pi)^{3/2} v_t^3}\right)\exp[-w^2/2v_t^2], \tag{6.137}$$

where $v_t^2 = T_e/m_e$ defines the electron thermal velocity.

The first-order equation is

$$\mathscr{C}[f_e^{(1)}] = \left(\frac{\partial}{\partial t} + w_\| \nabla_\| - \frac{e}{m_e} E_\| \frac{\partial}{\partial w_\|}\right) f_e^{(0)} \equiv \mathscr{L}[f_e^{(0)}], \tag{6.138}$$

a drift kinetic equation describing dynamics parallel to the magnetic field. The operator $\mathscr{L}$ is found by considering moving particles forced upon by the electric field in such a way that phase space volume is conserved. The characteristics of $\mathscr{L}$ are simply the particle trajectories.

The consistency conditions are the averages over particle number and kinetic energy, which yield the density and thermal energy since at lowest order there is no velocity (we assume that $u_\| \ll v_t$),

$$\int d^3w \; \mathscr{C}[f] = 0, \qquad \int d^3w \; \frac{m_e w^2}{2} \mathscr{C}[f] = 0. \tag{6.139}$$

These conditions are satisfied by arbitrary $f$, which is why we can apply them at each order. At zeroth order they provide constraints on $n_e$ and $T_e$,

$$\frac{\partial n_e}{\partial t} = 0 \qquad \frac{3}{2} n_e \frac{\partial T_e}{\partial t} = 0, \tag{6.140}$$

which simply say that $n_e$ and $T_e$ do not change on the collision time scale.

We now expand the equation to first order by inserting $f^{(0)}$ into $\mathscr{L}$ under the consistency constraints on $n_e$ and $T_e$,

$$\mathscr{C}[f_e^{(1)}] = w_\| \left[\frac{\nabla_\| n_e}{n_e} + \left(\frac{w^2}{2v_t^2} - \frac{3}{2}\right)\frac{\nabla_\| T_e}{T_e} + \frac{eE_\|}{T_e}\right] f_e^{(0)}. \tag{6.141}$$

The solution for $f$ at first order is trivial, since we know the eigenfunctions of the Lorentz operator are the Legendre polynomials, and the right side depends on $\zeta$ only through

$$w_\| = w\zeta = wP_1(\zeta). \tag{6.142}$$

Since the Legendre polynomials form an orthogonal basis for real functions on $[-1, 1]$, only the first eigenfunction has a nonzero coefficient. This means that

$$f_e^{(1)} = -\frac{1}{2\nu_p} \frac{w^3}{v_t^3} w_\| \left[\frac{\nabla_\| n_e}{n_e} + \left(\frac{w^2}{2v_t^2} - \frac{3}{2}\right)\frac{\nabla_\| T_e}{T_e} + \frac{eE_\|}{T_e}\right] f_e^{(0)}. \tag{6.143}$$

The formal step of inserting $f_e^{(1)}$ into $\mathscr{L}$ and applying the consistency constraints is simply a matter of finding the current and heat flux resulting from this $f_e^{(1)}$. These integrals are given by

$$J_\| = -\int d^3w \; e w_\| f_e^{(1)} \qquad q_{e\|} = \int d^3w \; \frac{mw^2}{2} w_\| f_e^{(1)} + \frac{5}{2} T_e \frac{J_\|}{e}, \tag{6.144}$$

subtracting off the advective part and assuming that $u_\parallel \ll v_t$. We find

$$J_\parallel = n_e e(u_\parallel - v_\parallel) = \eta^{-1} n_e e \frac{V_e^2}{\nu_e}\left[\frac{\nabla_\parallel p_e}{p_e} - \frac{eE_\parallel}{T_e} + \alpha\frac{\nabla_\parallel T_e}{T_e}\right], \tag{6.145}$$

for the current and

$$q_{e\parallel} = -\kappa n_e T_e \frac{V_e^2}{\nu_e}\frac{\nabla_\parallel T_e}{T_e} - \alpha\frac{T_e}{e}J_\parallel, \tag{6.146}$$

for the heat flux.

Placing these fluxes into the fluid equations for $n_e$ and $T_e$, and transforming back to the laboratory frame, we find

$$\frac{\partial n_e}{\partial t} + \nabla_\parallel n_e v_\parallel = 0, \tag{6.147}$$

$$\frac{3}{2}n_e\left(\frac{\partial}{\partial t} + v_\parallel \nabla_\parallel\right)T_e + n_e T_e \nabla_\parallel v_\parallel = -\nabla_\parallel q_{e\parallel}, \tag{6.148}$$

$$-n_e e E_\parallel = n_e e\frac{\partial A_\parallel}{\partial t} + n_e e\nabla_\parallel \phi = \nabla_\parallel p_e + \alpha n_e \nabla_\parallel T_e - n_e e \eta_\parallel J_\parallel, \tag{6.149}$$

where $J_\parallel$ and $q_{e\parallel}$ given by Eqs (6.145) and (6.146), $A_\parallel$ and $J_\parallel$ are related by Ampère's law, $v_\parallel$ is given in terms of $u_\parallel$ and $J_\parallel$ as

$$-\nabla^2 A_\parallel = \mu_0 J_\parallel \qquad v_\parallel = u_\parallel - \frac{J_\parallel}{n_e e}, \tag{6.150}$$

and $\eta_\parallel$ is the resistivity given by

$$\eta_\parallel = \eta\frac{m_e \nu_e}{n_e e^2}. \tag{6.151}$$

In a setting wherein only the parallel dissipation channels are treated, one can then add the perpendicular advection into the time derivative.

These relations are general, with constant coefficients $\eta$, $\kappa$, and $\alpha$ given by the specific model. They are also dependent on the charge state of the ions. For this simple Lorentz model we have

$$\eta = \frac{3\pi}{32} \qquad \kappa = \frac{128}{3\pi} = \frac{4}{\eta} \qquad \alpha = \frac{3}{2}, \tag{6.152}$$

for charge state $Z = +1$. For the more detailed, and standard, Braginskii model (1965) we have

$$\eta = 0.51 \qquad \kappa_e = 3.2 \qquad \kappa_i = 3.9 \qquad \alpha = 0.71, \tag{6.153}$$

with $\kappa_{e,i}$ for electrons and ions, also for charge state $Z = +1$, and we note that the definition of the thermal collision frequency

$$\nu_e = \frac{4\sqrt{2\pi}n_e\lambda e^4}{3m_e^{1/2}T_e^{3/2}}, \tag{6.154}$$

comes from that model (together with $\nu_i$ as given in Chapter 1). For different charge states the numerical constants change but the description does not.

It is interesting to note the presence of the cross terms represented by $\alpha$. This is called the thermoelectric effect, whereby a temperature gradient causes a current, and the reverse situation of a conductive heat flux caused by a current is guaranteed by Onsager symmetry. It is the result of the collision frequency's dependence on the particles' relative velocity, a feature of the Coulomb interaction. If the above calculation is repeated without the factor of $(v_t/w)^3$ in Eq. (6.134), the result is $\alpha = 0$. The physical origin of this effect is that particles imparting momentum to an ion at a given point originate from another point generally one mean free path away, and the mean free path is longer for faster particles. In the presence of a temperature gradient the faster particles therefore come from places where the temperature difference is greater, and hence they add more to the total impulse than would be the case if the distribution had the same Maxwellian everywhere. Faster particles are in a sense overcounted then, when adding up the impulses. Had we simply constructed the resistive and conductive dissipation terms from the usual simple considerations of a frictional drag and a finite but small mean free path in a temperature gradient, we would have missed these cross terms. The extra term proportional to $\alpha\nabla_\parallel T_e$ in Eqs (6.145) and (6.149) is also called the thermal force.

The above analysis is just an illustration of how this procedure is used to find the dissipative fluxes. A more realistic effort takes into account a finite ratio between $\nu_e$ and $\Omega_e$, and also includes collisions between like particles, but this requires a level of detail beyond the scope of this outline; the interested reader can consult the references in the Section Further Reading. The version presented here follows Hassam (1980), who was the one to point out that the electron inertia term is second order in $\omega/\nu_e$; that is, an arbitrary $v_\parallel - u_\parallel$ cannot appear to lowest order. Furthermore, several other important effects enter at second order along with the electron inertia. The standard among detailed versions is the one by Braginskii (1965), whose definition of $\nu_e$ is standard.

In the more realistic applications, one must face the fact that $\omega/\nu_e$ is almost never small enough to make these collisional models rigorously valid. Nevertheless, they form a useful starting point in understanding the physics of the dissipative fluxes. Extensions to the collisional fluid model which are designed to treat the more important kinetic effects will be discussed in Volume Two. One needs a time-dependent equation for the parallel heat fluxes as dynamical variables, so that the temperature and parallel heat flux of each species is treated at the same level as the density and parallel velocity. One also needs a model for Landau damping, a kinetic wave damping mechanism whose role here will be to provide a dissipation channel through the temperature dynamics.

### 6.10.1 Cross-species temperature differences

In this simplified analysis we have neglected the next correction, which is the one affecting the difference in temperature between the electrons and ions in a plasma. In many cases, especially laboratory discharge plasmas of low density, which are either of short duration or continually produced at a rate that can compete with collision processes, the ion temperature can be taken to be small, $T_i \ll T_e$, in a situation referred to as cold ions. In others, especially fusion plasmas where the energy density is as high as one can achieve, the opposite is true and the ion and electron temperatures are taken to be thermalised. If $T_i \ll T_e$ we refer to cold ions. On the other hand, if $T_i \sim T_e$, we refer to warm ions, and if it is much larger, then to hot ions.

The reason electron and ion temperatures can be different lies in the disparate mass ratio and the nature of the collision process. Collisions between like particles conserve energy and momentum, so they contribute to thermalisation of the species towards a Maxwellian distribution. But they do not change the temperature or the average directed velocity. Only cross-species particle collisions can change these. The details show that the small mass of the electrons allows an expansion in powers of the mass ratio, $\mu_e = m_e/m_i$, with the largest term given by a pitch angle scattering at constant energy. This yields the frictional drag giving resistivity and thermal conduction, as in the procedure just done above. Only the next term, a mass ratio correction, gives the thermalisation between the temperatures. So the temperature equilibration between electrons and ions proceeds at a rate given by $3\mu_e\nu_e$ times the temperature difference. Even in the collisional edge region of fusion plasmas, this is too weak to affect the dynamics, for which $\nu_e$ is comparable to the vorticity. It is usually taken to be effective in setting the equilibrium temperatures close to each other. Except for some exotic cases, the choice of $\tau_i = T_i/T_e = 1$ is reasonable for studying the physics. In astrophysical circumstances or laboratory discharge plasmas, where the time scales of equilibrium maintenance can be very short, then one can take $\tau_i = 0$ and revert to cold ions. It all depends on the situation. Generically then, we will take $\tau_i$ to be either unity or zero.

If transport processes and relaxation of a driven equilibrium (technically, not an equilibrium but a quasistatically evolving steady state) are followed, as in transport analysis of fusion experiments, then the thermal relaxation process is part of the set of relevant effects and is always treated. If several ions are present then their thermal relaxation proceeds on the scale of $\nu_i$, which although slower than $\nu_e$ is faster than $\mu_e\nu_e$. So the ions are usually assumed to be at the same temperature, while the electrons and average ion temperatures evolve separately. We will be treating the properties of the turbulence, leaving transport analysis, regimes, and phenomenology to other work. So the collisional effects which effectively contribute to dissipation of the dynamics, like resistivity and thermal conduction, and later parallel viscosity (related to isotropisation of the temperature within a species), are kept as dissipation processes, and these other higher-order effects are neglected.

There is an additional set of nonideal fluxes to treat, and while these do not represent dissipation, they affect the nature of nonlinear advection, especially in the momentum equation but also in the temperature. They are the diamagnetic fluxes, to be addressed next.

## 6.11 The moment approach—diamagnetic fluxes

The moment approach has the same basis as the Chapman–Enskog procedure, that of a distribution function in phase space. But instead of direct manipulation of $f$, a hierarchy of moments are built up. The quantities conserved in ideal fluid dynamics form the starting point,

$$n = \int d^3w\, f, \tag{6.155}$$

$$nm\mathbf{v} = \int d^3w\; m\mathbf{w}f, \tag{6.156}$$

$$\frac{3}{2}nT + nm\frac{v^2}{2} = \int d^3w\; m\frac{w^2}{2}f. \tag{6.157}$$

We find the equations these quantities satisfy by performing the same integrations on the kinetic equation, since they commute with $\partial/\partial t$. In doing this, however, we find that the advection term in the kinetic equation, $\mathbf{w}\cdot\nabla f$, introduces in each equation the moments of correspondingly higher order; these are the fluxes for the conserved quantities,

$$n\mathbf{v} = \int d^3w\; \mathbf{w}f, \tag{6.158}$$

$$\mathbf{P} + m\mathbf{v}\mathbf{v} = \int d^3w\; m\mathbf{w}\mathbf{w}f, \tag{6.159}$$

$$\mathbf{q} + \left(\frac{3}{2}nT + p + nm\frac{v^2}{2}\right)\mathbf{v} = \int d^3w\; m\frac{w^2}{2}\mathbf{w}f. \tag{6.160}$$

And then, writing the equations for these introduces quantities of still higher degree in $\mathbf{w}$ in the integrand. Here, a closure scheme is implemented, according to which one simply declares that moments at a certain level are functions of those already determined. In the simplest models, the pressure is assumed to be isotropic and the fourth moment is assumed to result from a Maxwellian. These last two statements imply the following relations:

$$p = \frac{1}{3}\int d^3w'\; mw'^2 f \qquad \mathbf{\Pi} = \int d^3w'\; m\left(\mathbf{w}'\mathbf{w}' - \frac{1}{3}w'^2\mathbf{g}\right)f = \mathbf{P} - p\mathbf{g}, \tag{6.161}$$

where

$$\mathbf{w}' = \mathbf{w} - \mathbf{v}, \tag{6.162}$$

and

$$R_{\|} = \int d^3w'\, m\frac{w'^2}{2}w_{\|}'^2 f = \frac{5}{2}\frac{pT}{m}. \tag{6.163}$$

For an ideal fluid, these moments are simply what results from using a Maxwellian for $f$, that is, $p = nT$ and $\mathbf{\Pi} = \mathbf{q} = 0$. The corrections to this situation then give the dissipative fluxes.

The term dissipative flux is often used for any correction that arises from the (small) departure of $f$ from the Maxwellian. However, in a magnetised plasma we find a further class of corrections which do not involve dissipation *per se* (that is, a process which increases entropy). These are the diamagnetic fluxes, and they arise from the finite gyroradius of the particles in the presence of inhomogeneities in the lower-order moments. From the pressure gradient we have a corresponding velocity, called the diamagnetic particle flux, or flow, which due to the opposite reaction of electrons and ions by this process to a pressure gradient gives rise to a diamagnetic current. From the temperature gradient we get a diamagnetic heat flux. And from the velocity gradient we get a diamagnetic momentum flux, historically called gyroviscosity because it has the form of a viscosity tensor. For dynamics much slower than the gyrofrequencies, $\omega \ll \Omega$ for each species, we can form a quasistatic balance between the spatial gradients, which provide force terms in the equations, and the Lorentz force, which provides a reaction. To lowest order in $\omega/\Omega$, and ultimately also the ratio between the gyroradius and the spatial scale of the gradient, we can use this balance to solve for the diamagnetic flux corresponding to a particular gradient.

The simplest example is the diamagnetic flow, which results in the same way the $\mathbf{E} \times \mathbf{B}$ velocity does in MHD: setting the sum of the relevant forces to zero. In a several fluids setting, we do not take the MHD approximation that $ne\mathbf{E} \gg \nabla p$ for each species. Instead, we set $\mathbf{E} = 0$ to isolate the diamagnetic part, and then set the magnetic Lorentz force equal to the pressure force. For example, for electrons we have

$$n_e e\mathbf{v} \times \mathbf{B} + \nabla p_e = 0, \tag{6.164}$$

which we solve for the diamagnetic flow,

$$\mathbf{v}_* = -\frac{1}{n_e e}\frac{1}{B^2}\mathbf{B} \times \nabla p_e. \tag{6.165}$$

The flow for the ions, $\mathbf{u}_*$, has similar form but opposite sign, and so we also have a diamagnetic current due to the gradient of the total pressure,

$$\mathbf{J}_* = \frac{1}{B^2}\mathbf{B} \times \nabla p, \tag{6.166}$$

valid for an arbitrary number of species. We note that this is the same as the perpendicular part of the current in an MHD equilibrium, and that this type of force balance consideration is what leads to the drifts description in low frequency motion.

We also note that the condition that this flow is small compared to the thermal velocity is a condition on the gyroradius:

$$\frac{\mathbf{v}_*}{V_e} = -\mu_e^{1/2}\frac{cT_e}{eB^2}\mathbf{B}\times\frac{\nabla p_e}{p_e} \sim \rho_e\nabla\log p_e. \tag{6.167}$$

The gyroradii must therefore be small compared to the perpendicular scale length, which means simply that these fluxes are only relevant to a magnetised plasma, and then to low frequency motion. This limitation applies to each of the diamagnetic fluxes, not only the flow.

As another example we can take the heat flux that arises from the temperature gradient, which we illustrate for ions of charge state $Ze$. We isolate the diamagnetic part by setting $\mathbf{E} = 0$ in the kinetic equation, and also assume that $m_i \mid \mathbf{u}_* \mid^2 \ll T_i$. Forming the third moment of the kinetic equation, we have

$$\nabla\cdot\int d^3w\, m_i\frac{w^2}{2}\mathbf{w}\mathbf{w}f = \frac{Ze}{m_i c}\int d^3w\, m_i\frac{w^2}{2}\mathbf{w}(\mathbf{w}\times\mathbf{B})\cdot\frac{\partial f}{\partial\mathbf{w}}. \tag{6.168}$$

We solve the right side via integration by parts, and following from $\rho_i\nabla \ll 1$ we set the Maxwellian into the left side. The fourth moment is evaluated as in Eq. (6.163) since for a Maxwellian it is an isotropic, diagonal tensor. We now have

$$\nabla\left(\frac{5}{2}\frac{p_i T_i}{m_i}\right) = \frac{Ze}{m_i}\left(\frac{5}{2}p_i\mathbf{u}_* + \mathbf{q}_{i\wedge}\right)\times\mathbf{B}. \tag{6.169}$$

By operation with $\mathbf{B}\times$ from the left, we find

$$\frac{5}{2}p_i\mathbf{u}_* + \mathbf{q}_{i\wedge} = \frac{1}{ZeB^2}\mathbf{B}\times\nabla\left(\frac{5}{2}p_i T_i\right), \tag{6.170}$$

and hence, using the result for $\mathbf{u}_*$, we have

$$\mathbf{q}_{i\wedge} = \frac{5}{2}\frac{p_i}{Ze}\frac{1}{B^2}\mathbf{B}\times\nabla T_i. \tag{6.171}$$

A similar result holds for the velocity field, but it is much more complicated. Following the method used to find $\mathbf{q}_{i\wedge}$ we set up the equation for the off-diagonal second moment and again balance the gradients involving the Maxwellian, $\nabla\mathbf{u}$ as a tensor, with the magnetic Lorentz force. After a symmetrising operation, this yields

$$p_i\mathbf{W}_s = \frac{Ze}{m_i c}(\mathbf{\Pi}^*\times\mathbf{B} - \mathbf{B}\times\mathbf{\Pi}^*), \tag{6.172}$$

where $\mathbf{\Pi}^*$ is the diamagnetic momentum flux, and $\mathbf{W}_s$ is the strain tensor given by

$$\mathbf{W}_s = \nabla\mathbf{u} + (\nabla\mathbf{u})^T - \frac{2}{3}(\nabla\cdot\mathbf{u})\mathbf{g}, \tag{6.173}$$

with the superscript denoting the transpose. Solution of this equation gives $\mathbf{\Pi}^*$, which is usually displayed in terms of its components perpendicular and parallel to the magnetic field,

$$\mathbf{\Pi}^*_{\perp\perp} = \frac{n_i m_i}{4} \frac{T_i}{ZeB^2}(\mathbf{B} \times \mathbf{W}_{\perp\perp} - \mathbf{W}_{\perp\perp} \times \mathbf{B}), \tag{6.174}$$

$$\mathbf{\Pi}^*_{\perp\parallel} = n_i m_i \frac{T_i}{ZeB^2}\mathbf{B} \times \mathbf{W}_{\perp\parallel}, \tag{6.175}$$

where $\mathbf{W}_{\perp\perp}$ and $W_{\perp\parallel}$ are the perp–perp and perp–parallel components of the $\mathbf{W}_s$ tensor. Due to the cross operations the parallel–parallel component is zero.

The moment approach is a viable alternative to the Chapman–Enskog procedure for calculating the dissipative transport equations for a plasma, and it can be shown that an expansion in a finite set of moments is equivalent to an expansion in a finite set of orthogonal polynomials in the Chapman–Enskog procedure (required for realistic collision operators). In both cases, the complications arise from the $w^{-3}$ dependence of the Coulomb interaction, which couples all moments. More details can be found in the references in the Section Further Reading.

The particle view of these diamagnetic fluxes is noteworthy. If we take a simple situation in which there is a temperature gradient, but the plasma is otherwise homogeneous and at rest, we isolate $\mathbf{q}_{i\wedge}$, illustrated in figure 6.2. Drawing the diagram such that the magnetic field is toward the observer and the gradient is toward the observer's left, the ions follow gyro orbits in a clockwise sense, and the electrons orbit counterclockwise. Examining all the particles which are incident on a given reference point in space, the particles arriving from the left have a higher

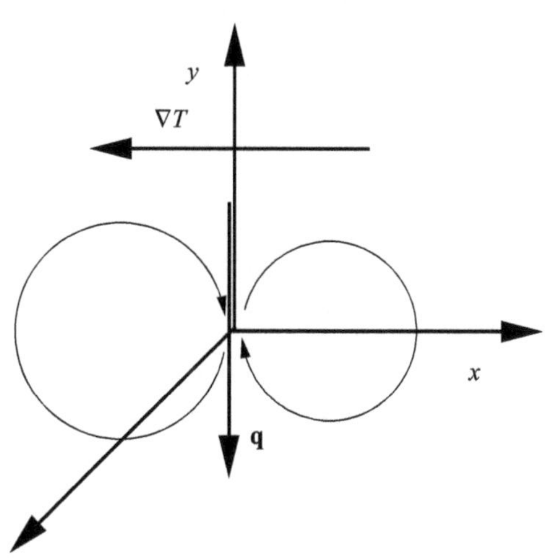


**Figure 6.2.** The diamagnetic heat flux. Gyromotion occurs in the $xy$-plane, perpendicular to the magnetic field, $\mathbf{B}$. For ions it is clockwise. A temperature gradient towards $-\nabla x$ results in a net heat flux $\mathbf{q}_{i\wedge}$ towards $-\nabla y$, generally towards $\mathbf{B} \times \nabla T/Ze$, due to the fact that the up-gradient particles are moving faster.

temperature than those arriving from the right. The faster ions are seen by the observer to move downward, and the faster electrons are seen to move upward. This gives rise to both a net flow and a net conductive flux, since due to the dependence of the particle gyroradius on velocity the faster of the particles on both sides have their gyrocentres farther away. So if the observer places a probe at the reference point, a net ion flow, a net current, and a net ion conductive heat flux will all be measured in the downward direction, and a net electron flow and conductive heat flux will be measured in the upward direction. However, no gyrocentres are in motion. The reconciliation with the measurement of a flow but no gyrocentres in motion, and therefore no actual transport of conserved quantities follows from a subtle result called the diamagnetic cancellation. This is the fact that all diamagnetic flux terms conspire to cancel each other out of the fluid equations to lowest order in the gyroradius, up to terms arising from any inhomogeneity in the magnetic field. For example, we have

$$\frac{3}{2}n_e\mathbf{v}_* \cdot \nabla T_e + p_e\nabla \cdot \mathbf{v}_* + \nabla \cdot \mathbf{q}_{e\wedge} = -\frac{1}{n_e e}\frac{1}{B^2}\mathbf{B} \times \nabla\left(\frac{5}{2}p_e T_e\right), \tag{6.176}$$

in the electron temperature equation which vanishes for homogeneous $\mathbf{B}$, so that $\mathbf{v}_*$ does not directly advect $T_e$. By the same result for ions, $\mathbf{u}_*$ does not directly advect $T_i$. With the velocities in the relative density equations, we have the simpler form

$$\mathbf{v}_* \cdot \nabla n_e + n_e\nabla \cdot \mathbf{v}_* = \nabla \cdot (n_e\mathbf{v}_*) = -\frac{1}{e}\frac{1}{B^2}\mathbf{B} \times \nabla p_e, \tag{6.177}$$

for the particle flux, and similarly for $n_i$ and $\mathbf{u}_*$. Similar results concerning $\mathbf{\Pi}^*$ prevent $\mathbf{u}_*$ from directly advecting $\mathbf{u}$, up to finite gyroradius corrections. Due to these cancellations, diamagnetic fluxes affect the fluid equations only through inhomogeneity in $\mathbf{B}$. Another approach to these effects, gyrokinetic theory, will be discussed in Volume Two.

The consideration of fluid motion in terms of net drifts leads naturally to the main subject of this work, namely, turbulence in a magnetised plasma. The main assumptions needed to contemplate the diamagnetic fluxes are the same as those needed to move from the fluid plasma model, and its MHD subset, to the fluid drift model: the gyroradius should be small compared to the perpendicular scale, and the frequencies of interest should be small compared to the gyrofrequency. We will further require the motion to be free of dynamical compression, and so the compressional Alfvén/magnetosonic waves form the parameter boundary. However, the main MHD assumption that the electric field should be larger than the pressure force is usually violated in this parameter regime, so our starting point will be the low frequency limit of the fluid plasma model. Under these restrictions the perpendicular dynamics is in force balance, which is the consideration used to derive fluid drift dynamics. That is the subject of the next Chapter.

## Further Reading

The derivation of Maxwell's equations from the Principle of Least Action is standard and appears in many texts, including L D Landau and E M Lifshitz

1975 *The Classical Theory of Fields* 4th edn (Oxford: Pergamon). For ideal fluids see Landau and Lifshitz 1972 *Fluid Dynamics* (Oxford: Pergamon).

High frequency wave properties of fluid electrodynamics is covered in many introductory texts of plasma physics, including S Ichimaru 1973 *Basic Principles of Plasma Physics* (London: Benjamin); N A Krall and A W Trivelpiece 1973 *Principles of Plasma Physics* (New York: McGraw-Hill); K Nishikawa and M Wakatani 2000 *Plasma Physics* new edition (Berlin: Springer); and K-H Spatschek 1990 *Theoretische Plasmaphysik* (Stuttgart: Teubner). Landau damping is covered in detail in these texts as well as those cited in Chapter 1.

Ideal magnetohydrodynamics also has an extensive literature. A good introductory text is J P Freidberg 1987 *Ideal Magnetohydrodynamics* (New York: Plenum Press). A good text for further study of the basic phenomena (especially reconnection and MHD turbulence) is D Biskamp 1993 *Nonlinear Magnetohydrodynamics* (Cambridge: Cambridge University Press). The standard text for astrophysical MHD applications is E N Parker 1979 *Cosmical Magnetic Fields* (Oxford: Clarendon Press). A tokamak-oriented text which includes the basic MHD problems as well as the more complicated phenomena one encounters is R B White 1989 *Theory of Tokamak Plasmas* (North Holland, Amsterdam).

References pertinent to magnetic helicity and cross helicity are given at the end of Chapter 2.

The Chapman–Enskog and moment hierarchy procedures fall under the rubric of collisional transport theory, reviewed by R Balescu 1988 *Transport Processes in Plasmas* 2 vols (Amsterdam: North Holland). This work has an entire Chapter on the mass ratio expansion, which makes the separation between cross-species momentum and energy relaxation very transparent. See also S Chapman and T G Cowling 1952 *The Mathematical Theory of Non-Uniform Gases* (Cambridge: Cambridge University Press).

The original references for the Chapman–Enskog procedure are S Chapman 'On the law of distribution of molecular velocities, and on the theory of viscosity and thermal conduction, in a nonuniform, simple, monatomic gas' *Philos. Trans. R. Soc. Lond. A* **216** (1916) 279; and D Enskog 1917 *Kinetische Theorie der Vorgänge in mässig verdünnten Gasen* (Uppsala: Almqvist and Wiskell).

Calculation of plasma resistivity due to electron–ion collisions is due to Robert S Cohen, Lyman Spitzer Jr and Paul McR Routly 'The electrical conductivity of an ionized gas' *Phys. Rev.* **80** (1950) 230; and the comprehensive calculation with thermal conduction and other transport coefficients was given by Lyman Spitzer Jr and Richard Härm 'Transport phenomena in a completely ionized gas' *Phys. Rev. B* **89** (1953) 977. For a plasma of several fluids, one of which is electrons, the standard reference is S I Braginskii 'Transport processes in a plasma' *Rev. Plasma Phys.* **1** (1965) 205. This work is the one which showed that temperature equilibration between electrons and ions is slow enough to make plasma transport a two-temperature process. The more introductory treatment presented in Section 6.10 follows A B Hassam 'Higher-order Chapman–Enskog theory for electrons' *Phys. Fluids* **23** (1980) 38, who pointed out that the electron inertia is formally an order down from resistive friction and therefore should not appear in the lowest order collisional regime.

For the moment approach the main reference is H Grad 'On the kinetic theory of rarefied gases' *Commun. Pure Appl. Math.* **2** (1949) 331. The most realistic collision operator is the one of Landau, reviewed by both Braginskii and Balescu, above. Common simplified models are the pitch angle scattering model of Lorentz, discussed by Balescu, or the damping/relaxation BGK model generalised by J Greene 'Improved Bhatnagar–Gross–Krook model of electron–ion collisions' *Phys. Fluids* **16** (1973) 2022.

The diamagnetic cancellation for temperature and heat flux was shown by S-T Tsai, F W Perkins and T H Stix 'Thermal conductivity and low frequency waves in collisional plasmas' *Phys. Fluids* **13** (1970) 2108; and the one for velocity and gyroviscosity was shown by F L Hinton and C W Horton Jr 'Amplitude limitation of a collisional drift wave instability' *Phys. Fluids* **14** (1971) 116.

The Poynting cancellation was given in B Scott 'The character of transport caused by ExB drift turbulence' *Phys. Plasmas* **10** (2003) 963.

**IOP** Publishing

Turbulence and Instabilities in Magnetised Plasmas, Volume 1
Fluid drift turbulence
**Bruce Scott**

# Chapter 7

# Fluid drift dynamics in a magnetised plasma

## 7.1 Introduction

We have seen in the previous Chapter that fluid plasma dynamics admits both compressional and shear Alfvén oscillations, through its MHD subset. In situations where the disturbances have very disparate length scales parallel and perpendicular to the background magnetic field, the frequencies of the corresponding Alfvén oscillations are also disparate, as both types of oscillation have simple dispersion relations describing propagation at the Alfvén speed. If the dynamics of a long, thin flux tube are to occur at low frequencies commensurate with the flux tube's shear Alfvén time, it is reasonable to expect that the compressional waves are not appreciably excited, and the motion remains dynamically incompressible. In such a situation, the perpendicular forces should remain in quasistatic balance, with finite time derivatives but without compressional waves. Fluid plasma dynamics under this restriction becomes fluid drift dynamics, with flows and thermal fluxes given in terms of associated stream functions which are scalar quantities among the dependent variables. This Chapter explains the qualitative nature of the drift approximation and the conditions under which it is expected to be valid. Perpendicular force balance is discussed with respect to the total current as well as the flows of each fluid species. Parallel dynamics is discussed in terms of the ordering on nonlinearities with an eventual eye toward exact energetic consistency in the final model. The finite ion inertia gives rise to the correction, which mediates the quasistatic evolution of the force balance. The polarisation drift of the ions, which carry the polarisation current, is needed to compensate any imbalances in the divergences of the other components of the current. Hence both the magnetic field and the current are exactly divergence free in the final formulation. Following this, a detailed derivation of the fluid drift equations for nonlinear disturbances on an inhomogeneous background is presented. Following that, the free energy theorem governing these equations is constructed, with discussion on the role of the background gradients as the drive process. Exact energetic consistency is thereby recovered.

 

Afterwards, brief discussions of two issues, stratification and quasineutrality, are given. Stratification refers to the difference between general nonlinearity and nonlinear turbulence on a background. In modern language, the general nonlinear set of equations is referred to as 'total-f' and the case of turbulence in the presence of gradients in either the thermal quantities, the flows, or the current is referred to as 'delta-f,' from the same distinction and derivation path in the corresponding kinetic theory. To do all of this requires the developments we will discuss later, in Volume Two. But we can make the same point herein using a simple one-dimensional fluid model with compression along the background magnetic field.

Finally, with the description of drift dynamics complete, we return to the concept of quasineutrality and give a direct demonstration of it within the regime of ExB motion in a magnetised plasma.

## 7.2 What the drift approximation is

We expect dynamical disturbances on a magnetised plasma background to consist of strongly anisotropic flux tubes, such that the scale of motion perpendicular to the background magnetic field, **B**, is several orders of magnitude smaller than the parallel scale, so that in terms of wavenumbers we have

$$k_\perp \gg k_\parallel. \tag{7.1}$$

By itself, this is called 'flute mode ordering' in the older literature. To justify the drifts we further need some statements on the frequencies.

We expect shear Alfvén dynamics to be important, so that the relevant frequencies are in the range of

$$\omega \sim k_\parallel v_A. \tag{7.2}$$

The reason is that the processes which enforce the flute mode structure in the first place enter through the electromagnetic electron responses. The combination with the flute mode ordering implies

$$\omega \ll k_\perp v_A. \tag{7.3}$$

If we further specify that

$$\omega \ll \Omega_i, \tag{7.4}$$

for all $k_\perp$ of interest, then it follows that the time scales of motion are much slower than those of compressional waves, and so these are not appreciably excited. That is, Eqs (7.3) and (7.4) are satisfied for any degree of freedom in the spectrum (here, $\Omega_i$ stated because it is the slowest of the $\Omega$'s). The drift approximation consists of a formalism which removes these fast time scales from the governing equations, leaving only the dynamics of interest.

The drift approximation is usually associated with two other inequalities in addition to Eq. (7.1). The plasma should be magnetised, which means that the spatial scale associated with the background gradient, $L_\perp$, be large compared to the drift scale, $\rho_s$,

$$L_\perp \, |\nabla \log T_e| \equiv 1 \qquad \rho_s \equiv \frac{c_s}{\Omega_i} \qquad \delta \equiv \frac{\rho_s}{L_\perp} \ll 1, \tag{7.5}$$

so that it is possible to have dynamics on scales comparable to $\rho_s$, which is local in the sense that the background can be taken as a set of constant parameters. The dynamical plasma beta should also be low,

$$\beta_e \equiv \frac{\mu_0 p_e}{B^2} = \frac{c_s^2}{v_A^2} \ll 1, \tag{7.6}$$

so that motion which strongly perturbs the gas pressure remains incompressible due to the much stronger magnetic pressure. We will see how these enter below. Both $\delta$ and $\beta_e$ are available as small parameters in a formal expansion.

These inequalities give rise to a quasistatically evolving force balance across magnetic field lines; quasistatic, since although not static it evolves without dynamical oscillations (compressional Alfvén or lower hybrid waves) involved in its restoration. Rather, the inertia gives rise to the small correction which continually acts to enforce the balance to lowest order in the small parameter $\delta$. The characteristics of that force balance are that the perpendicular and parallel dynamics largely separate, such that the perpendicular electric field is electrostatic and any disturbances in the magnetic field are perpendicular to the background field (i.e. of the shear Alfvén type).

The drift approximation works with both the several-fluids and single-fluid (MHD) versions of the fluid plasma dynamics model; in the latter case the model it gives rise to is usually called 'reduced MHD' in the literature. We will see that reduced MHD is wholly contained within fluid drift dynamics in the same sense that MHD is contained within the plasma dynamics model of several fluids.

## 7.3 Perpendicular force balance—diamagnetic current

To elucidate the character of the quasistatic force balance we start with the MHD force equation,

$$\rho \frac{d\mathbf{u}}{dt} = -\nabla p + \mathbf{J} \times \mathbf{B}, \tag{7.7}$$

where $d/dt$ is the advective derivative

$$\frac{d}{dt} = \frac{\partial}{\partial t} + \mathbf{u} \cdot \nabla, \tag{7.8}$$

for the one fluid. We formally solve for the perpendicular component of $\mathbf{J}$,

$$\mathbf{J}_\perp = \mathbf{J} - \frac{\mathbf{B}\mathbf{B} \cdot \mathbf{J}}{B^2} = \frac{1}{B^2}\mathbf{B} \times \nabla p + \frac{1}{B^2}\mathbf{B} \times \rho \frac{d\mathbf{u}}{dt}, \tag{7.9}$$

by operation with

$$\frac{1}{B^2}\mathbf{B}\times\,, \tag{7.10}$$

often called the drift operator. The quantity $B^{-1}\mathbf{B}\cdot\mathbf{J}$ in Eq. (7.9) defines $J_{\|}$, the parallel current.

Due to the smallness of the inertia we expect that to $O(1)$ the force balance holds,

$$\mathbf{J}_{\perp}^{(0)} = \mathbf{J}_{*} \equiv \frac{1}{B^2}\mathbf{B}\times\nabla p, \tag{7.11}$$

and then the inertial correction appears at $O(\delta)$,

$$\mathbf{J}_{\perp}^{(1)} = \frac{1}{B^2}\mathbf{B}\times\rho\frac{d\mathbf{u}}{dt}. \tag{7.12}$$

The smaller piece is called the polarisation current, since it emerges from the polarisation drift flows in each charged fluid, as we shall see shortly. The larger piece simply reflects approximate MHD equilibrium in the disturbances. As it holds during temporal changes in the pressure, the force balance is referred to as quasistatic.

## 7.4 Parallel dynamics–shear Alfvén nonlinearity

The parallel component of the MHD force equation is simply

$$\rho\frac{du_{\|}}{dt} = -\nabla_{\|}p, \tag{7.13}$$

since the contraction of the Lorentz force with **B** vanishes. Here it is important to note the approximation that the scale of the magnetic field is large enough, and the perturbations to it are small enough, that **B** may be commuted through the entire advective time derivative to form

$$u_{\|} = \frac{1}{B}\mathbf{B}\cdot\mathbf{u}. \tag{7.14}$$

The parallel gradient $\nabla_{\|}$ is given in terms of the total (background plus disturbances) field,

$$B\nabla_{\|} = \mathbf{B}\cdot\nabla + \tilde{\mathbf{B}}\cdot\nabla, \tag{7.15}$$

in which it is important to formulate $B\nabla_{\|}$ as an exact divergence at any level of approximation. It is at this point that we split the magnetic field between the background guide field **B**, which is assumed to be static, and the disturbances. Contraction of the Lorentz force by the total magnetic field is assumed to make it vanish, while in the definition of parallel components of a dependent variable the background field is assumed. Any differences are one order down in $\delta$ and therefore neglected.

Although the disturbances given by $\tilde{\mathbf{B}}$ are small, they combine with the perpendicular components of $\nabla$, denoted $\nabla_\perp$, which are by the same order (one order in $\delta$) much larger than $\nabla_\parallel$. We therefore have

$$\mathbf{B} \cdot \nabla \sim \tilde{\mathbf{B}} \cdot \nabla, \tag{7.16}$$

with the consequence that all three components of $\nabla_\parallel$ are of the same order. The disturbances are given by

$$\tilde{\mathbf{B}} = \nabla \cdot A_\parallel \mathbf{b} \qquad \text{hence} \quad \nabla_\parallel \approx -\frac{1}{B^2}\mathbf{B} \times \nabla A_\parallel, \tag{7.17}$$

in which the parallel component of the magnetic potential, $A_\parallel$, is the only one kept.

If the subtleties of what is kept and what is discarded seem capricious at first encounter, it is to be noted that they are decided by whatever is necessary to deliver exact energetic consistency in the final model. Under the drift approximation, all of $B$, $\mathbf{B}$ and $\mathbf{b}$ may be independently defined, subject to exact energetic consistency, which we will address in detail below.

## 7.5 Perpendicular force balance—fluid drifts

To obtain more insight into the consequences of quasistatic force balance, as well as to evaluate the polarisation drift, we turn to the equation of motion for each charged fluid separately. Assuming a one component plasma with singly-charged ions with $n_i = n_e$ and $m_e \ll m_i$, the equations of motion are

$$n_i m_i \frac{d\mathbf{u}}{dt} + \nabla \cdot \mathbf{\Pi}^* = -\nabla p_i + n_e e(\mathbf{E} + \mathbf{u} \times \mathbf{B}), \tag{7.18}$$

$$0 = -\nabla p_e - n_e e(\mathbf{E} + \mathbf{v} \times \mathbf{B}), \tag{7.19}$$

with $\mathbf{u}$ the ion flow and $\mathbf{v}$ the electron flow, and

$$\frac{d}{dt} = \frac{\partial}{\partial t} + \mathbf{u} \cdot \nabla, \tag{7.20}$$

the ion advective derivative. We retain the diamagnetic part of $\mathbf{\Pi}$ to cancel part of the advection, as noted in the previous Chapter. We assume *a priori* that the left side of the ion equation is one order smaller in $\delta$ than the right side.

Due to the frequency ordering (Eqs (7.2) and (7.3)), we assume that $\mathbf{E}_\perp$ is electrostatic, so that

$$\mathbf{E}_\perp = -\nabla_\perp \phi. \tag{7.21}$$

Neglecting inertia, the lowest order flows are then given in terms of $\phi$ and the pressures as stream functions,

$$\mathbf{u}_\perp^{(0)} = \frac{1}{B^2}\mathbf{B} \times \nabla\phi + \frac{1}{n_e e}\frac{1}{B^2}\mathbf{B} \times \nabla p_i, \tag{7.22}$$

$$\mathbf{v}_\perp = \frac{1}{B^2}\mathbf{B}\times\nabla\phi - \frac{1}{n_e e}\frac{1}{B^2}\mathbf{B}\times\nabla p_e, \tag{7.23}$$

where the perpendicular electron flow is finished because we are neglecting perpendicular electron inertia. In turn, the pieces of these flows are,

$$\mathbf{v}_E = \frac{1}{B^2}\mathbf{B}\times\nabla\phi \qquad \mathbf{v}_* = -\frac{1}{n_e e}\frac{1}{B^2}\mathbf{B}\times\nabla p_e \qquad \mathbf{u}_* = \frac{1}{n_e e}\frac{1}{B^2}\mathbf{B}\times\nabla p_i, \tag{7.24}$$

the ExB velocity, and the electron and ion diamagnetic velocities, respectively.

If we take these $O(1)$ flows and form the resulting lowest order current, we find

$$\mathbf{J}_\perp^{(0)} = n_e e(\mathbf{u}_\perp^{(0)} - \mathbf{v}_\perp) = \frac{1}{B^2}\mathbf{B}\times\nabla(p_i + p_e), \tag{7.25}$$

which is the same as Eq. (7.11). This is just the diamagnetic current $\mathbf{J}_*$, because the diamagnetic flows and fluxes are found by neglecting the inertial corrections. The ExB flows vanish in the current since they are the same for all charged fluids, and we have assumed a neutral plasma.

In computing the diamagnetic drift we have assumed that $n_i = n_e$ since there is one component. For several ion species, each one has the factor $1/n_e e$ replaced by $1/n_i Ze$, where $Z$ is the charge state.

## 7.6 The polarisation drift

We find the polarisation drift by inserting the lowest order $\mathbf{u}_\perp^{(0)}$ and inserting it into the corrections at $O(\delta)$,

$$\mathbf{u}_\perp^{(1)} = \frac{1}{n_e e}\left[n_i m_i \frac{1}{B^2}\mathbf{B}\times\frac{d\mathbf{u}_\perp^{(0)}}{dt} + \mathbf{\Pi}^*(\mathbf{u}_\perp^{(0)})\right] \qquad \mathbf{u}_p \equiv \frac{m_i}{e}\frac{1}{B^2}\mathbf{B}\times\frac{d_E}{dt}\mathbf{u}_\perp^{(0)}, \tag{7.26}$$

where the diamagnetic flow is cancelled by $\mathbf{\Pi}^*$ (evaluated with the lowest order $\mathbf{u}_\perp$), leaving perpendicular advection only by $\mathbf{v}_E$ as noted in the previous Chapter. Parallel advection is neglected due to the ordering

$$k_\| c_s \ll \omega \sim \mathbf{v}_E \cdot \nabla, \tag{7.27}$$

which we discuss further below. This leaves nonlinear advection under the drift approximation as advection solely by the ExB velocity. We then have

$$\frac{d_E}{dt} = \frac{\partial}{\partial t} + \mathbf{v}_E \cdot \nabla, \tag{7.28}$$

as the ExB advective derivative, which together with the nonlinear part of the parallel derivative are the only nonlinearities we eventually keep in these fluid drift models. We write it as $d_E/dt$ in this Chapter to distinguish it from the other forms of $d/dt$ we consider, such as the version with the total velocity $\mathbf{u}\cdot\nabla$.

In computing the polarisation drift we have assumed that $n_i = n_e$ since there is one component. For several ion species, each one has the factor $1/n_e e$ replaced by $1/n_i Ze$, where $Z$ is the charge state.

It remains to explain why indeed we must keep the polarisation drift at all. This is due to the fact that although $\mathbf{u}_p$ itself is one order down in $\delta$ compared to $\mathbf{v}_E$ and $\mathbf{u}_*$,

$$\mathbf{u}_p = \frac{m_i}{eB^2}\mathbf{B} \times \frac{d_E}{dt}\mathbf{u}_\perp = \frac{1}{\Omega_i}\mathbf{b} \times \frac{d_E}{dt}\mathbf{u}_\perp, \tag{7.29}$$

its divergence is not. We note now that the divergence of any expression of the form given by taking the drift operator to the gradient of a scalar vanishes unless the magnetic field is inhomogeneous. For example, the divergence of the ExB flow is

$$\nabla \cdot \frac{1}{B^2}\mathbf{B} \times \nabla\phi = \nabla \times \left(\frac{\mathbf{B}}{B^2}\right) \cdot \nabla\phi. \tag{7.30}$$

Although the divergence of $\mathbf{u}_*$ by itself has a contribution arising from $\nabla p_i \times \nabla n_e$, these pieces vanish in all situations through the appropriate diamagnetic cancellations. So the perpendicular flow and flux divergences enter only through the scale of the background magnetic field. By contrast, the divergence of the polarisation drift enters through the perpendicular scale of motion, e.g.

$$-\nabla \cdot \mathbf{J}_p = \nabla \cdot \frac{n_i m_i}{B^2}\frac{d_E}{dt}\nabla_\perp\phi \sim \frac{n_i m_i}{B^2}\frac{d_E}{dt}\nabla_\perp^2\phi, \tag{7.31}$$

in the case of the ExB contribution, for purposes of example. In terms of the ordering, this means

$$\nabla \cdot \mathbf{J}_* \sim \frac{\mathbf{J}_*}{L_B} \sim \frac{\delta^{-1}\mathbf{J}_p}{\delta^{-1}\Delta_\perp} \sim \frac{\mathbf{J}_p}{\Delta_\perp} \sim \nabla \cdot \mathbf{J}_p. \tag{7.32}$$

Effectively, the divergence operator drops the lowest order flows by one order in $\delta$, setting the divergences each at $O(\delta)$.

The same holds when comparing advection by $\mathbf{u}_\perp$ to the polarisation divergence. If the scale of motion is comparable to $\rho_s$ and the dynamical frequency is comparable to $c_s/L_\perp$, then advection of the density by the ExB velocity,

$$\mathbf{v}_E \cdot \nabla n_e \sim \frac{1}{B}\frac{\phi}{\rho_s}\frac{n_e}{L_\perp}, \tag{7.33}$$

and the divergence of the polarisation drift times the density,

$$n_e\nabla \cdot \mathbf{u}_p \sim \frac{n_e}{\rho_s}\frac{c_s/L_\perp}{\Omega_i}\frac{1}{B}\frac{\phi}{\rho_s} \sim \frac{n_e}{L_\perp}\frac{1}{B}\frac{\phi}{\rho_s}, \tag{7.34}$$

will be comparable, since $\rho_s\Omega_i = c_s$. This is how small-scale drift wave disturbances and their associated turbulence scale, so in general we must retain $\mathbf{u}_p$ when it appears under the divergence operator.

## 7.7 Drift ordering and 'delta-f'

The foregoing considerations, treated loosely and without formal justification, leads directly to the question of how we should set up the ordering in the expansion so as to make it systematic. As in the cases of the successively simpler models presented in the previous Chapter, this involves some assumptions about the space and time scales of interest. Loosely speaking, we wish to include frequencies, $\omega$, as fast as $c_s/L_\perp$ and space scales as small as $\rho_s$, and we wish to exclude the faster compressional or gyro frequencies. So in ordering the various terms in the equations we assume

$$\frac{\partial}{\partial t} \sim \omega \sim \frac{c_s}{L_\perp} \ll \Omega_i, \tag{7.35}$$

for the time derivative. Note that this inequality requires a magnetised background plasma, since $c_s/\Omega_i$ is just $\rho_s$.

The ordering will be done while keeping an eye on exact energetic consistency. While the fundamental fluid and fluid electrodynamic equations, as well as conventional MHD, have strict energy theorems such that long-time computations constitute well-posed problems, this is not always the case for fluid drift models, especially historically. What is necessary for turbulence is that the short dynamical time scale and the medium saturation time scale are well separated, and that one can hold a computation in statistically equilibrated saturation for indefinite time. For this we require an exact energy theorem for the nonlinear system at any level of approximation. Specifically, there must be a global quantity quadratic in all the dependent variables, plus some field energy contributions whose time derivative vanishes exactly, not just up to some order, when dissipation and drive terms are switched off. The global quantity is the 'free energy' for that system. Later, when considering kinetic systems, we will see that this really is the second variation of the thermodynamic free energy.

Since the dynamics of interest is turbulence and gradient-driven instabilities, we have some guidance concerning the amplitude of motion at a given perpendicular space scale, $\Delta_\perp$. Any disturbances should satisfy (e.g. for the density)

$$\nabla \tilde{n}_e \sim \nabla n_e \qquad \Longrightarrow \qquad \frac{\tilde{n}_e}{n_e} \sim \frac{\Delta_\perp}{L_\perp}. \tag{7.36}$$

If the space scale is global then we expect $O(1)$ disturbances. If the space scale is $\rho_s$ then we expect $O(\delta)$ disturbances. In either case, this ordering of the amplitudes is called 'mixing level' since the dynamics is assumed to be just strong enough to equilibrate the background gradient.

If $\Delta_\perp \sim L_\perp$ then the equations become very complicated due to the need to keep all the operations by the advective derivative, the drift operator, and the divergence in their proper order, especially when considering the polarisation drift. This would require the fully nonlinear 'total-f' form of the equations. By far the more important applications of fluid drift dynamics have been in situations where the dynamics is local: $\Delta_\perp \ll L_\perp$. Most models, and all of the ones we will discuss herein, will take this assumption, and work with the 'delta-f' form of the equations, in which all effects are linearised except for the two quadratic nonlinearities represented by ExB advection (Eqs (7.24) and (7.28)) and the magnetic nonlinearities contributing to the parallel derivative (Eqs (7.15)–(7.17)).

It is important to note that we do not thereby linearise the dynamics; we can still treat mixing level turbulence so long as it occurs at small scales. For mixing level turbulence the nonlinear ExB advection is ordered comparable to the linear contribution arising from advection of the background gradient by the ExB eddy disturbances. For example,

$$\mathbf{v}_E \cdot \nabla \tilde{n}_e \sim \mathbf{v}_E \cdot \nabla n_e. \tag{7.37}$$

Both in turn are ordered to be comparable to the time derivative, following the expectation that advection of the background, also called 'gradient drive,' is responsible for the dynamics. In general,

$$\frac{\partial}{\partial t} \sim \mathbf{v}_E \cdot \nabla \sim v_A \nabla_\parallel, \tag{7.38}$$

and we keep quadratic nonlinearities involving the appearances of $\tilde{\phi}$ and $\tilde{\mathbf{B}}_\perp$ in $\mathbf{v}_E \cdot \nabla$ and $\nabla_\parallel$, respectively. The dynamics is therefore allowed to be nonlinear.

We do not, however, keep direct nonlinear contributions to the coefficients of the various terms. For example, the smallness of the amplitudes coupled with the small scale of motion guarantees that

$$\nabla \cdot n_i \nabla_\perp \phi = n_i \nabla_\perp^2 \phi + \nabla_\perp n_i \cdot \nabla_\perp \phi = n_i \nabla_\perp^2 \phi + O(\delta), \tag{7.39}$$

so the appearance of the density in the polarisation divergences is merely as a coefficient. Similarly, we neglect direct advection by the parallel velocity, because

$$\nabla \cdot n_i u_\parallel \mathbf{b} = n_i \nabla \cdot u_\parallel \mathbf{b} + u_\parallel \nabla_\parallel n_i = n_i \nabla \cdot u_\parallel \mathbf{b} + O(\delta), \tag{7.40}$$

due to the fact that the background has no $O(1)$ variation along the field lines, and that for the parallel dynamics there is no scale disparity; the dynamics tends to involve the longest parallel wavelengths allowed by the geometry of the background (cf. Eq. (7.27)).

All of this amounts to a two-scale expansion: we are following nonlinear disturbances on a (at most) slowly varying background, not the global approach to equilibrium of that background. Collecting all of it together, we expect that the relative amplitudes of all the disturbances scale together,

$$\frac{e\tilde{\phi}}{T_e} \sim \frac{\tilde{n}_e}{n_e} \sim \frac{u_\parallel}{c_s} \sim \frac{\tilde{\mathbf{B}}_\perp}{B} \sim \delta = \frac{\rho_s}{L_\perp} \ll 1, \tag{7.41}$$

that the perpendicular scale of motion follows $\rho_s$ although it may be several $\rho_s$,

$$\rho_s \lesssim \Delta_\perp \ll L_\perp, \tag{7.42}$$

and that the time scale follows from advection or Alfvén wave motion,

$$\frac{\partial}{\partial t} \sim \mathbf{v}_E \cdot \nabla \sim \{v_A, c_s\}\nabla_\| \sim \frac{c_s}{L_\perp} \ll \Omega_i. \tag{7.43}$$

All inequalities in these expressions imply smallness by one order in $\delta$. In the literature, this combination of both space and time scale ordering is called 'drift ordering' or 'gyrokinetic ordering' since the same set of approximations is the justification of gyrokinetic theory, which we will encounter in Volume Two.

We also expect the dynamical plasma beta to be small,

$$\beta_e = \frac{\mu_0 p_e}{B^2} = \frac{c_s^2}{v_A^2} \ll 1, \tag{7.44}$$

because we are neglecting compression of the magnetic field while allowing the pressure to evolve arbitrarily. Although this does mean that the values of $c_s$ and $v_A$ in Eq. (7.41) are disparate, we are taking a 'maximal ordering' designed to retain both Alfvén and acoustic dynamics parallel to the magnetic field. This is because they involve different subsets of the dependent variables and also different parallel scales of motion. Both together are responsible for limiting the range of excited parallel wavelengths, as we will see in Chapter 11. Due to the subsidiary ordering on $\beta_e$, which is taken at each level of the expansion in $\delta$, we neglect the disturbances of $\mathbf{B}$ in the drift operator, $B^{-2}\mathbf{B}\times$.

## 7.8 Derivation of the fluid drift equations

The starting point is the fluid plasma dynamics model of the previous Chapter. We assume a single component plasma of singly-charged ions and electrons. As dissipative fluxes we retain the heat flux $\mathbf{q}$ and momentum flux $\mathbf{\Pi}$ for each fluid species. The only other dissipation process treated is resistive friction between the electrons and ions. Diamagnetic heat and momentum fluxes are retained for each species.

The outline of the derivation is the procedure involved by the drift approximation. The perpendicular electric field is taken as electrostatic, so that only the parallel component of the magnetic vector potential is retained. Perpendicular flows and diamagnetic heat and fluxes are given to $O(1)$ in terms of scalar dependent variables. The polarisation drift and diamagnetic momentum fluxes are retained as $O(\delta)$ corrections. The electron mass remains finite although the $m_e \ll m_i$ limit is taken to neglect small corrections to terms which would still exist for $m_e \to 0$. The formulae for the perpendicular flows and fluxes are inserted into the rest of the equations to form a closed set, which determines the set of scalar variables formed by the densities, temperatures, and parallel flows for each species and the electrostatic and parallel magnetic potentials. Closure is achieved by specifying formulae or additional equations for the parallel heat and momentum fluxes (those which represent true dissipation). The simplest form of this is the procedure given by the Chapman–Enskog expansion in the previous Chapter. The neglect of the perpendicular magnetic potential (equivalently, disturbances in the strength of the magnetic field) is checked *a posteriori* by calculating it from the diamagnetic and polarisation currents.

The first step is the electric field. Taking the perpendicular components as electrostatic, we have through Faraday's law,

$$E_\parallel = -\frac{\partial A_\parallel}{\partial t} - \nabla_\parallel \phi \qquad \mathbf{E}_\perp = -\nabla_\perp \phi, \tag{7.45}$$

where the scalar functions $\phi$ and $A_\parallel$ become two of the dynamical variables. This effectively replaces the vector equations for $\mathbf{E}$ and $\mathbf{B}$ with scalar equations for $\phi$ and $A_\parallel$.

The second step is specification of the lowest order perpendicular flows and fluxes; collectively, the drifts. The flows are found by solving the equations of motion by applying the drift operator and neglecting the inertia,

$$\mathbf{u}_\perp = \mathbf{v}_E + \mathbf{u}_* = \frac{1}{B^2}\mathbf{B} \times \nabla\phi + \frac{1}{n_e e}\frac{1}{B^2}\mathbf{B} \times \nabla p_i, \tag{7.46}$$

$$\mathbf{v}_\perp = \mathbf{v}_E + \mathbf{v}_* = \frac{1}{B^2}\mathbf{B} \times \nabla\phi - \frac{1}{n_e e}\frac{1}{B^2}\mathbf{B} \times \nabla p_e. \tag{7.47}$$

Here, the diamagnetic and dissipative momentum fluxes are ordered with the inertia. The perpendicular heat fluxes are found using the moment approach explained in the previous Chapter,

$$\mathbf{q}_{e\perp} = -\frac{5}{2}\frac{p_e}{e}\frac{1}{B^2}\mathbf{B} \times \nabla T_e \qquad \mathbf{q}_{i\perp} = \frac{5}{2}\frac{p_i}{e}\frac{1}{B^2}\mathbf{B} \times \nabla T_i. \tag{7.48}$$

The polarisation drift is found by inserting the lowest order $\mathbf{u}_\perp$ into the inertia, with the diamagnetic momentum flux cancelling out the advection of $\mathbf{u}_\perp$ by $\mathbf{u}_*$, leaving

$$\mathbf{u}_p = -\frac{m_i}{eB^2}\frac{d_E}{dt}\left(\nabla_\perp\phi + \frac{1}{n_e e}\nabla_\perp p_i\right). \tag{7.49}$$

Here, we have taken the correction to $O(\delta)$ only, neglecting $O(\delta^2)$ and $O(\beta_e\delta)$ corrections due to the variation of $\mathbf{B}$.

A subtle point regarding the parallel and perpendicular components of a vector arises when taking divergences. In general, $\mathbf{B}$ is not homogeneous, so we must be careful with the order of the operations. Using $\mathbf{u}$ as an example, what we have done is to split the perpendicular and parallel parts,

$$\mathbf{u} = \mathbf{u}_\perp + \mathbf{b}u_\parallel = (\mathbf{g} - \mathbf{bb}) \cdot \mathbf{u} + \mathbf{bb} \cdot \mathbf{u}, \tag{7.50}$$

where $\mathbf{g}$ is the metric tensor (all derivatives are in general covariant, and $\nabla\mathbf{g} = 0$; this is important when using curvilinear coordinates, especially when they are non-orthogonal). The divergence of $\mathbf{u}$ contains both parallel and perpendicular contributions,

$$\nabla \cdot \mathbf{u} = \nabla_\perp \cdot \mathbf{u}_\perp + \nabla \cdot \mathbf{b} u_\parallel. \tag{7.51}$$

Under drift ordering, the scale of $\mathbf{B}$ is larger than that of $\mathbf{u}_\perp$ in the perpendicular directions, so $\mathbf{B}$ may be commuted at will through the perpendicular gradient or the perpendicular divergence. The parallel scale, however, is the same for both, namely, the inverse of $k_\parallel$. Noting that $\nabla \cdot \mathbf{B} = 0$, the divergence of $\mathbf{b}$ is finite,

$$\nabla \cdot \mathbf{b} = \nabla \cdot \frac{\mathbf{B}}{B} = \mathbf{B} \cdot \nabla \frac{1}{B} = B\nabla_\parallel \frac{1}{B}, \tag{7.52}$$

and so the parallel gradient of $B$ enters the parallel divergence:

$$\nabla \cdot \mathbf{u} = \nabla_\perp \cdot \mathbf{u}_\perp + B\nabla_\parallel \frac{u_\parallel}{B}. \tag{7.53}$$

So as mentioned before, no matter how $\nabla_\parallel$ is approximated, the parallel divergence given by the form with $B\nabla_\parallel$ must be formulated (and, in a computation, discretised) as an exact divergence. We also require the divergence of both the equilibrium $\mathbf{B}$ and its fluctuation controlled by $A_\parallel$ to vanish exactly. The forms used are to be adjusted such that these criteria are satisfied at any level of approximation.

The parallel divergences appear in the equations for the scalar quantities; the parallel gradients appear in those for the parallel fluxes. In general we will call these 'state variables' and 'flux variables,' respectively. State variables are $\phi$ in addition to $n_e$, $T_e$, and $T_i$, and represent conserved quantities (charge, mass, energy) and at the same time serve as force potentials for the flux variables. The latter are the parallel velocities $v_\parallel$ and $u_\parallel$ for electrons and ions, respectively, and the parallel heat fluxes $q_{e\parallel}$ and $q_{i\parallel}$ for both species. The equations for state variables are divergence balances, while those for flux variables are force balances.

The density equation for a single component, singly-ionised, neutral plasma can be derived either from the ion or electron densities. Here, we take the electron one due to the ability to neglect the polarisation drift. The material continuity equation for electrons is

$$\frac{\partial n_e}{\partial t} + \nabla \cdot n_e \mathbf{v} = 0. \tag{7.54}$$

The velocity is split into parallel and perpendicular components,

$$\frac{\partial n_e}{\partial t} + \nabla \cdot n_e \mathbf{v}_\perp + \nabla \cdot n_e v_\parallel \mathbf{b} = 0. \tag{7.55}$$

The perpendicular velocity is further split into the ExB and diamagnetic parts, and the divergence of the ExB velocity is set together with the diamagnetic divergence,

$$\frac{\partial n_e}{\partial t} + \mathbf{v}_E \cdot \nabla n_e + n_e \nabla \cdot \mathbf{v}_E + \nabla \cdot n_e \mathbf{v}_* + B\nabla_\parallel \frac{n_e v_\parallel}{B} = 0. \tag{7.56}$$

Instead of the parallel electron velocity directly, we will retain the parallel current,

$$J_\parallel = n_e e(u_\parallel - v_\parallel), \tag{7.57}$$

as a dynamical variable. It is related to $A_\parallel$ through Ampère's law,

$$J_\parallel = -\mu_0^{-1}\nabla_\perp^2 A_\parallel, \tag{7.58}$$

where to $O(1)$ only the perpendicular components of $\nabla^2$ contribute. The particular choice of the divergence forms results from the diamagnetic cancellation discussed in the previous Chapter, according to which these fluxes enter only through divergence combinations which vanish unless **B** is inhomogeneous. The magnetic divergence operator is defined for any scalar $f$ as

$$\mathcal{K}(f) \equiv -\left(\nabla \times \frac{\mathbf{B}}{B^2}\right) \cdot \nabla f = -\nabla \cdot \frac{1}{B^2}\mathbf{B} \times \nabla f, \tag{7.59}$$

so the divergence of the diamagnetic material flux is

$$\nabla \cdot n_e \mathbf{v}_* = -\nabla \cdot \frac{1}{e}\frac{1}{B^2}\mathbf{B} \times \nabla p_e = \frac{1}{e}\mathcal{K}(p_e). \tag{7.60}$$

The divergence of $\mathbf{v}_E$ is also of this form,

$$\nabla \cdot \mathbf{v}_E = \nabla \cdot \frac{1}{B^2}\mathbf{B} \times \nabla\phi = -\mathcal{K}(\phi). \tag{7.61}$$

Finally, putting the ExB advection together with the time derivative, we have

$$\frac{d_E n_e}{dt} = \frac{1}{e}B\nabla_\parallel\frac{J_\parallel}{B} - n_e B\nabla_\parallel\frac{u_\parallel}{B} - \frac{1}{e}\mathcal{K}(p_e) + n_e\mathcal{K}(\phi). \tag{7.62}$$

This is the equation for the electron density, but due to quasineutrality it also governs the ions. Note that due to the ordering we can use $\nabla_\parallel J_\parallel$ or $n_e e\nabla_\parallel(u_\parallel - v_\parallel)$ interchangeably, as the contribution in this term from $\nabla_\parallel n_e$ is at higher order in $\delta$.

The role of the polarisation drift on the ions is next, and it is easiest to see it by considering the quasineutrality condition,

$$\nabla \cdot \mathbf{J} = 0. \tag{7.63}$$

As with the velocities, we separate the parallel and perpendicular parts, and in the perpendicular part we retain the polarisation drift of the ions,

$$-\nabla \cdot n_i e\mathbf{u}_p = \nabla \cdot J_\parallel \mathbf{b} + \nabla \cdot \mathbf{J}_* = B\nabla_\parallel\frac{J_\parallel}{B} + \nabla \cdot (n_i e\mathbf{u}_* - n_e e\mathbf{v}_*). \tag{7.64}$$

In this case, all of the subtleties have been taken care of in writing down the polarisation drift, in Eq. (7.49). Inserting the formulae for $\mathbf{u}_*$ and $\mathbf{v}_*$, we have

$$\nabla \cdot \frac{n_i m_i}{B^2}\frac{d_E}{dt}\left(\nabla_\perp\phi + \frac{1}{n_e e}\nabla_\perp p_i\right) = B\nabla_\parallel\frac{J_\parallel}{B} - \mathcal{K}(p_e + p_i). \tag{7.65}$$

We can see the relation between the electron and ion density equations, since the differences are all in these contributions to $\nabla \cdot \mathbf{J} = 0$. One can use Eqs (7.62) and (7.65) within drift ordering to find

$$\frac{d_E n_i}{dt} = -n_i B\nabla_\parallel \frac{u_\parallel}{B} + \frac{1}{e}\mathcal{K}(p_i) + n_i\mathcal{K}(\phi) - \nabla \cdot n_i \mathbf{u}_p. \tag{7.66}$$

Since $\mathbf{u}_p$ contains its own time derivative, the electron Eq. (7.62) is more suitable for computations. Using Eq. (7.65) to eliminate $\mathbf{u}_p$ simply recovers Eq. (7.62). We will use a similar manipulation to eliminate $\nabla \cdot \mathbf{u}_p$ in the ion temperature equation.

The temperatures are manipulated in similar fashion to the densities. For the electrons we start with

$$\frac{3}{2}n_e\left(\frac{\partial}{\partial t} + \mathbf{v} \cdot \nabla\right)T_e + p_e\nabla \cdot \mathbf{v} = -\nabla \cdot \mathbf{q}_e. \tag{7.67}$$

Explicitly neglecting advection by $v_\parallel$ and separating $\mathbf{v}_E$ from $\mathbf{v}_*$, we have

$$\begin{aligned} &\frac{3}{2}n_e\left(\frac{\partial}{\partial t} + \mathbf{v}_E \cdot \nabla\right)T_e + p_e\nabla \cdot \mathbf{v}_E + p_e\nabla \cdot v_\parallel \mathbf{b} \\ &= -\frac{3}{2}n_e\mathbf{v}_* \cdot \nabla T_e - p_e\nabla \cdot \mathbf{v}_* - \nabla \cdot \mathbf{q}_{e\perp} - \nabla \cdot q_{e\parallel}\mathbf{b}. \end{aligned} \tag{7.68}$$

The first three terms on the right side combine to form the diamagnetic cancellation for the heat flux, whereupon

$$\begin{aligned} &\frac{3}{2}n_e\left(\frac{\partial}{\partial t} + \mathbf{v}_E \cdot \nabla\right)T_e + p_e\nabla \cdot \mathbf{v}_E + p_e B\nabla_\parallel \frac{v_\parallel}{B} \\ &= -T_e\nabla \cdot n_e\mathbf{v}_* - p_e\nabla \cdot \frac{\mathbf{q}_{e\perp}}{p_e} - B\nabla_\parallel \frac{q_{e\parallel}}{B}. \end{aligned} \tag{7.69}$$

The parallel velocity is eliminated in favour of the parallel current,

$$\begin{aligned} \frac{3}{2}n_e\left(\frac{\partial}{\partial t} + \mathbf{v}_E \cdot \nabla\right)T_e = &\frac{T_e}{e}B\nabla_\parallel \frac{J_\parallel}{B} - p_e B\nabla_\parallel \frac{u_\parallel}{B} - B\nabla_\parallel \frac{q_{e\parallel}}{B} \\ &-p_e\nabla \cdot \mathbf{v}_E - T_e\nabla \cdot n_e\mathbf{v}_* - p_e\nabla \cdot \frac{\mathbf{q}_{e\perp}}{p_e}. \end{aligned} \tag{7.70}$$

Finally, we express the magnetic divergences in terms of the $\mathcal{K}$ operator,

$$\begin{aligned} \frac{3}{2}n_e\frac{d_E T_e}{dt} = &\frac{T_e}{e}B\nabla_\parallel \frac{J_\parallel}{B} - p_e B\nabla_\parallel \frac{u_\parallel}{B} - B\nabla_\parallel \frac{q_{e\parallel}}{B} \\ &-\frac{5}{2}\frac{p_e}{e}\mathcal{K}(T_e) - \frac{T_e}{e}\mathcal{K}(p_e) + p_e\mathcal{K}(\phi). \end{aligned} \tag{7.71}$$

Similar manipulation of the ion temperature equation leads to the intermediate form involving the polarisation drift,

$$\begin{aligned} \frac{3}{2}n_i\frac{d_E T_i}{dt} = &-p_i B\nabla_\parallel \frac{u_\parallel}{B} - p_i\nabla \cdot \mathbf{u}_p + \frac{5}{2}\frac{p_i}{e}\mathcal{K}(T_i) + \frac{T_i}{e}\mathcal{K}(p_i) \\ &+p_i\mathcal{K}(\phi) - B\nabla_\parallel \frac{q_{i\parallel}}{B}. \end{aligned} \tag{7.72}$$

We then use Eq. (7.65) to eliminate $\mathbf{u}_p$, leaving

$$\begin{aligned} \frac{3}{2}n_i\frac{d_E T_i}{dt} = \frac{T_i}{e}B\nabla_\| \frac{J_\|}{B} - p_i B\nabla_\| \frac{u_\|}{B} + \frac{5}{2}\frac{p_i}{e}\mathcal{K}(T_i) - \frac{T_i}{e}\mathcal{K}(p_e) \\ + p_i\mathcal{K}(\phi) - B\nabla_\| \frac{q_{i\|}}{B}. \end{aligned} \tag{7.73}$$

The curious appearance of $J_\|$ and $\mathcal{K}(p_e)$ in this equation is simply due to their relation to $\mathbf{u}_p$ under the divergences. The same applies to $p_e$ and $p_i$ under $\mathcal{K}$ in $\mathbf{J}_*$.

At this point we need the equations for the parallel velocities. For the ions we contract the equation of motion with the (total) magnetic field, to eliminate the magnetic Lorentz force,

$$\mathbf{B}\cdot\left[n_i m_i \frac{d_E}{dt}\mathbf{u} = -\nabla p_i + n_i e\mathbf{E} + n_i e\mathbf{u}\times\mathbf{B} - \nabla\cdot\left(\Pi_\|\mathbf{b}\right) - \mathbf{R}_{ei}\right], \tag{7.74}$$

where $\mathbf{R}_{ei} = R_{ei}\mathbf{b}$ is retained as parallel resistive friction, $\Pi_\|\mathbf{b} = \mathbf{bb}\cdot\mathbf{\Pi}$ is the parallel part of viscous dissipation, and the diamagnetic (gyroviscous) cancellation and drift ordering are invoked to write the advective derivative only with $\mathbf{v}_E$. Neglecting higher order corrections, we carry the contraction with $\mathbf{B}$ through $d_E/dt$ such that it operates only on $\mathbf{u}$ and substitute $n_i = n_e$ in the electric force,

$$n_i m_i \frac{d_E}{dt}u_\| = -\nabla_\| p_i + n_e e E_\| - B\nabla_\| \frac{\Pi_\|}{B} - R_{ei}. \tag{7.75}$$

By similar operation on the electron equation of motion we have

$$n_e m_e \frac{d_E}{dt}v_\| = -\nabla_\| p_e - n_e e E_\| + R_{ei}, \tag{7.76}$$

where we neglect electron viscosity. Adding the two and neglecting the electrons' contribution to the total inertia, we find the equation for $u_\|$,

$$n_i m_i \frac{d_E}{dt}u_\| = -\nabla_\|(p_e + p_i) - B\nabla_\| \frac{\Pi_\|}{B}, \tag{7.77}$$

to which $R_{ei}$ does not contribute since like $E_\|$ it cancels in the subtraction, conserving total momentum. Forming an equation for the relative drift velocity, $u_\| - v_\|$, we have

$$n_e m_e \frac{d_E}{dt}(u_\| - v_\|) = \nabla_\| p_e + n_e e E_\| - R_{ei} + O(\mu_e), \tag{7.78}$$

where as before $\mu_e = m_e/m_i$ is the electron ion mass ratio. Putting in Faraday's law (Eq. 7.45) for $E_\|$ and dropping the $O(\mu_e)$ corrections we have

$$n_e e\frac{\partial A_\|}{\partial t} + n_e m_e \frac{d_E}{dt}(u_\| - v_\|) = \nabla_\| p_e - n_e e\nabla_\|\phi - R_{ei}. \tag{7.79}$$

The first two terms on the right side represent the adiabatic response and the other terms represent the workings of inductivity, electron inertia, and resistive dissipation to oppose its statis force balance. Finally, under drift ordering the factor of $n_e$ appearing in the inertia may be commuted through the advective derivative, neglecting $O(\delta)$ corrections, to yield the Ohm's law,

$$n_e e\frac{\partial A_\parallel}{\partial t} + \frac{m_e}{e}\frac{d_E}{dt}J_\parallel = \nabla_\parallel p_e - n_e e\nabla_\parallel \phi - R_{ei}. \tag{7.80}$$

If we provide constitutive relations for the dissipation effects, $q_{e\parallel}$, $q_{i\parallel}$, $R_{ei}$, and $\Pi_\parallel$, in terms of the dependent variables $\{\phi, A_\parallel, n_e, T_e, T_i, u_\parallel\}$, with $J_\parallel$ already given by Ampère's law (Eq. 7.58) and $n_i$ by $n_e$, we then have a closed set. The simplest form of this (aside from neglecting thermal effects altogether and all dissipation except for resistive friction, as we will do for simplified, explanatory treatments) is the Chapman–Enskog formalism of the previous Chapter. In the literature these are sometimes called the reduced Braginskii equations, since a procedure very similar to that leading to reduced MHD is being applied to the two fluid Braginskii (1965) equations. That set amounts to the dynamical equations (Eqs (7.62), (7.65), (7.71), (7.73), (7.77), (7.80)), plus the constitutive relations

$$R_{ei} = n_e e\eta_\parallel J_\parallel - 0.71 n_e \nabla_\parallel T_e \qquad \eta_\parallel = 0.51\frac{m_e \nu_e}{n_e e^2}, \tag{7.81}$$

$$q_{e\parallel} = -0.71\frac{T_e}{e}J_\parallel - n_e\kappa_{\parallel e}\nabla_\parallel T_e \qquad \kappa_{\parallel e} = 3.2\frac{T_e}{m_e\nu_e}, \tag{7.82}$$

$$q_{i\parallel} = -n_i\kappa_{\parallel i}\nabla_\parallel T_i \qquad \kappa_{\parallel i} = 3.9\frac{T_i}{m_i\nu_i}, \tag{7.83}$$

$$\Pi_\parallel = -\mu_\parallel\nabla_\parallel u_\parallel \qquad \mu_\parallel = 0.96 n_i m_i\frac{T_i}{m_i\nu_i}, \tag{7.84}$$

and Ampère's law (Eq. 7.58) for $J_\parallel$. Note that the term $\nabla_\parallel u_\parallel$ in $\Pi_\parallel$ is not a divergence, but a gradient. If the divergence $B\nabla_\parallel(\Pi_\parallel/B)$ is multiplied by $u_\parallel$ and integrated over the volume, or spatial domain, then integrating $B\nabla_\parallel$ by parts and noting its formulation as a pure divergence, we obtain

$$\int d\mathcal{V}\; u_\parallel\, B\nabla_\parallel\frac{\Pi_\parallel}{B} = \int d\mathcal{V}\;(-\Pi_\parallel)\nabla_\parallel u_\parallel = \int d\mathcal{V}\;\mu_\parallel\left(\nabla_\parallel u_\parallel\right)^2, \tag{7.85}$$

which is a positive-definite quantity. All dissipation terms will have to produce such forms in the resulting energy theorem of the equations.

This set can be generalised by taking models for the coefficients $\eta_\parallel$, $\kappa_{\parallel e}$, $\kappa_{\parallel i}$, $\mu_\parallel$, which are more realistic in the weakly collisional regime, or by extending the model to retain $q_{e\parallel}$ and $q_{i\parallel}$ as dynamical variables. Electron viscosity ($\mu_{\parallel e}$) is usually neglected because the electron inertia itself only enters in the weakly collisional regime (in a Chapman–Enskog expansion it enters only at second order: cf. Hassam

1980). Perpendicular dissipative effects (scaling as $\rho_i^2\nu_i$ for the ions and $\rho_e^2\nu_e$ for the electrons) are neglected because they are typically very weak at $\rho_s$ scales in strongly hot, magnetised plasmas (since $\nu_i$ is usually much smaller than $c_s/L_\perp$, and $\mu_e\nu_e$ is smaller still).

The *a posteriori* check involving the neglect of $\mathbf{A}_\perp$ can be done by estimating $\nabla p_e \sim p_e(e/T_e)\nabla\phi$, and then using Ampère's law to obtain $\mathbf{A}_\perp$ from the dominant perpendicular current, $\mathbf{J}_*$, which is determined by $p = p_e + p_i$ via Eq. (7.25). We find that $\mathbf{J}_\perp \sim n_e e \mathbf{v}_E$. The effective compressibility of the magnetic field lines (perpendicular to $\mathbf{B}$) can then be determined by taking the curl of $\mathbf{A}_\perp$. One finds that the field strength disturbance $\tilde{B}/B$ is of order $\beta_e\delta$, in agreement with the simple postulate that $\nabla_\perp(p + B^2/2\mu_0)$ remains close to zero. Estimation of the magnitude ratio $A_\parallel/A_\perp$ is more complicated since it depends on estimation of $J_\parallel$, but if we assume that $J_\parallel/n_e e c_s \sim \delta$, then we find that since $\mathbf{v}_E/c_s \sim \delta$ we have $J_\parallel \sim J_\perp$ and hence $A_\parallel \sim A_\perp$. Ultimately, $\mathbf{A}$ follows $\mathbf{J}$ in the relative scaling of its components, we have the three pieces of $\mathbf{A}$ following the parallel, diamagnetic and polarisation currents in scaling as $\delta$, $\delta$, and $\delta^2$, respectively. They therefore contribute in similar magnitude to the divergence, $\nabla \cdot \mathbf{A}$, which like $\nabla \cdot \mathbf{J}$ vanishes. Therefore, while $\omega A_\parallel$ can compete with $\nabla_\parallel\phi$, the inductive perpendicular electric field $\omega\mathbf{A}_\perp$ is small compared to $\nabla_\perp\phi$ by the flute mode ordering factor $k_\parallel/k_\perp \sim \delta$. It is therefore a good approximation to neglect $\mathbf{A}_\perp$ in the perpendicular electric field in fluid drift dynamics whenever both $\beta_e$ and $\delta$ are small compared to unity.

## 7.9 Energetics of the fluid drift equations

The drift equations form an expansion of the fluid plasma model to lowest order in $\delta$, and as a result they only conserve total energy up to $O(\delta)$ corrections. To achieve the usual goal—expansion of the equations up to $O(\delta)$ corrections with an energy theorem up to $O(\delta^2)$ corrections—one has to do better than this. The fluid drift equations are customarily used in a different context, however. Rather than evolving a global system through $O(1)$ changes, one follows the dynamics of small scale, small amplitude disturbances, even if nonlinear, on a prescribed background which is at most slowly varying compared to the disturbances. In this case we are not interested in total energy conservation but in the conservation of fluctuation free energy. That is, the background itself appears as a set of parameters, with all the activity in the fluctuations which represent the departure from the equilibrium state. By equilibrium we do not mean force balance, but a true equilibrium represented by a minimisation of the thermodynamic free energy, which means maximum entropy for a given energy or minimum energy for a given entropy. The systematic derivation of this requires details of the kinetic theory. For now we construct it heuristically, noting that if there are terms representing waves ($\nabla_\parallel$) or compression ($\mathcal{K}$) which don't balance properly by forming up into total divergences, then we would have to go back and reconsider the derivation of the equations.

The gradient itself can appear as a forcing process, with drive terms, or be the result of a set of true source terms. The dissipation mechanisms appear as energetic sinks, or forcing processes which are negative-definite (positive-definite quantities preceded by a minus sign in the conservation law). In a global total-f system, only true sources can serve as positive forcing, while a dissipation mechanism is a transfer process between ordered motion (e.g. wave energy or fluid kinetic energy) and random motion (temperature). In a local system, a source maintains a gradient and a sink is an actual loss effect, since one does not follow the heating which results from a dissipation mechanism like friction or viscosity. Nevertheless, one does have an exact free energy theorem with forcing and dissipation.

In order to see why the local drift equations do not conserve global energy completely, consider how advection and compression work together in conserving and transferring thermal and kinetic energy (refer to Section 6.8 of the previous Chapter). In the total advective heat flux of $(5/2)p_i\mathbf{u}$, a factor of 3/2 of that comes from advection of $n_i$ and $T_i$, so far as the action of the divergence on $n_i$ and $T_i$ is concerned, but for the velocity divergence a factor of 3/2 comes from $\nabla \cdot n_i\mathbf{u}$ in the density equation and the other factor of 1 comes from $p_i\nabla \cdot \mathbf{u}$ in the temperature equation. So for each contribution to $\mathbf{u}$, both advection and compression are involved. But for $u_\|$ and $\mathbf{u}_p$, the advection pieces have been dropped, and so we cannot have complete conservation unless these are retained.

As far as fluctuation free energy is concerned, however, there are no problems. We must construct an energy-like quantity that is quadratic in the disturbances. The way to do this is to generalise the kinetic energy, which takes care of the velocities, to a free energy expression which incorporates the state variables, which collectively refer to those that appear under gradient operators as force terms for the velocities. The potential $\phi$ already appears through the ExB kinetic energy, and no additional term due to $\phi$ should appear, because of quasineutrality. The potential is therefore already taken care of. For the others, it suffices to take their squares and adjust the coefficients such that (1) free energy sinks such as $\eta_\| J_\|^2$ appear properly as negative-definite, and (2) compression and work terms such as $u_\|\nabla_\| p_i$ appear as transfer couplings and not as forcing processes. Denoting the disturbances with the tilde symbol and adjusting the units, we have

$$U = \frac{1}{2}n_i m_i\left[(\mathbf{v}_E + \mathbf{u}_*)^2 + \tilde{u}_\|^2\right] + \frac{\tilde{\mathbf{B}}^2}{2\mu_0} + \frac{1}{2}n_e m_e \frac{\tilde{J}_\|}{n_e e}^2 + \frac{p_e}{2}\frac{\tilde{n}_e^2}{n_e^2} + \frac{3}{2}\frac{p_e}{2}\frac{\tilde{T}_e^2}{T_e^2} + \frac{p_i}{2}\frac{\tilde{n}_e^2}{n_e^2} + \frac{3}{2}\frac{p_i}{2}\frac{\tilde{T}_i^2}{T_i^2}, \tag{7.86}$$

for the free energy density $U$, where $\mathbf{v}_E$ and $\mathbf{u}_*$ are given in terms of $\nabla_\perp\tilde{\phi}$ and $\nabla_\perp\tilde{p}_i$, with $\tilde{p}_i = n_i\tilde{T}_i + T_i\tilde{n}_e$ noting $\tilde{n}_i = \tilde{n}_e$, and $\tilde{\mathbf{B}}$ is given in terms of $\nabla_\perp\tilde{A}_\|$. Note that the density enters twice, once each for electrons and ions. The polarisation drift does not appear in the velocities, due to the drift ordering. Nevertheless, it is represented by the ExB energy and is used to derive the expression. Hence it has to be kept in the divergence balance for the current.

In keeping with the ordering, we find the time evolution of $U$ through the dependent variables, which given the expression for the free energy are formally the equations for the disturbances. There are two ways to treat background gradients. If they are to be done self consistently then we assume that the only variables present are the disturbances, then derive the equations as we did above, namely, the quantities under the gradient operators are the dependent variables, and all the coefficients multiplying them are background constant parameters. Then, the existence of gradients lies within the dependent variables themselves, which are maintained by a set of source terms. Alternatively, we may add the gradient drive terms wherever the nonlinearities, $\mathbf{v}_E \cdot \nabla$ or $\nabla_\parallel$, act upon any of the densities or temperatures. In the latter case we have

$$\mathbf{v}_E \cdot \nabla n_e = n_e \mathbf{v}_E \cdot \nabla \log n_e + \mathbf{v}_E \cdot \nabla \tilde{n}_e, \tag{7.87}$$

e.g. for the density, so that there are now two terms, one for the local disturbance and one for the background gradient. In either case, the free-standing coefficient $n_e$ is a background parameter. The easier method is the first one, and is used here. Hence, anywhere we find for example, $\nabla_\parallel T_e$, we express it as $\nabla_\parallel \tilde{T}_e$ to keep track of where the dependent variables go, while recalling that a drive term is implicitly included there and can be written separately at the end. Recall also that $\nabla_\parallel p_e$ is expressed as $T_e \nabla_\parallel \tilde{n}_e + n_e \nabla_\parallel \tilde{T}_e$, both pieces of which can contain a drive term.

Writing the equations explicitly in terms of the dependent variables, first for the state variables, we have

$$\nabla \cdot \frac{n_i m_i}{B^2} \frac{d_E}{dt} \nabla_\perp \tilde{W}_i = B \nabla_\parallel \frac{\tilde{J}_\parallel}{B} - \mathcal{K}(\tilde{p}_e + \tilde{p}_i), \tag{7.88}$$

$$\frac{d_E \tilde{n}_e}{dt} = \frac{1}{e} B \nabla_\parallel \frac{\tilde{J}_\parallel}{B} - n_e B \nabla_\parallel \frac{\tilde{u}_\parallel}{B} + n_e \mathcal{K}(\tilde{\phi}) - \frac{1}{e} \mathcal{K}(\tilde{p}_e), \tag{7.89}$$

$$\begin{aligned} \frac{3}{2} n_e \frac{d_E \tilde{T}_e}{dt} = \frac{T_e}{e} B \nabla_\parallel \frac{\tilde{J}_\parallel}{B} - p_e B \nabla_\parallel \frac{\tilde{u}_\parallel}{B} - B \nabla_\parallel \frac{\tilde{q}_{e\parallel}}{B} \\ + p_e \mathcal{K}(\tilde{\phi}) - \frac{T_e}{e} \mathcal{K}(\tilde{p}_e) - \frac{5}{2} \frac{p_e}{e} \mathcal{K}(\tilde{T}_e), \end{aligned} \tag{7.90}$$

$$\begin{aligned} \frac{3}{2} n_i \frac{d_E \tilde{T}_i}{dt} = \frac{T_i}{e} B \nabla_\parallel \frac{\tilde{J}_\parallel}{B} - p_i B \nabla_\parallel \frac{\tilde{u}_\parallel}{B} - B \nabla_\parallel \frac{\tilde{q}_{i\parallel}}{B} \\ + p_i \mathcal{K}(\tilde{\phi}) + \frac{T_i}{e} \mathcal{K}(\tilde{p}_e) + \frac{5}{2} \frac{p_i}{e} \mathcal{K}(\tilde{T}_i), \end{aligned} \tag{7.91}$$

where in Eq. (7.88) for the vorticity we define

$$\tilde{W}_i = \tilde{\phi} + \frac{1}{n_e e} \tilde{p}_i, \tag{7.92}$$

as the total ion velocity potential, including the diamagnetic part. In the last three of these equations a drive term is implicitly included in each ExB advective derivative, but not in the first, since second derivatives of such terms are ordered small.

For the flux variables we have

$$n_e e \frac{\partial \tilde{A}_\parallel}{\partial t} + \frac{m_e}{e} \frac{d_E}{dt} \tilde{J}_\parallel = \nabla_\parallel \tilde{p}_e - n_e e \nabla_\parallel \tilde{\phi} - \tilde{R}_{ei}, \tag{7.93}$$

$$n_i m_i \frac{d_E}{dt} \tilde{u}_\parallel = -\nabla_\parallel (\tilde{p}_e + \tilde{p}_i) - B \nabla_\parallel \frac{\tilde{\Pi}_\parallel}{B}. \tag{7.94}$$

The heat flux Eqs (7.82) and (7.83) become

$$\tilde{q}_{e\parallel} + 0.71 \frac{T_e}{e} \tilde{J}_\parallel = -n_e \kappa_{\parallel e} \nabla_\parallel \tilde{T}_e, \tag{7.95}$$

$$\tilde{q}_{i\parallel} = -n_i \kappa_{\parallel i} \nabla_\parallel \tilde{T}_i, \tag{7.96}$$

in the Braginskii model. In each of these equations, including the heat fluxes, a drive term is implicitly included in each nonlinear parallel derivative which acts upon a temperature or pressure.

In these equations, the background quantities are regarded as parameters unless they appear as gradient forcing terms under either $\mathbf{v}_E \cdot \nabla$ or $\nabla_\parallel$, and, of course, pressures under gradients appear with density and temperature pieces separately, e.g.

$$\mathcal{K}(\tilde{p}_e) = T_e \mathcal{K}(\tilde{n}_e) + n_e \mathcal{K}(\tilde{T}_e) \qquad \nabla_\parallel \tilde{p}_e = T_e \nabla_\parallel \tilde{n}_e + n_e \nabla_\parallel \tilde{T}_e. \tag{7.97}$$

This is referred to as linearised pressure. The quantities under the gradient operator are the dependent variables, and the background gradient drive terms if done separately, and the coefficients are background quantities. These coefficients are now constants, since treating them with stratification will preclude a closed free energy theorem, such that the model would not have a conservation law at all, neither global energy nor local free energy (see Section 7.11, below).

In forming the evolution equation for $U$, many terms form pairs which can be recast in the form of total divergences. These are transfer mechanisms, just like their similar counterparts in the global energy theorems discussed in previous Chapters. For example, multiplying Eq. (7.94) with $\tilde{u}_\parallel$, we obtain a piece involving $\tilde{u}_\parallel$ and $\nabla_\parallel \tilde{p}_e$,

$$n_i m_i \frac{d_E}{dt} \frac{\tilde{u}_\parallel^2}{2} = \cdots - \tilde{u}_\parallel \nabla_\parallel \tilde{p}_e \, . \tag{7.98}$$

In both the density and temperature equations, in each case multiplying with the relative fluctuations, we find the corresponding pieces which involve $\nabla_\parallel$ acting on $\tilde{u}_\parallel$,

$$n_e T_e \frac{d_E}{dt} \frac{\tilde{n}_e^2}{n_e^2} = \cdots - T_e \tilde{n}_e B \nabla_\parallel \frac{\tilde{u}_\parallel}{B}, \tag{7.99}$$

$$\frac{3}{2} n_e T_e \frac{d_E}{dt} \frac{\tilde{T}_e^2}{T_e^2} = \cdots - n_e \tilde{T}_e B \nabla_\parallel \frac{\tilde{u}_\parallel}{B}. \tag{7.100}$$

The right sides of these equations combine into the following total divergence,

$$\frac{\partial U}{\partial t} = \cdots - B\nabla_{\|}\frac{\tilde{p}_e\tilde{u}_{\|}}{B}. \tag{7.101}$$

Similarly, the $\mathbf{v}_E \cdot \nabla$ part of the advective derivative becomes a total divergence, because we are keeping only the lowest order, divergence-free part of it. For example, the ordering follows

$$\mathbf{v}_E \cdot \nabla\tilde{n}_e \sim \mathbf{v}_E \cdot \nabla n_e \sim n_e\nabla \cdot \mathbf{v}_E \gg \tilde{n}_e\nabla \cdot \mathbf{v}_E, \tag{7.102}$$

noting that the $n_e\nabla \cdot \mathbf{v}_E$ term is already accounted for in $\mathcal{K}(\tilde{\phi})$. The action of $\mathbf{v}_E \cdot \nabla$ on $U$ simply enters a divergence,

$$\frac{d_E U}{dt} = \frac{\partial U}{\partial t} + \mathbf{v}_E \cdot \nabla U = \frac{\partial U}{\partial t} + \nabla \cdot U\mathbf{v}_E + O(\delta). \tag{7.103}$$

The corrections of order $\delta$ are further dropped by treating the velocity in $\mathbf{v}_E \cdot \nabla$ as exactly incompressible, leaving the quasistatic compression given by $\mathcal{K}(\tilde{\phi})$ as a separate set of terms. Within this model, conservation by both $\mathbf{v}_E \cdot \nabla$ and $\mathcal{K}(\tilde{\phi})$ is exact.

The gradient drive terms, however, do not vanish in this fashion of integration over total divergences, but remain as forcing terms For example,

$$\frac{\partial U}{\partial t} = \cdots - \tilde{u}_{\|}\nabla_{\|}p_e = \cdots - \frac{\tilde{u}_{\|}}{B}\tilde{\mathbf{B}}\cdot\nabla p_e, \tag{7.104}$$

which simply means that a flow along a perturbed field line down a pressure gradient will lead to the driving of the parallel flow. The same holds for ExB advection of the background gradients. For example, the gradient forcing terms in Eq. (7.89) and (7.90) give rise to

$$\frac{\partial U}{\partial t} = \cdots - T_e\frac{\tilde{n}_e}{n_e}\mathbf{v}_E \cdot \nabla n_e - n_e\frac{3}{2}\frac{\tilde{T}_e}{T_e}\mathbf{v}_E \cdot \nabla T_e, \tag{7.105}$$

and so the transport of $\tilde{n}_e$ and $\tilde{T}_e$ by the ExB velocity down their respective gradients leads to the excitation of free energy in the form of just these fluctuations.

Similarly, a heat flux down a temperature gradient along a perturbed field line will drive the temperature disturbances. With the dissipative model of Eqs (7.82) and (7.83) this is somewhat subtle, since the contribution due to the gradient must be extracted from a second-order $\nabla_{\|}$ operation which is a divergence and a gradient. For the electrons this is manipulated as follows. Writing the contribution to the energy theorem as

$$\frac{\partial U}{\partial t} = \cdots - \frac{\tilde{T}_e}{T_e}B\nabla_{\|}\frac{\tilde{q}_{e\|}}{B}, \tag{7.106}$$

we add the heat flux Eq. (7.95) times $\tilde{q}_{e\|}/n_e T_e \kappa_{\|e}$, to obtain

$$\begin{aligned}\frac{\partial U}{\partial t} = \cdots &- B\nabla_{\|}\left(\frac{\tilde{T}_e}{T_e}\frac{\tilde{q}_{e\|}}{B}\right) - \frac{\tilde{q}_{e\|}}{\kappa_{\|e}n_e T_e}\left(\tilde{q}_{e\|} + 0.71\frac{T_e}{e}\tilde{J}_{\|}\right) \\ &- \tilde{q}_{e\|}\nabla_{\|}(\log T_e),\end{aligned} \tag{7.107}$$

where the first term is a total divergence and the last one is the gradient drive term for $\tilde{q}_{e\|}$. Now, we complete the square on the combination involving $\tilde{q}_{e\|}$ and $\tilde{J}_\|$ by considering the Ohm's law's contribution to the energy, whose relevant terms involving dissipation are the ones from $R_{ei}$ in Eq. (7.81),

$$\frac{\partial U}{\partial t} = \cdots - \eta_\| \left(\tilde{J}_\|\right)^2 + \frac{\tilde{J}_\|}{n_e e} 0.71 n_e \nabla_\| (\tilde{T}_e + T_e). \tag{7.108}$$

We then substitute for the term involving 0.71 using Eqs (7.81) and (7.82) and combine the two sets of terms neglecting the divergence term, to find

$$\frac{\partial U}{\partial t} = \cdots - \eta_\| \left(\tilde{J}_\|\right)^2 - \frac{p_e}{\kappa_{\|e}} \left(\frac{\tilde{q}_{e\|}}{p_e} + 0.71 \frac{\tilde{J}_\|}{n_e e}\right)^2 - \tilde{q}_{e\|} \nabla_\| (\log T_e). \tag{7.109}$$

We now have two negative-definite sinks, representing resistivity and thermal conduction, respectively, and the gradient drive term arising from conductive heat flows along perturbed field lines down the background temperature gradient.

In all cases, the corrections incurred by the operation of $\nabla_\|$ or $\mathbf{v}_E \cdot \nabla$ on the background quantity are down by one order in $\delta$ from the same operation on the disturbance; e.g.

$$\mathbf{v}_E \cdot \nabla \frac{\tilde{T}_e}{T_e} = \frac{\mathbf{v}_E \cdot \nabla \tilde{T}_e}{T_e} + \tilde{T}_e \mathbf{v}_E \cdot \nabla \frac{1}{T_e} = \frac{\mathbf{v}_E \cdot \nabla \tilde{T}_e}{T_e} + O(\delta). \tag{7.110}$$

Equations for the disturbances are scaled against the local parameters $n_e$, $T_e$, and $T_i$, and then the parameters are assumed to be constants except where operated upon by $\mathbf{v}_E \cdot \nabla$ or $\nabla_\|$, and then these gradients are taken to be separate drive terms, as done in Eqs (7.88)–(7.96). The energy theorem is then exact without these $O(\delta)$ corrections.

The combined energy theorem of Eqs (7.89)–(7.93) is for the total energy,

$$\mathcal{E} = \int d\mathcal{V}\, U. \tag{7.111}$$

Under the dissipative flux formulae of Eqs (7.81)–(7.84), the total free energy theorem is

$$\begin{aligned} \frac{\partial \mathcal{E}}{\partial t} = \int d\mathcal{V}\, \Big\{ &\Big[ -\tilde{q}_{e\|} \nabla_\| (\log T_e) - \tilde{q}_{i\|} \nabla_\| (\log T_i) - \tilde{u}_\| \nabla_\| (p_e + p_i) \\ &- (T_e + T_i) \frac{\tilde{n}_e}{n_e} \mathbf{v}_E \cdot \nabla n_e - \frac{3}{2} n_e \frac{\tilde{T}_e}{T_e} \mathbf{v}_E \cdot \nabla T_e - \frac{3}{2} n_e \frac{\tilde{T}_i}{T_i} \mathbf{v}_E \cdot \nabla T_i \Big] \\ &- \left[ \eta_\| \left(\tilde{J}_\|\right)^2 + \frac{p_e}{\kappa_{\|e}} \left(\frac{\tilde{q}_{e\|}}{p_e} + 0.71 \frac{\tilde{J}_\|}{n_e e}\right)^2 + \frac{p_i}{\kappa_{\|i}} \left(\frac{\tilde{q}_{i\|}}{p_i}\right)^2 + \mu_\| \left(\nabla_\| \tilde{u}_\|\right)^2 \right] \Big\}, \end{aligned} \tag{7.112}$$

for a localised system, in which the total divergences vanish under the integral. Note the scaling against $p_e$, especially for $\tilde{T}_i / T_e$. The first two lines on the right side give the forcing processes of indefinite sign, all potentially positive gradient drive terms, and the last line gives the sinks, all negative-definite and resulting from collisional dissipation. Drive terms due to $J_\parallel \nabla_\parallel (p_e + p_i)$ vanish due to the relationship between $\tilde{J}_\parallel$ and $\tilde{\mathbf{B}}$ via $\tilde{A}_\parallel$. We note the mutual cancellation of all $\mathscr{K}()$ terms, which proceeds similarly to the $\nabla_\parallel$ terms as both represent compression phenomena: $\mathscr{K}()$ is the quasistatic remnant of compression due to an inhomogeneous magnetic field, which is left behind after the drift approximation removes dynamical (wave-like) compression from the system. The operator $\mathscr{K}()$ is a first-order differential operator, which possesses the linearity property. One therefore ends with expressions of the form $\mathscr{K}(f)$, with $f$ a scalar, and these are total divergences, by Eq. (7.59).

## 7.10 Summary

To summarise this Chapter, we have defined the drift approximation as a parameter limit in which we can take $k_\parallel / k_\perp$, $\delta$, and $\beta_e$ as arbitrarily small, and the frequencies $\Omega_i$ and $k_\perp v_A$ as arbitrarily fast. This leads to quasistatic, dynamically incompressible perpendicular dynamics while retaining compressibility along the magnetic field, hence allowing frequencies as fast as $k_\parallel v_A$. It follows that the electric field splits into an electrostatic $\mathbf{E}_\perp$ and an electromagnetic $E_\parallel$. Due to the smallness of $\delta$ the dynamics is allowed to be nonlinear at small relative amplitude, with disturbances obeying $\tilde{n}_e / n_e \ll 1$ but $\nabla \tilde{n}_e \sim \nabla n_e$. All $O(\delta)$ corrections are neglected, except for the polarisation drift, which under divergences reverts to $O(1)$. The gradient operator on disturbances is always an order larger than on the background except for the first operation involving background gradients, so that although $\nabla \tilde{n}_e \sim \nabla n_e$ we have $\nabla n_i \cdot \nabla \phi \ll n_i \nabla_\perp^2 \phi$. Due to the flute mode ordering we can use $\nabla^2$ and $\nabla_\perp^2$ interchangeably. The parallel structure of $\mathbf{B}$ is retained since all parallel wavelengths are comparable. A local model of nonlinear, turbulent dynamics is derived under the drift approximation, and the model satisfies a fluctuation free energy theorem in which background gradients appear as drive processes (fluxes down gradients). Global energy is only conserved to first order in $\delta$, but the free energy is conserved exactly.

## 7.11 Delta-f versus total-f energetics

There are many instances in which one can use delta-f equations even for physical situations involving self-consistent gradients. In this case we rewrite the delta-f equations, Eqs (7.88)–(7.96), without the gradient drive terms. Boundaries would be appropriate for a confined system. The usual case is to have all dependent variables vanish on an outer magnetic surface, such that $\nabla_\parallel$ applied to any of them vanishes there. On an inner surface, the cross-field derivative would vanish such that the values themselves could float. Such a system would relax, as the large-scale gradient is converted into dynamics. A sink region at the outer boundary can contain the transport due to fluxes. Recall here that if the gradient drive is positive, it is implied that the down-gradient flux is positive. Without any external sources, the overall energy must indeed decay, as this outcome is required by the free energy theorem. The only alternative is complete stability, so that all dependent variable flux quantities decay to zero while the content of free energy relaxes to a finite value. For dynamics to cease the parallel gradient of all quantities would vanish, so that the only finite component remaining would be the one with constant values within the surfaces in which the magnetic field lies. If the field lines are open, then the only final state would be the one in which all dependent variables are zero at all locations. Evolution toward the final state would occur under conservation of free energy. The only possibility for a dissipative system to be in stationary statistical equilibrium is if it is driven by actual sources, so instead of equations with gradient drive terms, we would have the same equations with one or more sources of particles and energy.

The ultimate reason that such a driven/dissipative system with self-consistent gradients can be treated with delta-f equations is that all the energy transfer effects in a global energy theorem with fully stratified equations are the same as their counterpart free energy transfer effects in a self consistent but delta-f homogeneous free energy theorem. To see this we consider the simplest treatment of sound wave effects. Eqs (7.89) and (7.94) written without the electromagnetic field terms would be

$$\frac{\partial \tilde{p}_e}{\partial t} + p_e B\nabla_{\|}\frac{\tilde{u}_{\|}}{B} = 0, \tag{7.113}$$

$$\rho\frac{\partial}{\partial t}\tilde{u}_{\|} = -\nabla_{\|}\tilde{p}_e, \tag{7.114}$$

representing a simplified isothermal delta-f system in which $\rho$ and $p_e$ are constants. The free energy theorem would be

$$\frac{1}{2}\frac{\partial}{\partial t}\left(\frac{1}{p_e}\,|\,\tilde{p}_e\,|^2\right) + B\nabla_{\|}\frac{\tilde{p}_e\tilde{u}_{\|}}{B} = \tilde{u}_{\|}\nabla_{\|}\tilde{p}_e, \tag{7.115}$$

$$\frac{1}{2}\frac{\partial}{\partial t}\left(\rho\,|\,\tilde{u}_{\|}\,|^2\right) = -\tilde{u}_{\|}\nabla_{\|}\tilde{p}_e, \tag{7.116}$$

having multiplied by $\tilde{p}_e/p_e$ and $\tilde{u}_{\|}$, respectively. This version shows the respective transfer effects, which appear on the right-hand side of each equation.

Now we consider the same equations in a total-f system,

$$\frac{\partial p_e}{\partial t} + u_{\|}\nabla_{\|}p_e + \Gamma p_e B\nabla_{\|}\frac{u_{\|}}{B} = 0, \tag{7.117}$$

$$\rho\frac{\partial}{\partial t}u_{\|} = -\nabla_{\|}p_e, \tag{7.118}$$

in which $p_e$, $\rho$, and $u_{\|}$ are all variable. The continuity equation for $\rho$ is

$$\frac{\partial \rho}{\partial t} + B\nabla_{\|}\frac{\rho u_{\|}}{B} = 0. \tag{7.119}$$

The energy theorem this time,

$$\frac{\partial}{\partial t}\frac{p_e}{\Gamma - 1} + B\nabla_{\|}\left(\frac{p_e}{\Gamma - 1} + p_e\right)\frac{u_{\|}}{B} = u_{\|}\nabla_{\|}p_e, \tag{7.120}$$

$$\frac{\partial}{\partial t}\frac{\rho}{2}u_\|^2 + B\nabla_\|\frac{\rho}{2}u_\|^2\frac{u_\|}{B} = -u_\|\nabla_\| p_e, \tag{7.121}$$

is the global one, not a free energy one.

In fact, it is straightforward to show that the free energy is not conserved in the total-f system, and the global energy is not conserved in the delta-f system. However, we can see that the transfer processes between thermal free energy and parallel flow energy in the delta-f system are identical to those between thermal energy and parallel flow energy in the total-f system. In general, these energy and free energy transfer processes are the same in both models. That is why it is physically sensible to understand the physical processes of small-scale turbulence in a system with gradients by exploring them within a delta-f set of equations with self-consistent gradients and an exact free energy theorem. The latter piece of content is the critical one. Only when the pieces of a free energy system are identifiable with the corresponding pieces in a global energy system, and the transfer effects are the same ones, can we make this statement. The reason it is practicable is that with quadratic quantities involving divergence-free flow derivatives being the only non-linearities in the model, the numerical quality is much cleaner and one can be much more sure of the integrity of the results.

## 7.12 Quasineutrality in Drift Dynamics

The principal reason we have to keep the polarisation drift is that it is what maintains quasineutrality in the dynamics. Here we illustrate this explicitly using the formulation just developed. The first part is also valid for MHD since we leave the electric field **E** arbitrary and note that even MHD expects the plasma flow to be no larger than the ExB velocity $\mathbf{v}_E$ even in strongly compressible dynamics.

If in the formulation of MHD we were to keep the partial charge densities $n_j e_j$ for each species and leave them arbitrary, Eq. (7.7) would appear as

$$\rho\frac{d\mathbf{v}_E}{dt} = \rho_{ch}\mathbf{E} + \mathbf{J}\times\mathbf{B} - \nabla p \tag{7.122}$$

where for scaling purposes we insert $\mathbf{v}_E$ for $\mathbf{u}$ in each fluid. Since even in MHD we assume time scales below the gyrofrequency, the inertia term yields the polarisation current as an $O(\delta)$ correction when we solve for $\mathbf{J}$,

$$\mathbf{J} + \mathbf{J}_D = J_\|\mathbf{b} + \frac{1}{B^2}\mathbf{B}\times\nabla p + \frac{\rho}{B^2}\mathbf{B}\times\frac{d\mathbf{v}_E}{dt} + \rho_{ch}\mathbf{v}_E + \epsilon_0\frac{\partial\mathbf{E}}{\partial t} \tag{7.123}$$

where $\mathbf{J}_D$ is the Maxwell displacement current (last term on the right hand side). The divergence of the quantity on the left ($\mu_0^{-1}\nabla\times\mathbf{B}$) vanishes, yielding

$$\begin{aligned}\nabla\cdot\left[J_\|\mathbf{b} + \frac{1}{B^2}\mathbf{B}\times\nabla p + \frac{\rho}{B^2}\mathbf{B}\times\frac{d\mathbf{v}_E}{dt}\right]\\ +\frac{\partial}{\partial t}\left(\epsilon_0\nabla\cdot\mathbf{E}\right) + \nabla\cdot[(\epsilon_0\nabla\cdot\mathbf{E})\mathbf{v}_E] = 0\end{aligned} \tag{7.124}$$

using $\rho_{ch} = \epsilon_0(\nabla \cdot \mathbf{E})$ in the last term. Since $\mathbf{B} \times \mathbf{v}_E = \mathbf{E}$, the terms on the second line are small if $\epsilon_0 \ll \rho/B^2$, which makes this the condition for quasineutrality. The two terms came from $\mathbf{J}_D$ and $\rho_{ch}\mathbf{E}$, respectively. Neglecting these means respectively taking

$$\nabla \cdot \mathbf{J} \to 0 \qquad \rho_{ch} \to 0 \qquad \text{so long as} \quad \epsilon_0 \ll \frac{\rho}{B^2} \tag{7.125}$$

This forms the set of statements describing quasineutrality in the dynamics.

On the other hand, the existence of a rotational flow with the electrostatic potential as its stream function requires the existence of a nonvanishing charge density. Let us assume at this point that we have ExB vortices in a homogeneous magnetic field. Then, now again with $\mathbf{E} = -\nabla\phi$,

$$\nabla \times \mathbf{v}_E = \frac{\mathbf{b}}{B}\nabla_\perp^2\phi \tag{7.126}$$

hence the vorticity is proportional to the charge density. Let us further assume that the adiabatic response is active, holding disturbances in $\phi$ down to those in the electron density. Then, the requirement that the total charge density be small compared to the charge density of just the electrons becomes a space scale restriction:

$$\frac{\rho_{ch}}{n_e e} \ll \frac{\widetilde{n}_e}{n_e} \sim \frac{e\widetilde{\phi}}{T_e} \Rightarrow \frac{\epsilon_0}{n_e e}\nabla^2\widetilde{\phi} \ll \frac{e\widetilde{\phi}}{T_e} \Rightarrow \lambda_D^2\nabla_\perp^2 \ll 1 \tag{7.127}$$

where $\lambda_D$ is the Debye length defined in Chapter 1. If instead of $e\widetilde{\phi}/T_e$ we set the normalised charge density small compared to $(\rho_s^2\nabla_\perp^2)(e\widetilde{\phi}/T_e)$, the normalised vorticity, the requirement that $\lambda_D \ll \rho_s$ results. In order for ExB flow dynamics to remain quasineutral, we require

$$v_A \ll c \qquad \lambda_D \ll \rho_s \qquad \Omega_i \ll \omega_{pi} \tag{7.128}$$

all of which follow if $\epsilon_0 \ll \rho/B^2$, which once again is the fundamental requirement. Then, the dynamics below the Alfvén compressional frequency $k_\perp v_A$ also has to satisfy $k_\perp\lambda_D \ll 1$, which for $k_\perp\rho_s \sim 1$ or smaller follows from the fundamental requirement.

The inequalties listed in this section constitute the requirements for quasineutrality in the dynamics. But in general, all of them descend from the one statement that $\epsilon_0 \ll \rho/B^2$, namely that the vacuum polarisability (another way to say permittivity of free space) be small compared to the polarisability of the plasma.

## Further Reading

Drift dynamics was considered for linear instabilities in the late 1950s and early 1960s, for which one can see S S Moiseev and R Z Sagdeev 'On the Bohm diffusion coefficient' *Sov. Phys. JETP* **17** (1963) 515; but the procedure for nonlinear dynamics involving polarisation was first systematically worked out by F L

Hinton and C W Horton Jr 'Amplitude limitation of a collisional drift wave instability' *Phys. Fluids* **14** (1971) 116. Similar operations leading to reduced MHD were systematised by H R Strauss *Phys. Fluids* **20** (1976) 1354. The operations explicitly indicating the polarisation drift were given in a turbulence work by A Hasegawa and K Mima 'Stationary spectrum of strong turbulence in magnetized nonuniform plasma' *Phys. Rev. Lett.* **39** (1977) 205.

The concept of drifts in magnetised plasma physics is similar to the concept of quasigeostrophic flow in geophysical fluid dynamics. Here, the magnetic field ($\mathbf{B}$) provides the anchor, while there, the background vorticity of the rotation ($\mathbf{\Omega}$) does the same. The ion diamagnetic velocity here is analogous (with $\mathbf{B} \times \nabla p_i$) to the geostrophic flow there (with $\mathbf{\Omega} \times \nabla p$). The disparity in scales along or across the anchor vector is also used to derive the shallow water equations in geophysics. For more background see texts such as A Gill 1982 *Atmosphere-Ocean Dynamics* (New York: Academic Press); and J Pedlosky 1987 *Geophysical Fluid Dynamics* (Berlin: Springer).

The references to Braginskii (1965) and Hassam (1980) are given in the previous Chapter.

Systematic fluid drift equations involving self-consistent electron dynamics were developed by J F Drake and T M Antonsen Jr 'Nonlinear reduced fluid equations for toroidal plasmas' *Phys. Fluids* **27** (1984) 898; following earlier work by D A Monticello and R B White 'Nonlinear drift tearing modes' *Phys. Fluids* **23** (1980) 366. The local forms were developed by B Scott, A B Hassam and J F Drake 'Nonlinear evolution of drift-tearing modes' *Phys. Fluids* **28** (1985) 275; and by R E Waltz 'Numerical simulation of electromagnetic turbulence in tokamaks' *Phys. Fluids* **28** (1985) 577.

Fluctuation free energy was developed as a concept of small kinetic fluctuations on an MHD background, with a unified energy principle for the whole, by M D Kruskal and C R Oberman 'On the stability of plasma in static equilibrium' *Phys. Fluids* **1** (1958) 275. It was picked up as the fluctuation dissipation theorem for kinetic computations, such as in W W Lee and W M Tang *Phys. Fluids* **31** (1988) 612.

The use of the quadratic dependent variable as an invariant, as used here, is from M Wakatani and A Hasegawa *Phys. Fluids* **27** (1984) 611. Development of it as a method to diagnose the turbulence is from B Scott *J. Comput. Phys.* **78** (1988) 114; and *Phys. Rev. Lett.* **65** (1990) 3289; and *Phys. Fluids B* **4** (1992) 2468. It also appears in R E Waltz *Phys. Fluids B* **2** (1990) 2118.

The free energy as its second variation away from a thermodynamic equilibrium, as in M D Kruskal and C R Oberman *op. cit.*, is fundamental to the kinetic models to be treated in Volume Two.

**IOP** Publishing

Turbulence and Instabilities in Magnetised Plasmas, Volume 1
Fluid drift turbulence
**Bruce Scott**

# Chapter 8

# Parallel dynamics—Alfvén/sound waves

## 8.1 Introduction

The simplest way to examine the relationship between MHD and fluid drift dynamics, is to compare the properties of wave-like disturbances on a homogeneous background in both models. We encountered the sound waves and compressional and shear Alfvén waves in MHD in Chapter 6. Herein we study them under the fluid drift model in detail, especially the roles of the various terms in the free energy theorem for the system, and compare to its reduced MHD subset. In both cases, a sound wave is an oscillation of free energy between pressure disturbances and parallel (ion) mass flows, and a shear Alfvén wave is such an oscillation between magnetic field energy and (ion) ExB kinetic energy. The ions are referred to in parentheses, because the much less massive electrons contribute negligibly to the fluid kinetic energy. In the case of shear Alfvén waves, a substantial component of the energy which oscillates against the magnetic field can come from the pressure disturbances rather than the ExB flows. This is the case in which the conventional (single-fluid) MHD model breaks down, since under MHD the fluid pressure is not coupled to the magnetic field energy (equivalently, the electric current) at all in a homogeneous magnetic field. Alfvén waves whose parallel dynamics is coupled to the electron pressure in this fashion are called kinetic shear Alfvén waves, or simply kinetic Alfvén waves, due to their original derivation from kinetic theory. The excitation of a parallel current by the imbalance of the pressure and electrostatic forces is called an adiabatic response, since it tends towards maintenance of the state of adiabatic electrons, or force balance, so named due to its quasistatic nature on slower time scales.

The MHD model usually treats the most basic shear Alfvén waves as plane waves, as shown in figure 8.1. Under both reduced MHD and fluid drift dynamics, the basic mode is a torsional Alfvén wave, following from the description of incompressible velocity and magnetic field disturbances in terms of their stream functions, or potentials, as in figure 8.2. The basic motion is described by an

 

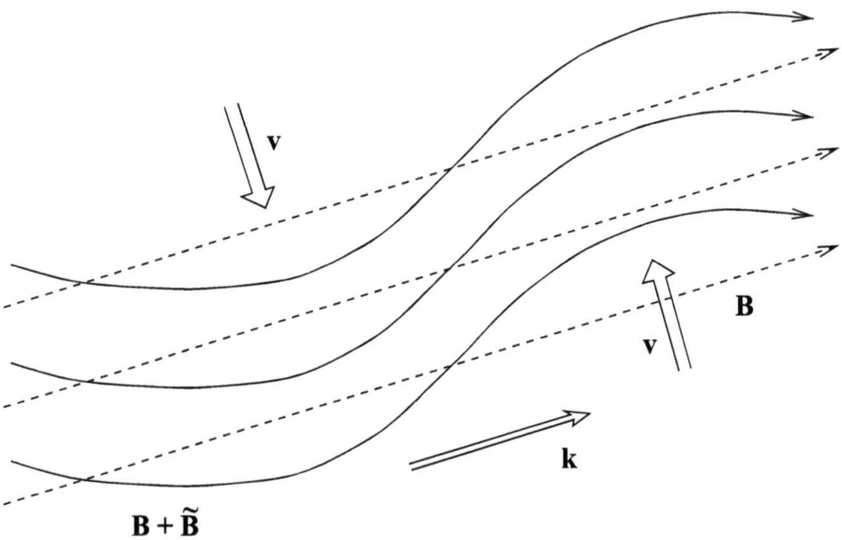


**Figure 8.1.** The conventional shear Alfvén wave under MHD, as a plane wave oscillation. The dashed lines show the undisturbed magnetic field lines, **B**. Under the influence of the plasma flow (**v**) across the magnetic field, the field lines are bent into the positions shown by the solid lines, $\mathbf{B} + \tilde{\mathbf{B}}$. The effect of magnetic tension to oppose this deformation leads to transverse wave propagation (**k**) along **B**.

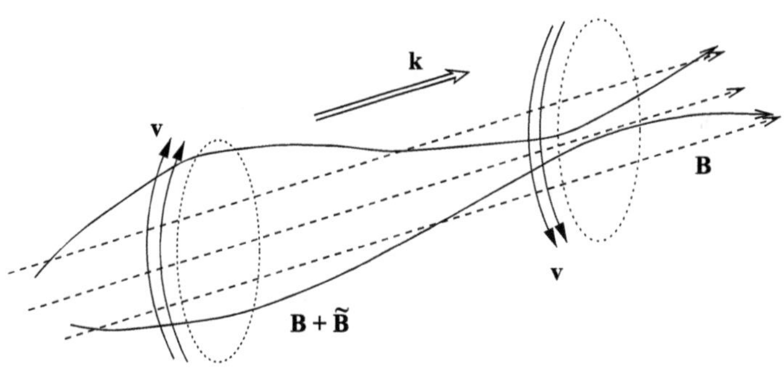


**Figure 8.2.** The same shear Alfvén wave oscillation, but localised to a flux tube with a finite radius in the perpendicular plane. The flow is now vortical, representing ExB eddy motion. The field lines are twisted, and again their curvature leads to magnetic tension which opposes the deformation. This is now a torsional Alfvén wave. Due to the finite perpendicular scale (inverse of $k_\perp$), the magnetic disturbance ($\tilde{\mathbf{B}}$) and ExB flow (**v**) may be replaced by their respective stream functions, $\tilde{A}_\parallel$ and $\tilde{\phi}$. The result is the representation in terms of potentials under drift dynamics or reduced MHD, rather than in terms of vector fields as under pure MHD.

oscillating ExB vortex which has different phases at different positions along the field lines. The perpendicular scale is always finite, never so large as the parallel scale along the background guide field. Drift dynamics departs from MHD by taking into account the coupling between this Alfvén dynamics and the electron pressure, as we develop in this Chapter.

## 8.2 The four-field fluid drift model

In this and the next few Chapters we will study the fundamental properties of the fluid drift dynamical model in qualitative form using the simplest possible model which still incorporates the necessary physics, that is, perpendicular ExB flows and parallel electron and ion dynamics, but neglecting the details of dissipation and some extra instabilities that go with the specific dynamics of the temperatures. The main point to understand is that this is a 'two-fluid' model, with the differing electron and ion parallel velocities being treated at the same level of consistency. This means that the parallel current participates in electron compressibility, with the polarisation current doing the same for the ions such that the current divergence vanishes. It has the consequence that disturbances in the electron pressure launch shear Alfvén wave oscillations with the energetic coupling $\tilde{J}_\parallel \leftrightarrow \tilde{p}_e$, called 'adiabatic compression,' serving as the main mediator. As this physics does not appear in conventional MHD, it makes sense to highlight it first, before moving on to more detail.

We assume a single-component plasma with singly-charged ions and equal electron and ion densities. The key feature of the two fluids which gives them differing responses to a fluctuating electromagnetic field is their disparate unit masses; the electron/ion mass ratio is small, so that the inertia of the electrons in response to the field can often be neglected. This means that the faster electron response will govern how the plasma responds to electric fields and currents, while the ion inertia will govern how the plasma responds to neutral flows. The principal examples of these are the shear Alfvén waves and sound waves, respectively. The dependent variables we need to keep are $\{\phi, p_e, A_\parallel, u_\parallel\}$, with $J_\parallel$ determined from $A_\parallel$ via Ampère's law. This is a situation in which the electrons provide the dynamics through their finite temperature (hence pressure), while the ions provide the inertia through their finite mass.

To find the model equations, we start with the dissipative drift dynamics model given in Sections 7.8 and 7.9 of the previous Chapter, and then set the temperature and heat flux disturbances to zero. We then eliminate the background gradient and magnetic divergence terms. The resulting two-fluid, four-field model for small-scale disturbances on a homogeneous background with resistive friction and ion viscosity appears as

$$\frac{d}{dt}\rho_s^2\nabla_\perp^2\frac{e\tilde{\phi}}{T_e} = c_s B\nabla_\parallel\frac{\tilde{J}_\parallel/nec_s}{B}, \tag{8.1}$$

$$\frac{d}{dt}\frac{\tilde{p}_e}{p_e} = \Gamma c_s B\nabla_\parallel\left(\frac{\tilde{J}_\parallel/nec_s}{B} - \frac{\tilde{u}_\parallel/c_s}{B}\right), \tag{8.2}$$

$$\frac{\partial}{\partial t}\frac{\tilde{A}_\parallel}{B\rho_s} + \mu_e\frac{d}{dt}\frac{\tilde{J}_\parallel}{nec_s} = c_s\nabla_\parallel\left(\frac{\tilde{p}_e}{p_e} - \frac{e\tilde{\phi}}{T_e}\right) - 0.51\mu_e\nu_e\frac{\tilde{J}_\parallel}{nec_s}, \tag{8.3}$$

$$\frac{d}{dt}\frac{\tilde{u}_\parallel}{c_s} = -c_s\nabla_\parallel\frac{\tilde{p}_e}{p_e} + B\nabla_\parallel\frac{\mu_\parallel}{B}\nabla_\parallel\frac{\tilde{u}_\parallel}{c_s}, \tag{8.4}$$

with Ampère's law

$$\frac{\tilde{J}_\parallel}{nec_s} = -\beta_e^{-1}\rho_s^2\nabla_\perp^2\frac{\tilde{A}_\parallel}{B\rho_s}. \tag{8.5}$$

The nonlinear advective derivative is now written solely with the ExB velocity, and the parallel derivative is with the magnetic flutter nonlinearities. These are given by

$$\frac{d}{dt} = \frac{\partial}{\partial t} + \mathbf{v}_E\cdot\nabla \qquad B\nabla_\parallel = (\mathbf{B} + \mathbf{B}_\perp)\cdot\nabla, \tag{8.6}$$

in which it is assumed that both $\mathbf{v}_E$ and $\mathbf{B}_\perp$ are divergence free. In models where $\mathbf{v}_E$ has a divergence, it is written separately in another term, as in the previous Chapter. The tilde symbol serves to distinguish fluctuating quantities from background parameters. Later, we will normalise the equations and then drop the tilde symbols for increased clarity.

The mass ratio and dynamical beta are the main parameters for the Alfvén responses. Assumed to be small, they are never completely negligible. Once more, they are given by

$$\mu_e = \frac{m_e}{m_i} \qquad \beta_e = \frac{c_s^2}{v_A^2} = \frac{\mu_0 p_e}{B^2}, \tag{8.7}$$

and for the purposes of modelling we simply take the parallel viscosity, $\mu_\parallel$, and ratio of specific heats, $\Gamma$ as free parameters. This is a convenient way to scale the equations, showing the symmetries among the various interactions. The nonlinear effects in $d/dt$ and $\nabla_\parallel$ are still generally present, and they took centre stage in Chapters 2–5, but for the purposes of studying waves, we will neglect them until they return in Chapter 11.

To isolate the properties of wave-like motion we will first neglect dissipation entirely. Then, considering the initial value problem with some specific numerical examples we will reintroduce resistive friction and ion viscosity, showing how they act differently on the various waves.

## 8.3 Wave-like motion

We saw in Chapter 6 how linearised wave oscillations on a homogeneous background are treated in the context of several different dynamical models, one of which was MHD. We now do the same for the simplified four-field fluid drift model. All we need do to the above set of equations is to eliminate the nonlinear terms by taking

$$\frac{d}{dt}\to\frac{\partial}{\partial t} \qquad \nabla_\parallel\to\frac{\partial}{\partial z}, \tag{8.8}$$

where $z$ is a coordinate along the guide field, a magnetic field given by $\mathbf{B} = B\nabla z$, in which $B$ is a constant. The perpendicular plane may be represented by $x$ and $y$, which are equivalent in the absence of gradients. The perpendicular operator $\nabla_\perp^2$ then becomes

$$\nabla_\perp^2 = \left(\frac{\partial^2}{\partial x^2} + \frac{\partial^2}{\partial y^2}\right). \tag{8.9}$$

These equations represent wave dynamics parallel to the magnetic field involving the state variables $\tilde{\phi}$ and $\tilde{p}_e$ and parallel flux variables $\tilde{A}_\parallel$ and $\tilde{u}_\parallel$, which are localised in the $xy$-plane such that

$$\nabla_\perp^2 \gg \nabla_\parallel^2, \tag{8.10}$$

and therefore (following the drift approximation) do not exhibit oscillatory wave dynamics perpendicular to the magnetic field.

A set of four equations for four dependent variables implies four distinct modes of oscillation—four eigenfunctions and their frequencies. In the case that both $\beta_e$ and $\mu_e$ are small, we can split them into two pairs of two waves each. For frequencies much faster than $c_s\nabla_\parallel$ we can neglect $u_\parallel$ and combine $\tilde{p}_e$ and $\tilde{\phi}$ to form the following set of equations for the two faster waves,

$$\frac{\partial}{\partial t}\rho_s^2\nabla_\perp^2\left(\frac{\tilde{p}_e}{p_e} - \frac{e\tilde{\phi}}{T_e}\right) = \left(1 - \Gamma\rho_s^2\nabla_\perp^2\right)c_s\nabla_\parallel\frac{\tilde{J}_\parallel}{nec_s}, \tag{8.11}$$

$$\frac{\partial}{\partial t}\frac{\tilde{A}_\parallel}{B\rho_s} + \mu_e\frac{\partial}{\partial t}\frac{\tilde{J}_\parallel}{nec_s} = c_s\nabla_\parallel\left(\frac{\tilde{p}_e}{p_e} - \frac{e\tilde{\phi}}{T_e}\right). \tag{8.12}$$

Using Eqs (8.3) and (8.5) to eliminate $\tilde{A}_\parallel$ and $\tilde{J}_\parallel$, we may further combine the two first-order equations into one that is second order,

$$\left(\beta_e - \mu_e\rho_s^2\nabla_\perp^2\right)\rho_s^2\nabla_\perp^2\frac{\partial^2}{\partial t^2}\left(\frac{\tilde{p}_e}{p_e} - \frac{e\tilde{\phi}}{T_e}\right) = \left(1 - \Gamma\rho_s^2\nabla_\perp^2\right)\rho_s^2\nabla_\perp^2 c_s^2\nabla_\parallel^2\left(\frac{\tilde{p}_e}{p_e} - \frac{e\tilde{\phi}}{T_e}\right). \tag{8.13}$$

We note that since in any degree of freedom the perpendicular wavenumber $k_\perp$ is always finite even if arbitrarily small, we can cancel the common factor of $\rho_s^2\nabla_\perp^2$ from both sides. This is now in the form of a wave equation, but the coefficients still depend on $\nabla_\perp^2$. What this means is that the speed of the wave in the parallel direction is dependent on the scale of motion in the perpendicular plane. Clearly, then, we still have a three-dimensional character to the dynamics: with $\tilde{\phi}$ and $\tilde{A}_\parallel$ serving as stream functions for the MHD disturbances $\mathbf{v}_E$ and $\mathbf{B}_\perp$, we have vortical ExB motion and magnetic fluctuations in the perpendicular plane, coupled together by currents which flow in the parallel direction. However, as shown in figure 8.3, we have the situation that this Alfvénic activity, represented by the pair ($\mathbf{v}_E$, $\mathbf{B}_\perp$), or ($\tilde{\phi}$, $\tilde{A}_\parallel$), is now coupled to the electron pressure disturbances, $\tilde{p}_e$. This is seen in Eqs (8.11)–(8.13), as it is the difference between $\tilde{p}_e/p_e$ and $e\tilde{\phi}/T_e$ to which $\tilde{J}_\parallel$ interacts, not just $e\tilde{\phi}/T_e$. The forces on

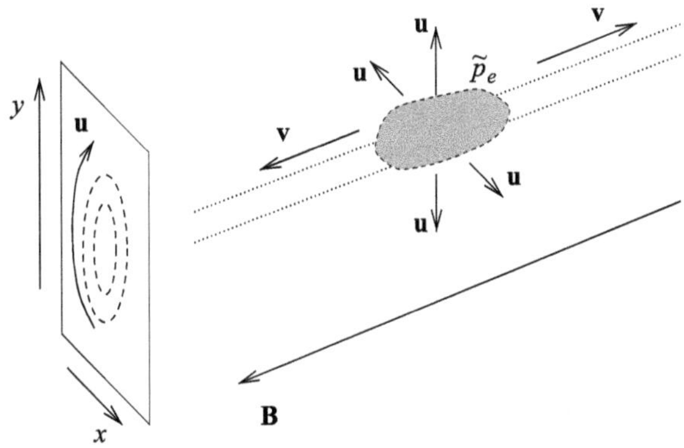


**Figure 8.3.** The kinetic shear Alfvén wave oscillation, localised to a flux tube in the $xy$-plane, caused by an initial electron pressure disturbance. The parallel pressure gradient gives rise to parallel electron flows (**v**) which are balanced by the ion polarisation drift (**u**). Both species have their ExB drifts, which do not cause a net current. As the electron current oscillates in the direction of **B**, the ion flow spirals into and out of the flux tube as a correction to the ExB flow, which is along the dashed circles drawn in the $xy$-plane. The response of the pressure disturbance depends on the perpendicular scale of the flux tube; if this scale is close to $\rho_s$, substantial parallel displacement of the pressure disturbance results. Under MHD or reduced MHD, there would be no response of the pressure to the Alfvén dynamics, and the pressure disturbance would not give rise to ExB flows. There would be no response to an initial pressure disturbance, other than sound waves.

the electrons are given by the pressure gradient as well as the electric field, and the imbalance drives the current. The stream function is now $e\tilde{\phi}/T_e - \tilde{p}_e/p_e$, for the perpendicular electron flow including the diamagnetic component, $\mathbf{v} = \mathbf{v}_E + \mathbf{v}_*$. The existence of $\tilde{J}_\parallel$ with the finite $\nabla_\parallel$, with which it is excited, drives both $\tilde{p}_e$ and $\tilde{\phi}$ so as to restore balance between the two forces. The polarisation current arises to balance $\tilde{J}_\parallel$, preserving $\nabla \cdot \mathbf{J} = 0$ overall.

If we Fourier decompose the disturbances in the perpendicular plane, we find that each component has its own wave speed. The wave equation becomes

$$\frac{\partial^2}{\partial t^2}\left(\frac{\tilde{p}_e}{p_e} - \frac{e\tilde{\phi}}{T_e}\right) = V_a^2\nabla_\parallel^2\left(\frac{\tilde{p}_e}{p_e} - \frac{e\tilde{\phi}}{T_e}\right), \tag{8.14}$$

with the wave speed $V_a$ given by

$$V_a^2 = \frac{1 + \Gamma k_\perp^2\rho_s^2}{\beta_e + \mu_e k_\perp^2\rho_s^2}\, c_s^2 = \frac{1 + \Gamma k_\perp^2\rho_s^2}{1 + k_\perp^2 d_e^2}\, v_A^2, \tag{8.15}$$

where $d_e$ is the collisionless skin depth given by

$$d_e^2 = \frac{c^2}{\omega_{pe}^2} = \frac{m_e}{\mu_0 n e^2} = \frac{\mu_e}{\beta_e}\rho_s^2. \tag{8.16}$$

If we take the limit $k_\perp^2\rho_s^2 \to 0$, we find propagation at the Alfvén speed

$$\lim_{k_\perp^2 \to 0} V_a = v_A. \tag{8.17}$$

This limit corresponds to the single-fluid MHD torsional shear Alfvén waves, for which all scales propagate at $v_A$. We note the possibility of $k_\perp^2\rho_s^2 \to 0$ within the drift approximation, so long as we still satisfy $\nabla_\perp^2 \gg \nabla_\parallel^2$ and hence that parallel (shear) Alfvén motion is arbitrarily slow compared to perpendicular (compressional) Alfvén motion. If, on the other hand, we take $k_\perp^2\rho_s^2 \to \infty$, we find propagation at the electron thermal speed

$$\lim_{k_\perp^2 \to \infty} V_a = V_e, \tag{8.18}$$

corresponding to electrostatic dynamics with electromagnetic induction ($\tilde{A}_\parallel$) now negligible compared to electron inertia ($\tilde{J}_\parallel$).

The dependence of the wave propagation parallel to the magnetic field on the scale of motion perpendicular to the field is apparent in this dispersion relation, as shown in figure 8.4. Propagation along the field lines is non-dispersive, with $\omega = k_\parallel V_a$ for all $k_\parallel$ and $V_a$ dependent only on parameters other than $k_\parallel$. However, the propagation velocity $V_a$ is always between the two limits given by the Alfvén velocity, $v_A$, and the electron thermal velocity, $V_e$. These two velocities represent the limits of the fastest and slowest speeds at which kinetic shear Alfvén waves propagate, but which is the faster depends on whether $\beta_e$ is larger or smaller than $\mu_e$. The waves with the largest perpendicular scales propagate at $v_A$, reflecting ideal MHD behaviour, with $\tilde{\phi}$ and $\tilde{A}_\parallel$ interacting such that $E_\parallel = 0$, that is, the wave is an oscillation between the inductive and static parts of the electric field, or similarly between the parallel and polarisation currents. At the smallest perpendicular scales the waves propagate at $V_e$, reflecting electrostatic dynamics between $\tilde{p}_e$ and $\tilde{J}_\parallel$. The existence of this electrostatic limit is possible only within the drift model, as it results from the coupling between the electron pressure and the Alfvénic activity.

The corners of the effective dispersion relation $V_a(k_\perp)$ are of interest. The figure is drawn for $\beta_e/\mu_e = 10^{\pm 2}$ for the two cases. For $\beta_e < \mu_e$ the first corner left to right is at

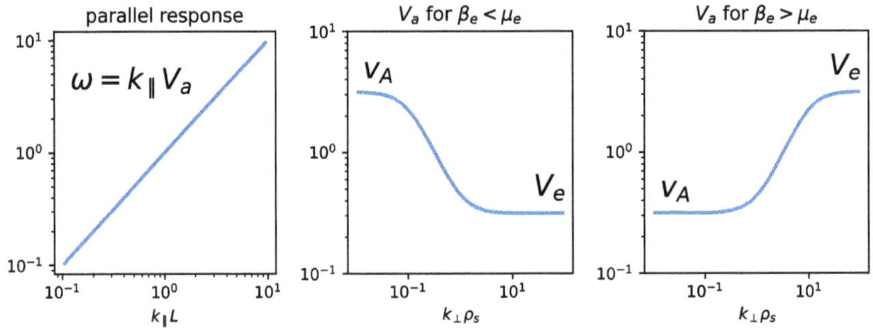


**Figure 8.4.** The kinetic shear Alfvén wave dispersion relation, showing $\omega = \pm k_\parallel V_a$ with the propagation velocity, $V_a$, as a function of the inverse perpendicular scale, $k_\perp$. For $\beta_e > \mu_e$ $V_a$ rises with $k_\perp$, and for $\beta_e < \mu_e$ it falls with rising $k_\perp$. In both cases, the large-scale limit is the Alfvén speed, $v_A$, reflecting ideal MHD between $\tilde{\phi}$ and $\tilde{A}_\parallel$, and the small-scale limit is the electron thermal speed, $V_e$, reflecting electrostatic dynamics between $\tilde{p}_e$ and $\tilde{J}_\parallel$. The speed of parallel electron dynamics is always between these two limits.

$k_\perp d_e = 1$ and the second is at $k_\perp \rho_s = 1$, and $v_A$ is higher than $V_e$ by a factor of ten here. For $\beta_e > \mu_e$ the first corner left to right is at $k_\perp \rho_s = 1$ and now it is the second at $k_\perp d_e = 1$, and $v_A$ is now lower than $V_e$ by a factor of ten here. In each case, the ratio $\rho_s / d_e$ is the same factor $\beta_e / \mu_e = 10^{\pm 2}$ as for $V_e / v_A$ for the velocities.

This is a different sort of dispersion relation than one usually sees. More conventional would be a variation of $V_a$ upon $k_\parallel$ in some form. But here we have $V_a$ dependent not upon $k_\parallel$ but upon $k_\perp$, which one can find in fluid dynamics in some cases (e.g. gravity waves in the shallow water equations). This dependence is a fundamental property of electromagnetic responses mediated by electron inertia in a magnetised plasma. Obviously, if electron inertia were neglected, in the more common assumption of massless electrons, one would not see all of these features, but at most the turn-up of $V_a$ with $k_\perp \rho_s$ as seen for $\beta_e > \mu_e$. On the other hand, the electrostatic limit sees only the ever increasing speed of the response as $k_\perp$ goes toward zero. Both are pathological, unless both inductivity and electron inertia are kept.

The two perpendicular scales, $\rho_s$ and $d_e$, therefore place limits on whether the single-fluid MHD model with massless electrons is applicable to the treatment of parallel dynamics. The skin depth merely brings in finite electron inertia to control the propagation speed, but the drift scale brings in the part of the dynamics which involves the electron thermodynamics through $\tilde{p}_e$. If $k_\perp^2 \rho_s^2$ is finite, then the interaction between the electron pressure and the compression along the guide field is important in distinguishing electron and ion dynamics. Plasma pressure does not couple to shear Alfvén dynamics in single-fluid MHD, but in general it does so.

The other two waves in the four-field model are found by assuming slower frequencies, such that the two kinetic shear Alfvén waves are not excited. We find this limit by assuming the state often referred to as 'adiabatic electrons'

$$\frac{\tilde{p}_e}{p_e} = \frac{e\tilde{\phi}}{T_e}, \tag{8.19}$$

which we encountered in Chapter 4. It results from the demand that the eigenfunctions of the kinetic shear Alfvén waves are at zero amplitude, hence that both sides of Eqs (8.11)–(8.13) vanish. The name 'adiabatic' refers to the quasistatic evolution of the electric fields, assuming much slower dynamics than the Alfvénic responses, with the electrons therefore progressing through successive states of force balance. We subtract Eq. (8.1) from Eq. (8.2) and insert Eq. (8.19) to find

$$\left(1 - \rho_s^2 \nabla_\perp^2\right) \frac{\partial}{\partial t} \frac{\tilde{p}_e}{p_e} = -\Gamma c_s \nabla_\parallel \frac{\tilde{u}_\parallel}{c_s}, \tag{8.20}$$

$$\frac{\partial}{\partial t} \frac{\tilde{u}_\parallel}{c_s} = -c_s \nabla_\parallel \frac{\tilde{p}_e}{p_e}, \tag{8.21}$$

having also assumed that $\tilde{J}_\parallel \to 0$ due to its coupling to departures from Eq. (8.19) via the kinetic shear Alfvén dynamics. Performing the Fourier decomposition of the $xy$-plane and combining the two equations into one, we find another wave equation,

$$\left(1 + \Gamma k_\perp^2 \rho_s^2\right) \frac{\partial^2}{\partial t^2} \frac{\tilde{p}_e}{p_e} = \Gamma c_s^2 \nabla_\parallel^2 \frac{\tilde{p}_e}{p_e}. \tag{8.22}$$

Propagation of these sound waves occurs at the velocity $V_s$ given by

$$V_s^2 = \frac{\Gamma}{1 + \Gamma k_\perp^2 \rho_s^2} c_s^2. \tag{8.23}$$

This reveals $c_s$ as the isothermal, cold-ion sound speed, since for isothermal electrons we would have $\Gamma = 1$. In this limit compression of the electron flow along the field lines still occurs, but only as a result of the finite polarisation current which the factor $k_\perp^2 \rho_s^2$ represents.

In the single-fluid MHD limit ($k_\perp^2 \to 0$) the sound waves and shear Alfvén waves propagate with velocities given by $V_s$ and $v_A$, respectively, with $c_s$ the isothermal limit of $V_s$. When $k_\perp^2 \rho_s^2$ becomes finite, both velocities are altered. The sound waves slow down with $k_\perp^2 \rho_s^2$, while the kinetic Alfvén waves go towards propagation at $V_e$ as both $k_\perp^2 \rho_s^2$ and $k_\perp^2 d_e^2$ become larger than unity (note that $V_e^2$ is $\beta_e/\mu_e$ times $v_A^2$, as $\rho_s^2$ is $\beta_e/\mu_e$ times $d_e^2$). In all cases, these two pairs of waves form a complete set of eigenmodes for the four-field model on a homogeneous background. The propagation speed in the parallel direction depends on $k_\perp$ rather than $k_\parallel$, giving the dispersion an interesting character: for propagation of waves within a thin flux tube, differing Fourier modes propagate at differing speeds, so that an initial structure will 'dephase' if its perpendicular extent is comparable to $\rho_s$. It is important to note that while the mathematics is suggestive of longitudinal wave propagation—scalar variables depending on $z$ and propagating along $\nabla z$—the ExB fluid motion and the magnetic disturbances have directions in the perpendicular plane. The scalar variables $\phi$ and $A_\parallel$ are only useful in describing shear Alfvén propagation if their perpendicular extent is limited; i.e. we require $\nabla_\perp^2 \gg \nabla_\parallel^2$. But that is the usual situation.

On fast time scales, with $\tau \sim (k_\parallel V_a)^{-1}$, the sound waves appear as zero frequency modes. The two quantities that only change slowly are $\tilde{u}_\parallel$, which is clear, and another quantity given by

$$\frac{\partial}{\partial t}\left(\frac{\tilde{p}_e}{p_e} - \Gamma \rho_s^2 \nabla_\perp^2 \frac{e\tilde{\phi}}{T_e}\right) = 0, \tag{8.24}$$

essentially representing pressure continuity including the ion polarisation drift. This is the third equation we find if we set $\tilde{u}_\parallel$ to zero in Eqs (8.1)–(8.4). Although $\tilde{p}_e$ and $\tilde{\phi}$ vary rapidly, through their difference $\tilde{p}_e - \tilde{\phi}$, the pressure and ExB vorticity always vary together. What this represents is a variation of the density (and temperature) in response to the divergence in the current carried by one of the species. For electrons it is the parallel divergence, $\nabla_\parallel \tilde{J}_\parallel$. For ions it is the polarisation divergence, which balances $\tilde{J}_\parallel$ such that the total $\nabla \cdot \mathbf{J}$ remains zero. For both species the divergences force the density and temperature equally—even if the amount is small by the factor $\rho_s^2\nabla_\perp^2$ the effect on the charge density is non-negligible and represents the vorticity/current pair of variables which is equivalent to the electrostatic/-magnetic potential or (ExB-)velocity/magnetic field pair of variables in the familiar MHD shear- (in this case torsional-) Alfvén wave. We will return to this point when discussing the transport of density/pressure by the electric current, below.

## 8.4 Energetics, dissipation

The four-field model has a relatively simple free energy theorem in the absence of background gradients. Its construction follows the same procedure as in Chapters 3 and 7. The total free energy density is

$$\frac{U_T}{p_e} = \frac{\mathbf{v}_E^2 + \tilde{u}_\parallel^2}{2c_s^2} + \frac{\tilde{\mathbf{B}}^2}{2\mu_0 p_e} + \frac{\mu_e}{2}\left|\frac{\tilde{J}_\parallel}{nec_s}\right|^2 + \frac{1}{2\Gamma}\left|\frac{\tilde{p}_e}{p_e}\right|^2. \tag{8.25}$$

Explicitly substituting the definitions of $\mathbf{v}_E$ and $\tilde{\mathbf{B}}$,

$$\mathbf{v}_E = \frac{1}{B}\nabla z \times \nabla\tilde{\phi} \qquad \tilde{\mathbf{B}} = -\nabla z \times \nabla\tilde{A}_\parallel, \tag{8.26}$$

this becomes

$$\begin{aligned}\frac{U_T}{p_e} = \frac{1}{2}\left|\rho_s\nabla_\perp\frac{e\tilde{\phi}}{T_e}\right|^2 + \frac{1}{2}\left|\frac{\tilde{u}_\parallel}{c_s}\right|^2 + \frac{1}{2}\beta_e^{-1}\left|\rho_s\nabla_\perp\frac{\tilde{A}_\parallel}{B\rho_s}\right|^2 \\ + \frac{\mu_e}{2}\left|\frac{\tilde{J}_\parallel}{nec_s}\right|^2 + \frac{1}{2\Gamma}\left|\frac{\tilde{p}_e}{p_e}\right|^2,\end{aligned} \tag{8.27}$$

in the language of fluid drift dynamics in physical units.

For purposes of clarity we simply re-express these scaled units as normalised quantities, and drop the tilde symbol. Since we are really doing isothermal dynamics, at this point we equate electron pressure with electron density through

$$\tilde{p}_e = T_e\tilde{n}_e \qquad \Gamma = 1, \tag{8.28}$$

also setting the ratio of specific heats to unity. The normalised quantities are then

$$n = \frac{\tilde{n}_e}{n_e} \qquad \phi = \frac{e\tilde{\phi}}{T_e}, \tag{8.29}$$

for the state variables, and

$$u = \frac{\tilde{u}_\parallel}{c_s} \qquad J = \frac{\tilde{J}_\parallel}{n_e e c_s} \qquad A = \frac{\tilde{A}_\parallel}{B\rho_s}, \tag{8.30}$$

for the flux variables.

The equations are normalised in terms of a periodicity length in the parallel direction given by $L_{\|} = 2\pi/K_{\|}$, regarded as the effective field line connection length. The unit of normalised time is set to $(K_{\|}c_s)^{-1}$, since in the neglect of background gradients $\rho_s$ and $d_e$ are the only important perpendicular scales for localised dynamics, i.e. there is no $L_{\perp}$ until we consider the background gradients later on. The dissipation goes over into normalised parameters denoted

$$\nu = \nu_e(K_{\|}c_s)^{-1} \qquad \eta = 0.51\mu_e\nu \qquad X = \mu_{\|}(K_{\|}c_s)^{-2}, \tag{8.31}$$

where we quote the input parameters in terms of $\nu$ and $X$ in this Chapter. The scaling with $T_e$ is given by the adiabatic response, and the scaling with $c_s$ is given by the acoustic dynamics, which results when the electrons are adiabatic. We become familiar with the wave-like character of the underlying responses. This is why we do not use an Alfvén normalisation as in pure MHD cases, let alone anything based on a collisional scale.

The linearised equations are

$$\frac{\partial\varpi}{\partial t} = \nabla_{\|}J, \tag{8.32}$$

$$\frac{\partial n}{\partial t} = \nabla_{\|}(J - u), \tag{8.33}$$

$$\frac{\partial}{\partial t}(A + \mu_e J) = \nabla_{\|}(n - \phi) - \eta\, J, \tag{8.34}$$

$$\frac{\partial u}{\partial t} = -\nabla_{\|}n + X\nabla_{\|}^2, \tag{8.35}$$

with $\nabla_{\|} = \partial/\partial z$ and with field equations

$$\varpi = \nabla_{\perp}^2\phi \qquad J = -\beta_e^{-1}\nabla_{\perp}^2 A. \tag{8.36}$$

For the free energy density we find

$$U_T = \frac{1}{2}\left(|\nabla_{\perp}\phi|^2 + n^2 + \beta_e^{-1}|\nabla_{\perp}A|^2 + \mu_e J^2 + u^2\right), \tag{8.37}$$

which satisfies

$$\frac{\partial U_T}{\partial t} + \nabla\cdot(\cdots) = -\eta\, J^2 - X\left|\nabla_{\|}u\right|^2. \tag{8.38}$$

The total free energy is the integral of $U_T$ over the spatial domain,

$$E_T = \mathscr{V}^{-1}\int d\mathscr{V}\; U_T, \tag{8.39}$$

defined as an average in computational cases (i.e. $E_T$ instead of the global total $\mathscr{E}$). In a periodic domain the divergence terms vanish under the integral. That is, the free energy decays according to resistive and viscous dissipation only, and there are no sources.

We can examine the various transfer effects by looking at the pieces of $\partial U_T/\partial t$ together. Variously, we find them by multiplying the equation for $\varpi$ with $-\phi$, the equation for $n$ with $n$, the equation for $A$ and $J$ with $J$, and the equation for $u$ with $u$. We note that the effective electron inertia is actually only one piece, due to Ampère's law,

$$\frac{\partial}{\partial t}(A + \mu_e J) = \left(1 - \frac{\mu_e}{\beta_e}\nabla_\perp^2\right)\frac{\partial A}{\partial t}. \tag{8.40}$$

These pieces of the free energy theorem appear as

$$\frac{\partial}{\partial t}\frac{1}{2} | \nabla_\perp \phi |^2 + \nabla \cdot \left(-\phi \nabla_\perp \frac{\partial \phi}{\partial t}\right) + \nabla_\parallel(\phi J) = J \nabla_\parallel \phi \tag{8.41}$$

$$\frac{\partial}{\partial t}\frac{n^2}{2} + \nabla_\parallel(nu - nJ) = u \nabla_\parallel n - J \nabla_\parallel n \tag{8.42}$$

$$\frac{\partial}{\partial t}\frac{1}{2}\beta_e^{-1} | \nabla_\perp A |^2 + \frac{\partial}{\partial t}\frac{1}{2}\mu_e J^2 + \nabla \cdot \left(\beta_e^{-1}\frac{\partial A}{\partial t}\nabla_\perp A\right) = J\nabla_\parallel n - J\nabla_\parallel \phi - \eta J^2, \tag{8.43}$$

$$\frac{\partial}{\partial t}\frac{u^2}{2} + \nabla_\parallel\left(-X\nabla_\parallel u\right) = -u\nabla_\parallel n - X\left| \nabla_\parallel u \right|^2, \tag{8.44}$$

including several terms which appear as total divergences. Terms appearing in two different equations with opposite sign represent transfer mechanisms, while terms appearing only once represent sources and sinks. The divergence terms represent transport of free energy due to fluxes along field lines, and wave fluxes (polarisation and induction) across them.

The four pieces of $E_T$ given in Eq. (8.39) are defined by

$$E_e = \mathcal{V}^{-1}\int d\mathcal{V}\, \frac{1}{2}\left(-\phi\varpi\right) \qquad E_b = \mathcal{V}^{-1}\int d\mathcal{V}\, \frac{1}{2}J\left(A + \mu_e J\right) \tag{8.45}$$

$$E_n = \mathcal{V}^{-1}\int d\mathcal{V}\, \frac{1}{2}n^2 \qquad E_u = \mathcal{V}^{-1}\int d\mathcal{V}\, \frac{1}{2}u^2 \tag{8.46}$$

for ExB, Ohmic, thermal, and parallel flow parts. The magnetic energy is the part of $E_b$ due to $JA$ and the parallel electron kinetic energy is the part due to $J^2$. Integrations of $\nabla_\perp^2$ by parts in $\varpi$ and $J$ in $E_e$ and $E_b$ yield the ExB and magnetic disturbance energies, respectively, using the field Eqs (8.36). The free energy sink rates are

$$\Gamma_c = \frac{1}{2E_T}\mathcal{V}^{-1}\int d\mathcal{V}\, \eta\, J^2 \qquad \Gamma_s = \frac{1}{2E_T}\mathcal{V}^{-1}\int d\mathcal{V}\, X\left| \nabla_\parallel u \right|^2 \tag{8.47}$$

defined (positive) as energetic damping rates. The growth rate and energetic error for this model is given by

$$\Gamma_T = \frac{1}{2E_T}\frac{\partial E_T}{\partial t} \qquad \Gamma_E = \Gamma_T - \Gamma_c - \Gamma_s. \tag{8.48}$$

These are used to elucidate computational examples given below.

There are three pathways of free energy coupling, all of which go through the terms with one factor of $\nabla_{\parallel}$. Two of them appear also in single-fluid MHD: the shear Alfvén coupling between $\phi$ and $J$ over $J\nabla_{\parallel}\phi$ and the sound wave coupling between $n$ and $u$ over $u\nabla_{\parallel}n$. As noted above, there is no direct coupling between $\phi$ or $J$ and $n$ in single-fluid MHD, due to the absence of $\nabla_{\parallel}n$ in Ohm's law and $\nabla_{\parallel}J$ in the pressure equation. But in drift dynamics these two terms create a third coupling, between $n$ and $J$ over $J\nabla_{\parallel}n$, which we can refer to as the kinetic shear Alfvén transfer mechanism. As $\phi$ and $J$ are already coupled through the basic shear Alfvén dynamics, this ultimately represents a coupling through the parallel electron dynamics between $\phi$ and $n$. This effective coupling mechanism is the one which mediates the adiabatic response, as noted in the Introduction. In terms of transport of thermodynamic quantities by the ExB velocity, this represents a coupling between the principal transported quantity and the advective velocity doing the transporting. As we will see in later Chapters, this gives fluid drift turbulence some very important distinctions when compared to such simple concepts as random mixing. Simply put, this kinetic shear Alfvén coupling is the hallmark feature of fluid drift dynamics in general, and drift-wave physics in particular.

## 8.5 Transient responses to a disturbance

We can further emphasise the kinetic shear Alfvén coupling between pressure and ExB flows by considering the transient response of the plasma to an initial pressure disturbance. We assume that $\beta_e \ll 1$ and $\mu_e \ll 1$, while $\nu$ and $X$ are finite. Our usual scenario for this is an initial disturbance in the electron density, localised in all three dimensions. The charge density is zero so the ions and electrons have the same initial disturbance. Due to the finite background electron temperature, this also constitutes an electron pressure disturbance, providing the initial force potential to launch the dynamics.

The initial disturbance $n(x, y, z)$ is set up such that $\nabla_{\perp}^2 \gg \nabla_{\parallel}^2$; that is, it is a flute mode disturbance localised in all three dimensions, with most of the dynamics occurring in a thin flux tube of radius $\Delta_{\perp}$ somewhat larger than $\rho_s$ and length $\Delta_{\parallel}$, dimensions basically set by those of the initial $n$. We set this up such that $n$ is the only nonzero disturbance; the other three dependent variables start at zero. A simple example is

$$n(x, y, z, t = 0) = a_0 \exp[-(x^2 + y^2)/\Delta_{\perp}^2]\exp\left[-z^2/\Delta_{\parallel}^2\right], \tag{8.49}$$

where $a_0$ is the initial amplitude. For a periodic domain the disturbance is considered well localised if $K_{\parallel}\Delta_{\parallel}$ is much larger than unity. On the other hand, if the scales are set by $\Delta_{\parallel}$ itself, it is reasonable to set $K_{\parallel}\Delta_{\parallel}$ to unity, redefining the parallel scale to that of the initial disturbance, for the linear response problem.

In single-fluid MHD the only response would be sound waves: $n$ forces $u$ and $u$ re-forces $n$. There would be no Alfvénic activity, as $\phi$ and $J$ would remain zero for all time. But what actually happens is that the finite pressure gradient forces the electron fluid such that $J$ rises away from zero, starting the shear Alfvén activity. It is important to note that with the initial finite $n$ we always get shear Alfvén waves, no matter what the parameters; basically, no matter how small $\rho_s^2\nabla_\perp^2$ is. If we are in the MHD regime because $\Delta_\perp \gg \rho_s$, the electron compression represented by $\nabla_\parallel J$ is a negligible force on $n$. However, the finite $n$ always launches the Alfvén waves, whose amplitude is given by the adiabatic response: $e\tilde{\phi}/T_e \sim \tilde{n}_e/n_e$ in physical units, or $\phi \sim n$ here. In this regime $n$ is not significantly affected and the end state after dissipation of the Alfvén dynamics is then the adiabatic electron state. Anything slower than the Alfvénic time scale would then proceed with adiabatic or near-adiabatic electrons. This is among the reasons that such ExB drift dynamics below the ion gyrofrequency in a magnetised plasma is always outside the validity of the single-fluid MHD equations.

Dissipation of the Alfvénic activity proceeds through resistivity ($\eta$) in the four-field model, while the sound wave response is damped by the parallel viscosity ($X$). The viscous damping rate is simply $X\nabla_\parallel^2$. The resistive damping rate is complicated by the presence of both terms on the left-hand side of Eq. (8.34). But if the damping can be treated as a perturbative correction to the wave frequency (small enough $\eta_\parallel$), the result for the damping rate of magnetic energy is

$$\gamma_b = -\frac{1}{2}\frac{\eta K^2}{\beta_e + \mu_e K^2}, \tag{8.50}$$

where $K = k_\perp\rho_s$, also assuming separation from the sound wave dynamics. The damping rate scales roughly as $k_\perp^2 d_e^2$ times $\nu_e$ in physical units (with $\eta = 0.51\mu_e\nu_e$, and then the factor of $\mu_e/\beta_e$ times $\rho_s^2$ giving $d_e^2$). If oscillations are forced at a given frequency $\omega$, the quantity $d_e(\nu_e/\omega)^{1/2}$ is often referred to as the collisional skin depth.

## 8.6 Numerical examples

We illustrate the combined action of kinetic shear Alfvén waves and sound waves by taking two example sets of parameters, and looking at large and small perpendicular scale responses for each. Large and small scales are taken to be $K = k_\perp\rho_s = 0.1$ and 1.0, respectively, for single Fourier components in the perpendicular plane. A value of $k_\perp\rho_s = 0.1$ might not sound like large scale, but since $K$ enters only through its square for these responses, it gives percent-level corrections, so that the large-scale regime is already reached. The parallel direction is assumed to be periodic on a length $2\pi/K_\parallel$. The initial state for the Fourier component $k_\perp$ is given by

$$n(\mathbf{k}_\perp) = \exp\left[-(K_\parallel z)^2\right], \tag{8.51}$$

and we solve Eqs (8.32)–(8.35) with $\nabla_\perp^2$ replaced by $-K^2$ and with $\Gamma = 1$. Since in this model $\nabla_\perp$ always appears in the form of $\nabla_\perp^2$, all coefficients of the equations are real and we can therefore use (e.g. for the electron density) $n$ and $n(\mathbf{k}_\perp)$ interchangeably, so long as we recall that we are keeping only one Fourier component.

For this study we discretise the equations on a uniform grid of 128 grid points in the $z$-direction with step size $h_z = 2\pi/128K_\parallel$, and set the time step to $10^{-4}$, somewhat smaller than $0.2h_z/V_e$, where $V_e = \mu_e^{-1/2}$ in these units. The numerical scheme is given by fourth-order centred differences for $\nabla_\parallel = \partial/\partial z$ and the stiffly stable time step described in Chapter 3.

We take two large-scale cases, with $K = 0.1$, showing what is nominally the MHD regime. A warm plasma case is defined such that $\beta_e > \mu_e$ and $\nu_e > K_\parallel v_A$ but not by much, so that the corrections enter appreciably. This parameter regime is appropriate to the outer regions of magnetically confined fusion plasma experiments. The nominal case is given by

$$\beta_e = 5 \times 10^{-4} \qquad \mu_e = 2.72 \times 10^{-4} \qquad \nu = 1000 \qquad X = 0.1. \tag{8.52}$$

A hot plasma case is defined such that $\beta_e > \mu_e$ by a reasonable margin with collisions at most a small correction. This parameter regime is appropriate to the core regions of magnetically confined fusion plasma experiments, and also to most natural situations where a fluid description can be used. With $K^2$ also small, it is usually considered as the nearly ideal MHD regime. The nominal case is given by

$$\beta_e = 10^{-2} \qquad \mu_e = 2.72 \times 10^{-4} \qquad \nu = 1.0 \qquad X = 0.1. \tag{8.53}$$

In both cases the value of $\mu_e$ reflects deuterium ions, which is a fusion application. To get the same cases for proton ions, replace the 2.72 factor with 5.45. The main differences between these two cases are the relative strength of the dissipation and the ratio $\beta_e/\mu_e$, which is assumed by the traditional MHD picture to be large but is not always so.

The resulting transient activity for the two cases is shown in figures 8.5–8.7. In each figure, the free energy pieces in Eqs (8.41)–(8.44) are shown in the left column and the total free energy, the resistive and sound wave damping rates, and the total damping rate and energetic error are shown in the central column. The right column has the amplitudes at $z = 0$ of $\phi$, $n$, $h = n - \phi$, and $\varpi$, respectively. The error ($\Gamma_E$, blue dashed at centre bottom), is very small in all cases.

On the fast time scale (to $t = 2$) for the hot case we find the initial Alfvénic response, with the ExB and magnetic energies rising to peak values within about $\Delta t = 0.3$. The complicated waveforms reflect the excitation of several $k_\parallel$ values. The resistive damping follows the magnetic energy, through $J$. The sound wave damping follows $u$. On the slow time scale (to $t = 20$) for the same case we find very weak damping of the Alfvén activity, due to the relatively high $\beta_e$ and small $K$. We find the sound wave activity damping actually faster, with the total energy damping unsteadily due to the oscillation of $u$. The amplitude of $h$ barely damps at all on these time scales. The collisional dissipation ($\Gamma_c$) rises with time because since the total energy ($E_T$) is falling, the slow damping of the magnetic energy becomes larger relative to the energy content.

For the warm plasma case (to $t = 2$) we find the faster Alfvénic activity damping resistively, while sound waves with adiabatic electrons survive for longer periods,

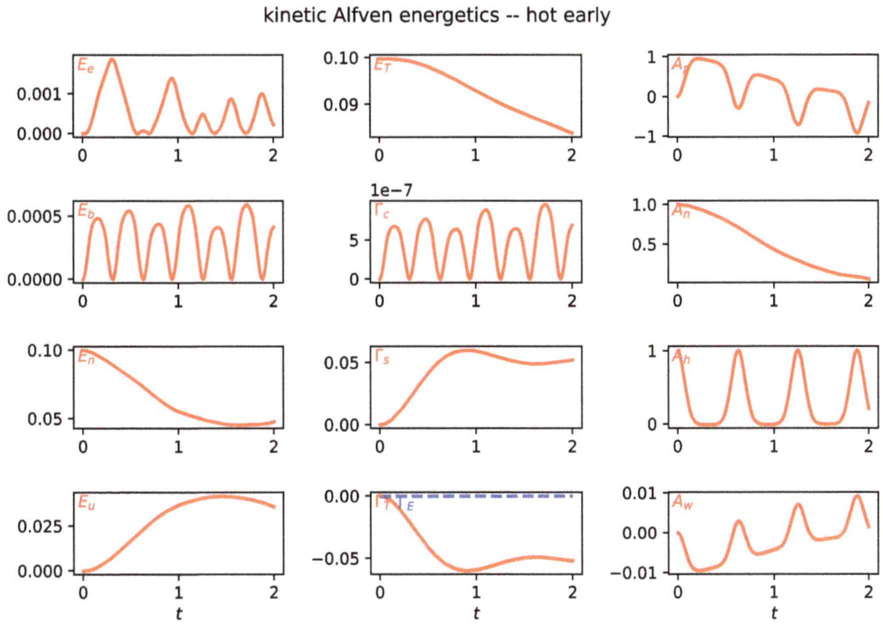


**Figure 8.5.** Time evolution of the energetic quantities in a kinetic shear Alfvén wave oscillation, for the hot plasma case with $K = 0.1$, on the Alfvén time scale. The potential and magnetic field oscillate together, and the sound waves begin to oscillate and damp. The quantities $E_e$, $E_b$, $E_n$, and $E_u$ give the contributions of $\phi$, $A$, $n$, and $u$ to the total energy ($E_T$), respectively. The damping rates $\Gamma_c$ and $\Gamma_s$ (as positive quantities) are due to parallel resistivity and viscosity, respectively. The total growth rate ($\Gamma_T$), negative for damping, is much larger in absolute than the error ($\Gamma_E$). The amplitudes are given at $z = 0$, not the integrated averages as in turbulence cases. Here, $\phi$ oscillates with the Alfvén response but in a slower envelope follows $n$ with the sound wave response. The amplitude of the nonadiabatic density component $h = n - \phi$ is the same as $n$ or $\phi$, such that the condition $\nabla p_e \ll n_e e \nabla\phi$ for MHD validity is never satisfied. The vorticity ($A_w$) is smaller due to the large perpendicular scale.

still viscously damped. The $K = 1.0$ cases are not shown, but they are similar except for one important fact: the oscillation between $n$ and the Alfvénic activity is stronger, with the result that $n$ is actually transported by this activity. We display this for the hot plasma case in the next Section.

In all cases we find that most of the damping in terms of energy is through parallel viscosity acting upon the sound waves ($\Gamma_s$). This is damping on $u$ and by extension on $n$. The Alfvén dynamics, however, proceeds independently of this, damped only through the resistivity. The overall damping rate follows the sound waves since with $K < 1.0$ the ExB flow energy is smaller than the pressure disturbance free energy even with similar amplitudes in $\phi$ and $n$. For the hot plasma case, this Alfvén damping rate is very small, due to the largeness of $\beta_e/\mu_e$, especially for $K = 0.1$ as the damping scales as $K$ squared. But here the energetic content in the Alfvén dynamics is very small.

The amplitude of the oscillations follows from the adiabatic response. These are measured at $z = 0$ by averaging the values at the two neighbouring grid points ($z = 0$ falls between points). Although falling and rising again mostly through the acoustic response, the potential follows it on the slow time scale, and the difference $h = n - \phi$

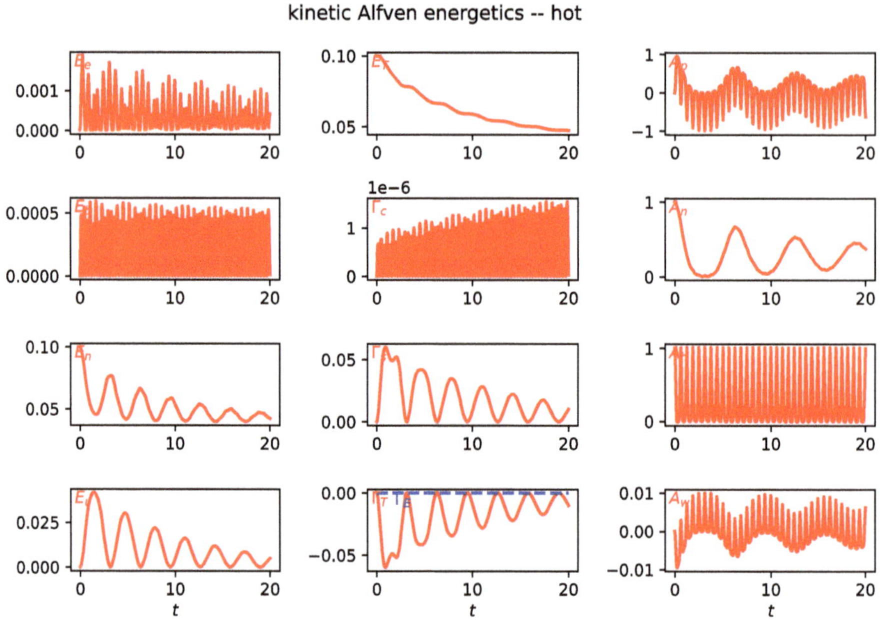


**Figure 8.6.** Time evolution of the energetic quantities in a kinetic shear Alfvén wave oscillation, for the hot plasma case with $K = 0.1$, as in figure 8.5, but on the sound wave time scale. The pressure and parallel ion velocity oscillate together as Alfvén waves proceed through many periods. Essentially all of the damping is due to the viscosity.

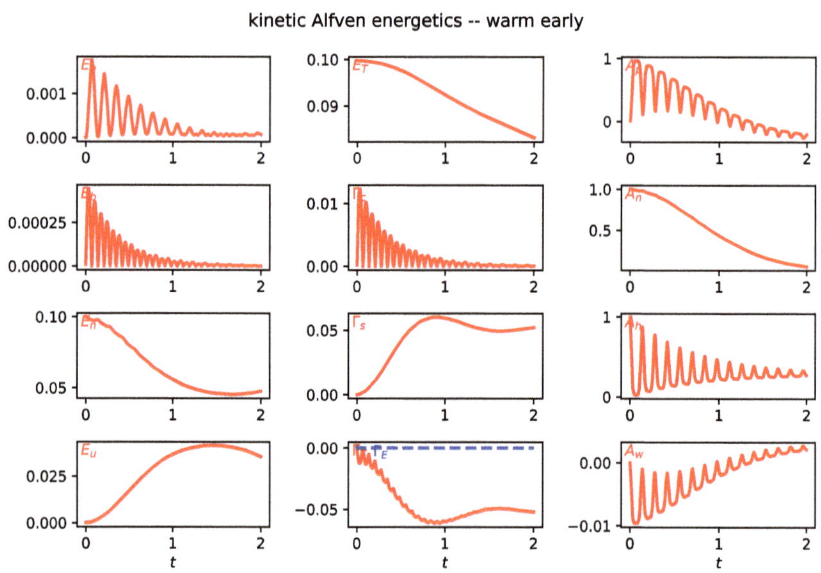


**Figure 8.7.** Time evolution of the energetic quantities in a kinetic shear Alfvén wave oscillation, for the warm plasma case with $K = 0.1$, on the Alfvén time scale. Although even here most of the damping is due to the viscosity, the resistivity is what damps the Alfvénic activity. $E_e$ and $A_p$ remain finite, as the sound waves evolve with adiabatic electrons, but $E_b$ drops away to zero.

oscillates with an amplitude of unity as the wave damps slowly (here, $K^2$ times $\mu_e/\beta_e$ is a little larger than $10^{-4}$). It should noted that it is impossible to get this result from the traditional MHD model without setting $\phi(t = 0)$ precisely to ensure it, since the MHD model does not contain the adiabatic response physics.

## 8.7 Energetics and decay rates

The energetics of the shear Alfvén wave oscillations deserves special attention. All four components are involved, and it is an important test of the numerics that this be properly demonstrated. Afterward, the decay rates will be checked. We have already seen in Chapter 4 that the dissipation of the parallel current by resistivity is an enabler for driven/dissipative turbulence. Checking the decay rates for homogeneous resistive shear Alfvén waves may be regarded as very basic, but this result will serve as a control test for the same exercise when going to a sheared magnetic field or to models with more detail in their physics. This is always the first thing to be tested any time one produces a new numerical scheme or a computation of a new model, for dynamics in magnetised plasmas. This part of the physics is always at the centre of any dynamics in turbulence in magnetised plasmas.

We take the hot plasma case from the previous Section. The dissipation channels are switched off by setting $\nu$ and $X$ to zero. Run to $t = 20$, the energy pieces and the total are given in figure 8.8. The dark line gives the total energy, which is conserved to high accuracy in the absence of dissipation. The four pieces $E_e$, $E_b$, $E_n$, and $E_u$, defined by in Section 8.4, are given separately. Since the amplitude of the wave is $\phi \sim n$ and $K = 0.1$ the ExB energy content and correspondingly also the magnetic energy is small. Most of the energetics is in the sound wave oscillation since this case is deep into the MHD regime. The same case with $K = 0.3$ is given in figure 8.9. Now all four pieces are visible and the Alfvén energy exchanges are noticeably faster. The same case with $K = 0.7$ is given in figure 8.10. Now the Alfvén exchanges are as large as the sound wave exchanges and are even faster. Under conditions of good energy conservation, the characteristics of the dependence of the shear Alfvén wave oscillation parallel to the guide field on the scale of motion across it are clear.

The shear Alfvén wave oscillation damping rate $\gamma_b$ is given in Eq. (8.50), independent of normalisation since physically $\gamma_b$ and $\nu_e$ are in the same units. The decay for the hot plasma case with $K = 0.3$ and a larger $\nu = 10$ is highlighted in figure 8.11. The magnetic energy $E_b$ is oscillatory between zero and a set of temporally local maxima whose envelope slowly decays. The maxima are taken into a secondary time series with $(t, 0.5 \log E_b)$ values, and then a linear fit is done. The slope gives the magnetic amplitude growth rate, $\gamma_b$, which is a decay rate when negative. The sound wave oscillation is seen superposed, so the envelope curve is noisy on a small amplitude. For this set of maxima, the decay rate is finitely calculable down to a few times $10^{-3}$, which is smaller than is relevant in turbulent cases. The value found here is $\gamma_b = -6.23 \times 10^{-3} \pm 9.57 \times 10^{-4}$, giving a fairly marginal case but one which is still well determined.

The scaling of $\gamma_b$ with respect to $\nu$ and $K$ is shown in figure 8.12. The integrity of the scaling holds down to the uncertainty of about $\gamma_b = 10^{-3}$ for these values—longer

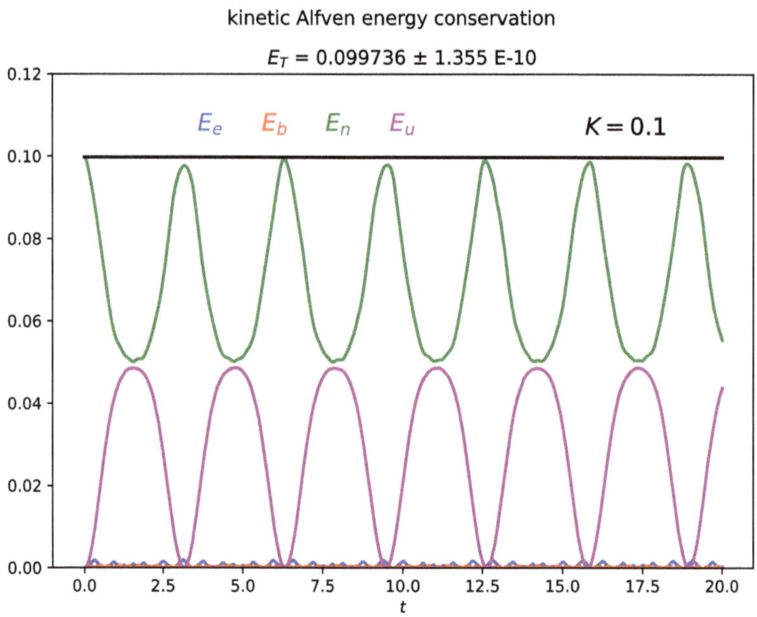


**Figure 8.8.** Energetics of the kinetic shear Alfvén wave oscillation, for the hot plasma case with $K = 0.1$ and all dissipation channels set to zero. The dark line and the top label give the accuracy of total energy conservation. The ExB, magnetic, thermal, and parallel momentum free energy channels are given by $E_e$, $E_b$, $E_n$, and $E_u$, respectively, colour coded as labelled. The energy content in the ExB and magnetic channels is small because the amplitudes of $\phi$ and $n$ are comparable but the energy carries the factor of $K^2$ in comparison. Several sound wave oscillations are visible.

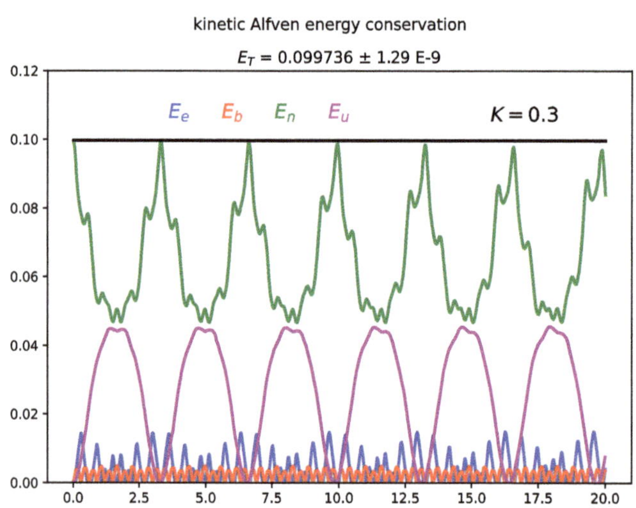


**Figure 8.9.** Energetics of the kinetic shear Alfvén wave oscillation, for the hot plasma case with $K = 0.3$ and all dissipation channels set to zero. Labels are as in figure 8.8. The energy content in the ExB and magnetic channels is visible since the $K^2$ factor is an order of magnitude larger. Due to the thermal coupling channel $n \leftrightarrow J$, which is responsible for the second term in $\beta_e + \mu_e K^2$, the shear Alfvén component is faster in this case.

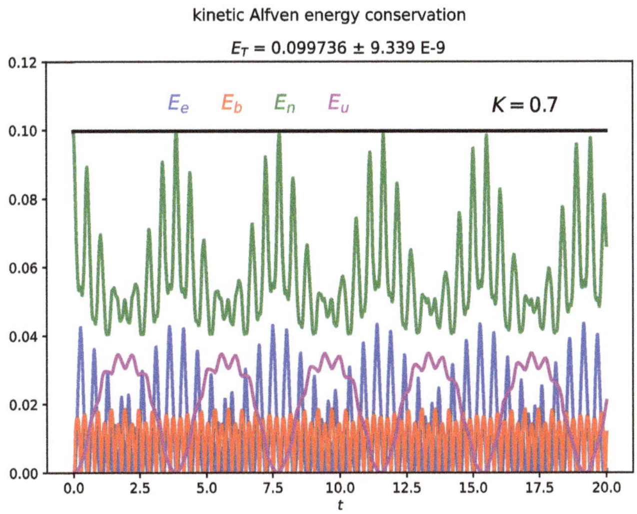


**Figure 8.10.** Energetics of the kinetic shear Alfvén wave oscillation, for the hot plasma case with $K = 0.7$ and all dissipation channels set to zero. Labels are as in figures 8.8 and 8.9. The shear Alfvén component is faster still, and its energy content is about half of the total by time average.

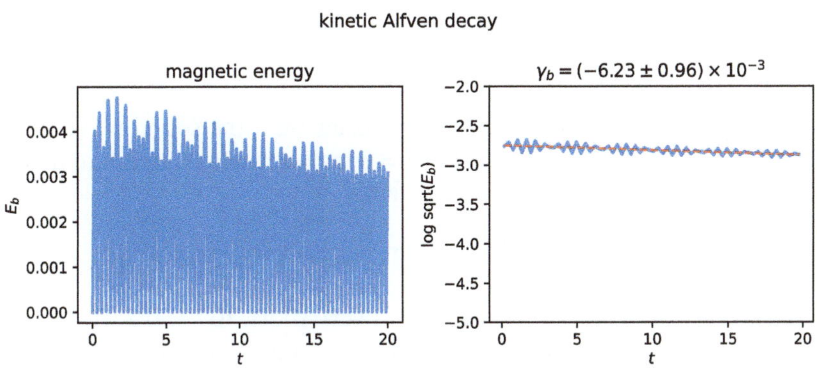


**Figure 8.11.** Decay of the kinetic shear Alfvén wave oscillation, for the hot plasma case with $K = 0.3$ and $\nu = 10$ in contrast to the base case in figure 8.5. The magnetic energy oscillates on both the shear Alfvén and acoustic oscillation frequencies as it decays. The peaks in $E_b(t)$ are sampled and fit to a log-linear decay line to give the growth rate $\gamma_b$ and its fitting uncertainty, with $\gamma < 0$ giving a decay rate.

time series are necessary in that case. For larger $\nu$ or $K$ the damping rate is too fast for this diagnostic, as we run out of a reasonable number of peaks, and more frequently sampled and shorter time series would be necessary. In each frame, the markers show the cases computed, and the solid line gives the comparison to Eq. (8.50), showing that the perturbative assumption on the decay versus the oscillation is well justified.

This exercise serves two purposes: the control case for the same study with more detail in physics or geometry, and a test on the numerics. We hold it in reserve for later versions as they arise.

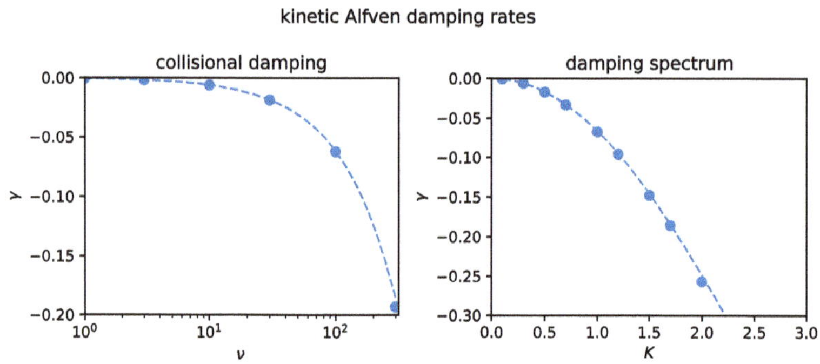


**Figure 8.12.** Scaling of the kinetic shear Alfvén wave damping rate with $\nu$ and $K$, for the hot plasma case around the values in figure 8.11. Markers show the cases computed, and the solid line gives the comparison to Eq. (8.50). Within the limits given by the finiteness of the diagnostic, the decay rate follows the perturbative scaling.

## 8.8 Thermal transport by the current

An interesting question posed by the foregoing results is the prospect that disturbances in the plasma density and pressure may be spread out along the magnetic field lines by this Alfvénic activity rather than by the sound waves. In MHD this process does not exist, since the plasma thermodynamical quantities behave in parallel dynamics just like a fluid. We can illustrate this contrast very well by taking the hot plasma case and restricting toward a pure kinetic shear Alfvén case by considering transient time scales much shorter than the sound wave transit. To watch a coherent wave packet depart the $z = 0$ region, we take an initial disturbance more tightly localised along the field line, with

$$p_e(\mathbf{k}_\perp) = \exp\left[-(K_\| z/0.3)^2\right] \qquad \text{or} \qquad n = \exp[-(z/0.3)^2], \tag{8.54}$$

with $n$ in normalised units. The run length is very short, to $t = 0.2$, just enough for the oscillation to spread out all the way to the periodicity length along the field lines. The resulting evolution for $K = 0.1$ is shown in figure 8.13, giving trace lines for the form of $\phi$, $n$, and $J$ in each time frame. Apart from the fact that it is a pressure disturbance doing the initial launching, we find the familiar shear Alfvén waves of MHD, with negligible effect on $n$. If $\rho_s$ takes a typical hot confined plasma value of 1 mm, and we consider the initial density disturbance as a pellet added to the background, the value $K = 0.1$ corresponds to a pellet diameter of about 3 cm. For $K = 0.3$ (pellet diameter 1 cm) we have a marginal case; the density/pressure disturbance changes only slightly, but we see a visible pulse in it which follows $\phi(\mathbf{k}_\perp)$, in other words, the Alfvén waves. Finally, the result for $K = 1.0$, corresponding to a small but not tiny pellet diameter of 3 mm, is displayed in figure 8.14. The lines for $J$ go off-scale since $\phi$ and $n$ are highlighted. We find about half the density disturbance taking direct part in the Alfvénic activity. The fraction scales roughly as $K^2/(1 + K^2)$, following from the absence of a $\nabla_\perp^2$ factor between $n$ and $J$ in Eqs (8.33) and (8.34). The wave is slightly faster due to the coupling between $J$ and $n$,

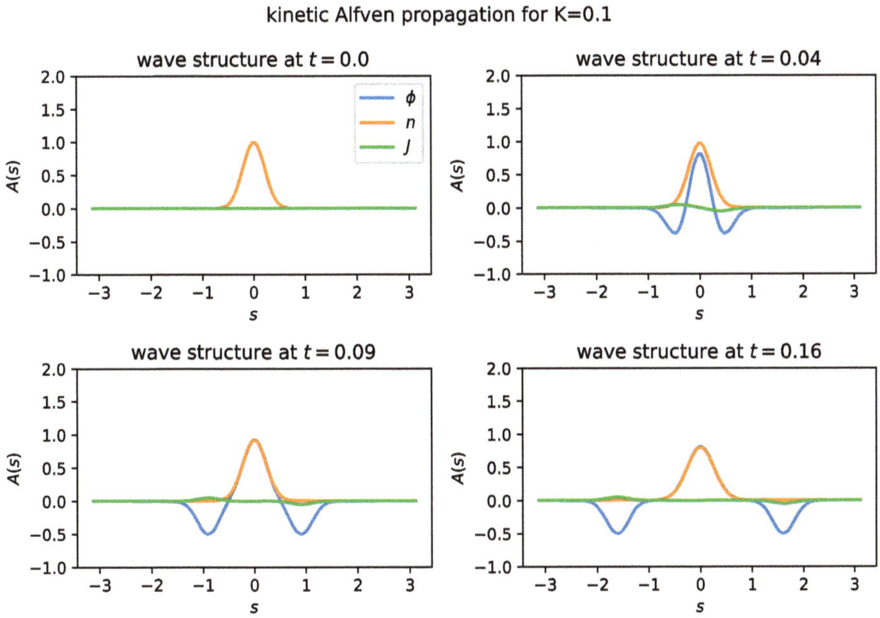


**Figure 8.13.** Wave motion in a kinetic shear Alfvén wave oscillation, for the hot plasma case with $K = 0.1$. The wave is excited by $n$, but most of the motion involves $\phi$.

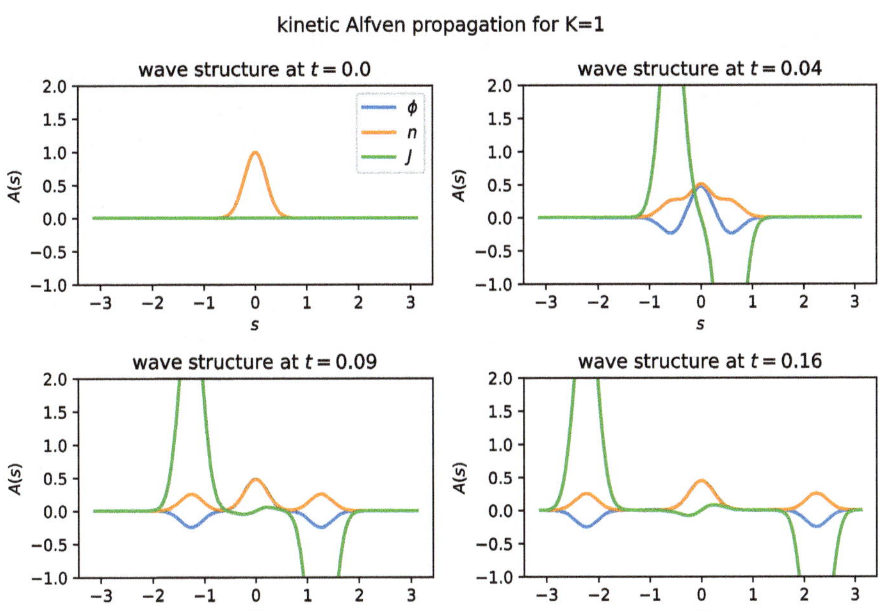


**Figure 8.14.** Wave motion in a kinetic shear Alfvén wave oscillation, for the hot plasma case with $K = 1.0$. The wave is excited by $n$, and now due to the thinness of the flux tubes about half of the pressure disturbance is involved in the Alfvén wave along with $\phi$. Note also that the wave speed is a little faster than the case in figure 8.13, due to the adiabatic coupling.

affecting the shorter perpendicular scales more strongly. On the short time scale represented in these two figures, the re-introduction of the sound waves was found to have negligible effect.

In the case with $K^2 \ll 1$, the response is closer to being MHD-like, but the manifestation of the adiabatic coupling is that the amplitude of $\phi$ is the same as for the initial $n$. Moreover, $\phi$ conforms to $n$ after the wave packets have departed. This is how the adiabatic response works for open or sheared magnetic field lines where no standing wave is formed. Even in the nominal MHD regime, in a magnetised plasma, a density disturbance launches an Alfvén wave transient, and the end state, especially in the zone where the wave was launched, is that of adiabatic electrons.

In the case with $K^2$ close to unity, the density/pressure spreads out via Alfvén wave activity, not sound waves. This is made possible by how small amounts of charge enter and leave the pieces of the flux tube at various positions along the $z$-axis. For a localised disturbance of the form in Eq. (8.49), we can consider the single flux tube it occupies. At $z = 0$ the initial response is a parallel current according to which electrons leave the region in both directions along $\mathbf{z}$. The parallel current divergence causes $\nabla \cdot \mathbf{E}$ to rise since the net charge density would tend to rise, but since the ion polarisability $\rho/B^2$ is much larger than $\epsilon_0$ the dynamics remains quasineutral. With the resulting polarisation current, the ions spiral out of the region in the $xy$-plane. While both species undergo circular ExB motion the ions have a finite radial slip, which is their polarisation drift keeping the total $\nabla \cdot \mathbf{J} = 0$. At positions on the $z$-axis at distance from $z = 0$ (up and down the magnetic field lines from the initial disturbance; consult figure 8.3), what happens is that electrons enter the region along the field lines ($J$) while ions spiral radially inward. This is the Alfvén activity represented by $\varpi$ and $J$ (hence $\phi$ and $A$, or $\mathbf{v}_E$ and $\mathbf{B}_\perp$) both responding to $n$. The mass flow follows the fate of $n = T_e\tilde{n}_e$, which is proportional to $\tilde{\rho} = m_i\tilde{n}_e$. If $\Delta_\perp > \rho_s$, then the mass flow follows sound waves, while if $\Delta_\perp < \rho_s$ it follows from the redistribution of ions due to the polarisation drift; i.e. ultimately from the kinetic Alfvén waves. All of this is decided by the relative sizes of the effective coefficients in front of the time derivatives of $\phi$ and $n$ in Eqs (8.32) and (8.33).

We note that there is no response at all to a pressure disturbance under the reduced MHD model neglecting sound waves, whose equations are found by neglecting $u$ and the two terms involved in the kinetic shear Alfvén coupling mechanism, leaving

$$\frac{\partial \varpi}{\partial t} = \nabla_\parallel J, \tag{8.55}$$

$$\frac{\partial n}{\partial t} = 0, \tag{8.56}$$

$$\frac{\partial A}{\partial t} + \mu_e \frac{\partial J}{\partial t} = -\nabla_\parallel \phi - \eta J. \tag{8.57}$$

If the current or electrostatic potential are not initially excited, there is no evolution in this system at all. Given the results just illustrated we therefore conclude that the responses to pressure disturbances are incorrectly predicted by conventional MHD at all scales, not just those near $k_\perp\rho_s \sim 1$.

## 8.9 Effects of temperature dynamics

We are deferring the discussion of temperature effects in general until later, because of some additional complications that require expansion of the scope of the model equations. However, we note that in the context of homogeneous parallel dynamics, the character of the wave dynamics does not change as a result of adding the temperatures. Rather, energy equipartition and the dissipation mechanisms are affected, which restricts the overall effect to the level of details: quantitative effects on both the wave speeds and the damping rates of kinetic shear Alfvén and sound waves. Without the details of dissipation, the electron temperature and the parallel heat flux satisfy equations which are similar to those for the density and current, and due to the separation $V_e \gg V_i \sim c_s$, the ion parallel dynamics does not enter appreciably. Even the finite gyroradius corrections only end in rendering the Alfvén dynamics to the electrostatic electron thermal regime below the collisionless skin depth. In the hot plasma regime, electron Landau damping takes over from resistivity as the main dissipation mechanism. This will be treated in detail in Volume Two.

The main point for now is that for $k_\perp\rho_i$ up to unity there are no qualitative effects on kinetic shear Alfvén waves and sound waves due to the physics of the temperatures. In particular, the transport of density/pressure by the Alfvén dynamics for $K \lesssim 1$ survives.

## 8.10 Summary

The drift approximation removes the compressional Alfvén waves of MHD but leaves the shear Alfvén waves alone. Applied solely to the MHD model, the drift approximation becomes the reduced MHD model, in which the shear Alfvén waves are mathematically described by $(\phi, A_\parallel)$ or as well $(\varpi, J_\parallel)$, instead of $(\mathbf{v}_E, \mathbf{B}_\perp)$, although their physics is the same. When two-fluid behaviour is retained, the drift approximation leads to the fluid drift dynamics model, which takes the pressure gradient into account in the electron force balance, forming the adiabatic response of the electrons. To conserve energy, the parallel current is retained in the parallel thermodynamics of the electrons, with the combination $(\tilde{u}_\parallel - \tilde{J}_\parallel/ne)$ describing the parallel electron motion. Correspondingly, the combination $\mathbf{v}_E + \mathbf{v}_*$, for which the scaled potential/pressure difference is the stream function, describes the perpendicular electron motion. The pressure forces and the difference between the electron and ion fluid velocities therefore go together, and the Alfvénic activity follows the total electron velocity, not the ion velocity or the ExB drift. Under drift dynamics, the electrons have an adiabatic response, and therefore disturbances in the electron pressure launch Alfvén waves. If $\beta_e \ll 1$, the resulting kinetic shear Alfvén waves oscillate on relatively fast time scales, while the sound wave activity is slower. The electron inertia is instrumental in preserving finite limits of the wave speed at either very low beta, $\beta_e \ll \mu_e$, or for $k_\perp\rho_s > 1$, perpendicular scales smaller than the drift scale. Given moderate resistivity, the Alfvénic activity damps, leaving the sound waves to oscillate with adiabatic electrons and dissipate viscously. If the magnetic background is inhomogeneous, $\nabla_\perp^2$ and $k_\parallel$ do not generally commute, but the physical character remains the same. If the collisionality is very low, then we have to treat the temperatures and their associated kinetic dissipation mechanisms.

## Further Reading

Kinetic shear Alfvén waves were described by A B Mikhailovskii 1974 *Theory of Plasma Instabilities* (New York: Consultants Bureau). They were described using kinetic theory by A Hasegawa and L Chen 'Kinetic processes in plasma heating by resonant mode conversion of Alfvén wave' *Phys. Fluids* **19** (1976) 1924. Kinetic shear Alfvén waves as transients to instabilities and turbulence were discussed in these references, and by K T Tsang, J C Whitson, J D Callen, P J Catto and J Smith 'Drift Alfvén waves in tokamaks' *Phys. Rev. Lett.* **41** (1978) 557; and S Camargo, B Scott and D Biskamp 'The influence of magnetic fluctuations on collisional drift-wave turbulence' *Phys. Plasmas* **3** (1996) 3912. The wave energetics was discussed by B Scott 'Three-dimensional computation of drift Alfvén turbulence' *Plasma Phys. Control. Fusion* **39** (1997) 1635.

**IOP** Publishing

Turbulence and Instabilities in Magnetised Plasmas, Volume 1
Fluid drift turbulence
**Bruce Scott**

# Chapter 9

# Perpendicular dynamics—drift waves

## 9.1 Introduction

In studying the part of fluid drift dynamics which involves wave propagation parallel to the magnetic field, we have already encountered one very subtle part of the dynamics which takes place in the plane locally perpendicular to the magnetic field. This is the polarisation drift. When other perpendicular drifts, with the exception of the ExB velocity, are absent, the polarisation drift is responsible for allowing parallel currents which can have a divergence. Charge neutrality is then maintained by a balance between the divergences of these two currents, with the reaction of the polarisation drift to the other divergences controlling the evolution of the perpendicular electric field, and hence the ExB velocity.

In this Chapter we are also concerned by the effects of advection by the ExB velocity. In a homogeneous background the ExB advection is a purely nonlinear effect given by advection of a disturbance by a velocity whose stream function is also a disturbance. When there is a background gradient, however, we encounter linear forcing terms arising from the advection of the background gradient by the ExB disturbances. Flows down the gradient of a given quantity drive disturbances in that same quantity (e.g. the pressure). The interaction between this perpendicular ExB advection and the self-consistent adiabatic response parallel to the magnetic field forms the dynamical structure which we call the drift wave. Magnetic disturbances in the presence of the gradient also invoke drive terms, which can have de- or stabilising effects according to how the various components of the dynamics are relatively phased. Drift wave turbulence is what transpires when the advection is strong enough that the nonlinear and linear effects are comparable; we treated an introduction to this in Chapter 4. Herein, we treat drift waves as linear modes of oscillation, or as instabilities, which occur in a magnetised plasma in which there is a pressure gradient and a finite parallel scale. When unstable, they give a forcing mechanism to the turbulence as studied in Chapter 4.

 

## 9.2 ExB advection in a gradient—the drift frequency

The simplest process by which advection in a background gradient drives the associated disturbance is given simply by the model in which the total ExB advection of that quantity is zero:

$$\frac{dp_e}{dt} = \frac{\partial p_e}{\partial t} + \mathbf{v}_E \cdot \nabla p_e = 0. \tag{9.1}$$

If we linearise this equation, replacing $p_e$ by the background $p_e$ and the disturbance $\tilde{p}_e$, we neglect the action of $\mathbf{v}_E \cdot \nabla$ on $\tilde{p}_e$, and note that with a fixed background we advance only the disturbance. This says that

$$\frac{\partial \tilde{p}_e}{\partial t} = -\mathbf{v}_E \cdot \nabla p_e, \tag{9.2}$$

or in keeping with our procedure with $p_e$ a background parameter,

$$\frac{1}{p_e}\frac{\partial \tilde{p}_e}{\partial t} = -\mathbf{v}_E \cdot \nabla \log p_e. \tag{9.3}$$

A sketch of the process is shown in figure 9.1. We set a right-handed coordinate system $\{x, y, z\}$ such that

$$\mathbf{B} = B\nabla z \qquad \nabla \log p_e \equiv -\frac{1}{L_p}\mathbf{v}_E \cdot \nabla x, \tag{9.4}$$

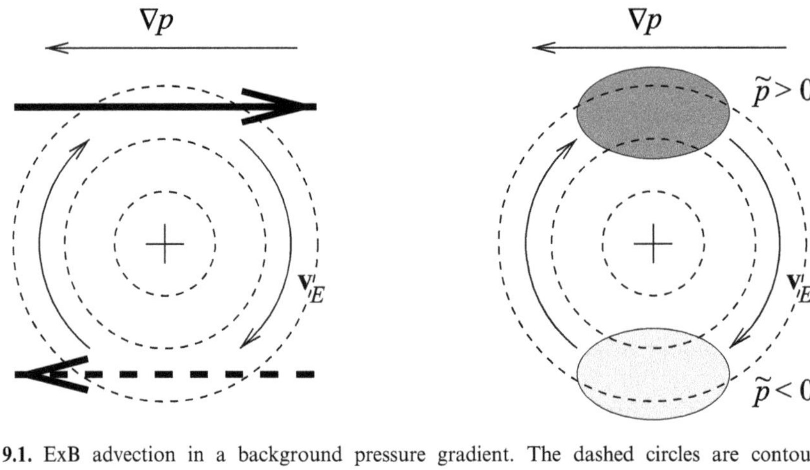


**Figure 9.1.** ExB advection in a background pressure gradient. The dashed circles are contours of $\tilde{\phi}$, representing a localised ExB flow vortex, clockwise around the potential maximum. The ExB flow, $\mathbf{v}_E$, is shown by the solid arrow arcs, and where it is down or up the gradient it is shown (left frame) by the thick solid and dashed arrows, respectively. These flows excite the pressure disturbances, $\tilde{p}_e$, positive where the flow is down-gradient, and negative where the flow is up-gradient (right frame). The resulting $\tilde{p}_e$ is driven on the leading edge of the $\tilde{\phi}$ maximum, in the drift direction (towards the top of the picture). For a vortex wave train, the pressure is therefore driven with a phase ahead of the potential. Spatially, the magnetic field **B** is out of the plane, toward the observer.

with $B$ and $L_p$ constants. The ExB velocity in the $x$-direction is

$$\mathbf{v}_E \cdot \nabla x = \frac{1}{B^2}\mathbf{B} \times \nabla\tilde{\phi}\cdot\nabla x = -\frac{1}{B}\frac{\partial\tilde{\phi}}{\partial y}, \tag{9.5}$$

which is what distinguishes the two perpendicular dimensions, establishing $y$ as a preferred direction. Scaling the disturbances in terms of $\tilde{p}_e/p_e$ and $e\tilde{\phi}/T_e$ (electron pressure, temperature, and fundamental charge), we have the evolution of the pressure disturbance,

$$\frac{\partial}{\partial t}\frac{\tilde{p}_e}{p_e} = -\frac{T_e}{eBL_p}\frac{\partial}{\partial y}\frac{e\tilde{\phi}}{T_e}. \tag{9.6}$$

This is the usual expression in any situation where the adiabatic response of the electrons is relevant, tending to keep $e\tilde{\phi}/T_e$ close to $\tilde{p}_e/p_e$. It leads to the typical time scale associated with drift dynamics in which the source of free energy is forced by the background gradient. One expression of the definition of $\rho_s$, besides as an ion inertial gyroradius $c_s/\Omega_i$, is as the scale which combines with the sound speed to produce $T_e/eB$ as a diffusion coefficient; i.e.

$$\rho_s c_s = \frac{T_e}{eB} \qquad \text{since} \quad c_s^2 = \frac{T_e}{m_i} \quad \text{and} \quad \rho_s^2 = \frac{T_e m_i}{e^2B^2}, \tag{9.7}$$

in which the drift scale $\rho_s$ and sound speed $c_s$ appear with the electron temperature in combination with the ion mass, reflecting electron mobility and ion inertia as before. Combining the factor of $\rho_s$ with the $y$-derivative, we obtain the background drift frequency, $c_s/L_p$. For a particular wavenumber $k_y$, we have the frequency with the associated wavelength $2\pi/k_y$,

$$\omega_* = \frac{T_e k_y}{eBL_p} = \frac{k_y\rho_s c_s}{L_p}. \tag{9.8}$$

This is called the 'diamagnetic frequency,' due to the fact that the wave propagation velocity multiplying $k_y$ is simply the diamagnetic drift velocity given by

$$v_D = \frac{T_e}{eBL_p} = \frac{\rho_s c_s}{L_p} = |\,\mathbf{v}_*\,|. \tag{9.9}$$

Motion of this type represents a drift in the $y$-direction, which is the electron drift direction, $-\mathbf{B} \times \nabla p_e$, given by the orientation of the $\mathbf{v}_*$ due to the background gradient. This is the diamagnetic drift from Chapter 6.

## 9.3 Drift waves—the very simplest model

The foregoing considerations give rise to the picture of drift waves one most often finds in textbooks. We know that there is an adiabatic response in the electrons parallel to the magnetic field, but we discard the details and simply assume that it takes place on a much faster time scale than that of the diamagnetic drift. We also have to assume that the effects of parallel ion dynamics are negligible, and so the diamagnetic drift must in turn take place much faster than sound wave time scales. We therefore assume a nonvanishing parallel wavenumber, $k_\parallel$, and demand that

$$k_\parallel V_a \gg \omega \sim \omega_* \gg k_\parallel V_s, \tag{9.10}$$

where $V_a$ and $V_s$ are the generalised Alfvén and sound speeds given by

$$V_a^2 = c_s^2 \frac{1 + k_\perp^2 \rho_s^2}{\beta_e + \mu_e k_\perp^2 \rho_s^2} \qquad V_s^2 = c_s^2 \frac{1}{1 + k_\perp^2 \rho_s^2}, \tag{9.11}$$

encountered in the previous Chapter. We note that $V_a$ is used here rather than the electron thermal velocity $V_e$ as written in some texts ($V_e^2 = T_e/m_e$ is the high-$k_\perp$ limit of $V_a^2$). The relevant velocity is $V_e$ only when the adiabatic response is purely electrostatic, mediated by electron inertia. This would require $v_A > V_e$, or $\beta_e < \mu_e k_\perp^2 \rho_s^2$, equivalently $k_\perp^2 d_e^2 > 1$, which does enter relevant cases but only at the short wavelength end of the spectrum. In any case, however, if we can say that the time scale for the adiabatic response is arbitrarily fast, *and* the parallel gradient does not vanish, we may replace further consideration of the parallel dynamics with the condition that the electrons be adiabatic:

$$\frac{e\tilde{\phi}}{T_e} = \frac{\tilde{p}_e}{p_e} = \frac{\tilde{n}_e}{n_e} \qquad \text{(adiabatic electrons).} \tag{9.12}$$

The last equality is due to the fact that the parallel electron dynamics also relaxes the temperature disturbances, and if it is arbitrarily fast then $\tilde{T}_e$ will go to zero. At this point, electron pressure and density become synonymous, as in the four-field model with $\Gamma = 1$. Specification of the details of the parallel dynamics is very vague, no more involved than the simple arguments just given. Among other things, the polarisation drift of the ions is neglected, and quasineutrality is guaranteed by having the electrons adiabatic, so that not only is the current divergence free, it is also zero.

Advection of the background density by ExB disturbances under these conditions leads to the very simple dispersion relation,

$$\omega = \omega_* = k_y v_D, \tag{9.13}$$

which we find by substituting the adiabatic state, Eq. (9.12), into the advection equation, Eq. (9.6), yielding

$$\left(\frac{\partial}{\partial t} + v_D \frac{\partial}{\partial y}\right) \frac{e\tilde{\phi}}{T_e} = 0, \tag{9.14}$$

and taking the appropriate Fourier transforms in space and time. The physical picture of this is very simple. The disturbances have spatial variation in at least two directions, $y$ and $z$, the drift and parallel directions, respectively. At this level of simplicity, any $x$-dependence has no effect. Even the $z$-dependence makes no difference, except for the prescription that the ordering in Eq. (9.10) be maintained. The simplest case is a sinusoidal wave in the $y$-direction, a sketch of which is shown in figure 9.2. At all places where there is $y$-variation in $\tilde{\phi}$, the pressure disturbance is driven. On the 'leading edge' of the wave, the ExB flow in the $x$-direction is positive, that is, down the gradient, and $\tilde{p}_e$ is driven positive there. Since the adiabatic response is effectively instantaneous, $\tilde{\phi}$ follows $\tilde{p}_e$ and the structure simply propagates in the $y$-direction with the diamagnetic drift velocity. It is important to note that this propagation does not result from a proper wave equation; Eq. (9.14) is only a first order equation, so there is no counterpart propagating in the opposite direction. An initial disturbance simply moves unidirectionally.

## 9.4 Drift waves—polarisation and dispersion

There are two principal features of the very basic drift wave model: there is no instability, because the frequency is purely real, and there is no dispersion, because the propagation velocity is independent of the wavelength. Systems which have this property are called 'self similar' because there is no preferred scale of motion as long as the assumptions and restrictions leading up to the model are satisfied. However, we have already seen that in the parallel dynamics, both for the adiabatic (kinetic shear Alfvén) and sound wave responses, the wave speeds for *parallel* motion depend on the *perpendicular* wavenumbers. This is still true for purely adiabatic sound waves, so we ought to watch out for its emergence for drift waves. This is done by taking the polarisation current into account. Here, we assume that the parallel current is not zero, although it is still negligible enough that we can assume adiabatic electrons. But the small but finite parallel divergence now requires the finite polarisation current.

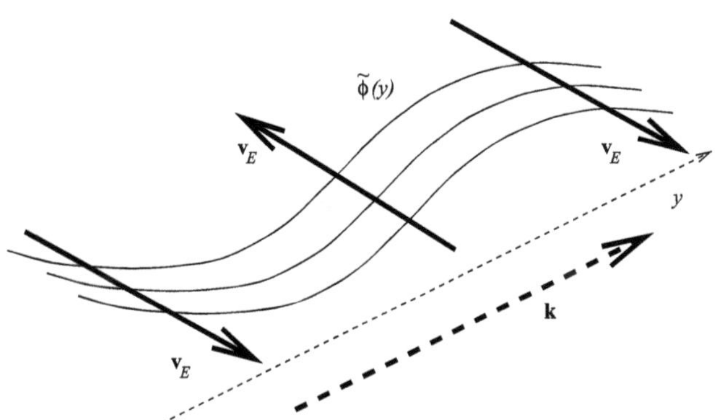


**Figure 9.2.** Sketch of a simple one-dimensional drift wave. The implied vertical axis is $\tilde{\phi}$, and the sinusoidal curves represent a wave train $\tilde{\phi}(y)$. The ExB flow, $\mathbf{v}_E$, is up or down the pressure gradient according to the direction of the thick solid arrows. Flow to the right is down-gradient, and there the pressure disturbance tends to be excited positive. But the arbitrarily fast adiabatic response ties $\tilde{\phi}$ back to $\tilde{p}_e$, giving a drift propagation in the $y$-direction, with $\mathbf{k} = k_y \nabla y$. Spatially, $\mathbf{B}$ is in the direction of the vertical axis, that is 'up' in this orientation.

This is a second version of drift waves one can encounter in textbooks: we start with the equation for the ion density in a plasma with a density gradient but a homogeneous magnetic field. With **B** constant we have $\nabla \cdot \mathbf{v}_E = 0$ and no diamagnetic flux effects, but there is still the polarisation drift to watch out for. We have

$$\frac{\partial n_i}{\partial t} + \mathbf{v}_E \cdot \nabla n_i + \nabla \cdot n_i \mathbf{u}_p = 0, \tag{9.15}$$

where now the polarisation drift, given by

$$n_i \mathbf{u}_p = -\frac{n_i m_i}{eB^2}\left(\frac{\partial}{\partial t} + \mathbf{v}_E \cdot \nabla\right)\nabla_\perp \phi, \tag{9.16}$$

(here neglecting the parallel ion motion), breaks the situation of pure ExB advection. Linearising these equations and applying drift ordering, by which the factors in Eq. (9.16) are taken as constant, we find

$$\frac{\partial}{\partial t}\frac{\tilde{n}_i}{n_i} + v_D \frac{\partial}{\partial y}\frac{e\tilde{\phi}}{T_e} - \rho_s^2 \nabla_\perp^2 \frac{\partial}{\partial t}\frac{e\tilde{\phi}}{T_e} = 0. \tag{9.17}$$

Applying quasineutrality, we replace $\tilde{n}_i$ by $\tilde{n}_e$, that is, in the disturbances as well as the background. Finally, we prescribe adiabatic electrons, Eq. (9.12), in this case also, to find

$$\left[\left(1 - \rho_s^2 \nabla_\perp^2\right)\frac{\partial}{\partial t} + v_D \frac{\partial}{\partial y}\right]\frac{e\tilde{\phi}}{T_e} = 0. \tag{9.18}$$

The only difference between this and Eq. (9.14) is the appearance of the factor of $\rho_s^2 \nabla_\perp^2$. If we again take the appropriate Fourier transforms in space and time, we find the modification to the dispersion relation of Eq. (9.13) caused by the polarisation drift,

$$\omega = \frac{\omega_*}{1 + k_\perp^2 \rho_s^2} = \frac{c_s}{L_p}\frac{k_y \rho_s}{1 + k_\perp^2 \rho_s^2} \equiv \omega_L. \tag{9.19}$$

This is the origin of the name of $\rho_s$ as the 'drift wave dispersion scale.' At this scale the propagation starts to slow down, with $\omega_L$ giving the basic drift wave frequency and representing the slowing of the time scale as $k_\perp \rho_s$ approaches unity. Otherwise the physical picture does not change from that of the previous Section; we have simply been slightly more careful in setting up the model. If $\rho_s$ is arbitrarily small, we recover the simplest self-similar model.

In both of these simple models, we still have the problem that the result describes only pure propagation; there are no energetics or instability. For each Fourier component **k** there is a reference frame one can transform to, using $x' = x - V_\mathbf{k} t$ for appropriate velocity $V_\mathbf{k}$, in which even $\partial/\partial t$ vanishes. Nonlinearly there is no magnetic flutter with electrostatic responses and no ExB nonlinearity since $\mathbf{v}_E \cdot \nabla\phi$ (hence $[\phi, \phi]$; see Chapter 4) vanishes. This means that there is no concept of energetics. Even the dissipation-free transfer phenomena require a more careful description of the dynamics than simple specification of adiabatic electrons; we have already encountered some of this in contemplating the kinetic shear Alfvén waves. Fortunately, having done that already, it is a simple step to arrive at a complete description of drift waves, not only how their structure is established and maintained, but also how free energy is tapped from the background and then transferred through the dynamical system until it ends at dissipation. All we have to do is to add the extra phenomenon of ExB nonlinearity advection in a gradient to what we have already done in deriving the kinetic shear Alfvén waves.

## 9.5 Drift waves—self-consistent dynamics

### 9.5.1 Perpendicular forcing, parallel response

In order to understand the drift wave structure, one has to investigate how it is initially established. The adiabatic electron state has to be maintained, even if it is the initial state. More generally, the departures from the adiabatic state must be found by considering the self-consistent electron dynamics following from the adiabatic response. We have already seen, however, that adiabatic electrons are the end state of dissipative Alfvén activity, regardless of how the system is initialised. So we know that the time scale on which the adiabatic state is itself established is that on which the Alfvén waves die out. Here, we investigate the ExB advection and kinetic shear Alfvén waves interacting together, and show that this is the mechanism by which drift waves are set up and maintained. We simply add the extra terms arising from the background gradient to the linearised four-field model we have already seen in Section 8.2 of the previous Chapter. There are two such terms, one from the ExB advection of the gradient,

$$\mathbf{v}_E \cdot \nabla \log p_e = -\frac{1}{L_p}\frac{1}{B^2}\mathbf{B} \times \nabla\tilde{\phi}\cdot\nabla x = \frac{c_s}{L_p}\rho_s\frac{\partial}{\partial y}\frac{e\tilde{\phi}}{T_e}, \tag{9.20}$$

and one from the parallel derivative of the quantity with a gradient,

$$\frac{1}{B}\mathbf{B}_\perp \cdot \nabla \log p_e = \frac{1}{L_p}\frac{1}{B^2}\mathbf{B} \times \nabla\tilde{A}_\parallel \cdot \nabla x = -\frac{c_s}{L_p}\rho_s\frac{\partial}{\partial y}\frac{\tilde{A}_\parallel}{B\rho_s}. \tag{9.21}$$

If we add these two terms to the linearised four-field model where appropriate, we have

$$\frac{\partial}{\partial t}\rho_s^2\nabla_\perp^2\frac{e\tilde{\phi}}{T_e} = c_s B\nabla_\parallel\frac{\tilde{J}_\parallel/nec_s}{B}, \tag{9.22}$$

$$\frac{\partial}{\partial t}\frac{\tilde{p}_e}{p_e} + \frac{c_s}{L_p}\rho_s\frac{\partial}{\partial y}\frac{e\tilde{\phi}}{T_e} = \Gamma c_s B\nabla_\parallel\left(\frac{\tilde{J}_\parallel/nec_s}{B} - \frac{\tilde{u}_\parallel/c_s}{B}\right), \tag{9.23}$$

$$\frac{\partial}{\partial t}\frac{\tilde{A}_\parallel}{B\rho_s} + \mu_e\frac{\partial}{\partial t}\frac{\tilde{J}_\parallel}{nec_s} = c_s\nabla_\parallel\left(\frac{\tilde{p}_e}{p_e} - \frac{e\tilde{\phi}}{T_e}\right) - \frac{c_s}{L_p}\rho_s\frac{\partial}{\partial y}\frac{\tilde{A}_\parallel}{B\rho_s} - 0.51\mu_e\nu_e\frac{\tilde{J}_\parallel}{nec_s}, \tag{9.24}$$

$$\frac{\partial}{\partial t}\frac{\tilde{u}_\parallel}{c_s} = -c_s\nabla_\parallel\frac{\tilde{p}_e}{p_e} + \frac{c_s}{L_p}\rho_s\frac{\partial}{\partial y}\frac{\tilde{A}_\parallel}{B\rho_s} + \mu_\parallel\nabla_\parallel^2\frac{\tilde{u}_\parallel}{c_s}, \tag{9.25}$$

with Ampère's law

$$\frac{\tilde{J}_\parallel}{nec_s} = -\beta_e^{-1}\rho_s^2\nabla_\perp^2\frac{\tilde{A}_\parallel}{B\rho_s}, \tag{9.26}$$

in the same language as before.

We note that the effective frequency represented by the gradient forcing terms is simply

$$\frac{c_s}{L_p}\rho_s\frac{\partial}{\partial y} \to \frac{c_s}{L_p}k_y\rho_s = \omega_*, \tag{9.27}$$

that is, the diamagnetic frequency defined in Eq. (9.8). So these terms add $\omega_*$ to the mix of frequencies which also includes $k_\parallel V_a$ and $k_\parallel V_s$, the two-fluid shear Alfvén and sound wave frequencies, and $\nu_e$, the electron collision frequency representing resistive dissipation of the waves, the main channel to dissipation along with the ion viscosity. This is the first time we encounter the basic competition between parallel constraining and perpendicular excitation in a three-dimensional setting in a magnetised plasma. It is related to the role of the coupling parameter in the two-dimensional drift wave turbulence model, seen in Chapter 4. In this study of drift wave dynamics, we see its origin.

### 9.5.2 Normalised linear equations

These equations can be simply normalised in the way suggested by the scaled units, but we now have a choice to resolve, namely, that of the parallel versus perpendicular scale. This provides the opportunity to motivate the normalisation scheme for drift wave dynamics in its entirety. The first step is the choice between $L_\parallel$ and $L_\perp$ for the system scale for the time. On traditional grounds we choose $L_\perp$ as the one $\omega_*$ depends on. We choose it independently of $L_p$ in order to be able to turn the gradient on and off in a computation. Time is therefore normalised against $c_s/L_\perp$. The state variables, $\tilde{\phi}$ and $\tilde{p}_e$, serving as stream functions and parallel force potentials, are normalised in terms of the scaled units suggested by the adiabatic response, appropriate to any situation in which that response controls the amplitude of $\tilde{\phi}$. The flux variables, $\tilde{A}_\parallel$ and $\tilde{u}_\parallel$, and also $\tilde{J}_\parallel$, are normalised in terms of $c_s$. The parallel gradient is still normalised against the wavenumber of the parallel scale, $K_\parallel = 2\pi/L_\parallel$, and an extra factor of the ratio $K_\parallel L_\perp$ is folded into the flux variables and will appear in the dimensionless parameters of the system, reflecting the interaction between the parallel and perpendicular dynamics. The factor of $\beta_e$ can be placed either in Ampère's law (normalising $A_\parallel$ against $A\,B\rho_s$), as done here, or in Faraday's law ($A_\parallel$ against $A\,B\rho_s\beta_e$), as in some earlier papers. Finally, $\nabla_\perp$ is normalised against $\rho_s$ when acting on a dependent variable.

The normalised quantities are

$$\phi = \frac{e\tilde{\phi}}{T_e} \quad n = \frac{\tilde{p}_e}{p_e} \quad A = \frac{\tilde{A}_\parallel}{B\rho_s}(K_\parallel L_\perp) \quad u = \frac{\tilde{u}_\parallel}{c_s}(K_\parallel L_\perp) \quad J = \frac{\tilde{J}_\parallel}{n_e e c_s}(K_\parallel L_\perp). \tag{9.28}$$

Since the model is isothermal we equate $\tilde{p}_e/p_e$ with $\tilde{n}_e/n_e$ and set $\Gamma = 1$ when considering the normalised equations, as we did for the wave response cases.

The normalised and linearised equations appear as

$$\frac{\partial \varpi}{\partial t} = \nabla_\parallel J, \tag{9.29}$$

$$\frac{\partial n}{\partial t} = -\omega_n \frac{\partial \phi}{\partial y} + \nabla_\parallel (J - u), \tag{9.30}$$

$$\frac{\partial}{\partial t}(A + \hat{\mu} J) = \nabla_\parallel (n - \phi) - \omega_n \frac{\partial A}{\partial y} - C\,J, \tag{9.31}$$

$$\hat{\epsilon}\frac{\partial u}{\partial t} = -\nabla_\parallel n + \omega_n \frac{\partial A}{\partial y} + X\nabla_\parallel^2 u, \tag{9.32}$$

with field equations

$$\varpi = \nabla_\perp^2 \phi \qquad J = -\hat{\beta}^{-1}\nabla_\perp^2 A, \tag{9.33}$$

forming the same system as in Chapter 8 except for the addition of the gradient drive terms.

The main parameters are the ones controlling the kinetic shear Alfvén dynamics, that is, the adiabatic response,

$$\hat{\beta} = \beta_e (K_\parallel L_\perp)^{-2} \qquad \hat{\mu} = \mu_e (K_\parallel L_\perp)^{-2} \qquad C = \frac{0.51\nu_e}{c_s/L_\perp}\hat{\mu}, \tag{9.34}$$

where we note the scaling against the parallel to perpendicular scale ratio for the background. Replacing $\beta_e$ and $\mu_e$, the parameters $\hat{\beta}$ and $\hat{\mu}$, respectively, give the effective inertia of the adiabatic response on drift wave time scales. The collisionality $C$ is the normalised parallel resistivity controlling the shear Alfvén damping rate in these units and serving as the basic collisionality for drift waves. The two parameters $\omega_n$ and $\hat{\epsilon}$ are free switches. Their nominal values are

$$\omega_n = \left| L_\perp \nabla \log p_e \right| \qquad \hat{\epsilon} = (K_\parallel L_\perp)^{-2} \qquad X = \frac{\mu_\parallel}{c_s L_\perp}, \tag{9.35}$$

but we can set them arbitrarily, to turn the gradient on and off or to speed up the sound waves in a computation, respectively.

### 9.5.3 Limits under conventional ordering

The parameters controlling the parallel wave speeds determine the competition between the ExB perpendicular dynamics and the generalised parallel dynamics. They may also be written as

$$\hat{\beta} = \left(\frac{c_s/L_\perp}{K_\| v_A}\right)^2 \qquad \hat{\mu} = \left(\frac{c_s/L_\perp}{K_\| V_e}\right)^2 \qquad \hat{\epsilon} = \left(\frac{c_s/L_\perp}{K_\| c_s}\right)^2. \tag{9.36}$$

Traditional drift ordering in terms of these parameters is

$$(\hat{\beta},\, \hat{\mu},\, C) \ll 1 \ll \hat{\epsilon}, \tag{9.37}$$

reflecting an almost adiabatic balance in the electrons and the neglect of sound waves for the ions. If we take the square root of all the terms and express the reciprocal of the inequality, and then replace '1' by $k_y\rho_s$, we recover Eq. (9.10).

To concentrate on the drift wave dynamics we will neglect sound wave dynamics, but before doing so there is an interesting relationship between the two to take note of. In a homogeneous plasma under the four-field model, one has four eigenmodes for each Fourier component, which we saw in the previous Chapter. These are the kinetic shear Alfvén and sound waves, one pair of each. When the background gradient is introduced, one of the sound waves gets transformed into a drift wave. We can see this by subtracting Eqs (9.30) and (9.29) to find,

$$\frac{\partial}{\partial t}(n - \varpi) + \omega_n \frac{\partial \phi}{\partial y} = -\nabla_\| u. \tag{9.38}$$

This is the same as Eq. (9.17), except for the sound wave term on the right side. We can estimate the size of the sound wave term using Eq. (9.32), to find

$$\frac{\partial}{\partial t}(n - \varpi) + \omega_n \frac{\partial \phi}{\partial y} \sim \Gamma \nabla_\|^2 n \left(\hat{\epsilon}\frac{\partial}{\partial t}\right)^{-1}. \tag{9.39}$$

With everything besides $\hat{\epsilon}$ of order unity, we now see that the sound wave physics is preempted by the drift waves, so long as we retain flute mode ordering, $\hat{\epsilon}^{-1} \ll 1$. This is why the sound waves are neglected in the conventional linear analysis.

If we keep the sound wave term, apply adiabatic electrons (Eq. 9.12), and Fourier transform time and $\nabla_\|$ as before, we find

$$(1 + k_\perp^2\rho_s^2)\omega^2 - \omega\omega_* - k_\|^2 c_s^2 = 0, \tag{9.40}$$

in physical units, whose solution is

$$\omega = \frac{\omega_L}{2} \pm \left(\frac{\omega_L^2}{4} + k_\|^2 V_s^2\right)^{1/2}, \tag{9.41}$$

with $\omega_L$ and $V_s$ given by Eqs (9.19) and (9.11), respectively. This exercise makes the sound wave preemption clear if $\omega_* \gg k_\| c_s$, and sound waves with a correction by $\omega_*$ if the reverse. In normalised units replace $\omega_*$ with $k_y\rho_s\omega_n$ and $k_\parallel^2c_s^2$ by $\hat{\epsilon}$ for $k_\parallel = K_\parallel$.

If we drop the sound wave wave term and set the gradient parameter $\omega_n$ to zero, both terms in Eq. (9.41) drop out and we recover the zero frequency mode in the pure kinetic shear Alfvén system, which we saw in the previous Chapter:

$$\frac{\partial}{\partial t}\left(\frac{\tilde{p}_e}{p_e} - \Gamma\rho_s^2\nabla_\perp^2\frac{e\tilde{\phi}}{T_e}\right) = 0, \tag{9.42}$$

in physical units, or

$$\frac{\partial}{\partial t}(n - \varpi) = 0, \tag{9.43}$$

with $\Gamma = 1$ when normalised. This zero frequency mode becomes part of the sound wave system on its slower frequencies. With the gradient present, however, it becomes the drift wave described in Eqs (9.17) and (9.18), if the electrons are adiabatic.

### 9.5.4 Self consistency: drift-Alfvén modes

We now allow the electrons to respond self consistently, through the interaction between the parallel and perpendicular dynamics. As we saw in the previous Chapter, the parallel dynamics tends towards an adiabatic state, with $n = \phi$. But the ExB advection tends to excite $n$ at extremal locations of $\mathbf{v}_E \cdot \nabla x$, that is, the $x$-component of the ExB velocity (Eq. 9.5), not $\phi$. The actual state of the electrons, the relationship between $n$ and $\phi$, is determined from these differing tendencies. Any difference of the actual response from $n = \phi$ will give the corrections to the adiabatic dispersion relation, and hence any instability.

In fact, Eq. (9.38) is the starting point for discussing drift waves at this level. Assuming $\omega^2 \gg k_\parallel^2V_s^2$ we drop the sound wave term and concentrate on the faster dynamics. We may define a 'nonadiabatic pressure' by combining the terms in Eq. (9.18) on the left side and everything else on the right. We define

$$\frac{\tilde{h}_e}{n_e} = \frac{\tilde{p}_e}{p_e} - \frac{e\tilde{\phi}}{T_e} \qquad \text{normalised to} \quad h = n - \phi, \tag{9.44}$$

which is the state variable for the kinetic shear Alfvén waves, and rewrite Eq. (9.38) as

$$(1 + k_\perp^2\rho_s^2)\frac{\partial\phi}{\partial t} - \omega_n\frac{\partial\phi}{\partial y} = -\frac{\partial h}{\partial t}. \tag{9.45}$$

If we Fourier transform this in time and space we have

$$\omega(1 + k_\perp^2) - \omega_n k_y = -\omega(h_\mathbf{k}/\phi_\mathbf{k}), \tag{9.46}$$

for a given Fourier component $\mathbf{k} = (k_x, k_y, k_\parallel)$, and frequency $\omega$, noting that in normalised units $\omega_* \rightarrow \omega_n k_y$. The nonadiabatic response on the right side is determined by the Ohm's law (Eq. 9.31), which we can express by Fourier transforming and placing all terms except the nonadiabatic forcing on the left side,

$$(\omega - \omega_*)A_\mathbf{k} + (\hat{\mu}\omega + iC)J_\mathbf{k} = -k_\parallel h_\mathbf{k}. \tag{9.47}$$

Substituting $k_\perp^2 A_\mathbf{k} = \hat{\beta} J_\mathbf{k}$ from Ampère's law (Eq. 9.33), we find

$$[\hat{\beta}(\omega - \omega_*) + (\hat{\mu}\omega + iC)k_\perp^2]J_\mathbf{k} = -k_\perp^2 k_\parallel h_\mathbf{k}. \tag{9.48}$$

We write the Fourier transformed vorticity equation as

$$\omega k_\perp^2 \phi_\mathbf{k} = k_\parallel J_\mathbf{k}. \tag{9.49}$$

This is general except for the neglect of the sound waves.

Use Eq. (9.49) to eliminate $J_\mathbf{k}$ in Eq. (9.48) in favour of $\phi_\mathbf{k}$ to find

$$k_\parallel^2 h_\mathbf{k} = -\omega[\hat{\beta}(\omega - \omega_*) + (\hat{\mu}\omega + iC)k_\perp^2]\phi_\mathbf{k}. \tag{9.50}$$

This is the equation for the self-consistent response. In the weakly nonadiabatic limit ($h_\mathbf{k} \ll \phi_\mathbf{k}$), the principal mechanism for an imaginary part in the response, hence a phase shift, is the resistivity, which gives a correction proportional to $-iCk_\perp^2/k_\parallel^2$ and hence a positive phase shift. Here we can see why it is important that $k_\parallel$ not vanish. If it could do so, the mode could not be determined in this fashion, for there would be no response path from $\tilde{p}_e$ back to $\tilde{\phi}$ and we would have to look for additional physics or different types of modes, which we will do in later Chapters. On the other hand, if $k_\parallel$ is relatively large (parallel wavelength much shorter than $L_\parallel$), then the adiabatic response is fast enough to render the departures of $\omega$ from the adiabatic drift $\omega_L$ negligible (Eq. 9.19, or $\omega_L \rightarrow \omega_n k_y/(1 + k_\perp^2)$ in normalised units). So the most important Fourier components in a homogeneous magnetic field are those with $k_\parallel/K_\parallel$ of order unity or at least close to zero, from the standpoint of instability or dissipation.

We also see how the $y$-direction is made distinct as a preferred orientation. So we can no longer use $K = k_\perp \rho_s$ as a parameter, unless we say that $k_x = 0$. For many applications, however, this is what is done, returning the parameters set to the usual one while setting $\omega_n$ and $\Gamma$ to unity.

Finally, Eqs (9.46) and (9.50) may be combined to form

$$\omega - \omega_L = \frac{\omega^2}{k_\parallel^2(1 + k_\perp^2)}[\hat{\beta}(\omega - \omega_*) + (\hat{\mu}\omega + iC)k_\perp^2]. \tag{9.51}$$

This is the version of the dispersion relation for the drift Alfvén case neglecting sound waves. Alternatively we may form the same expression as a correction for small $k_\perp^2$,

$$(\omega - \omega_n k_y)(\hat{\beta}\omega^2 - k_\parallel^2)\gamma + k_\perp^2\omega\left[(\hat{\mu}\omega^2 - k_\parallel^2) + iC\omega\right] = 0, \tag{9.52}$$

by setting all terms containing $k_\perp^2$ apart from the others. This shows the large-scale regime most commonly associated with MHD. In MHD the $\omega_* = \omega_n k_y$ term is assumed to be negligible itself, leaving the shear Alfvén waves ($\hat{\beta}\omega^2 = k_\parallel^2$) and a zero frequency mode which becomes the sound wave pair if the sound waves are reintroduced. But that latter picture is usually not valid because in a magnetised plasma we usually have $\omega_* \gg K_\parallel c_s$ for the energetically active part of the wavenumber spectrum. It is interesting to note that all departure from the simple waves $\omega = \omega_*$ and $\omega = \pm k_\parallel v_A$ is dependent upon a finite $k_\perp \rho_s$, which is what Eq. (9.52) shows.

## 9.6 Dissipation: phase shifts and energetics

### 9.6.1 Excitation of the phase shift

The role of the phase shift in drift wave free energy access is heuristically shown in figure 9.3. By examination of Eq. (9.46) we note that instability or dissipation, represented by a finite imaginary part to the mode frequency, is possible only if the nonadiabatic response, $\tilde{h}_e$, is out of phase with $\tilde{\phi}$. More generally, there must be a phase shift of $\tilde{p}_e$ ahead of $\tilde{\phi}$ in order for there to be an instability. To see this, we can assume we already know that there is such a phase shift, $\alpha$, in the normalised Fourier components,

$$n_{\mathbf{k}} = \phi_{\mathbf{k}} \exp(-i\alpha) \approx \phi_{\mathbf{k}}(1 - i\alpha), \tag{9.53}$$

with the latter form valid if $\alpha \ll 1$. Then, we find from Eq. (9.46) that

$$\omega(1 + k_\perp^2) - \omega_* = -\omega(-i\alpha) = i\alpha\omega. \tag{9.54}$$

If $\alpha \ll 1$ we may solve this by expansion,

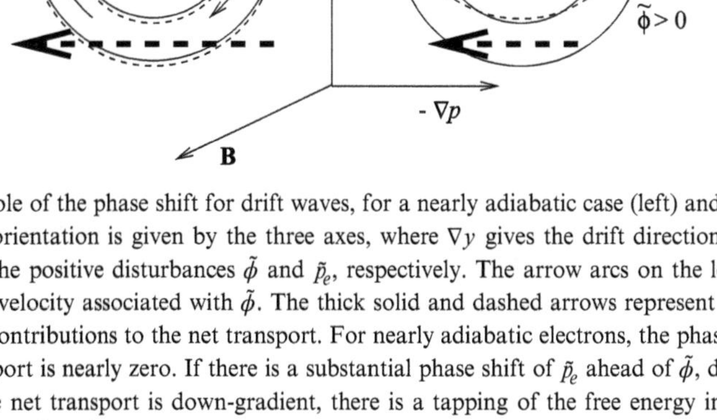


**Figure 9.3.** The role of the phase shift for drift waves, for a nearly adiabatic case (left) and a more collisional one (right). The orientation is given by the three axes, where $\nabla y$ gives the drift direction. Solid and dotted circles represent the positive disturbances $\tilde{\phi}$ and $\tilde{p}_e$, respectively. The arrow arcs on the left figure show the sense of the ExB velocity associated with $\tilde{\phi}$. The thick solid and dashed arrows represent the down-gradient and up-gradient contributions to the net transport. For nearly adiabatic electrons, the phase shift is negligible and the net transport is nearly zero. If there is a substantial phase shift of $\tilde{p}_e$ ahead of $\tilde{\phi}$, due to collisions for example, then the net transport is down-gradient, there is a tapping of the free energy in the gradient, and hence growth of the disturbances.

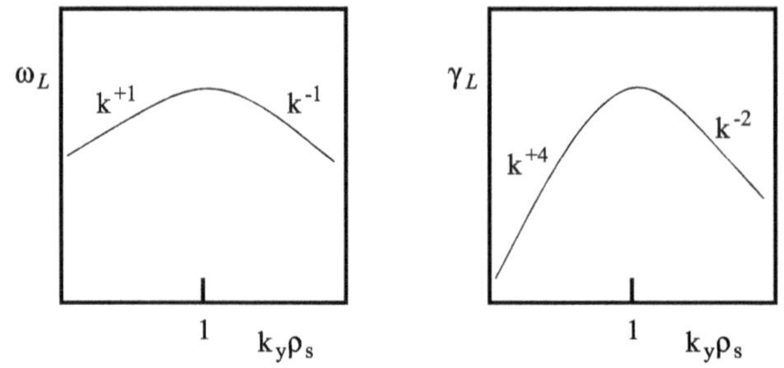


**Figure 9.4.** The simple collisional drift wave dispersion relation, showing the mode frequency, $\omega_L$, and growth rate, $\gamma_L$, as a function of the drift wavenumber, $k_y$. The phase shift is caused by resistive friction acting as a drag on the adiabatic response. In the four-field model for finite collisionality, $C > 1$, all scales yield $\gamma_L > 0$, that is, instability, with the fastest growth rates at $k_y\rho_s = 1$. The maximum frequency is $c_s/2L_p$. The maximum growth rate is proportional to the phase shift, which in this case is proportional to $C$.

$$\omega = \omega_L + i\alpha\frac{\omega_L}{1 + k_\perp^2}, \tag{9.55}$$

with $\omega_* \to \omega_n k_y$ in normalised units. This simple sketch of the run of drift wave frequency and growth rate with $k_\perp^2$ is shown in figure 9.4. The two sides $k_y < 1$ and $k_y > 1$ (in $\rho_s$ units) follow the simple power laws as shown, and are joined in the middle by a rollover. The two sides are symmetric for $\omega$, but for $\gamma$ the low-$k_y$ side is steeper following the greater difficulty of the polarisation current to maintain a finite parallel current when the factor of $k_\perp^2$ multiplying the time derivative for the vorticity becomes small.

Now, we examine Eq. (9.50). Assuming that $k_\parallel^2 = O(1)$ and that $(\hat{\beta}, \hat{\mu}, C) \ll 1$, we find that $h_\mathbf{k} \ll \phi_\mathbf{k}$ and hence that the correction on the left side of Eq. (9.46) is indeed small. If $C = 0$, that is, there is no dissipation, then this correction is purely real, that is, $h_\mathbf{k}$ differs from $\phi_\mathbf{k}$ in amplitude but not in phase. This means that inertia by itself amounts purely to a real frequency shift and hence cannot cause instability. In an exceptional limit, given by large $\hat{\mu}$ and small $\hat{\beta}$ and $C$, the electron inertia is the largest term controlling the current responses and is enough to cause reactive instabilities (addressed below). The other limit is the classic case of the collisional drift wave: assume that $C$ is much larger than $\hat{\beta}$ or $\hat{\mu}$, so that $J_\mathbf{k}$ is a purely resistive response to $h_\mathbf{k}$. If this is true we find that

$$h_\mathbf{k} = -iC\frac{\omega k_\perp^2}{k_\parallel^2}\phi_\mathbf{k} \qquad \Longrightarrow \qquad \alpha = C\frac{\omega k_\perp^2}{k_\parallel^2}, \tag{9.56}$$

which is equivalent to the simplest model of a phase shift induced by resistive drag in the adiabatic response.

Dissipation therefore leads to instability, with waves propagating at $\omega = \omega_L$ but growing at a rate proportional to $C$. In the simple case of pure collisional drift waves with $C \ll 1$, we find a growth rate spectrum given by

$$\gamma = \alpha\frac{\omega_L}{1 + k_\perp^2} = C\frac{\omega_L^2}{k_\parallel^2}\frac{k_\perp^2}{1 + k_\perp^2} = C\frac{\omega_n^2}{k_\parallel^2}\frac{k_y^2}{(1 + k_\perp^2)^2}\frac{k_\perp^2}{1 + k_\perp^2}. \tag{9.57}$$

For $C \ll 1$ and $k_\parallel$ quantised at integer multiples of $K_\parallel$, the fastest growing components are at $k_\parallel = \pm K_\parallel$ and $k_x = 0$ and $k_y\rho_s = 1$, or $k_\parallel = \pm 1$ and $k_y = 1$ in normalised units. There is no instability for $k_\parallel = 0$ since $\tilde{h}_e$ does not excite $\tilde{J}_\parallel$ and $\tilde{J}_\parallel$ does not excite $\tilde{\phi}$ then. We also see how all of $C$, $\omega_*$, and the finite $k_\perp\rho_s$ are necessary for the existence of a finite phase shift.

### 9.6.2 Self-consistent energetics

The phase shift is related to the net transport down the background gradient caused by the disturbances, and therefore to the tapping of the free energy represented by the gradient. We build the energy theorem for drift waves in exactly the same fashion as for kinetic shear Alfvén waves, except that now in addition to the transfer effects and sinks we have the gradient forcing which enters through the extra terms arising from Eqs (9.20) and (9.21). Following the procedure of the previous Chapter, we multiply Eqs (9.29)–(9.32) by $-\phi$, $n$, $J$, and $u$, respectively, and apply the field Eqs (9.33) to convert the $\nabla_\perp^2$ terms to positive-definite quantities and divergences. For the free energy density we find

$$U_T = \frac{1}{2}\left( | \nabla_\perp \phi |^2 + n^2 + \hat{\beta}_e^{-1} | \nabla_\perp A |^2 + \hat{\mu} J^2 + \hat{\epsilon} u^2 \right), \tag{9.58}$$

which satisfies

$$\frac{\partial U_T}{\partial t} + \nabla \cdot (\cdots) = -\omega_n n \frac{\partial \phi}{\partial y} + \omega_n (u - J) \frac{\partial A}{\partial y} - C\, J^2 - X \left| \nabla_\parallel u \right|^2. \tag{9.59}$$

These are the same forms as in the previous Chapter, except for the normalisation, and here the two forcing terms due to the gradient. The total free energy is the average of this over the spatial domain,

$$E_T = \mathscr{V}^{-1} \int d\mathscr{V}\ U_T, \tag{9.60}$$

in which the divergence terms are assumed to vanish. The forcing term involving $J$ also vanishes overall because it can be formed into a total divergence,

$$J\frac{\partial A}{\partial y} = -\hat{\beta}^{-1}(\nabla_\perp^2 A)\frac{\partial A}{\partial y} = \frac{\partial}{\partial y}\frac{1}{2}\hat{\beta}^{-1} | \nabla_\perp A | - \nabla \cdot \left( \hat{\beta}^{-1} \frac{\partial A}{\partial y} \nabla_\perp A \right). \tag{9.61}$$

This leaves two forcing terms expressed as a drive if they are positive,

$$\Gamma_n = \frac{1}{2E_T} \mathscr{V}^{-1} \int d\mathscr{V} \left( -\omega_n n \frac{\partial \phi}{\partial y} \right) \qquad \Gamma_v = \frac{1}{2E_T} \mathscr{V}^{-1} \int d\mathscr{V} \left( \omega_n u \frac{\partial A}{\partial y} \right), \tag{9.62}$$

to go with the two sink terms, defined as positive-definite quantities,

$$\Gamma_c = \frac{1}{2E_T} \mathscr{V}^{-1} \int d\mathscr{V}\ (CJ^2) \qquad \Gamma_s = \frac{1}{2E_T} \mathscr{V}^{-1} \int d\mathscr{V} \left( X \left| \nabla_\parallel u \right|^2 \right), \tag{9.63}$$

which we had before, in the new normalisation. The free energy theorem can be expressed as

$$\frac{1}{2E_T}\frac{\partial E_T}{\partial t} \equiv \Gamma_T = \Gamma_n + \Gamma_v - \Gamma_c - \Gamma_s - \Gamma_w, \tag{9.64}$$

showing how the drive via the gradient can be balanced in general by resistive and viscous dissipation, and in a turbulence computation by $\Gamma_w$, the subgrid dissipation, which for the 3D models of later Chapters is defined therein. The difference between the left- and right-hand sides of this in a computation gives the error,

$$\Gamma_E = \Gamma_T - (\Gamma_n + \Gamma_v - \Gamma_c - \Gamma_s - \Gamma_w), \tag{9.65}$$

which as in the 2D turbulence models of Chapter 3 is used to check the integrity of the results.

We will use this later to diagnose the energetics of turbulence, but for now we apply it to the linear system, where it describes the energetics of either an instability or decay of a stable mode. Applying the Fourier transforms in space but not in time, any of these quadratic integral quantities become a sum over all **k**, and in a linear system they separate into each individual **k** which grow or decay on their own. Each positive-definite quantity becomes an amplitude of a complex quantity (this includes the sink terms, which are always positive-definite preceded by a minus sign), and each forcing term becomes the real part of a complex quantity. Transfer terms are also like this, for example

$$\mathcal{V}^{-1}\int d\mathcal{V}\; J_\parallel \nabla_\parallel \phi = \sum_{\mathbf{k}} 2\,\mathrm{Re}\left[J_{\mathbf{k}}^* i k_\parallel \phi_{\mathbf{k}}\right], \tag{9.66}$$

that is, the average becomes a sum over **k**, both positive and negative. In these terms, the pieces of the free energy theorem including the forcing and dissipation terms appear as

$$\frac{\partial}{\partial t} k_\perp^2 \mid \phi_{\mathbf{k}} \mid^2 = 2\,\mathrm{Re}\left[J_{\mathbf{k}}^* i k_\parallel \phi_{\mathbf{k}}\right], \tag{9.67}$$

$$\frac{\partial}{\partial t}\frac{1}{\hat{\beta}}(1 + \hat{\mu} k_\perp^2) k_\perp^2 \mid A_{\mathbf{k}} \mid^2 = 2\,\mathrm{Re}\left[J_{\mathbf{k}}^* i k_\parallel n_{\mathbf{k}} - J_{\mathbf{k}}^* i k_\parallel \phi_{\mathbf{k}} - C \mid J_{\mathbf{k}} \mid^2\right], \tag{9.68}$$

$$\frac{\partial}{\partial t} \mid n_{\mathbf{k}} \mid^2 = 2\,\mathrm{Re}\left[-J_{\mathbf{k}}^* i k_\parallel n_{\mathbf{k}} + u_{\mathbf{k}}^* i k_\parallel n_{\mathbf{k}} - \omega_n n_{\mathbf{k}}^* i k_y \phi_{\mathbf{k}}\right], \tag{9.69}$$

$$\frac{\partial}{\partial t}\hat{\epsilon} \mid u_{\mathbf{k}} \mid^2 = 2\,\mathrm{Re}\left[-u_{\mathbf{k}}^* i k_\parallel n_{\mathbf{k}} + \omega_n u_{\mathbf{k}}^* i k_y A_{\mathbf{k}} - X k_\parallel^2 \mid u_{\mathbf{k}} \mid^2\right], \tag{9.70}$$

noting the application of the condition that the physical disturbances be real-valued, for example, $\phi_{-\mathbf{k}} = \phi_{\mathbf{k}}^*$, and that the electromagnetic forcing in Ohm's law, given to be proportional to the real part of $\omega_n J_{\mathbf{k}}^* i k_y A_{\mathbf{k}}$, vanishes since $J_{\mathbf{k}} = \hat{\beta}^{-1} k_\perp^2 A_{\mathbf{k}}$ (proportional with a real coefficient). The transfer effects and sinks are the same as those in the homogeneous case. In these terms we write the contribution of Fourier mode $\mathbf{k}$ to the total energy $E_T$ as $E_{\mathbf{k}}$. The transfer channels are

$$T_a = \frac{1}{E_{\mathbf{k}}} \mathrm{Re}\left[J_{\mathbf{k}}^* ik_{\|}\phi_{\mathbf{k}}\right], \tag{9.71}$$

giving the Alfvén (MHD) transfer from $J$ to $\phi$,

$$T_n = \frac{1}{E_{\mathbf{k}}} \mathrm{Re}\left[J_{\mathbf{k}}^* ik_{\|}n_{\mathbf{k}}\right], \tag{9.72}$$

giving the adiabatic compressional transfer from $n$ to $J$, and

$$T_s = \frac{1}{E_{\mathbf{k}}} \mathrm{Re}\left[-u_{\mathbf{k}}^* ik_{\|}n_{\mathbf{k}}\right], \tag{9.73}$$

giving the acoustic (sound wave) compressional transfer from $n$ to $u$, respectively, all expressed as relative transfer rates. The dissipation terms are

$$\Gamma_c = \frac{1}{E_{\mathbf{k}}} C \mid J_{\mathbf{k}}|^2 \qquad \Gamma_s = \frac{1}{E_{\mathbf{k}}} X k_{\|}^2 \mid u_{\mathbf{k}}|^2, \tag{9.74}$$

giving the resistive and viscous dissipation processes. The gradient forcing can be identified as the terms multiplied by $\omega_n$,

$$\Gamma_n = \frac{1}{E_{\mathbf{k}}} \mathrm{Re}\left[-\omega_n n_{\mathbf{k}}^* ik_y \phi_{\mathbf{k}}\right] \qquad \Gamma_v = \frac{1}{E_{\mathbf{k}}} \mathrm{Re}\left[\omega_n u_{\mathbf{k}}^* ik_y A_{\mathbf{k}}\right]. \tag{9.75}$$

The first of these is the ExB gradient drive, and the second is the sound wave magnetic drive. Positive forcing is a drive and negative is a damping, and the indefinite sign of these terms reflects the contingency of the relative phase shift, which is a result not an input. Almost all interesting cases have $\Gamma_v \ll \Gamma_n$. As with the kinetic shear Alfvén wave model, the resistivity acts only on the adiabatic response (finite $J$), and the viscosity acts only on the sound waves ($u$). The total energy for this $\mathbf{k}$ is given by

$$E_{\mathbf{k}} = k_\perp^2 \mid \phi_{\mathbf{k}}|^2 + \mid n_{\mathbf{k}}|^2 + \frac{1}{\hat{\beta}}\left(1 + \frac{\hat{\mu}}{\hat{\beta}} k_\perp^2\right) k_\perp^2 \mid A_{\mathbf{k}}|^2 + \hat{\epsilon} \mid u_{\mathbf{k}}|^2, \tag{9.76}$$

representing ExB, thermal, electromagnetic, and sound wave free energy, respectively. The forcing terms are normalised in terms of a total growth rate,

$$\gamma = \frac{1}{2E_{\mathbf{k}}} \frac{\partial E_{\mathbf{k}}}{\partial t} = \Gamma_n + \Gamma_v - \Gamma_c - \Gamma_s, \tag{9.77}$$

which in this case is simply the linear growth rate. The mode frequency is measured from the phase of the disturbance,

$$\omega = \frac{\partial \alpha}{\partial t} \qquad \text{where} \quad \alpha = \mathrm{Im} \log n_{\mathbf{k}}, \tag{9.78}$$

correcting for the branch cut at $\pm\pi$.

The principal forcing is the ExB gradient drive, $\Gamma_n$. If we go back to the spatial representation and take note of Eq. (9.5), we can recast it as

$$\Gamma_n = \frac{1}{2E_T}\mathscr{V}^{-1}\int d\mathscr{V}\ (\omega_n n v_E^x) \qquad \text{where} \qquad v_E^x = -\frac{\partial\phi}{\partial y}, \tag{9.79}$$

in normalised units. Similarly, the magnetic gradient forcing can be recast as

$$\Gamma_v = \frac{1}{2E_T}\mathscr{V}^{-1}\int d\mathscr{V}\ (\omega_n u\tilde{B}^x) \qquad \text{where} \qquad \tilde{B}^x = \frac{\partial A}{\partial y}. \tag{9.80}$$

In both cases, the forcing as a drive is given by net down-gradient transport times the gradient strength. We can see where the phase shifts are relevant by returning to the Fourier representation for $\Gamma_n$, and then substituting $\tilde{p}_{e\mathbf{k}} = (1 - i\alpha)\tilde{\phi}_{\mathbf{k}}$, to find

$$\Gamma_n = \frac{1}{E_{\mathbf{k}}}\omega_n k_y\ \alpha\ |\ \tilde{\phi}_{\mathbf{k}}|^2. \tag{9.81}$$

If the details of the adiabatic response give rise to a positive phase shift, then we have positive net down-gradient transport and a growing disturbance.

This is the origin of the coupling and drive mechanisms in the turbulence studied in Chapter 4. It is the most basic version of turbulence driven by an instability in a magnetised plasma.

## 9.7 Alfvénic transients

### 9.7.1 The initial value problem for drift waves

We have four equations, with the sound waves, but so far have looked at only one mode at the given $(k_x, k_y, k_\parallel)$. In a real initial value problem we cannot simply pick out the eigenmode of choice but must allow the dynamical system to evolve from an initial state. That is, we are not allowed to Fourier transform time.

As in the cases we considered for wave dynamics in the previous Chapter, the linear nature of the problem allows restriction to a single spatial Fourier component, so it is sufficient to start with the appropriate periodic initial state. The evolution is described by that of each linear eigenmode, of which there are three if we neglect the sound waves. We can find all three of these modes by combining the methods above to find the drift wave and those of the previous Chapter to find the two kinetic shear Alfvén waves. Neglecting $u$, we may manipulate Eqs (9.29)–(9.31) into the following form for the Fourier components,

$$\left(\frac{\partial}{\partial t} + i\frac{\omega_n k_y}{1 + k_\perp^2}\right)\phi_{\mathbf{k}} = -\frac{1}{1 + k_\perp^2}\frac{\partial h_{\mathbf{k}}}{\partial t}, \tag{9.82}$$

$$\left(\frac{\partial}{\partial t} \pm iV_a k_\parallel\right)[h_{\mathbf{k}} \mp V_a(A_{\mathbf{k}} + \hat{\mu}J_{\mathbf{k}})] = -i\omega_n k_y(\phi_{\mathbf{k}} \mp V_a A_{\mathbf{k}}) \pm V_a C J_{\mathbf{k}}, \tag{9.83}$$

where $V_a$ given by $V_a^2 = (1 + k_\perp^2)/(\hat{\beta} + \hat{\mu}k_\perp^2)$ is the kinetic shear Alfvén phase velocity in these units (compare to Eq. 9.11). The first equation is the same as Eq. (9.46), which gives the drift wave with $\tilde{h}_e$ as a correction, as we have seen. The second combines the two adiabatic response equations, in which the conventional ordering $\partial/\partial t \sim \hat{\beta}^{-1/2} \gg 1$ leaves the $\omega_* = \omega_n k_y$ terms on the right as corrections, and the $C$ term for resistive damping. This latter pair of equations describes the two kinetic shear Alfvén waves to which the drift wave is coupled. Together, these are the three eigenmodes in terms of which the initial state evolves. Since $\phi_\mathbf{k}$ starts at zero, all the modes are excited (including the drift mode, since we may replace $\phi_\mathbf{k}$ by $n_\mathbf{k}$ in its eigenmode equation, adjusting the correction on the right side accordingly). If $k_\perp^2 \ll 1$ in addition to $(\hat{\beta}, \hat{\mu}, C) \ll 1$, the evolution is relatively simple: the Alfvén waves' zero frequency mode, $n_\mathbf{k} - k_\perp^2\phi_\mathbf{k}$, does not evolve on the short time scale, $\tau \sim \hat{\beta}^{1/2}$, on which the two Alfvén transients propagate and dissipate. On the slower time scale of $\tau \sim (\omega_n k_y)^{-1}$, the nearly adiabatic structure is well established and the drift wave begins to propagate. At later times the transients are long passed, leaving the drift mode which propagates with $\omega \approx \omega_L$ and slowly grows on the time scale of $\tau \sim (\omega_L \alpha)^{-1}$. This is the eventual solution, justifying the simpler procedure of finding the sole mode with a positive growth rate. At very much later times, the drift mode interacts with the remaining sound wave, which is damped in the four-field model through parallel viscosity. From an initial pressure disturbance given by a density disturbance times a constant temperature, therefore, the Alfvén transients are launched and then dissipate, and the drift mode structure forms. Afterward, evolution is slower growth of the drift Alfvén mode of instability. We see the significance of the scenario in the previous Chapter, letting the Alfvén dynamics arise from an initial density pellet. It is a physically realisable mechanism for forming these structures.

## 9.8 Numerical examples

It is instructive to take the same numerical examples we did for kinetic shear Alfvén and sound waves, and include the background gradient effects, and with them the drift wave. We take a single $k_\parallel = 1$, and set $k_x = 0$ for simplicity, and examine the spectrum of modes of various $k_y$. Then, we can proceed as before, with the parameter $K = k_y\rho_s$ set in normalised units, and $\varpi_\mathbf{k} = -K^2\phi_\mathbf{k}$ for example. We write $k_\parallel$ explicitly, although it is set to unity in normalised units, with $-\pi < z < \pi$ as before.

As in the wave examples, each is followed from an initial disturbance in the electron density until the final mode structure is established and the growth rate and mode frequency can be measured. The sound waves are also included, so that we solve the same problem as in the wave responses case, with the finite $\omega_* = \omega_n k_y \rho_s$ being the only addition, reducing to $\omega_n K$ in normalised units. The equations to be solved are the four ordinary differential equations,

$$\frac{\partial \phi_\mathbf{k}}{\partial t} = -ik_\parallel A_\mathbf{k}, \qquad (9.84)$$

$$\frac{\partial n_{\mathbf{k}}}{\partial t} = -i\omega_n K\phi_{\mathbf{k}} + ik_{\parallel}(K^2 A_{\mathbf{k}} - u_{\mathbf{k}}), \tag{9.85}$$

$$\left(1 + \hat{\beta}^{-1}\hat{\mu}K^2\right)\frac{\partial A_{\mathbf{k}}}{\partial t} = ik_{\parallel}\left(n_{\mathbf{k}} - \phi_{\mathbf{k}}\right) - i\omega_n K A_{\mathbf{k}} - \hat{\beta}^{-1}CK^2 A_{\mathbf{k}}, \tag{9.86}$$

$$\hat{\epsilon}\frac{\partial u_{\mathbf{k}}}{\partial t} = -ik_{\parallel}n_{\mathbf{k}} + i\omega_n K A_{\mathbf{k}} - Xk_{\parallel}^2 u_{\mathbf{k}}, \tag{9.87}$$

for the four dependent variables ($\phi_{\mathbf{k}}$, $n_{\mathbf{k}}$, $A_{\mathbf{k}}$, $u_{\mathbf{k}}$), which follow from Eqs (9.31)–(9.32). The initial state for all cases is $n_{\mathbf{k}} = 1.0$ and $\phi_{\mathbf{k}} = A_{\mathbf{k}} = u_{\mathbf{k}} = 0$, so the dynamics is launched from an initial density/pressure disturbance, like a pellet, as done before for the kinetic shear Alfvén cases. This is a very simple task easily taken care of with the timestep scheme we have been using (Chapter 3). The timestep in all cases is nominally $\tau = 0.001$. The sampling rate for measurements is one every 0.02 time units, and averages given are over the last one-third of a sample.

The warm plasma case is given by

$$\hat{\beta} = 5 \qquad \hat{\mu} = 2.7 \qquad \nu = 10 \qquad X = 10 \qquad \hat{\epsilon} = 10^4, \tag{9.88}$$

where $\nu = \nu_e L_\perp / c_s$, reflecting a scale ratio of $K_\parallel L_\perp = 10^{-2}$ and giving a value of $C = 13.9$ for the collisionality. The hot plasma case is given by

$$\hat{\beta} = 1 \qquad \hat{\mu} = 0.027 \qquad \nu = 0.1 \qquad X = 1 \qquad \hat{\epsilon} = 10^2, \tag{9.89}$$

reflecting a scale ratio of $K_\parallel L_\perp = 10^{-1}$, and giving a value of $C = 0.139$ for the collisionality. In both cases the value of $\hat{\mu}$ reflects deuterium ions, and the numerical value of $C$ reflects 0.51 times $\hat{\mu}$ times the choice for $\nu_e L_\perp / c_s$. These are the same as the two cases from the previous Chapter, with the scale ratio $K_\parallel L_\perp$ folded into the other parameters. We also examine a near-electrostatic case, given by the warm plasma parameters but with $\hat{\beta} = 0.1$ set arbitrarily.

The warm plasma case with $K = 0.2$ is exemplary for the transient phase and establishment of the drift wave mode structure, starting from the initial pressure disturbance. The mode frequency and growth rate finish at $\omega = 0.186$ and $\gamma = 0.0213$, and the phase shift is $\alpha = 0.119$, reflecting a robust drift Alfvén instability with nevertheless small enough phase shift to be consistent with drift wave mode structure. The nominal resistive damping rate $(1/2)CK^2/(\hat{\beta} + \hat{\mu}K^2)$ is about 0.054. In figure 9.5, we see the transient phase, to $t = 25$, too short for the resistive damping to have been completely effective, but long enough for about two kinetic shear Alfvén periods. The amplitudes of $n_{\mathbf{k}}$ and $\phi_{\mathbf{k}}$ are already comparable, reflecting the adiabatic response. Sound waves are negligible on this time scale. The ExB gradient drive rate $\Gamma_n$ oscillates, showing that the mode structure has not found itself. The mode frequency, however, has found its eventual value and is settling there. The emergence of the drift wave from this transient state is shown in figure 9.6, with the evolution to $t = 125$. Here we find all pieces of the electron dynamics ($\Gamma_n$ and $\Gamma_c$), as well as the mode frequency and growth rate, pinned to their eventual values. Only the sound wave component has not settled, but it is of negligible impact. For all

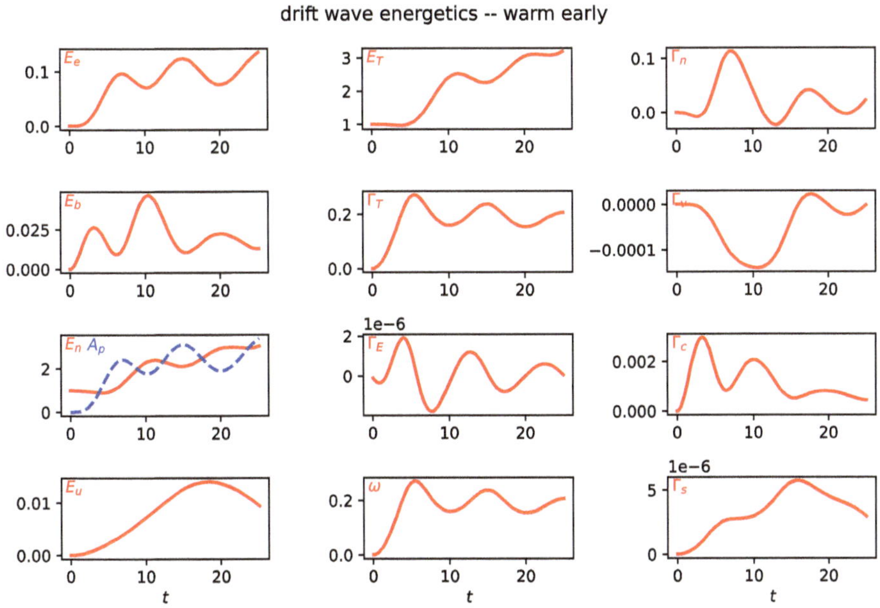


**Figure 9.5.** Transient response to the initial pressure disturbance, for the warm plasma case with $K = 0.2$. The Alfvénic transients are excited and then $\phi_{\mathbf{k}}$ and $n_{\mathbf{k}}$ oscillate about each other at similar amplitude. This is just long enough to show the first part of the associated sound wave oscillation during the same phase. The quantities $E_e$, $E_b$, $E_n$, and $E_u$ give the contributions of $\phi_{\mathbf{k}}$, $A_{\mathbf{k}}$, $n_{\mathbf{k}}$, and $u_{\mathbf{k}}$ to the total energy ($E_T$), respectively. The amplitude $A_p$ is given together with $E_n$, showing how the amplitudes of $\phi_{\mathbf{k}}$ and $n_{\mathbf{k}}$ come to track each other, having started with $\phi_{\mathbf{k}} = 0$. The sink rates $\Gamma_c$ and $\Gamma_s$ are due to resistivity and viscosity, and the drive rates $\Gamma_n$ and $\Gamma_v$ give the ExB and magnetic parts of the pressure gradient forcing, respectively. The total growth rate ($\Gamma_T$) is much larger than the error ($\Gamma_E$). The instantaneous mode frequency, $\omega$, is also shown.

intents and purposes, the drift wave structure is well established, and the disturbance is growing as an instability.

The hot plasma case is quite different. Here, $\hat{\beta} \gg \hat{\mu}$ and furthermore $C$ is small enough to make $\hat{\beta}\omega_* \gg CK^2$ unless $K$ is unrealistically large. Or with $K$ reasonable $C$ would have to become unrealistically large (as a rate, $\nu_e \gg k_\| v_A$). This is a reactive case, where we could more or less set $C$ to zero. If that is done, however, the Alfvén oscillations persist forever and the mode structure never settles. For this value of $C = 0.139$ the mode does settle but takes a long time to do so. The nonadiabatic effects are small on average, but the temporal excursions are large and mode frequency information is not accessible. These properties are shown in figure 9.7, with the early phase $t < 80$, in which we can see that the total energy and ExB drive rate oscillate robustly about a value much closer to zero than the standard deviation. The total energy does drop with time, in an overall sense. The later phases are shown in figure 9.8. This is a stable case, which is due to the lack of robust collisionality.

If we set a much higher collisionality we obtain a robust eigenmode albeit one with a very low growth rate. Setting $\nu = 10^2$ so that $\nu_e$ is $10^2$ times $k_\| v_A$ in physical units, we have a case in the traditional MHD regime. All the Alfvén activity dies out

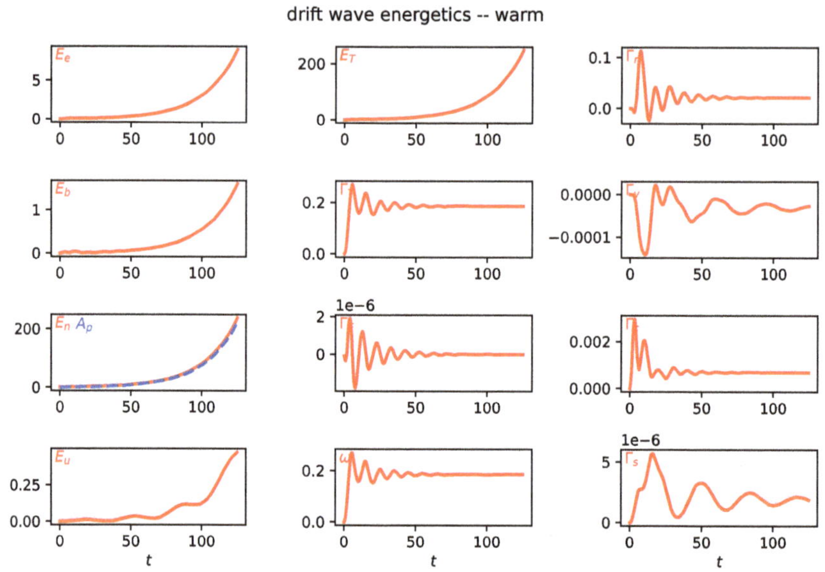


**Figure 9.6.** Development of the drift wave structure from the initial pressure disturbance, for the warm plasma case with $K = 0.2$. The Alfvénic transients are excited and then decay on a time scale of about $\Delta t = 50$. Both the mode frequency and growth rate settle out on this same scale, as the drift wave emerges from the initial state of all four eigenmodes. The sound wave energy and damping are very small, due to the small $K_{\|}L_{\perp}$; a companion case run without the sound waves gave nearly identical results for the other quantities. The only important free energy drive is the ExB transport ($\Gamma_n$), and the only significant sink is resistivity ($\Gamma_c$), although the small $K^2$ keeps $\Gamma_c \ll \Gamma_n \approx \gamma$. Note how the small magnetic flutter drive ($\Gamma_v$) is associated with the sound waves.

very early, and we end with a nearly adiabatic drift wave at late times. The phase shift is measured at $\alpha = 0.0131$ and the growth rate is $6.02 \times 10^{-4}$, taken over $1667 < t < 5000$. The mode frequency $\omega$ is 0.233, which is higher than the nominal electrostatic adiabatic value of 0.192. The evolution up to $t = 500$ is shown in figure 9.9. Many astrophysical situations are in this regime, where one usually assumes MHD is ideal due to a high, not low, frequency, but with a value of $k_{\perp}\rho_s$ sufficiently small that the resistive diffusion rate given by $(\eta_{\|}/\mu_0)k_{\perp}^2$ is negligible (the coefficient $\eta_{\|}/\mu_0$ is $0.51\mu_e\nu_e/\beta_e$ in these units). In those cases, however, drift instabilities are too small to be of interest. It is a laboratory plasma problem mostly due to the finite $\rho_s/L_{\perp}$ exhibited there.

In this simplified model, hot plasma cases with low enough collisionality can become reactive. Even more pathological is the case with $\hat{\mu} \gg 1$ and $\hat{\beta}$, $C$ both small. The difficulties with such cases typically point to the need for additional physics. We will discuss reactive dynamics further in Section 9.10.

Generally, the mode frequency slows down as the structure becomes more nonadiabatic, and the growth rate is also reduced. The reduction of $\omega$ and $\gamma$ below $\omega_L$ becomes evident as $K$ increases for the electrostatic cases, shown in figure 9.10, for $\nu = 1$ and 10, hence $C = 1.39$ and 13.9, chosen to keep the resistivity dominant in controlling the response of the current, leaving the dynamics in the collisional

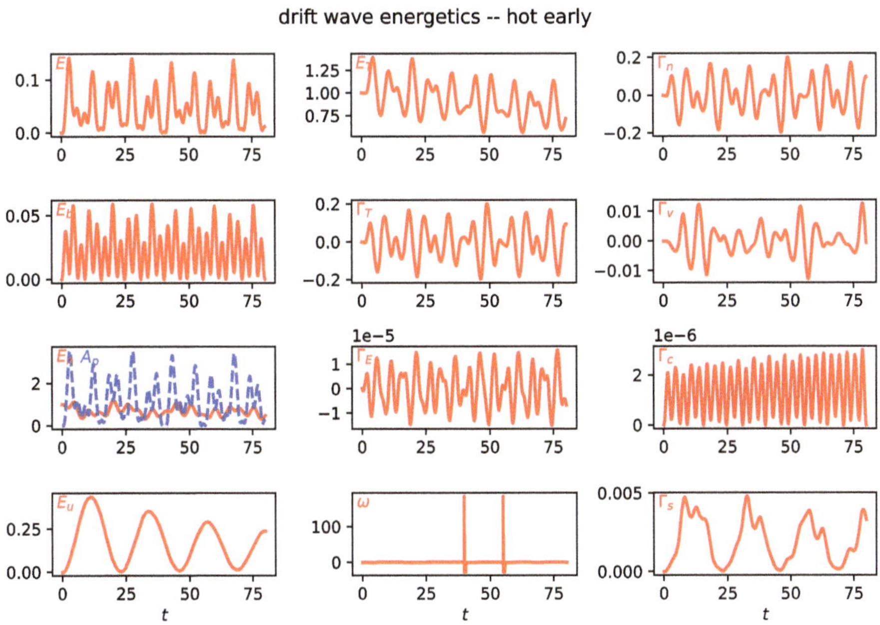


**Figure 9.7.** Initial development of the drift wave structure from the initial pressure disturbance, for the hot plasma case with $K = 0.2$. The Alfvénic transients are excited but then do not decay effectively, since the collisionality is so low. Oscillations in the total energy ($E_T$) and gradient drive rate ($\Gamma_n$) are large. There is a discernible long-term drop in $E_T$, but one caused by the sound wave damping (figure 9.8). The diagnostics cannot discern the mode frequency among the oscillations.

regime. The larger value is enough to also ensure a strong nonadiabatic response, with growth rates above 0.1 in the last half decade before $K = 1$. Nevertheless, the overall behaviour does follow the simple near-adiabatic model, defined by the regime in which the corrections on the right sides of Eqs (9.85) and (9.86) are small.

Turning to the finite-beta warm plasma case, we find that the finite electromagnetic behaviour has minimal effect, as seen in figure 9.11. Although $\hat{\beta} > \hat{\mu}$, for $\nu = 10$ we still have $CK^2 > \hat{\beta}$ for all but the longest wavelength range. The growth rates are lowered by the perturbative effect of the magnetic flutter effect ($\tilde{\mathbf{B}} \cdot \nabla p_e$) on the phase shifts, as we will explore in the next Section. But for this value of $\hat{\beta}/C$ the effect is not large. One finds more sensitivity when $\nu = 1$ is set, and $C$ is reduced by ten. The growth rates near the spectrum peak are reduced more noticeably.

## 9.9 Drift Alfvén waves—the magnetic flutter effect

When the electromagnetic effects entering through $\tilde{A}_\parallel$, hence the finite $\beta_e$, start to have significant effect on the drift wave, the drift wave becomes a drift Alfvén wave, since the particular properties of the Alfvénic transients as such start to alter the structure of the drift wave mode, and therefore the mode frequency and growth rate. The place in the dynamics this happens is in the response of the parallel current $\tilde{J}_\parallel$ to

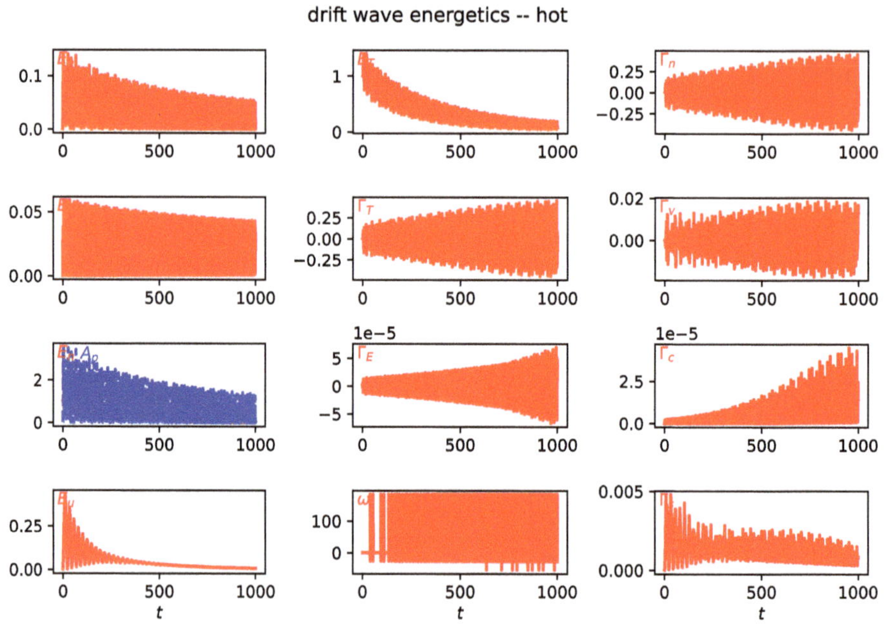


**Figure 9.8.** Long-term development of the drift wave structure from the initial pressure disturbance, for the hot plasma case with $K = 0.2$. The Alfvénic transients remain active after the viscosity has eliminated the sound waves. Nevertheless, long-term decay of the total energy ($E_T$) in a stable state is established. Despite the noisiness, the energetic error ($\Gamma_E$) is small. These are the features of a $C \ll 1$ case in the four-field model.

its force potential $\tilde{h}_e$. If the situation is collisional and electrostatic, the response is dissipative, with $\tilde{J}_\parallel$ remaining $\pi/2$ out of phase with $\tilde{h}_e$. In this limit the parallel dynamics acts as a resistive diffusion of the force potential along the magnetic field. When the magnetic effects enter, this phase relation alters, with effect on the phase shift $\tilde{h}_e \leftrightarrow \tilde{\phi}$ of the drift wave.

There are two electromagnetic effects to consider. One is induction, represented by $\partial \tilde{A}_\parallel/\partial t$ in the Ohm's law. The other is magnetic flutter, arising from the parallel gradient of the background electron pressure, $\tilde{\mathbf{B}} \cdot \nabla p_e$, also in the Ohm's law. This comes from the fact that the disturbances to the magnetic field lines can give them a down-gradient component. The causal chain for this is the dependence of variation of the drift wave state variables in the drift direction ($\partial \tilde{h}_e/\partial y$), so that the current they generate has the same spatial phase, due to the linearity of the electrostatic, dissipative part of the Ohm's law. Ampère's law, which is also linear, therefore yields a magnetic potential, $\tilde{A}_\parallel$, with the same $y$-dependence as $\tilde{h}_e$. This in turn yields a disturbance in the magnetic field which is in the $x$-direction, down or up the gradient, depending on the sign.

Each of these effects carries with it a particular perpendicular scale. These are the collisional and collisionless skin depths, $d_e$ and $d_{ec}$ respectively, given by

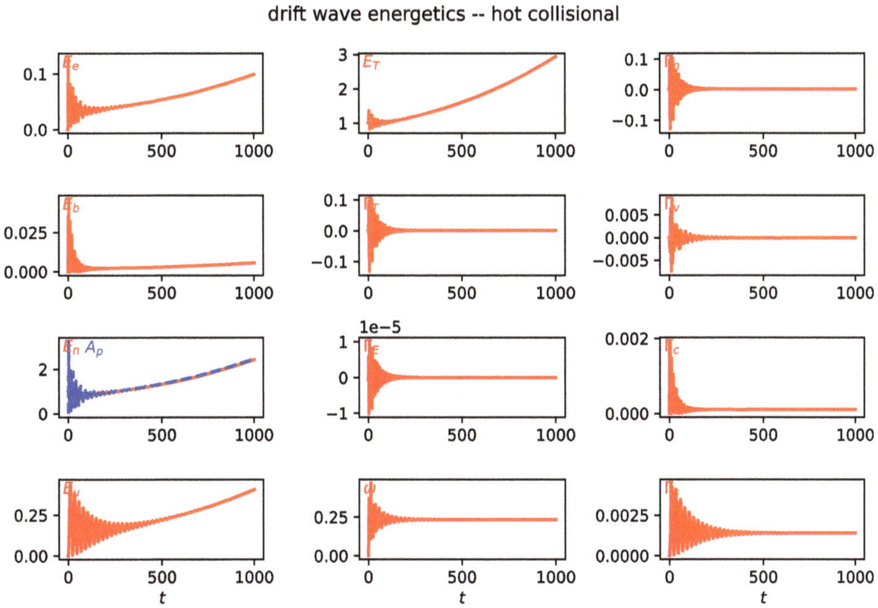


**Figure 9.9.** Development of the drift wave structure from the initial pressure disturbance, for a collisional hot plasma case with $\nu = 10^2$. The Alfvénic transients now die out effectively, leaving a slowly growing adiabatic drift wave ($A_p = E_n$ closely) behind. The mode structure takes a long time to settle due to the low growth rate compared to the Alfvén frequency. Nevertheless, the energetic error ($\Gamma_E$) remains small, and even the mode frequency ($\omega$) is now discernible.

$$d_e^2 = \frac{\mu_e}{\beta_e}\rho_s^2 \qquad\qquad d_{ec}^2 = \frac{0.51\mu_e\nu_e}{\beta_e\omega_*}\rho_s^2, \tag{9.90}$$

in $d_{ec}$ with $\omega \approx \omega_*$, in physical units (note also that $\hat{\beta}/\hat{\mu} = \beta_e/\mu_e$). The first follows from the competition between electron inertia ($\partial\tilde{J}_\parallel/\partial t$) and magnetic induction ($\beta_e\omega$); the second follows from the competition between resistivity and either the wave frequency or with magnetic flutter ($\beta_e\omega_*$). When both of these scales are smaller than the perpendicular scale of motion, these finite $\beta_e$ effects significantly influence the response of $\tilde{J}_\parallel$ to $\tilde{h}_e$.

The electromagnetic effects have opposite tendencies, reflected by the entrance of $\hat{\beta}$ in the response equation for the Fourier components, Eq. (9.50). The physical explanation for this, in the regime that the finite $\hat{\beta}$ effects are weak, is that when a current starts to flow, the field line is tipped from its equilibrium position towards the direction of the gradient. The resulting pressure force opposes the induced current, leaving the resistive dynamics unperturbed if $\omega = \omega_*$. But this is the real frequency part of $\omega - \omega_*$, which doesn't directly affect the phase shift anyway. Eq. (9.50) may be rewritten as

$$k_\parallel^2 h_{\mathbf{k}} = \omega[\hat{\beta}(\omega_* - \omega) - (\hat{\mu}\omega + iC)K^2]\phi_{\mathbf{k}}, \tag{9.91}$$

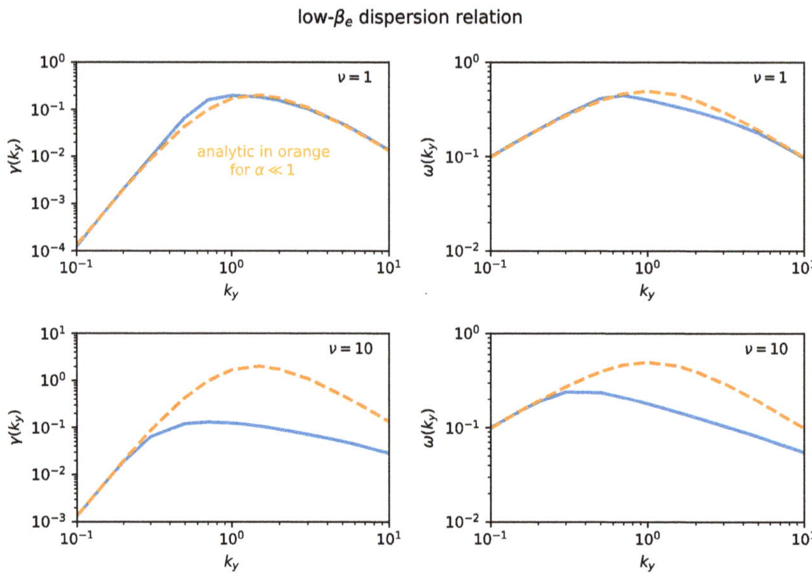


**Figure 9.10.** Spectra of the growth rates (left) and mode frequency (right), for two sets of electrostatic cases. The near-adiabatic model is closely followed for the $\nu = 1$ case ($C = 1.39$) but departures for $k_y > 0.3$ in the $\nu = 10$ case ($C = 13.9$) reflecting the nonadiabatic response become substantial. The spectrum for $\gamma$ is still much steeper on the long-wave side of the peak.

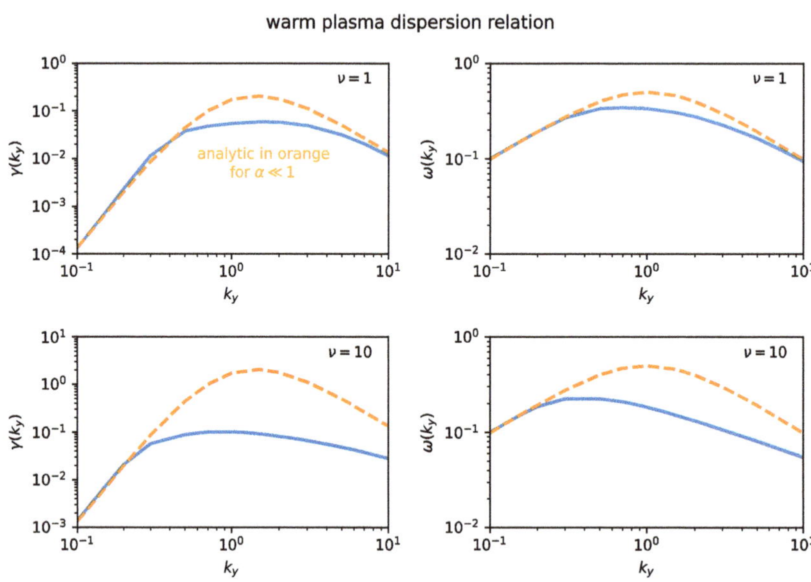


**Figure 9.11.** Spectra of the growth rates (left) and mode frequency (right), for two sets of the warm plasma case. The difference to figure 9.10 for $\nu = 10$ is minimal, with the growth rates very slightly lower in the peak range. The effect of finite inductivity ($\hat{\beta} = 5$ rather than 0.1) is felt more for $\nu = 1$, which is $C = 1.39$ here. The effect on the growth rate $\gamma$ in the peak range is the largest.

with $K = k_\perp \rho_s$. Setting the factor of $\omega$ to $\omega_L$, assuming that $\tilde{h}_e$ is a correction, we find that the imaginary part of the response $\tilde{h}_{e\mathbf{k}}/\tilde{\phi}_\mathbf{k}$ is affected only by $C$, that is, dissipation. But especially when the growth rate is substantial, there is also an imaginary correction due to $\gamma$ multiplied by $\hat{\beta}(\omega_* - \omega)$. If $\omega_* > \omega$, the usual case even for the isothermal model, the imaginary part of the finite $\hat{\beta}$ correction actually subtracts from the phase shift represented by $C$. We may rewrite Eq. (9.91) in terms of the (real) mode frequency and growth rate, obtaining

$$k_\parallel^2 h_\mathbf{k} = (\omega_R + i\gamma)[\hat{\beta}(\omega_* - \omega_R) - (\hat{\mu}\omega_R + iC)K^2]\tilde{\phi}_\mathbf{k}, \tag{9.92}$$

where $\omega_R$, $\gamma$, and $\omega_*$ are all real coefficients. The correction due to $\hat{\beta}\omega_*\gamma$ is obviously stabilising since it acts to reduce the phase shift in the self-consistent electron response. If $\omega_R$ is close to $\omega_*$, the magnetic induction and flutter effects cancel, but if (as usual) $\omega_R < \omega_*$, then the net effect is stabilisation by magnetic flutter. We note as well the destabilising tendency of the electron inertia, $\hat{\mu}$, which has the same physical role as the magnetic induction.

We can find if this is the actual mechanism for the finite $\hat{\beta}$ stabilisation we saw marginally in the warm plasma case but clearly in the hot plasma case, by setting $C$ to something intermediate between the two cases and varying $\hat{\beta}$. We do this for the model equations unaltered (the same $\omega_*$ in the magnetic flutter effect as in the ExB gradient drive), for the flutter effect doubled, and then for the flutter effect removed entirely. Varying the flutter effect in this fashion amounts to placing a constant coefficient in front of the two $\omega_n$ terms in Eqs (9.86) and (9.87), setting it to 1, 2, and 0, respectively. We start with the warm plasma case with $\nu = 1$, reducing $C$ to 1.39. The inductivity $\hat{\beta}$ is swept from 0.1 to $10^2$, which is $\beta_e$ from $10^{-5}$ to $10^{-2}$ for this case. The point at which $\beta_e = \mu_e$ is at $\hat{\beta} = 2.72$, right in the middle. The results are shown in figure 9.12, and it is clear that the lack of a strong $\hat{\beta}$ effect for the nominal case is due to the cancellation between the magnetic induction and flutter effects, because when we double the flutter effect we find a clear stabilisation. With no flutter we find a nominal destabilisation at first, but then at $\hat{\beta}K \sim 1$ the mode jumps into a reactive instability with large phase shift, discussed in the next Section. The finding is that magnetic induction is destabilising, and magnetic flutter is stabilising. We note that this cannot be explained by the effects on the real frequency, which does not exhibit strong changes with which we could explain the effects on $\gamma$. What does change with the magnetic flutter effect, however, is the drift wave phase shift. We therefore find that the magnetic flutter effect does its work indirectly, on the phase shift of a growing instability, rather than directly through the real frequency effects associated with the skin depths.

## 9.10 Reactive instabilities

In the previous two Sections we found another curious effect. This is that when the electron inertial effect in the adiabatic response is large enough we find instabilities which do not depend on dissipation. We referred to this briefly in Section 9.6, after Eq. (9.55), that is, that the effect of the real part of the self-consistent response

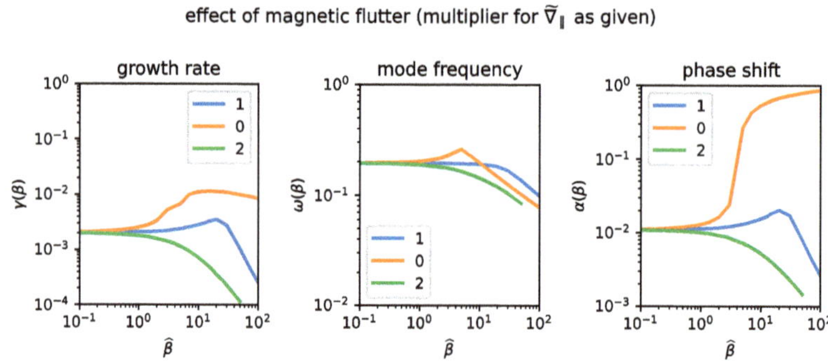


**Figure 9.12.** Scalings of the mode frequency, growth rates, and phase shifts, versus inductivity ($\hat{\beta}$), for the warm plasma case with $K = 0.2$ and $\nu = 1$. Note that $\beta_e = 10^{-4}\hat{\beta}$ for these parameters. The curves are labelled according to the artificial multiplier of the magnetic flutter effect ($\tilde{\nabla}_\parallel$), with '1' nominal. The role of the flutter effect is to cancel the magnetic induction in the Ohm's law, with the induction destabilising and the flutter effect stabilising. If the flutter effect is doubled, a clear stabilising tendency results, explained by the phase shifts rather than the mode frequencies, as in the text. For $\hat{\beta}K$ in the vicinity of unity, the case without magnetic flutter jumps to a new mode with different properties (reactive).

cannot cause instability if it is a correction. However, with substantial $\hat{\mu}$ for $K = k_y\rho_s$ near unity (with $k_x = 0$), we start to find reactive instabilities, termed as such because they are not catalysed by dissipation. These can take over the dynamics even for finite but small $C$, for which there becomes a threshold in $K$. We can set all the parameters as in the warm plasma case as done above, but now set $\hat{\beta} = 0.1$ and $\hat{\mu} = 10$ but $C = 0.1$. For $K = 0.2$ we still find a reasonable collisional drift wave, but very weak, with $\alpha = 0.004$ and $\gamma = 7.8 \times 10^{-4}$. This is almost completely adiabatic. However, raising $K$ finds $\alpha = 0.0177$ and $\gamma = 0.00479$ at $K = 0.3$, and then $\alpha = 0.597$ and $\gamma = 0.21$ at $K = 0.5$. Two intermediate points at $K = 0.35$ and 0.4 show $\alpha = 0.042$ and $\gamma = 0.013$ and $\alpha = 0.241$ and $\gamma = 0.090$, respectively. There is a sharp jump at what we can call a threshold value of $K > 0.35$ in this case. That this is a reactive instability is confirmed by setting $\nu = C = 0$ and doing these cases again. For $K = 0.4$ and 0.5 the effect is minimal and we still obtain a very robust instability with well-defined mode structure and mode frequency, while for $K = 0.35$ the mode is oscillatory and essentially stable. So the cases with $\nu = 0.1$ are showing a smoothed transition.

This threshold is roughly at the point where $K > \hat{\mu}^{-1/2}$, which corresponds to $\omega_* > K_\parallel V_e$ in physical units. At this point the fluid dynamics effectively competes with the parallel electron dynamics along the field line and access to the free energy in the gradient is no longer impeded by the electron motion. The behaviour is roughly the same as shown in figure 9.12 for the case with magnetic flutter turned off. Then, we can get the reactive modes through inductivity. In physical cases, though, they depend on electron inertia.

Reactive instabilities are similar to such textbook ideal fluid results as the Kelvin–Helmholtz and firehose instabilities. They are part of a system which is perfectly conservative, and in simple examples the equation to be solved involves Hermitian operators whose eigenvalues are purely real. These eigenvalues are squared frequencies, so that when they are negative the eigenmode becomes a

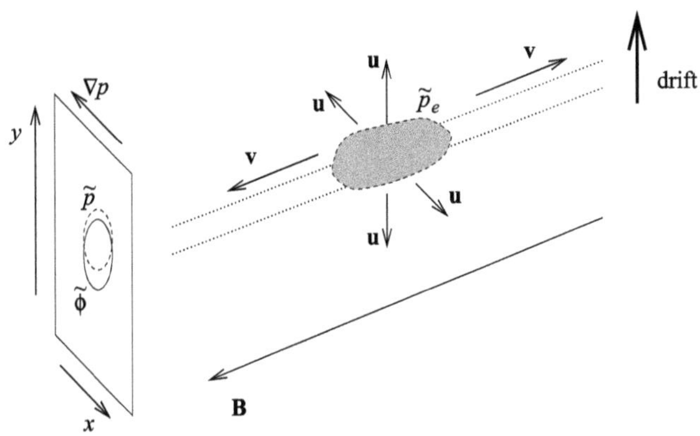


**Figure 9.13.** The drift wave structure, localised to a flux tube in the $xy$-plane, caused by an initial electron pressure disturbance, showing how it is formed by the kinetic shear Alfvén wave oscillation. The adiabatic response consists of the pressure disturbance ($\tilde{p}_e$) launching the Alfvénic activity, tends to force $\tilde{\phi}$ towards $\tilde{p}_e$, with the polarisation (ions, **u**) and parallel (electrons, **v**) currents balancing to maintain quasineutrality. When a background gradient is present, the ExB vortex in the $xy$-plane represented by $\tilde{\phi}$ (contour given by the solid circle) continually excites $\tilde{p}_e$ (contour given by the dashed circle) where the ExB flow is down-gradient, that is, phase-shifted in the forward drift direction ($y$). The combination of these two effects results in a net drift of the entire structure in the $y$-direction. When the Alfvénic transients have damped away, the result is a drift wave.

purely growing instability. In drift Alfvén dynamics, however, we have the cross-field gradient acting on the waves, so the frequencies we find are nearly always complex. We have four eigenmodes, but they do not come in pairs with equal and opposite frequencies. We saw this due to the finite $\omega_*$ even in the simplified case in Eq. 9.41. For these reactive instabilities we have nonzero real mode frequencies as large as $\omega_*$ and larger than the growth rates.

For fluid dynamics these are physically real instabilities, but for a plasma they usually signify that we might be missing some important physics. For drift waves that is indeed the case. Our exemplary case is an instability caused by parallel dynamics of a finite electron mass, and that is a sign that we need to extend the dynamical model to incorporate Landau damping, an expressly kinetic effect. We will not go further to address such effects in this Chapter, but it is useful to keep the general concept in mind. In this case it means we cannot use the four-field model for low collisionality (even if the reactive instability is removed by the magnetic flutter effect), but must include the temperature dynamics, especially for the electrons. This will be revisited in Volume Two after the thermal and kinetic physics needed to treat it has been introduced.

## 9.11 Mode structure

A sketch of the basic process is shown in figure 9.13, which repeats the kinetic shear Alfvén structure of figure 8.3 of Chapter 8, but then adds the background gradient. The initial adiabatic response is the same, and the rest of the dynamics sets in more slowly. The adiabatic response tends to keep $\tilde{\phi}$ close to $\tilde{p}_e$ as for the kinetic shear Alfvén waves, but now the advection of the gradient also tends to push $\tilde{p}_e$ ahead of $\tilde{\phi}$, as in figure 9.1. The resulting structure, the drift wave, reflects the combination of these two processes, and drifts in the $y$-direction. If the phase shift is not small, the simple models of Sections 9.3 and 9.4, and even the small phase shift expansion of Section 9.5, break down, and we have to consider the general problem. The initial value problem as done above, especially the early responses (figure 9.5) elucidates the basic physics.

## 9.12 Summary

Drift waves represent a fundamental interaction between parallel and perpendicular dynamics which occurs in a magnetised plasma with an electron pressure gradient. Advection of the background gradient liberates free energy, which is then dissipated through the adiabatic response. The advection effect tends to drive the phase shift of the pressure disturbance ahead of the electrostatic potential disturbance by a quarter wavelength, while the adiabatic response tends to reduce it to zero. The phase shift and free energy tapping, and net transport, all go together. In the most idealised case the kinetic shear Alfvén oscillations, and their dissipation, are arbitrarily fast, so that the electrons are almost adiabatic, and the phase shift and growth rate are small. For interesting parameters, however, the corrections rapidly become too large for this implied expansion to be valid, and a computation which involves the damped transients is necessary. It is possible to do an eigenvalue analysis to find the maximum growth rate, but the physics of the drift wave dynamics is more transparent in the initial value problem for which the instability has to evolve. The initial state of a pressure disturbance is exemplary in showing the relationship of drift waves to kinetic shear Alfvén waves. By the same token, one could not study drift waves with an MHD model, since the coupling of Alfvénic dynamics to the electron pressure would then be missing. Drift waves are dissipative instabilities, requiring either resistivity or kinetic dissipation effects on the parallel electron dynamics, so that if the resistivity is too small one has to keep additional physics involving the electron temperature, which is deferred to Volume Two. When the dynamical beta, $\beta_e$, is large enough, the adiabatic response becomes electromagnetic, and the drift wave becomes a drift Alfvén wave. Magnetic induction, part of the effective inertia of the adiabatic response, is generally destabilising, and magnetic flutter, which is a parallel gradient of the background along disturbed field lines, is generally stabilising, in both cases due to indirect effects on the drift wave phase shift. The details of stability or instability then follow from the relative size of these two effects. Drift wave stability in general is sensitive to the details of the cancellation between the mode frequency and the diamagnetic frequency.

## Further Reading

Drift wave instabilities have an extensive literature, most of which is concerned on the details introduced by particular magnetic geometric or electron kinetic effects. The original reference for the isothermal collisional model is S S Moiseev and R Z Sagdeev 'On the Bohm diffusion coefficient' *Sov. Phys. JETP* **17** (1963) 515. The collisionless model used to be called the universal instability. Two reviews of the linear work are W M Tang 'Microinstability theory in tokamaks' *Nucl. Fusion* **18** (1978) 1089; and P C Liewer 'Measurements of microturbulence in tokamaks and comparisons with theories of turbulence and anomalous transport' *Nucl. Fusion* **25** (1985) 543.

The electromagnetic generalisation, usually called the drift Alfvén wave, goes back to L V Mikhailovskaya and A B Mikhailovskii *Sov. Phys. JETP* **18** (1964) 1077; treated by A B Mikhailovskii 1974 *Theory of Plasma Instabilities* (New York: Consultants Bureau). See also J T Tang and N C Luhmann Jr 'Destabilization of hydromagnetic drift-Alfvén waves in a finite pressure, collisional plasma' *Phys. Fluids* **19** (1976) 1935, for a very clear exposition. The tokamak version was given by K T Tsang, J C Whitson, J D Callen, P J Catto and J Smith 'Drift Alfvén waves in tokamaks' *Phys. Rev. Lett.* **41** (1978) 557.

The relationship of the drift wave to the transient electron parallel dynamics was treated by S Camargo, B Scott and D Biskamp 'The influence of magnetic fluctuations on collisional drift-wave turbulence' *Phys. Plasmas* **3** (1996) 3912; B Scott 'Three-dimensional computation of drift Alfvén turbulence' *Plasma Phys. Control. Fusion* **39** (1997) 1635.

See also the Further Reading Sections of neighbouring Chapters.

**IOP** Publishing

Turbulence and Instabilities in Magnetised Plasmas, Volume 1

Fluid drift turbulence

**Bruce Scott**

# Chapter 10

# Mode structure diagnostics

## 10.1 Introduction

In Chapter 2 the basic properties of two-dimensional turbulence were worked out and in Chapters 3–5 the statistical properties and constraint by forcing were introduced. A few basic diagnostics, mainly energetics and spectra, were introduced to elucidate the main points. The dual cascade and energy transfer properties can be demonstrated in this fashion. But once we start with coupling mechanisms which may or may not lead to drive of the turbulence, and basic properties of wave or instability mode structure, we are on to concepts whose activity in a turbulent state require demonstration with something qualitatively objective. There are several channels of forcing and several types of instability potentially present, but the dominant ones will leave their typical mode structure still present in the turbulence. For example, how adiabatic or hydrodynamic is the electron response in the turbulence, or whether one or the other type of instability is still measurably present in fully developed turbulence, is something which should not be left to arbitrary interpretation.

Even for the cascade dynamics and the local energy transfer processes which are involved, the state of affairs is not obvious in the more heuristic diagnostics, such as the turbulent transport flux caused by net ExB advection, or disturbance amplitudes, visualisations, or even the spatial scale of motion. A detailed diagnostic was necessary to show that the cascade tendencies in the flow energy and the thermal free energy are simultaneously present.

In general we will find that more than one type of eigenmode is potentially present in the overall turbulent system; in some cases there will be as many as four different eigenmode types. Whether the turbulence is actually more of the character of one of these eigenmodes than of the others can only be decided by detailed diagnosis. Herein, we introduce the basic statistical diagnostics of disturbances in turbulence, a nonlinear, nonequilibrium state which does not lend itself to exact analytical solution, or even modelling beyond very heuristic considerations. Together, these

doi:10.1088/978-0-7503-2504-2ch10  

diagnostics show the fundamental character of the turbulence: how its different pieces are coupled, how it transfers energy from drive to dissipation. This experience is useful and necessary in sorting out the dynamical character of turbulence in the more complicated situations we will find, especially after magnetic geometry and electron and ion temperature dynamics are introduced.

In this Chapter we will use the notation $\tilde{\phi}$ and $\phi$ interchangeably, the former always for quantities in physical units, the latter always in the normalised units of Chapter 4, as the two-dimensional dissipative coupling model is used in the examples. When introducing concepts in a general way, the physical units are used, and when showing computational examples, the normalised units are used.

## 10.2 Temporal diagnostics

The first diagnostic of use is the average saturated amplitude of the disturbances and the average transport they cause. It is important to define what we mean by an average in this context. If we measure the transport, for example, as a function of time, then we can say that it is saturated if the average measured for it in some interval of time is statistically equal to that measured over some other interval of time. This interval should be long compared to the dynamical time scale of individual vortices or other disturbances, but short compared to the time we would expect the background to evolve as a result of the transport. In a local computation such as the ones we have dealt with so far, the background is either held fixed, or due to the boundary conditions (periodic, Neumann, or Dirichlet in the $x$-direction) the average profile gradient is held fixed. We may therefore take arbitrarily long intervals, limited by reasonable computational resources. The shortest interval we should use is that given by the longer of the eddy turnover time or the correlation time of the turbulence. The eddy turnover time is the typical dynamical time scale on which energy is transferred between similar spatial scales; one can define it in terms of the average vorticity. The correlation time is that on which two signals displaced in time begin to lose relation to each other; for this one can take the width of the correlation function in terms of the time displacement. In practice we must use a time rather longer than either of these in order to obtain a good statistical sample.

As an example we can use the ExB kinetic energy, $E_e$, which is determined by $\phi$, the electrostatic potential. It is defined as an average over the spatial domain,

$$E_e = \mathcal{V}^{-1} \int d\mathcal{V} \, \frac{1}{2} \frac{\rho}{B^2} \mid \nabla_\perp \tilde{\phi} \mid^2 = \mathcal{V}^{-1} \int d\mathcal{V} \left( -\frac{1}{2} \tilde{\phi} \varpi \right), \tag{10.1}$$

where the vorticity variable is defined as

$$\varpi = \nabla \cdot \frac{\rho}{B^2} \nabla_\perp \tilde{\phi}. \tag{10.2}$$

The volume element $d\mathcal{V}$ is defined by the problem, so that in the two-dimensional cases we have been using, we have

$$d\mathcal{V} = dx\, dy \qquad \int d\mathcal{V} = \int dx \int dy = \int d^2x, \tag{10.3}$$

where $x$ and $y$ are the two coordinates defining the perpendicular plane, and the domain of integration is the domain size $L_xL_y$. Given a constant volume element, the integrations are simple averages over grid points. If not, then the volume element is included in a weighted average. So we have

$$\begin{aligned} E_e(t) &= \frac{1}{N_xN_y}\sum_{\text{grid}}\frac{1}{2}\frac{\rho}{B^2}\,|\,\nabla_\perp\tilde{\phi}(x,\,y,\,t)|^2 \\ &= \frac{1}{N_xN_y}\sum_{\text{grid}}\frac{1}{2}\Big(-\,\tilde{\phi}(x,\,y,\,t)\varpi(x,\,y,\,t)\Big), \end{aligned} \tag{10.4}$$

in a two-dimensional model, with the grid and time dependences written explicitly. In sums over grid points, which ones to sum is given by the spatial dependences of the right-hand side. This becomes clearer when we look at spectra.

Such quantities are measured at each of a set of time points, usually equidistant, with a sampling interval and a sample length. The average and standard deviation of $E_e$ over the interval are expressed as

$$E_e = \int_{t\in\mathcal{T}}\frac{dt}{\mathcal{T}}\,E_e(t) \qquad \Delta E_e = \left[\int_{t\in\mathcal{T}}\frac{dt}{\mathcal{T}}\,(E_e(t)\,-\,E_e)^2\right]^{1/2}, \tag{10.5}$$

where $\mathcal{T}$ is the extent of the time interval, and $t\in\mathcal{T}$ denotes integration over the interval (this label is shown here only). With the sample, the averaged integrals are replaced by sums of the samples divided by the number of samples. The quantity $E_e$ is statistically stationary if for differing intervals of fixed $\mathcal{T}$ the standard deviation of the measures of $E_e$ is less than the standard deviation $\Delta E_e$ itself. The turbulence is temporally saturated if not only the average transport but also the average amplitude of each dependent variable is statistically stationary by this definition.

A computation of turbulence involving several dependent variables yields correspondingly several time series, and therefore a wealth of information even simply in these time traces. It would not do justice to such a dynamical system if understanding of it were limited to the average transport or the average amplitude of a single dependent variable. The first step beyond that is the ratio between two of these amplitudes. If the density and electrostatic potential amplitudes were always equal, for example, we would have a strong indication that the electrons might be adiabatic. If these amplitudes differ such that the nonadiabatic part of the density disturbances, $\tilde{h}_e = \tilde{n}_e - (n_e e/T_e)\tilde{\phi}$, or $h = n - \phi$ in normalised units, is comparable by amplitude to $\tilde{n}_e$ or in these units

$$h \sim n \sim \phi \quad \Longrightarrow \quad \text{‘nonadiabatic cases’}, \tag{10.6}$$

the electrons are surely quite strongly nonadiabatic. By convention we use the half-squared amplitudes as diagnostics, since some of them also appear as parts of the free energy. For example, the time trace of the half-squared amplitude of $\tilde{\phi}$ is given by

$$A_p(t) = \mathcal{V}^{-1}\int d\mathcal{V}\,\frac{1}{2}\left|\frac{e\tilde{\phi}}{T_e}\right|^2 \qquad \text{or} \qquad \frac{1}{L_xL_y}\int d^2x\,\frac{1}{2}\phi^2. \tag{10.7}$$

In a similar manner, we have $A_n$ and $A_h$ for $n$ and $h$ in normalised units, and $A_w$ for the vorticity. The transport flux of electrons due to net ExB advection of the density disturbances is given by

$$F_e = \mathcal{V}^{-1} \int d\mathcal{V}\ \tilde{n}_e v_E^x \qquad v_E^x = [\tilde{\phi}, x] \qquad \text{or} \qquad \frac{1}{L_x L_y} \int d^2x\ n v_E^x, \tag{10.8}$$

where $v_E^x$ is the $x$-component of the ExB velocity, and the bracket is defined separately with normalised or physical units. These time traces of the amplitudes and energetics were shown in figures 4.13 and 4.14 from Chapter 4, for the cases $D = 0.1$ and $D = 0.01$ of the two-dimensional dissipative coupling model. Here we show the amplitudes again, together with the transport. The most basic indication of the drift-wave character of the turbulence is that the amplitudes of $\tilde{n}_e$ and $\tilde{\phi}$ always track each other, at least moderately closely, even when the dynamics of $\tilde{h}_e$ is strong enough to lead to substantial vorticity, energetics, and transport. Note that $A_n$ is also $E_n$, the thermal free energy, so the latter is used in any context involving energy or energetics.

In this context, substantial transport means greater than an averaged particle flux density of greater than $n_e v_D$ in physical units, or unity in the local normalised units defined in Chapter 4. Here, we have $F_e \sim 3$ in these units.

The vorticity by itself is an important quantity, since it measures the eddy turnover rate of the turbulence. The vorticity of an ExB flow is defined as

$$\Omega_E = \frac{1}{B}\nabla_\perp^2 \phi, \tag{10.9}$$

and has units of frequency. If we consider an eddy of radius $r$ turning rigidly so that the velocity at radius $r$ is $v$, then the vorticity is $v/r$ and the turnover time is $2\pi r/v$. Hence by definition, the RMS vorticity is $2\pi$ divided by the eddy turnover time. Given the half-squared amplitude $A_w$, the RMS vorticity is $(2A_w)^{1/2}$. A value of unity for this, or $A_w = 1/2$, indicates $\Omega_E \sim c_s/L_\perp$. The result in figure 10.1 is an average $A_w$ of 2.92 (times $c_s/L_\perp$) over the interval $200 < t < 400$ (times $L_\perp/c_s$), which indicates robust drift wave turbulence. Nevertheless, it is smaller than the ion gyrofrequency $\Omega_i$ by the drift parameter $\delta_0 = \rho_s/L_\perp$, so the drift approximations remain valid.

## 10.3 Spectral diagnostics

Besides the amplitudes, we also want to know information about the basic scale of motion: where most of the free energy resides, whether the large or small scales are important even if not dominant. Moreover, the dynamical character of the turbulence may vary with the scale of motion. The simplest way to get this is through the amplitude spectrum of the disturbances. As for the averaged amplitudes themselves, these spectra will tell us more if we compare them for the various dependent variables in a given computation. Even if the boundary conditions are not periodic, the spectrum can still be measured. It is done as if there were a Fourier decomposition. The easiest one to use for an introduction is the $y$-direction, which unlike time and usually the radial direction, is periodic (in a coordinate system

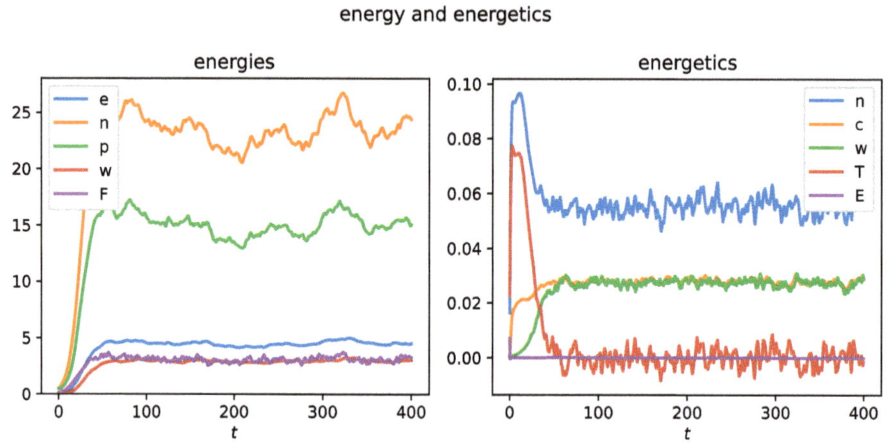


**Figure 10.1.** Time traces of the half-squared amplitudes of $\tilde{n}_e$, $\tilde{\phi}$, $\tilde{h}_e$, and $\varpi$, respectively, labelled by 'n', 'p', 'h', 'w', and of $F_e$, the ExB flux of electrons, labelled 'F', for the case with $D = 0.1$, as in figures 4.13 and 4.14 from Chapter 4. These quantities are averaged over the spatial domain. The substantial turbulent flux and vorticity reflects strong nonadiabatic electron dynamics, even though the adiabatic response forces $\tilde{n}_e$ and $\tilde{\phi}$ to similar size. The energetics show the gradient drive rate $\Gamma_n$ and the sink rates due to resistive ($\Gamma_c$) and subgrid ($\Gamma_w$) dissipation, labelled by their respective subscript. The total ('T') and error ('E') are also shown, with the latter too small to see. The growth is positive until saturation whereafter it finds a statistical zero.

aligned to the field lines of a sheared magnetic field, the parallel coordinate is not strictly periodic either, cf. Chapter 12). But before we get into this any deeper we have to discuss normalisation of the Fourier transforms.

### 10.3.1 Normalisation of the Fourier transform

The Fourier decomposition as we use it assumes that for a function of the periodic coordinate $y$, we have for any scalar quantity $f$,

$$f(y) = \sum_{l=-n}^{n} f_l \exp[i\, k_y\, y] \qquad k_y = l\, k_0 \qquad f(k_y) = f_l\,, \tag{10.10}$$

where $k_0$ is the fundamental wavenumber on a periodicity length of $2\pi/k_0$, and $k_y$ is the wavenumber, and $l$ is the mode number, and in this instance $n$ gives the range of Fourier components, or modes for short. The grid is assumed to be equidistant, given that all the coordinate gradients as used herein are expected to be constant. If the number of grid points is $N$, then the number of modes, positive or negative, is $n = N/2$. The Fourier amplitudes are complex in general. The condition that $f(y)$ be real-valued is

$$f(-k_y) = f^*(k_y). \tag{10.11}$$

The inverse transform is whatever is needed to return the original $f(y)$ from the decomposition into $f(k_y)$ and vice versa, and is given by

$$f(k_y) = \oint \frac{k_0\, dy}{2\pi} \exp[-i\, k_y\, y] f(y), \tag{10.12}$$

namely, the integral is given as an average. Some definitions, especially in the computer literature, are opposite to this. In our case, we take the average when going to $k_y$-space, not when going to $y$-space. So we have to be careful to divide by the number of points when finding $f(k_y)$ but not when finding $f(y)$.

We nearly always express a spectrum in terms of the positive mode side, due to the reality condition. If the energy is half the average of the square, then the spectral energy is half the sum over the complex square of the modes, since now

$$\sum_{j=1}^{N} f^2(y_j) = \sum_{l=-n}^{n} \left|f(k_y)\right|^2, \tag{10.13}$$

where $j$ is the grid point index. If the reality condition is used, then the information from the negative modes is summed into the positive modes, which now have twice the energy, but only for $l > 0$. In measuring a spectrum we usually do not include $k_y = 0$ anyway, so then the spectrum is simply the nonzero modes in a real-to-complex finite Fourier transform (FFT, in the RFFT variant). In some computer implementations this average is already done and in others it is not. One has to know which way that goes.

### 10.3.2 Spectra in the 2D domain

We are usually interested in the perpendicular spectrum, either the wavenumber spectrum in $y \leftrightarrow k_y$ or the isotropic spectrum in $\{xy\} \leftrightarrow k_\perp$, where $k_\perp$ is the perpendicular wavenumber, with the binning described below. In a three-dimensional setting these are functions also of $s$, where we use $s$ for the parallel coordinate generally and $z$ specifically only when it is the Cartesian one with no transformations (cf. Chapter 12).

For the wavenumber spectrum we have the one-dimensional decomposition,

$$f(x, y, s, t) = \sum_l f_l(x, s, t) \exp[i\, k_y\, y], \tag{10.14}$$

where $\{xys\}$ are the three-dimensional coordinates, and the mode range in the RFFT is now understood. The half-squared amplitude is given by

$$A_f(t) = \mathscr{V}^{-1} \int d\mathscr{V}\, \frac{1}{2} f^2(x, y, s, t) = \left(N_x\, N_y\, N_s\right)^{-1} \sum_{\text{grid}} \frac{1}{2} f^2(x, y, s, t), \tag{10.15}$$

where the $N$'s are the grid point counts in each dimension and the sum is over grid points. The wavenumber spectrum of the half-squared amplitude is given by

$$A_f(k_y, t) = (N_x\, N_s)^{-1} \sum_{\text{grid}} \left|f(k_y, x, s, t)\right|^2, \tag{10.16}$$

where if $l = 0$ it is multiplied by 1/2 but not otherwise. Then, the sum given by

$$\sum_{l=0}^{n} A_f(k_y, t) = A_f(t), \tag{10.17}$$

agrees with the spatial average (this has to be checked thoroughly within any computational package).

In two dimensions, if the proper RFFT2D is used, the considerations are the same. The representation is

$$f(x, y, s, t) = \sum_l \sum_m f_{lm}(s, t) \exp[i(k_x x + k_y y)] \qquad k_\perp = (k_x^2 + k_y^2)^{1/2}, \tag{10.18}$$

where $\{ml\}$ are the mode ranges for $k_x$ and $k_y$ on their respective periodicity lengths. If the $x$-direction is periodic then the same procedure is used, except that in the real-to-complex FFT the mode range is 0 to $N_y/2$ for $k_y$ and $[-N_x/2, N_x/2]$ for $k_x$ (note that $y \to k_y$ is always done first). In a Dirichlet domain one can use sine transforms, but one can also use a windowing given by

$$f(k_\perp, s, t) = \oint \frac{dx}{L_x} \oint \frac{dy}{L_y} \exp[-i(k_x x + k_y y)] f(x, y, s, t)\, w_f(x), \tag{10.19}$$

where the windowing function is given by

$$w_f = 1 - \exp[-100(x + L_x/2)^2] - \exp[-100(x - L_x/2)^2]. \tag{10.20}$$

It screens off the edges so that the same periodic numerics can be used for measurement even though not in the computation. This cannot be used as a transformation in a computation, but it is sufficient for measurement.

This $f(k_\perp, s, t)$ is the wavenumber spectrum in two dimensions, taken with the appropriate RFFT2D. We then have to turn that into the isotropic spectrum, which is one-dimensional. The isotropic perpendicular spectrum is given by

$$A_f(k) = (N_s)^{-1} \sum_{\text{grid}} \sum_{\mathbf{k} \in \mathscr{K}} |f(k_\perp, s, t)|^2, \tag{10.21}$$

where

$$\mathbf{k} \in \mathscr{K} \quad \text{if} \quad k - \frac{1}{2}k_0 < \left(k_x^2 + k_y^2\right)^{1/2} < k + \frac{1}{2}k_0, \tag{10.22}$$

as in Eq. (3.36) from Chapter 3. The fundamental wavenumber $k_0$ in the isotropic perpendicular spectrum is defined as $2\pi$ divided by the shorter of $L_x$ or $L_y$ since this is what constrains the largest isotropic eddy size. The average over $s$ is present for three-dimensional cases. Done this way, the equivalence between the spatial and spectral amplitudes remains.

These quantities are measured time by time, but in a turbulent process they fluctuate rapidly. Hence the averaging of the wavenumber spectrum over the other coordinates and the isotropic spectrum over the third coordinate if present (this is why they look better for the 3D cases). We usually want to average them over time, particularly for nonlinear energy transfer processes. For saturated turbulence the interval can be arbitrarily long. For decaying cases or during the transition to nonlinearity from a linear start, the interval is shorter but in any case the average over only ten snapshot steps is better than the instantaneous version. The time average of any of these is simply over a set of snapshot steps over a given time interval $\mathscr{T}$, as discussed above,

$$A_p(k_y) = \int \frac{dt}{\mathscr{T}} A_p(k_y, t), \tag{10.23}$$

for the wavenumber spectrum of an amplitude.

Spectra involving different variables, as in the ExB transport of electrons, are measured in similar fashion, as in

$$\begin{aligned} F_e(k_y, t) &= (N_x N_s)^{-1} \sum_{\text{grid}} \tilde{n}_e^*(k_y, x, s, t) v_E^x(k_y, x, s, t) \\ F_e(k_y) &= \int \frac{dt}{\mathscr{T}} F_e(k_y, t), \end{aligned} \tag{10.24}$$

for the wavenumber spectrum of the ExB transport of electrons in a three-dimensional case. In a transfer spectrum in the density equation, $v_E^x$ would just be replaced by another quantity such as a parallel gradient or a nonlinearity.

These last forms, by averaging over as much of the statistical fluctuation as possible, tell us the most about the mode structure.

### 10.3.3 In drift wave turbulence

As an example, we can look at the relationship among the variables $\phi$, $n$, $h = n - \phi$, and $\varpi$ in 2D drift wave turbulence. The respective amplitudes are then $A_p$, $A_n$, $A_h$, and $A_w$. For moderate levels of dissipative coupling the electrons are close to adiabatic at the larger scales but less so at the smaller scales. This tells us that the nonlinear advection of the vorticity, $[\phi, \varpi]$, is more able at smaller scales where the factors of $k_\perp$ are larger to compete with the dissipative coupling, $Dh$, since $D$ is independent of $k_y$ in this model. This is also reflected in $A_h(k_y)$, which follows $A_n$ at small scales where the direct cascade sends enough free energy into $A_n$ so that $A_n > A_p$. At the larger scales, though, we have $A_h \ll A_n \approx A_p$, the signature of nearly adiabatic electrons. Due to these characteristics, we also find that the transport spectrum has a peak displaced to the short wavelength side of most of the free energy. These results are shown in the left frame of figure 10.2.

It is also interesting to note the spectrum of the vorticity, $A_w(k_y)$, in this figure. As expected due to the fact that it contains more factors of $k_\perp$ than $A_p$, the vorticity has a flatter spectrum than $A_p$ or $A_n$ for the state variables. This is important to keep in mind, as the nonlinear vorticity dynamics is important to drift wave turbulence overall, and therefore spatial resolution of the density or electrostatic potential is not sufficient to ensure an adequately resolved computation. In this model $\omega$ and $h$ have very similar dynamics, due to the constancy of the coupling mechanism, and exhibit similar spectra. Finally, the fact that the high-$k_y$ drop of $A_w$ proceeds right to the last

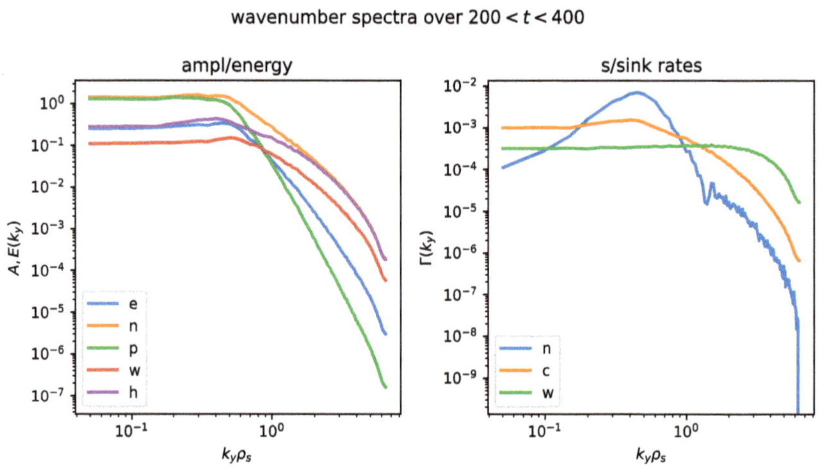


**Figure 10.2.** (left) Wavenumber spectra of $\tilde{n}_e$, $\tilde{\phi}$, $\tilde{h}_e$, and $\varpi$ in normalised units, respectively, labelled by 'n', 'p', 'h', 'w', and of the ExB energy, $E_e$, labelled by 'e', and (right) of the source/sink rates due to gradient drive ('n') and resistive and subgrid dissipation ('c' and 'w'). The flux, not shown, has the same shape as the gradient drive. These quantities are averaged over time and the $x$-direction. The characteristics of note are that $\tilde{\phi}$ and $\tilde{n}_e$ track each other at low $k_y$, with $\tilde{h}_e$ small, but that the small scales see mostly density disturbances, with $\tilde{\phi}$ small. For weak to moderate coupling, $\tilde{h}_e$ and $\varpi$ also tend to track each other. This follows from the competition between the dissipative coupling and a strongly $k_\perp$-dependent excitation mechanism, which is dependent on the polarisation nonlinearity (ExB advection of the vorticity). The resulting transport peaks at intermediate wavelength. Note the peakedness of the drive and relative flatness of the sinks (these are $k_y$ spectra, not $k_\perp$).

mode indicates a well-behaved result (i.e. there is no sharp upward turn). The strength of the vorticity nonlinearity at small scales guarantees pressure against the capability of the numerical scheme. In electromagnetic cases the spectrum of $J_\|$ also has this role and both should be shown, as was the case for many MHD turbulence papers in the past.

## 10.4 Energetics

Beyond the amplitudes and spectra of the disturbances we may apply these same diagnostics to the terms corresponding to the major processes in the energetics: each step in the pathway from drive to dissipation. Drift wave energetics takes the pathway

$$\text{gradient drive} \rightarrow \tilde{p}_e \leftrightarrow \tilde{J}_\| \rightarrow \tilde{\phi} \rightarrow \tilde{J}_\| \rightarrow \text{dissipation}, \qquad (10.25)$$

where we write $\tilde{p}_e$ generically to cover both $\tilde{n}_e$ and $\tilde{T}_e$ in the fluctuations of the dependent variables in turbulence.

The path through $\tilde{\phi}$ is necessary to get the first of these arrows, which represents the drive, simply the product of the gradient and the transport. The second arrow is the adiabatic transfer (adiabatic coupling, compressional along field lines), given by $\tilde{J}_\|\nabla_\|\tilde{p}_e$. The third and fourth arrows represent the more familiar MHD Alfvén transfer, $\tilde{J}_\|\nabla_\|\tilde{\phi}$. Both these transfers can go either way as they are of indefinite sign and represent a coupling. In the dissipative coupling model the same process is simplified by

$$\text{gradient drive} \to n \to \phi \to n \to \text{dissipation}, \tag{10.26}$$

through the common term $Dh$ in each of two equations. Finally, the dissipation is simply $\eta_{\|}\tilde{J}_{\|}^2$ in the four-field resistive models with no temperature dynamics. In a drift wave computation like the ones shown so far this transfer pathway is all there is between $\tilde{p}_e$ and $\tilde{\phi}$, but in more general situations other channels are available. Diagnosis of the energetics is important for proper understanding of the dynamics in such situations.

The energetics of the dissipative coupling model is given by

$$-\phi\,\frac{\partial \varpi}{\partial t} = -\phi\big(-[\phi,\, \varpi] - Dh - \mathscr{D}(\varpi)\big), \tag{10.27}$$

$$n\,\frac{\partial n}{\partial t} = n\big(-[\phi,\, n] + \omega_n v_E^x - Dh - \mathscr{D}(n)\big), \tag{10.28}$$

in normalised units, where $\omega_n$ is the gradient strength. The first term on the right-hand side in each equation is the ExB advection, conserving energy separately but giving the nonlinear transfer. The $\omega_n n v_E^x$ term gives the gradient drive as flow down the gradient in the presence of a positive fluctuation amplitude. The dissipative coupling term in the first equation gives transfer in from the thermal free energy and is the only available overall input. The dissipative coupling term in the second equation is a combination of transfer out and the resistive sink, arranged with the density in the energetics (lacking a separate equation for the current). The last term in each equation is the subgrid dissipation. These various processes are divided by twice the energy, expressing them as forcing rates which can be compared to growth rates of linear instabilities. Hence, we have

$$\Gamma_T = \frac{1}{2E_T}\frac{\partial E_T}{\partial t} = \Gamma_n - \Gamma_c - \Gamma_w, \tag{10.29}$$

after integrating the above over the domain and dividing by $2E_T$. In this expression, the transfer terms cancel out. But we wish to know how the transfer processes work, so we measure them as well. The transfer terms are defined such that positive values indicate energy into the given range in the spectrum.

The gradient drive, resistive dissipation, and subgrid dissipation are given by

$$\Gamma_n = \frac{1}{2E_T}\mathscr{V}^{-1}\int d\mathscr{V}\,[\omega_n n v_E^x], \tag{10.30}$$

$$\Gamma_c = \frac{1}{2E_T}\mathscr{V}^{-1}\int d\mathscr{V}\,[Dh^2], \tag{10.31}$$

$$\Gamma_w = \frac{1}{2E_T}\mathscr{V}^{-1}\int d\mathscr{V}\,[n\mathscr{D}(n) - \phi\mathscr{D}(\varpi)], \tag{10.32}$$

respectively, where $E_T$ is the total free energy given by

$$E_T = \mathscr{V}^{-1} \int d\mathscr{V} \, \frac{1}{2}\left[ | \nabla_\perp \phi |^2 + n^2 \right], \tag{10.33}$$

noting that the gradient drive is proportional to the transport times the gradient strength. In evaluations we can compute $E_T$ using

$$E_T = \mathscr{V}^{-1} \int d\mathscr{V} \, \frac{1}{2}\left[ (-\phi\varpi) + n^2 \right], \tag{10.34}$$

as the difference is an integral over a divergence which is assumed to vanish due to the boundary conditions.

The spectra of the drive and dissipation are then given by

$$\Gamma_n(k_y) = \frac{1}{E_T} \int \frac{dt}{\mathscr{T}} \int \frac{dx}{L_x} \, \mathrm{Re}\left[ n^*(k_y, x, t)(\omega_n v_E^x)(k_y, x, t) \right], \tag{10.35}$$

$$\Gamma_c(k_y) = \frac{1}{E_T} \int \frac{dt}{\mathscr{T}} \int \frac{dx}{L_x} \, \mathrm{Re}\left[ h^*(k_y, x, t)(Dh)(k_y, x, t) \right], \tag{10.36}$$

$$\begin{aligned} \Gamma_w(k_y) = \frac{1}{E_T} \int \frac{dt}{\mathscr{T}} \int \frac{dx}{L_x} \\ \mathrm{Re}\left[ n^*(k_y, x, t)(\mathscr{D}(n))(k_y, x, t) - \phi^*(k_y, x, t)(\mathscr{D}(\varpi))(k_y, x, t) \right], \end{aligned} \tag{10.37}$$

where the relevant term in each case is computed and then fed to the RFFT as just another scalar quantity. The factors in the integration are due to the normalisation of the RFFT (see above). These are given in a positive sense in the right frame of figure 10.2, above.

In figure 10.3, the drive and dissipation processes are shown together with the transfer processes as signed quantities. The spectrum of a transfer process is given by the same multiplier ($-\phi$ or $n$) times the quantity in question, in the same manner as just done, but now we preserve the sign in the equation. So for the nonlinear transfer within the ExB and thermal free energy we have

$$T_E(k_y) = \frac{1}{E_T} \int \frac{dt}{\mathscr{T}} \int \frac{dx}{L_x} \, \mathrm{Re}\left[ \phi^*(k_y, x, t)([\phi, \varpi])(k_y, x, t) \right], \tag{10.38}$$

$$T_n(k_y) = \frac{1}{E_T} \int \frac{dt}{\mathscr{T}} \int \frac{dx}{L_x} \, \mathrm{Re}\left[ -n^*(k_y, x, t)([\phi, n])(k_y, x, t) \right]. \tag{10.39}$$

The spectra of the dissipative coupling transfer in the ExB and thermal free energy, the latter combined with the resistive dissipation, are given by

$$T_p(k_y) = \frac{1}{E_T} \int \frac{dt}{\mathscr{T}} \int \frac{dx}{L_x} \, \mathrm{Re}\left[ \phi^*(k_y, x, t)(Dh)(k_y, x, t) \right], \tag{10.40}$$

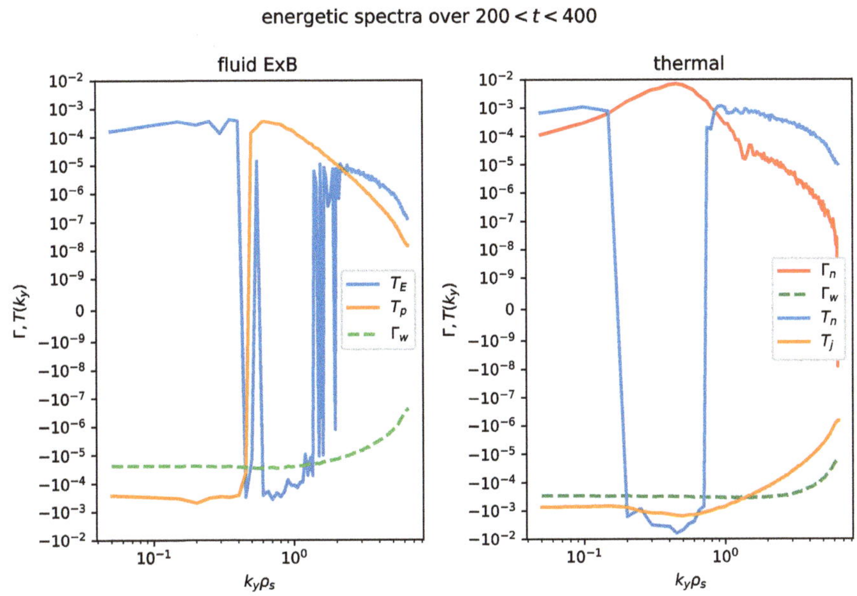


**Figure 10.3.** ExB fluid (left) and thermal (right) energetics for the dissipative coupling case with $D = 0.1$. The curves are in symmetric-log form, as some of them are of indeterminate sign. The dashed curves are subgrid dissipation in each equation (both labelled $\Gamma_w$). The solid curves are the gradient drive ($\Gamma_n$), and the nonlinear ExB transfer ($T_E$ and $T_n$) and separate effect of dissipative coupling ($T_p$ and $T_j$), in each equation, respectively. The gradient drive supports overall transfer to the ranges outside the peak, in the density equation, and general inverse ExB transfer. The ExB energetics is supported by transfer in from the current, modelled as $\phi(Dh)/2E_T$ in normalised units (the two orange curves). The dissipative transfer spectrum is positive at high $k_y$ but negative at low $k_y$, balancing the inverse ExB energy cascade. The direct cascade in thermal free energy dumps significant energy into the smallest scales, where it is balanced by both dissipation channels. The transfer proceeds in a direction given by the excess of $n$ over $\phi$ or vice versa.

$$T_j(k_y) = \frac{1}{E_T}\int \frac{dt}{\mathcal{T}} \int \frac{dx}{L_x} \operatorname{Re}\left[-n^*(k_y, x, t)(Dh)(k_y, x, t)\right]. \qquad (10.41)$$

The gradient drive and subgrid dissipation spectra are also shown. The latter is now shown to be negative-definite and split according to each energy component.

The driver for all the energetics is $\Gamma_n$, determined by the ExB transport flux down the gradient. The nonlinear transfer $T_n$ reacts to this. There is a positive signal also at long wavelength, not to be confused with large scale since the low-$k_y$ component includes all $k_x$, including small scales. The transfer out to the ExB energy combined with the resistive dissipation is shown by $T_j$, with the subgrid dissipation giving a sink in both components. The ExB turbulence exists solely due to coupling to another quantity which is supported by the drive. In this case $T_p$ is the only process transferring energy into the ExB fluid energy component. The transfer in is all at short wavelength, and it supports the cascade process to longer wavelengths. At long wavelength this net input is balanced not by subgrid dissipation but by transfer back out to the thermal free energy, through what is essentially the parallel current.

The cascade dynamics proceeds in both directions, which tends to result in the spectrum of the dissipation to be flatter than that of the drive, as the energy balance of the spectral ranges on both sides of the drive peak adjust to the nonlinear transfer. The driven modes have a particular structure, but the randomness of the cascade phenomenon results in the creation of modes of arbitrary structure at lower and higher $k$. This scattering is very effective as it involves all possible three-wave triplets in the entire system. All the damped eigenmodes are thereby kept alive, and they do most of the dissipation. Much of the transfer in or out of each component also results from the nonlinear ExB transfer within the component.

The cascade dynamics ensures that there is always activity at low and high $k_y$ outside the main driven range. The linear coupling proceeds in the direction given by whatever excess in the one or the other variables results from the cascades. Even in this simplified model there are several processes acting at once, and with the turbulence in statistical saturation, all of them have to be in balance. The short correlation time we will see in the next Section shows that the nonlinear interactions are mostly incoherent, even if the linear coupling by itself would produce coherent eigenmodes.

## 10.5 Correlations

In some cases it is important to know the correlation time and lengths, especially when comparing with the predictions of a simple model based on scaling a set of equations or on the typical scales of a linear instability. Analytical models may make assumptions about the scales based on isotropy as well as the wavelength and growth rate of the strongest instability. Measuring the correlation length in both the $x$- and $y$-directions indicates how isotropic the turbulence is, and these lengths and the correlation time may be compared to the wavelength and the growth time of the linear instability. Measuring the correlation time indicates how important the incoherent part of the dynamics is to the whole.

In each case the correlation length or time is given by the $1/e$ width of the corresponding correlation function. We take the time as an example. The normalised temporal correlation function $F(\tau)$ for any scalar quantity $f$ is given by

$$F(\tau) = A(\tau)/A(0) \qquad A(\tau) = \int \frac{dt}{\mathcal{T}} \, \mathcal{V}^{-1} \int d\mathcal{V} \, f(t)f(t+\tau), \tag{10.42}$$

automatically normalised, with each integral actually an averaged sum over the time interval $\mathcal{T}$. If $\mathcal{T}$ is long enough then $F(\tau)$ is symmetric about zero. The curve is normally smooth enough that $F(\tau)$ is equal to $e^{-1}$ only for two values of $\tau$ in a positive/negative pair with roughly equal magnitudes. The correlation time $\tau_c$ may be defined as the average of these two magnitudes.

The correlation lengths may be measured in a similar fashion, but if the coordinate domain is periodic the integrals may be replaced by sums over the spectra. For example, using a wavenumber spectrum, the normalised spatial correlation function in the $y$-direction is

$$F(\Delta y) = A(\Delta y)/A(0) \qquad A(\Delta y) = \sum_l |f_l|^2 \cos(k_y \Delta y), \tag{10.43}$$

where $l$ is the mode number. This $F(\Delta y)$ is obviously both symmetric about zero and periodic. It is expected that $\lambda_y$ is significantly smaller than $L_y$ so that the periodicity artefact does not affect the measurement of $\lambda_y$ for a turbulence case. Similar considerations apply for the $x$-direction whether or not it is periodic.

To improve the statistics, the integrals in each of these correlation functions for one direction, including time, are also averaged over the others. For example, $f(x, y, t)f(x + \Delta x, y, t)$ in a two-dimensional case is averaged over $(x, y, t)$ rather than just $t$. In a three-dimensional computation it may be averaged over $(x, y, s, t)$ or some subdomain, especially if the geometric coefficients depend on the position along the magnetic field lines.

In figure 10.4 the spatial correlation functions for $n$ in the dissipative coupling model with $D = 0.1$ are shown for the each independent variable $(x, y, t)$, averaged over volume and time. Since simple rollover (also called cshift) functions for arrays are very fast, these are used instead of the cosine formula for $x$ and $y$. In figure 10.5 the temporal correlation functions for $n$ and $\varpi$ are shown. For these, a series $f(y, t)$ measured at $x = 0$ is taken for each variable, sampled at $\Delta t = 0.2$ for 1024 samples, for all $N_y = 256$ $y$-points. The range of $\tau$ is $[-255, 255]$ times $\Delta t$, and the integration interval $\mathcal{T}$ is the central 512 points, so there are no sample boundary issues. The correlation lengths are $\lambda_x = 2.60$ and $\lambda_y = 2.44$ for the density, and the correlation times are $\tau_c = 2.95$ for the density and 2.07 for the vorticity. Scatter in these values is a few 0.01 among runs with different initial states, that is, the realisation limit is comparable to the grid scale. This feature of $\lambda_x > \lambda_y$, even slightly, is common in this model, in which the linear drive process favours modes with $k_x < k_y$ which are relatively extended down the gradient. We will see this change as the models get more complicated. The correlation time is short, reflecting the dominance of small scales in the dynamics. In three-dimensional models the scales will get slightly larger

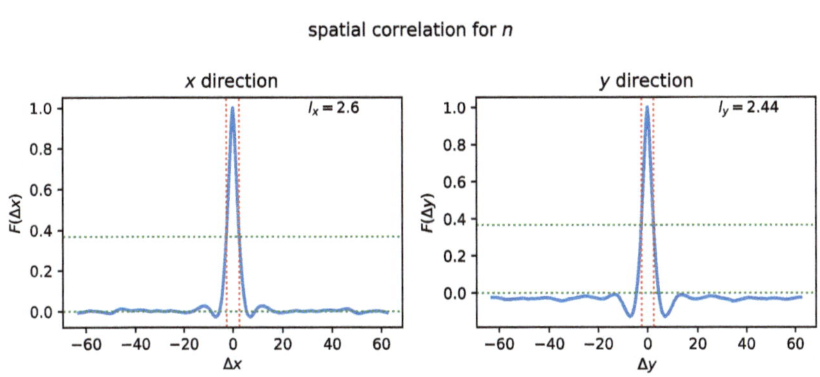


**Figure 10.4.** Spatial correlation functions for $\tilde{n}_e$ in each case averaged over volume and time over $200 < t < 400$. The horizontal dashed line is drawn at the level $1/e$, and the vertical lines then give the corresponding correlation length. The correlation lengths are much shorter than the domain size. The negative dips in $F(\Delta y)$ are vestiges of the wave-like signal seen in the linear phase, below.

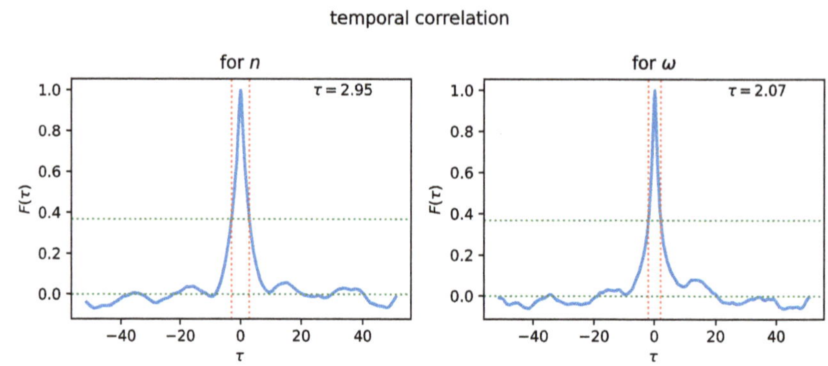


**Figure 10.5.** Temporal correlation functions for $\tilde{n}_e$ and $\omega$, averaged over $y$ and integrated over $250 < t < 350$ so as to avoid overlap using the $200 < t < 400$ time sample. The horizontal dotted line is drawn at the level $1/e$, and the vertical lines then give the corresponding correlation time. The correlation times are much shorter than the duration of the saturation phase.

but otherwise the tendency to scale with $\rho_s$ and $c_s/L_\perp$ will remain. With the results on instabilities and these correlation results in mind, we generally find that the correlation time in turbulence is much shorter than the growth time of an instability and shorter still compared with the time is takes a linear eigenfunction to settle.

## 10.6 Linear growth phase versus turbulence

In a linear or quasi-linear computation, each of which assume linear mode structure even while allowing more general time dependence, these damped eigenmodes are absent at late times, and all one sees is the structure of the most (or only) unstable eigenmode. In turbulence, by contrast, all degrees of freedom are maintained at finite amplitude. It is therefore of interest to examine some of this mode structure information during a well-defined linear growth phase starting with the same random-phase density field with $\phi = n$ as usual, but at an RMS amplitude of $10^{-6}$, which is low enough to ensure negligible nonlinear effect. The case $D = 0.1$ is the same, and only the starting amplitude is different.

The time traces as in figure 10.1 are shown with the same format in figure 10.6. We see now that the starting amplitude is just barely low enough. The drive and dissipation rates rise until the strongest instability crystallises out. It is a linear instability which remains so long as the ExB advection process is too small for effect. The growth rate can be read from the run of $\Gamma_T$ which together with the drive and sink rates $\Gamma_{n,c,w}$ serves as a control against the corresponding rates in the turbulence. The linear growth phase lasts until about $t = 110$ when the drive rate begins to drop. Initial saturation occurs when $\Gamma_T$ first crosses zero, but statistical saturation, when all curves are stationary, is only after $t = 200$, after which this case and the one with nonlinear start are convergent. The main note of interest is (1) the subgrid dissipation ('w'), initially negligible, rises to collect the nonlinear energetics in the turbulence, and (2) the drive rate itself is much lower in the turbulence than during the growth phase. Nonlinear drive rates in general are a lot lower than corresponding instabilities. Saturation in turbulence is defined by the statistical zero given by a

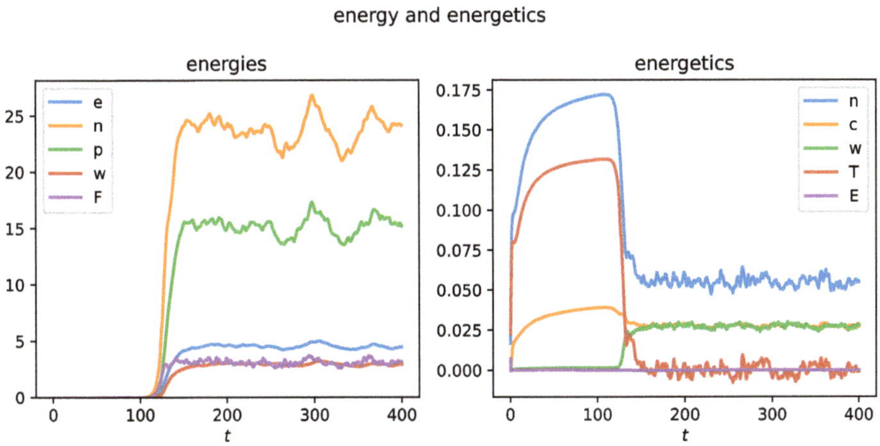


**Figure 10.6.** Time traces of the energetics for the linear growth phase of a companion run of the case with $D = 0.1$, with an initial amplitude $10^{-6}$ rather than the $10^0$ nonlinear start. The format is the same as in figure 10.1, with which it is compared. The linear growth and drive rates are much larger than the drive rate in the turbulence, to which this case converges in time. Initial saturation is defined as the time at which $\Gamma_T$ first drops to zero. Transition to turbulence is very rapid.

time range in which the mean value of $\Gamma_T$ is at least one, preferably two, orders of magnitude smaller than the standard deviation. As we will see, this is why three-dimensional cases in particular have to be run for so long.

Next is the wavenumber spectra, shown with spatial morphology of the vorticity in figure 10.7, at three time points, one early enough in the linear phase to see the peak, one in the turbulence, and one in the middle near the time of initial saturation. The linear phase exhibits eddy flows at a given wavelength extended down the gradient. This extension increases right up until the moment of initial saturation. One sees the rest of the spectrum rising to the level of the peak, and during this phase all modes outside the peak rise together—it is only in fully developed turbulence later on that the cascade tendencies we have discussed impose themselves. There is not much change in the morphology at initial saturation, just that the stable modes acquire enough energy to stop the growth. Only then does the flow advection property of the nonlinearities start, and then the morphology changes. In the fully developed turbulence, the spectrum has fleshed out and the vorticity begins to exhibit its nonlinear structure, which in this case involves the spatial intermittency seen in the last time point.

Lastly, the spatial correlation functions for $n$ at $t = 100$, near the end of the linear growth phase, for this case are shown in figure 10.8. The format is the same as in figure 10.4, with which one can compare. The difference between the linear structure and the turbulence is directly given. The linear state is a wave train, which progressively becomes clearer as time goes on—in the absence of the nonlinear processes. This evolution is still underway when the linear growth phase is stopped by initial saturation, which is why we can say the initial amplitude is just barely low enough. The correlation length in the $x$-direction increases slowly, here 9.13, and is

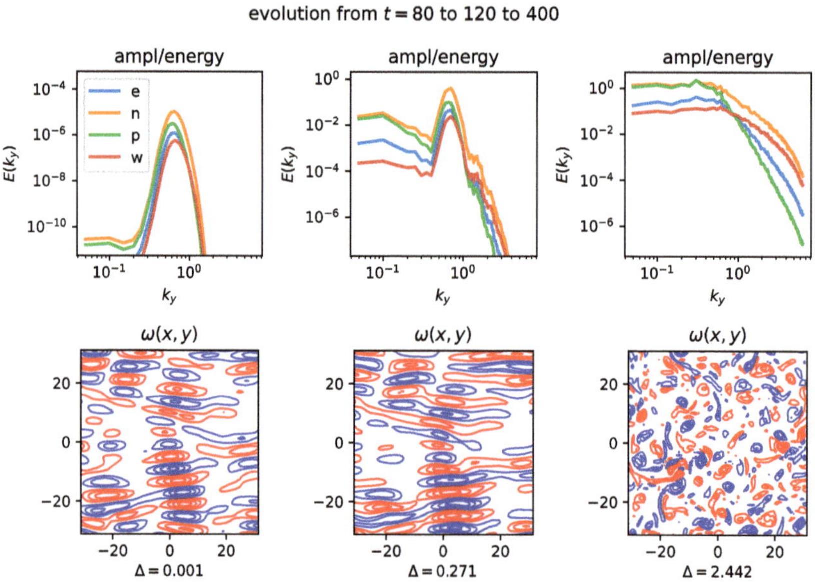


**Figure 10.7.** Wavenumber spectra of the amplitudes (top row), with the same format as in figure 10.2, and spatial morphology of the vorticity (bottom) at three times, during the linear growth phase (left), just when nonlinearities start, at initial saturation (centre), and later in the fully developed turbulence (right). Once all the energetic processes find their long-term balances and the spectrum has fleshed out, statistical saturation and the emergence of nonlinear mode structure defines the nature of the turbulence.

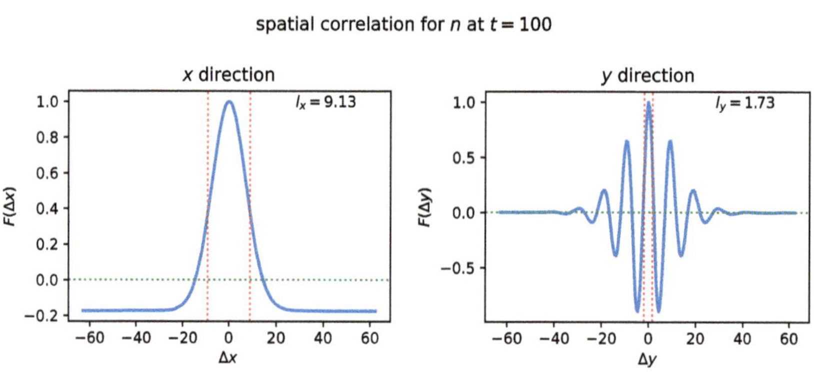


**Figure 10.8.** Spatial correlation functions for $\tilde{n}_e$ for the case with $D = 0.1$ taken at $t = 100$, during the linear growth phase of this companion run. The format is the same as in figure 10.4, for the fully developed turbulence, with which it is compared. In the $x$-direction the correlation length is much longer than in the $y$-direction or in the turbulence, and in the $y$-direction the correlation function partially shows a sinusoidal character.

maximum in the following frame at $t = 110$ before collapsing down to the level it has in the turbulence. In the $y$-direction the correlation length is not only lower (it is a short wavelength instability), but the correlation function itself has a different shape. For a perfect wave train the correlation function would also be sinusoidal, since $\cos(y + \Delta y) = \cos y \cos \Delta y - \sin y \sin \Delta y$, which when multiplied by $\cos y$ and averaged and normalised yields $F(\Delta y) = \cos \Delta y$. This is also why $F(\Delta y)$ has those negative lobes outside the first zero in figure 10.4. It is a vestige of the linear state. So again here we do not see any sign of the mode structure in turbulence until that phase is fully developed. Initial saturation is not enough.

All in all, the mode structure in fully developed turbulence is completely different from the same measurements during the linear growth phase. We will find that this holds for practically any case of turbulence in a magnetised plasma in a three-dimensional setting, which is why doing this now is relevant. We have a simplified model which can be run by anyone with a personal laptop—high performance computing is not necessary to learn the basic phenomenology of this type of turbulence (by contrast especially to three-dimensional convection in fluids!). Not only is the look and feel different, the statistical properties and the morphology are measured to be different.

Moreover, in turbulence one finds signatures in the energetics which not only do not exist for the linearised version of the same case, but also could not possibly exist. The energy transfer into the ExB part of the system cannot possibly be negative in a linear mechanism such as dissipative coupling, as that is the only way to get energy into this component in the first place. The direction of transfer through a linear mechanism can only be reversed by sufficient nonlinear redistribution in some part of the spectrum. This is what we see in figure 10.3. Hence in turbulence even this model shows energetic pathways that cannot possibly result from a system described by a set of intrinsically unstable modes.

Quasi-linear modelling, which takes the properties of unstable linear eigenmodes into an expression of the turbulence, does not take this type of process into account. Hence we will not treat that in any detail within this work. To understand turbulence one has to understand the underlying energetics and the processes that give it its basic description in terms of signs and spectrum. It is common for the signatures of the dominant linear instability to be different from those of the turbulence, for which is it therefore not a reliable guide.

## 10.7 Randomness

One of the things to understand about turbulence is that it is a nonperiodic phenomenon involving an effectively arbitrarily large number of degrees of freedom. Both correlation times and lengths are expected to be limited in some sense, and the above examples show that they are. How random the phenomenon is can be characterised by looking at the distribution of amplitudes as well as their averages. The RMS amplitude of a signal is the standard deviation relative to the mean. For disturbance amplitudes sampled over the entire spatial domain the mean is usually very close to zero, but even if in a more limited sampling the mean may be subtracted away and then the signal normalised to its standard deviation,

$$f = N^{-1}\sum_i f_i \qquad \Delta f = \left[N^{-1}\sum (f_i - f)^2\right]^{1/2} \qquad A_i = \frac{f_i - f}{\Delta f}. \tag{10.44}$$

For quantities which have a significant nonzero mean (comparable to or larger than the standard deviation), we still measure these by subtracting off the mean value before normalising to the deviation. This allows comparison of the shape of the distribution to a shifted Gaussian in general cases.

We then measure the distribution of amplitudes of $A_i$; typically four standard deviations about zero are enough. The one-dimensional space with $A$ as a coordinate is split up into equal bins over the interval $[-4, 4]$ for an odd number of bins evaluated as cell-centred volumes in this phase space (i.e. one is centred at the origin). For each point in the sample, the bin into which $A_i$ falls is incremented by one, and then all bins are divided by the number of points. The resulting histogram is shown as $P(A)$.

For a random sampling the distribution $P(A)$ is given by a Gaussian in $A$, whose definition is $(2\pi)^{-1/2}e^{-A^2/2}$. One can also find $P(A)$ for a perfectly coherent sine wave by sampling it at equal intervals; the result is a double hump curve with a minimum at zero and maxima at the wave amplitude. This is far enough removed from the results in turbulence that we will not have to show it. There is also a model for a passively advected quantity in a random flow field, according to which $P(A)$ should have more of a Lorentzian character, which has significantly extended tails compared to a Gaussian due to the tendency towards spatial intermittency (the passive scalar distribution is given by $c[1 + kA^2]^{-\gamma}$, with $c = 1/1.38\,624\sqrt{\pi}$, $k = 0.08$, and $\gamma = 1 + 1/2k$; see the Further Reading Section, below). The amplitude distributions for $\phi$, $n$, $h$, $\varpi$, and the two ExB velocity components $v_E^x$ and $v_E^y$ are shown in figure 10.9, for the case with $D = 0.1$. Since this case was done in a periodic domain, the mean in each case is actually zero. The distributions for $\phi$ and $n$ are indistinguishable from Gaussian. They are definitively different from a double-peaked coherent distribution, as befits the short correlation time and lengths. However, the distributions for $h$, then $v_E^x$ and $v_E^y$, and especially $\varpi$ have increasingly flatter tails tending toward the passive scalar distribution. This reflects the increasing power of the turbulent dynamics for variables which track smaller scales.

The transport dynamics shows somewhat more interesting behaviour. In figure 10.10 we display the distributions of the transport, $nv_E^x$, both pointwise and as an average over surfaces of constant $x$ (later, this will be seen as a 'flux surface average'). Both the pieces in the transport, $n$ and $v_E^x$, are closer to Gaussian, but the transport is not. This largely reflects the increased probability of smaller amplitudes in the product of Gaussian quantities, but the distribution of the transport is also skewed towards positive values. The favouring of small amplitudes is compensated for in the average, which is much closer to Gaussian. However, even the average is skewed, as if one rotated the curve slightly to the left. Relative to the mean, the negative values are slightly favoured by the peak, but the tail is flatter to the positive side. The zero itself is shifted by the average so that the curve is centred. The statistics for the pointwise case is $F_e = 3.01 \pm 17.70$ and $\langle F_e \rangle = 3.01 \pm 1.19$ for the average. The skewness in both $F_e$ and $\langle F_e \rangle$ is mostly the shift away from zero, although the curve is tilted slightly to the negative side with a wider tail to the positive side (this feature would be more extreme in strongly driven convection, for example). It results from the energetic favouring (through the gradient drive) of any

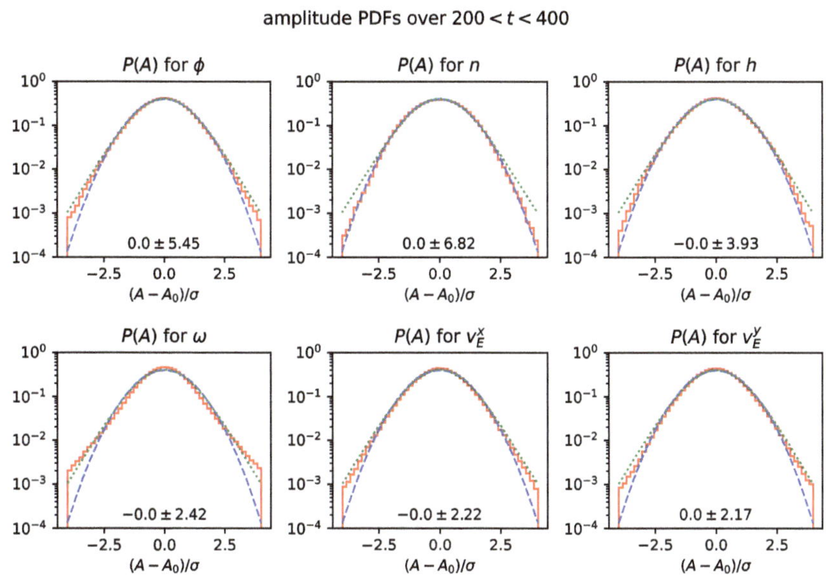


**Figure 10.9.** (top row) Amplitude probability distributions of $\phi$ (left), $n$ (centre), and $h$ (right), for the case with $D = 0.1$. (bottom row) The same for $\varpi$ (left), $v_E^x$ (centre), and $v_E^y$ (right). The distributions of $\phi$ and $n$ do not differ significantly from Gaussian (blue dashed curve). But from $h$ to $v_E^{x,y}$ and especially $\varpi$ the tails get flatter and the distributions are closer to the passive scalar model (green dotted curve).

disturbances which happen to find themselves aligned with down-gradient flows of the same sign. These ones tap the free energy and do the transport. A nonzero net transport reflects their preeminence. The same signal for the more adiabatic cases shows little skewness, following from the comparatively weaker gradient drive.

## 10.8 Cross coherence

Randomness in a single amplitude may mask a significant degree of correlation between the amplitudes of two different quantities, each of which may have a Gaussian distribution. The most obvious case is that of adiabatic electrons in otherwise random turbulence: both $\tilde{n}_e$ and $\tilde{\phi}$ have a near-Gaussian distribution but they are perfectly correlated because they are held equal to each other by arbitrarily strong cross coupling. We may quantify this through two measurements. One is the cross coherence. Each sampled quantity is normalised to its standard deviation and zero mean as shown above. Then, each point in the sample contains two numbers, the normalised amplitudes of each quantity, and may be placed in a two-dimensional space with these two amplitudes, call them $A_1$ and $A_2$, as the coordinate axes. If the two quantities are perfectly correlated the points lie on the line $A_2 = A_1$, or if they are anticorrelated they lie on the line $A_2 = -A_1$. If they are perfectly uncorrelated then the points will be scattered with a density independent of the angle about the coordinate origin, $A_2 = A_1 = 0$.

We measure this as a sort of two-dimensional histogram. The space $(A_1, A_2)$ is binned in the same way as the one-dimensional axis $A$ for single quantity

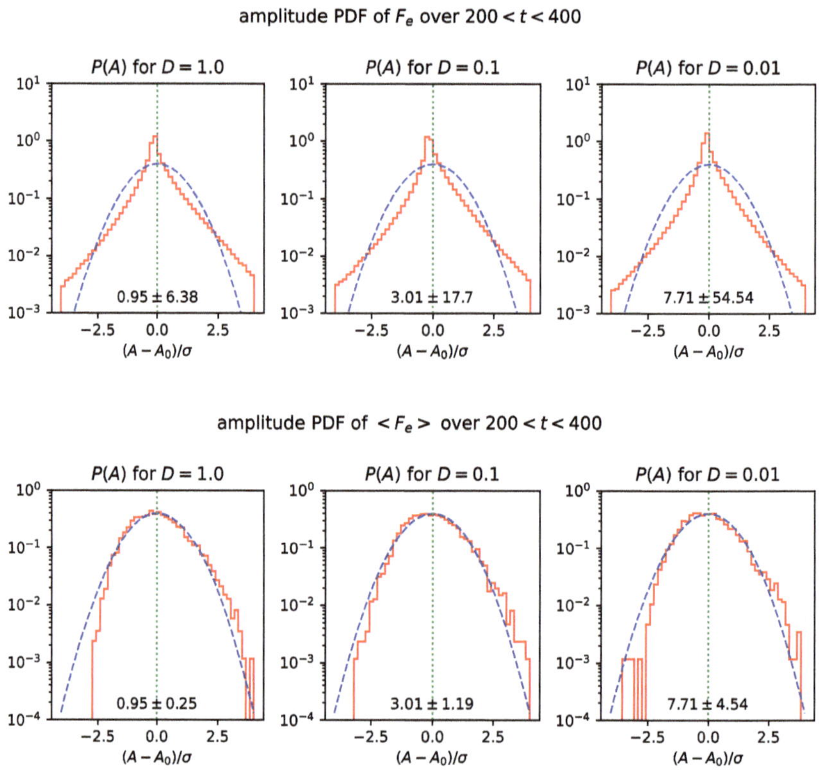


**Figure 10.10.** (top row) Amplitude probability distributions of the pointwise turbulent ExB transport flux ($F_e$), for the cases with $D = 1.0$, 0.1, and 0.01. (bottom row) The same for the surface-averaged flux ($\langle F_e \rangle$). The pointwise flux distribution is peaked at small values, while the average is closer to Gaussian. The standard deviation is larger than the mean for the pointwise distributions. The mean values for $\langle F_e \rangle$ are actually larger than their standard deviation, but the curves are shown as such to compare their shapes to the Gaussian (blue) curves.

distributions, and then each point is placed into the bin in which $A_1$ and $A_2$ fall for that point, and the corresponding bin number incremented. Contours of $P(A_1, A_2)$ are drawn in figure 10.11 for the cases with $D = 1.0$, $D = 0.1$, and $D = 0.01$ in the dissipative coupling model. These are ellipses oriented along the line $A_2 = A_1$, reflecting a cross coherence intermediate between the two extremes of perfect correlation and perfect relative randomness. The ellipses are typically thin for drift wave turbulence, reflecting the tendency towards an incompletely adiabatic response. As we will see in later Chapters, the ellipses become circles when the coupling between $\tilde{p}_e$ and $\tilde{J}_\parallel$ is left out, which reflects a character of passive advection.

The second measure is the relative phase between the two Fourier components for each given wavenumber. Taking the Fourier decomposition in the $y$-direction, for both $\tilde{n}_e$ and $\tilde{\phi}$ for example, we define the relative phase as

$$\alpha_l(\tilde{n}_e, \tilde{\phi}) = \mathrm{Im} \log \left( \tilde{n}_{e\,l}^{*} \tilde{\phi}_l \right), \tag{10.45}$$

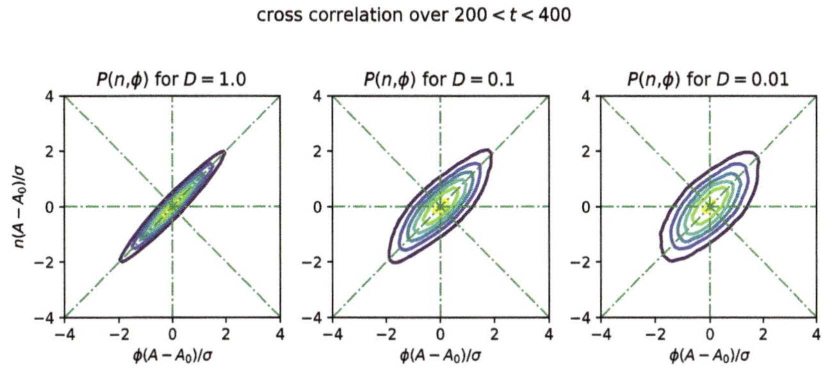


**Figure 10.11.** Cross coherence between $\tilde{n}_e$ and $\tilde{\phi}$, for the cases with $D = 1.0$, 0.1, and 0.01. These forms reflect the general trend of the adiabatic response with coupling strength, with even weak values showing significant departures from the randomised state, which would produce circular contours here. Cases with $D > 1$ show strong adiabatic coupling, with the normalised difference $\tilde{h}_e$ small everywhere, producing contours on the line from bottom left to top right.

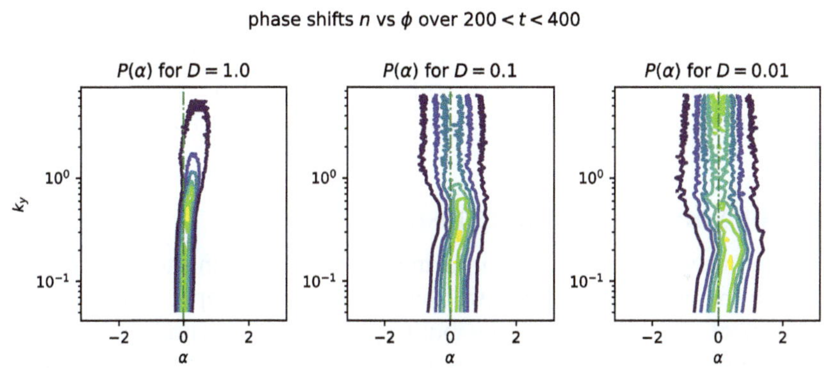


**Figure 10.12.** Phase shift distributions between $\tilde{n}_e$ and $\tilde{\phi}$, for the cases with $D = 1.0$, 0.1, and 0.01. For $D = 1$ and larger the phase shifts are close to zero for all wavelengths ($2\pi/k_y$). For $D = 0.01$ the adiabatic response is still present, however, despite the overly strong turbulence ($\tilde{n}_e/n_e$ as large as $\delta_0^{-1}$). For $D < 1$ the range of strongest drive, indicated by the spectra in figure 10.2, produces slightly larger phase shifts than the rest of the spectrum, but is still closer to zero than to $\pi/2$. A convective instability maximising growth potential would have the flows going down the gradient coinciding with a density fluctuation relative maximum. This would give $\alpha$ in the vicinity of $\pi/2$.

assuming that $\tilde{n}_{el} = a_l\tilde{\phi}e^{-i\alpha_l}$ with both $a_l$ and $\alpha_l$ real. We then assemble the distribution $P(\alpha_l)$ as a one-dimensional histogram over the interval $[-\pi, \pi]$ for each wavenumber $l$. Contours of $P(\alpha_l)$ are drawn in figure 10.12 for the cases with $D = 1.0$, $D = 0.1$, and $D = 0.01$ in the dissipative coupling model. These do better than the simple averages to reflect the character of the phase shift distribution. For adiabatic electrons $P(\alpha_l)$ would be a delta function at zero for each $l$, while a perfectly uncorrelated distribution would be a flat line whose level is the inverse number of bins in the histogram. Perfectly driven linear instabilities would show a narrow distribution at $\pi/2$, reflecting a down-gradient flow through a disturbance with positive amplitude. As we see, for intermediately nonadiabatic drift wave turbulence we find that the distribution is still quite well localised near zero even though it is shifted to positive values of $\alpha$. Most of the signal for $P(\alpha)$ is well below $\pi/2$, and this may be taken as a signature for drift wave mode structure.

## 10.9 Interscale transfer

We have already seen the nonlinear transfer, but we mention it here again as is it part of the energetics. In addition to the transfer terms such as $\tilde{p}_e \leftrightarrow \tilde{J}_\|$, which for electrostatic turbulence is a linear process at each $k$, we have the interscale transfers $T(k_\perp \leftarrow k_\perp')$ for each quantity, reflecting the scattering in wavenumber space by the nonlinear ExB advection. This builds a picture of drift wave energetics which involves longer and shorter wavelengths as well as the $\tilde{\phi}$ and $\tilde{p}_e$ disturbances (in the dissipatively coupled model, $\phi$ and $n$). The simultaneous, oppositely directed nonlinear transfer dynamics in the ExB energy $E_e$ and the thermal free energy $E_n$ is compensated by the linear transfer (cf. figure 10.3). The flow of energy is therefore circular, from low-$k_\perp$ $E_n$ to high-$k_\perp$ $E_n$ through the direct cascade of turbulent mixing, then through the linear kinetic shear Alfvén transfer to high-$k_\perp$ $E_e$, to low-$k_\perp$ $E_e$ through the inverse energy cascade of vortex merger, and then through the linear transfer back to low-$k_\perp$ $E_n$. At all scales the gradient drive and nonadiabatic (Alfvénic) dissipation are operative, with the intermediate range of positive net drive determined self consistently by the dynamics. A diagram of this is shown in figure 10.13.

## 10.10 Three-dimensional diagnostics

We will have new mode structure information when moving to three dimensions, starting with the next Chapter. We will have parallel envelopes, which are averages over $(x, y, t)$ of quadratic quantities built from variables with $(x, y, s, t)$ dependence. These will be the same quantities as above, e.g. $A_p(s)$, just with $s$ dependence. We will also have a parallel wavenumber spectrum, analogous to $k_y$, in which a parallel Fourier mode is defined as

$$f(k_\|, x, y, t) = \int \frac{ds}{L_s} \exp[-ik_\| s] f(x, y, s, t), \tag{10.46}$$

for any scalar, which can be a variable or as in the case of energetics a nonlinear advection bracket. This assumes that the coordinate $s$ follows the field lines. The boundary conditions will preclude using this in computations, but we will only use it for measurement. More details on these are left to the next few Chapters.

## 10.11 Summary—mode structure in turbulence

The basic properties of drift wave turbulence, as measured through the mode structure, are as follows. It is a process of incompressible, two-dimensional ExB drift turbulence occurring in drift planes, locally perpendicular to the background magnetic field, with the drift planes coupled together by the parallel dynamics, especially the adiabatic response of the electrons. The ExB flow disturbances follow

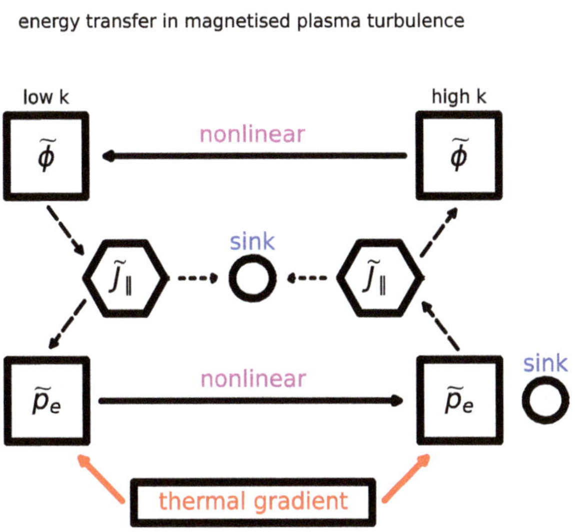


**Figure 10.13.** Diagram of the principal dynamics of energy transfer in magnetised plasma turbulence. The thermal gradient drives electron pressure disturbances ($\tilde{p}_e$) and the parallel current ($\tilde{J}_\parallel$) is dissipative at all scales. Nonlinear transfer via unsteady ExB advection is between scales, shown with solid lines. The scale transfer is mostly through $k_\perp$ even in three dimensions, since the guide field makes the parallel dynamics wave-like. Linear transfer is mostly through the adiabatic response, directed as $\tilde{p}_e \rightarrow \tilde{J}_\parallel \rightarrow \tilde{\phi}$ into the ExB energy at small scales and vice versa at large scales. The parallel current acts as a mediator between the two state variables ($\tilde{p}_e$, $\tilde{\phi}$). Due to the bi-directional nonlinear transfer, the sink spectrum is wider than the drive spectrum, with most of the energetics in an intermediate spectral range. Where this range falls depends on the relative strengths of all the interactions, self consistently, eventually determined by the parameters. The entire spectrum is energetically tied together, and reacts to changes as a unit.

a Eulerian equation and the thermodynamic quantities they transport would attempt to follow passive scalar dynamics. The basic character of this part of the dynamics is a turbulent local cascade process. This description is shown by the energy transfer spectra, which measure the interscale exchanges to occur within about a half decade of wavenumber space, and by the short space and time correlations, and by the fact that the instantaneous spectra are noisier than the average spectra, which are always smooth. This ExB drift plane turbulence is acted upon by dissipative coupling strongly enough that the potential and pressure disturbances, or more generally the flow eddies and the quantities they transport, are always highly correlated. The coupling is, however, not generally so strong that the electrons are adiabatic; rather, the amplitude of $\tilde{h}_e$ is usually a sizeable fraction of the amplitude of $\tilde{n}_e$. Drift wave turbulence is therefore well described as a competition between ExB turbulence in the drift plane, excited by the background gradient, and dissipative coupling, treated in a simplified manner as a relaxation, but proceeding generally through finitely resistive kinetic shear Alfvén wave responses along the magnetic field. The dual ExB/pressure cascade is reflected in the relative flatness of the sink spectra compared with the drive spectra. The vorticity spectrum in $k_y$ is flatter than that of any other advected quantity, while in $k_\perp$ it is collected at shorter wavelength than the energy.

The statistics is close to Gaussian, showing that (1) the system is turbulent and (2) the passive scalar character of the thermodynamic quantities is felt only in the energy transfer properties, that is, the parallel coupling is strong enough that the thermodynamic quantities do not depart statistically from the ExB flow disturbances. The unified nature of the spectrum is seen in the phase shift distributions, specifically in the absence of a range in wavenumber space in which the influence of the coupling is lost.

Finally, it is worth emphasising that this information comes from a detailed energetic and statistical diagnosis of the turbulence, not from superficial information such as linear growth rates, the saturated transport scaling, or visualisations of the disturbances in any of the dependent variables. What is important to understanding at this stage is the qualitative character of the dynamics, not the quantitative results, and not how things look to the eye. Magnetised plasma turbulence is a complicated system which does not lend itself well to simple, heuristic analytical modelling, except perhaps in an ad hoc explanatory mode. But we can get the necessary understanding just as well with this fully developed package of diagnostics, which has the advantage of operating on actual nonlinear solutions to the equations as they stand.

## Further Reading

This Chapter concentrates on energetics because for gradient-driven systems the key question is how the turbulence maintains itself against dissipation, and because the free energy source is in the gradient of a thermodynamic state variable while it is liberated by flow eddies, and the energy transfer channel from transported quantity back to flow stream function is of central importance. Cross correlation among the variables is the most useful diagnostic. The cross coherence diagnostic is standard for more than one variable and is covered in the texts cited below. The phase shift diagnostic was developed for drift wave turbulence and used in B Scott *Plasma Phys. Control. Fusion* **39** (1997) 471. The energy transfer diagnostics originally come from neutral fluid turbulence, e.g. used to demonstrate local cascade dynamics in three-dimensional incompressible turbulence, in A J Domaradzki *Phys. Fluids* **31** (1988) 2747. It was expanded for drift wave turbulence in S Camargo, D Biskamp and B Scott *Phys. Plasmas* **2** (1995) 48; and *Phys. Plasmas* **3** (1996) 3912. The passive scalar distribution referred to in the Section on Randomness is in equation (14) and figure 1 of Ya G Sinai and V Yakhot 'Limiting probability distributions of a passive scalar in a random velocity field' *Phys. Rev. Lett.* **63** (1989) 1962. For an arbitrarily long time the distribution was expected to acquire a logarithmic singularity at zero, which is the signature of strong intermittency in which the fluctuations in the advected quantity do not fill space. See also I Goldhirsch and A Yakhot 'Probability distribution of a passive scalar: A soluble example' *Phys. Fluids A* **2** (1990) 1303.

A large volume of literature exists on the statistical characterisation of a single fluctuating variable, and much of statistical diagnosis of turbulence actually falls into this area. For drift wave turbulence see the review by W Horton *Rev. Mod. Phys.* **7**(3)

(1999). For experimental data on disturbances in fusion plasmas see E Sánchez, *et al.* *Phys. Plasmas* **7** (2000) 1408. Standard techniques are treated in many texts on turbulence, including U Frisch 1995 *Turbulence: the Legacy of A.N. Kolomogorov* (Cambridge: Cambridge University Press); M Lesieur 1997 *Turbulence in Fluids* 3rd edn (Dordrecht: Kluwer); and A S Monin and A M Yaglom 1975 *Statistical Fluid Mechanics: Mechanics of Turbulence* vol 2 (Cambridge, USA: MIT Press).

**IOP** Publishing

Turbulence and Instabilities in Magnetised Plasmas, Volume 1
Fluid drift turbulence
**Bruce Scott**

# Chapter 11

## Three-dimensional drift wave turbulence

### 11.1 Introduction

Having explored drift wave turbulence in the two-dimensional perpendicular plane (also called the drift plane), the Alfvénic responses parallel to the guiding magnetic field, and the drift wave instability mode structure, we allow variations in the third dimension to participate. In principle, three-dimensional (3D) turbulence in the hydrodynamic sense (dominant nonlinearity in all three dimensions) would be possible. However, all we have to do is to allow more three dimensionality in the beginning than will result in the final turbulent state. This will be enough to understand how strong variation in the direction along the guide field is relaxed through the dissipation channels (sound waves through viscosity, and Alfvén waves through resistivity). Moreover, we will find some energetic processes that transfer free energy in the third dimension as well as within the perpendicular wavenumber spectrum. Although the $k_\parallel \ll k_\perp$ flute mode ordering is still assumed, incoherent nonlinearity in the drift plane can excite variation in the parallel direction by broadening the parallel wavenumber spectrum. By allowing all of these processes to occur, and then measuring the transfer energetics to see which ones remain active, and then finding the resulting three-dimensional spectrum, we can find out whether the turbulence tends to remain or becomes more three-dimensional, or whether it tends to align to the guide field and become purely two-dimensional. This intrinsic physics is to be worked out before considering spatial variation of the background magnetic field or a wider set of physical processes. Sorting these questions out is the purpose of this Chapter on the fundamental nature of three-dimensional drift wave (drift-Alfvén) turbulence.

### 11.2 Drift Alfvén model and energetics

The set of equations for this purpose is a union of the sets used to study parallel responses and drift waves, and homogeneous two-dimensional four-field fluid drift

doi:10.1088/978-0-7503-2504-2ch11  

turbulence, in all cases electromagnetic. In the same language as in the previous two Chapters, we have

$$\frac{d}{dt}\rho_s^2\nabla_\perp^2\frac{e\tilde{\phi}}{T_e} = c_s B\nabla_\parallel\frac{\tilde{J}_\parallel/nec_s}{B}, \tag{11.1}$$

$$\frac{\partial}{\partial t}\frac{\tilde{A}_\parallel}{B\rho_s} + \mu_e\frac{d}{dt}\frac{\tilde{J}_\parallel}{nec_s} = c_s\nabla_\parallel\left(\frac{\tilde{p}_e}{p_e} - \frac{e\tilde{\phi}}{T_e}\right) - \frac{c_s}{L_p}\rho_s\frac{\partial}{\partial y}\frac{\tilde{A}_\parallel}{B\rho_s} - 0.51\mu_e\nu_e\frac{\tilde{J}_\parallel}{nec_s}, \tag{11.2}$$

$$\frac{d}{dt}\frac{\tilde{p}_e}{p_e} + \frac{c_s}{L_p}\rho_s\frac{\partial}{\partial y}\frac{e\tilde{\phi}}{T_e} = \Gamma c_s B\nabla_\parallel\left(\frac{\tilde{J}_\parallel/nec_s}{B} - \frac{\tilde{u}_\parallel/c_s}{B}\right), \tag{11.3}$$

$$\frac{d}{dt}\frac{\tilde{u}_\parallel}{c_s} = -c_s\nabla_\parallel\frac{\tilde{p}_e}{p_e} + \frac{c_s}{L_p}\rho_s\frac{\partial}{\partial y}\frac{\tilde{A}_\parallel}{B\rho_s} + B\nabla_\parallel\frac{\mu_\parallel}{B}\nabla_\parallel\frac{\tilde{u}_\parallel}{c_s}, \tag{11.4}$$

with Ampère's law

$$\frac{\tilde{J}_\parallel}{nec_s} = -\beta_e^{-1}\rho_s^2\nabla_\perp^2\frac{\tilde{A}_\parallel}{B\rho_s}. \tag{11.5}$$

These are in physical units, with the dependent variables given dimensionless combinations. The dependent variables are denoted with the tilde, and all other quantities are constant parameters of the background. The only change to Chapter 9 is to replace the partial time derivative with the nonlinear ExB advective one and the linear parallel gradient with the nonlinear one admitting the magnetic flutter effects. So we have the terms involving the field variables in

$$\frac{d}{dt} = \frac{\partial}{\partial t} + \frac{1}{B}\mathbf{b}\times\nabla\tilde{\phi}\cdot\nabla \quad B\nabla_\parallel = B^z\frac{\partial}{\partial z} + B^y\frac{\partial}{\partial y} - \mathbf{b}\times\nabla\tilde{A}_\parallel\cdot\nabla, \tag{11.6}$$

as the set of nonlinearities being added. The magnetic induction term $\partial A/\partial t$ retains the pure partial derivative since it is a field term and part of the parallel electric field (as derived in Chapter 7).

These superscripts denote contravariant components of the vector, which are defined by

$$B^{(x,\,y,\,z)} = \mathbf{B}\cdot\nabla(x,\,y,\,z), \tag{11.7}$$

component by component, as defined by the gradients of the coordinates. The dot product is called a contraction—component pairs are multiplied and then summed. This is a small bit of tensor mathematics, but for now we won't need any more of that. The coordinates are orthogonal: the gradients of any two different ones contracted together yield zero. In this geometry the coordinate gradients are straight, so that we focus on the physics and delay expansion of the mathematical generality until later.

The component of the background magnetic field in the $z$-direction, $B^z$, is assumed to be constant (equal to $B$), and a small $y$-component, $B^y$, which can have $x$-dependence, is added. We assume $B^y \ll B^z \approx B$ and therefore evaluate $\mathbf{b} = \nabla z$ and $B^z = B = B_0$, the reference magnetic field value, and $B^y(x)/B^z$ as a small component allowing for magnetic shear. The coordinates $xyz$ satisfy $\nabla x \times \nabla y \cdot \nabla z = 1$, forming a right-handed system. The above defines the $z$ coordinate as mostly parallel, and the $xy$-plane as the drift plane. The background gradient, defined by a finite value for $1/L_p$, is in the $x$-direction. The above combination then defines the $y$-direction as the drift direction and surfaces of constant $x$ as magnetic surfaces of the background guide field (i.e. $B^x$ is assumed to be zero). The $x$-direction is the direction across the surfaces in which the magnetic field lies ($B^x = 0$) and is thought of as the down-gradient or the cross-surface direction.

The physical parameters are the sound speed $c_s$ and the drift scale $\rho_s$ determined by $B_0$ and the electron temperature $T_e$ and ion mass $m_i$ via

$$c_s^2 = \frac{T_e}{m_i} \quad \rho_s^2 = \frac{m_i T_e}{e^2 B_0^2}. \tag{11.8}$$

The effective frequency represented by the gradient forcing terms is

$$\frac{c_s}{L_p}\rho_s \frac{\partial}{\partial y} \rightarrow \frac{c_s}{L_p} k_y \rho_s = k_y v_D = \omega_*, \tag{11.9}$$

where $v_D$ is the diamagnetic drift velocity and $\omega_*$ is the diamagnetic drift frequency for each $k_y$, the wavenumber of each component in the drift direction. The mass ratio and dynamical beta, assumed to be small, are

$$\mu_e = \frac{m_e}{M_i} \quad \beta_e = \frac{c_s^2}{v_A^2} = \frac{\mu_0 p_e}{B_0^2}. \tag{11.10}$$

The nominal Alfvén speed and electron thermal speed are then

$$v_A = \beta_e^{-1/2} c_s \quad V_e = \mu_e^{-1/2} c_s. \tag{11.11}$$

This is the minimal three-dimensional superset of the four-field model turbulence of Chapter 3 and the drift-Alfvén gradient-driven response system of Chapter 9.

### 11.2.1 Normalised nonlinear equations

In the normalisation scheme we will normalise to the perpendicular scale $L_\perp = L_p$ and leave its ratio to the parallel scale $L_\parallel = 2\pi/K_\parallel$ as a parameter. The language of Chapter 3 is used for the turbulence, which we marry to the linear coupling and forcing terms. The language of Chapter 9 is used for the linear terms and therefore the normalisation. So the factor $K_\parallel L_\perp$ is folded into the normalisation and appears in the dimensionless parameters reflecting the interaction between the parallel and perpendicular dynamics. The perpendicular gradient $\nabla_\perp$ is normalised against $L_\perp$ when acting on a dependent variable. The time derivative is normalised to $c_s/L_\perp$ and the parallel gradient to $K_\parallel$. The normalised quantities are

$$\phi = \frac{e\tilde{\phi}}{T_e} \quad n = \frac{\tilde{p}_e}{p_e} \quad A = \frac{\tilde{A}_\parallel}{B\rho_s}(K_\parallel L_\perp) \quad u = \frac{\tilde{u}_\parallel}{c_s}(K_\parallel L_\perp) \quad J = \frac{\tilde{J}_\parallel}{n_e e c_s}(K_\parallel L_\perp), \tag{11.12}$$

as before. Also as before, the model is isothermal, and so we equate $\tilde{p}_e/p_e$ with $\tilde{n}_e/n_e$ and set the ratio of specific heats to unity.

The spatial Poisson bracket is defined and normalised as

$$[a, b] = \frac{1}{B}\nabla z \cdot \nabla a \times \nabla b \to \delta_0\left(\frac{\partial a}{\partial x}\frac{\partial b}{\partial y} - \frac{\partial b}{\partial x}\frac{\partial a}{\partial y}\right), \tag{11.13}$$

for scalars $a$ and $b$, and the perpendicular Laplacian normalises to

$$\rho_s^2\nabla_\perp^2 \to \nabla_\perp^2 = \delta_0^2\left(\frac{\partial^2}{\partial x^2} + \frac{\partial^2}{\partial y^2}\right), \tag{11.14}$$

namely, we are putting the drift parameter $\delta_0 = \rho_s/L_\perp$ into these operators rather than into the dependent variables, with $xy$ normalised to $L_\perp$ rather than $\rho_s$. The ExB advective derivative and the parallel gradient then normalise to

$$\frac{d}{dt} = \frac{\partial}{\partial t} + [\phi, \,] \quad \nabla_\parallel = \frac{\partial}{\partial z} + \alpha(x)\frac{\partial}{\partial y} + [-A, \,] \quad \alpha(x) = \frac{B^y}{B_0}, \tag{11.15}$$

where $\alpha$ is the field line pitch. The parallel gradient is then re-expressed as

$$\nabla_\parallel = \frac{\partial}{\partial z} + [-\psi_T, \,] \quad \text{where} \quad \psi_T = \psi(x) + A \quad \text{and} \quad \frac{\partial \psi}{\partial x} = -\alpha. \tag{11.16}$$

The gradient drive terms are done in the same way, so that they are folded into both $\nabla_\parallel$ and $d/dt$ with, for example,

$$\nabla_\parallel n - \frac{\partial A}{\partial y} = \nabla_\parallel(n_0 + n) = \frac{\partial n}{\partial z} + [-\psi, (n_0 + n)] \quad \text{where} \quad n_0 = -\omega_n x, \tag{11.17}$$

with $\omega_n$, nominally unity, serving as the gradient drive parameter.

Then we have as nonlinear, normalised equations

$$\frac{\partial \varpi}{\partial t} + [\phi, \varpi] = \nabla_\parallel J, \tag{11.18}$$

$$\frac{\partial}{\partial t}(A + \hat{\mu}J) + \hat{\mu}[\phi, J] = \nabla_\parallel(n_0 + n - \phi) - C\,J, \tag{11.19}$$

$$\frac{\partial n}{\partial t} + [\phi, n_0 + n] = \nabla_\parallel(J - u), \tag{11.20}$$

$$\hat{\varepsilon}\frac{\partial u}{\partial t} + \hat{\varepsilon}[\phi, u] = -\nabla_\parallel(n_0 + n + P), \tag{11.21}$$

for the dependent variable evolution equations, and

$$\varpi = \nabla_\perp^2\phi \quad J = -\hat{\beta}^{-1}\nabla_\perp^2 A \quad P = -X\nabla_\parallel u_\parallel, \tag{11.22}$$

for the auxiliaries and field equations, in which $P$ is the viscous correction to the pressure and $X$ is the parallel viscous diffusivity. The parameters controlling the kinetic shear Alfvén dynamics, that is, the adiabatic response, are

$$\hat{\beta} = \beta_e(K_\| L_\perp)^{-2} \quad \hat{\mu} = \mu_e(K_\| L_\perp)^{-2} \quad C = \frac{0.51\nu_e}{c_s/L_\perp}\hat{\mu}. \tag{11.23}$$

We also have

$$\delta_0 = \frac{\rho_s}{L_\perp} \quad \omega_n = |L_\perp \nabla \log n_e| \quad \hat{\varepsilon} = (K_\| L_\perp)^{-2} \quad X = \frac{\mu_\|}{c_s L_\perp}, \tag{11.24}$$

controlling the drift wave and acoustic part of the dynamics. These can be set arbitrarily as noted in Chapter 9. One can compare Eqs (11.18)–(11.21) with Eqs (3.121)–(3.124) from Chapter 3, noting the similar structure and different language.

### 11.2.2 Energetics

The free energy theorem for the normalised nonlinear equations is formed in the same way as for the linear case in Chapter 9, since the nonlinear $d/dt$ and $\nabla_\|$ are still linear in its arguments. The following relations still hold:

$$\phi\frac{d\varpi}{dt} = \phi\frac{\partial\varpi}{\partial t} + [\phi, \phi\varpi] \quad n\frac{dn}{dt} = \frac{1}{2}\frac{\partial}{\partial t}n^2 + \frac{1}{2}[\phi, n^2], \tag{11.25}$$

for the ExB advection, and

$$\phi\nabla_\| J + J\nabla_\|\phi = \nabla_\|(\phi J) \quad u\nabla_\| X\nabla_\| u = \nabla_\|(uX\nabla_\| u) - X(\nabla_\| u)^2, \tag{11.26}$$

for the parallel gradient, for example, as the reader can show. The cross-quantity energetics in detail is given by

$$-\phi\frac{d\varpi}{dt} + \nabla_\|(\phi J) = J\nabla_\|\phi, \tag{11.27}$$

$$J\frac{\partial A}{\partial t} + \hat{\mu}J\frac{dJ}{dt} = J\nabla_\| n - J\nabla_\|\phi - \omega_n J b^x - C\,J^2, \tag{11.28}$$

$$n\frac{dn}{dt} + \nabla_\|[n(u - J)] = -J\nabla_\| n + u\nabla_\| n + \omega_n n v_E^x, \tag{11.29}$$

$$\hat{\varepsilon}u\frac{du}{dt} + \nabla_\|(uX\nabla_\| u) = -u\nabla_\| n + \omega_n u b^x - X(\nabla_\| u)^2, \tag{11.30}$$

where the down-gradient field perturbations are

$$v_E^x = [\phi, x] = -\delta_0\frac{\partial\phi}{\partial y} \quad b^x = [-A, x] = \delta_0\frac{\partial A}{\partial y}, \tag{11.31}$$

as in the linear cases. The transfer effects on the right side are the same as before, except that $\nabla_\parallel$ is nonlinear and therefore the parallel dynamics now transfers energy between both quantities and scales. The ExB nonlinear advective transfer is within quantities, contained within the $d/dt$ terms, but between scales. The terms with $\omega_n = -\partial n_0/\partial x$ represent the gradient drive depending on the sign of correlation (with the $Jb^x$ one averaging to zero over finite regions). The positive-definite terms with minus signs represent the resistive ($C$) and viscous ($X$) dissipation acting as sinks, along with the subgrid dissipation. Summing these together, the total free energy theorem is

$$\begin{aligned}
&\frac{1}{2}\frac{\partial}{\partial t}\left(u_E^2 + \hat{\beta}^{-1}B_\perp^2 + n^2 + \hat{\varepsilon}u^2 + \hat{\mu}J^2\right)\\
&- \nabla \cdot \left(\phi\delta_0^2\nabla_\perp\frac{\partial\phi}{\partial t} + \hat{\beta}^{-1}\frac{\partial A}{\partial t}\delta_0^2\nabla_\perp A\right)\\
&+\left[\phi, -\phi\varpi + \frac{1}{2}(n^2 + \hat{\varepsilon}u^2 + \hat{\mu}J^2)\right]\\
&+ \nabla_\parallel(\phi J + nu - nJ - uX\nabla_\parallel u)\\
&= \omega_n(nv_E^x + ub^x - Jb^x) - CJ^2 - X(\nabla_\parallel u)^2,
\end{aligned} \tag{11.32}$$

where

$$u_E^2 = \delta_0^2\,|\nabla_\perp\phi|^2 \quad B_\perp^2 = \delta_0^2\,|\nabla_\perp A|^2, \tag{11.33}$$

give the squared ExB and magnetic amplitudes in the turbulence. To this is added the subgrid dissipation, treated next.

### 11.2.3 Subgrid dissipation

In computations the subgrid dissipation treatment is extended as well to the third dimension. As in the 2D cases it is applied to the same quantities which are directly advected by the ExB velocity. The need for it in turbulence to contain the cascade dynamics was shown in Chapter 5. This structure with the subgrid dissipation is given by

$$\frac{d}{dt} = \frac{\partial}{\partial t} + [\phi, ] + D() \quad D() = \nu_\perp(\nabla_\perp^2)^2 + \nu_\parallel\left(\frac{\partial^2}{\partial z^2}\right)^2. \tag{11.34}$$

As in the 2D turbulence models of Chapter 3, the dissipation is tied to the grid resolution, using

$$\nu_\perp = \nu_0\frac{h_\perp^4}{16} \quad \nu_\parallel = \nu_0\frac{h_z^4}{16}, \tag{11.35}$$

assuming $h_x = h_y = h_\perp$, namely, isotropic grid cells in the perpendicular plane. Nominally, $\nu_0$ is set to 1.0, but the cases driven by the gradient ($\omega_n = 1$) with periodic boundary conditions require larger values, to be discussed as they occur. The energetic dissipation is given by

$$\Gamma_w = \frac{1}{2E_T}\mathscr{V}^{-1}\int d\mathscr{V}\left(-\phi\, D(\varpi) + n\, D(n) + \hat{\mu}J\, D(J) + \hat{\epsilon}u\, D(u)\right), \qquad (11.36)$$

and it is measured along with the other terms $\Gamma_n$, $\Gamma_v$, $\Gamma_c$, $\Gamma_s$, and the total, $\Gamma_T$, and its error, $\Gamma_E$, which are defined the same way as in Eqs (9.62)–(9.65) from Chapter 9.

### 11.2.4 Computational Courant limit

With more processes of various types being represented in the equations we have to be more systematic about the check involved in setting the time step in computational cases. The limit for any scheme is set by the consideration that no signal should proceed all the way across a grid cell volume in any direction within one time step. The details bring in a numerical constant given by the properties of the scheme. In our case the value is about 0.2, which translates to a similar effort as in a Runge–Kutta fourth-order, 4-step scheme where the value is about 0.8. The stability analysis which gives these limits is covered in textbooks on numerical methods (see Further Reading).

The Alfvén speed along the part of $B^y$ due to $\psi$ as well as the nonlinearities in $xy$ and the guide field in $z$, together with the ExB velocity in $xy$, enter the check on the Courant limit. In this usage the effective Alfvén speed is the larger of $v_A$ and $V_e$, or $\hat{\beta}^{-1/2}$ and $\hat{\mu}^{-1/2}$ in normalised units. All spatial step sizes must be large enough, or the time step $\tau$ small enough, that

$$0.2\,\tau^{-1} > {}_{\max}\frac{v_E^i}{h_i} \quad \text{with} \quad i \in \{xy\}, \qquad (11.37)$$

for the ExB dynamics, and

$$0.2\,\tau^{-1} > {}_{\max}\frac{V_a}{h_i B_0} b^i \quad \text{with} \quad i \in \{xyz\}, \qquad (11.38)$$

for the Alfvén dynamics, otherwise the time step has to be reduced. In the ExB dynamics the derivatives of $\phi$ compute $v_E^i$ and and in the Alfvén dynamics both $b^z = 1$ and the derivatives of $\psi_T$ (see Eq. (11.16)) are involved, hence both the linear and nonlinear terms. The components are the contravariant ones involved in the $\mathbf{v}\cdot\nabla$ or $\mathbf{b}\cdot\nabla$ contraction operations. This is checked in trial runs for several cases within a suite and then a constant value for $\tau$, which satisfies all of them, is used for the suite.

## 11.3 Periodic cases

For periodic cases we set up a square domain in the drift plane and more points in the $z$-direction than we will eventually need. This is to allow for relaxation into saturation with less variation along the guide field than allowed for by the grid. The same perpendicular plane as in the 2D drift wave cases of Chapter 4 is used. The time step is set as just mentioned. Periodic boundary conditions in the $x$-direction carry a number of issues when finite magnetic shear is considered, as we will address later on in this Chapter. So for now we set the magnetic shear, hence $B^y$, to zero and continue with the straight guide field. The first set of cases will be on a square perpendicular domain with $L_x = L_y = 1$ as in the 2D cases. Then we address the aspect ratio issue with $L_y/L_x \neq 1$, and after that, with the 3D structure becoming clear, we will return to finite magnetic shear.

The boundaries are periodic in all three dimensions. The periodicity is given by

$$f(x + L_x, y, z) = f(x, y + L_y, z) = f(x, y, z + L_\|) = f(x, y, z), \tag{11.39}$$

where in normalised units the periodicity length in $z$ is $2\pi$. As in the 2D cases this is implemented using two guard cells at each boundary, loaded once per time step, so that the spatial derivatives can be taken at all active points without further ado.

The step schemes are the 3rd-order stiffly stable (Karniadakis) scheme for time, the 4th-order bracket (Arakawa) scheme for the $xy$-plane, both as in Chapter 3, and the 4th-order centered differences for $\partial/\partial z$ as in Chapter 8. At the end of each time step $\phi$ from $\varpi$ and $A$, $J$ from $A + \hat{\mu}J$ are solved for as in Chapter 3.

Initially we try three cases built around the warm plasma case from Chapters 8 and 9, with differing domain sizes and resolution. The small case has $\delta_0 = 0.016$, which means $L_\perp = 62.5\rho_s$, with $-\pi < z < \pi$ in normalised units, and a grid of $64 \times 64 \times 64$ in the three coordinates. The medium case has $\delta = 0.008$, or twice as large in units of $\rho_s$, and a grid of $128 \times 128 \times 128$. The large case doubles the grid resolution of the medium case with a grid of $256 \times 256 \times 256$. In each of these the parameters are

$$\hat{\beta} = 5 \quad \hat{\mu} = 2.7 \quad C = 13.9 \quad X = 10 \quad \hat{\epsilon} = 10^4, \tag{11.40}$$

reflecting a scale ratio of $K_\| L_\perp = 10^{-2}$. The value of $\hat{\mu}$ reflects deuterium ions. The value of $C$ reflects $\nu_e = 10c_s/L_\perp$. The magnetic shear is set to zero, so that $\psi = 0$ in Eq. (11.16). The small and medium cases have approximately $k_0\rho_s = 0.1$ and 0.05 for the fundamental wavenumber in units of $\rho_s$, for comparisons to the corresponding 2D cases in Chapter 4. The time step is set to $\tau = 0.002$, which passes the Courant test for all cases unless the flow pathologies addressed below emerge.

The initial state is formed starting with a 2D random-phase density field determined by

$$n_{\mathbf{k}} = A_0\, k \left(1 + (k/k_1)^6\right)^{-1} \exp(i\Theta), \tag{11.41}$$

with the expression involving $k$ giving the spectrum and $\Theta$ a random number on $[0, 2\pi]$ giving the phases. We choose $k_1 = 0.3$, which starts most of the energy at long wavelength. The constant $A_0$ chosen such that the RMS amplitude is set to $\delta_0$. This is Fourier transformed back to $xy$-space and labelled $n_{2D}$. It is then multiplied by an envelope function in $z$ so that

$$n(x, y, z) = n_{2D}(x, y)\exp\left[-(z/2.5)^2\right] \tag{11.42}$$

gives the 3D initialisation. The vorticity is initialised finite, with $\varpi = 0.1\delta_0 n$. The initial potential field $\phi(x, y, z)$ then results from the field equation (Eq. 11.22).

### 11.3.1 Decaying cases

The first cases are set to decay from the nonlinear start as was done for the 2D model systems in Chapter 3. Hence, we start with $\omega_n$ set to zero. The nonlinearities are all in the perpendicular plane, essentially the same as in the four-field two-fluid MHD case at the end of that Chapter. The new element is the variation along the guide field, and the coupling between the density $n$ and ExB potential $\phi$ mediated by the current $J$ in the adiabatic response. This is quantified by measuring the parallel wavenumber spectra in the same way as the $k_y$-spectra in 2D, except now we Fourier decompose in $z$,

$$n(x, y, z) = \sum n_m(x, y)\exp(imz) \quad m = k_\parallel / K_\parallel, \tag{11.43}$$

and form the spectra by averaging over $x$ and $y$ for each $m$, for example,

$$E_n(m) = \mathcal{A}^{-1} \oint\oint dx\, dy\, \mathrm{Re}[n_m^*(x, y)\, n_m(x, y)] \quad \mathcal{A} = \oint\oint dx\, dy \tag{11.44}$$

for the thermal free energy due to $n$. Both positive and negative mode numbers are kept. They are displayed on a symmetric-log horizontal scale, so both sides of the spectrum and the zero are shown. The case with the $256^3$ grid is displayed.

The time traces for the free energy components, dissipation (sink) rates, and transfer rates are given in figure 11.1. The energy components are

$$E_e = -\frac{1}{2}\phi\varpi \quad E_b = \frac{1}{2}J(A + \hat{\mu}J) \quad E_n = \frac{1}{2}n^2 \quad E_u = \frac{1}{2}\hat{\epsilon}u^2, \tag{11.45}$$

all in volume average. The $E_b$ is a combination of the magnetic energy $E_B = JA/2$ and electron parallel kinetic energy $E_j = \hat{\mu}J^2/2$ and is referred to interchangeably as total magnetic energy or Ohmic energy. Note that $-\phi\varpi$ is equivalent to $|\nabla_\perp\phi|^2$ and $JA$ to $|\nabla_\perp A|^2$ in volume average. The sink rates $\Gamma_c$, $\Gamma_s$, and $\Gamma_w$ are given above. The transfer rates at this point are

$$T_j = J\nabla_\parallel\phi \quad T_a = J\nabla_\parallel n \quad T_s = u\nabla_\parallel n, \tag{11.46}$$

all in volume average and normalised by the $1/2E_T$ factor along with the $\Gamma$'s. The $T_j$ is the Alfvénic transfer from $E_b$ to $E_e$, and $T_a$ is the adiabatic transfer from $E_n$ to $E_b$. The two together constitute the three-dimensional version of the adiabatic coupling we have been referring to. Defined positive, they denote eventual transfer from $E_n$ to $E_e$, maintaining the turbulent ExB flow. The $T_s$ is the acoustic transfer from $E_n$ to $E_u$, the more familiar sink channel for thermal free energy towards viscous dissipation. We will elaborate on these effects, especially their spectra, in the next few Chapters.

There is an initial Alfvén response phase, as the density fluctuation field launches kinetic shear Alfvén waves which dissipate the part of $h = n - \phi$ with finite $k_\parallel$ through the $J$, as in Chapter 8. This is over by about $t = 50$, as shown in the first

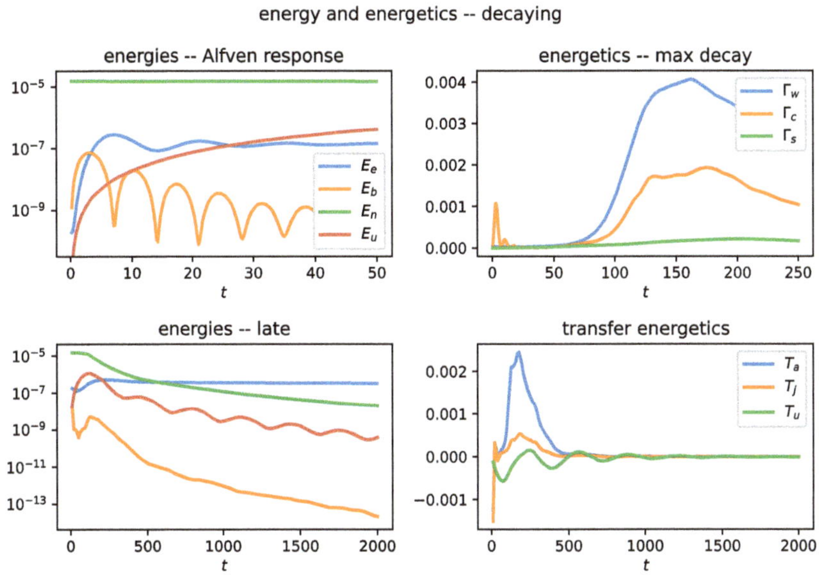


**Figure 11.1.** Temporal evolution of the free energy pieces with time on three time scales (ExB: $E_e$, magnetic: $E_b$, thermal: $E_n$, sound: $E_u$), all in volume average, for the decaying case with square domain. (top left) Relaxation of the initial Alfvén transients, with secular rise of $E_e$, $E_b$, and $E_u$ at the expense of $E_n$. (top right) The intermediate phase, with most of the decay mediated by the parallel dynamics, as the decay of $E_n$ is accelerated by coupling to the other pieces. (bottom left) The entire evolution, most of which is the late phase during which the 2D dynamics is dominant. (bottom right) Evolution of the three transfer channels (adiabatic: $T_a$, Alfvénic: $T_j$, acoustic: $T_s$), all as rates (volume average, normalised to the total energy). These are signed such that positive $T_a$ and $T_j$ reflects transfer from $E_n$ through $E_b$ to $E_e$, and negative $T_s$ reflects transfer from $E_n$ to $E_u$. The small relative amplitude of $E_b$ is significant, reflecting efficient throughput with small content in the dynamics of the current.

frame. The intermediate phase has most of the energy still in $E_n$ since most of the spectrum is on the $k_\perp\rho_s < 1$ side. As the other three pieces decay, the evolution of the ExB energy is very slow as in the 2D fluid cases with the other components as passive scalars. The last frame shows the transfer channels in the parallel dynamics. Much of the thermal free energy in $E_n$ spills out mostly through the adiabatic ($T_a$) but also through the acoustic ($T_s$) channels. Hence, the main decay event is much sharper in time than is the case in two dimensions. It does not occur purely dissipatively, although the energetic dissipation ($\Gamma_c$ and $\Gamma_s$) is finite. The energy first transfers out of $E_n$ and then into $E_b$ and $E_u$, which then dissipate. The ExB kinetic energy content $E_e$ increases, and the reason is found by measuring the Alfvén transfer $T_j$. Since both $T_a$ and $T_j$ are positive, free energy is being taken out of $E_n$ and transferred to $E_e$ through $\nabla_\parallel J$. The transfer process is very efficient, shown by the fact that the energy content $E_b$ in both the magnetic energy $E_B$ and electron parallel kinetic energy $E_j$, which are comparable, remains much smaller than either $E_n$ or $E_e$. Transfer into the sound wave energy $E_u$ due to $T_s$ is smaller than either $T_a$ or $T_j$ but persists for longer, so that $E_n$ is removed about equally over time through both adiabatic and sound wave channels. Even in the absence of gradient drive, this is basic drift wave physics in the energetic processing. We will find that this set of effects underlies practically anything we study. Even if the drive mechanism is something else, the nonlinear drift wave energetics is always involved in mediating saturation or decay of this type of turbulence. At late times the content in $E_b$ and $E_u$ is very small and most of the energy is in the $k_{\|} = 0$ flute mode. The energetic error starts in the $10^{-6}$ range during the initial Alfvén transient response, and it drops to below $10^{-8}$ at later times.

The scale transfer dynamics is shown by the isotropic perpendicular spectra and the parallel spectra, especially during the peak of this drift wave decay phase. These are shown averaged over the $150 < t < 200$ range, in figures 11.2 and 11.3, respectively. Along with the energy components, the half-squared amplitudes $A_p$, $A_w$, and $A_j$ for $\phi$, $\varpi$, and $J$, are given in the spectra, the Γ's are given in the energetics, and the nonlinear transfer rates are given on the right-hand side. Most of the changes seen in figure 11.1 occur within this time range. The vorticity is within the decade centred upon $k_{\perp}\rho_s = 1$, before this range preferentially decays at late times. The $k_{\|}$-spectra show the dominant $m = 0$ flute mode but the width of the vorticity, which carries the most finite $m \gg 1$ activity, is largest during this time at

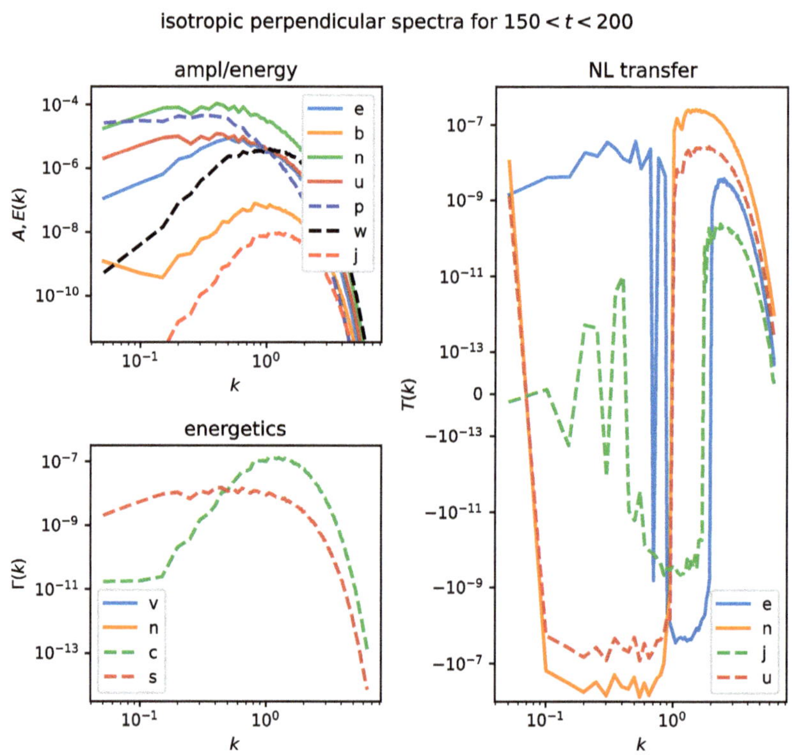


**Figure 11.2.** Isotropic perpendicular spectra for the decaying case with square domain during $150 < t < 200$, roughly the phase of maximum energetics. The labels are 'e', 'b', 'n', and 'u' for $E_e$, $E_b$, $E_n$, and $E_u$, with 'p', 'w', and 'j' for $A_p$, $A_w$, and $A_j$, respectively. In the energetics the labels are for each of the Γ's as drive and sink rates. The vorticity and resistive dissipation peak near the $\rho_s$ scale, with the viscous dissipation is flatter and smaller. The inter-scale transfer due to advection of $\varpi$, $n$, $\hat{\mu}J$, and $\hat{\epsilon}u$, is labelled 'e', 'n', 'j', and 'u', respectively. It is from large scales to small in the thermal free energy ('n') but inverse in the ExB energy ('e'), as in the 2D cases. Transfer in the parallel channels ('j' and 'u') has the same signature as for $E_n$ but is subdominant.

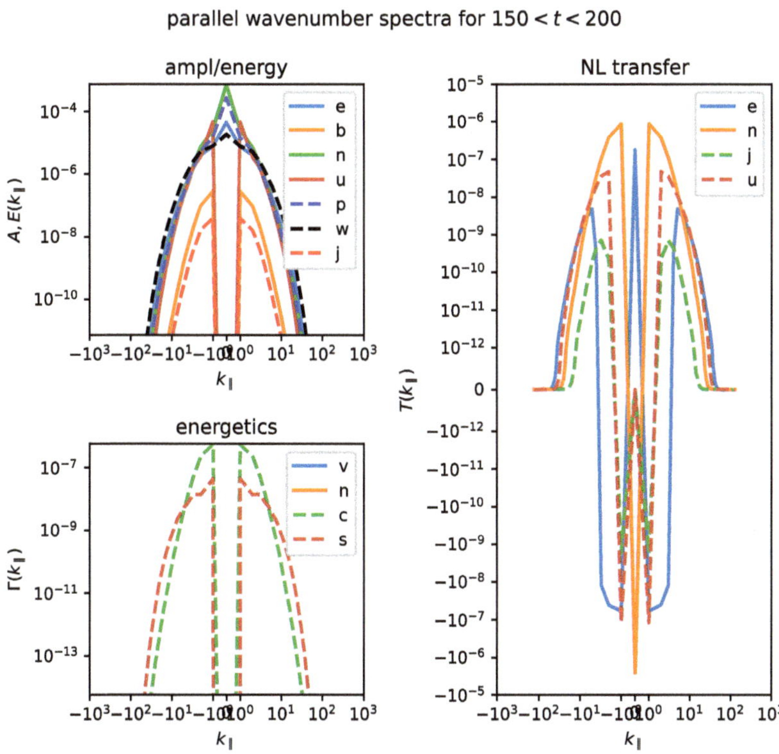


**Figure 11.3.** Parallel wavenumber spectra for the decaying case with square domain during $150 < t < 200$, roughly the phase of maximum energetics. The format and labels are the same as in the previous figure. The flute mode at $k_\parallel = 0$ is dominant, especially in $\phi$ ('p') and $n$ ('n'), while the widest spectrum is in the vorticity ('w'). The width is determined by the dissipation (bottom left), and does not extend effectively to the maximum value on this grid (128). The inter-scale transfer (right) is out of the flute mode into all others in the thermal free energy ('n') but for the ExB energy ('e') it is into the flute mode, as well as into short wavelengths from the low but finite $k_\parallel$ region. This is what drives the flute mode most effectively in all of these cases, driven or decaying.

about $m \sim 10$, with essentially all activity down to the small-amplitude tails within $m \sim 30$, namely, much less than the allowed maximum of 128 for this grid. At late times, the $k_\parallel$-spectrum narrows with practically all the energy in the flute mode, and the $k_\perp$-spectrum shows the small-scale component decaying away the faster, while the largest-scale $\phi$ component remains. The ExB energy peak at about $k_\perp \rho_s \sim 0.2$ is the largest piece for about $t > 600$, and it slowly shifts to larger scale during the rest of the evolution. The spectral transfer is shown in both these figures, for the perpendicular and parallel scales, respectively. The perpendicular transfer shows the same tendency as in the 2D models, with the ExB energy transfer robustly inverse, towards larger scales, and the thermal free energy transfer direct, towards smaller scales. The parallel transfer also shows this, with the thermal transfer out of the flute mode ($k_\parallel = 0$) and into all the others, and the ExB transfer out of the lower range into the higher but, importantly in light of what is to be shown below, also into the flute mode.

The spatial morphology during the decay phase is shown in figure 11.4. The turbulence forms its initial structure with the density pieces being churned up by the ExB eddies, which have the look and feel of a fluid. At later times, as shown in figure 11.5, after most of the drift wave energy transfer has taken place, the flow structure has more of the large-scale isolated vortex type of the unforced 2D cases, with the vorticity forming the same type of sheet structures and then isolated cells we saw before. This is because of the dominance of the flute mode, so that most of the later evolution is in fact a 2D phenomenon.

Later, we will see this large-scale structure disappear when the magnetic shear is increased away from zero. But for now it is something we have to investigate. We also have the fact that this high resolution is necessary for the 3D decaying cases. Figure 11.6 shows the decay rate for the cases with $N_x$ = 64, 128, 256, and a control test with $N_x$ = 512. The grid is given this node count in all three dimensions for all but the last case, which has $N_z$ = 256 in the parallel direction. Although the first two cases look more or less convergent, they are not convergent on the correct result, which has the rapid decay phase separate from the initial Alfvén transient. After that initial phase, the decay rate drops back to zero until the nonlinear couplings establish themselves, and then it peaks sharply around $t$ = 150. The two lower resolution levels wash these two phases together. Moreover, the isolated vortex character is only found with the higher resolution. This emphasises the need to check this once again for any new set of cases. The behaviour seen here will change in the driven cases and then again when magnetic shear is introduced.

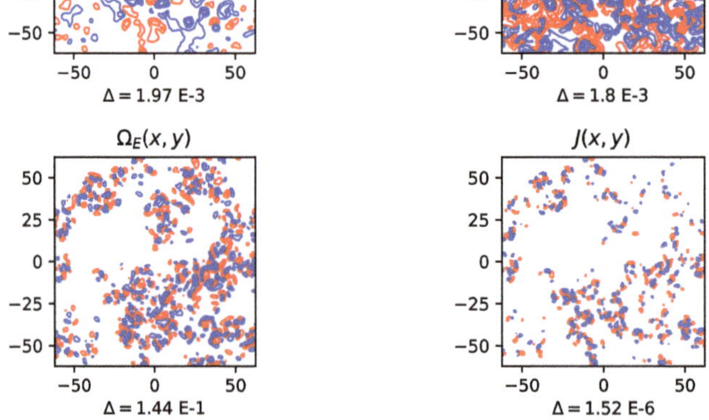


**Figure 11.4.** Spatial morphology of the decaying case with square domain, at $t$ = 180, during the phase of maximum energetics. The turbulence is well mixed across scales, and somewhat spatially intermittent as it spreads along the magnetic field lines.

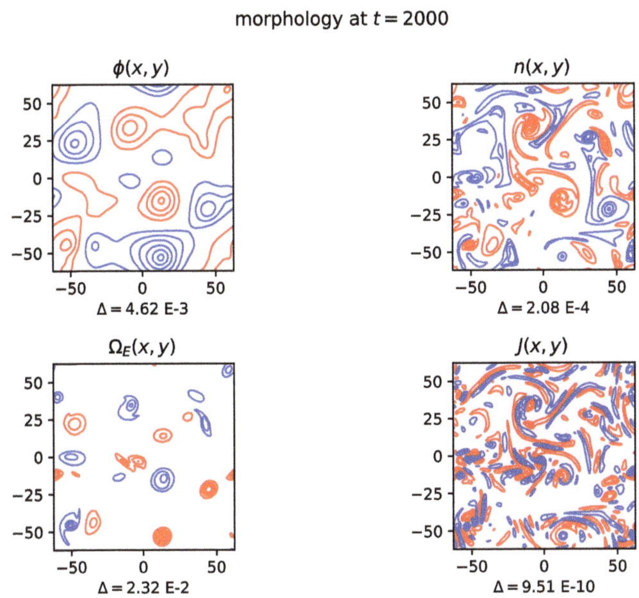


**Figure 11.5.** Spatial morphology of the decaying case with square domain, at $t = 2000$, during the late phase in which most of the energy is in the ExB motion. The turbulence shows the isolated vortices structure seen in the 2D hydrodynamic decaying case (Chapter 4, Section 4.3.1). By this time, the flute mode ($k_\| = 0$) has almost all the energy.

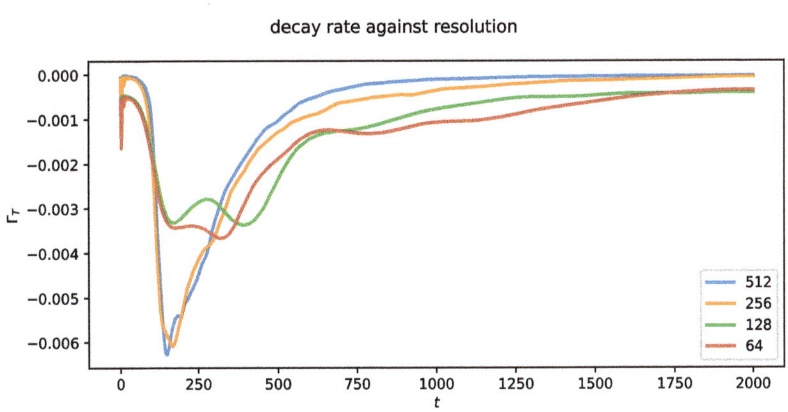


**Figure 11.6.** Dependence of the decay rate ($\Gamma_T < 0$) on spatial resolution. The $N_x = 256$ case is the lowest which is convergent, as the $N_x = 512$ case shows. The strength of the decay and shortness of its phase is missed by coarser cases, which dissipate the smaller scales and shorter wavelengths more strongly. Note that by themselves the two coarsest cases might give the impression of approximate convergence.

### 11.3.2 Driven cases

We next consider the driven cases, with $\omega_n = 1$ turned back on. Concentrating again on the case with $N_x = 256$, we find very violent turbulence and the necessity to reduce the time step to $\tau = 0.001$ accordingly. We also have to increase the

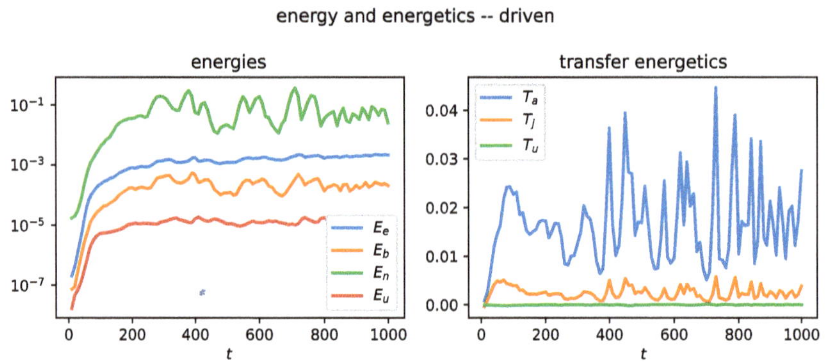


**Figure 11.7.** Temporal evolution of the free energy pieces with time for the driven case with periodic square domain (left). Evolution of the three transfer channels (right), with the same labelling as in figure 11.1. Note the bursty character of especially $E_n$ and $T_a$, on the logarithmic scale. The signs of $T_a$ and $T_j$ indicate adiabatic response energetically from $n$ to $\phi$ through $J$, with $T_s$ indicating the acoustic coupling energetically from $n$ to $u$. With the drive in $E_n$, the adiabatic coupling transfer to $E_b$ is the largest, with a remainder going into $E_e$. The small relative amplitude of $E_b$ reflects efficient throughput with small content in the dynamics of the current.

perpendicular and parallel subgrid dissipation to $\nu_\perp = 10$ and $\nu_\| = 10$, about which more later.

The time traces (figure 11.7) show the turbulence saturating initially but then growing to higher amplitude, showing a strongly bursting character in time. The isotropic perpendicular spectra (figure 11.8) show that the reason for this is a condensation of much of the energy in the largest available scale, where there are fewer neighbouring degrees of freedom to effectively decorrelate a structure. Since this is not seen in the corresponding decaying case, the continuously applied gradient drive is the cause. The spectrum of the gradient drive rate, mostly due to the ExB advection component $\Gamma_n$, show the dominance of the lowest-$k_\perp$ component. Since the $k_x = \pm 1$ and $k_y = \pm 1$ mode in normalised units corresponds to the scale of periodicity in both directions, it is called the 'box mode,' noting that any higher-$k_\perp$ mode, with either $k_x$ or $k_y$ equal to 2 or more in magnitude, is binned into mode two or higher in the isotropic spectrum. So the box mode is seen to become the dominant component in any of these cases (the same morphology is seen in all of them). The parallel spectra (figure 11.9) show that the lowest-$k_\|$ component, namely zero, the flute mode, becomes dominant at soon as saturation is reached. The energy and the transport flux both peak in the box mode in $k_\perp$ and flute mode in $k_\|$ simultaneously. The vorticity, on the other hand, is mostly at small scale in $k_\perp$ and has a wide and flat spectrum in $k_\|$. Both of these features remain with any available resolution. The spatial morphology of this box mode is shown in figure 11.10. Superficially similar to the 2D hydrodynamic case with gradient drive (Chapter 4, Section 4.3.3), the difference lies in the fact that the couplings allow the energy in $n$ to transfer into $\phi$, first through the parallel (adiabatic) response, then through the inverse ExB vorticity transfer (in the $k_\|$-spectrum) into the flute mode, a specifically 3D phenomenon.

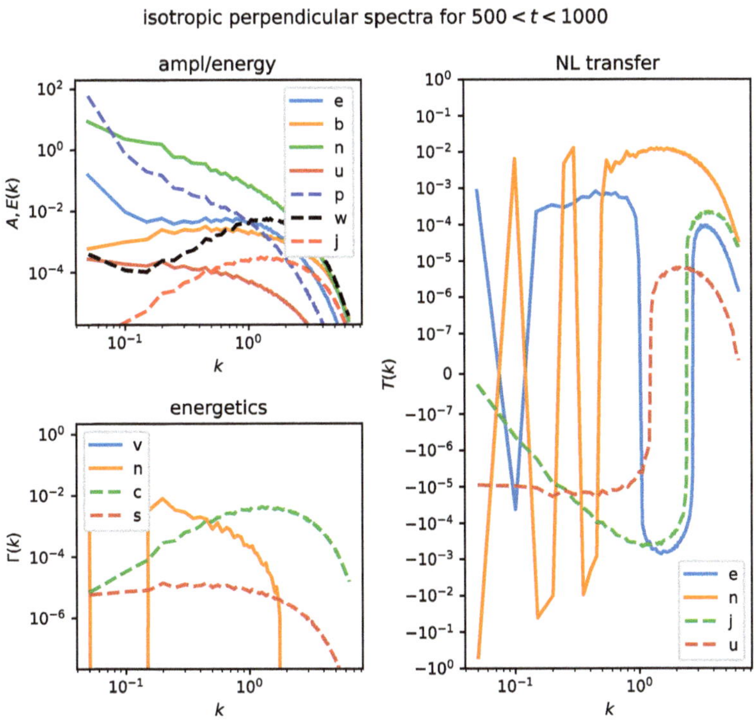


**Figure 11.8.** Isotropic perpendicular spectra for the driven case with periodic square domain during $500 < t < 1000$, roughly the phase of saturation. The format and labels are the same as in figure 11.2. The small-scale structure is similar to the decaying case, but the gradient drive ensures energetic dominance of the largest scales. The inter-scale transfer shows transfer from large scales to small in $E_n$ but inverse in $E_e$, as in the 2D cases. The direct transfer in $E_n$ is less pronounced than in the decaying case, due to the intermittency of the drive and its domination by the largest-scale mode.

We find very strong turbulence in these $1 \times 1$ cases, all dominated by the box/flute mode. Since this is manifestly unrealistic, we now investigate further by relaxing the restrictions, first of the square domain, and then of the periodic boundary conditions. Since the turbulence pumps its energy into the largest mode, the restriction of periodicity on any given scale determines the result, and that is not realistic. We consider the perpendicular domain aspect ratio $L_y/L_x$ first.

## 11.4 Aspect ratio

The computational domain in any situation with gradients satisfying $K_\parallel L_\perp \ll 1$ reflects a thin atmosphere situation, in which the direction of the pressure gradient corresponds to the shortest scale among system sizes in the $xyz$ directions. As a result it is reasonable to consider aspect ratios $L_y/L_x$ significantly greater than unity in turbulence driven by gradients. For convection in an atmosphere this is well known. The terrestrial atmosphere is an extreme case (height smaller than north–south zonal band, in turn smaller than the entire extent of longitude), but even with convection in a shell comprising the outer one-third of the radius, the aspect ratio is still of order10.

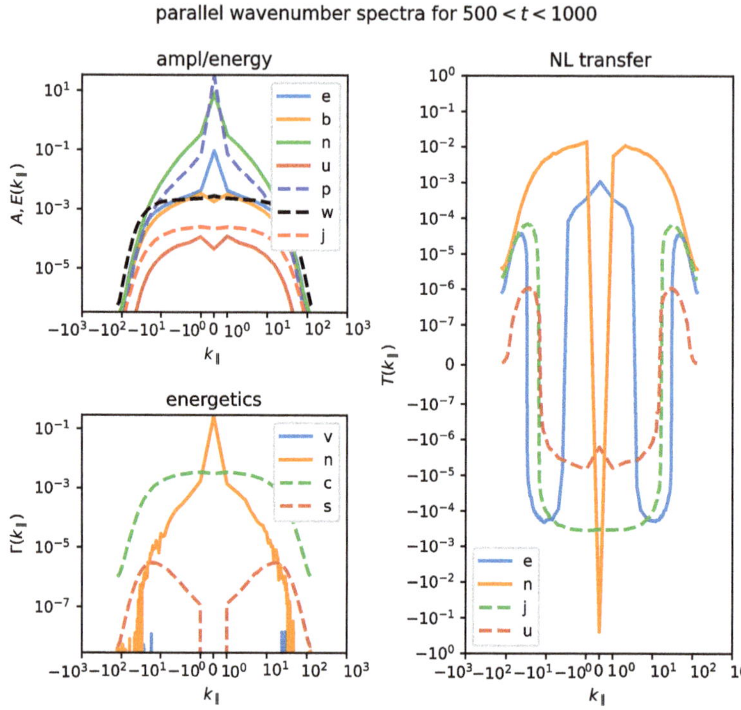


**Figure 11.9.** Parallel wavenumber spectra for the driven case with periodic square domain during $500 < t < 1000$, roughly the phase of saturation. The format and labels are the same as in figure 11.3. The structure is similar to the decaying case, but due to the gradient drive the flute mode at $k_{\|} = 0$ is even stronger. The widest spectrum is still in the vorticity. The energetics reflects gradient drive dominantly in the flute mode, with the dissipation channels extending to the maximum value on this grid (128). The flute mode is net positive, and all others are net negative. The inter-scale transfer is the same as in the decaying case, just stronger due to the gradient drive. The transfer through the vorticity nonlinearity is into the flute mode, and the transfer through advection out of it maintains the rest of the spectrum.

Values of order 10 or greater are usual as well in the outer regions of confined plasmas. So we have to study this effect on the large-scale 'box-mode' structures we found in the $1 \times 1$ square-domain cases.

We take the small and medium cases as above, with the 64 or 128 grid node counts in $x$, unchanged, and simply extend the $y$-domain by a factor of four, enough for a down-gradient flow with width $L_x$ to flow through twice (the reason for this is the Kelvin–Helmholtz shear flow instability cutoff mentioned below, after we see the results). Hence, we will have $L_y/L_x = 4$ for all subsequent cases. The small case now has a grid of $64 \times 256 \times 64$ and the medium case has $128 \times 512 \times 128$. We are holding $L_x$ and $\rho_s$ constant in physical units, therefore these cases have the same $\delta_0$ values as the corresponding square-domain cases, but here $L_y$ is increased from 1 to 4, yielding a $4 \times 1$ long-domain case. This gives more of a thin atmosphere character to the domain.

Now, instead of the large-scale 'box-mode' structures we saw in all the square-domain periodic cases, we find a dominant large-scale down-gradient flow, with

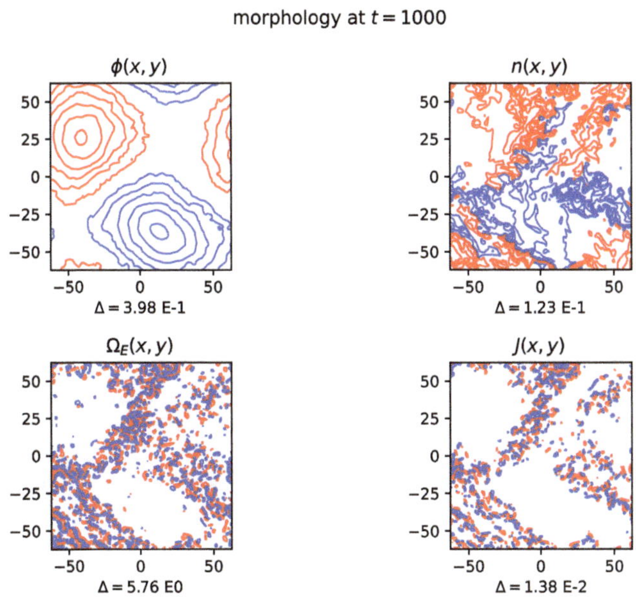


**Figure 11.10.** Spatial morphology at $t = 1000$ of the driven case with periodic square domain. The box mode is most evident in $\phi$, with $n$ showing thin sheets, and the vorticity and current showing small-scale structure. This is very strong nonlinear dynamics, limited only by the domain size, but as dominated by essentially one mode, it is not really turbulence, despite the fact that different quantities peak at different places in the spectrum.

thickness sufficient to flow through the extent of the domain in the $x$-direction without breaking up. The resulting amplitudes are so large it is found necessary to increase the subgrid dissipation to $\nu_0 = 10$ to contain the cascades. The time traces in figure 11.11 show much stronger dynamics and more burstiness. When the transport becomes largest a sharp event in the ExB and magnetic energies is seen, namely, in the Alfvén dynamics. The isotropic perpendicular spectra in figure 11.12 and especially the parallel spectra in figure 11.13 show that this strong flow leads to very wide-spectrum dissipation through the action of resistivity on the Alfvén dynamics, whose vorticity spectrum fills the available spectral range. The flow itself is seen clearly in figure 11.14, showing the spatial morphology at $t = 1000$. This flow is not found in the square-domain cases because the periodicity length in the $y$-direction in those cases is too short. That is, in a square domain a flow in the $x$-direction cannot become wide enough in the $y$-direction to become Kelvin–Helmholtz stable. So, driven by the gradient and sustained by the nonlinear dynamics it becomes dominant and is only limited by finite-$k_{\parallel}$ Alfvén dynamics with a very unsteady coupling.

The main lesson of these periodic-domain studies is the effect of the aspect ratio on flows. This is a property of the Kelvin–Helmholtz instability, one of the main elementary instabilities of hydrodynamics covered in the text by Chandrasekhar (see Further Reading). The condition for instability is an extremum in the main vorticity, by the Rayleigh theorem. In any version of it, whether the flow shear, or main vorticity, is piecewise constant or given a localised profile, the instability exists only at long wavelength, down to a cutoff given by $k_{\perp}\Delta_V < \lambda_c$, where the constant $\lambda_c$,

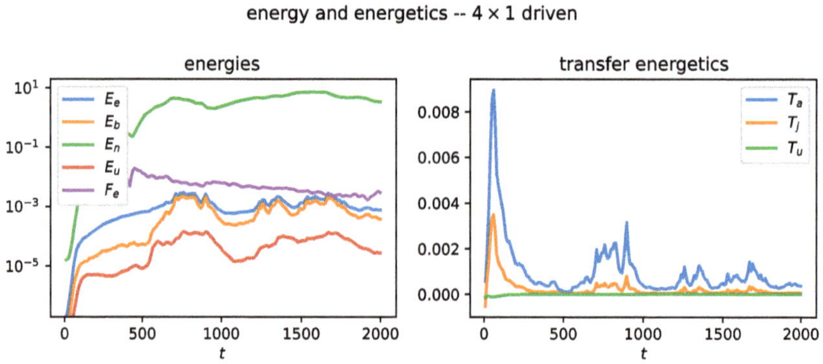


**Figure 11.11.** Temporal evolution of the free energy pieces and transfer energetics with time, for the driven case with periodic 4 × 1 long domain. The labelling and format is as in figure 11.1, with the turbulent ExB flux ($F_e$) also given. This is much stronger dynamics than in the square domain—compare with figure 11.7 and note the different scale. The energies in the parallel channels (magnetic and acoustic) are very unsteady. This burstiness is reflected in the three transfer channels.

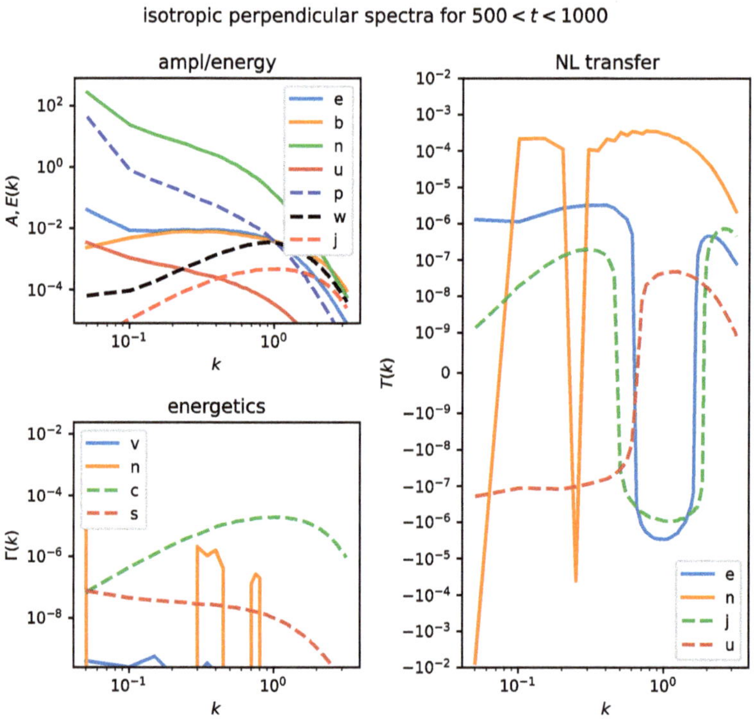


**Figure 11.12.** Isotropic perpendicular spectra for the driven case with periodic 4 × 1 long domain during $500 < t < 1000$, roughly the phase of saturation. The labelling and format are the same as in figure 11.2. The structure is similar to the square-domain driven case, but the dynamics is even stronger, indicated by the weak turn-down in the vorticity and dissipation spectra at high $k_\perp$. The inter-scale transfer spectra are essentially the same as before. Here as well, the dominance by the flow leaves the long wavelength side in a situation with intermittent statistics.

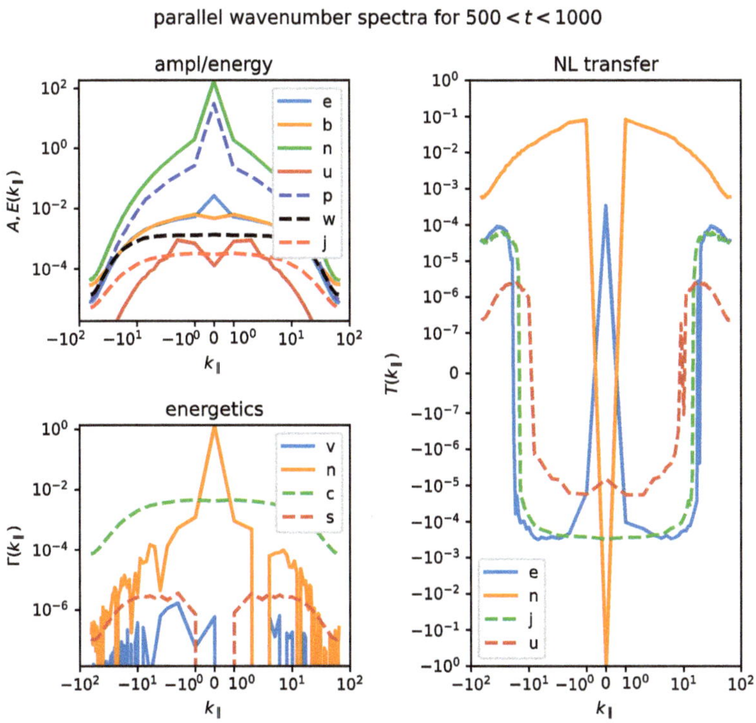


**Figure 11.13.** Parallel wavenumber spectra for the driven case with periodic 4 × 1 long domain during $500 < t < 1000$, roughly the phase of saturation. The labelling and format are the same as in figure 11.3. The structure is similar to square-domain cases, with the flute mode dominant. The increased strength of the dynamics results in very wide vorticity and dissipation spectra. This signature persists with increased resolution, indicating that this is not a well saturated case of turbulence, more an energetic blowout sustained only by the constant external gradient drive.

of order unity or a little smaller, is a weak function of the shape of the main vorticity envelope. Here, $\Delta_V$ is the scale of the flow shear and $k_\perp$ is the wavenumber of any perturbation of the flow (the wavelength is in the direction of the flow and the displacements of perturbation are across the flow). For a flow in the $x$-direction, $k_\perp$ is $k_x$ and the flow width $\Delta_V$ is in the $y$-direction. So to make $\Delta_V$ wide enough to keep any available $k_x$ stable requires that $\Delta_V$ is comparable to the domain size $L_x$, which is possible only if $L_y$ is significantly greater than $L_x$. We will cover this in more detail in Chapter 16, on the interaction between turbulence and flows, where the Reynolds stress and its properties, and the energetics of flow/turbulence exchange are covered. The simple picture of it is that for $k_\perp \Delta_V \gg 1$ the structures are tilted with the flow and for $k_\perp \Delta_V \ll 1$ the structures are rolled against the flow so that the energetics for drive (transfer from flow to eddies) is favourable. The reason for the critical aspect ratio of four is that down-gradient flows of width greater than the 'length' in $L_x$ first become allowed, and their pathology has to be faced. Basically, turbulence driven by gradients in a channel has to allow long wavelengths in the drift direction, perpendicular to both $\nabla p$ and $\mathbf{B}$ in a magnetised plasma.

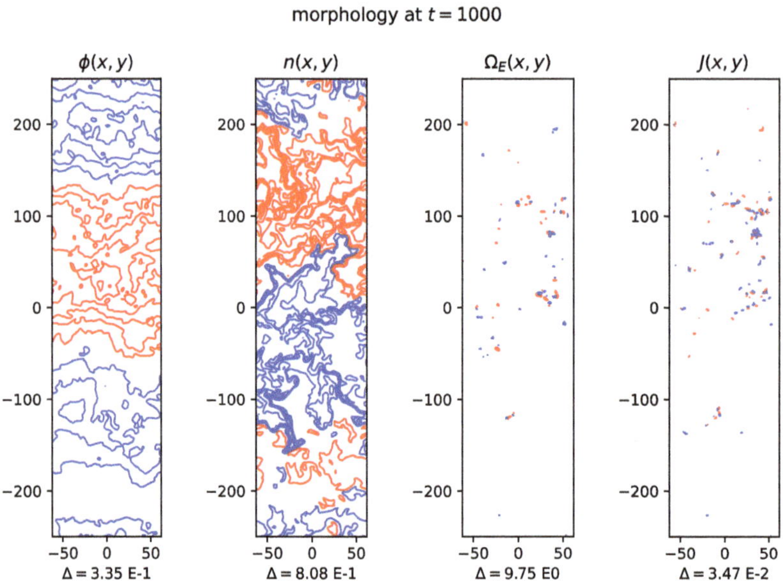


**Figure 11.14.** Spatial morphology at $t$ = 1000 for the driven case with periodic 4 × 1 long domain. Now the large-scale flows in the $x$-direction are the dominant component. They are not smooth or steady, but temporally bursty with some small-scale structure. These flows are disallowed in the square domain, which does not have sufficient space in the $y$-direction for a Kelvin–Helmholtz stable $x$-direction to occur.

The down-gradient flows are found to extend through the entire domain for sufficient $L_y/L_x$, in this case a ratio of only four was enough. Flows down a gradient that go forever are not realistic in most physical cases. It is necessary at least to force them to turn around as they must do in a thin atmosphere case anyway. To do this we have to turn to bounded cases, considered separately.

## 11.5 Bounded cases

A bounded case is defined as one in which the values of all the dependent variables are fixed to zero at one or more boundaries. In our case we do this at both boundaries in $x$ while leaving both $yz$ coordinates periodic, according to

$$f(\pm L_x/2, y, z) = 0 \quad f(x, y + L_y, z) = f(x, y, z + L_\parallel) = f(x, y, z), \tag{11.47}$$

where in normalised units the periodicity length in $z$ is $2\pi$. The $x$-direction is loaded once per time step with guard cells as before, but now the $x = \pm L_x/2$ points are taken to fall at cell centres, leftward of the first active point $x_1$ and rightward of the last active point $x_n$, where there are $n$ points on the grid. Then the boundary cells are copied across the boundaries according to negative reflection, e.g.

$$f(x_0) = -f(x_1) \quad f(x_{-1}) = -f(x_2), \tag{11.48}$$

at the left boundary. Then, up to the fourth-order operations on the grid, the assumed value halfway between, e.g. $f(x_0)$ and $f(x_1)$, is zero. The $yz$-planes at each

$x$ are 2D-tori in which the field lines lie ($B^x = 0$) in the equilibrium, and they are called flux surfaces. There will be more on that in the next Chapter.

Once we consider Dirichlet boundaries in $x$, we have to add to the subgrid dissipation a finite diffusion or damping in a layer at $x = \pm L_x/2$ on all the dependent variables. As with the subgrid dissipation, this is applied to whichever quantity is nonlinearly advected by the ExB velocity. In Eqs (11.18)–(11.21) these are $\varpi$, $\hat{\mu}J$, $n$, and $\hat{\epsilon}u$, respectively. The layer is a narrow Gaussian centred upon the boundary location, with the relevant half being that which is within the domain. This structure with the subgrid and boundary dissipation together is given by

$$\frac{d}{dt} = \frac{\partial}{\partial t} + [\phi, \,] + D() \quad D() = \nu_\perp\left(\nabla_\perp^2\right)^2 + \nu_\parallel\left(\frac{\partial^2}{\partial z^2}\right)^2 + s_f p_f, \tag{11.49}$$

where the boundary dissipation at the edges of the domain in $x$ is done as a damping layer with amplitude $s_f$ and profile function given by

$$p_f(x) = \exp\left[-100\left(\frac{x}{L_x} + \frac{1}{2}\right)^2\right] + \exp\left[-100\left(\frac{x}{L_x} - \frac{1}{2}\right)^2\right]. \tag{11.50}$$

This is nominally a damping layer with coefficient $s_f = 1$ and $1/e$ half-width $L_x/10$ centred on each boundary. The dissipation is usually sufficient to keep the nonlinear vorticity dynamics away from the boundary and also maintain the background gradient (i.e. keep the dependent variable $n$ from forming a structure of the form $\omega_n x$ to cancel out $n_0$ which forms the gradient). The boundary dissipation is incorporated into the overall subgrid dissipation operator $D$, so that the energetics of it is still described by $\Gamma_w$ in Eq. (11.36), with the new $D$.

We return to the long-domain case, now as a bounded case with the boundary dissipation. The $4 \times 1$ case for $N_x = 128$ is shown. The time traces (figure 11.15) show well saturated turbulence which is now without prominent bursts. For this reason we are able to reduce the subgrid dissipation back to $\nu_0 = 1$. The amplitude is still very large (compare with $\delta_0^2 = 2.56 \times 10^{-4}$). The transfer channels show a robust initial growth and saturation phase which then settles into a statistically equilibrated phase without large excursions. Although we still have the very large amplitude, this is more of what one expects from driven dissipative turbulence. Large amplitude no longer means order unity, just rather larger than order $\delta_0$.

The isotropic perpendicular spectra (figure 11.16) show the same structure as in the periodic, driven cases. The largest scales are still dominant in $\phi$ and $n$ while the vorticity $\varpi$ and parallel flux variables ($J$ and $u$) are peaked in the decade around the $\rho_s$ scale (top right). The gradient drive ($\Gamma_n$ and $\Gamma_v$) descends down from the largest scale, while the resistive ($\Gamma_c$) and viscous ($\Gamma_s$) dissipation are spread out with the former following the vorticity (bottom right). The nonlinear transfer shows a robust inverse tendency in the ExB nonlinearity, while the thermal free energy is mostly out of the largest scale into all the others.

The parallel spectra (figure 11.17) show the dominant flute mode in $\phi$ and $n$, with the other quantities spread widely out (top left, denoted as before). These are

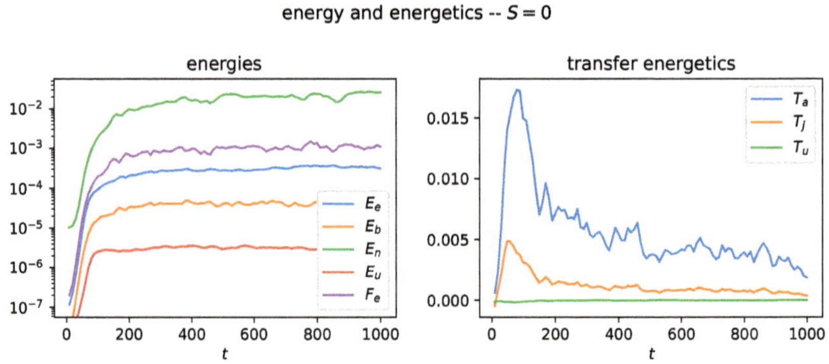


**Figure 11.15.** Temporal evolution of the free energy pieces and transfer energetics with time, as in figures 11.1 and 11.11, for the driven case with bounded $4 \times 1$ long domain. This is also the zero-shear $S = 0$ case. The dynamics is now much more regular, with the turbulence rising out of the initial state to saturate after about $t = 500$. With the periodic throughput flow in the $x$-direction now eliminated by boundary conditions, large bursts are no longer prominent, even though the dynamics is still strong. (right) The transfer dynamics is more regular, largest in the growth phase, then settling into saturation. The signal of adiabatic transfer from $n$ to $J$ and Alfvénic transfer from $J$ to $\phi$, which maintains the ExB energy, is very clear.

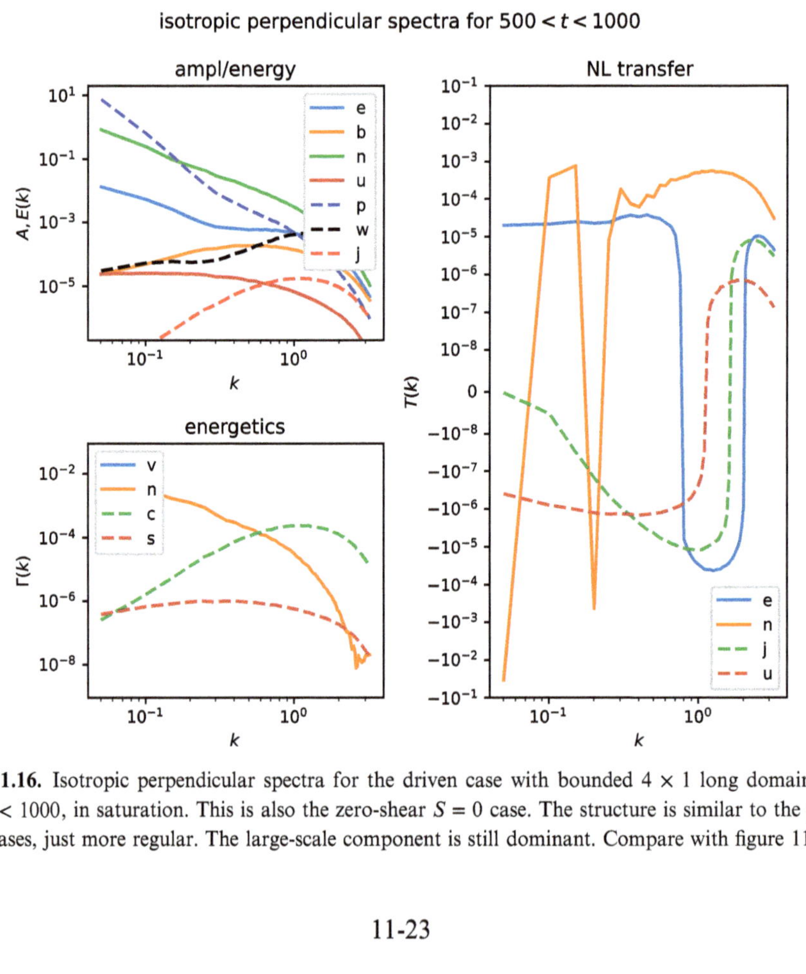


**Figure 11.16.** Isotropic perpendicular spectra for the driven case with bounded $4 \times 1$ long domain during $500 < t < 1000$, in saturation. This is also the zero-shear $S = 0$ case. The structure is similar to the periodic driven cases, just more regular. The large-scale component is still dominant. Compare with figure 11.12.

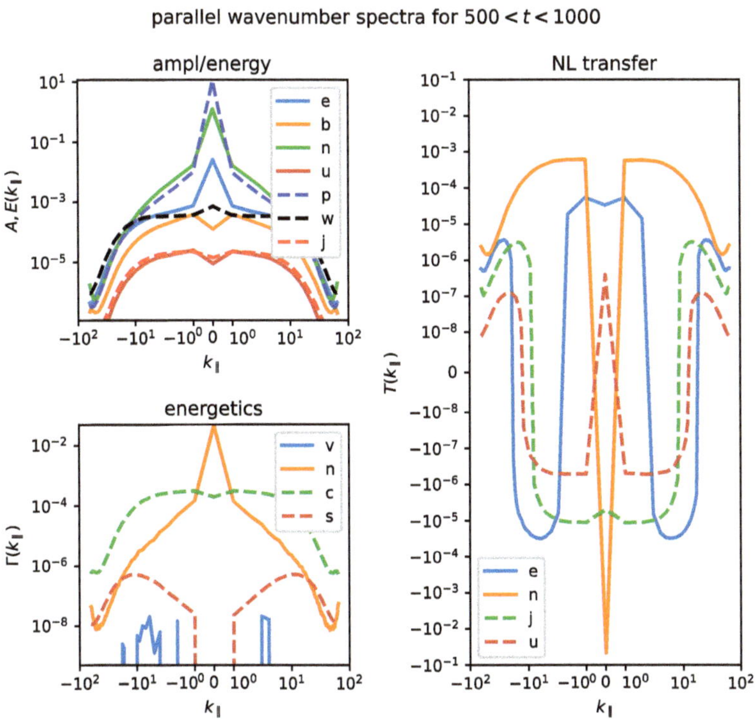


**Figure 11.17.** Parallel wavenumber spectra for the driven case with bounded 4 × 1 long domain during $500 < t < 1000$, in saturation. This is also the zero-shear $S = 0$ case. The structure is similar to the periodic driven cases, just more regular. The flute mode component is still dominant. The nonlinear transfer into the flute mode in the ExB vorticity nonlinearity and out of it in the ExB advective nonlinearity is prominent. Compare with figure 11.13.

reflected in the flute mode drive and the wide-spectrum dissipation channels (bottom left). The nonlinear transfer in the parallel domain is the same as before, except the statistics are better with the same number (50) of samples. We still have the result that the ExB vorticity nonlinearity feeds the flute mode region while the ExB advective nonlinearity depletes it into the rest of the spectrum.

The spatial morphology (figure 11.18) shows how the ExB dynamics tends to build one large vortex ($\phi$), which since it represents a flute mode can be thought of as a tube or rope structure aligned with the magnetic field in three dimensions. Large-scale structure is formed in $n$ as well, but the rest of it is the small-scale component which excites the parallel dynamics, feeding the flow eddies at all scales through the parallel transfer channels (figure 11.15, bottom right), which then feed the flute mode through the nonlinear dynamics. In this manner, the large-scale structure maintains itself. But the down-gradient flow all the way through the domain is now disallowed.

Finally, in our chain of decreasing pathology, we turn on the magnetic shear and watch it take this large-scale structure away, leaving the turbulence in the form which will occupy us for the remainder of the work.

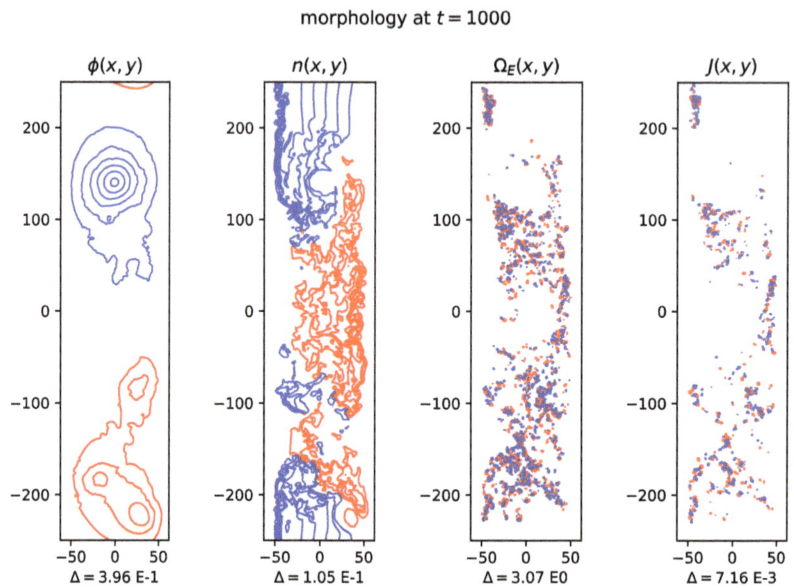


**Figure 11.18.** Spatial morphology at $t = 1000$ for the bounded, driven case with $4 \times 1$ long domain. This is also the zero-shear $S = 0$ control case. Large eddies form ($\phi$), rather than unimpeded throughput flows in the $x$-direction, which are now stopped by forcing them to turn around at the boundary.

## 11.6 Cases with magnetic shear

The magnetic field is now allowed to be sheared, by which is meant that the small component $B^y \ll B^z = B_0$ has $x$-dependence, $B^y(x)$, vanishing at $x = 0$ since without loss of generality we can choose the coordinates $xyz$ such that the origin picks up an arbitrary field direction, and we can set $\mathbf{B} = B_0 \nabla z$ there and choose $x$ to be the perpendicular direction in which the direction of the field varies. We then set $\nabla y = \nabla z \times \nabla x$. The shear measures the rate with distance in $x$ with which the direction of the field varies. This defines the shear length $L_s$, given by

$$L_s^{-1} = \left| \frac{1}{B_0} \frac{\partial B^y}{\partial x} \right|, \tag{11.51}$$

which is defined positive. A dimensionless shear, $S$, which can have either sign, is then defined in terms of $K_\|$, so that

$$S\, K_\| = \frac{1}{B_0} \frac{\partial B^y}{\partial x}. \tag{11.52}$$

The shear is thereby defined in terms of the parallel periodicity length. The shear length in terms of $S$ is $L_s = \left| S\, K_\| \right|^{-1}$.

In a layer-localised situation with a domain size $L_x$ much smaller than the shear length, we may assume the above statement that

$$B^y \ll B^z = B_0 \quad \text{hence} \quad B^y = B_0\,(S\,K_{\|})x, \tag{11.53}$$

where the shear is now given by a dimensionless parameter. In this initial investigation of the shear, we will start with the $S = 0$ cases as given above and then introduce it with values less than unity. More severe shear in a computation requires changes to the coordinate system, which we will introduce in the next Chapter. For now, the purpose is to show its effect on the large-scale $k_{\|} = 0$ flute mode structures we have just encountered.

Magnetic shear introduces a complication into the boundary conditions for periodic cases. Having already shown why periodicity constraints in the direction of the pressure gradient are undesirable, it is a short step to continue with that here. For reasons of MHD equilibrium in the background geometry, addressed in the next Chapter, the pressure gradient and the variation of the direction of the magnetic field lines are in the same direction, in this case the $x$-direction. The vectors $\nabla p$ and $\nabla(B^y/B^z)$ are collinear. If the $x$-direction is periodic, the structures on one side of the domain must match over to the other. The pressure gradient is consistent with that, only having the difficulty of allowing the $k_x = 0$ down-gradient flows. But the magnetic shear and periodic boundary conditions might have to be arranged such that a structure extending in the $x$-direction can overlap itself in a different location in $y$. Moreover, this $y$-displacement also depends upon the position along the field line. This means that the shear length should be tied to the periodicity length if the structure can overlap itself. We also have the choice, somewhat arbitrary but which might affect the physics, of whether to shift the periodicity in the $x$-direction like this or not (we could also ignore that and set $x$ periodic independently of the other coordinates as before). The arrangement might prefer one type of structure to the other at the largest scale. So, without further ado, we restrict to Dirichlet boundaries for the dependent variables in the $x$-direction as in the bounded cases above. We will stick to that for the remainder of the work.

We take the same bounded case as just above, which now serves as the $S = 0$ control case. The shear is gradually turned on, and the effects on the mode structure are measured. We also find a transport scaling with magnetic shear, which in turn forms a control case for the transformations of the coordinate geometry done in the next Chapter. For these cases we are able to revert to using the $\tau = 0.002$ time step. The parameters are the same as above (Eq. 11.40). The shear is varied through

$$S = \{0, 0.1, 0.2, 0.3, 0.5\}. \tag{11.54}$$

We display the details of the $S = 0.5$ case and the variation of the mode structure and transport scaling for the others.

The time traces are given in figure 11.19 for this case. We now find fairly smooth saturation as we saw in the 2D dissipative drift wave cases. The turbulence is robust, but there are no strong events which give the impression of explosions. The saturated amplitude and turbulent ExB flux level is commensurate with the $\delta_0^2$ scaling expected for the drift ordering (delta-f) assumptions behind the equations. The transfer rates through the parallel dynamics are larger in the beginning, then relax into lower levels as the turbulence saturates. The direction of the transfer indicates a robust adiabatic response wherein free energy driven by the gradient into $n$ passed through $J$ and then into $\phi$.

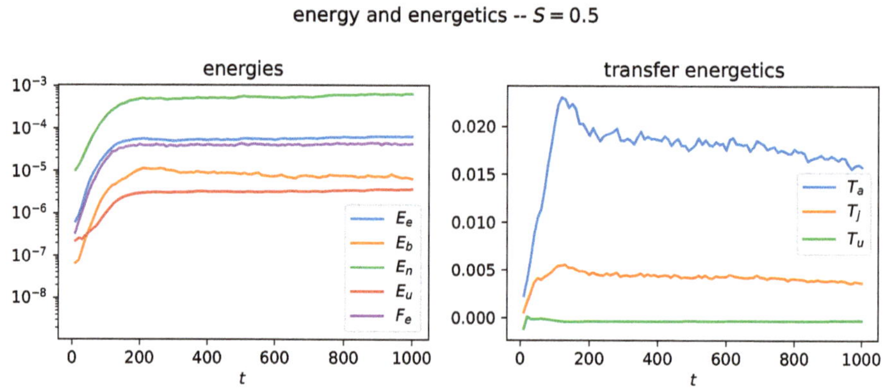


**Figure 11.19.** Temporal evolution of the free energy pieces and transfer energetics with time, as in figure 11.15. The dynamics is as regular as for $S = 0$, but saturates at a much lower level on about the same time scale. The transfer dynamics is similar to the $S = 0$ case but appears stronger since it is given as a rate. The signal of adiabatic transfer from $n$ to $J$ and Alfvénic transfer from $J$ to $\phi$ remains.

The isotropic perpendicular spectra (figure 11.20) show that the large-scale dominance seen in the cases with $S = 0$ is now absent. The spectra of the free energy, the transport, and the vorticity and dissipation are all fairly wide, with peaks slightly shifted relative to one another. These are roughly at $k_\perp \rho_s = 0.3$, then 0.5, and then closer to 1 for this case. It is important to note that there is no single dominant scale, rather, the entire spectrum is involved in the dynamics. The inter-scale transfers in the ExB vorticity nonlinearity towards larger scale, and in the advective ExB nonlinearity towards smaller scale, are very evident and statistically well formed, with these 50 samples. One reason is that since the large scales are no longer dominant, all of the dynamics is efficiently mutually decorrelated since there are many available degrees of freedom at neighbouring $k_\perp$ to the mode ranges well above the lowest.

The parallel spectra (figure 11.21) show that the dominant flute mode in $\phi$ and $n$ seen for $S = 0$ is now absent, even though the amplitudes still peak at $k_\parallel = 0$. The vorticity and dissipation spectra are still broad, indicating robustly mixed turbulence. The nonlinear transfer remains in the same character as in all the other cases, with the $k_\parallel = 0$ region still driven nonlinearly through the vorticity and depleted through the ExB turbulent advection.

The spatial morphology (figure 11.22) shows that any identifiable large-scale eddy motion is absent. One finds turbulent structure in the range of one to ten times the drift scale, with individual structures emerging and then becoming decorrelated by the vorticity. Any hint of coherent $x$-direction flow, or of coherent vortices at later times, is absent. One finds a flow layer, but in the $y$-direction. Flows of this type will be addressed in Chapter 16.

One issue in these diagnostics is in the parallel spectra. In these coordinates, which due to the shear no longer follow the magnetic field lines, a spectrum in $k_z$ is not the same thing as a spectrum in $k_\parallel$, although we are casting is this way. The result of this is that with $B^y k_y$ finite, the $k_\parallel = 0$ range involves fairly finite levels of $B^z k_z$ to

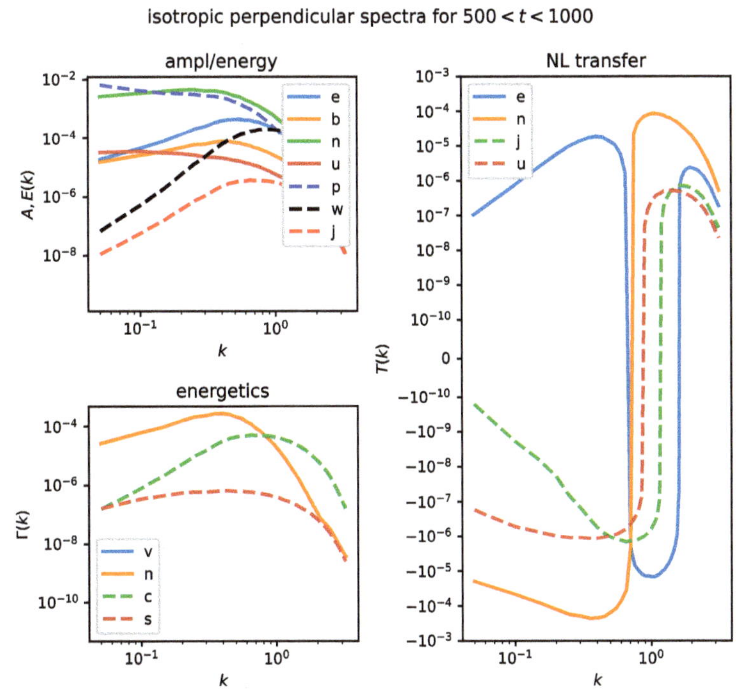


**Figure 11.20.** Isotropic perpendicular spectra for the $S = 0.5$ case, during $500 < t < 1000$, in saturation. The large-scale structure seen before is no longer present, and the energy peaks at only somewhat larger scale than the vorticity ($k_{\perp}\rho_s \sim 0.3$ versus 1.0, respectively). The gradient drive peaks in between the energy and vorticity peaks, with the dissipation in a similar form to the $S = 0$ case. The inter-scale transfer in the vorticity and advective nonlinearities has the now-familiar signature, but is statistically much cleaner (this reflects the larger role of the smaller scales, where the vorticity redistributes faster). Compare with figures 11.12 and 11.16.

compensate. We will repair this in the next Chapter with the introduction of field-aligned coordinates, and then a parallel spectrum will mean the same thing independently of the value of magnetic shear.

### 11.6.1 Mode structure changes with shear

The introduction of the magnetic shear away from zero is seen to eliminate the large-scale box mode and even the large-scale vortices in a bounded system with anisotropic domain sizes in the perpendicular place. We examine this in more detail with three of the diagnostics, each presented as a series of increasing values of magnetic shear.

The parallel wavenumber spectrum come first. The dominant flute mode structure is apparent in the $S = 0$ case and largely eliminated for $S = 0.5$, as seen in comparison between figures 11.17 and 11.21. As $S$ is gradually increased, we progress from the cases $S = 0.1$ to 0.3, still a relatively small value, as shown in figure 11.23. The flute mode peak is seen to subside gradually as $S$ increases, and is no longer the dominant component even for $S = 0.3$, and even more so for $S = 0.5$ as shown above.

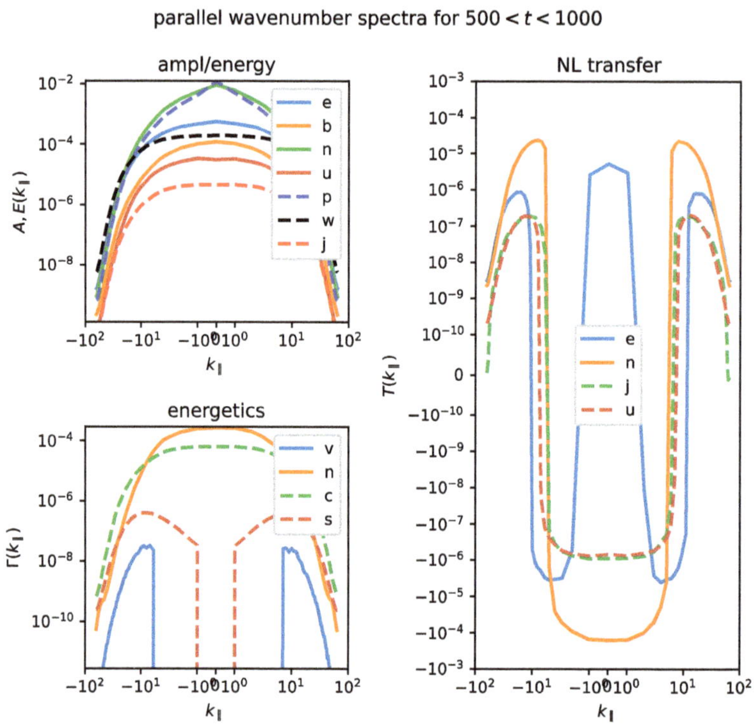


**Figure 11.21.** Parallel wavenumber spectra for the $S = 0.5$ case during $500 < t < 1000$, in saturation. Compare with figure 11.17. The flute mode has been nearly eliminated, and $\phi$ and $n$ and the gradient drive have broader spectra. The signatures in the vorticity and the resistive and viscous dissipation are more similar to the $S = 0$ cases. The magnetic flutter at higher $k_\parallel$ is positive but small. The nonlinear transfer into the flute mode region through the vorticity and out of it through the turbulent advection is prominent. Transfer in the parallel channels remains subdominant. These signals are broadened by being measured in the $z$-direction and not exactly parallel to the background magnetic field (see text). This is fixed in the next Chapter when field-aligned magnetic coordinates are used, for which these finite-$S$ cases serve as an overall control case.

Next are the cross coherence and phase shift distributions introduced in Chapter 10, shown in figure 11.24. As $S$ is gradually increased, we progress from a hydrodynamic mode structure signature to a drift wave signature. For weak shear the state variables $n$ and $\phi$ are uncorrelated, and long wavelength phase shift distribution peaks at $\alpha = \pi/2$, which gives the maximum drive. Positive $n$ is in phase with positive $v_E^x$ and therefore $\Gamma_n$ proportional to $\sin\alpha$ is most efficient. This is the signature of a hydrodynamic mode structure. As $S$ increases, this structure gradually morphs into one with strong correlation between $n$ and $\phi$ and with phase shifts still positive but closer to zero. Moreover, the distributions for every $k_y$ in the spectrum are very similar, which reflects the entire spectrum acting as a unit. This is the signature of the drift wave mode structure.

The main finding of this test by which $S$ increases from 0 to 0.3 is that magnetic shear eliminates the large-scale flute mode (low $k_\perp$ and practically zero $k_\parallel$), and replaces hydrodynamic character with drift wave mode structure. The vision of

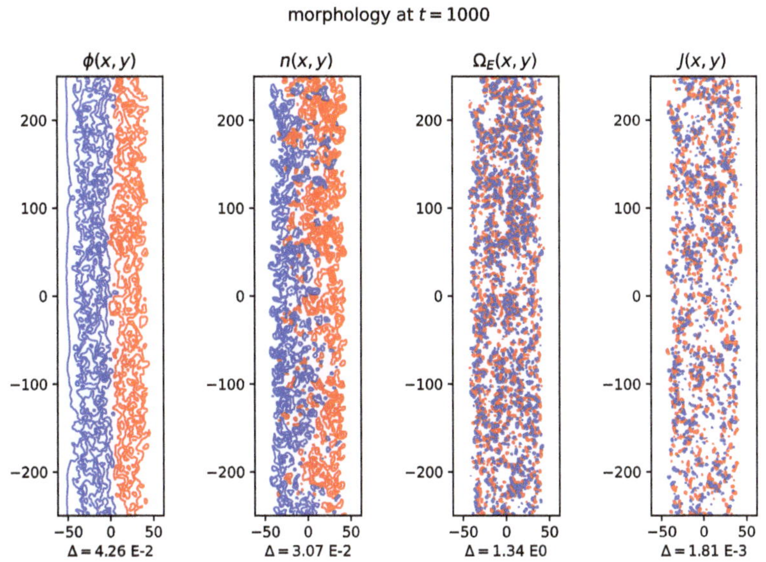


**Figure 11.22.** Spatial morphology at $t = 1000$ for the bounded, driven case with magnetic shear ($S = 0.5$). The large-scale eddies are now absent, and the range of scales is so wide as to impede visibility.

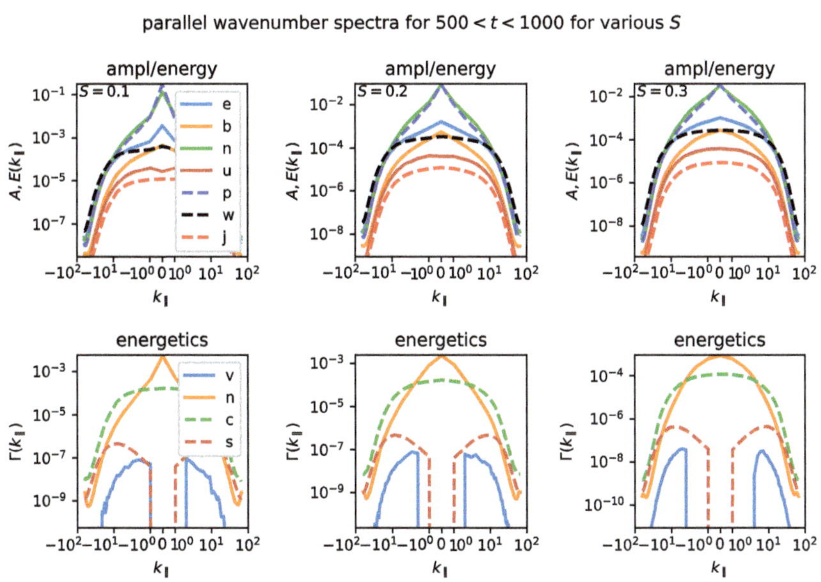


**Figure 11.23.** Series of parallel spectra, amplitudes (top) and energetics (bottom), for increasing magnetic shear, $S = 0.1$ to 0.3, still a relatively small value. The flute mode peak subsides as $S$ is increased.

dissipative drift wave turbulence only exists under this condition. As practically all systems with steep gradients, for which this type of steady drive is relevant, are both bounded and have finite magnetic shear, we expect more of an obviously magnetised plasma look and feel of the turbulence rather than a more (magneto-)hydrodynamic

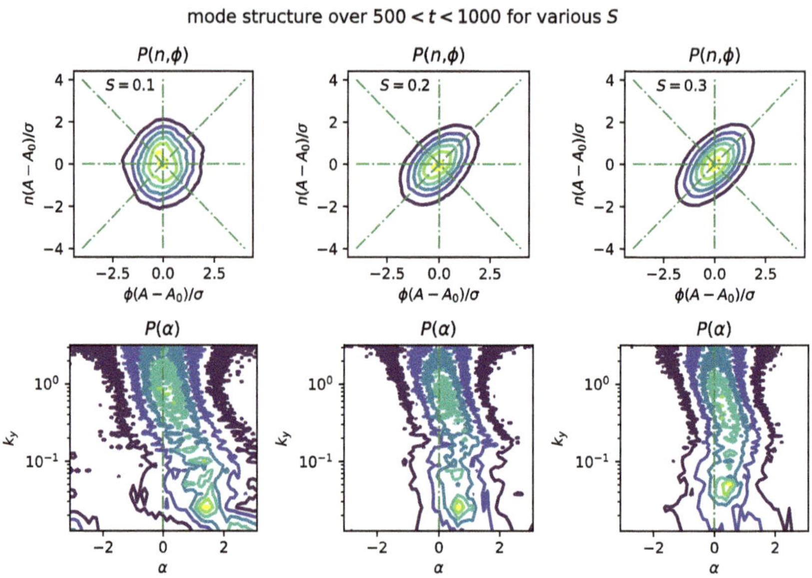


**Figure 11.24.** Cross coherence (top), and phase shift distributions (bottom) between $n$ and $\phi$, for increasing magnetic shear, $S = 0.1$ to 0.3, still a relatively small value. Compare with figures 10.11 and 10.12 from Chapter 10. The mode structure transitions from the hydrodynamic to the drift wave signatures within this shear range.

one. Gradient-driven turbulence in magnetised plasmas usually exhibits this type of mode structure. We will investigate this again as more complications in the magnetic geometry are incorporated over the next few Chapters.

The introduction of magnetic shear also reduces the amplitude of the turbulence from a level approaching unity down to the level commensurate with $\delta_0 = \rho_s/L_\perp$, which we expect from the mixing level when the dominant scale reduced from system size $L_\perp$ towards the $\rho_s$ scale. The scaling with magnetic shear for four representative quantities is shown in figure 11.25. The four quantities are the turbulent ExB flux $nv_E^x$, the enstrophy $\varpi^2/2$, the thermal free energy $n^2/2$, and the combination of magnetic and parallel electron flow energy $J(A + \hat{\mu}J)/2$, all in volume average. As $S = 1$ is approached, these quantities are an order of magnitude or less than they were with $S = 0$. The enstrophy changes the least, since as the turbulence reduces in scale the increase of $k_\perp$ compensates the reduction of the amplitude. So with finite shear we now have the small-scale turbulence for which we built this system of model equations. This is finally drift-Alfvén turbulence in three dimensions.

## 11.7 On pathology

It is important to realise when naïvely chosen simplified models become pathological in a computational realisation. For a turbulence computation, especially a decaying one, the choice of Cartesian geometry seems obvious. When in a magnetised plasma, setting $L_\parallel \gg L_\perp$ is also a clear choice even if less obvious when one starts from a *tabula rasa* situation. In decaying turbulence there was no obvious pathology. The scale of motion does grow, but during the decay phase the dependence on initial

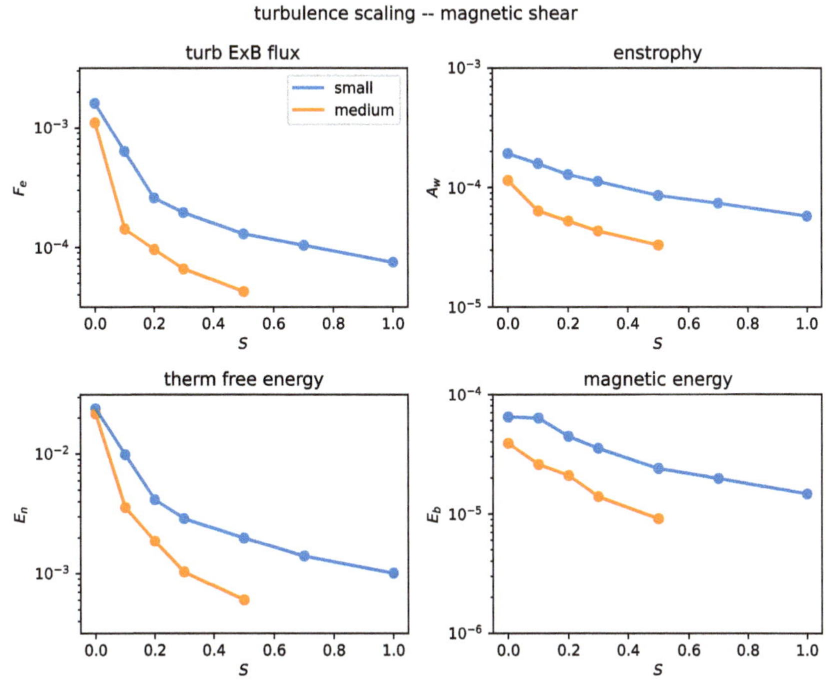


**Figure 11.25.** Scaling of the turbulent ExB transport flux ($F_e$), the enstrophy ($A_w$), the thermal free energy ($E_n$), and the total magnetic, or Ohmic, energy ($E_b$), all in volume average, against the magnetic shear ($S$). The orange points are for the $N_x = 128$ cases with $\delta_0 = 0.008$, and the blue points are for the $N_x = 64$ cases with $\delta_0 = 0.016$. The difference in the flux roughly reflects the expected $\delta_0^2$ scaling for the higher $S$.

conditions allows a reasonable initial choice of spectrum such that the domain-size box mode does not emerge until very much later, and then it can be avoided with finite run time. However, if a background gradient is present, one has to think more about setting up the scenario. If $L_\| \gg L_\perp$ and at the same time $L_\perp$ is the gradient scale length, first the box mode and then the forever flow in the direction of the gradient becomes an unrealistic pathology. The choice of periodic boundary conditions in allowing either the one or the other to exist then emerges as pathological in the first place, and the scenario has to be changed.

A three-dimensional setting with straight magnetic geometry and steep gradients is itself not physically realistic. The existence of a domain-sized result shows that the expectation of turbulence sufficiently local that such a homogeneous domain can be used to model it is thereby falsified. A 3D model with no magnetic shear and turbulence driven by gradients is not realistic unless the drive persists for a short time, after which the turbulence decays. That indeed is then the same scenario of decay out of an arbitrary initial state. At best, an external or initial source of activity which then decays is reasonable if that is the physical situation being considered. But that is the model addressed by the decaying cases, which do deliver realistic structure. In reasonable situations in nature, a confined sheet structure with a gradient is accompanied by magnetic shear, and two-dimensional periodicity on the magnetic surfaces implies an MHD equilibrium with closed magnetic surfaces, usually containing currents, which we treat in the next three Chapters. Systems in nature in which a gradient drives dynamics in a sheet layer are usually open at the ends of the field lines, or as in the case of a solar loop bounded at these ends. In these, the properties created by magnetic shear on closed surfaces in eliminating the $k_\parallel = 0$ flute mode component are then replaced by the conditions at the ends of the field lines, be they open or bounded, and then whether the flute mode is allowed or not depends on the situation. We will treat such open field line cases in Chapter 15, after the toroidal cases and flows have been addressed.

## 11.8 Summary

The basic character of 3D magnetised plasma turbulence can be summarised in a few points. The dynamics does tend to line up along the magnetic field, due both to the adiabatic response coupling $\phi$ back to $n$, which assists the nonlinear decay processes as well as the linear dissipation channels such as resistivity and viscosity (and, later, thermal conduction). In the absence of drive, the parallel wavenumber spectrum narrows until at later times the flute mode component is most of what is left. The vorticity has the widest spectrum of all the dependent variables. Next, driven cases with no magnetic shear produce large-scale flute mode activity with strong bursts acting to excite the decay channels. But due to the nonlinear tendency to collect the ExB flow energy in the largest scales—the condensation property—the large-scale flute mode dominates the dynamics, even if due to the 3D coupling and nonlinear transfer channels the dynamics is much more violent than in the same perpendicular domain as a hydrodynamic 2D model. These cases always demand more resolution than one can give them, and one ends with the same quandary as in 3D homogeneous fluid convection, which has always been among the high performance computational challenges.

The introduction of magnetic shear across two-dimensional surfaces which are periodic in both dimensions and in which the field lines lie introduces a property called 'field line connection'—the flute mode component is eliminated because all available degrees of freedom in which flows down the gradient, hence with any finite derivative within the surfaces, also have a finite parallel gradient. Once the shear is introduced, the cross correlation between $\phi$ and $n$, which is the signature in measurements that the adiabatic response is active, emerges and then the entire spectrum involves energetic coupling by currents and the narrow phase shift distributions that go with it. This is then the main change which establishes this kind of turbulence: three-dimensional magnetised plasma turbulence with identifiable drift wave mode structure. Field line connection acts to guarantee finite $k_\parallel$ for every degree of freedom such that the adiabatic response always acts. The background guide field also limits the $k_\parallel$ spectrum enough to confirm the flute mode ordering, with $k_\parallel/K_\parallel$ no larger than a few tens for most of the energetics.

Drift wave turbulence with magnetic shear forms the central topic of the next Chapter. Field line connection is an essential element and is derived and quantified therein.

## Further Reading

Basic numerical methods including Courant limits for both diffusion and advection problems are treated in R D Richtmeyer and K D Morton 1967 *Difference Methods for Initial-Value Problems* (New York: Wiley). The references containing the numerical scheme used herein were given at the end of Chapter 3.

The phenomenon of condensation of fluid energy into the largest scales in two dimensional fluid turbulence is treated by A Hasegawa 1985 Self-organization processes in continuous media *Adv. Phys.* **34** 1. For drift wave turbulence in three dimensions, see A Hasegawa and M Wakatani 1987 Self-organization of electrostatic turbulence in a cylindrical plasma' *Phys. Rev. Lett* **59** 1581, and S B Korsholm and P K Michelsen and V Naulin 1999 Resistive drift wave turbulence in a three-dimensional geometry' *Phys. Plasmas* **6** 2401.

Hydrodynamic stability with highly detailed treatment of the Kelvin–Helmholtz instability is treated in the text S Chandrasekhar 1961 *Hydrodynamic and Hydromagnetic Stability* (Oxford: Oxford University Press). There will be more on this in Chapter 16.

Three-dimensional drift wave turbulence with dissipative coupling and the adiabatic response was given by H Sugama, M Wakatani and A Hasegawa *Phys. Fluids* **31** (1988) 1601. More references are given at the end of the next Chapter.

**IOP** Publishing

Turbulence and Instabilities in Magnetised Plasmas, Volume 1
Fluid drift turbulence
**Bruce Scott**

# Chapter 12

# Drift wave turbulence in a sheared magnetic field

## 12.1 Introduction

The last phase of the previous Chapter introduced magnetic shear, and we found that the scales of motion were restricted so as to make three-dimensional computations tractable. The magnetic field was essentially straight, but the existence of periodicity in the parallel direction turns the system into one which has toroidal topology. In fact, toroidal topology is defined by a set of finite surfaces which are periodic in every dimension within the surface. So in our case we have one coordinate in the plane perpendicular to the field, in which the pitch of the magnetic field lines varies. It is also assumed to be the direction of the gradient. We call this the radial coordinate even if it is just a finite layer. For orientation, the direction down the gradient is taken as the positive radial direction. The direction parallel to the field and the direction perpendicular to the field, which is also a symmetry coordinate (no dependence of the equilibrium system on this coordinate), form a plane. At each location in the radial direction, the flux lines of the magnetic field lie within this plane. As a result, these planes are called flux surfaces.

This is a fundamental effect with large consequences to the dynamics: magnetic shear on closed flux surfaces. In a magnetically confined plasma in equilibrium, the magnetic field lines lie within these surfaces. The magnetic field is further generally sheared, which means that any regular, simply connected region in the drift plane becomes deformed by shearing if each point in it is transported along the magnetic field. The magnetic field has this property if the surfaces it lies in are shaped or the plasma lying in it carries a current. Generally, any finite pressure equilibrium guarantees the presence of currents, due to the necessary magnetic forces. Drift wave dynamics and turbulence are most relevant in systems with magnetic field lines which lie in closed surfaces, whose properties are the reason that we have been implicitly assuming that the parallel wavenumber $k_{\|}$ does not vanish. The finite parallel gradient, $\nabla_{\|}$, is the path to dissipation through the kinetic shear Alfvén channel. So whether its unperturbed form with $k_{\|}$ is in general allowed by the

doi:10.1088/978-0-7503-2504-2ch12

geometry to vanish or not is important. It is a theorem of topology that if a vector field should lie within closed surfaces with no boundaries (e.g. a sphere or torus) and is everywhere nonvanishing, then the domain must be toroidal. For a magnetic field in three-dimensional Euclidean space this means that the surfaces must be two-dimensional tori, namely, the flux surfaces referred to above. The point is not about the details of the curved geometry, but only about how it is connected.

There are two very important results of this geometric change to the dynamics of drift waves. First, drift waves as such become linearly stable. In the simple linear models of Chapter 9, we found that for most parameters the linear modes had substantial growth rates. But when $k_{\|}$ becomes a secular function of the position across flux surfaces, which is the effect of shear, we find that the eigenmodes are all stable. Second, the turbulence can persist nevertheless, if it is initialised at finite amplitude. It sustains itself by altering the linear mode structure into something which can support much stronger nonadiabatic dynamics, with both larger drive and dissipation rates. The basic dynamics of this is already present in the simple dissipative coupling model of Chapter 4, but there it is overshadowed by the fact that the linear modes can grow by themselves. In a sheared magnetic field, these nonlinear interactions become the reason we find turbulence at all. Stabilisation of drift waves by magnetic shear and nonlinear destabilisation, or self-sustainment, of drift wave turbulence form the focus of this Chapter.

## 12.2 Field line connection and magnetic shear

If the background is an MHD equilibrium, we may use pressure as the coordinate tracking displacements across flux surfaces. The basic condition of MHD equilibrium with no mass flows is that the right side of the MHD force equation vanishes, or (cf. Chapter 6)

$$\mathbf{J} \times \mathbf{B} = \nabla p. \tag{12.1}$$

If we take the dot product of this with either $\mathbf{J}$ or $\mathbf{B}$, we find that both of these vector fields must lie in surfaces of constant $p$. This defines the direction across flux surfaces as the direction down the pressure gradient. The obvious choice of coordinate directions is what we have been using: $\nabla p$, $\mathbf{B}$, and their cross product.

The magnetic field is in general sheared, and what this means is that its direction is a function of the coordinate across flux surfaces. The very simplest model of this situation is the three-dimensional periodic slab, which treats toroidal topology and magnetic shear, but no other complication of closed flux surfaces. We did this in the previous Chapter. It is the combination of periodicity and shear which carries the important result that the parallel gradient does not in general vanish. In three dimensions, we model the magnetic field and background pressure as

$$L_p \nabla \log p = -\nabla x \qquad \mathbf{B} = B_0(\nabla z + \alpha \nabla y) \quad \text{with} \quad \alpha = \alpha(x), \tag{12.2}$$

in a domain with the $y$ and $z$-coordinates periodic, reflecting the topology of the $yz$-surfaces, which are the flux surfaces. The contravariant components of $\mathbf{B}$ are given by $B^i = \mathbf{B} \cdot \nabla x^i$ for $i \in \{xyz\}$, which are

$$B^x = 0 \qquad B^y = B_0\alpha(x) \qquad B^z = B_0, \tag{12.3}$$

where $\alpha$ is the field line pitch. The pressure scale length is given by $L_p$, which is also the characteristic domain size in the $x$-direction. The periodicity lengths are $L_y = 2\pi/k_0$ and $L_\parallel = 2\pi/K_\parallel$, with $k_0$ and $K_\parallel$ the fundamental wavenumbers in $y$ and $z$ respectively, so that

$$f(x, y + L_y, z) = f(x, y, z + L_\parallel) = f(x, y, z). \tag{12.4}$$

The shear length is given by $L_s$, and is defined by

$$L_s^{-1} = \frac{\partial\alpha}{\partial x}. \tag{12.5}$$

We will model this with $L_s$ a constant parameter, so that $S$ is a normalised shear defined by

$$L_s^{-1} = S\,K_\parallel. \tag{12.6}$$

In a local computation, $L_p$, $k_0$, $K_\parallel$, $B_0$, and $S$ are all constant parameters.

The unperturbed (equilibrium) parallel gradient of any scalar quantity $f$ is given by

$$\frac{1}{B_0}\mathbf{B}\cdot\nabla f = \frac{\partial f}{\partial z} + \frac{x}{L_s}\frac{\partial f}{\partial y} = \frac{\partial f}{\partial z} + SxK_\parallel\frac{\partial f}{\partial y}. \tag{12.7}$$

The periodicity within the surfaces leads to a Fourier representation for $f$,

$$f(x, y, z) = \sum_l \sum_m f_{lm}(x)\exp\left(ilk_0y + imK_\parallel z\right). \tag{12.8}$$

The equilibrium parallel gradient can be evaluated in terms of a set of wavenumbers,

$$\begin{aligned}\frac{1}{B_0}\mathbf{B}\cdot\nabla f &= \sum_l \sum_m iK_\parallel(m + lk_0Sx)f_{lm}(x)\exp\left(ilk_0y + imK_\parallel z\right)\\ &= \sum_l \sum_m ik_\parallel f_{lm}(x)\exp\left(ilk_0y + imK_\parallel z\right),\end{aligned} \tag{12.9}$$

with $k_\parallel$ defined as such for each $lm$-mode. Due to the finite domain size and 2D toroidal topology, the set of available parallel wavenumbers,

$$k_\parallel = K_\parallel(m + l\,k_0Sx), \tag{12.10}$$

with $l$ and $m$ integers, is discrete. The parallel gradient can only vanish if either the Fourier component $f_{lm}$ itself vanishes, or if $l$ and $m$ are chosen such that

$$m + l\,k_0Sx = 0, \tag{12.11}$$

where the quantity multiplying $l$ can be called a field line pitch parameter. Flux surfaces on which $k_\parallel(l, m)$ vanishes are called rational surfaces, and flux surfaces on which $k_\parallel(l, m)$ can vanish for $l = 1$ are called main rational surfaces.

There is a finite spacing between these surfaces. For each $l$, neighbouring rational surfaces are given at the locations near a given rational surface for mode $(l,m)$, where modes $(l,\ m \pm 1)$ have their rational surfaces. The spacing is therefore

$$\Delta_{rs}(k_y) = \frac{1}{lk_0 S} = \frac{1}{Sk_y}, \tag{12.12}$$

for each Fourier mode $k_y$ in the $y$-direction, which is always assumed to be in a periodic domain. This distance is given solely by the shear and the wavenumber in the drift direction.

Obviously, if the pitch parameter $\alpha(x)$ is irrational then this particular $k_\parallel$ cannot vanish, and neither can any other, so long as $l$, hence $k_y = lk_0$, does not vanish. That is, not all surfaces are rational. If there is finite shear, then the pitch parameter is in general irrational because it is continuously varying, and the set of rational numbers on the real line is a set of measure zero. This is why the assumption that $k_\parallel$ and hence $\nabla_\parallel$ does not vanish for typical situations is a good one, so long as the background magnetic field lies in closed surfaces and there is finite shear. This combined property of periodicity and magnetic shear is called field line connection. The computational resolution issues it raises are addressed below.

## 12.3 The 2D sheared slab model

A simple way to treat magnetic shear is to take into account the field line connection property, and examine the dynamics in the neighbourhood of a major rational surface of a closed domain with toroidal topology. The radial coordinate $x$ gives the distance across magnetic flux surfaces from this rational surface, the $y$-coordinate is perpendicular to the magnetic field within the $x = 0$ surface, and the third coordinate is taken along the magnetic field at the $x = 0$ surface. In a 2D model we take variation in the third coordinate to be zero but allow parallel gradients to exist in the $y$-direction, with the magnetic field component $\mathbf{B} \cdot \nabla y \equiv B^y$ having a radial dependence which crosses zero at $x = 0$. We therefore have

$$\mathbf{B} \cdot \nabla f = B_0 \frac{x}{L_s} \frac{\partial f}{\partial y}, \tag{12.13}$$

for any scalar quantity $f = f(x,\ y)$. The corresponding parallel wavenumbers are therefore tied to those in the drift direction,

$$k_\parallel = \frac{x}{L_s} k_y. \tag{12.14}$$

The parallel wavenumber is, however, a valid statistical parameter distinct from $k_y$, as due to the ability of disturbances to alter their extent in the $x$-direction, the $k_\parallel$ spectrum can change independently of the $k_y$ spectrum. Additionally, this secular dependence of the geometry on $x$ now forbids periodic boundary conditions in the $x$-direction. This follows from the experience in the previous Chapter, inserting magnetic shear into the simple three-dimensional cases.

Other than these changes in the geometry, the basic dynamics we will examine is that of the preceding Chapters, drift Alfvén linear modes and turbulence within the simplest possible geometrical model. The equations and normalisation scheme are the same as before,

$$\frac{\partial \varpi}{\partial t} + [\phi, \varpi] = \nabla_{\parallel} J, \tag{12.15}$$

$$\frac{\partial}{\partial t}(A + \hat{\mu} J) + \hat{\mu}[\phi, J] = \nabla_{\parallel}(n_0 + n - \phi) - C\, J, \tag{12.16}$$

$$\frac{\partial n}{\partial t} + [\phi, n_0 + n] = \nabla_{\parallel}(J - u), \tag{12.17}$$

$$\hat{\varepsilon}\frac{\partial u}{\partial t} + \hat{\varepsilon}[\phi, u] = -\nabla_{\parallel}(n_0 + n + P), \tag{12.18}$$

for the dependent variable evolution equations, and

$$\varpi = \nabla_{\perp}^2 \phi \qquad J = -\hat{\beta}^{-1}\nabla_{\perp}^2 A \qquad P = -X\nabla_{\parallel} u_{\parallel}, \tag{12.19}$$

for the auxiliaries and field equations, in which $P$ is the viscous correction to the pressure and $X$ is the parallel viscous diffusivity. The density $n$ is also the pressure as it carries the background temperature with it. For the 2D sheared-slab case, the nonlinear parallel gradient is

$$\nabla_{\parallel} = [-\psi_T, \ ] \quad \text{where} \quad \psi_T = \psi(x) + A \quad \text{and} \quad \psi = -\frac{1}{2}Sx^2, \tag{12.20}$$

with $Sx$ controlling $k_{\parallel}$ and $A$ the magnetic flutter. The normalisation of $z$ in terms of $K_{\parallel}$ and $y$ in terms of $L_{\perp}$ results in the form of $\psi$ in terms of $S$ and $x$ only. The parameters controlling the kinetic shear Alfvén dynamics, that is, the adiabatic response, are

$$\hat{\beta} = \beta_e(K_{\parallel}L_{\perp})^{-2} \quad \hat{\mu} = \mu_e(K_{\parallel}L_{\perp})^{-2} \quad C = \frac{0.51\nu_e}{c_s/L_{\perp}}\hat{\mu}, \tag{12.21}$$

reflecting the competition between perpendicular drift wave excitation and parallel shear Alfvén responses and resistive dissipation. We also have

$$n_0 = -\omega_n x \quad \omega_n = \left| L_{\perp}\nabla \log p_e \right| \quad \hat{\varepsilon} = (K_{\parallel}L_{\perp})^{-2} \quad X = \frac{\mu_{\parallel}}{c_s L_{\perp}}, \tag{12.22}$$

controlling the acoustic part of the dynamics. These can be set arbitrarily as noted in Chapter 9.

The Laplacian and spatial bracket are

$$\nabla_{\perp}^2 = \delta_0^2\left(\frac{\partial^2}{\partial x^2} + \frac{\partial^2}{\partial y^2}\right) \qquad [a, b] = \delta_0\left(\frac{\partial a}{\partial x}\frac{\partial b}{\partial y} - \frac{\partial b}{\partial x}\frac{\partial a}{\partial y}\right) \qquad \delta_0 = \frac{\rho_s}{L_{\perp}}, \tag{12.23}$$

for arbitrary scalars $a$ and $b$, and $\delta_0$ is the drift parameter. Dependent variables are functions of $(x,y)$, with $y$ periodic on an interval of $L_y$. The domain in $x$ is localised to $-L_x/2 < x < L_x/2$, and all dependent variables vanish for $x = \pm L_x/2$, as in the Dirichlet cases from the previous Chapter.

## 12.4 Linear stability of electrostatic drift waves

An important result of field line connection in closed surfaces is that the simple drift waves we have been studying are all linearly stable. We recall that if $k_\parallel^2$ may be taken as a constant in the dissipative coupling model, then every Fourier component in the entire set is linearly unstable—the imaginary part of the frequency eigenvalue is positive. We saw this in Chapter 4. There have been several attempts in the past to prove linear stability of the sheared slab to drift waves, but this problem cannot be solved exactly. The solutions in the literature are asymptotic or numerical, and most are approximate. Here we will recover this result with the well-defined initial value problem. It will prove especially important to do this without resorting to eigenmodes, because for many situations there is no localised eigenmode which keeps its structure. Due to the secular $x$-dependence of $k_\parallel$, the initial disturbance launches a wave-like component whose propagation velocity is partly in the $x$-direction, and therefore some of the activity will propagate away from the rational surface.

One point should be kept in mind. We are spending time with a two-dimensional model while a three-dimensional one is generally more germane. However, the linear eigenmodes of a sheared slab are the same in both two and three dimensions, since they are only active in the vicinity of their rational surfaces, and which plane we take in the space of Fourier components in $z$ (the set of $m$'s) is arbitrary. We can therefore set $m = 0$ without loss of generality for the linear problem.

The definition of the problem is as above, with the nonlinear terms set to zero. We have a single Fourier component $k_y$, defined for $0 < t < \infty$ and $-L_x/2 < x < L_x/2$, subject to the boundary condition that all dependent variables vanish for $x = \pm L_x/2$. The dependent variables are $\varpi$, $J$, $n$, and $u$, with field variables $\phi$ and $A$, all taking complex values, satisfying the set of equations given by

$$\frac{\partial \varpi}{\partial t} = ik_\parallel J, \tag{12.24}$$

$$\frac{\partial}{\partial t}(A + \hat{\mu} J) = -i\omega_* A + ik_\parallel(n - \phi) - C\,J, \tag{12.25}$$

$$\frac{\partial n}{\partial t} = -i\omega_* \phi + ik_\parallel(J - u), \tag{12.26}$$

$$\hat{\varepsilon}\frac{\partial u}{\partial t} = i\omega_* A - ik_\parallel n - Xk_\parallel^2 u, \tag{12.27}$$

with field equations

$$\delta_0^2\left(\frac{\partial^2\phi}{\partial x^2} - k_y^2\phi\right) = \varpi \qquad \delta_0^2\left(\frac{\partial^2 A}{\partial x^2} - k_y^2 A\right) = -\hat{\beta}J, \tag{12.28}$$

where

$$k_\parallel = Sk_y x \qquad \omega_* = \omega_n k_y \delta_0, \tag{12.29}$$

with $\delta_0$, $k_y$, $\omega_n$, $\hat{\beta}$, $\hat{\mu}$, $C$, $S$, and $X$ constant parameters. The linear system is run without subgrid and boundary dissipation operator $D()$, as there is no cascade to work with. Any damping is due to $X$ and $C$.

There is no longer any grid in the $y$-direction so we may quote the wavenumber and dimensions in units of $\rho_s$, which is simpler for a linear model. So we will have $K = k_y\delta_0$ which is $k_y\rho_s$ in normalised units, and $x_L = L_x/2\delta_0$ which is the radial domain half-width in units of $\rho_s$. With only one $k_y$ there are no domain aspect ratio considerations.

There is a pathology to work out before we move on to the nonlinear situation. This is that the sound wave resonance point must be within the domain size, or

$$\mathrm{Max}\left(k_\parallel\right) = Sx_L k_y > \hat{\varepsilon}^{1/2}. \tag{12.30}$$

So an $\hat{\varepsilon}$ too high or $x_L$ too low will not satisfy this. The waves start in the region around the rational surface at $x = 0$ and then are seen to propagate away from it, toward the radial boundaries in both directions. If the waves interact with the boundaries, then this propagation dynamics is interfered with, and the wave becomes a trapped mode which can grow.

We can see the beginnings of this in the adiabatic limit in which there is no growth. If all three of $\hat{\beta}$, $\hat{\mu}$, and $C$ vanish, we have the adiabatic limit with $J = 0$ and $n = \phi$, whose equations are

$$\frac{\partial}{\partial t}(\phi - \varpi) = -i\omega_*\phi - ik_\parallel u, \tag{12.31}$$

$$\hat{\varepsilon}\frac{\partial u}{\partial t} = -ik_\parallel\phi - Xk_\parallel^2 u, \tag{12.32}$$

with field equations and other considerations as before. In each time step, $\phi - \varpi$ and $u$ are advanced, and then the Helmholtz equation represented by $\phi - \varpi$ is solved for $\phi$. The dynamics is started with a localised Gaussian of $1/e$ half-width $x_L/3$ for $\phi - \varpi$, which is solved for $\phi$ in the same way. The parameters held fixed are $K = 0.1$ and $\omega_n = S = 1$. The grid consists of 256 equidistant nodes in a domain bounded by $x_L = 50$. The time step is 0.001. To illustrate the radial propagation situation, we will vary $\hat{\varepsilon}$ and $X$ away from $\hat{\varepsilon} = 100$ and $X = 1$, serving as nominal values.

What happens is shown in figure 12.1. The initial disturbance, localised near the rational surface and forming a wave train in the $y$-direction, spreads out in the $x$-direction, losing amplitude such that the total free energy remains the same. When

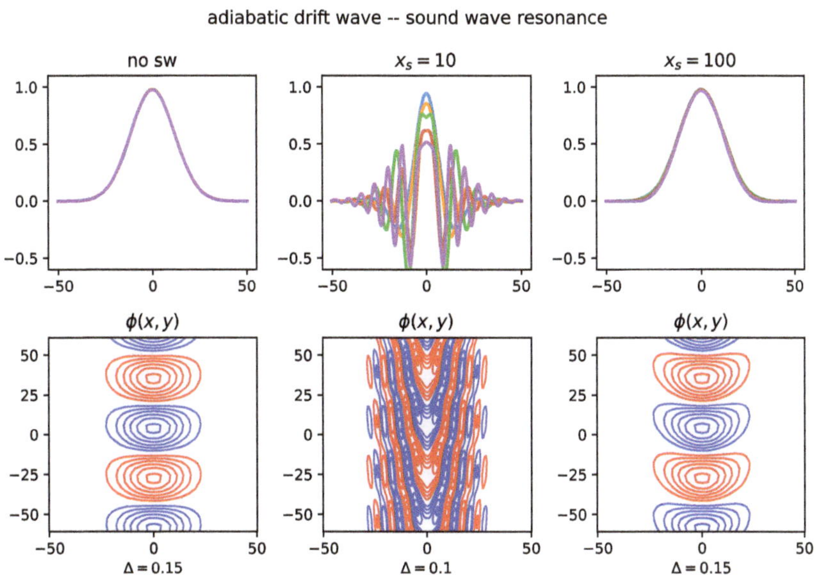


**Figure 12.1.** (left to right) Spread of adiabatic linear drift waves away from the rational surface, for the case with no sound waves, and for $x_s = 10$ and 100. (top row) Envelope lines for $n$ are shown at intervals of $\Delta t = 20$, to $t = 100$. (bottom row) Spatial structure of $\phi$ at $t = 100$. Only the middle case shows the sound wave resonance, as for $x_s = 100$ it is outside the domain.

the disturbances reach positions at which $k_{\|}c_s > \omega_*$, which in normalised units becomes

$$x_s = \frac{\omega_*}{k_{\|}'c_s} = \frac{k_y\rho_s}{Sk_y\hat{\epsilon}^{-1/2}} \rightarrow S^{-1}\hat{\epsilon}^{1/2}, \tag{12.33}$$

the wave energy goes over into sound waves. For a domain of $x_L = 50$ this turnover is only reached before hitting the boundary for cases with $\hat{\epsilon} < 2500$ if $S = 1$. If the boundary distance would be taken $x_L \rightarrow \infty$, then this spreading would continue indefinitely. With finite $x_L$ it is halted only when the disturbances hit the limit $x = \pm x_L$. The spatial manifestation of this is the formation of the meniscus shapes as shown for the $x_s = 10$ case in the bottom row of figure 12.1. The other cases, $x_s = \infty$ as a control and $x_s = 100$ show no coupling to the sound waves. The latter case has its $x_s$ outside of the domain.

As another control we show the same thing for larger grids with 1024 nodes in $x$ and $x_L = 200$, holding the resolution fixed. The cases with $x_s = 10$, 32, and 100 are all shown in figure 12.2, and here we see the sound wave resonance for all three cases. So it is just a matter of having the $x = \pm x_s$ point within the domain to be able to see this. The meniscus shapes form and then continuously develop, producing ever sharper structure in the x-direction, as a result of two space derivatives on the time derivative, balancing $x^2$ for the given $k_y$. The structure is constrained by the properties of this feature of the differential equation.

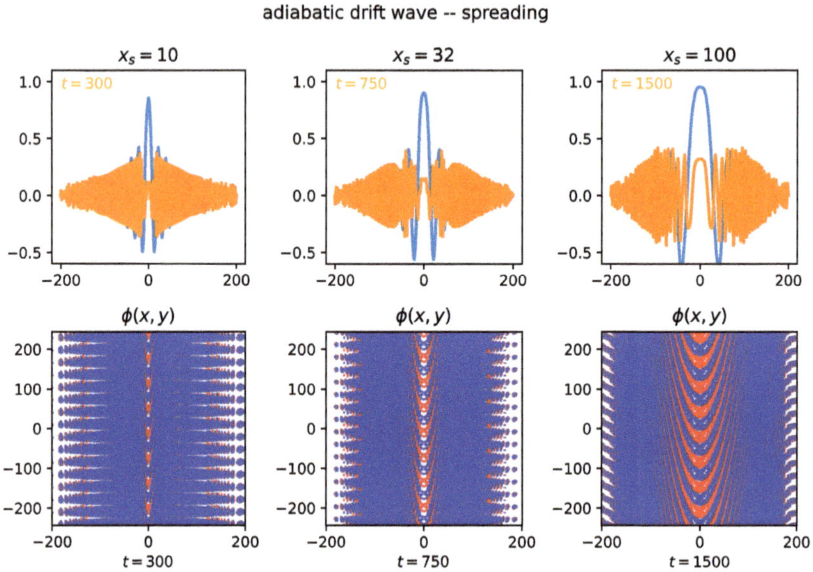


**Figure 12.2.** (left to right) Spread of adiabatic linear drift waves away from the rational surface, for $x_s = 10$, 32, and 100, for a larger domain. (top row) Envelope lines for $n$ are shown at an early time and at the time given. (bottom) Spatial structure of $\phi$ at the time given. The sound wave resonance occurs but does not stop the spreading away from $x = 0$.

The resonance with the sound waves does not by itself limit the spreading. The sound waves themselves continue until they either pile up at the computational boundary or they are dissipated. A simple way to do the dissipation is to reintroduce $X$. We find that if we set $X = 1$ for all three cases, the spreading is halted near $x = \pm x_s$ and then the run is valid until striation due to the weak dependence of the propagation speed in the $y$-direction causes the structure to become unresolved. The cases with $X = 1$ are shown in figure 12.3. This is an effective way to contain the dynamics to the part of the possible domain in $x$ actually carried in the computations. Both $\hat{\varepsilon}$, that is, the ratio $K_{\|}L_{\perp}$, and the shear $S$ come into consideration. If we choose $\hat{\varepsilon} = 10^4$ as in the nominal warm plasma case, but with $x_L = 50$, then no value of $X$ helps because the resonance itself is outside the domain. We duly find the waves hitting the boundaries for $\hat{\varepsilon} = 10^4$ and $X = 10$ or 100 and therefore set $\hat{\varepsilon} = 100$ and $X = 1$ artificially for the rest of this demonstration.

To see the situation for the drift Alfvén cases with finite energetics, we do the same study as in the previous figure with a warm plasma case, varying $\hat{\varepsilon}$ at $X = 1$ and $x_L = 200$ while holding other parameters fixed. They are set to

$$\omega_n = 1 \quad S = 1 \quad K = 0.1, \quad \hat{\beta} = 5 \quad \hat{\mu} = 2.72 \quad C = 13.9, \tag{12.34}$$

with $C$ reflecting a choice of $\nu_e L_{\perp}/c_s = 10$. We start with the same localised Gaussian, but for $n$ only, with all other variables set to zero, and allow the dynamics to evolve naturally. The drift wave structure establishes quickly as in the simple drift Alfvén cases of Chapter 9. The rest of the evolution proceeds mostly as in the adiabatic case

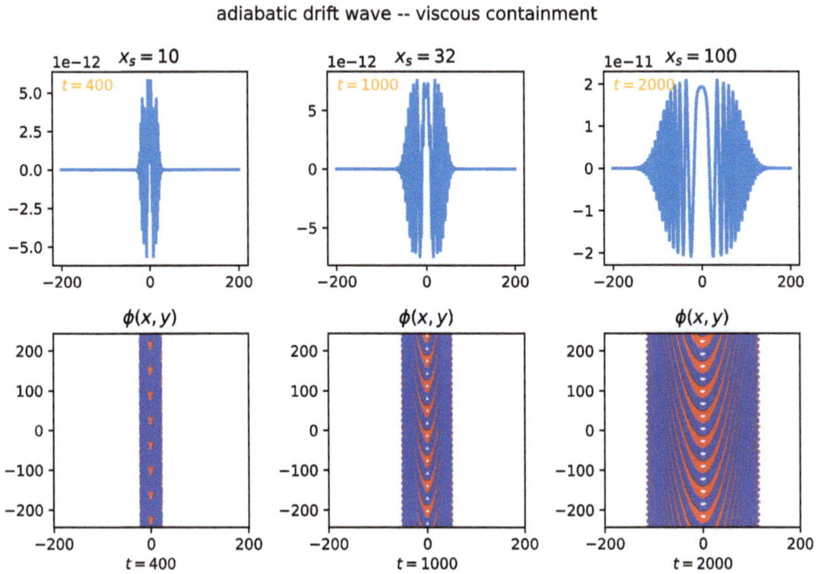


**Figure 12.3.** (left to right) Spread of adiabatic linear drift waves away from the rational surface, for $x_s = 10$, 32, and 100, with $X = 1$ set for each. (left) Envelope lines for $n$ are shown at intervals of $\Delta t = 10$, to $t = 100$. (right) Spatial structure of $\phi$ at $t = 100$.

and is shown in figure 12.4. Localisation interior to the boundaries is found, but there is a tendency to vacate the innermost region. Then, a new disturbance train can form there which is transiently decoupled from the outer regions. During a short phase the disturbances can grow again, until they too spread out in $x$, interact with the sound waves, and are dissipated by the parallel viscosity. This can cause an ambiguous result for indefinite time, especially if $\hat{\varepsilon} = 10^4$ and $x_s$ is too far away from the $x = 0$ surface.

The disturbances are now nonadiabatic and therefore potentially able to tap enough free energy from the background gradient in order to grow. For slow growth and strong collisionality ($C > \hat{\mu}$) the resistivity is the largest effect controlling the response of $J$ to $h = n - \phi$. There is a channel centred upon $x = 0$ in which $\omega_*$ is large enough to overcome the tendency of the resistivity to damp $J$ and hence $h$. The width of this layer is given by

$$\Delta_D^2 = \frac{0.51\omega_* \nu_e}{\left(k_\parallel' V_e\right)^2} \quad \rightarrow \quad \Delta_D = \left(\frac{C}{k_y}\right)^{1/2}, \tag{12.35}$$

with the second value the one in normalised units (note we usually have $\Delta_D < x_s$). This is called the hydrodynamic layer, since within it the dissipative coupling is smaller than the wave frequency, and outside of it the electron response becomes adiabatic. The waves tap the free energy well, within this channel, but then they spread to the adiabatic region $|x| \sim x_s > \Delta_D$. This is why spike structures quasi-periodically form and dissipate near $x = 0$. The eventual fate of the wave is given by

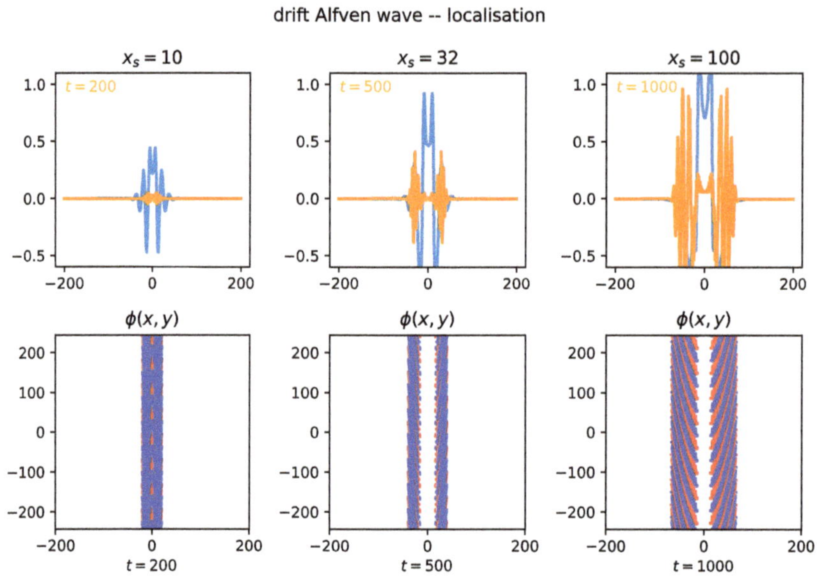


**Figure 12.4.** (left to right) Spread of electromagnetic linear drift Alfvén waves away from the rational surface, for $x_s = 10$, 32, and 100, with $X = 1$ set for each. (left) Envelope lines for $n$ are shown at intervals of $\Delta t = 10$, to $t = 100$. (right) Spatial structure of $\phi$ at $t = 100$.

the competition between this dynamics and the quasi-adiabatic sound wave coupling going on further out. Here as well, if $x_s$ is not interior to the domain, the resonance effect will not work and the result becomes a trapped mode which grows.

This has illustrated a very important property of numerical systems when no physically growing eigenmode is present: the grid often contains its own eigenmodes which exist solely due to its discrete nature. Stability computations done by other means also include some form of spurious mode elimination; one uses a semi-analytical variational principle to assume a mode of regular structure, or in an eigenvalue computation one will assume a certain symmetry at the rational surface and specify that wave energy should be outgoing at $x = \pm x_L$. In other words, even for analytical calculations asymptotic methods are used. Arbitrary resolution at $x = 0$ and arbitrarily large $x_L$ are assumed. In some computational cases the elimination of spurious modes depends on the form of the grid spacing. Generally, it is important to use all the diagnostics at our disposal when undertaking numerical investigations of such delicate systems as linear drift waves, so as to distinguish physical from spurious effects.

## 12.5 Magnetic shear in 3D—field-aligned coordinates

We now return to a more realistic three-dimensional situation to test the stability properties in the case of strong magnetic shear. In the previous Chapter we approached this situation from a different angle: the existing structure for the case of $S = 0$, for which all $k_\parallel \neq 0$ modes are unstable, was shown to become constrained by the gradual introduction of magnetic shear. In realistic cases we have $S = 1$ or possibly higher. For Cartesian coordinates with high resolution for the perpendicular dynamics, we had to have it in the third coordinate as well, to cover the possibility $k_{\parallel} = 0$ for each $(k_y, k_z)$ pair on the grid. In regions where $B^y(x)$ departs from zero, we need more grid nodes in $z$ to allow $B^y/h_y$ and $B^z/h_z$ to mutually cancel. This leaves very large $k_{\parallel}$ values on the grid, giving a very small Courant limit for the time step, comparable to the maximum value of $k_{\parallel}v_A$ or $k_{\parallel}V_e$ anywhere on the grid. If one degree of freedom causes trouble with this, the turbulence will find it. This is what we faced even for $S$ up to 0.5 in the previous Chapter, where a rather small time step was needed, even for cases without extreme amplitudes. For the large domain cases we would have had to reduce the time step even more, so we could have taken $S$ to higher values only with a large increase of computational expense. On the other hand, the actual $k_{\parallel}$ spectrum by itself only requires several tens of grid nodes, and then the limit is given only by the range of $k_{\parallel}$ values the physics actually needs to resolve. This is the plan behind the development of field-aligned coordinates.

Another manifestation of strong shear is that for substantial $S$ there are a large number of rational surfaces within the domain in the x-direction for short wavelengths in the $y$-direction. Recall the double-periodicity in Eq. (12.4), which has the result that rational surfaces are given by Eq. (12.11), and their spacing is given by Eq. (12.12). If an eddy in the turbulence is approximately isotropic and has a width $d$, then its wavenumber in the drift direction is $\pi/d$. If this is inserted into Eq. (12.12), then we find

$$\Delta_{rs} < d \qquad \text{if} \quad S > \pi^{-1}. \tag{12.36}$$

With typical values in a sheared magnetic field of $S = O(1)$, we have the scaling of the spacing as $1/k_y$ for each $k_y$, so that a single isotropic ($k_x \sim k_y$) eddy always overlaps several surfaces (the test number is $\pi$ times $S$). This makes field line connection relevant to the physics of the turbulence and therefore its spatial resolution demands computations.

The only remedy for these problems is to align the coordinates to the magnetic field, not at only one flux surface, but everywhere. The resolution problem is caused by the fact that the parallel gradient involves two coordinates, since two contravariant components of $\mathbf{B}$ are nonzero. What we need is a coordinate transformation which removes this problem by leaving only one nonzero $B^\mu = \mathbf{B} \cdot \nabla x^\mu$ among the $x^\mu \in \{x, y, s\}$, where $s$ becomes the label of the parallel coordinate as distinct from $z$, the angle-like coordinate which remains periodic. It will become very important to note that pure periodicity in the $s$ coordinate will not exist. The simplest transformation which does this is

$$y_k = y + Sx(z - z_k) \quad y = y_k - Sx(s - s_k) \quad \text{with} \quad s = z = s, \tag{12.37}$$

where the back transformation is also given, and $x$ and $s = z$ remain unchanged. The quantity $s_k = z_k$ is a constant which sets the reference point of the transform to a given location in $s = z$. The meaning of this reference point is that the transformed coordinate cells are square only at the reference point; the field-aligned coordinate system is generally nonorthogonal because of the shear. The inverse metric coefficients for the untransformed coordinate system are given by

$$g^{xx} = 1 \quad g^{yy} = 1 \quad g^{zz} = 1 \quad \text{others} = 0. \tag{12.38}$$

For the field-aligned system they are given by

$$g^{xx} = 1 \quad g_k^{xy} = S(s - s_k) \quad g_k^{yy} = 1 + S^2(s - s_k)^2 + S^2x^2, \tag{12.39}$$

$$g^{xs} = 0 \quad g_k^{ys} = Sx \quad g^{ss} = 1. \tag{12.40}$$

We evaluate these at $x = 0$ due to the drift ordering, leaving

$$g^{xx} = 1 \quad g_k^{xy} = S(s - s_k) \quad g_k^{yy} = 1 + S^2(s - s_k)^2 \quad g^{ss} = 1 \quad \text{others} = 0. \tag{12.41}$$

The element $g^{xy}{}_k$ tells us that the coordinate cells are indeed square only at $s = s_k$. The usefulness of this is that each member of this family of transformations is equivalent in the sense that it covers the whole space. So we can use the one referenced at a given location in $s$ for evaluation of the perpendicular drift dynamics, keeping care when taking parallel derivatives to include the shift in $y$ introduced by the fact that for two members of the family of transformations, the $s_k$ values, and hence the $y_k$-coordinate reference, are different.

One can verify that this coordinate system is indeed aligned by finding the contravariant components of $\mathbf{B}$:

$$B^x = 0 \quad B_k^y = 0 \quad B_s = B_0. \tag{12.42}$$

Note that $B^y{}_k$ vanishes not only at $x = 0$ or $s = s_k$, but everywhere. Due to the fact that only one of the $B^\mu$ is nonvanishing, the parallel gradient involves only that coordinate:

$$\mathbf{B} \cdot \nabla f = B^\mu \frac{\partial f}{\partial x^\mu} = B^s \frac{\partial f}{\partial s}. \tag{12.43}$$

The flute mode condition of drift ordering shows that derivatives with respect to $s$ are small, since the parallel and perpendicular scales of motion are defined by

$$\nabla_\parallel^2 = B^{-2}(\mathbf{B} \cdot \nabla)^2 = \frac{\partial^2}{\partial s^2} \quad \nabla_\perp^2 = g^{xx}\frac{\partial^2}{\partial x^2} + 2g_k^{xy}\frac{\partial^2}{\partial x \partial y} + g_k^{yy}\frac{\partial^2}{\partial y^2}. \tag{12.44}$$

The metric coefficients, given in Eq. (12.41), are all of order unity. So if $k_\parallel \ll k_\perp$ then it follows that $\partial/\partial s$ is much smaller than the other derivatives,

$$\frac{\partial}{\partial s} \ll \frac{\partial}{\partial x} \sim \frac{\partial}{\partial y}. \tag{12.45}$$

We are back to the plan which led to the two-dimensional model, and have generalised it to three dimensions. In contrast to the high resolution in all three dimensions we needed before, to cover $k_\parallel = 0$ for each $(k_y, k_z)$ pair on the grid, the third dimension now only has to cover the finite-$k_\parallel$ spectrum itself, so that we need significantly fewer nodes in $s$ than we did in $z$ before. As a result we can increase the time step and hence run a computation of similar expense for a much longer physical time.

This transformation affects the boundary conditions. Although the flux surface itself is still physically periodic in both $y$ and $z$, the fact that we have replaced $y$ with $y_k$ means that perfect periodicity in $s$ is lost, since the $y$ locations are different at different locations in terms of $y_k$ along $s$, i.e. the field line. In the untransformed coordinate system we have

$$f(x, y + L_y, z) = f(x, y, z) \qquad f(x, y, z + L_\parallel) = f(x, y, z), \tag{12.46}$$

for periodicity in the $y$ and $z$-directions, respectively. To express these conditions in the field-aligned coordinate system we apply the transformation in Eq. (12.37) to them, obtaining

$$f(x, y_k + L_y, s) = f(x, y_k, s) \qquad f(x, y_k + 2\pi Sx, s + L_\parallel) = f(x, y_k, s). \tag{12.47}$$

Periodicity in $y$ is maintained, since the differing meanings of $y$ and $y_k$ appear only when we vary $s$. We still have

$$\left.\frac{\partial}{\partial y}\right|_s = \left.\frac{\partial}{\partial y}\right|_z, \tag{12.48}$$

but in the other angle we have

$$\left.\frac{\partial}{\partial s}\right|_y = \left.\frac{\partial}{\partial z}\right|_y - Sx\left.\frac{\partial}{\partial y}\right|_z, \tag{12.49}$$

which has the following consequence.

The boundary condition in $s$ becomes one of what we can call pseudo-periodicity, since when we apply the Fourier decomposition in $y_k$ (allowed because we still have periodicity in $y$), we obtain a phase-shifted periodicity in $s$:

$$f(x, k_y, s + L_\parallel) = f(x, k_y, s)e^{-2\pi i S k_y x}, \tag{12.50}$$

which we obtain by multiplying Eq. (12.47) by $\exp(-ik_y y)$ and integrating over $y$ (taking care not to confuse $y_k$, the $k$th member of the family of field-aligning transformations, with the drift direction wavenumber $k_y$). The phase factor in Eq. (12.50) shows where the rational surfaces are: the locations where $Sk_y x$ is an integer. On these rational surfaces we have true periodicity in $s$ as well as $y$, and so $k_\parallel$ can vanish even for finite $k_y$. Elsewhere, however, $k_\parallel$ cannot vanish for finite $k_y$, and so we have the conclusion that $k_\parallel$ is in general finite on closed magnetic surfaces. This recovers the field line connection property of shear and periodicity, once more, by which $k_\parallel$ is in general finite on topologically toroidal surfaces when there is magnetic shear. This property is why we get drift wave dynamics, as shown in the study in the previous Chapter varying weak shear away from zero.

The shifting transformation implied by a nonzero $s_k$ in Eq. (12.37) affects the boundary conditions and the parallel derivatives. All of the above equations hold, but we want to express them for differing members of the family of transforms for different locations in $s$. At $s = s_1$ we use $s_k = s_1$, while at $s = s_2$ we use $s_k = s_2$. In general, the $k$ value will track the $s$-node being used, and parallel derivatives and the parallel boundary conditions involve differing $s$-nodes. The parallel derivative for a given $s_k$ is given by

$$2h_s\frac{\partial f}{\partial s}(x,\, y_k,\, s_k) = f(x,\, y_k,\, s_k + h_s) - f(x,\, y_k,\, s_k - h_s), \tag{12.51}$$

where $h_s$ is the node spacing in $s$. Setting $s_{k\pm1} - s_k = h_s$, we re-express this derivative such that each location in $s$ uses the field-aligned transform referenced there,

$$2h_s\frac{\partial f}{\partial s}(x,\, y_k,\, s_k) = f(x,\, y_{k+1} - \Delta^+,\, s_{k+1}) - f(x,\, y_{k-1} - \Delta^-,\, s_{k-1}), \tag{12.52}$$

with the shifts given by

$$\Delta^\pm = \pm Sxh_s. \tag{12.53}$$

In discrete numerics, this is equivalent to taking the shift implied in Eq. (12.50) over an entire periodicity length $L_\|$, and distributing it among all the $s$-nodes. The parallel boundary condition becomes

$$f(x,\, y_{k+N},\, s_{k+N}) = f(x,\, y_k,\, s_k), \tag{12.54}$$

where $N$ is the number of $s$-nodes, and therefore $s_{k+N} - s_k$ is $L_\|$. Each location along a field line is now treated numerically equivalently, which is in order with the property of the slab geometry that all locations along the field line are equivalent.

We encounter another numerical resolution problem with the field line connection, especially with linear modes. The parallel boundary condition in Eq. (12.50) is what defines field line connection in this model, and it prescribes a phase shift that each Fourier component in $k_y$ must have at displacements of one connection length in $s$. The distance between neighbouring rational surfaces is given by $1/k_yS$. On the grid, the $k_y$ for the shortest wavelength is $\pi/h_y$. The distance to the next rational surface for this mode is $h_y/\pi S$, which is less than $h_x$ itself. In that case, the minimum parallel gradient imposed by the field line connection is not felt at all, and there arises a grid mode with the lowest $k_y$, which as a result doesn't notice the shear. This mode therefore grows and is effectively a linear instability with $k_\| = 0$, namely, a flute mode. In the previous Chapter we saw that the $S = 0$ case was powerfully unstable, leading to very strong turbulence. The introduction of magnetic shear eliminated the large-scale flute mode vortex with $\phi \gg n$ in normalised units, bringing $\phi$ down to the level of $n$ as the long wavelength adiabatic response became effective. However, ever stronger shear runs the grid out of resolution, and then the instability returns. The consequence of these grid modes in turbulence is rapid growth of small-scale clutter which corrupts the solution.

To fix this, we need small enough $h_x/h_y$ for neighbouring points to see the field line connection property for the shortest wavelength, and the criterion is

$$h_y > h_x\, \pi\, S, \tag{12.55}$$

so for $S = 1$ a grid ratio of $h_y/h_x = 4$ should be sufficient, giving the necessary resolution ratio of 4 points in $x$ for each one in $y$.

For cases which start turbulent and always remain turbulent the increase of $x$-resolution is not necessary, since the small-scale vorticity is enough to overwhelm this type of instability. In other words, the turbulence turns out to be insensitive to these grid modes, since they are coherent structures with no or almost no $s$-dependence and as such are destroyed by the turbulence faster than they can grow. But proof of nonlinear instability does require increased $x$-resolution, and in extreme cases a convergence study must be done in the $x$-direction to ensure that the conclusion is robust.

## 12.6 Self-sustained drift wave turbulence

We use this coordinate treatment to address three-dimensional sheared slab drift Alfvén turbulence under the four-field model. The same procedure as in the previous Chapter for cases with finite shear and bounded in the $x$-direction is used again here, just with differences in the coordinate system. We need to find the correspondence in the linear stability results for strong shear just done above and the existence of robust turbulence in cases with magnetic shear done in the previous Chapter. Either the turbulence results are incorrect or the linear stability does not hold in turbulence. To prove which way that decision falls requires effective control cases. We will find indeed that finite nonlinearity sustains the turbulence by itself in the face of linear stability. This Section will show how we find out whether and especially through which physical mechanism that becomes a real result.

The same Eqs (12.15)–(12.19), which are the same as in the previous Chapter, are used. The main difference in the expressions is the three-dimensional nonlinear parallel gradient given by

$$\nabla_{\parallel} = \frac{\partial}{\partial s} + [-A, \ ]. \tag{12.56}$$

Note that the piece $\psi$ due to $S$ does not appear in $\nabla_{\parallel}$ as it does for Cartesian geometry, since the contribution to $\nabla_{\parallel}$ due to $\nabla\psi$ has been folded into the map $(y, z) \leftrightarrow (y_k, s)$ in the field-aligned coordinates. The computational grid is $256 \times 256 \times 32$ in $\{xys\}$, in a domain with $L_x = 1$ and $L_x = 4$ and $L_s = 2\pi$ in units of $L_{\perp}$ and $1/K_{\parallel}$, respectively. Hence the aspect ratio is $L_y/L_x = 4$ and the resolution ratio is $h_y/h_x = 4$ as well. The parameters which are kept constant for these cases are

$$\delta_0 = 0.015 \quad S = 1 \quad \omega_n = 1 \quad \hat{\epsilon} = 100 \quad X = 1, \tag{12.57}$$

around which $\hat{\beta}$, $\hat{\mu}$ and $C$ are varied.

The energetic drive and damping rates, growth rate, and error are the same as in Chapters 9 and 11, except that the subgrid dissipation is aligned to the field in the same way as the grid. We now have

$$D() = \nu_0 \left( \frac{h_x^2}{4} \frac{\partial^2}{\partial x^2} + \frac{h_y^2}{4} \frac{\partial^2}{\partial y^2} \right)^2 + \nu_0 \left( \frac{h_s^2}{4} \frac{\partial^2}{\partial s^2} \right)^2 + s_f p_f, \tag{12.58}$$

with the hyperdiffusion tied to the resolution in each coordinate. The boundary dissipation $s_f p_f$ is done as in Eq. (11.49). The coefficients are set to $\nu_0 = 1.0$ and $s_f = 0.1$, sufficient to contain the turbulence for these cases. The energetics is the same as in Eqs (9.62)–(9.65) for the physical terms and in Eq. (11.36) for the subgrid dissipation, evaluated with $D$ as just defined.

The initial state is set up with the same steps as in Eqs (11.41) and (11.42) from Chapter 11, with the 2D random-phase density field at $s = 0$, but with the envelope structure directly in $s$, along the field, rather than the unaligned $z$-coordinate. The coordinate system with $y_0 = y + Sxz$, anchored at $z = 0$ is used everywhere for this purpose (see Eq. (12.37)). The parallel envelope is expressed as

$$n(x, y_0, s) = n_{2D}(x, y_0) \exp[-(s/2.5)^2]. \tag{12.59}$$

This state is then shifted from $n(x, y, z)$ back onto $n(x, y_k, s)$ in the field-aligned coordinate system using the $y_k$ at each $s = s_k$ whose $g^{xy}{}_k$ metric element vanishes there, and then the run is started. As the drift wave mode structure forms due to the (Alfvénic) adiabatic response, the fluctuations spread out along the field lines, and due to the shear the interactions at $s = \pm\pi$ for various $xy$ are incoherent and there is no standing-wave transient.

The first set of cases we look at is the electromagnetic model, with $\hat{\beta} = \hat{\mu} = 1$ and $C = 1$. This is a low collisionality case with a finite time dependence in the adiabatic response. Several runs are compared, differing only in the initial amplitude, from deeply linear to nonlinear. The range of amplitudes tested is

$$A_0 = 10^{-8}, \quad 10^{-4}, \quad 10^{-2}, \quad 10^{-1}, \quad 3 \times 10^{-1} \quad 10^{0}, \quad 10^{1}, \tag{12.60}$$

which holds everything constant across cases except for the degree of nonlinearity. The extra case with $A_0 = 0.3$ is exemplary in showing how the nonlinear instability becomes established. Due to the slow initial evolution, it is run to $t = 4000$ while a run to 1000 is sufficient for all the others. The time traces of energetic quantities are shown in figure 12.5. The ExB vorticity is $\Omega_E$ in units of $c_s/L_\perp$, which is given by $\delta_0^{-1}\varpi$ in this normalisation. The RMS ExB vorticity is given for all the cases, and the energetics for the $A_0 = 0.3$ case. We find that for $A_0 = 0.1$ and lower, each case decays in a manner independent of the amplitude, indicating linear stability with insignificant nonlinear perturbation. For $A_0 = 1$ and higher, each case saturates quickly and converges to the same amplitude. The extra case with $A_0 = 0.3$ is transitional, initially decaying, finding itself, growing back to nonlinear amplitude, and then saturating after $t = 1000$ to a value convergent with the other nonlinear cases. The development of the spectra in this case is shown later on, below. Its energetics shows the small value of $\Gamma_n$ initially, with $\Gamma_T < 0$ indicating net damping. From $300 < t < 500$ the net drive begins to rise, and then the raggedness of the curve for $t > 800$ indicates visible turbulence. At this point the RMS ExB vorticity has reached a value above about 0.3, which is faster than any reasonable growth or damping rate for these modes. Between about $500 < t < 1000$ the nonlinear growth rate $\Gamma_T$ is net positive. As the turbulence picks up, the resistive dissipation $\Gamma_c$ and also

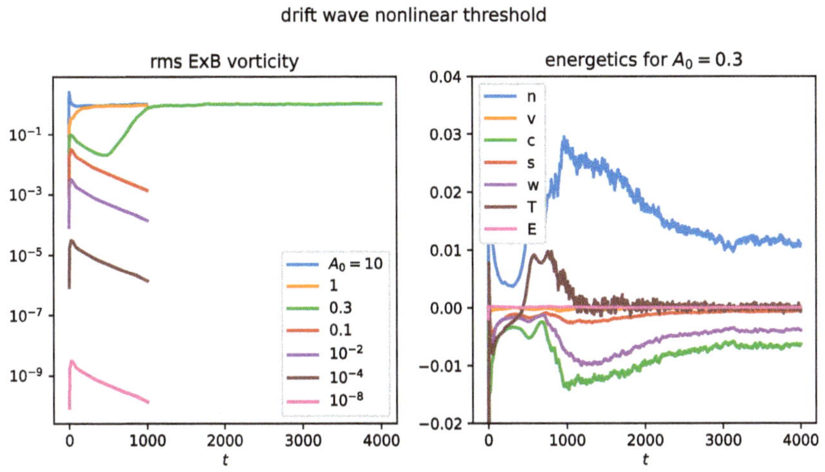


**Figure 12.5.** (left) Time traces of the RMS ExB vorticity, for various initial amplitudes denoted by $A_0$. The basic signature of nonlinear instability, the cases decay if $A_0$ is too small, and saturate convergently otherwise. The run of the $A_0 = 0.3$ case indicates the threshold—the amplitude decays until nonlinear correlations can self sustain the dynamics. The energetics (right, labels according to Γ subscript) indicate that the ExB gradient advection drive ($\Gamma_n$) increases first, and then the resistive dissipation ($\Gamma_c$) and subgrid dissipation ($\Gamma_w$) rise to compensate. The magnetic flutter drive ($\Gamma_v$) and the viscous dissipation ($\Gamma_s$) remain small. The total nonlinear growth rate ($\Gamma_T$) saturates at zero, and the error ($\Gamma_E$) remains small.

the subgrid dissipation $\Gamma_w$ rise in response. At later times we have $\Gamma_n \approx \Gamma_c + \Gamma_w$ in approximate balance, with $\Gamma_c$ the largest sink, with all others small. Measured over $3000 < t < 4000$ we have $\Gamma_n = 0.011 \pm 6 \times 10^{-4}$ for the drive rate, and $\Gamma_T = (0.6 \pm 2.7) \times 10^{-4}$, i.e. essentially zero, for the growth rate. The error $\Gamma_E = (2 \pm 0.3) \times 10^{-5}$ is finite but about three orders of magnitude smaller than the total drive and dissipation rates.

It remains for us to find out if this result is just due to a fortunate combination of parameters. So we take the following parameter sets,

$$\begin{array}{lll} \hat{\beta} = 0.15 & \hat{\mu} = 0.15 & C = 0.918 \\ \hat{\beta} = 0.15 & \hat{\mu} = 1 & C = 1.02 \\ \hat{\beta} = 1 & \hat{\mu} = 0.15 & C = 0.918 \\ \hat{\beta} = 1 & \hat{\mu} = 0.15 & C = 0.153 \\ \hat{\beta} = 1 & \hat{\mu} = 1 & C = 1.02 \\ \hat{\beta} = 5 & \hat{\mu} = 2.72 & C = 2.77, \end{array} \tag{12.61}$$

giving a reasonable range of parameter combinations controlling the adiabatic response of the electrons.

Each of these is run with both $A_0 = 1$ and $A_0 = 10^{-4}$ to test the effect of nonlinearity at finite amplitude as in the above cases. The time traces are shown together in figure 12.6, with the same format as in figure 12.5. The lines run together, which one is which is not important—only that all but one of the nonlinear starts

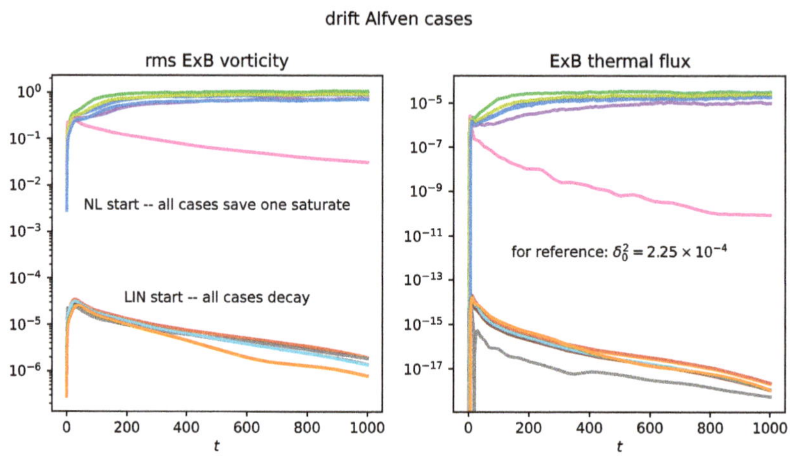


**Figure 12.6.** Time traces of the RMS ExB vorticity (left) and the turbulent ExB flux (right), for various cases all with initial amplitude $A_0 = 1$ for a nonlinear start or $A_0 = 10^{-4}$ for a linear start. The nonlinear cases saturate to robust nonlinear amplitude, except for the one with $C < 1$ (see text). The linear-start cases all decay, showing the same result as in figure 12.5.

saturates at reasonable amplitude, and all the linear starts decay. The one case which did not find robust saturation is the one with $C = 0.153$, which indicates that $C$ in the vicinity of unity or higher is required to find this nonlinear physics. The finding is that finite electron-channel dissipation and the ExB vorticity nonlinearity out of a nonlinear start are the necessary conditions for the nonlinear instability, while their combination is sufficient.

Taken together, these results constitute the signature of nonlinear instability. Among the cases, only the initial amplitude is different, and therefore it is finite amplitude which causes the turbulence to find saturated amplitude. This cannot be a result of grid modes, as otherwise we would see eventual growth in the cases with the $A_0 = 10^{-4}$ linear start. This first result is the stable, damped nature of the disturbances in the linear case, and the basic existence of the nonlinear instability for the nonlinear case. The second result is the convergence of all the saturated cases for a given parameter set, independent of initial conditions. The third result is the increase of the nonlinear drive rate directly caused by finite vorticity. These cases are sufficient to demonstrate the nonlinear instability for the three-dimensional model including the $k_y = 0$ component; in other words, realistic dynamics. One sees the importance of the control cases with linear start in establishing this conclusion.

### 12.6.1 Role of nonlinear polarisation

The nonlinearities can be investigated as to which one causes these changes away from the linear result. With some hindsight, the following is how to do it for the ExB vorticity nonlinearity $[\phi, \varpi]$. The reason for focus on this nonlinearity is that it represents the nonlinear part of the polarisation current. It is the dynamics which allows access of the free energy source represented by the background gradient. The polarisation current opposes the parallel current divergence, supporting a finite level of the parallel current in the overall balance. If the parallel current is zero then the product of $k_{\parallel}$ and the nonadiabatic response $h = n - \phi$ must be zero. Any component of the density or potential representing turbulent eddies which are mostly isotropic must have a finite $k_{\parallel}$, which follows from the field line connection property of the geometry. Hence, to have a robust $h$ especially at long wavelength (where $\omega_* \ll c_s/L_{\perp}$) we require something to support $J$. The agent enabling this is $[\phi, \varpi]$ representing the nonlinear polarisation current. All the other nonlinearities are referred to as the ExB advective nonlinearity, with the main one being $[\phi, n]$ in terms of energetic effect. These provide the saturation of the turbulence.

The following describes a control test study against this proposition. Start with the standard case with $A_0 = 1.0$ which saturates quickly. Then, create a 'vorticity only' case by suppressing all the ExB advective nonlinearities (leave out the nonlinear $[\phi, \ ]$ terms except for $[\phi, \varpi]$, while leaving $[\phi, n_0]$ in to provide the gradient drive). As an additional control, create a 'no vorticity' case by suppressing only the ExB vorticity nonlinearity (leave out $[\phi, \varpi]$). The time traces for RMS ExB vorticity for these three are shown in figure 12.7. It can be clearly seen that while the nominal case saturates, the 'no vorticity' case decays almost as if it were linear, and the 'vorticity only' case grows to much larger amplitude than the 'nominal' case (in which no changes are made). This indicates not only that the ExB vorticity nonlinearity is the cause of the nonlinear instability, but also that the other nonlinearities are most of what causes the saturation. In the nominal case, both nonlinearities are simultaneously active, and one increases the activity in the vorticity and current dynamics, while the other acts more like turbulent mixing and causes saturation by redistributing the dynamics among all available degrees of freedom, including all the dissipatively stable components. Examination of the isotropic spectra for the 'vorticity only' case, in figure 12.8, shows that the turbulence collects at a larger scale (top left) than is usual, and is more due to the larger gradient drive (bottom left, $\Gamma_n$) than the nonlinear transfer (right) through the ExB vorticity (blue), which is not different from what is seen in the nominal case. This indicates that the absence of the ExB advective nonlinearity leads to less decorrelation of the driving structures than is nominal, so that they can take over the dynamics. It also indicates that the dynamics in a case assuming linear responses to the existing turbulence is very different from the nominal. The responses in turbulence with the adiabatic response are therefore fundamentally nonlinear.

With the vorticity nonlinearity enabling the turbulence generally, we examine some details in the energetics which show how it acts. We take the three cases shown in figure 12.7, and plot the isotropic spectra of the energetic transfer channels involving $\nabla_{\parallel} J$ in the equations for $\varpi$ and $n$, which quantifies the adiabatic compression in the parallel current and hence the adiabatic response. These are shown in figure 12.9, with the ExB energetics on the left side and the thermal free energetics on the right side. The definitions of the effects, or channels, are

$$\text{Alfvén} = \frac{1}{2E_T} \mathscr{V}^{-1} \int d\mathscr{V} \, (-\phi) \nabla_{\parallel} J \quad \text{adiabatic} = \frac{1}{2E_T} \mathscr{V}^{-1} \int d\mathscr{V} \, (n) \nabla_{\parallel} J, \qquad (12.62)$$

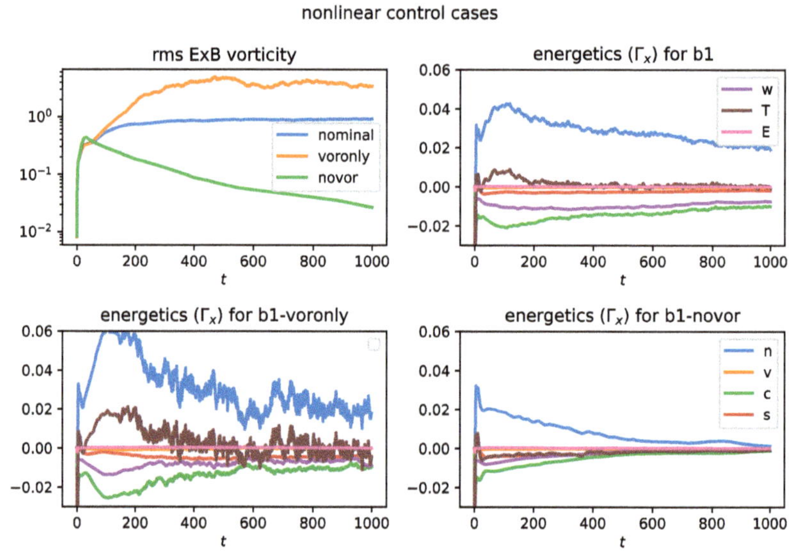


**Figure 12.7.** (top left) Time traces of the RMS ExB vorticity, for cases with $A_0 = 1$, with all terms in the equations ('nominal'), with only the ExB vorticity nonlinearity ('voronly'), and with all terms except the ExB vorticity nonlinearity ('novor'). The nominal case is the one with sufficient amplitude to exhibit nonlinear instability and saturation. Absence of the ExB vorticity nonlinearity leads to decay as in the near-linear cases with small $A_0$. Presence of it alone leads to very large amplitude. This identifies the ExB vorticity nonlinearity as the principal cause. The energetics in the 'nominal' (top right), 'voronly' (bottom left), and 'novor' (bottom right) collectively show that the gradient drive is stronger when the ExB vorticity nonlinearity is present and stronger still when it acts alone (labels are split for clarity). Coupled with the tendency of it to collect the ExB energy into the larger scales in a dissipation-free manner, this indicates that the nonlinear dynamics is a result of vorticity scattering exciting the free energy access, opposing the parallel couplings which act as a clamp on it. The total nonlinear growth rate ($\Gamma_T$) falls to near zero in all cases, while the magnetic flutter drive ($\Gamma_{\tilde{v}}$) and the error ($\Gamma_E$) remain small.

as the Alfvén and adiabatic transfer rates, respectively, with positive values indicating the rate of energy transfer into each quantity. Note that the actual amplitude of the turbulence is normalised away in these measurements as in those for the Γ's giving the drive and dissipation pieces. The nominal and 'voronly' cases show comparable rates even though the amplitude of the latter is about one order of magnitude larger. There is more of an emphasis on the larger scales in the 'voronly' case, essentially reflecting the lack of the nonlinear transfer in the ExB advective nonlinearity which would remove thermal free energy back to smaller scales. In the 'novor' case the ExB vorticity nonlinearity is absent, and without it the rate of transfer into the ExB energy through $\nabla_{\|}J$ is much smaller. The result is that the 'novor' case decays while the other two reach and maintain robust amplitude.

The role of the ExB vorticity nonlinearity is to open up the range of accessible states with nonadiabatic electrons, namely, larger possible values of the parallel current. Taking the spectra for the nominal case as a benchmark, we find that the linear dynamics acting by itself, or even in the 'novor' case, produces very weak nonadiabaticity. Indeed, whatever nonadiabaticity is left in the linear case results

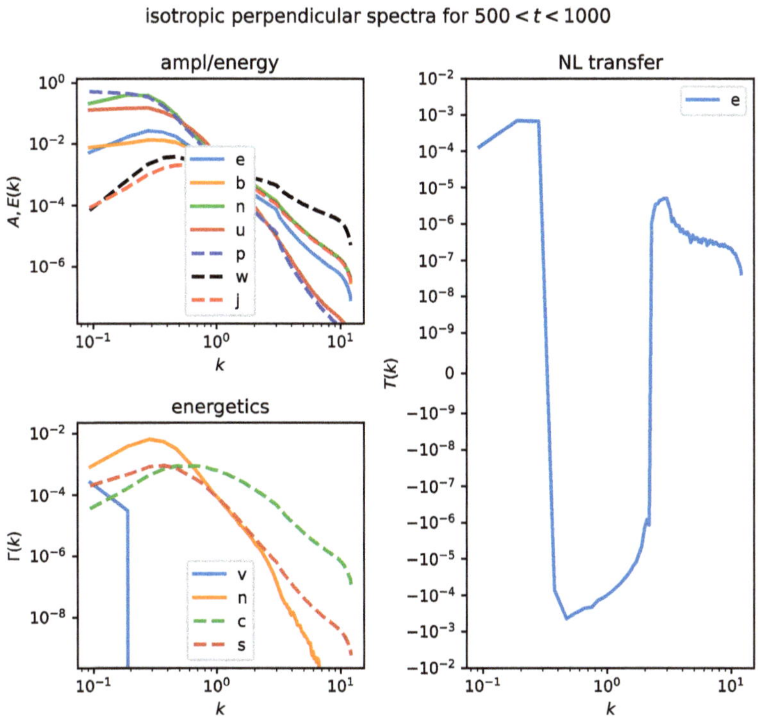


**Figure 12.8.** Isotropic perpendicular spectra for the $A_0 = 1$ case 'voronly' during $500 < t < 1000$, in a phase close to saturation. The labelling and format are as in the spectra shown in the previous Chapter. The amplitudes (top left) and gradient drive (bottom left, 'n') are much more concentrated at large scales than in the nominal case. The nonlinear ExB vorticity transfer (right) reflects this shift but is otherwise not different from nominal. The other ExB transfer channels are absent in the 'voronly' case.

solely from the linear time derivative of the vorticity. The vorticity nonlinearity produces very strong nonadiabaticity by contrast, in either the nominal or 'voronly' cases, and then enables and maintains turbulence. These are the nonlinear and linear polarisation current effects, and the whole of this has shown that the nonlinear polarisation current is very much stronger than its linear counterpart in producing electron nonadiabaticity. This is why we have a nonlinear instability.

The vorticity nonlinearity has a special structure, due to the double conservation properties (energy, enstrophy) that lead to its inverse cascade tendency. In a set of triads in the equations for the larger scale components which represent incoming energy from smaller scales, the extra factors of $k_\perp$ in $[\phi, \varpi]$ are larger than the $k_\perp$ for the larger scales (cf. Chapter 2). The two beat waves driving a particular $k_\perp$ in three wave coupling triplets with $\mathbf{k} + \mathbf{k}' + \mathbf{k}'' = 0$ tend to come from higher $k_\perp'$ rather than lower, and this gives larger than usual factors of $k_\perp'$ and $k_\perp''$ in comparison to the three wave interactions in the other nonlinearities, as well as two additional factors of either $k_\perp'$ or $k_\perp''$ in $\varpi$ itself. This extra large nonlinear effect arises precisely where it is needed to enhance the amplitude of the nonadiabatic response $h = n - \phi$ by expanding the set of accessible nonadiabatic

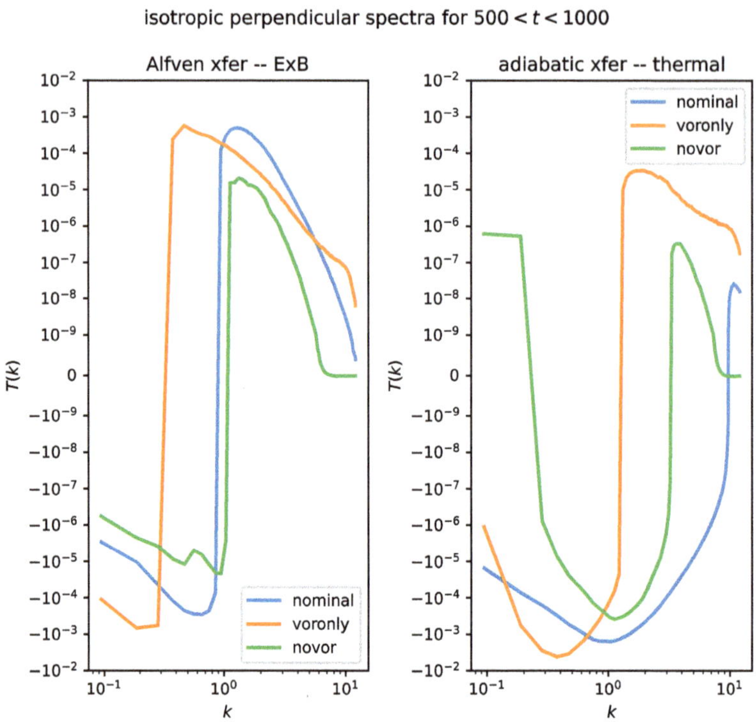


**Figure 12.9.** Isotropic spectra of the energy transfer involving the parallel current divergence for cases with $A_0 = 1$, with all terms in the equations ('nominal'), with only the ExB vorticity nonlinearity ('voronly'), and with all terms except the ExB vorticity nonlinearity ('novor'). The ExB energetics (Alfvénic) is on the left and the thermal free energetics (adiabatic) is on the right. The 'voronly' case shows comparable transfer rate to the nominal case, with the spectrum shifted towards a larger scale. The 'novor' case shows much weaker excitation of the vorticity through the current. The presence of $[\phi, \varpi]$ enables the nonlinear instability by exciting a stronger energy transfer channel through the parallel current, and its absence leads to weak energetics and no nonlinear instability. This is the effect of nonlinear polarisation on drift wave turbulence.

states, whether larger $k_\parallel^2$ for a given $h$ or a larger $h$ for a given $k_\parallel^2$. With larger $h$ possible, there is more gradient drive ($\Gamma_n$), and with larger $k_\parallel^2$ there is more dissipation ($\Gamma_c$). Both changes occur in turbulence in this fashion, in contrast to the linear waves. As a result, the central features of drift wave turbulence are something one cannot tell from linear drift wave analysis.

Taken together, these tests show that the physics of self-sustained drift wave turbulence as a nonlinear instability is a result of nonlinear polarisation supporting a robust parallel current, therefore robust nonadiabatic response, and therefore larger background gradient drive. This is the fundamental manifestation of the competition between perpendicular and parallel dynamics—turbulence versus coupling. The ability of the parallel gradient to wax and wane according to how much it is excited by the turbulence is needed to represent this, and the full freedom of $k_\parallel$ away from both $k_x$ and $k_y$ in the three-dimensional model is needed for full realism.

## 12.7 Magnetic shear and drift wave mode structure

The mode structure of nonlinear drift wave turbulence is similar in both two- and three-dimensional models. Using the 2D dissipative coupling cases from Chapter 10 as a control, we list herein the important features of three-dimensional drift wave mode structure, noting the comparisons to and contrasts against the structure found in the 2D cases.

The case from which these measurements come is one with robust, electromagnetic turbulence, with parameters set as in the warm plasma case, with

$$\hat{\beta} = 5 \quad \hat{\mu} = 2.72 \quad C = 13.9 \quad \hat{\epsilon} = 10\,000 \quad X = 100 \quad \delta_0 = 0.015, \tag{12.63}$$

and $S = \omega_n = 1$, with the mass ratio $\hat{\mu}/\hat{\epsilon}$ reflecting deuterium ions and the parallel viscosity set to $\mu_\parallel = c_s/K_\parallel$ which more realistically reflects the weakly collisional case for the ions. The grid is $64 \times 256 \times 32$ in $\{x, y, s\}$, and the time step is 0.01. The domain is $L_x = 1$ and $L_y = 4$ giving $L_x/\rho_s$ of about 67. The case runs to $t = 4000$, and averages are taken over the last $\Delta t = 3000$. The ExB transport flux averages to $F_e = (1.03 \pm 0.059) \times 10^{-4}$, translating to 0.46 in units of $\delta_0^2 c_s$, the local benchmark.

The development of the isotropic spectra of the amplitudes and energetics as the nonlinear setting is established is shown in figure 12.10 at $t = 350$ and in figure 12.11 at $t = 500$, for the case above starting with the initial amplitude of $A_0 = 0.3$. The eventual structure is initially not present and only develops as the vorticity nonlinearity becomes effective. At later times, we see the structure in figure 12.12 giving the same spectra for the nominal case in saturation. The amplitudes of $\phi$ and $n$ peak towards large scale, the vorticity peaks near $\rho_s$, and the gradient drive $\Gamma_n$ is spread out mostly within the nominal $0.1 < k_\perp\rho_s < 1$ drift wave turbulence range. The energy transfer through ExB nonlinearities are towards small scale in every channel except the ExB energy which is robustly inverse. The transfer channels in ion and electron inertia, through $u$ and $J$, respectively, are subdominant. Hence the ExB advective nonlinearity is mostly in the state variable, as $[\phi, n]$ in this model.

The parallel wavenumber spectra are shown in figure 12.13. The amplitudes and drive rates all peak at $k_\parallel = 0$, with the vorticity having the flattest spectrum, supporting relatively flat dissipation spectra. The energy transfer through ExB nonlinearities are out of the flute mode region into the wider ranges in the 'nonadiabatic' channels (thermal free energy, and magnetic and electron parallel flow energy), and back into the flute mode range for the ExB and sound wave channels. The transfer channels in ion and electron inertia are subdominant. The structure of the amplitude spectra is the manifestation of the various coupling and constraint channels in the parallel dynamics. The nonadiabatic response $h$ occupies a smaller range than $n$ or $\phi$, each in turn narrower than the vorticity for which the mixing rate is largest.

The parallel envelopes of amplitudes and transport are shown in figure 12.14, mainly for completeness, although the fact that they are independent of $s$ apart from statistical variation reflects a non-transient, saturated turbulent state. These serve as control tests against the toroidal geometry effects as tested in Chapter 14. It is necessary to apply the shifting transformations discussed at the end of Section 12.6

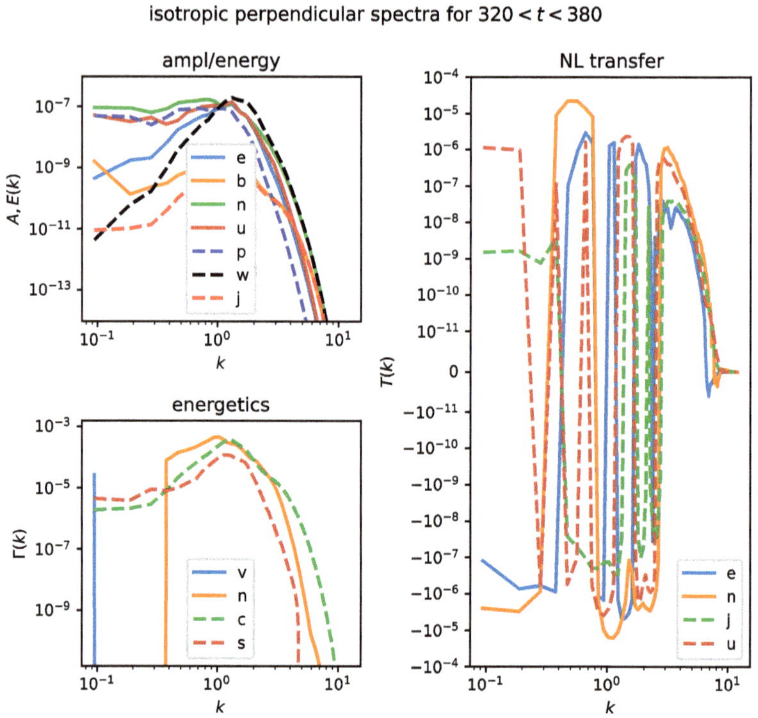


**Figure 12.10.** Isotropic perpendicular spectra for the $A_0 = 0.3$ case during $320 < t < 380$, just as the nonlinear energetics begins to establish itself. Note the time in figure 12.5. The energy and energetics all still peak where the initial state was set, and the transfer spectra have not become established yet. In this phase the energy is still decaying. The labelling and format are as in the spectra shown in Chapter 11 (esp. figure 11.2).

to get this result in a computation with field-aligned coordinates, as otherwise the strong grid deformation due to the variation of especially $g^{yy}{}_k$ with $s$ would lead to spurious structure in the $s$-direction.

The cross coherence and phase shift distributions are shown in figure 12.15. The density and potential are well correlated, and the phase shift distributions are narrow and closer to zero than to $\pi/2$. The signature of drift wave turbulence here is the near independence of the phase shift distribution with $k_y$, reflective of the tendency of the entire spectrum to act as a unit.

The amplitude probability distributions are shown in figure 12.16. These exhibit the form intermediate between Gaussian and passive scalar benchmarks, with especially the vorticity showing the wider tails of the passive scalar distribution. This is a manifestation of the small-scale sheared structure seen next.

The spatial form of the disturbances in shown in figure 12.17 for the latest snapshot. The difference in structure between $n$ and $\phi$ on the one hand and $\varpi$ and $J$ on the other is the main point. Here, $\varpi$ is represented by $\Omega_E$, the actual ExB vorticity. Several different scales with different morphology are involved collectively in a turbulent state which does not merely reflect dynamics at one dominant scale. We also see a flow layer in the $y$-independent component of $\phi$ in these pictures.

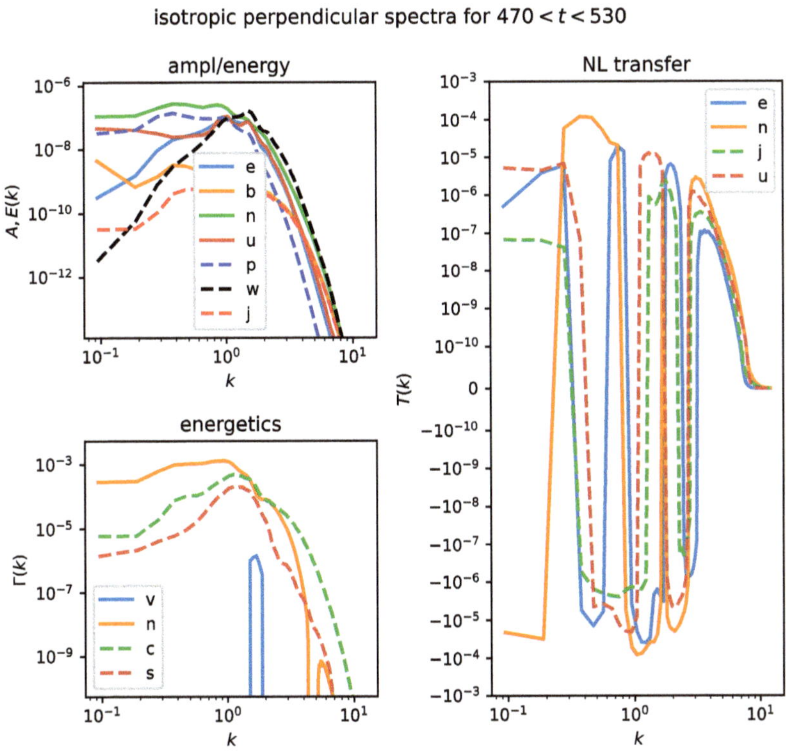


**Figure 12.11.** Isotropic perpendicular spectra for the $A_0 = 0.3$ case during $470 < t < 530$, as the amplitude begins to rise after the initial drop. Compare with the previous figure and note the time in figure 12.5. The nonlinear growth phase has just begun, assisted by a transient peak in the nonlinear advective transfer in the range around $k_\perp \rho_s = 0.3$, with the larger scales beginning to fill out.

We leave further treatment of these flow layers to Chapter 16, in which they take centre stage.

The three-dimensional morphology is shown in figure 12.18. There are 32 drift planes in all, and 8 are shown. One sees a general tendency to follow the field lines, but since disturbances are born at all locations in $s$ with equal probability it is moderately obscured. The profiles, which are not simply the $k_y = 0$ part but the $k_y = k_\parallel = 0$ flux surface averages, have been subtracted out in order to highlight the disturbances.

### 12.7.1 Details of the energetics

Besides time traces of the energetic quantities giving the drive and dissipation (source and sink) effects and a few of the transfer channels, the isotropic spectra of the energetics are very revealing. The effect of the ExB vorticity nonlinearity on the transfer of energy into the ExB part of the system through the parallel current has already been seen. Herein, we detail these for each of the four equations, with the state variables and flux variables grouped separately. When computing spectra one

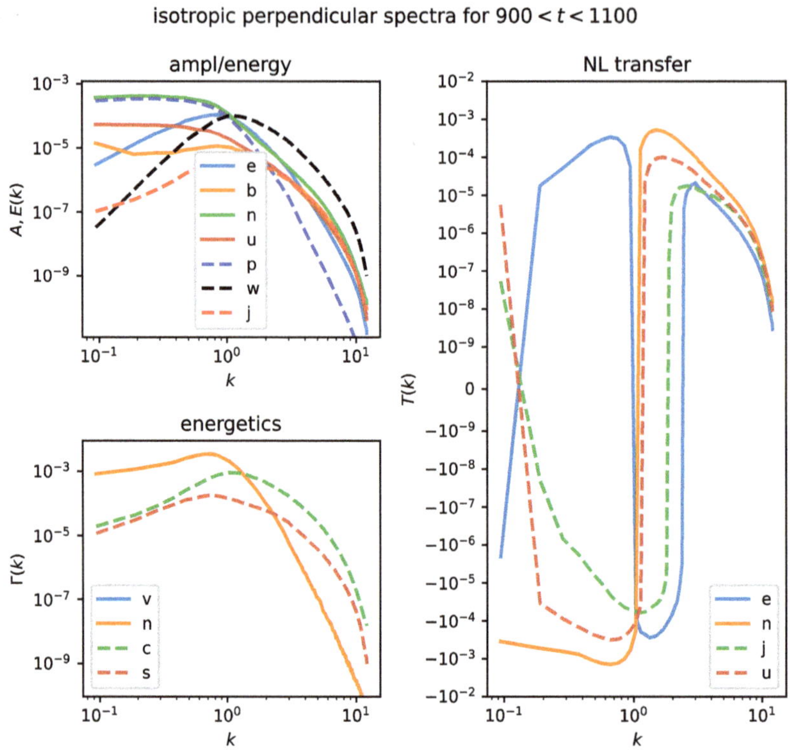


**Figure 12.12.** Isotropic perpendicular spectra for the nominal case, with $\hat{\beta} = 5$ and $C = 13.9$, averaged over $2000 < t < 4000$, in saturation. This is the same structure as for the $A_0 = 0.3$ case during $900 < t < 1100$, as it enters the saturated phase. Compare with figures 12.10 and 12.11. The amplitudes, gradient drive, and vorticity and resistive dissipation peak at large, intermediate, and small scales, respectively. The transfer due to the ExB vorticity nonlinearity towards large scale and the ExB advective nonlinearity towards small scale, together with the adiabatic response, ties the spectrum together as a unit, so that there is no single dominant mode. This is basic drift wave mode structure.

should recall that the transport term itself, into which two transfer effects combine, e.g.

$$\phi\nabla_{\parallel}J + J\nabla_{\parallel}\phi = \nabla_{\parallel}(\phi J), \tag{12.64}$$

has a nontrivial spectrum. Hence, when doing these the state variable terms are left as combinations of a state variable multiplying a flux variable divergence, and the flux variable terms are kept as combinations of a flux variable multiplying a state variable gradient.

As in Chapter 11, the cross-quantity energetics in detail is given by

$$-\phi\frac{\partial\varpi}{\partial t} = -\phi\left(-[\phi, \varpi] + \nabla_{\parallel}J\right), \tag{12.65}$$

$$n\frac{\partial n}{\partial t} = n\left(-[\phi, n] + \nabla_{\parallel}J - \nabla_{\parallel}u + \omega_n n v_E^x\right), \tag{12.66}$$

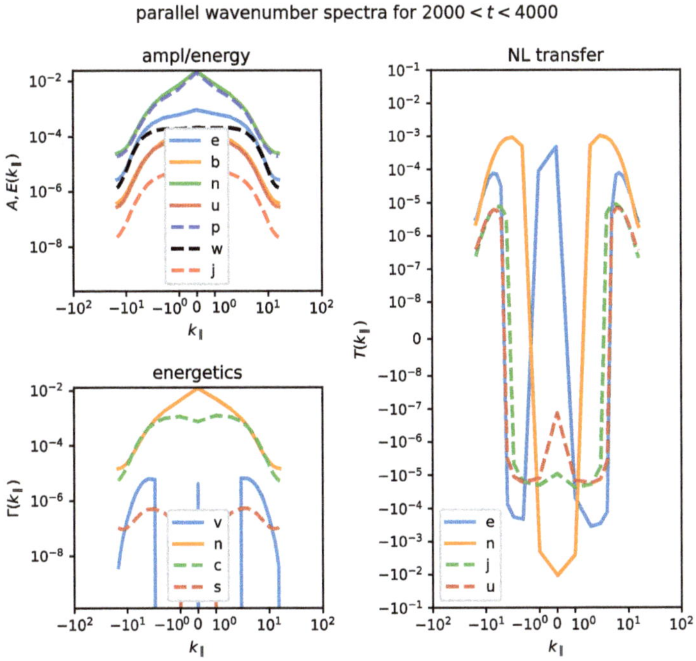


**Figure 12.13.** Parallel wavenumber spectra for the case with $\hat{\beta} = 5$ and $C = 13.9$ averaged over $2000 < t < 4000$, in saturation. The labelling and format are as in the parallel spectra shown in Chapter 11 (esp. figure 11.3). These are now with the proper $k_{\parallel} = b^s k_s$, not the $k_{\parallel} \approx b^z k_z$ as before. The peak of the amplitudes and drive at $k_{\parallel} = 0$ reflects the dominance of this region, but is not a spike reflective of a separate mode. The vorticity ('w') has the widest spectrum, while $\phi$ and $n$ ('p' and 'n') follow each other, and the resistive dissipation ('c') is slightly lower at $k_{\parallel} = 0$ and slightly higher in the mid-range $k_{\parallel} < 10K_{\parallel}$, than the gradient drive ('n'). The nonlinear transfer of ExB energy ('e') is into the $k_{\parallel} = 0$ region and the convective transfer in the other channels ('nju') is out of it into all other modes, with more of the finite-$k_{\parallel}$ region a net sink for the parallel channels ('c' and 's'), and with the thermal free energy ('n') the dominant channel. This is a basic drift wave mode structure.

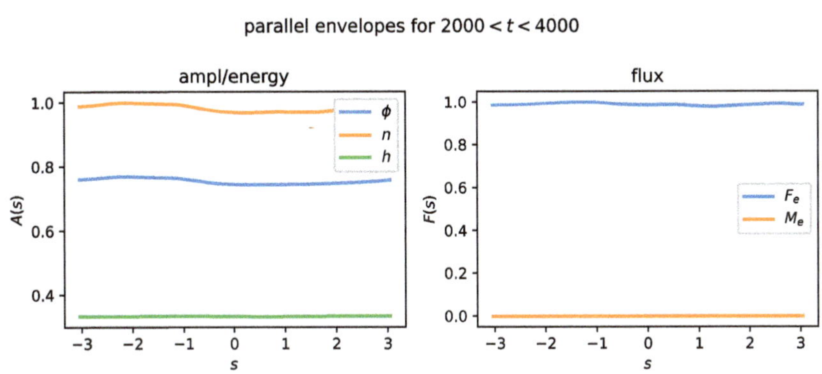


**Figure 12.14.** Parallel envelopes of the squared amplitudes (left) and turbulent ExB transport flux (right), which are averages over $x$ and $y$ and time $2000 < t < 4000$ of each quantity. As expected, the curves are very flat. This serves as a control case, not only of the field-aligned coordinate system but also for the toroidal cases addressed in later Chapters.

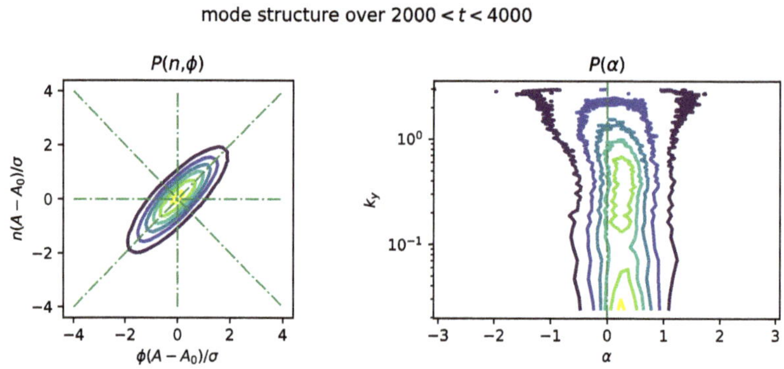


**Figure 12.15.** Cross coherence (left), and phase shift distributions (right) between $n$ and $\phi$, measured over $2000 < t < 4000$. Compare with the 2D cases of figures 10.11 and 10.12 from Chapter 10, and figure 11.24 from Chapter 11. The same basic mode structure is exhibited in the 2D dissipative drift wave and 3D drift Alfvén cases.

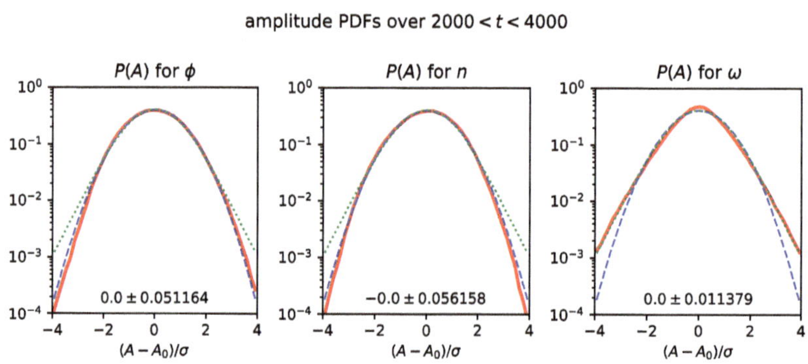


**Figure 12.16.** Amplitude probability distributions of $\phi$ (left), $n$ (centre), and $\varpi$ (right), measured over $2000 < t < 4000$. The blue (dashed) and green (dotted) curves show Gaussian and passive scalar distributions for comparison. The vorticity especially follows the latter closely. This is essentially the same signature of fully developed turbulence found in the 2D cases of figure 10.9 from Chapter 10.

for the state variables, and

$$J \frac{\partial}{\partial t}(A + \hat{\mu} J) = J\big(-\hat{\mu}[\phi, J] + \nabla_{\|} n - \nabla_{\|}\phi - \omega_n b^x - C\, J\big), \tag{12.67}$$

$$u \frac{\partial}{\partial t}(\hat{\epsilon} u) = u\big(-\nabla_{\|} n + \omega_n b^x + \nabla_{\|}(X \nabla_{\|} u)\big), \tag{12.68}$$

for the flux variables, where the down-gradient field perturbations are

$$v_E^x = -\frac{\partial \phi}{\partial y} \qquad b^x = \frac{\partial A}{\partial y}, \tag{12.69}$$

and the subgrid dissipation is not written. Each of these is plotted as an isotropic spectrum in $k_\perp$ reflecting $k_x$ and $k_y$, averaged over $s$. The quantities in large

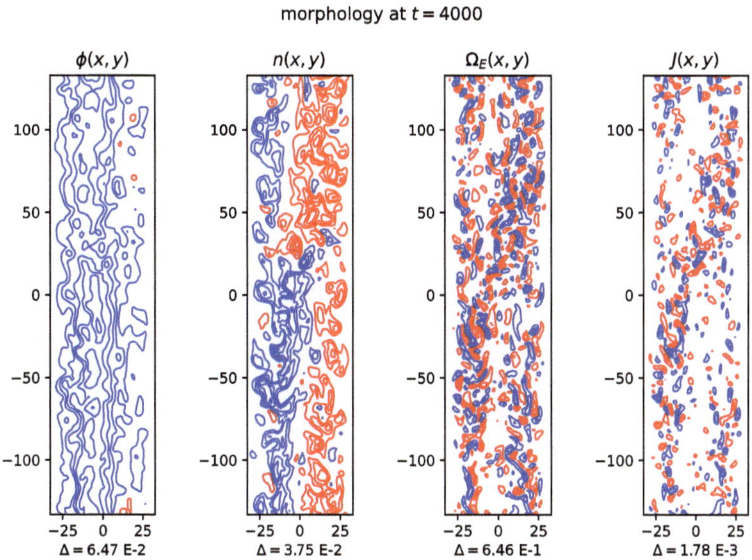


**Figure 12.17.** Spatial morphology at $t$ = 4000 of the state variables, potential, and current. These show ExB eddies in strong turbulence, with $\phi$ and $n$ vaguely similar at the eddy scale, with the potential of a domain-scale flow superposed. The vorticity ($\Omega_E = \delta_0^{-1}\varpi$) and current ($J$) form small-scale structures which are vigorously sheared. The interval indicates that the maximum $n$ is about 0.2 times the background value, and the maximum vorticity is about 3.5 times $c_s/L_\perp$, indicating vigorous turbulence. Different quantities at different scales is a manifestation of self-consistent dynamics involving the entire spectrum.

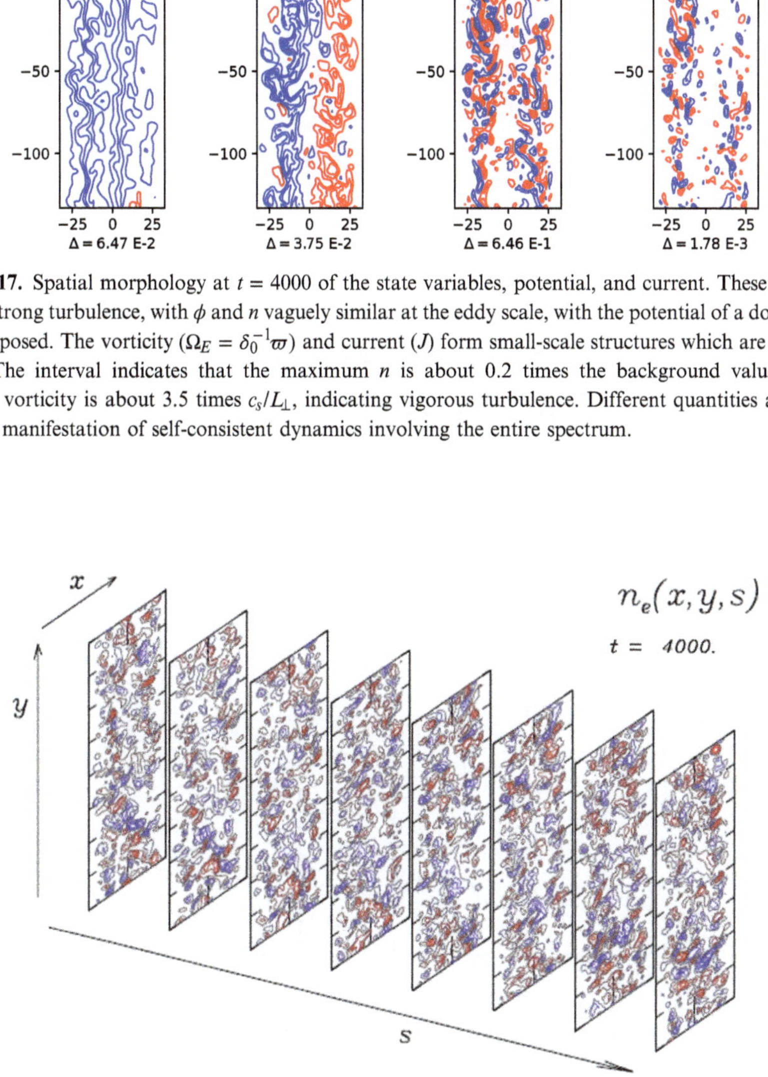


**Figure 12.18.** Three-dimensional structure of $n$ for the representative case. Of the 32 planes in $s$ in the computation, 8 are shown. Despite the finite width of the parallel wavenumber spectrum (figure 12.13) the tendency to follow field lines is vaguely visible. Positive contours are in red; negative are in blue.

parentheses are calculated as a scalar $f(x, y)$ and then 2D Fourier transformed into $f(k_x, k_y)$, and then the quantities multiplying them, for example $\phi$, are separately transformed. These form binary products whose spectra are evaluated as the real part of $\phi^* f$ in the example, at each $k_x$ and $k_y$, at each $s$. These are then binned into $k_\perp$ as described in Eq. (3.36) of Chapter 3. The result is divided by the total energy (twice the real part divided by twice the energy), to evaluate it as a sample of a transfer rate into the given $k_\perp$ for the given piece of the free energy (e.g. $-\phi\varpi$ or $n^2$, divided by two—in this case the volume average divides out). Each sample is then summed into a total over each $s$ and each snapshot time. The resulting spectrum is then a transfer rate in units of $c_s/L_\perp$ averaged over time.

The isotropic spectra of the energetics for the state variables are shown in figure 12.19. These are for the ExB energy $E_e$ on the left and thermal free energy $E_n$ on the right. The definitions of the channels are

$$
\begin{aligned}
T_e &= -\phi(-[\phi, \varpi]) \qquad T_j = -\phi\left(\nabla_\| J\right) \\
T_n &= n(-[\phi, n]) \qquad T_a = n\left(\nabla_\| J\right) \qquad T_s = n\left(-\nabla_\| u\right),
\end{aligned}
\tag{12.70}
$$

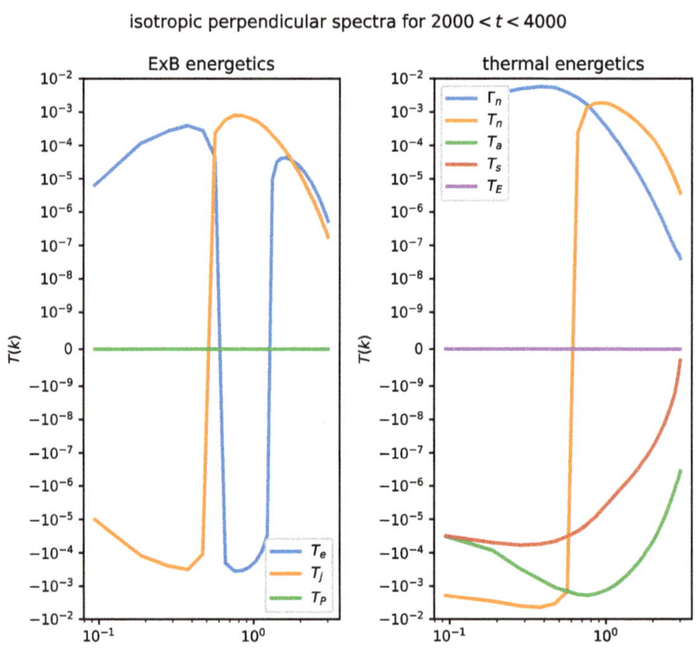


**Figure 12.19.** State variable energetics, for the ExB (left) and thermal free energy (right). Labels and definitions are as given in the text. The ExB energy has no sources, but is supported by incoming transfer from the current via Alfvén transfer at small scale, and the ExB vorticity nonlinearity at large scale. The thermal free energy is supported by the gradient drive at all scales, balanced mostly by nonlinear advective transfer at larger scale and adiabatic transfer to the current at smaller scale. The large scale signature, especially in $T_j$, would not be possible for a linear instability involving these equations.

along with $\Gamma_n$ as the ExB gradient drive for $E_n$. All of these are defined in volume average as with $\Gamma_n$, and the spectra are computed in the same way, as in Eq. (12.62). The nonlinear transfer channels $T_e$ and $T_n$ are the same ones encountered in Chapter 4 (the label $T_E$ was used there; here we label it $T_e$, not to be confused with the electron temperature). The two other effects, $T_P$ and $T_E$, are toroidal magnetic compression effects we will encounter in the next Chapter. In considering the ExB energy we note there are no sources and, apart from the subgrid dissipation not shown, no sinks.

For the thermal free energy, $\Gamma_n$ is mostly a source, and again there are no physical sinks. The main drive of all the dynamics is this particular source, which is prominently large at all scales except the dissipation range, loosely for $k_\perp\rho_s > 1$. This is balanced at larger scale mostly by nonlinear ExB advective transfer ($T_n$) to smaller, and secondarily by adiabatic compressional transfer ($T_a$) out of this energy piece into that of the current. The transfer out into the sound waves by acoustic compression ($T_s$) is small in comparison. At smaller scale the adiabatic transfer channel balances the nonlinear transfer incoming from larger scales, and can be said to be excited to the level it has by the nonlinearity.

The only way for energy to get into the ExB piece at all is Alfvénic transfer ($T_j$) in from the current. This is mostly at smaller scale, while it is back into the current at larger scale. The robust ExB vorticity nonlinearity ($T_e$), transfers energy within this piece from small scale to large, and the main balance is between this effect and the Alfvénic transfer. As we have already seen, the absence of the ExB vorticity nonlinearity leads to much smaller Alfvénic transfer, so that it can be said that the nonlinearity itself drives this transfer. This gives the turbulence a mechanism to sustain itself.

The isotropic spectra of the energetics for the flux variables are shown in figure 12.20. These are for the magnetic and parallel electron flow energy $E_b$ (in combination) on the left and parallel ion flow (sound wave) energy $E_u$ on the right. The definitions of the channels are

$$\begin{aligned} T_j &= J(-\hat{\mu}[\phi, J]) \qquad & T_a &= J\left(\nabla_\parallel n\right) \qquad & T_A &= J\left(-\nabla_\parallel \phi\right) \\ T_u &= u(-[\phi, u]) \qquad & T_s &= u\left(-\nabla_\parallel n\right), & & \end{aligned} \tag{12.71}$$

along with $\Gamma_v$ (two pieces, with $J$ and $u$, for $E_b$ and $E_u$, respectively) as the magnetic flutter gradient drive for $E_b$, and the resistive ($\Gamma_c$) and viscous ($\Gamma_s$) sinks for $E_b$ and $E_u$, respectively. In both cases the magnetic flutter gradient drive is labelled $\Gamma_v$. The two effects appear on separate graphs and so there is no confusion. The resistive dissipation curve is the lowest one in the left frame.

In considering the Ohmic energy $E_b$, the magnetic flutter gradient drive averages to zero ($J$ and $b^x$ being related through Ampère's law), and is small everywhere. In the acoustic energy $E_u$ it is simply small. The sinks $\Gamma_c$ and $\Gamma_s$ are negative definite. The main sources for both pieces are the incoming transfer through parallel dynamics, adiabatic $T_a$ and Alfvénic $T_A$ transfer for the Ohmic side and sound wave transfer $T_s$ for the acoustic side. The adiabatic transfer into the Ohmic energy and acoustic transfer into the acoustic energy are positive everywhere and constitute

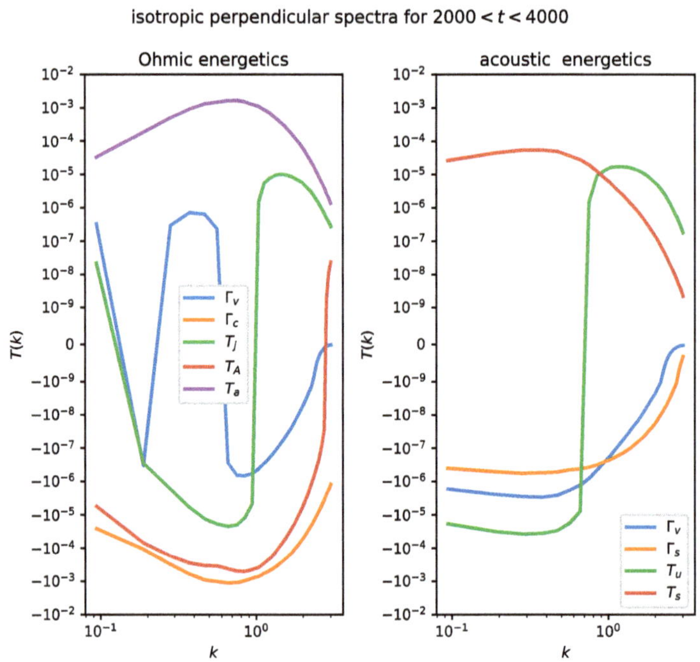


**Figure 12.20.** Flux variable energetics, for the electron (Ohmic, left) and ion channels (acoustic, right). Labels and definitions are as given in the text. Sinks are signed and therefore negative. The Ohmic (including magnetic) energy is supported at all scales by adiabatic transfer incoming from the thermal free energy, balancing both resistive dissipation and Alfvénic transfer to the ExB energy. The acoustic energy is supported mostly by transfer from the thermal free energy, balancing nonlinear transfer at large scale and mostly viscous dissipation at small.

the main source effect for these pieces. The Alfvénic transfer to the ExB energy and the resistive dissipation constitute the main sink for Ohmic energy. Nonlinear transfer is significant only in the acoustic energy where it transfers from large to small scale, balancing the incoming acoustic transfer at large scale, and the viscous dissipation at small scale. This is also the main effect limiting the size of the parallel gradient in the adiabatic part of the system.

Taken together, this gives a picture of a unified, self-consistent spectrum of nonlinear dynamics. It is the same one which was shown in figure 10.13 from Chapter 10. The energy transfer is mostly circular: start with thermal free energy at low-$k_\perp$ and follow how it is thrown to small scale, transfers into the Ohmic energy, then into the ExB energy, through both adiabatic and Alfvénic coupling, then back to large ExB eddy scales, then through Alfvénic coupling to the current and then adiabatic coupling out of it, and thereby back to the thermal free energy. The net drive due mostly to ExB advection of the background gradient ($\omega_n v_E^x$) is needed to support all of this, but the result of it is that the RMS ExB vorticity is much larger than the net drive by at least one order of magnitude. The typical level of RMS $\varpi$ is order few times $\delta_0$ and the drive is order few times $\delta_0^2$, translating into an ExB vorticity $\Omega_E$ of order few times $c_s/L_\perp$ and a drive rate or order few times $v_D/L_\perp$ in physical units. This falsifies any model which assumes that a nonlinear decorrelation rate balances a linear drive.

## 12.8 Electromagnetic effects

Simple quantitative scaling is not the way to understand the physics of a turbulent system, since as we have seen the physical character can persist in the presence of widely varying quantitative variation. An example is the energy transfer dynamics just studied, as well as the same for the 2D models in Chapter 4. Simply looking at total energy levels or even total transfer would have obscured the fact that each individual transfer mechanism keeps its form throughout the parameter range. But we do need to know the quantitative information after all, and in this case we are going to find out something very interesting. The magnetic induction and flutter effects work against each other in a similar way to the linear cases seen in Chapter 9. This makes the induction parameter $\hat{\beta}$ an interesting one to scale against. The collisionality $C$ is next, and then the magnetic shear, and then follows the control cases against magnetic flutter and parallel resolution. The reason for the latter is to test the proposition that relatively few grid nodes are needed in the parallel coordinate when aligning the grid to the magnetic field, and this despite the case that convergence is found for maximum values of $k_\parallel c_s$ rather smaller than the RMS ExB vorticity.

For this purpose we set up the grid and geometry, with a grid resolution of $64 \times 256 \times 32$ in $\{x, y, s\}$, in the field-aligned coordinate system with Dirichlet boundary conditions in $x$. The nominal case is the one listed above, with parameters $\hat{\beta} = 5$, $\hat{\mu} = 2.72$, and $C = 13.9$ controlling the adiabatic response. The other parameters, $\hat{\varepsilon} = 10\,000$, $S = 1$, and $\delta_0 = 0.015$, and $\omega_n = 1$ and $X = 1$ are kept fixed.

The main scalings are those against $\hat{\beta}$ and $C$. The variation against $\hat{\beta}$ between 1 and 50 is carried for three values of $C$ on either side of unity. These are shown in figure 12.21, in which the measured quantities are the turbulent ExB transport flux across the field lines, $F_e = nv_E^x$, the enstrophy, $A_w = \varpi^2/2$, the thermal free energy, $E_n = n^2/2$, and the Ohmic energy (magnetic and parallel electron flow), $E_b = J(A + \hat{\mu}J)/2$, all in volume average. The benchmark level for these is $\delta_0^2 = 2.25 \times 10^{-4}$ for comparison. The higher collisionality cases all show robust turbulence. Increasing inductivity is net stabilising in general, with complete stability setting for higher $\hat{\beta}$ at weaker collisionality. Cases with $F_e < 10^{-6}$ are declared stable, though at larger inductivity ($\hat{\beta} \gg 10$) with the nominal values of $\hat{\mu}$ and $\hat{\varepsilon}$ the linear decay is indeterminate, and values in the range $10^{-9}$ and $10^{-6}$ are found. But for those cases no nonlinearly relevant results emerge.

The next test takes the $\hat{\beta} = 5$ case and sweeps $C$ between the two extremes to find the threshold. These are shown in figure 12.22, where a threshold is found such that the cases are stable for about $C < 1$. The turbulent flux is roughly proportional to $C$ for the cases above the threshold. An additional test lowering viscosity to $X = 1$ for the four cases around the threshold showed less than 10% effects on the four quantities and no change to the threshold, so the stability properties are not affected by viscous dissipation.

The effect of magnetic flutter on the scaling with beta reflects the cancellation between induction ($\partial A/\partial t$) and magnetic flutter ($[-A,\ ]$ in $\nabla_\parallel$), which was noted for

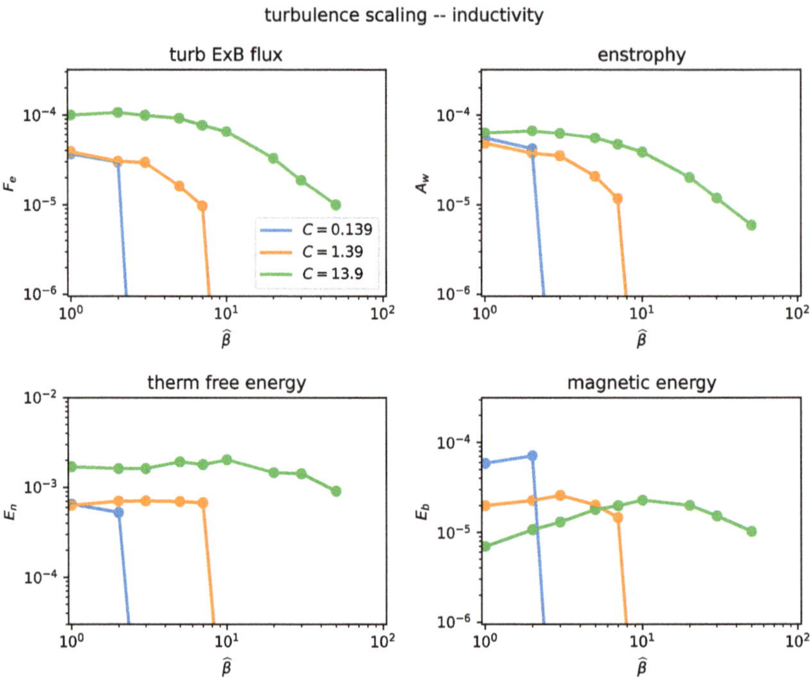


**Figure 12.21.** Scaling of the turbulent ExB transport flux ($F_e$), the enstrophy ($A_w$), the thermal free energy ($E_n$), and the Ohmic energy ($E_b$), all in volume average, against magnetic inductivity $\hat{\beta}$, for various $C = 0.139$ 1.39, and 13.9. If the finite $\hat{\beta}$ does not stabilise the turbulence the scaling is weakly falling with inductivity. Values below $10^{-6}$ are declared stable (see text).

the linear drift wave cases in Chapter 9. The scaling against $\hat{\beta}$ for the cases with $C = 13.9$ is compared with the same cases run without the gradient drive due to magnetic flutter. In these, the only difference is that the overall parallel gradient term $[-A, (n + n_0)]$ term is replaced by $[-A, n]$, which means that the nonlinear dynamics in all respects is left intact. The result is shown in figure 12.23. The turbulence is now found to rise with inductivity, becoming very strong for $\hat{\beta} \sim 10$. Hence, it is solely the terms proportional to $\omega_* b^x$ in the parallel dynamics that maintain the low amplitude and eventually cause stabilisation with $\hat{\beta}$.

Two control cases are now given. First is a test of the coordinate geometry and the differing numerical implementation of the parallel gradient. In the previous Chapter we had the linear part of $\nabla_\|$ involving derivatives in both $y$ and $z$, with $B^y$ proportional to $Sx$ times $B^z$, with purely Cartesian coordinate geometry. In the present case, the coordinates $y$ and $z$ are combined into $s$ in such a way that $B^y$ now vanishes and the effect of it is incorporated into the way in which the field lines map coordinate volumes from plane to plane in the $s$-direction. It is of interest to know what effect, if any, this has on the solution. So we re-run the small cases (with the $64 \times 256$ grid in $x$ and $y$) involved in the magnetic shear scaling of the previous Chapter in the field-aligned coordinate geometry. The parameters are

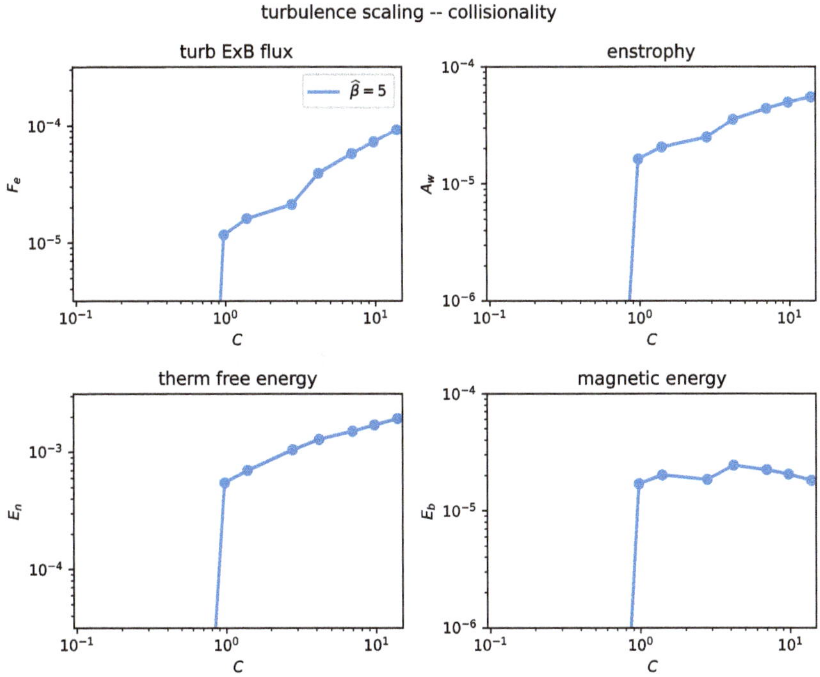


**Figure 12.22.** Scaling of the turbulent ExB transport flux ($F_e$), the enstrophy ($A_w$), the thermal free energy ($E_n$), and the Ohmic energy ($E_b$), all in volume average, against collisionality $C$. For this value $\hat{\beta} = 5$ the turbulence is stabilised for $C < 1$. The turbulent flux $F_e$ is roughly proportional to $C$ above the threshold. An additional test lowering viscosity to $X = 1$ for the four cases around the threshold showed less than 10% effects on the four quantities and no change to the threshold, so it is not due to viscous dissipation.

$$\hat{\beta} = 5 \quad \hat{\mu} = 2.7 \quad C = 13.9 \quad X = 10 \quad \hat{\epsilon} = 10^4. \tag{12.72}$$

Here, the shear is taken to larger values which is made possible by the field-aligned geometry. The result is shown in figure 12.24. We find the same dependence in both sets of cases. The previous, for $S$ up to 1, is compared within the wider range here, and we find fairly close convergence.

Finally, it is of interest to know how valid it is to coarsen the grid in the parallel coordinate direction to only $N_s = 32$ given the field-aligned geometry. The grid in this direction is refined by steps of a factor of two, four times, up to the highest value $N_s = 512$. Since the subgrid dissipation is tied to the resolution, the 512 case has a dissipation which is $16^4$ times smaller than the 32 case, a difference of slightly less than five orders of magnitude. The result, shown in figure 12.25, is indeed largely independent of $N_s$, although a small rise in especially vorticity with increasing resolution is found. But this test shows that we are allowed to use $N_s = 32$ to enable the large number of cases required to separate the physical effects, and the result is largely convergent at a level of $k_{\|}c_s$ rather smaller than the nominal vorticity level of $c_s/L_{\perp}$, since at $N_s = 32$ and $\hat{\epsilon} = 10^4$ the maximum value of $k_{\|}L_{\perp}$ is 0.16. Moreover, no special effect is seen in the last two cases, where the maximum value of $k_{\|}c_s$ on the

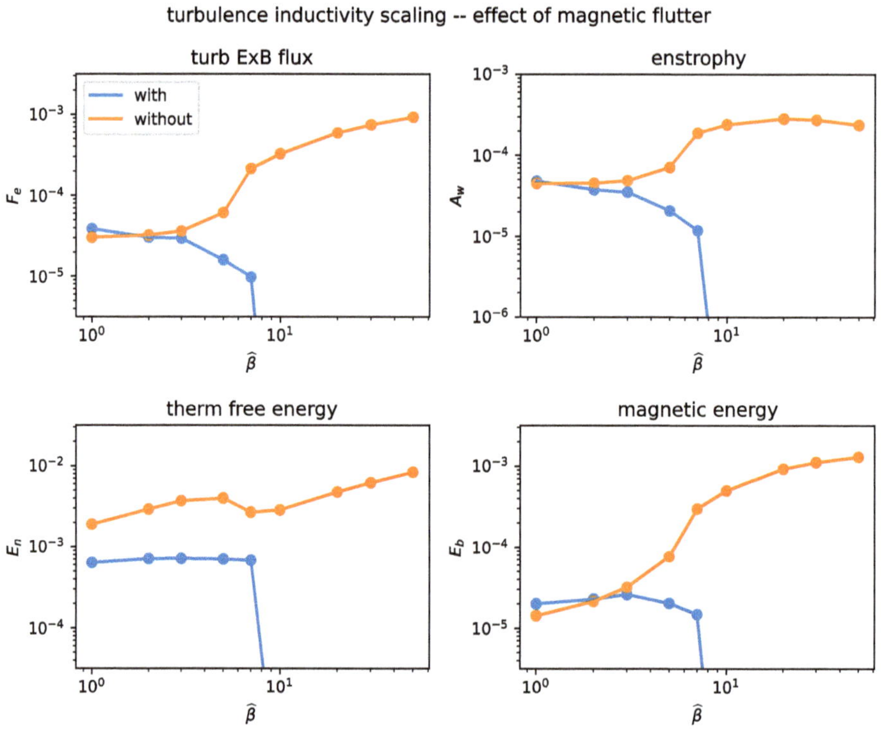


**Figure 12.23.** Scaling of the turbulent ExB transport flux ($F_e$), the enstrophy ($A_w$), the thermal free energy ($E_n$), and the Ohmic energy ($E_b$), all in volume average, against magnetic inductivity $\hat{\beta}$, for the $C = 1.39$ cases. The curves are labelled as to with or without the linear magnetic flutter term $\nabla_{\|} n_0$ in the Ohm's law. The magnetic flutter effect is responsible for the decrease of turbulence strength with inductivity, eventually causing stability for large values. On the other hand, inductivity by itself would lead to stronger turbulence due to more delay in the adiabatic response.

grid passes $c_s/L_{\perp}$. Hence we find no special effects of the nominal sound wave resonance in these three-dimensional cases, with a property which caused trouble for linear cases dropping away when considering turbulence.

## 12.9 Contingent role of linear stability

The first conclusion of this Chapter is that linear drift waves are stabilised by magnetic shear and the eigenmode structure it introduces in the direction across flux surfaces. The mechanism is field line connection, which depends on the combined effect of periodicity (toroidal topology) and magnetic shear. If either of these are absent then the linear eigenmodes are unstable again.

The second conclusion is that turbulence can exist in the absence of any linear instabilities. It sustains itself by changing the basic structure of the linear eigenmodes, so that much more robust nonadiabatic, dissipative electron activity and therefore background gradient drive are possible for a given set of parameters. The agent responsible for this change is the ExB vorticity nonlinearity, which represents

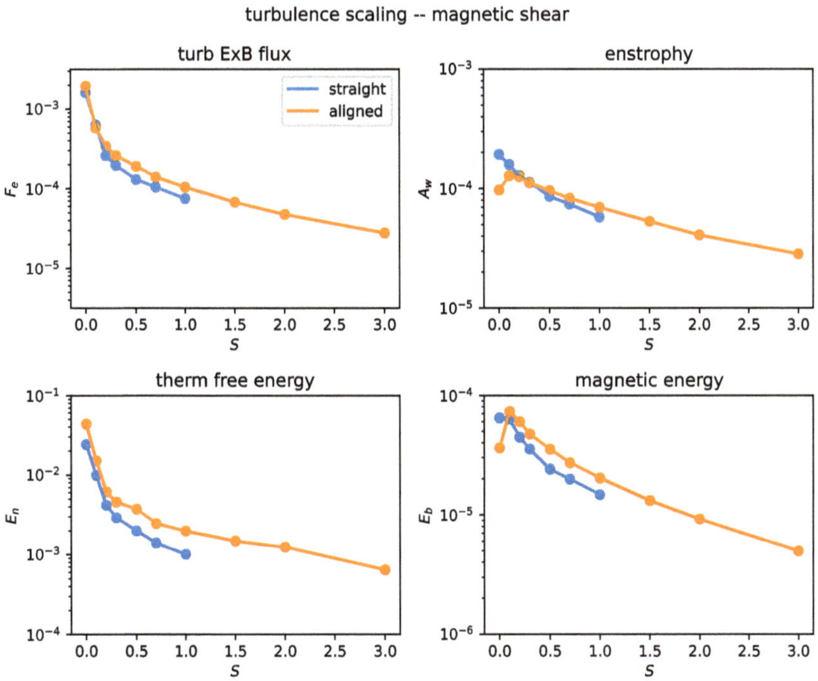


**Figure 12.24.** Scaling of the turbulent ExB transport flux ($F_e$), the enstrophy ($A_w$), the thermal free energy ($E_n$), and the Ohmic energy ($E_b$), all in volume average, against the magnetic shear ($S$). The basic parameters are the same. The red points are for the field-aligned coordinate system, and the blue points are for the $N_x = 64$ cases with $\delta_0 = 0.016$. The difference is that the field-aligned coordinate system has the same $k_\|$ spectrum everywhere in the $x$-domain, while otherwise it is dependent on $k_z$ and $k_y$ to cancel to form $k_\| = 0$, and $B^y$ is dependent on $x$. In the field-aligned coordinate system $B^y = 0$ and $k_\|$ follows only $B^s$.

self advection of ExB eddies or equivalently the nonlinear polarisation drift. This nonlinearity has the property of an incoherent scattering, which expands the space of available states without appreciable diffusivity. The other nonlinearities are mostly stabilising and act more conventionally as turbulent mixing diffusion. The fundamental cause of this difference between the nonlinearities is their differing cascade character; inverse for the ExB energy, direct for everything else.

The implication of these results is that linear stability analysis is not authoritative in determining whether turbulence will result, or if it does what its mode structure or transport scaling will be. For practical results one must turn to self-consistent computation. It may well be possible to toy-model this nonlinear instability given its properties as revealed by those computations, but the basic requirements of self consistency (at least a two-field model is needed) and a statistical analysis (with the differing character of the nonlinearities) make that a difficult prospect, one whose predictive value is questionable. The computations by themselves are sufficient to understand the physics.

Linear mode structure is too constrained; oscillations in the solution depend too much on the order of the system of equations and the factors of $x^2$ that appear.

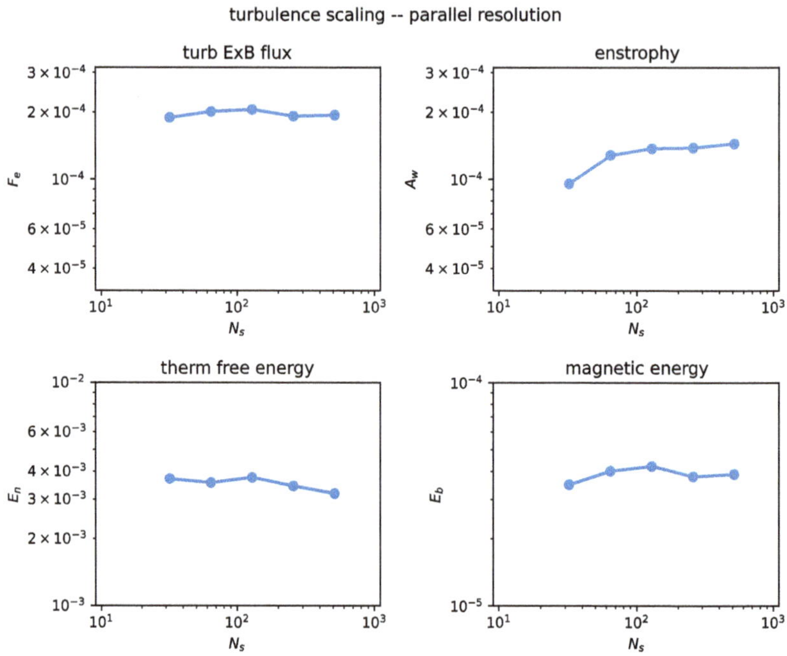


**Figure 12.25.** Scaling of the turbulent ExB transport flux ($F_e$), the enstrophy ($A_w$), the thermal free energy ($E_n$), and the Ohmic energy ($E_b$), all in volume average, against the grid resolution in the $s$-direction ($N_s$). Powers of two between 32 and 512 are used. The parallel subgrid dissipation scales as $k_\parallel^4$, so that a range of 65536 exists between lowest to highest. The basic parameters are the same. One does not have to extend $k_\parallel$ to $k_\parallel c_s > c_s/L_\perp$ to achieve convergence.

The presence of a randomising agent like turbulence breaks these constraints and allows a wider accessibility to the space of possible degrees of freedom.

## 12.10 Summary

The study of drift waves arose out of electrostatic linear analysis in the 1960s of what was then called the universal mode, as it seemed a low frequency flute mode type of perturbation was always unstable in the presence of a background gradient. The flute mode character coupled with a strong adiabatic response defined the mode structure, so that turbulence studied in simple laboratory devices was called drift wave turbulence. Equations combining the relevant features with the basic model of Eulerian 2D turbulence were developed, and these became mature at the end of the 1970s, with the Hasegawa–Mima model, which had the necessary invariants (conserved quantities whose densities are quadratic in the dependent variables). At the same time, consideration of magnetic shear led to the finding that the linear modes were stabilised. Also at the same time, investigation of the Alfvén drift instability, also called drift Alfvén modes, was underway. If all three of these are combined into one, drift Alfvén turbulence results. Since the mode structure is the defining property in turbulence, one can call this drift wave turbulence even for electromagnetic cases, which differ only in how the phase shifts are regulated.

The various nomenclature in early studies was due to what one thought should be emphasised: the Alfvén wave, the adiabatic response, the magnetic shear, or the turbulence. Today we realise that this is all a unified whole.

Drift wave turbulence results as such when the magnetic shear and other properties of the geometry prevent the existence of pure flute mode eddy disturbances ($k_{\parallel} = 0$ with $k_y$ finite). We saw the shear eliminate these in the previous Chapter. Herein, we saw the drift wave turbulence for larger values of magnetic shear and the involvement of the parallel wavenumber spectrum. Especially, we saw once and for all the result that the dissipation processes, namely, how the turbulence saturates not how it is driven, limit the $k_{\parallel}$ spectrum to a finite width such that $k_{\parallel} \ll k_{\perp}$ is satisfied with reasonable margin in general, even for the largest perpendicular scales of the turbulence. This justifies the main assumptions behind the setup: the three-dimensional turbulence is really a two-plus-one dynamical system, with turbulence in the plane perpendicular to the background guide field and wave-like but dissipative responses along it. The resulting dynamics in a magnetised plasma is distinct from the usual distinction between two- and three-dimensional turbulence one studies in other situations, in which there is no dominant guide field, or if there is one it is considered to be straight. Hence, also the restriction to the low frequency ($\omega \ll k_{\perp} v_A, \Omega_i$) and low dynamical beta ($p_e \ll B^2/\mu_0$) regime in the parameters. Despite robust turbulence, the dynamics is still too weak to compete against its constraint by the background magnetic field.

The issue of flows has arisen and their consideration in detail is necessary. Two geometric situations remain to be worked out first, though. The next two Chapters will treat the issues of curved magnetic field lines, which introduce the quasistatic compressional effects and with them some additional energy transfer channels, even at low frequency, and of open magnetic field lines, which relax the field line connection situation but replaced it with effective coupling among the dependent variables due to the outgoing wave (or 'exit flux') boundary conditions. After that will come a Chapter specifically devoted to the energetics of the interaction with flows, both those set by the background equilibrium and those self consistently introduced by the dynamics.

## Further Reading

The topology theorem referred to in the Introduction is called the hairy-ball or hedgehog theorem. A vector field cannot lie in a sphere without vanishing somewhere if the dimension is even. The simple version is due to Poincaré 'Sur les courbes définies par les équations différentielle' *J. Math. Pures Appl.* **4** (1885) 167. It is covered by Paul Renteln 2013 *Manifolds, Tensors, and Forms: An Introduction for Mathematicians and Physicists* (Cambridge: Cambridge Univ. Press).

'Shear damping' of linear drift waves became known in the 1960s and quickly became a central theme in magnetic fusion research. The earlier work is reviewed by W M Tang *Nucl. Fusion* **18** (1978) 1089. Linear drift waves in sheared slab geometry were definitively found stable by D W Ross and S M Mahajan *Phys. Rev. Lett.* **40** (1978) 324; and by K T Tsang, P J Catto, J C Whitson and J Smith *Phys. Rev. Lett.*

**40** (1978) 327. The Alfvén drift instability was studied directly as a manifestation of two-fluid MHD, as covered in the monograph by A B Mikhailovskii 1974 *Theory of Plasma Instabilities* (New York: Consultants Bureau). An early version of it was given by A B Mikhailovskii and L I Rudakov 'The stability of a spatially inhomogeneous plasma in a magnetic field' *Sov. Phys. JETP* **17** (1963) 621.

Immediately afterward, the delicate nature of the linear stability problem, and its sensitivity to very small-scale structure across the rational surface, was noted by S P Hirshman and K Molvig, who proposed the nonlinear instability in *Phys. Rev. Lett.* **42** (1979) 648. In the adiabatic sheared slab model with drift and sound waves, with an external drive parameter, D Biskamp and M Walter found the mode structure changes which eventually led to the nonlinear instability, in 'Suppression of shear damping in drift wave turbulence' Phys. Lett. A **109** (1985) 34. The analytical model in that work merits further study.

Self-sustained turbulence in the self-consistent dissipative drift wave system was demonstrated by B Scott 'Self-sustained collisional drift-wave turbulence in a sheared magnetic field' *Phys. Rev. Lett.* **65** (1990) 3289; and the mode structure changes and the importance of the simultaneous action of the two nonlinearities were detailed by B Scott, H Biglari, P W Terry and P H Diamond 'Self-organization in sheared drift-wave turbulence' *Phys. Fluids B* **3** (1991) 51; and by B Scott 'The mechanism of self-sustainment in collisional drift wave turbulence' *Phys. Fluids B* **4** (1992) 2468. Three-dimensional mode structure was presented by B Scott 'Three-dimensional computation of drift Alfvén turbulence' *Plasma Phys. Control. Fusion* **39** (1997) 1635; and parts of the results of this Chapter demonstrating the three-dimensional nonlinear instability in ExB shear flows and electromagnetic gyrofluid turbulence *Phys. Plasmas* **7** (2000) 1845; and in The nonlinear drift wave instability and its role in tokamak edge turbulence *New J. Phys.* **4** (2002) 52.

Drift wave computation in general is reviewed by W Horton *Rev. Mod. Phys.* **7** (1999), without having much to say about the nonlinear instability.

Field-aligned coordinates were proposed by K V Roberts and J B Taylor *Phys. Fluids* **8** (1965) 315; and expanded for toroidal flux tube models by S C Cowley, R M Kulsrud and R N Sudan 'Considerations of ion-temperature-gradient-driven turbulence' *Phys. Fluids B* **3** (1991) 2767; and by M A Beer, S C Cowley and G W Hammett 'Field-aligned coordinates for nonlinear simulations of tokamak turbulence' *Phys. Plasmas* **2** (1995) 2687. The global consistency requirement was noted by shear, periodicity, and plasma ballooning modes J W Connor, R J Hastie and J B Taylor *Phys. Rev. Lett.* **40** (1978) 396; and was first properly worked out for flux tube models by B Scott 'Global consistency for thin flux tube treatments of toroidal geometry' *Phys. Plasmas* **5** (1998) 2334. For the global version, see R L Dewar and A H Glasser 'Ballooning mode spectrum in general toroidal systems' *Phys. Fluids* **26** (1983) 3038; as well as B Scott 'Shifted metric procedure for flux tube treatments of toroidal geometry: avoiding grid deformation' *Phys. Plasmas* **8** (2001) 334, in which the shift operation between successive perpendicular planes is systematically worked out.

**IOP** Publishing

Turbulence and Instabilities in Magnetised Plasmas, Volume 1
Fluid drift turbulence
**Bruce Scott**

# Chapter 13

## MHD interchange turbulence

### 13.1 Introduction

In some situations the drift wave paradigm we have been developing does not apply. We have seen the important dependence of the dynamics upon the adiabatic response of the electrons, and of this in turn upon finite parallel wavelengths. It then follows that if in our physical situation parallel dynamics has no role, then something else must be invoked to explain the dynamics we see. This can happen if the parallel length scale is arbitrarily long, so that not only sound waves but also kinetic shear Alfvén waves are too slow to affect the perpendicular dynamics. It can also happen if the parallel dynamics is arbitrarily fast but the magnetic geometry allows a mode of oscillation with $k_{\|} = 0$ for finite drift wavelength. The effects of the flute mode were seen in Chapter 11. If allowed by the magnetic geometry this mode dominates the dynamics even if three-dimensional energy transfer processes are active. The flute mode is dynamically similar to the case of 2D Eulerian hydrodynamics with a passively advected scalar. The dynamics is inherently at large scale, and a background gradient in the latter quantity then leads to effective large-scale mixing.

If the magnetic field is inhomogeneous, the addition of the magnetic divergences gives rise to what is known as interchange dynamics. In this case the ExB flow has a finite divergence, so that motion away from the centre of curvature of the magnetic field lines leads to expansion of the fluid elements. The interchange of these elements leads to free energy release if that direction is also down a pressure gradient. The situation is similar to convection in a fluid, where an entropy gradient plays the same role. In the plasma case there is next to no threshold, but otherwise the dynamics is similar and in the 2D case is described by almost identical equations. The instability is called an interchange mode and it leads to interchange turbulence. If the magnetic curvature and the pressure gradient line up, one refers to unfavourable curvature and to the coupling through the magnetic divergence terms as interchange forcing. It is a conservative coupling mechanism, which creates an alternative pathway for

doi:10.1088/978-0-7503-2504-2ch13

energetic exchange between pressure disturbances and ExB flows, which is how it gives rise to these modes.

The relationship and competition between the mechanisms behind these interchange modes and drift waves is sufficiently subtle so that we will leave it to Chapter 14. For now, the interchange dynamics takes centre stage. We first study it in the simpler two-dimensional model, and compare it with the Rayleigh–Taylor forcing for neutral fluids. By contrast to that latter case, the interchange forcing conserves energy exactly and so forms a coupling mechanism, rather than serving as a potential energy source as gravity would do, even if the mechanism for destabilisation is very similar. We next study the three-dimensional version under the reduced MHD (RMHD) model, which we get by dropping the adiabatic coupling while keeping the magnetic divergences and hence the interchange forcing.

If the interchange forcing varies more or less sinusoidally in the direction along the magnetic field, then the basic 2D interchange mode is only excited where the gradients of the pressure and the magnetic field strength line up. In a toroidal magnetic field, the gradient of the magnetic field strength and the radius of curvature are closely equivalent, so these are commonly called curvature effects. The modes they give rise to are called ballooning modes, because the region on which they are excited is localised to the outboard midplane of the torus, or where the alignment just described takes place. We need to understand the physics of these modes and their associated turbulence before moving on to comparisons or considering their physics in competition with other free energy mechanisms.

The model magnetic geometries discussed in this and Chapter 14 are very simple. We require consideration only of magnetic shear and curvature. More realistic geometry and device modelling falls outside of our scope. The basic methods are the same however, and this should serve as a useful introduction.

## 13.2 Magnetic divergences and the interchange model

We review the origin of the diamagnetic flux and ExB divergences and the new terms they introduce to the basic conservation laws, this time with specific attention to how they build the physical model for interchange dynamics. When the magnetic field is inhomogeneous, the drift operator is no longer divergence free. The drift operator is

$$\frac{1}{B^2}\mathbf{B}\times\nabla, \tag{13.1}$$

which acts on scalar quantities. Acting on the electrostatic potential, it gives the ExB velocity, and acting on the total pressure it gives the diamagnetic current,

$$\mathbf{v}_E = \frac{1}{B^2}\mathbf{B}\times\nabla\phi \qquad \mathbf{J}_* = \frac{1}{B^2}\mathbf{B}\times\nabla p. \tag{13.2}$$

The divergence of the drift operator, which we will call the curvature operator, is defined as

$$\mathcal{K} \equiv -\nabla \cdot \frac{1}{B^2}\mathbf{B} \times \nabla = -\left(\nabla \times \frac{\mathbf{B}}{B^2}\right) \cdot \nabla. \tag{13.3}$$

As the second form shows, $\mathcal{K}$ is only nonzero if the magnetic field is inhomogeneous, and then these drifts have finite divergences.

### 13.2.1 Considering the MHD equilibrium

The simplest model magnetic field which gives rise to these magnetic divergence terms is a straight field with a perpendicular gradient in the field strength,

$$\mathbf{B} = B_0\left(1 - \frac{x}{L_B}\right)\nabla s, \tag{13.4}$$

in the same $\{x, y, s\}$ coordinate system we have been using. If this is the only variation (i.e. $\mathbf{B} \cdot \nabla\mathbf{B}$ vanishes), we require for MHD equilibrium that

$$\nabla\left(p + \frac{B^2}{2\mu_0}\right) = 0 \qquad \text{hence} \qquad \frac{\partial}{\partial x}\log B = -\frac{\mu_0 p}{B^2}\frac{\partial}{\partial x}\log p, \tag{13.5}$$

noting that $\beta = 2\mu_0 p/B^2$ is the plasma beta given in the first Chapter. In other words, the gradients of the pressure and the magnetic field cannot be in the same direction unless the equilibrium is determined by other means, namely, curvature. The opposite limit is no magnetic gradient but only curvature. Suppose

$$\nabla p = \frac{1}{\mu_0}\mathbf{B} \cdot \nabla\mathbf{B} \approx \frac{B^2}{\mu_0}\mathbf{b} \cdot \nabla\mathbf{b}. \tag{13.6}$$

Now we have the curvature and the pressure gradient in the same direction, and the model for $\mathcal{K}$ becomes

$$\mathcal{K} = -\left(\nabla \times \frac{\mathbf{B}}{B^2}\right) \cdot \nabla = -\frac{1}{B}(\nabla \times \mathbf{b}) \cdot \nabla \approx \frac{1}{B}\mathbf{b} \times (-\mathbf{b} \cdot \nabla\mathbf{b}) \cdot \nabla, \tag{13.7}$$

using the vector identity $\nabla(b^2/2) = \mathbf{b} \cdot \nabla\mathbf{b} + \mathbf{b} \times \nabla \times \mathbf{b}$ with the left side zero in this case and neglecting the parallel component. The quantity $\mathbf{b} \cdot \nabla\mathbf{b}$ is the magnetic curvature vector, whose magnitude is the inverse radius of curvature and whose direction is toward the centre of curvature. Hence, these terms are called 'curvature terms' and appear in the equations defined in terms of a curvature radius $R_c$. In a low beta tokamak, the curvature and magnetic field gradient are the same, and the curvature radius is the toroidal major radius $R$, so the curvature operator can be approximated by

$$\begin{aligned} \mathcal{K} &= -\frac{1}{B}\mathbf{b} \times (\nabla \log B + \mathbf{b} \cdot \nabla\mathbf{b}) \cdot \nabla \\ &\approx \frac{1}{B}\mathbf{b} \times (2\nabla \log R) \cdot \nabla \rightarrow \frac{R_0}{B_0}\nabla\varphi \cdot (2\nabla \log R) \times \nabla, \end{aligned} \tag{13.8}$$

where we take $\mathbf{b} = R_0\nabla\varphi$ with $\varphi$ the geometric toroidal angle, while always remembering that $\mathcal{K}$ must be a pure divergence. It is only if $\mathbf{B}\cdot\nabla\mathbf{B}$ nearly balances $\nabla B^2/2$—the magnetic field is nearly force free—or $\mathbf{B}\cdot\nabla\mathbf{B}$ in fact balances $\nabla p$, that $\nabla p$ can be free of the constraint in Eq. (13.5). In other words, the equilibrium is confined by currents, either external (entering via boundary conditions on a force-free case) or internal. In a low beta case the magnetic field is nearly force free with the balance between the magnetic gradient and the curvature. As we will see, the interchange forcing is stabilising unless $\nabla p$ and $\nabla B$ or the curvature vector $\mathbf{b}\cdot\nabla\mathbf{b}$ are in the same direction. Interchange turbulence only enters in low beta equilibria under such cases.

We therefore don't want to use the simple slab-gradient model in Eq. (13.4), but the simple curvature model in Eq. (13.7), using $\mathbf{b}\cdot\nabla\mathbf{b}$ instead of $\nabla\log B$ as its basis. One can think of a magnetic scale length which is given by $L_B = R_0/2$. Again, the parameter has to be a constant so that the operator is a divergence.

### 13.2.2 The curvature terms

In the same spirit as for the background pressure, we may neglect the variation of $B$ under drift ordering everywhere except where its gradient makes a term nonzero; in this case, the curvature terms. The two vector fields $\mathbf{v}_E$ and $\mathbf{J}_*$ written above acquire finite divergences through this variation of $\mathbf{B}$,

$$\nabla\cdot\mathbf{v}_E = -\mathcal{K}(\phi) \qquad \nabla\cdot\mathbf{J}_* = -\mathcal{K}(p). \tag{13.9}$$

If we scale the units of the fluctuations $\tilde{\phi}$ and $\tilde{p}$ as disturbances, we find for the ExB velocity divergence, the scale of the curvature forcing is

$$\nabla\cdot\mathbf{v}_E = -\mathcal{K}(\tilde{\phi}) \to \frac{2T_e}{eB_0R_0}(\nabla)\left(\frac{e\tilde{\phi}}{T_e}\right), \tag{13.10}$$

where $(\nabla)$ denotes derivatives and carries the scale of the wavenumber. This is compared with $T_e/eB_0L_\perp$ in the gradient forcing, so it is usually ordered small in an incompressible MHD model. Nevertheless, it is required for energy conservation (see below). Similarly, the divergence of the diamagnetic current is

$$\frac{1}{n_e e}\nabla\cdot\mathbf{J}_* = -\frac{1}{n_e e}\mathcal{K}(\tilde{p}) \to \frac{2T_e}{eB_0R_0}(\nabla)\left(\frac{\tilde{p}}{p_e}\right), \tag{13.11}$$

which enters the vorticity equation as a significant forcing effect since even if $L_\perp/R_0$ is small, it carries one fewer factor of $k_\perp\rho_s$ than the polarisation, in normalised units. In a model with cold ions, the total pressure becomes $p \to p_e$, and its fluctuation becomes $\tilde{p} \to \tilde{p}_e$. In an isothermal model, we have $\tilde{p}_e = T_e\tilde{n}_e$ and $\Gamma = 1$ for the ratio of specific heats.

In a four-field model, these curvature terms go into the vorticity and pressure equations, where the latter is the same as the electron density equation. We have

$$\frac{\partial \tilde{\varpi}}{\partial t} = \cdots - \mathscr{K}(\tilde{p}), \tag{13.12}$$

$$\frac{\partial \tilde{p}_e}{\partial t} = \cdots + p_e \mathscr{K}(\tilde{\phi}) - \frac{T_e}{e} \mathscr{K}(\tilde{p}_e), \tag{13.13}$$

as additions, assuming $p = nT_e$ and $\tilde{p} = \tilde{p}_e = T_e \tilde{n}_e$, with the terms in the pressure equation representing ExB and diamagnetic compression, respectively, recalling that the divergence of the diamagnetic flow is part of the continuity dynamics in a two-fluid model, but not in an MHD one. These are written with $\tilde{p}$ and $\tilde{p}_e$ separately, since if $T_i \neq 0$ then $\tilde{p} = \tilde{p}_e + \tilde{p}_i$, which for energetic consistency reasons has also to bring in the warm-ion polarisation at the same time. We will address this in Chapter 14. Hereafter, $\tilde{p}_i \to 0$ and $\tilde{p} \to \tilde{p}_e$.

## 13.3 Interchange energetics

These new terms give rise to an energetic coupling between $\tilde{p}_e$ and $\tilde{\phi}$ which the purely hydrodynamic drift wave model lacks. The interesting property is that the coupling is reactive; that is, it is dissipation free. Interchange dynamics follows from the quasistatic compression of perpendicular flows, which although dynamically incompressible (since we still have $\omega \ll k_\perp v_A$), have a finite divergence through the drift motion in the inhomogeneous magnetic field. This compression mechanism must conserve energy during their transfer of energy between $\tilde{\phi}$ (kinetic energy of ExB motion) and $\tilde{p}_e$ (thermal free energy), in the same way that parallel sound waves do. The methods are the same as in previous Chapters: integral quantities are cast as volume averages, with the volume integration represented by $d\mathscr{V}$ which is $dx\, dy$ in a Cartesian 2D model or $\sqrt{g}\, dx\, dy\, ds$ in a 3D model with coordinate system $\{x, y, s\}$ and volume element $\sqrt{g}$ given by the inverse of the Jacobian $\nabla x \times \nabla y \cdot \nabla s$ which is positive.

Multiplying Eq. (13.12) with $\tilde{p} \to \tilde{p}_e$ by $-\tilde{\phi}$ and Eq. (13.13) by $\tilde{p}_e / p_e$, and averaging over the volume, we find

$$\begin{aligned} \frac{\partial \mathscr{V}^{-1}}{\partial t} \int d\mathscr{V}\, \frac{1}{2} \frac{\rho}{B^2} |\nabla_\perp \tilde{\phi}|^2 &= \cdots + \mathscr{V}^{-1} \int d\mathscr{V}\, \tilde{\phi} \mathscr{K}(\tilde{p}_e) \\ &= \cdots + \mathscr{V}^{-1} \int d\mathscr{V}\, \mathscr{K}(\tilde{p}_e \tilde{\phi}) \\ &\quad - \mathscr{V}^{-1} \int d\mathscr{V}\, \tilde{p}_e \mathscr{K}(\tilde{\phi}), \end{aligned} \tag{13.14}$$

$$\frac{\partial}{\partial t} \mathscr{V}^{-1} \int d\mathscr{V}\, \frac{1}{2} \frac{1}{p_e} |\tilde{p}_e|^2 = \cdots + \mathscr{V}^{-1} \int d\mathscr{V} \left( \tilde{p}_e \mathscr{K}(\tilde{\phi}) - \frac{T_e}{n_e e} \tilde{p}_e \mathscr{K}(\tilde{p}_e) \right), \tag{13.15}$$

in addition to all the other effects we have covered. The last term is a pure divergence and transfers within the thermal free energy between degrees of freedom. The equal and opposite terms represent the interchange coupling. The ExB compression in the thermal free energy balances the diamagnetic curvature forcing in the ExB vorticity, so that it is a conservative coupling. The addition to the total energy theorem is given by

$$\begin{aligned}&\frac{\partial}{\partial t}\mathcal{V}^{-1}\int d\mathcal{V}\,\frac{1}{2}\left(\frac{\rho}{B^2}|\nabla_\perp\tilde{\phi}|^2+\frac{1}{p_e}|\tilde{p}_e|^2\right)\\&\quad=\cdots+\mathcal{V}^{-1}\int d\mathcal{V}\left(\mathcal{K}(\tilde{p}_e\tilde{\phi})-\frac{1}{2}\frac{T_e}{n_e e}\mathcal{K}(\tilde{p}_e^2)\right)=\cdots+0,\end{aligned}\tag{13.16}$$

namely, it is unchanged except for subdominant free energy transport terms in the local form. So we have all the sources and sinks as before, but the interchange effects integrate to zero.

Note the importance of casting $\mathcal{K}$ as a pure divergence in any nonlinear model. Only then will the energetic consistency be maintained.

The free energy conservation law is the same in 3D as in any drift Alfvén model or in 2D as in any drift wave or hydrodynamic model, except for the addition of extra coupling processes in the energetics. The curvature coupling, representing either ExB or diamagnetic compression, always conserves free energy. In a pure 2D interchange model, the energy theorem is the same as that for the hydrodynamic limit of two-dimensional drift waves, including the gradient drive, except we have the interchange coupling mechanism instead of the adiabatic coupling in the drift wave case. In the absence of parallel dynamics, there is no dissipation in the model, so instabilities tend to grow until the free energy source is depleted nonlinearly. Nevertheless, we will find situations in which the mode structure and energetics can be measured.

## 13.4 The 2D interchange model

Using the basic model given by Eq. (13.8), in a 2D setting, the curvature operator becomes

$$\mathcal{K}=\frac{1}{B}\nabla s\cdot(2\nabla\log R)\times\nabla\rightarrow\frac{2}{R_0}\frac{1}{B_0}\nabla s\cdot\nabla x\times\nabla\rightarrow\frac{2}{B_0R_0}\frac{\partial}{\partial y},\tag{13.17}$$

where $B_0$ and $R_0$ are constants.

The simplest model for interchange dynamics consists of the four-field model we have been studying, with the curvature terms but without the parallel dynamics. The parallel flows, $\tilde{u}_\parallel$ and $\tilde{J}_\parallel$ are therefore set to zero, leaving only $\tilde{p}_e$ and $\tilde{\phi}$ as dependent variables. With $\tilde{u}_\parallel=\tilde{J}_\parallel=0$ we have the hydrodynamic limit of the perpendicular dynamics, with addition of the curvature terms. The variables $\tilde{\phi}$ and $\tilde{p}_e$ are normalised to $\phi$ and $n$ with vorticity $\varpi$ as in Eqs (4.15) and (4.16) from Chapter 4. There, we have dissipative coupling terms; here, the curvature terms. The normalised equations appear as

$$\frac{\partial\varpi}{\partial t}+[\phi,\varpi]=-\omega_B\frac{\partial n}{\partial y},\tag{13.18}$$

$$\frac{\partial n}{\partial t} + [\phi, n + n_0] = \omega_B \frac{\partial \phi}{\partial y}, \tag{13.19}$$

$$\nabla_\perp^2 \phi = \varpi, \tag{13.20}$$

in the same scheme as in Chapter 4 with the simplest 2D drift wave turbulence model. In the 2D case, the local geometry is Cartesian and the Laplacian and spatial brackets keep their simple structure

$$\nabla_\perp^2 = \frac{\partial^2}{\partial x^2} + \frac{\partial^2}{\partial y^2} \qquad [a, b] = \frac{\partial a}{\partial x}\frac{\partial b}{\partial y} - \frac{\partial b}{\partial x}\frac{\partial a}{\partial y}, \tag{13.21}$$

for arbitrary scalars $a$ and $b$, as previously. For the 3D toroidal model, below, we will have to modify this.

In a reduced MHD model we do not keep $\mathcal{K}(n)$ in the equation for $n$. The term $\mathcal{K}(\phi)$ is ExB compression, while the diamagnetic compression denoted by $\mathcal{K}(n)$ is absent, a consequence of the MHD ordering which assumes $n \ll \phi$ and $k_\perp \rho_s \ll 1$ in these units. In the vorticity equation, $\mathcal{K}(n)$ is the main forcing term which exchanges free energy conservatively with ExB compression in the pressure equation. The 2D interchange model is the same as a Eulerian fluid with a passive scalar, with the addition of the curvature coupling terms. The profile scale given by $L_\perp$ and the pressure gradient are given by

$$n_0 = -\omega_n x \qquad \omega_n = \left|L_\perp \nabla \log p_e\right| = \frac{L_\perp}{L_p}, \tag{13.22}$$

where $L_p$ is the pressure scale length. The new parameter is the interchange forcing, entering through $\mathcal{K}$, with controlling parameter

$$\omega_B = 2L_\perp \left|\nabla \log R\right|, \tag{13.23}$$

where $R = L_B/2$ is the radius of curvature. The curvature operator for this model is then

$$\mathcal{K} = \omega_B \frac{\partial}{\partial y}. \tag{13.24}$$

In a computation of two-dimensional interchange dynamics, the only parameters are $\omega_n$, usually set to unity, and $\omega_B$, which for a low beta plasma is smaller. We additionally have the system size, since the disturbance perpendicular length scale is normalised to $\rho_s$.

## 13.5 The ideal interchange mode

We here introduce the ideal interchange mode as a linear instability. The linearised equations are those of Eqs (13.18) and (13.19) without the nonlinear terms,

$$\frac{\partial}{\partial t}\nabla_\perp^2\phi = -\omega_B\frac{\partial n}{\partial y}, \tag{13.25}$$

$$\frac{\partial n}{\partial t} = -\omega_n\frac{\partial \phi}{\partial y} + \omega_B\frac{\partial \phi}{\partial y}. \tag{13.26}$$

Taking a single Fourier component,

$$n(x, y, t) = n\exp[i(k_x x + k_y y - \omega t)], \tag{13.27}$$

where $n$ is now the complex amplitude, we solve Eq. (13.26) for $n$ in terms of $\phi$,

$$n = \frac{\omega_* - \omega_c}{\omega}\phi, \tag{13.28}$$

where $\omega_* = \omega_n k_y$ and $\omega_c = \omega_B k_y$ are the diamagnetic and interchange drift frequencies. Using this we can eliminate $n$ in Eq. (13.25) and then solve it for $\omega$,

$$\omega k_\perp^2\omega = -\omega_c(\omega_* - \omega_c), \tag{13.29}$$

Solving for $\omega$, we find

$$\gamma^2 = -\omega^2 = \frac{k_y^2}{k_\perp^2}\omega_B(\omega_n - \omega_B) \rightarrow \omega_B\omega_n, \tag{13.30}$$

with the latter limit originally taken in MHD models which assume incompressible flow in continuity equations. This is a purely growing mode if $\omega_n > \omega_B > 0$. That is, the gradient of $B$ must be aligned with that of $p$, but it must not be too large (we note, however, that if $\omega_B > \omega_n$ then the low beta assumption of the equilibrium would break down). The maximum value of this growth rate for finite $k_y$ is found for $k_x = 0$, namely, a structure known as a 'streamer,' extended in the direction of forcing. The value is known as the ideal interchange growth rate,

$$\gamma_I = \frac{c_s}{(L_p L_B)^{1/2}}, \tag{13.31}$$

in physical units. This is the classic interchange instability of MHD, and we see that an instability results for arbitrarily large scale.

The incompressible MHD limit is particularly similar to the classic Rayleigh–Taylor instability of a stratified fluid under gravity, since the main forcing term for the flows themselves has the same form. For the governing equations we take the two-dimensional Boussinesq ideal fluid model, which starts with the ideal fluid equations but then neglects compressional effects everywhere except where the density disturbance multiplies gravity,

$$\rho\left(\frac{\partial \mathbf{v}}{\partial t} + \mathbf{v}\cdot\nabla\mathbf{v}\right) = -\nabla p + \tilde{\rho}g\nabla x, \tag{13.32}$$

$$\frac{\partial p}{\partial t} + \mathbf{v} \cdot \nabla p = 0, \tag{13.33}$$

with $\rho$ the mass density, $\tilde{\rho}$ its disturbance, and $g\nabla x$ the gravity vector. The gravity vector is down the background gradient, in the $x$-direction, and motion is in the $xy$-plane. Taking the stream function and vorticity defined as

$$\mathbf{v} = \hat{\mathbf{z}} \times \nabla\phi \qquad \Omega = \hat{\mathbf{z}} \cdot \nabla \times \mathbf{v} = \nabla_\perp^2 \phi, \tag{13.34}$$

with $\hat{\mathbf{z}} = \nabla x \times \nabla y$ the normal to the plane, we find

$$\left(\frac{\partial}{\partial t} + \mathbf{v} \cdot \nabla\right)\Omega = \hat{\mathbf{z}} \cdot \nabla \times (\tilde{\rho} g \nabla x) = -\hat{\mathbf{z}} \times g\nabla x \cdot \nabla\tilde{\rho}. \tag{13.35}$$

The linearised form of this replaces $\tilde{\rho}$ by $(\rho/\Gamma p)\tilde{p}$, where $\Gamma$ is the ratio of specific heats, finding

$$\frac{\partial}{\partial t}\nabla_\perp^2\phi = -\frac{g}{\Gamma p}\frac{\partial \tilde{p}}{\partial y} \qquad \frac{\partial \tilde{p}}{\partial t} = -\frac{p}{L_p}\frac{\partial \phi}{\partial y}. \tag{13.36}$$

This analogy has led to the moniker 'g-mode' turbulence for the 2D interchange model. Even the three-dimensional case is sometimes called toroidal g-mode or resistive g-mode turbulence.

In the vorticity equation, the forcing by gravity in the Boussinesq fluid model is the same as the forcing by curvature in the plasma interchange model, and in the absence of compression in the latter model the two are basically the same. However, gravity functions as a potential energy source in the Boussinesq fluid model, while in the plasma interchange model the interchange forcing is a conservative energy transfer mechanism. Hence, the analogy, while useful for initial understanding, should not be overdone. What actually takes place in the plasma interchange model is that the pressure disturbance in a magnetised plasma gives rise to a diamagnetic current, and then in the case with magnetic curvature this current has a nonzero divergence. Charges would build up but in this case of low frequency dynamics what does build up is ExB vorticity, as the polarisation current compensates the divergence of the diamagnetic current. This ExB vorticity represents flow both down and up the gradient, but if $\nabla p \cdot (\mathbf{b} \cdot \nabla\mathbf{b} + \nabla B) > 0$ the down-gradient flow is through a pressure enhancement, so there is net transport. In other words, the phase shift of the pressure ahead of the electrostatic potential in the electron drift direction is $\pi/2$, or the ideal value for an instability. The reason for the energetic difference is that gravity forms a true potential energy, while compression is a conservative transfer process.

## 13.6 2D interchange turbulence

We now investigate interchange turbulence in a two-dimensional setting. A simple case will suffice to show the issues with saturation. We start with a naïve approach and take $\omega_n = 1$ and $\omega_B = 0.1$, and an initial state given by a finite train of localised pressure disturbances with

$$n = \phi = A_0 \exp\left(-\frac{x^2 + 0.1y^2}{\Delta_\perp^2}\right)\cos(\pi y/\Delta_\perp), \tag{13.37}$$

with initial amplitude and scale set to $A_0 = 10^{-8}$ and $\Delta_\perp = 6$, and $\varpi$ given by Eq. (13.20). This is an adiabatic start as in Chapter 4, except for the spatial localisation. We let the domain size be given by a minimum wavenumber $K = 0.1$, with $L_x = L_y = 2\pi/K$ on a 64 × 64 grid. The time step is set to $\tau = 0.001$ due to the large amplitudes to be faced. The subgrid dissipation is given by

$$\frac{\partial \varpi}{\partial t} = \cdots - D(\varpi), \tag{13.38}$$

$$\frac{\partial n}{\partial t} = \cdots = D(n), \tag{13.39}$$

as in Chapter 3, with $D = \nu_0(h^4/16)(\nabla_\perp^2)^2$, setting $\nu_0 = 1$ in anticipation of the same strong nonlinear transfer of $n$ to small scale experienced in the driven hydrodynamic cases. The energetic quantities and amplitudes, with the gradient drive rate given by

$$\Gamma_+ = \omega_n \frac{1}{2E_T} \mathcal{V}^{-1} \int d\mathcal{V}\ n v_E^x \qquad v_E^x = -\frac{\partial \phi}{\partial y}, \tag{13.40}$$

are as in the drift wave cases of Chapter 4. The subgrid dissipation rate is

$$\Gamma_- = \frac{1}{2E_T} \mathcal{V}^{-1} \int d\mathcal{V}\ \nu_0 \frac{h^4}{16}\left(|\nabla_\perp \varpi|^2 + |n|^2\right), \tag{13.41}$$

with the error given by the difference between $\Gamma_T = (1/2E_T)\partial E_T/\partial t$ and $\Gamma_+ - \Gamma_-$ as before (Eq. (3.47) from Chapter 3). The derivatives in the curvature terms are done with standard fourth-order centred differences. The numerical methods (time step, spatial discretisation, subgrid dissipation) are otherwise the same as in Chapter 3.

The first case illustrates the basic problem with pure interchange turbulence in a periodic domain with a constant external gradient drive. After forming up as the curvature forcing gives $n$ and $\phi$ the same spatial form up to their phase shift, the disturbance grows until nonlinear amplitude is reached. The time traces for this case are shown in figure 13.1. On this scale the free energy pieces are not seen during the linear phase, but its nature can be seen from the total growth rate ($\Gamma_T$), which is given by the gradient drive rate less the subgrid dissipation. The drive rate is constant until the nonlinear phase is entered. It then drops to zero as the instability begins to saturate, but the ensuing phase is so violent that one cannot speak of statistically equilibrated turbulence in a saturated state. The ExB eddy transport flux ($F_e$) begins and ends at low levels but is unrealistically large in between. This is more of a disruption than turbulence, in fact. The error ($\Gamma_E$), is small outside of this phase but large enough during it that one sees the need for more spatial resolution to completely capture the dynamics. The amplitude of the potential ($A_p$) and vorticity ($A_w$) are unrealistically large, but smooth at later times. This indicates that the energy has gone into flows, which we will see momentarily.

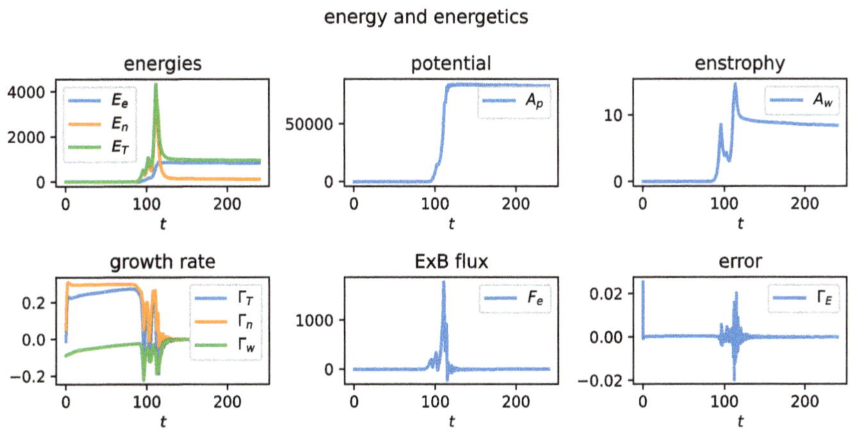


**Figure 13.1.** Time evolution of the 2D interchange model, for the case with external gradient drive in a 1 × 1 domain. The energy pieces (top left) are the total ($E_T$), thermal free ($E_n$), and ExB flow ($E_e$) energy, respectively. The total growth rate (bottom left, $\Gamma_T$) is due to gradient drive ($\Gamma_n$), and subgrid dissipation ($\Gamma_w$). The other curves are the potential amplitude ($A_p$), the enstrophy ($A_w$), the turbulent ExB transport flux ($F_e$), and the energy error ($\Gamma_E$). The linear growth phase proceeds until the amplitude reaches mixing level ($\tilde{p}_e/p_e \sim 1/k_y L_\perp$ in physical units), after which a short and explosive nonlinear phase yields overly large amplitude before throwing all the energy into flows. Simple cases like this one always end with most of the energy in flows, either in the $x$ (radial) or $y$ (zonal) direction.

The amplitude spectra for $\phi$, $n$, and $\varpi$ are given with the spatial morphology of $\phi$ during the initial transition into nonlinearity, in figure 13.2. The early spectra follow the initialisation, and the instability spreads out in the direction of forcing ($x$), growing in place in the linear phase. A completely developed linear eigenmode of a single Fourier component would fill out the $y$-domain, but before that happens the nonlinear interactions start.

The first manifestation of nonlinearity is the visible motion of these streamer structures in their direction of forcing. The instability has a positive flow down the gradient through a positive density disturbance, giving the energetically favourable situation for growth (positive $\Gamma_+ = \Gamma_n$), so when nonlinear advection becomes visible it slips the structures to left and right[1].

The nonlinearities spread the spectrum out, and then each eigenmode accesses free energy on its own, leading to the rapid saturation seen in the time traces as $\Gamma_T$ initially falls to zero. Then, as the dynamics becomes nonlinear, the initial spatial forms are destroyed by the emerging turbulence. Transition to full nonlinearity is very rapid, essentially skipping a saturated phase. Either all the energy condenses into a flow in the $y$-direction, which we call a zonal flow (see below), or into a flow in the $x$-direction, a radial flow. In the latter case the dynamics does not saturate, namely, the disturbances grow to large amplitude and the run is stopped. The wavenumber spectra ($k_y$) show a growing mode range, and then at the onset of nonlinearity the long-wavelength $\phi$ amplitude greatly increases and the spectrum of $n$ spreads out. Both effects are very fast, and the subgrid dissipation does not stop the growth.

[1] The shift of positive to the right and negative to the left is clearer in $n$, but $\phi$ is shown to emphasise the nonlinear formation of the box mode.

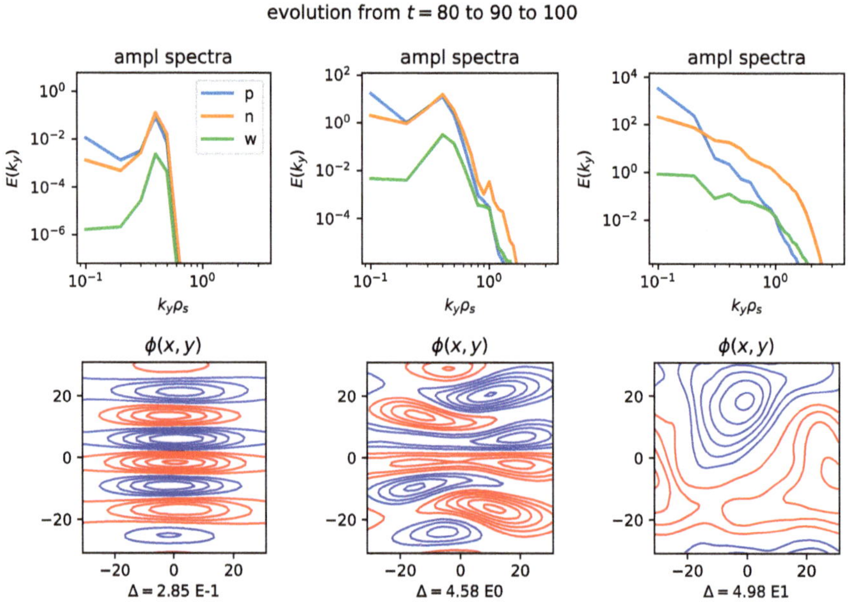


**Figure 13.2.** Transition from linear growth (left) to nonlinearity (centre) and then turbulence (right), in the wavenumber ($k_y$) amplitude spectra of $\phi$ (top row, 'p'), $n$ ('n'), and $\varpi$ ('w'), and in the spatial morphology of $\phi(x, y)$ with contours drawn red where positive and blue where negative, with interval given by the $\Delta$ values. The instability grows in place, then the first manifestation of nonlinearity is the visible shift to right and left of maxima and minima, respectively, as advection sets in, and then in turbulence large-scale eddies form with $\phi > n$ in the amplitudes, also shown by the low-$k_y$ part of the spectra. The dynamics driving large-scale radial flows always leads to the 'box mode' in two-dimensional cases.

Which fate one finds depends on the domain aspect ratio, much like the situation of the 3D drift wave flute modes noted in Chapter 11. This $1 \times 1$ case, whose spatial morphology at later time is given in figure 13.3, shows the same box mode and flow problem we saw there. If we run the same case in a $4 \times 1$ domain, with a $64 \times 256$ grid in $xy$, respectively, and with scales given by $K = 0.025$ and $L_y = 4L_x = 2\pi/K$, the linear stage is the same, but the nonlinear one is very different. The rise of $\phi$ at low $k$ is much more pronounced, and a flow layer runs through the short extent of the domain in the $x$-direction, for which especially in two dimensions there is no saturation mechanism. We see this flow and associated density and vorticity disturbances in figure 13.5. When the flow becomes wide enough in the $y$-direction, there is no possibility of Kelvin–Helmholtz destabilisation to break up, and the flow strengthens until the run has to be stopped. If we then run it in a $1 \times 4$ domain, with a $256 \times 64$ grid in $xy$, respectively, and with scales given by $K = 0.1$ and $L_x = 4L_y$, the linear stage is the same, but in nonlinearity one finds transient turbulence which proceeds until the eddies fill the $x$-domain and then the energy immediately transfers to a flow, as depicted in figure 13.4. The flow layer is very stable because once again its width is as large or larger than the periodicity length in its direction. For indefinitely growing cases, the level of unphysicality is reached when the relative amplitude of the disturbance reaches unity in physical units, while the model has

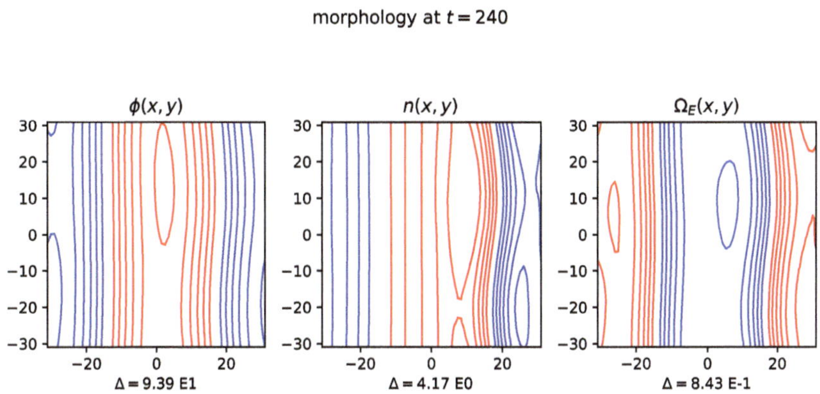


**Figure 13.3.** Spatial morphology of the $1 \times 1$ case at $t = 240$, the end of the run. All of the energy has gone into a flow in the $y$-direction and evolution has mostly stopped.

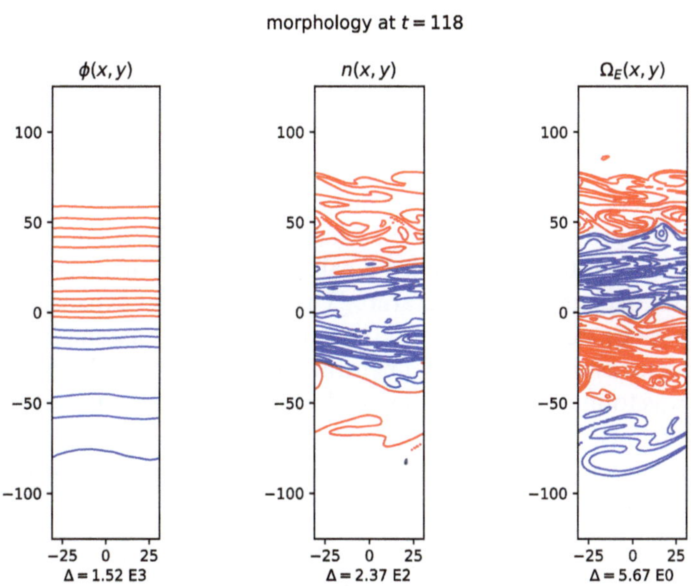


**Figure 13.4.** Spatial morphology of the $4 \times 1$ case before $t = 120$, when the run is stopped since it does not saturate. This time, all of the energy has gone into a flow in the $x$-direction which grows indefinitely since it is in the direction of the effective forcing.

assumed it is at mixing level. This is measured by the RMS value of $n$. In this 2D model, mixing level is normalised to unity, so the validity is lost when the amplitude reaches 64, the level of $L_\perp / \rho_s$ with $L_\perp$ equal to the smaller of $L_x$ or $L_y$ in each of these cases. With the $1 \times 4$ case (reverse, with $256 \times 64$ and $K = 0.1$, so that the long dimension is in the $x$-direction), the flow goes through the short extent, this time in the $y$-direction, and the drive stops since $\nabla y$ is not the direction which liberates free energy.

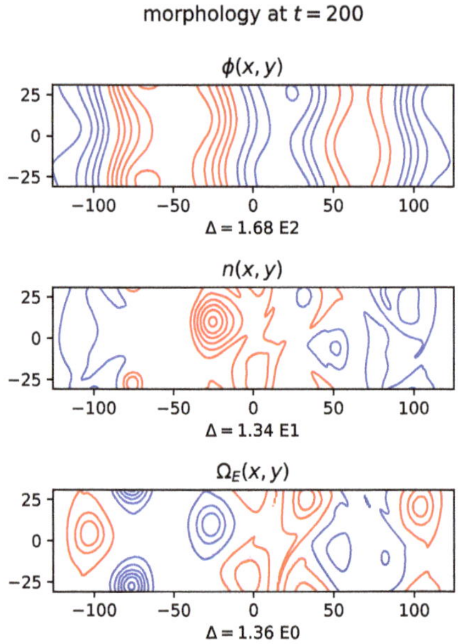


**Figure 13.5.** Spatial morphology of the 1 × 4 case at $t = 240$, the end of the run. Again, all of the energy has gone into a flow in the $y$-direction and evolution has mostly stopped.

This 2D model can produce a realistic result if we approach it transiently. Instead of a background gradient drive which never stops, we can initialise it with a localised layer of finite density, called a roll, which represents a pressure layer due to the constant background temperature. There is no direct expansion of the width of the layer since compressibility is controlled by the magnetic field, not the pressure, in contrast to the case of a neutral fluid. For this study the initial state in Eq. (13.37) was found to produce too much symmetry, with late-time flow patterns which look artificial. So we revert to the 2D random-phase flow field used in Eq. (11.41) from Chapter 11. This is referred to as $n_{2D}$, and to it is added a large-scale layer, so that

$$n = n_{1D} + n_{2D} \qquad n_{1D} = 16\cos(2\pi x/L_x) \qquad \varpi = \phi = 0. \tag{13.42}$$

The other variables are initialised at zero, and $\omega_n$ is set to zero. The value of 16 gives a relative amplitude of 1/4 for the roll in physical units. The initial fluctuation amplitude is set to $A_0 = 1$ to proceed directly to turbulence. The ultimate reservoir of free energy is the density roll that we initialised, and it represents thermal free energy in the density multiplied by the background temperature (specifically, the square of the relative density fluctuation amplitude, multiplied by 1/2 and then by the background pressure). Evolution out of this state proceeds as in a decaying case, which is relatively free of pathology.

The time traces for this case are shown in figure 13.6. With the initial density disturbances acted upon by curvature forcing (magnetic compression of the diamagnetic current), the instability develops with the ExB eddies in the same

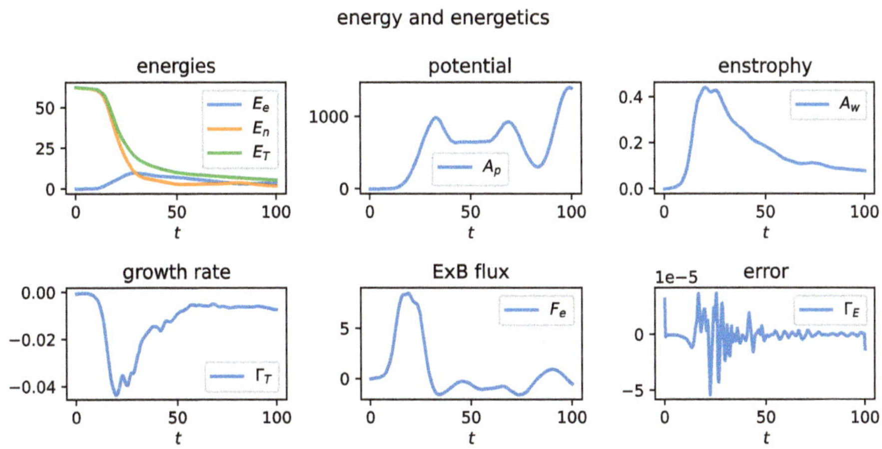


**Figure 13.6.** Time evolution of the density roll case, wherein the drive reservoir is in the initial state of the dependent variables, and instead of an external drive, the instability manifests as a transfer of free energy from the $k_y = 0$ component of $n$ into the $k_y \neq 0$ components of all the dependent variables. The energy pieces (top left) are the total ($E_T$), thermal free ($E_n$), and ExB flow ($E_e$) energy, respectively. The total growth rate (bottom left, $\Gamma_T$) is due solely to subgrid dissipation ($\Gamma_w$, not shown). The other curves are the potential amplitude ($A_p$), the enstrophy ($A_w$), the turbulent ExB transport flux ($F_e$), and the energy error ($\Gamma_E$). The linear growth phase proceeds until the amplitude reaches mixing level ($\tilde{p}_e/p_e \sim 1/k_y L_\perp$ in physical units), after which the developing turbulence destroys the density roll, which serves as its free energy source.

form as the disturbances. The initial energy ($E_T$) is all in the thermal component, but the ExB energy rises to become visible on the same graph in the $10 < t < 20$ range. Peak turbulence, tracked by the enstrophy ($A_w$) and the ExB eddy transport flux ($F_e$), is in the range around $t = 20$, with the dissipation (equal to the total growth rate $\Gamma_T$) following. The error ($\Gamma_E$) remains small throughout.

The amplitude spectra for $\phi$, $n$, and $\varpi$ are given with the spatial morphology of $\phi$ during the development of this phase, in figure 13.7. The early spectra follow the initialisation, and the instability forms on the outer side of the roll, namely, where $\nabla p$ and $\nabla \log B$ align. On the inner side, where they are anti-aligned, no instability forms. As the nonlinear phase is entered, the ExB eddies form a component at longer wavelength as in the hydrodynamic cases of Chapter 4, which begins to take over the nonlinear dynamics. This becomes more intense in the later phase, near peak turbulence as just noted. The low-$k_y$ amplitude of $\phi$ is much larger, and the spectrum of $n$ has broadened.

The spatial morphology in the entire domain is shown for $\phi$, $n$, and $\varpi$ together for these three times, in figures 13.8, 13.9, and 13.10, respectively. Note how the structures of the instability are initially present only on the outer side of the roll. Then, turbulence forms as a finite range of scales is more obvious. Finally, the turbulence overwhelms the density roll, and the entire dynamics decays, much more slowly at later times as seen in the time traces.

The fully developed turbulent state after the destruction of the density roll fills the entire domain, and then the interchange turbulence forms sheet-like structures in $n$

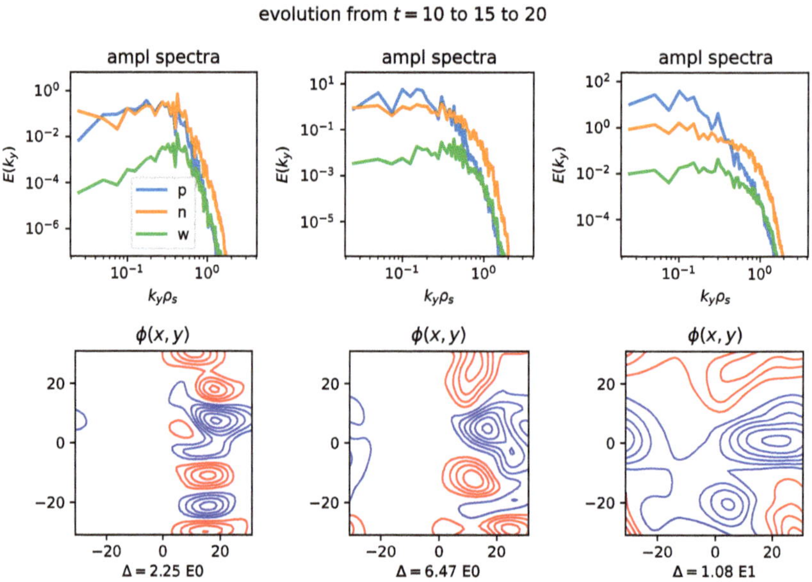


**Figure 13.7.** Development of the instabilities into turbulence, at $t = 10$ (left), $t = 15$ (centre), and $t = 20$ (right), highlighted by the wavenumber ($k_y$) amplitude spectra of $\phi$ (top row, 'p'), $n$ ('n'), and $\varpi$ ('w'), and in the spatial morphology of $\phi(x, y)$ with contours drawn red where positive and blue where negative, with the interval given by the Δ values (only the central 1/4 of the $y$-domain is shown). The spectrum of the growing instability is still given by the initial state, which was given a nonlinear initial amplitude. The instability forms on the outer side of the roll ($x > 0$), where the gradient lines up with the curvature vector (both inward). The ExB eddies form turbulence, which is fully developed by the later time at which the low-$k_y$ component in $\phi$ is prominent. Note the change in scale between the latter two times.

and $\varpi$ as in the hydrodynamic cases of Chapter 4. Before that happens, the peculiar morphology of the interchange turbulence is the mushroom cloud appearance of the pressure/density disturbances as the flows driving them form fronts. Seen in motion, it looks like 'boiling' namely convection.

## 13.7 Radial flows versus zonal flows

The flows generated by the turbulence are part of the dynamics and as we have seen their mechanisms conserve free energy properly. Even in the under-resolved violent phase of the cases with continuous external drive with no boundaries in the $x$-direction, the error was still small compared with the violence of the drive. As seen, we either see flows in the $x$- or the $y$-direction depending on what type of largest-scale disturbances are allowed by the geometry.

The general parlance for the down-gradient flows which drive interchange turbulence is 'radial flows' since the physical systems in which convecting flows are usually discussed are gravitationally confined spherical stars or spherical-shell or thin-layer atmospheres, and flows from core regions towards edge regions in a cylindrical or toroidal plasma. These radial flows appear explicitly in the energy theorem (Eq. 13.16), as the term $v_E^x$ multiplying $n$ and the gradient strength $\omega_n$. In a

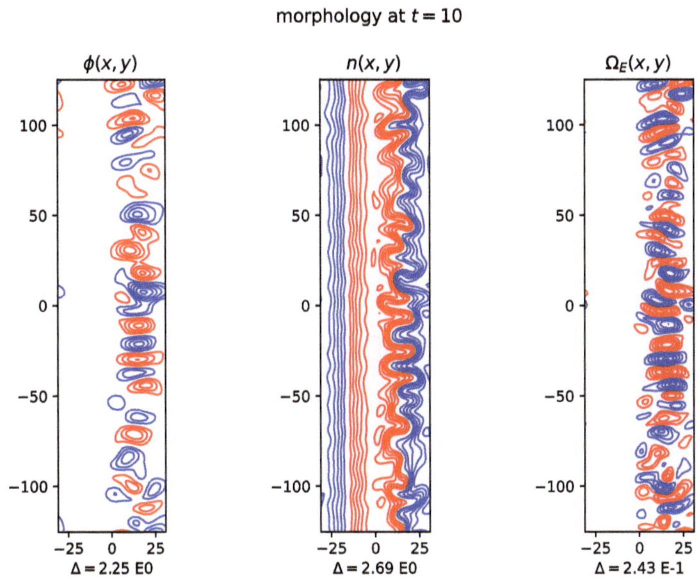


**Figure 13.8.** Spatial morphology of the density roll case at $t = 10$, when the instability is fully formed but before the spectrum has started to change away from the form in the initial state. The roll in the density is very visible, but at this time so are the disturbances. The activity is almost all on the outward side of the roll where $\nabla p$ and $\nabla \log B$ are aligned. On the inward side, the disturbances are locally stable.

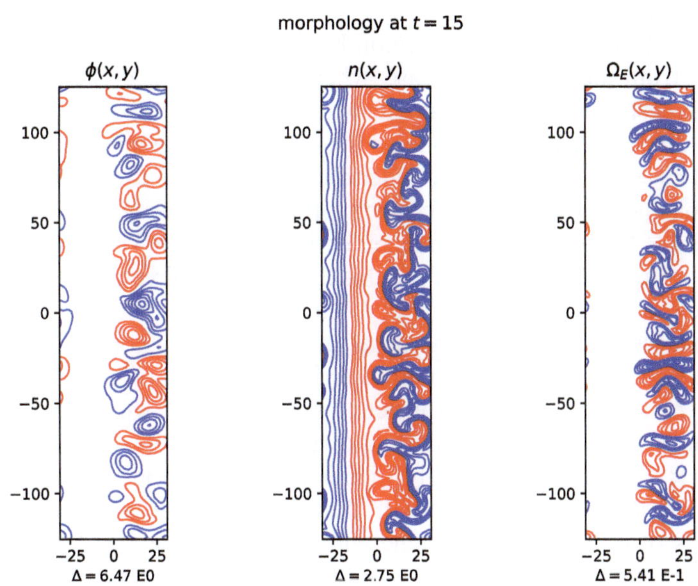


**Figure 13.9.** Spatial morphology of the density roll case at $t = 15$, in the transition to full nonlinearity. Most of the activity is still on the outward side of the roll where $\nabla p$ and $\nabla \log B$ are aligned.

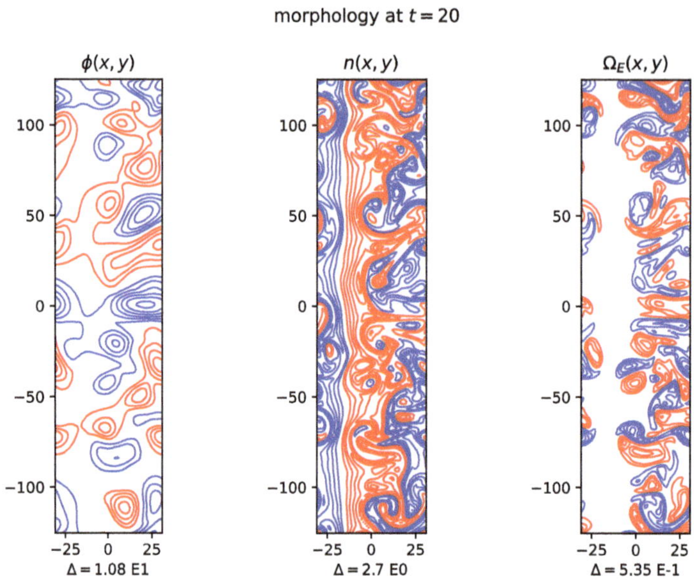


**Figure 13.10.** Spatial morphology of the density roll case at $t = 20$, the time of peak enstrophy, dissipation, and transport. The turbulence now involves a range of scales and is about to destroy the density roll.

nonlinear system we have simultaneously the tendencies to drive flows, in either direction, because they represent the largest-scale degree of freedom for the ExB energy, which has a tendency to put its energy into the largest scales. We get the flow either in the $x$-direction as a radial flow, or in the $y$-direction as a 'zonal flow,' so named due to the definition of the region between two flux surfaces as a 'zone' (the name comes from the geophysical case, wherein the globe is divided into latitudinal zones and an east–west flow of indefinite extent is called a zonal flow). Both types of flow are nonlinearly driven, and one or the other will take control.

Still with cases not involving boundedness in the radial direction, we find either that the turbulence grows without bound, with a radial flow going through the short extent in the $4 \times 1$ case, or that it shuts off entirely in the $1 \times 4$ case, with a zonal flow also going through the short extent but this time in the $y$-direction. The radial flow in the $4 \times 1$ case is not simply the case of a linear instability which does not saturate. The disturbances do experience the onset of saturation as nonlinearity starts, near $t = 100$ in the above cases, but the nonlinear interactions serve merely to dump most of the free energy into the long wavelengths, which take over when they are wide enough, which only the $4 \times 1$ domain allows (see below).

In the $1 \times 1$ case and especially in the $1 \times 4$ case the result is a final state dominated by a zonal, not a radial, flow, now in the $y$-direction. For some combinations of $\omega_n$ and $\omega_B$ either one will win out in the $1 \times 1$ domain, but the zonal flow always wins in the $1 \times 4$ domain. The turbulence grows and rapidly saturates as in the other cases but now the nonlinear interactions place energy into the eigenmode with $k_x = 1$ and $k_y = 0$, representing the zonal flow, which tends to block transport in the $x$-direction. We do in fact see $\Gamma_+$ dropping to zero in these cases. What remains is a nonlinear state with all the energy in the zonal component, dynamically aligned ($n$ and $\varpi$ functions of $\phi$) such that the nonlinearities are zero and that the net drive is zero. The initial break up and end states are seen in the first three figures.

So for computations done in a $1 \times 1$ or especially a $1 \times 4$ doubly periodic domain, the zonal flows will emerge nonlinearly to shut the turbulence and associated transport off. But this result, as it happens, is for most situations of physical interest entirely spurious.

### 13.7.1 Aspect ratio effects for interchange flows

The fate of these flows highlights the issue of the aspect ratio of the computational domain in the two dimensions which cover the turbulence ($x$ and $y$). If the domain is square, that is, the aspect ratio is unity, the radial flows drive the system but the nonlinear transfer dynamics will drive the zonal flows strongly enough for these to shear apart and shut off the eddies with dominantly radial flow, such that a radial flow does not emerge and the turbulence is essentially aborted. But we must remember the properties of the Kelvin–Helmholtz instability when discussing the fate of flows in realistic systems. The key feature of a periodic square domain is that a radial flow can only be half as wide as it is long. Going back to the outline of the Kelvin–Helmholtz problem, we note that this radial flow can be broken up by secondary disturbances, or those that grow on the primary disturbance which is the radial flow. For a radial flow in a square domain, the longest possible wavelength of a secondary disturbance is larger than the width of the widest possible radial flow. In this case the widest radial flow is mode $(k_x, k_y) = (0, 1)$, which has a half width from peak to zero of $\Delta = L_y/4$, and the most dangerous secondary disturbance has $k_x = 2\pi/L_x$. We find that $k_x\Delta = \pi/2$, and for a sinusoidal flow profile this is sufficient for the Kelvin–Helmholtz instability (the second derivative enters with two factors of $\pi$). On the other hand, the same sketch with $L_y/L_x = 4$ yields $k_x\Delta = 2\pi$, a case which is Kelvin–Helmholtz stable.

In physical situations with convective type flows causing the instabilities and driving the disturbances, the zonal/radial aspect ratio is always large. In the terrestrial atmosphere, for example, the north–south extent of a reasonable mid-latitude zone is about 20 degrees latitude, which at $a/3$ is much shorter than the longitudinal distance ($\pi a$ even at 60 degrees), where $a$ is the planetary radius. The situation is similar for the Jovian planets or for the Sun. These aspect ratios are larger than 10. In a cylindrical or toroidal plasma, the zonal periodicity length is at least as large as the minor radius, while the profile scale length in the important regions is much smaller. In order to ascertain the relative importance of radial and zonal flows for any type of instability or turbulence we wish to apply to such situations, we have to set the domain aspect ratio at least qualitatively commensurate with that of the actual situation in the real world. Most situations of relevance for this type of turbulence are thin-layer or thin-atmosphere cases, in which the aspect ratio corresponding to $L_y/L_x$ is large. We will discuss the details of the Kelvin–Helmholtz problem when considering flows in Chapter 16.

## 13.8 The mode structure of interchange turbulence

For purposes of comparison we need to establish what the mode structure of interchange turbulence is, and how to distinguish it from drift wave turbulence. A simple summary is sufficient. The mode structure is taken from the density roll case above, as it has proved sufficient to do the averages over the outer half of the domain, in which the disturbances are unstable, and over the phase of time during which peak turbulence is shown by the time traces in figure 13.6.

The energetics has been given in the time traces. The growth of the instability is due either to the background gradient drive or to transfer from the initial density roll, which during the linear phase—in which there is no coupling between $k_y \neq 0$ modes at small amplitude—largely amount to the same thing.

The cross coherence and phase shift distributions are shown in figure 13.11. These show a lack of cross coherence and a very wide phase shift distribution for each wavelength centred about the maximum driving value of $\pi/2$, largely as in the hydrodynamic cases with $D \to 0$ and forming a clear contrast to the pure drift wave cases discussed in Chapter 10. This is also seen in the differences in amplitude between $n$ and $\phi$ during the various phases.

With linear transfer minimal compared with the linear drive, the energetics for interchange turbulence is very simple: gradient drive at long wavelength, enough coupling to $\phi$ with the same spectrum to maintain the ExB flow eddies, and direct cascade through $n$ to arbitrarily small scales where the disturbances are dissipated. This is very like the classic cases of fluid turbulence driven by stirring at the largest scales, and due to the absence of strong, adiabatic coupling it is very unlike drift wave turbulence. We will see more of this in the three-dimensional cases, below.

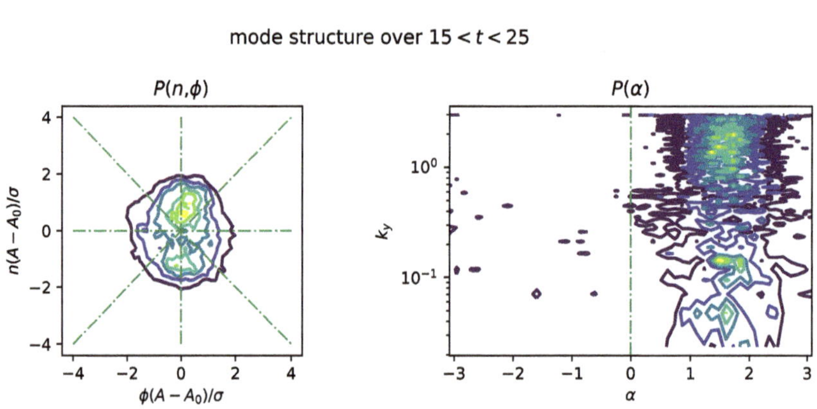


**Figure 13.11.** Cross coherence and phase shift distributions between $n$ and $\phi$, for the density roll case during $15 < t < 25$, the phase of maximum turbulent activity. The spatial averaging is over the outer half of the domain. This can be compared with the mode structure control cases in figures 10.11 and 10.12 from Chapter 10. The correlation between $n$ and $\phi$ is small and the phase shifts are centred on $\pi/2$ for maximum drive by the background gradient. These forms reflect the absence of an adiabatic response and represent hydrodynamic mode structure.

## 13.9 A simple model of a toroidal magnetic field

We now turn to three-dimensional interchange turbulence, in a model geometry which we can use to treat either slab geometry with interchange effects or a magnetic field lying in closed, toroidal surfaces. Toroidal topology and field line connection in a sheared magnetic field lying in surfaces of toroidal topology was already discussed in Chapter 12. Herein, we address the elements of toroidal curvature.

The simplest model of a set of nested toroidal surfaces is set of nested circular tori. Start with a 2D Cartesian surface with coordinates $RZ$, horizontal and vertical, respectively, a half-infinite plane wherein $R = 0$ is the left-hand boundary and the other three boundaries are at infinity. Add to this a geometric angle, $\varphi$, whose domain is one revolution. These have the domain

$$0 < R < \infty \qquad -\infty < Z < \infty \qquad -\pi < \varphi < \pi, \tag{13.43}$$

and the sense, Jacobian, and volume element

$$\nabla R \times \nabla Z \cdot \nabla\varphi = +\frac{1}{R} \qquad \sqrt{g} = R. \tag{13.44}$$

The surface $Z = 0$ is the midplane, and the $RZ$-plane is the poloidal plane. An arbitrary axisymmetric, divergence-free vector field can be written

$$\mathbf{B} = I\nabla\varphi + \nabla\psi \times \nabla\varphi, \tag{13.45}$$

with $I$ and $\psi$ scalar fields on the $RZ$-plane. With Ampère's law, the current is

$$\mu_0\mathbf{J} = \nabla I \times \nabla\varphi - (\Delta^*\psi)\nabla\varphi, \tag{13.46}$$

where the 'delta-star' operator is

$$\Delta^* = R^2\nabla \cdot \frac{1}{R^2}\nabla = R\frac{\partial}{\partial R}\frac{1}{R}\frac{\partial}{\partial R} + \frac{\partial^2}{\partial Z^2}. \tag{13.47}$$

The MHD equilibrium equation, contracted onto $\nabla\psi$ with $\nabla\psi \cdot \nabla\varphi = 0$, is

$$(\mathbf{J} \times \mathbf{B} - \nabla p) \cdot \nabla\psi = -\frac{|\nabla\psi|^2}{\mu_0 R^2}\left(I\frac{\partial I}{\partial\psi} + \Delta^*\psi + \mu_0 R^2\frac{\partial p}{\partial\psi}\right) = 0. \tag{13.48}$$

That the quantity in parentheses must vanish defines the Grad–Shafranov equation, after the authors who published it. That $I = I(\psi)$ results from axisymmetry, whence $\mathbf{J} \times \mathbf{B} \cdot \nabla\varphi$ vanishes. That $p = p(\psi)$ results from contracting the MHD equilibrium equation onto $\mathbf{B}$. That $\Delta^*\psi$ is a scalar field itself, $\mu_0 R^2\mathbf{J} \cdot \nabla\varphi$ indicates that if $\psi$ is given a boundary surface—a contour in the poloidal plane—on which it is zero, it is monotonic on the plane interior to the boundary. The extremum is called the magnetic axis. These considerations are why we can set up a model with nested surfaces. Furthermore, with $\beta \ll 1$ we have $I$ constant and $p$ negligible, and with the toroidal/poloidal aspect ratio large we may take $R$ constant in this equation, and then specify a model within a concentric circular geometry.

So now we may assume a model geometry with nested circular tori. The intersection of each surface with the poloidal plane is a circle. Let the boundary surface be defined by the circle with minor radius $r = a$, and the centre at $R = R_0$ and $Z = 0$. Each surface is defined with a circle with $0 < r < a$ centred at the same place. Then in the above terms,

$$I = I_0 = \text{constant} \qquad \psi = \psi(r) \qquad r\nabla\theta = R\nabla\varphi \times \nabla r, \tag{13.49}$$

also defines $\theta$, the geometric poloidal angle. Placing the branch cut $\theta = \pm\pi$ on the inboard midplane ($Z = 0$ with $R < R_0$), we define simple flux coordinates with domain and sense

$$0 < r < a \qquad -\pi < \theta < \pi \qquad -\pi < \varphi < \pi \qquad \nabla r \times \nabla\theta \cdot \nabla\varphi = \frac{1}{rR}. \tag{13.50}$$

Then, $\theta = 0$ defines the outboard midplane ($Z = 0$ with $R > R_0$). This is one way to cover the 2D toroidal surfaces and the radial direction separating them.

The pitch of the magnetic field lines $q$ is defined by the relative sizes of $I$ and $\nabla\psi$. Specifically, $q$ is defined as the number of toroidal turns a field line takes for one poloidal cycle around the surface. This is given by the average of the ratio of the toroidal and poloidal contravariant components of the magnetic field. Since it is an average, it is a 'flux function,' namely, a function of the flux surface label, for which $\psi$ is used fundamentally, and then if a radial coordinate $r = r(\psi)$ is defined in some way, then $q = q(r)$. In the tokamak magnetic field, $q$ is formally defined by

$$q = \oint \frac{d\theta}{\mathbf{B}\cdot\nabla\theta}\frac{I}{2\pi R^2}, \tag{13.51}$$

where in this context $\theta$ is any well-defined poloidal coordinate. The factor of $I/R^2$ is $\mathbf{B}\cdot\nabla\varphi$, the toroidal contravariant component. The integral does the full cycle, and the normalisation is common in the numerator and denominator. Coordinates for which this ratio $B^\varphi/B^\theta = q(\psi)$ is itself a flux function are called straight field line coordinates, but we will not pursue that here. It is sufficient for our purposes to note that in the circular tori model we have

$$q^{-1} = -\frac{R_0^2}{I}\frac{1}{rR_0}\frac{\partial\psi}{\partial r} \approx -\frac{1}{rB_0}\frac{\partial\psi}{\partial r}, \tag{13.52}$$

neglecting the structure in $\theta$. In a tokamak $q > 1$ at $r = 0$, rising to values near 3–4 times that at $r = a$, largely due to toroidal MHD stability constraints.

A magnetic field of this form has a field strength which is generally proportional to $R^{-1}$. In general it is determined by

$$R^2B^2 = I^2 + |\nabla\psi|^2 = I_0^2 + O(\beta, \epsilon^2/q^2), \tag{13.53}$$

where $\epsilon = r/R_0$ is the surface aspect ratio. A 'conventional tokamak' is defined as the case when both $\beta$ and $(\epsilon/q)^2$ are of order a few percent. At this stage we may take $I = I_0$ and then $B = I_0/R$. We then have

$$\nabla\log B = -\nabla\log R \approx \mathbf{b}\cdot\nabla\mathbf{b}, \tag{13.54}$$

and hence the validity of the simple curvature model in Eq. (13.8). The field gradient $\nabla B$ is therefore directed along $-\nabla R$, that is, towards the symmetry axis. The radius of curvature is the same, given by $R$ with curvature towards the symmetry axis. This is why interchange turbulence is often referred to as curvature driven, since the curvature and field gradient are synonymous, as they usually also are for a cylinder with a purely azimuthal magnetic field $(-\hat{\mathbf{z}} \times \nabla\psi)$.

The toroidal field geometry has a property which is different from a cylinder or slab; however, the interchange forcing is in the direction of the major radius but the pressure force is in the direction of the minor radius. These line up so as to cause the interchange instability on the outboard midplane of the torus, but on the inboard midplane they are anti-aligned, so that the curvature effects are stabilising. The magnetic field connects these two regions, so that on the same field lines there are locations where the interchange mechanism for instability is positive and others where it is negative. It is interesting to note that a negative interchange effect by itself does not imply damping, just neutral stability—from Eq. (13.29) one finds a real eigenmode frequency and therefore neutrally stable oscillation. If there is no magnetic shear to complicate the situation, we will have strong instability.

For the computational model, a thin flux tube is built from a small region in the drift plane on the outboard midplane side and then extended in each direction until the inboard midplane is reached. There, a matching condition is applied. The way to do this is to establish the geometry globally and then restrict to a flux tube by neglecting the $r$-dependence of all geometrical coefficients, taking the toroidal domain to be periodic on an integral fraction of the $2\pi$ cycle, and then applying flute mode ordering on the angle derivatives.

The coordinate system $r\theta\varphi$ may be used for this. The counterpart of Eq. (13.8) is to assume that the spatial brackets involve the structure

$$\nabla\varphi \cdot \nabla a \times \nabla b = \frac{1}{rR}\left(\frac{\partial a}{\partial r}\frac{\partial b}{\partial \theta} - \frac{\partial b}{\partial r}\frac{\partial a}{\partial \theta}\right), \tag{13.55}$$

for two scalars $a$ and $b$, with $rR$ the volume element preserving the pure divergence nature. Then apply Eq. (13.17) to this case, so that

$$\mathcal{K} = \frac{R_0}{B_0}\nabla\varphi \cdot \nabla(2\log R) \times \nabla, \tag{13.56}$$

with the factor of $R_0$ for dimensionality ($R_0\nabla\varphi$ is close to a unit vector).

For a spatial bracket between two dependent variables, such as $[\phi, n]$, we have the same structure, except that we have to watch out for the flute mode ordering on both derivatives, which arises in the field aligning. Beforehand, we have

$$[a, b] = \frac{R_0}{B_0}\nabla\varphi \cdot \nabla a \times \nabla b, \tag{13.57}$$

namely, the same structure as for the curvature. This is the same as the slab form except for the replacement of $\nabla s$ by $R_0\nabla\varphi$.

We then derive field-aligned coordinates the same way we did for the slab case in Chapter 12. Using $q$, we have

$$B^{\varphi} = \frac{I}{R^2} \qquad B^{\theta} = -\frac{1}{rR}\frac{\partial \psi}{\partial r} = \frac{I}{qR^2}, \tag{13.58}$$

and of course $B^r = 0$ as constant radius defines the flux surface. Taking

$$\xi = q\theta - \zeta \qquad s = \theta, \tag{13.59}$$

with $r$ unchanged and $q = q(r)$, we have the global version of toroidal field-aligned coordinates. It is important to note that at this point we still have global cover in a one-to-one and onto sense. This yields

$$B^{\xi} = 0 \qquad \left.\frac{\partial}{\partial \xi}\right|_{r,s} = -\left.\frac{\partial}{\partial \varphi}\right|_{r,\theta}. \tag{13.60}$$

Both properties represent field aligning with toroidal symmetry allowing eventual truncation of the toroidal domain. We also have

$$\nabla r \times \nabla \xi \cdot \nabla s = \frac{1}{rR}, \tag{13.61}$$

namely, an unchanged Jacobian. Most of the rest follows our previous effort, with the periodicity constraints

$$f(r, \theta + 2\pi, \varphi) = f(r, \theta, \varphi + 2\pi) = f(r, \theta, \varphi), \tag{13.62}$$

mapping to

$$f(r, \xi + 2\pi, s) = f(r, \xi + 2\pi q, s + 2\pi) = f(r, \xi, s), \tag{13.63}$$

namely, the parallel boundary condition is the same, also with

$$f(r, \xi, s) = \sum_l f_l(r, s)\, e^{il\xi} \qquad f_l(r, s + 2\pi) = f_l(r, s)\, e^{-2\pi i l q}. \tag{13.64}$$

With $q$ in general irrational, this limits the minimum average $k_{\|}$ to the vicinity of $K_{\|} = 1/qR_0$ for all degrees of freedom. The connection length is $2\pi q R_0$, namely one traversal of the poloidal domain measured along the field line, from the inboard midplane around to the bottom of the torus to the outboard midplane to the top of the torus continuing around until the inboard midplane is reached again. Since the periodicity constraint holds everywhere, one cannot extend the flux tube beyond this periodicity length without violating global consistency.

For the curvature operator we need only to set it up for $r\theta\varphi$ and then transform coordinates to

$$x = r - r_0 \qquad y = \frac{r_0}{q_0}\xi, \tag{13.65}$$

with $r = r_0$ a reference flux surface radius, $q_0 = q(r_0)$ like $r_0$ a constant, and $s$ unchanged. We also expand $q$ about $r = r_0$ to obtain

$$q = q_0 + q'x = q_0\left(1 + S\frac{x}{r_0}\right), \tag{13.66}$$

where $S$ is the normalised shear given by

$$S = \left.\frac{d \log q}{d \log r}\right|_{r=r_0}. \tag{13.67}$$

The wavenumber representation becomes

$$f(x, y, s) = \sum_k f_k(x, s)\, e^{iky} \qquad k = l\frac{q_0}{r_0}, \tag{13.68}$$

and the periodicity constraint becomes

$$f_k(x, s + 2\pi) = f_k(x, s)\, e^{-2\pi i S x k}, \tag{13.69}$$

which is the same as in Eq. (12.50) from Chapter 12. This establishes that the full length of the $y$-domain is $2\pi r_0/q_0$, which can be truncated by taking the minimum $l$ to be an integer different from unity, but which indicates how much space there is for an anisotropic domain in the perpendicular plane.

Putting this into the curvature operator, we have the global version

$$\mathcal{K} = \frac{R_0/B_0}{rR}[2 \log R, \quad ]_{r\theta}, \tag{13.70}$$

which we transform to the local field-aligned one,

$$\mathcal{K} = \frac{R_0}{rRB_0}\left([2 \log R, x]_{r\theta}\frac{\partial}{\partial x} + [2 \log R, y]_{r\theta}\frac{\partial}{\partial y}\right), \tag{13.71}$$

while restricting to flute mode ordering with the neglect of the $s$-derivative. With

$$\log R \approx \frac{r}{R_0} \cos\theta, \tag{13.72}$$

this yields

$$\mathcal{K} = \frac{2}{l_0}\left(\sin s\frac{\partial}{\partial x} + (\cos s + Ss \sin s)\frac{\partial}{\partial y}\right), \tag{13.73}$$

where in keeping with the local transformation the coefficients evaluated at $r = r_0$. This is another way to derive the shear $S$ for the toroidal flux tube.

For a shifted flux tube we apply the transformation analogous to Eq. (12.37) from Chapter 12, taking

$$x' = x \qquad y' = y - Ss_k \qquad s' = s, \tag{13.74}$$

which gives

$$\frac{\partial}{\partial x} = \frac{\partial}{\partial x'} - Ss_k\frac{\partial}{\partial y'} \qquad \frac{\partial}{\partial y} = \frac{\partial}{\partial y'}. \tag{13.75}$$

The effect of this on the field line connection part yields the same treatment as before, with Eq. (12.52) describing how to take $\partial/\partial s$ and Eq. (12.54) how to set the guard cells at the ends of the $s$-domain, both from Chapter 12. All the statements on rational surface spacing and resolution mentioned there also apply here. Truncation of the full toroidal domain to the flux tube is nothing more than setting the fundamental wavenumber to an integer multiple of $q_0/r_0$, which is the lowest $k_y$ available on the flux surface. Note that for aspect ratio purposes we now know, from Eq. (13.68), that the full length of the $y$-domain is $2\pi r_0/q_0$, which for any reasonable model is a lot larger than $L_\perp$, which is about how large the $x$-domain should be.

Applying the shifting transformation to the curvature operator, we have

$$\mathcal{K} = \frac{2}{I_0}\left(\sin s\frac{\partial}{\partial x'} + (\cos s + S(s - s_k)\sin s)\frac{\partial}{\partial y'}\right). \tag{13.76}$$

Evaluated at $s = s_k$ and dropping the primes, we have

$$\mathcal{K}|_{s=s_k} = \frac{2}{I_0}\left(\sin s\frac{\partial}{\partial x} + \cos s\frac{\partial}{\partial y}\right). \tag{13.77}$$

The dependence of these coefficients on $s$ while taking derivatives in $x$ or $y$ describes the rotation of the curvature vector about the field line away from the radial ($x$) direction as one moves away from the $s = 0$ position at the outboard midplane. At $s = 0$ it is directed as in the two-dimensional model in Eq. (13.24). As a function of position along $s$ it varies sinusoidally with a wavelength of $2\pi qR_0$. Working in a field-aligned coordinate system, it is a simple matter to jump back and forth between a slab and a toroidal model, simply by setting $\omega_B$ to zero or something finite, in this case $2L_\perp/R_0$ in normalised units multiplying the angle coefficients and derivatives in the parentheses.

For the spatial drift brackets including the nonlinearities, we have derivatives only in $x$ and $y$, so that

$$[a, b] = \frac{R_0}{B_0}(\nabla\varphi \cdot \nabla x \times \nabla y)\left(\frac{\partial a}{\partial x}\frac{\partial b}{\partial y} - \frac{\partial b}{\partial x}\frac{\partial a}{\partial y}\right). \tag{13.78}$$

Given that the magnetic field can be written as

$$\mathbf{B} = \nabla\chi \times \nabla y = \chi_x \nabla x \times \nabla y, \tag{13.79}$$

where $\chi = \chi(x)$ and $\chi_x = \partial\chi/\partial x$, since $B^x = B^y = 0$. This is called a Clebsch representation, and field-aligned coordinates are those chosen such that $\mathbf{B}$ is proportional to $\nabla x \times \nabla y$ times a flux function. The simple model for it is that if $x$ and $y$ are in units of distance, then $\chi_x = B_0$ and we have

$$[a, b] = \frac{1}{B_0}\left(\frac{\partial a}{\partial x}\frac{\partial b}{\partial y} - \frac{\partial b}{\partial x}\frac{\partial a}{\partial y}\right), \tag{13.80}$$

with $B_0 \to 1$ in normalised units. These terms are the same in a toroidal flux tube or Cartesian or sheared-slab geometry.

Since the other operations yielded the same model, this curvature operator and the associated terms form the only addition to the field-aligned slab model we need, to address the same type of turbulence, whether reduced MHD or drift Alfvén turbulence, in toroidal geometry.

## 13.10 The ballooning mode

The ballooning mode is one of the fundamental MHD instabilities in toroidal geometry. The original derivation was within what is known as the energy principle of MHD stability. For now we show how to derive this instability within the framework we already have. The traditional MHD approach does tell us, however, what to keep. Within MHD (which is not reduced MHD), the set of pressure-driven modes are within the low frequency regime, namely, the same as what we used to derive the two-fluid drift equations. But since MHD itself contains the compressional magnetosonic waves, the assumptions are taken within the analysis rather than at the start to derive the model. We are in a situation of similar ideas, except that the compressional waves were dispensed with the way the equations were derived (keeping only $A_\parallel$ in the variables), so that we can keep exact energetic consistency and still treat gradients, the adiabatic response, electron inertia, and dissipation. Having done all the hard work, it is a short step to derive the relevant equation.

The difference between a ballooning mode and an interchange mode is simply that the curvature forcing is not homogeneous in the domain of the parallel coordinate. Instead, for the toroidal case, it is sinusoidal, localised such that maximum drive is at $\theta = 0$ in the toroidal coordinates, or $s = 0$ in the flux tube. Away from this position the curvature vector $\mathbf{b} \cdot \nabla\mathbf{b}$, always in the direction towards the symmetry axis, rotates away from the direction of $\nabla p$ towards the poloidal direction. At the same time, the shear of the field lines deforms a structure approximately following the field lines, such that a growing mode should be localised in $s$, in this simple model symmetrically about $s = 0$. So the activity balloons to the outboard side of the torus and is small on the inboard side of the torus. Hence the name of the mode.

The basic model for these modes is the same as reduced MHD in two dimensions, extended to three by restoring the parallel dynamics and the geometric structure of the perpendicular Laplacian and the curvature operator along the field lines. Sound waves, electron inertia, and any two-fluid phenomena are neglected. The behaviour of the current is then the ideal MHD shear Alfvén response to the interchange forcing. Starting with the four-field model we have been using, these approximations yield linear equations

$$\frac{\partial \varpi}{\partial t} = \nabla_{\|} J_{\|} - \mathscr{K}(p), \tag{13.81}$$

$$\frac{\partial A_{\|}}{\partial t} = -\nabla_{\|}\phi, \tag{13.82}$$

$$\frac{\partial p}{\partial t} = -\frac{p_0}{BL_p}\frac{\partial \phi}{\partial y}, \tag{13.83}$$

with

$$\frac{\rho}{B^2}\nabla_{\perp}^2\phi = \varpi \qquad - \nabla_{\perp}^2 A_{\|} = \mu_0 J_{\|}, \tag{13.84}$$

for dependent variables $\varpi$, $J_{\|}$, and $p$, field quantities $\phi$ and $A_{\|}$, and every other quantity a constant background parameter. The geometric quantities are

$$\nabla_{\|} = K_{\|}\frac{\partial}{\partial s} \qquad \nabla_{\perp}^2 = \frac{\partial^2}{\partial x^2} + 2Ss\frac{\partial^2}{\partial x \partial y} + (1 + S^2 s^2)\frac{\partial^2}{\partial y^2}, \tag{13.85}$$

with $\mathscr{K}$ given in Eq. (13.73). The coordinate system is field aligned but not shifted. The structure of the metric and curvature operator forms a well, localised near $s = 0$ where the mode is expected to live. If we take the time derivative of the vorticity equation and substitute $J_{\|}$ and $p$ from the other equations, and assume $\partial/\partial x$ is zero (looking for $k_x = 0$ modes) we find

$$\frac{\rho}{B^2}\frac{\partial^2}{\partial t^2}\nabla_{\perp}^2\phi = \frac{1}{\mu_0}\nabla_{\|}\nabla_{\perp}^2\nabla_{\|}\phi + \frac{p_0}{BL_p}\mathscr{K}^y\frac{\partial^2\phi}{\partial y^2}, \tag{13.86}$$

where

$$\mathscr{K}^y = \frac{2}{I_0}(\cos s + Ss\sin s). \tag{13.87}$$

In two dimensions, $\nabla_{\|}$ is zero and $\mathscr{K}^y$ reduces to $2/I_0$, and we recover Eqs (13.25) and (13.26). If we now Fourier transform $y$ and time, the common factor of $k_y^2$ drops, leaving

$$K_{\|}^2\frac{\partial}{\partial s}g^{yy}\frac{\partial \phi}{\partial s} = \left(\frac{\gamma^2}{v_A^2}g^{yy} - \frac{2\mu_0 p_0}{B^2 R L_p}\mathscr{K}^y\right)\phi, \tag{13.88}$$

where

$$g^{yy} = 1 + S^2 s^2 \qquad \mathscr{K}^y = \cos s + Ss\sin s, \tag{13.89}$$

give the geometric structure. This is called the 'ballooning equation' and it is the same as the one derived from the MHD energy principle under the assumptions of incompressibility and low plasma beta. If $\partial/\partial s$ and $s$ are set to zero in this equation, the 2D result in Eq. (13.31) is recovered with $L_B$ replaced by $R/2$.

To understand this 'ballooning mode' we can simplify heuristically by taking $s = 0$ but setting $\nabla_\parallel \to ik_\parallel$, with the factors $g^{yy}$ and $\mathcal{K}^y$ reducing to unity, so that a finite parallel wavelength is assumed within the language of the 2D interchange model. The ballooning equation can then be solved simply, yielding

$$\gamma^2 = \gamma_I^2 - k_\parallel^2 v_A^2. \tag{13.90}$$

The parameter $\gamma_I$ is the ideal interchange growth rate in Eq. (13.31). Instability results when $\gamma^2 > 0$. If $k_\parallel = 0$ then we revert to the two-dimensional ideal interchange instability. This is opposed by the restoring force effect of the Alfvén response, leading to a threshold in the pressure gradient in the case that $k_\parallel^2$ is set by the geometry. In a toroidal magnetic field $k_\parallel^2$ usually scales as $(qR)^{-2}$, and $2\mu_0 p_0 / B^2 L_p$ becomes $-\nabla\beta$, yielding the standard ideal ballooning parameter for the pressure gradient,

$$\alpha_M = -q^2 R \nabla\beta, \tag{13.91}$$

which in terms of the parameters of the four-field model becomes

$$\alpha_M = \hat{\beta}\omega_B, \tag{13.92}$$

with a critical value of unity in the simple case. However, if there is no magnetic shear, then $k_\parallel$ is allowed to vanish, and we find instability and turbulence at all $\hat{\beta}$ for $\omega_B > 0$.

In the ballooning equation we have to find the $\phi$ which balances the three terms and yields a result for $\gamma$. There is a simple variational method by which one can get an estimate. Multiply the ballooning mode equation (Eq. 13.88) by the eigenmode and integrate the outermost $\partial/\partial s$ term by parts to find

$$\frac{\partial}{\partial s} K_\parallel^2 g^{yy} \phi \frac{\partial \phi}{\partial s} - K_\parallel^2 g^{yy} \left| \frac{\partial \phi}{\partial s} \right|^2 = \left( \frac{\gamma^2}{v_A^2} g^{yy} - \frac{2\mu_0 p_0}{B^2 R L_p} \mathcal{K}^y \right) |\phi|^2. \tag{13.93}$$

Note that we can assume the eigenmode is real, since there are no complex factors in the equation, even if $\gamma^2$ can also be negative. Now integrate over an arbitrary extent in $ds$ assuming the mode vanishes at some distance way from $s = 0$. Note that this neglects field line connection, but ballooning analysis is an approximation in which the spatial variation of the parallel boundary condition with $x$ is ordered with the global scale and therefore neglected. Then we have the integral condition

$$\frac{\gamma^2}{v_A^2} \int_{-\infty}^{\infty} ds\, g^{yy} |\phi|^2 = \frac{2\mu_0 p_0}{B^2 R L_p} \int_{-\infty}^{\infty} ds\, \mathcal{K}^y |\phi|^2 - K_\parallel^2 \int_{-\infty}^{\infty} ds\, g^{yy} \left| \frac{\partial \phi}{\partial s} \right|^2, \tag{13.94}$$

which is basically the same as Eq. (13.90) except for numerical factors, in which $g^{yy}$ and $\mathcal{K}^y$ act as shape factors. Now we can put in a trial function of Gaussian form for $\phi$ with $1/e$ width given by $\Delta$ in the $s$-direction. The resulting $\gamma^2$ is then given as a function of $\Delta$, and then $\Delta$ is varied until $\gamma^2$ is maximum. This is an estimate for the growth rate. To do better than this requires numerical techniques, which we do with the initial value problem.

What we want to do is test recovery of the standard result, which is $\alpha_M$ given by Eq. (13.91) with a multiplier of $0.66S$ giving the dependence upon the magnetic shear. So we set up a time-dependent initial value problem as for the wave analysis in Chapter 8. Instead of periodicity in $s$ we set up a domain with a multiple of $2\pi qR$ such that the mode will indeed decay at reasonable large separation from $s = 0$, which is the bottom of the well given by the forms of $g^{yy}$ and $\mathcal{K}^y$. We choose, somewhat arbitrarily and with some hindsight, to allow

$$-7\pi < s < 7\pi \qquad \text{with} \quad \phi(\pm 7\pi) = 0, \tag{13.95}$$

and evolve Eqs (13.81)–(13.83) in time, with Eqs (13.84) and (13.89) as auxiliaries. We assume $k_y\rho_s = 0.1$ and $k_x = 0$ for the maximum growth, using the fact that the result should be independent of $k_y$. The initial state is $p = 10^{-10}p_0 \exp(-s^2)$ with $\varpi = J_\| = 0$. There are 128 equidistant grid nodes in $s$ and the time step is 0.01. Runs are taken to $t = 1000$, and the energetic quantities are sampled every 0.2, for 20 000 samples. Statistics are taken over the last 4000 samples. Either of two results is found. Either the time dependence oscillates and a clear eigenmode is not found, or a clear eigenmode is found and the growth rate settles to a result level with very small time variation. We evaluate this by measuring the mean and standard deviation of $\gamma$ in the same way as for any other time trace. The only two parameters to vary are $\alpha_M$ and $S$, so we take a given $S$ and sweep $\alpha_M$ (through $\beta\omega_B$ in the equations), looking for the threshold. The results for $S = 1$ and $S = 2$ are shown in figure 13.12, which shows that we indeed recover

$$\alpha_M > 0.66\,S \qquad \text{for instability}, \tag{13.96}$$

as the relevant threshold condition, with the agreement best for higher shear as the 0.66 line is an asymptotic limit. For a typical case, $\alpha = 0.9$ for $S = 1$ the growth rate was $\Gamma_T = 0.056\,6 \pm 1.45 \times 10^{-15}$ and the error was $\Gamma_E = 4.7 \times 10^{-6} \pm 1.39 \times 10^{-15}$,

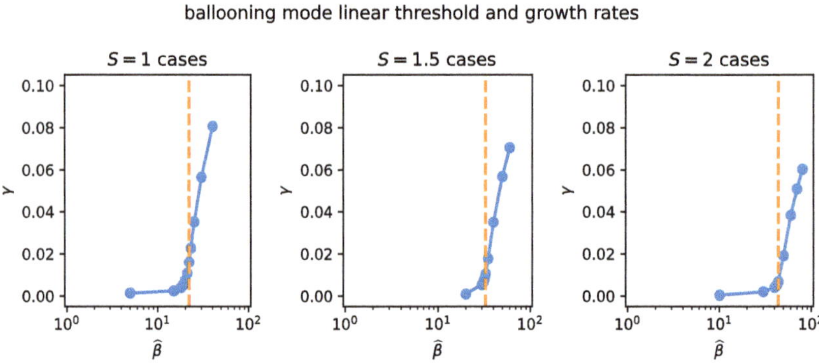


**Figure 13.12.** Ideal MHD ballooning mode growth rates as a function of $\hat{\beta}$ for three values of $S$, the magnetic shear. The MHD parameter $\alpha_M$ is $\hat{\beta}\omega_B$, and $\omega_B = 0.03$. The nominal threshold of $\alpha_M = 0.66\,S$ is indicated by the vertical dashed lines.

and the fact that in the incompressible model the curvature term enters as an extra drive term is taken into account. (In the analytical model it is ordered small, and in this case it is small by $\omega_B = 0.03$ compared with $\omega_n = 1$, so the effect on the growth rate is not large. We are mindful of the fact that in a turbulence computation that is not good enough and we will do these with full energetic consistency in the equations.) This is a useful test, not only on the numerics but also, once we get the result, on the ability of the simple ballooning model to capture the results we find in three dimensions, both linear and nonlinear. We turn to that next.

## 13.11 Three dimensions—ballooning mode turbulence

For the three-dimensional case of what changes from interchange to ballooning mode turbulence in a toroidal magnetic field, we start with the four-field drift Alfvén model of Chapters 11 and 12, delete the adiabatic and diamagnetic compression physics, and add the toroidal curvature forcing. This produces a reduced MHD, denoted RMHD, model which we use to study the mode structure and energetics. Besides study of the 3D interchange and ballooning mode physics, we also form a control case, comparable to the drift Alfvén case of Chapter 12, necessary to discern the physics of three-dimensional turbulence in a magnetised plasma in which both mechanisms are present. We will perform that study in Chapter 14. We need to lay down the ballooning mode turbulence case first.

Since we do not treat any complications of toroidal geometry beyond this version of $\mathcal{K}$, we may merely add the necessary curvature terms as in Eqs (13.18) and (13.19) to the three-dimensional drift Alfvén model in slab geometry given in Chapter 12. In keeping with the RMHD model, we leave out the adiabatic coupling mechanism ($\nabla_\parallel n$ in the equation for $J$ and $\nabla_\parallel J$ in the equation for $n$). We keep the electron inertia as before, to cover the case when both induction and resistivity are small. The equations are otherwise as in Eqs (12.15)–(12.19) from Chapter 12, under the same normalisation scheme and normalised parameters. They are

$$\frac{\partial \varpi}{\partial t} + [\phi, \varpi] = \nabla_\parallel J - \mathcal{K}(n), \tag{13.97}$$

$$\frac{\partial}{\partial t}(A + \hat{\mu} J) + \hat{\mu}[\phi, J] = -\nabla_\parallel \phi - C\, J, \tag{13.98}$$

$$\frac{\partial n}{\partial t} + [\phi, n_0 + n] = -\nabla_\parallel u + \mathcal{K}(\phi), \tag{13.99}$$

$$\hat{\epsilon}\frac{\partial u}{\partial t} + \hat{\epsilon}[\phi, u] = -\nabla_\parallel(n_0 + n + P), \tag{13.100}$$

for the dependent variable evolution equations, and

$$\varpi = \nabla_\perp^2 \phi \qquad J = -\hat{\beta}^{-1}\nabla_\perp^2 A \qquad P = -X\nabla_\parallel u_\parallel, \tag{13.101}$$

for the auxiliaries and field equations, in which $P$ is the viscous correction to the pressure and $X$ is the parallel viscous diffusivity. The nonlinear parallel gradient is

$$\nabla_\parallel = \frac{\partial}{\partial s} + [-A, ], \tag{13.102}$$

and the curvature operator is

$$\mathcal{K} = \omega_B\left(\sin s\frac{\partial}{\partial x} + \cos s\frac{\partial}{\partial y}\right) \qquad \text{where nominally} \quad \omega_B = 2\frac{L_\perp}{R_0}, \tag{13.103}$$

always evaluated at $s = s_k$ (as with the brackets) in the shifted coordinate system. The coefficient $\omega_B$ may be set independently.

### 13.11.1 Resistive ballooning modes and transition to turbulence

The above equations can be simply linearised to recover the ballooning result. To do that, neglect both ExB and acoustic compression in the equation for $n$, leaving just the advection on the left side, and also dispensing with $u$ and any of its dynamics. Then, delete the nonlinear terms, and evaluate $[\phi, n_0] = \omega_n(\partial\phi/\partial y)$, where the factors with physical units multiplying it are normalised out. Now, if finite $\hat{\mu}$ and $C$ are set to zero, the set of ballooning mode primitive equations are recovered (Eqs (13.81–13.83)). Then, putting the ballooning mode expressions for $\nabla_\parallel$, $\nabla_\perp^2$, and $\mathcal{K}$ and setting $k_x = 0$ recovers the ballooning equation (Eq. 13.88). One thing remains to mention before we go to the turbulence. That is, that with finite electron inertia $\hat{\mu}$ present, we can just as easily set the induction to zero ($\hat{\beta} = 0$) and replace it with the electron inertia. This simply puts a factor of $d_e^2\nabla_\perp^2$ on $A$ wherever it occurs. The effective substitution is

$$\frac{\partial A}{\partial t} \to d_e^2\nabla_\perp^2\frac{\partial A}{\partial t} \qquad \text{and} \qquad \nabla_\parallel\nabla_\perp^2\nabla_\parallel\phi \to d_e^{-2}\nabla_\parallel^2\phi. \tag{13.104}$$

The same analysis leading to Eq. (13.88) then yields

$$K_\parallel^2\frac{\partial}{\partial s}\frac{\partial\phi}{\partial s} = k_y^2 d_e^2\left(\frac{\gamma^2}{v_A^2}g^{yy} - \frac{2\mu_0 p_0}{B^2 R L_p}\mathcal{K}^y\right)\phi. \tag{13.105}$$

There are two differences. First, the $\partial J/\partial t$ term now yields $-\nabla_\parallel\phi$ with coefficients but no factor of $\nabla_\perp^2$, which is why no $g^{yy}$ factor appears on the left-hand side, and more importantly that the result is no longer independent of $k_y$. The most unstable set of modes is now the highest $k_y$ allowed by the resolution, since for $k_y d_e \gg 1$ we simply recover the ideal interchange result, up to a point. That is, the mode can simply spread out until the $k_\parallel^2$ term from both $K_\parallel$ and $\partial/\partial s$ is as small as $(k_y d_e)^2$ is large. If we pull out the factor $v_A^2$ and use the fact that $d_e/v_A = \rho_s/V_e$ we find that the result is independent of $\beta_e$ but instead depends on the electron mass.

We can use this to test the numerics of the nonlinear model in the linear regime by choosing negligible $C$ and starting with initial amplitude $A_0 = 10^{-10}$, varying $\hat{\beta}$

below the nominal ballooning regime and then increasing past it. In the linear regime there is no aspect ratio consideration, so we can preserve $L_x = 1$ and set $L_y = 0.25$, with the nominal grid at $256 \times 64$ in the $xy$-plane, yielding $h_x = h_y$ or a nominal resolution ratio of unity. Then, progressively increase $N_x$ from 256 to 4096, up to $h_y/h_x = 16$, to provide overly ample resolution of the field line connection. With $\delta_0 = 0.015$ the maximum $k_y\rho_s$ is about 12. For this capture of the linear regime, we set the subgrid dissipation in $x$ and $y$ to zero. With this, we find a small-scale instability growing at about $\gamma = 0.11$, or somewhat more slowly than $\gamma_I$ which is 0.173 in normalised units (i.e. in terms of $c_s/L_\perp$). The beginnings of the transition to nonlinearity are shown in figure 13.13. The small-scale instability is shown on the left side, as streamer flows and a peak in the spectrum as it grows. The central frames are taken when the structures begin visibly changing shape, with the corresponding spectral form showing the introduction of scales outside the peak region. The right side shows the eddy flows associated with the break up of the linear forms, and the spectrum shifting to longer wavelength. The run is stopped when the limited resolution in the $y$-direction and the lack of perpendicular subgrid dissipation cause it to lose integrity. This shows that when the ExB nonlinearities become active,

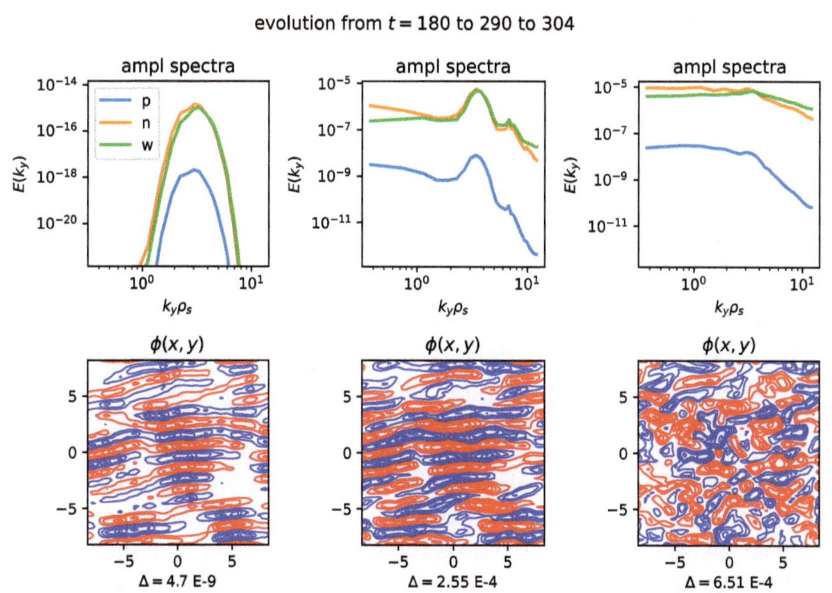


**Figure 13.13.** Transition out of the inertial ballooning mode instability to nonlinearity and turbulence. With a grid of $4096 \times 64 \times 32$ in $xys$ the grid is over-resolved in $x$ to guarantee resolution of the field line connection in the linear phase, but with too small a domain in $y$ to really capture the turbulence at late times—just enough to capture the beginnings of nonlinearity. Here, 1/4 of the $x$-domain (not $y$) is shown. Drift direction wavenumber ($k_y$) spectra of the amplitudes of $\phi$ (top, 'p'), $n$ ('n'), and $\varpi$ ('w'), and spatial morphology of $\phi$ in the drift plane (bottom), for $t = 180$ in the linear growth phase, $t = 290$ in the transition to nonlinearity, and $t = 304$ at the beginning of the turbulence phase in which the spectra are close to their eventual statistical equilibrated forms. The transition is very rapid. First the linear structures begin to move, and then they are wiped out by the turbulence.

the small-scale structure associated with this instability is wiped out and the turbulence assumes a different behaviour in which much of the energy is at larger scale.

The electron inertial instability just discussed becomes what is termed a 'resistive ballooning mode' when the collisional dissipation (resistivity) is turned back on and assumes values for which $\nu_e > c_s/L_\perp$, faster than any mode of interest. At this point the electron inertia is subdominant against resistivity, which for $\alpha_M$ below its critical value controls instability. There exists an unstable mode for any finite value of $C$, which we know having just seen the one for $C \to 0$. The scale of instability increases with larger $C$, such that it is near $\rho_s$ for $C = 1$ in these units. For $C = 13.9$, the middle of the range of collisionality we have been concerned with, the instability peak is in the range $0.5 < k_\perp\rho_s < 1$ so that we can examine this transition from instability to nonlinearity with the same resolution as we did with the nonlinear drift wave instability. The parameters for this case are

$$\begin{aligned} \hat{\beta} &= 1 \qquad \hat{\mu} = 2.72 \qquad C = 13.9 \\ \hat{\epsilon} &= 10\,000 \qquad X = 100 \qquad \delta_0 = 0.015, \end{aligned} \tag{13.106}$$

and $S = \omega_n = 1$, which except for $\hat{\beta}$ reduced to 1 from 5 is the same case with which we examined drift wave mode structure. The grid is $256 \times 256 \times 32$ in $xys$, with the same domain size of $L_x = L_y/4 = 1$. The resolution ratio $h_y/h_x = 4$ is set to guard against any difficulties in resolving field line connection in the linear regime. The lower $\hat{\beta}$ is chosen to highlight the shift to larger scale in the transition from instability to turbulence, and to remain well away from the influence of the ideal ballooning regime that enters with higher $\hat{\beta}$, which we will explore later on. This is a case that can display these effects while remaining easily resolved computationally.

The same set of transition to nonlinearity and then to turbulence is shown in figure 13.14. The instability grows fastest just longward of $k_y\rho_s = 1$ and when the transition to turbulence comes it is very rapid. First, the individual structures, growing in place, begin to move, then to break up, and then they suddenly lose their form as the entire spectrum reaches nonlinear amplitude together. In the onset phase the longer and shorter wavelength ranges grow faster than the instability itself as the latter throws energy into the neighbouring regions. Before the saturated phase is established the longer wavelengths are driven through both nonlinearities. Only after the nonlinear phase is fully entered does the spectral transfer take the form of direct transfer through $[\phi, n]$ and inverse through $[\phi, \varpi]$. At $t = 250$ the range of the instability loses energy to all the other modes and the ExB advective nonlinearity is the strongest channel especially out of the unstable range to the larger scales. This is shown by the isotropic spectra at that time, shown in figure 13.15. By $t = 300$ the familiar pattern is set up, even before the spectra has equilibrated (after about $t = 350$). The rest of the mode structure is shown in the next subsection, in which the cases are run from a nonlinear start on a conventional grid with isotropic grid cells in the $xy$-plane. Along the way we will get a resolution check against this case with four times as much in the $x$-direction.

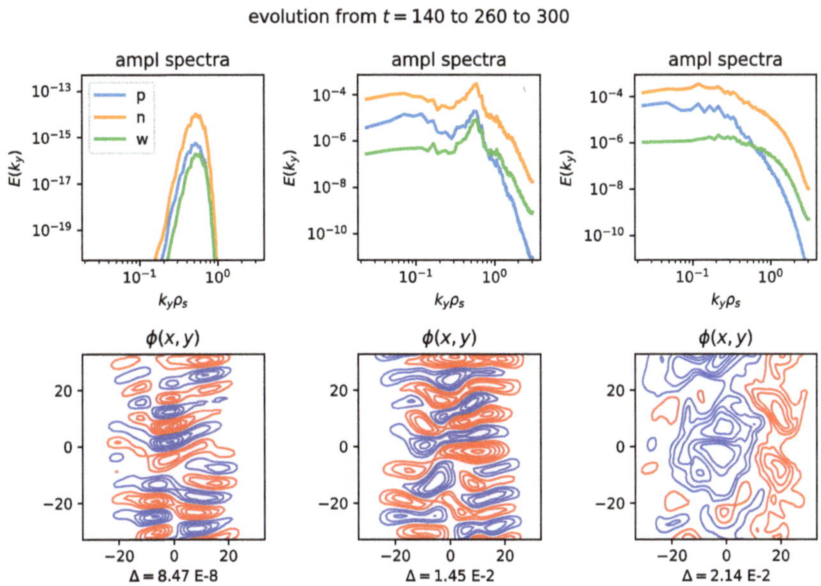


**Figure 13.14.** Transition out of the resistive ballooning mode instability to nonlinearity and turbulence. Fully resolved case with 256 × 256 × 32 grid, with enough collisionality ($C = 13.9$) to get the scale of the instability above $\rho_s$ (see text). Here, 1/4 of the $y$-domain (not $x$) is shown. Drift direction wavenumber ($k_y$) spectra of the amplitudes of $\phi$ (top, 'p'), $n$ ('n'), and $\varpi$ ('w'), and spatial morphology of $\phi$ in the drift plane (bottom), for $t = 140$ in the linear growth phase, $t = 260$ in the transition to nonlinearity, and $t = 300$ at the beginning of the turbulence phase in which the spectra are close to their eventual statistical equilibrated forms. The transition is very rapid. First the linear structures begin to move, and then they are wiped out by the turbulence.

## 13.12 Curvature forcing and ballooning mode structure

Concentrating now on the turbulence, we start with the same nonlinear initial state as in Section 12.6 of Chapter 12, and run the same set of cases as therein. The parameters controlling the electrons for the nominal case are

$$\hat{\beta} = 5 \qquad \hat{\mu} = 2.72 \qquad C = 13.9, \tag{13.107}$$

which is a case with robust resistive and electromagnetic responses. When $\hat{\beta} = 5$ or higher we have $\omega_* > (\eta_\parallel/\mu_0)k_\perp^2$ and $k_\perp d_e < 1$ everywhere in the spectrum at these parameters. The scale ratio and domain size parameters are

$$\hat{\epsilon} = 10\,000 \qquad \delta_0 = 0.015 \qquad L_y/L_x = 4. \tag{13.108}$$

The nominal grid is 64 × 256 × 32 in $xys$, and the time step is $\tau = 0.01$. The node count in $s$ is raised and the time step dropped if any of the cases show a need for it (the energetic error and Courant limit are the indicators). Dirichlet $x$-boundaries are used in $x$ with a damping layer set up as in the bounded cases of Chapter 11. We will run these same cases with the adiabatic response and diamagnetic compression turned back on in Chapter 14.

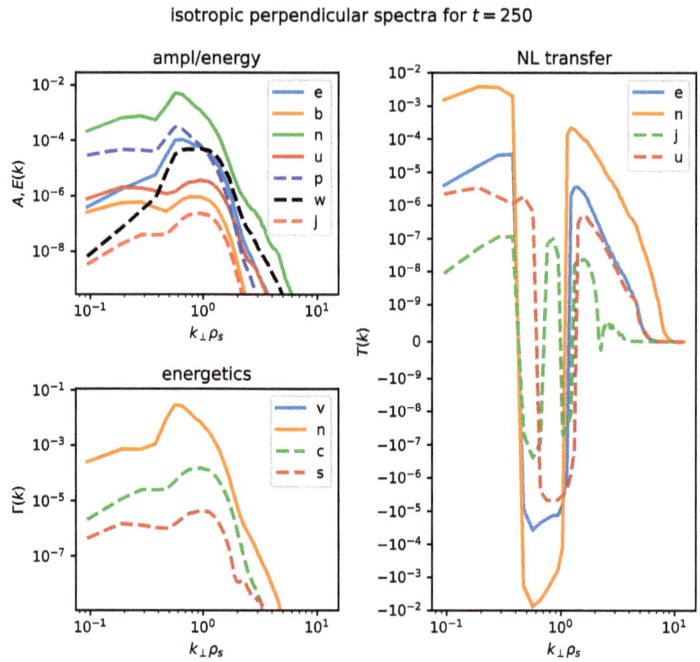


**Figure 13.15.** Isotropic perpendicular spectra for the $\hat{\beta} = 1$ and $C = 13.9$ case at $t = 250$, during the transition to nonlinearity. The labelling and format are as in figure 11.2 or 12.12 from Chapters 11 and 12. The amplitude peaks (top left) and the gradient drive (bottom left) still peak where the instability does. The ExB spectral transfer (right) show transfer out of the instability scale into all the others, especially the larger scale, for both the vorticity ('e') and advective ('n') nonlinearities. The cascade dynamics in the turbulence are not yet established.

The turbulence quickly rises to a high level and saturates after about $t = 200$. The runs are all taken to $t = 4000$, with the mode structure diagnostics averaged over $2000 < t < 4000$. The case with $C = 13.9$ and $\hat{\beta} = 5$ is referred to as nominal and is showcased. The case with $\hat{\beta} = 1$ has the same parameters as used to study the transition from the instability by itself, to nonlinearity and turbulence. The coarser resolution in the $x$-direction ($N_x = 256$ there and 64 here) gives isotropic grid cells as appropriate to strong turbulence. We also obtain a control test against the use of $h_y/h_x = 4$, which resolves the field line connection at the shortest wavelengths, so see if that is necessary when in turbulence. We will do this again each time we add physics.

The time traces of the amplitude and energetics of the nominal case are shown in figure 13.16. This result (top frames) is compared with the drift wave model from Chapter 12 (bottom), run at the same parameters, $\hat{\beta} = 5$ and $C = 13.9$. The speed of the development and saturation into the driven turbulence is evident. The clearest difference between the results of the two models, one without the curvature forcing and one without the adiabatic coupling, is that the ratio of enstrophy ($A_w$, denoted 'w') to thermal free energy ($E_n$, denoted 'n') is much larger in the drift Alfvén case in which the component of $\phi$ hence $\varpi$ at smaller scales is held up by the adiabatic coupling. This will be of central importance to the comparisons between drift wave

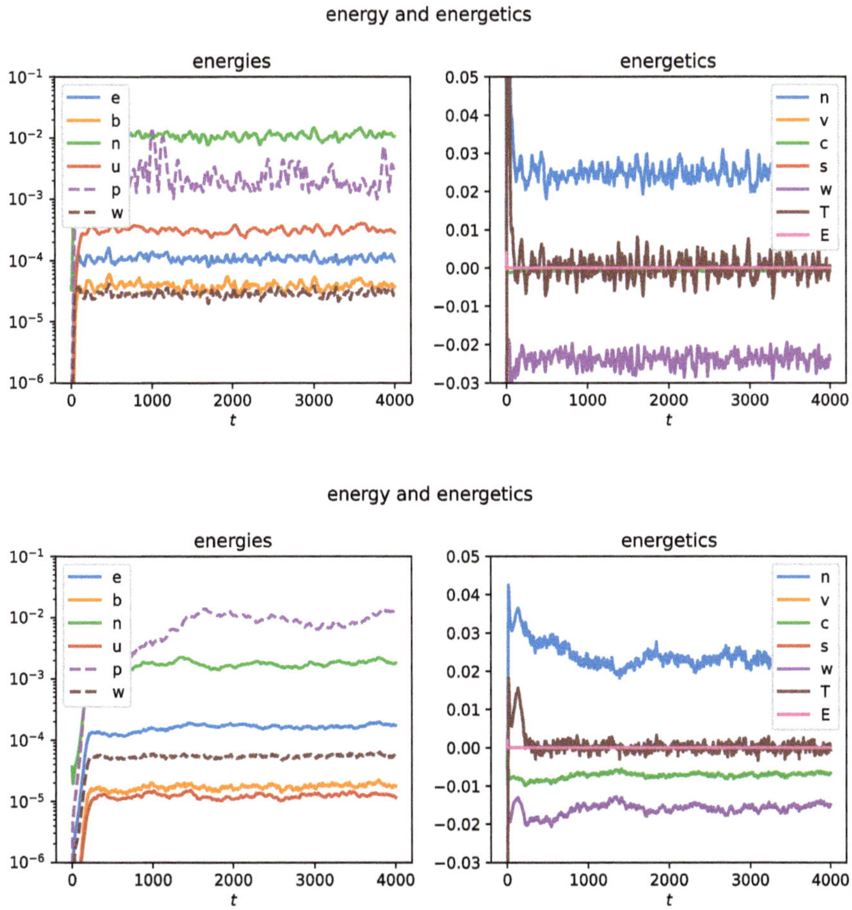


**Figure 13.16.** Time traces of (left) the free energy components ('ebnu' for $E_e$, $E_b$, $E_n$, $E_u$, respectively) and amplitudes ('pw' for $A_p$ and $A_w$), and (right) energetic drive and dissipation ('nvcsw' for, $\Gamma_n$, $\Gamma_v$, $\Gamma_c$, $\Gamma_s$, and $\Gamma_w$, respectively, with the sinks signed negative), nonlinear growth rate ($\Gamma_T$), and error ($\Gamma_E$). The top frames are for the toroidal RMHD case and the bottom ones are for the drift Alfvén sheared-slab case, both with the same parameters, $\hat{\beta} = 5$ and $C = 13.9$. The largeness of $A_p$ in the slab case is due to flows, addressed in Chapter 16, while in the RMHD case $E_e$, $E_b$, and $A_p$ are very small. The resistive dissipation ($\Gamma_c$) is appreciable only in the slab case. The standout difference is that the ratio of enstrophy ($A_w$, the lower of the two dashed curves in each case) to thermal free energy ($E_n$) is much larger in the drift Alfvén case. This will be of central importance to the analysis taken in Chapter 14.

turbulence forced by curvature and the resistive ballooning results, addressed in Chapter 14.

The isotropic spectra of the amplitude and energetics are shown in figure 13.17. This can be compared with the same case within the drift Alfvén model shown in figure 12.12 from Chapter 12. More energy (thermal free energy $E_n$ is denoted by 'n') is collected at larger scale, with the vorticity (enstrophy $A_w$ is denoted by 'w') still at the smaller scales but at the lower amplitude. The nonlinear transfer tendencies in the ExB vorticity ('e') and advective ('n') nonlinearities are the same but the one for

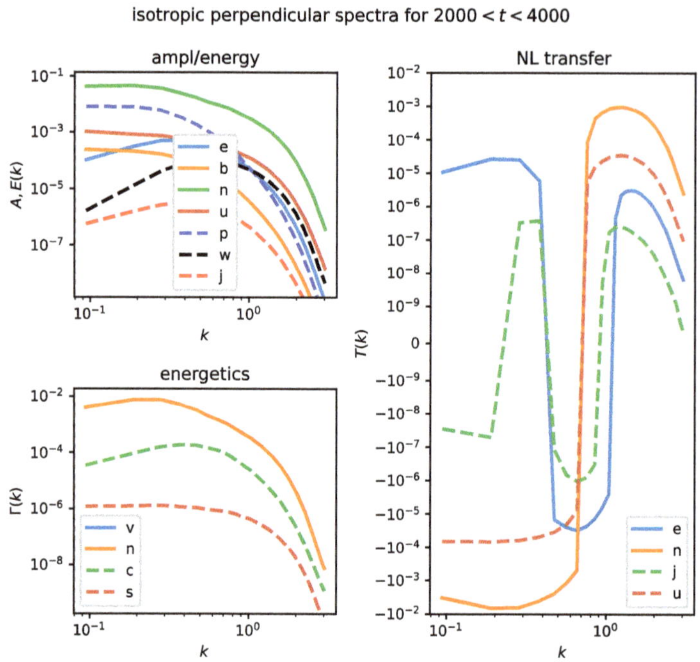


**Figure 13.17.** Isotropic perpendicular spectra for the nominal case during $2000 < t < 4000$, in the saturated phase. Contrast with figure 12.12 from Chapter 12, and with figure 13.15 for the linear growth phase. The spectra are weighted more at larger scale, with the vorticity (top left, 'w') still near $k_\perp \rho_s \sim 1$ but at lower amplitude. The nonlinear transfer (right) has the same directionality but the ExB vorticity nonlinearity ('e') is smaller than the advective ones for $E_n$, ('n'). Nonlinear transfer in the parallel channels ('j' and 'u') is smaller.

vorticity is smaller. The drive is much more at larger scale than in the drive Alfvén model. The weaker enstrophy transfer is due to the weaker enstrophy in general, so that the large-scale thermal free energy is less efficiently depleted. Together with the more direct large-scale drive, this leads to the much larger free energy content. The turbulent transport fluxes are compared later on, below.

The parallel wavenumber spectra are shown in figure 13.18. The peak in amplitudes and drive is at the $k_\parallel = 0$ flute mode range, but the forms are wider for the state variables but narrower for the vorticity. This reflects the absence of the adiabatic response in constraining $n$ and maintaining the strong vorticity non-linearities as in the drift Alfvén case of figure 12.13 from Chapter 12. The lack of ExB vorticity nonlinearity drive for the flute mode range is noteworthy. In the ballooning mode case this region is depleted by all the nonlinearities, transferring the energy to all the modes in the rest of the spectrum.

The parallel envelopes of amplitudes and transport are shown in figure 13.19. The action of the Alfvénic parallel dynamics on $\phi$ rather than $n - \phi$ is apparent in the much smaller amplitude and flatter envelope of $\phi$. By contrast, $n$ is very strongly ballooned, and so is the gradient drive. We will see this change in Chapter 14 when both drift wave and ballooning mode physics are present together. This diagnostic is

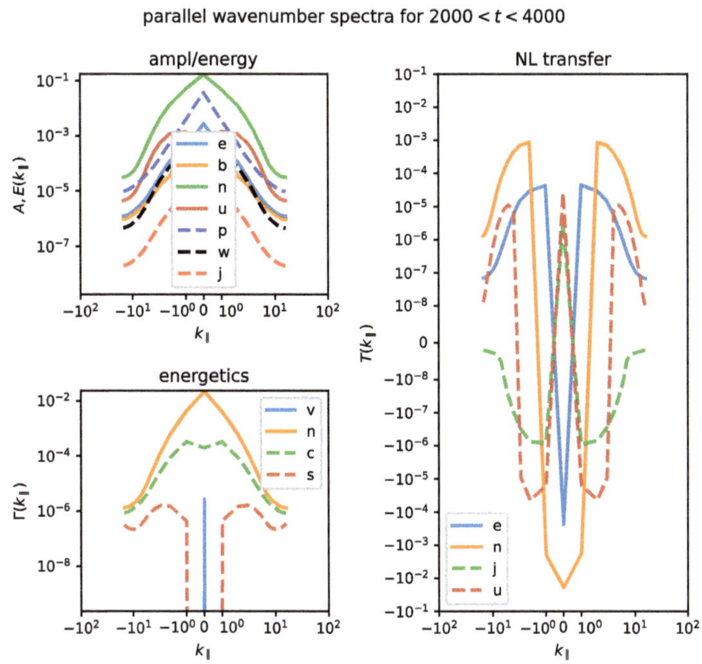


**Figure 13.18.** Parallel wavenumber spectra for the nominal case during $2000 < t < 4000$, in the saturated phase. See the contrast against figure 12.13 from Chapter 12. The spectra of both energy and energetics are narrower (left side). Note especially that the vorticity ('w') is narrower than the sound waves ('u'), which have a larger role as a sink than in the drift Alfvén case. The flute mode component is not driven by the ExB vorticity nonlinearity (right side, 'e').

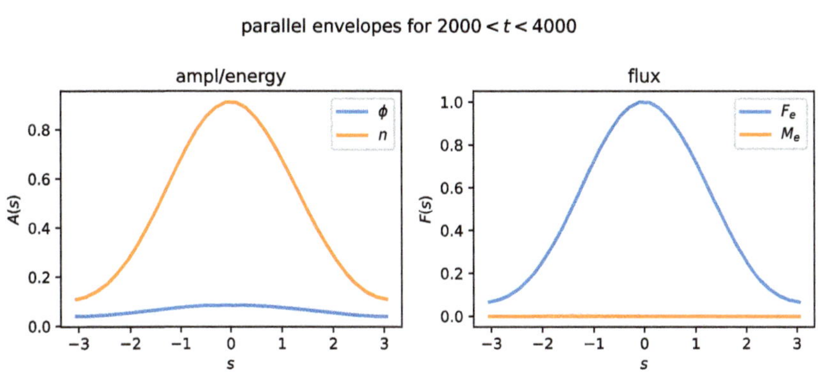


**Figure 13.19.** Parallel amplitude envelopes (left) and turbulent ExB transport flux (right), which are averages over $x$ and $y$ and time $2000 < t < 4000$ of each quantity. With the interchange forcing positive on the outboard side the curves balloon there. See the contrast against figure 12.14 from Chapter 12. Note especially that the density (left, 'n') is larger than the potential ('p'), with the difference $h = n - \phi$ essentially given by $n$. This is a result of the Alfvén dynamics constraining $\phi$ rather than $h$, and it violates the assumptions according to which the adiabatic response is neglected in an MHD model. This changes only for unrealistically large $C$.

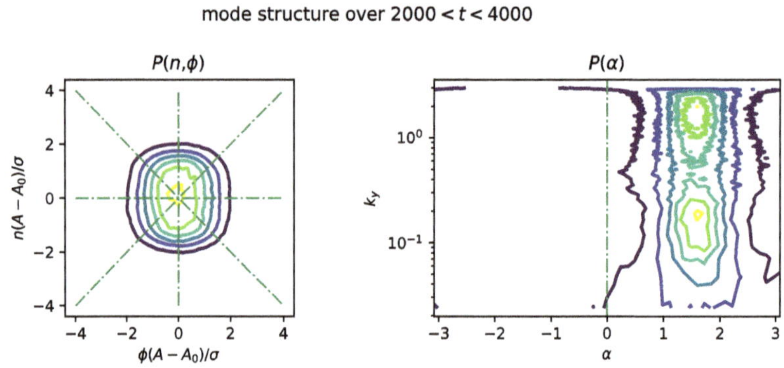


**Figure 13.20.** Cross coherence (left), and phase shift distributions (right) between $n$ and $\phi$, which one can compare with figure 12.15 from Chapter 12. The correlation between $n$ and $\phi$ is small and the phase shifts are centred on $\pi/2$ for maximum drive by the background gradient. These forms reflect the absence of an adiabatic response, by contrast to drift wave turbulence.

the clearest in showing how the RMHD model falsifies its basic assumption of $n \ll \phi$ *a posteriori*.

The cross coherence and phase shift distributions are shown in figure 13.20, which one can compare with the 2D case in figure 13.11, and to figure 12.15 from Chapter 12. Instead of lining up in response to the adiabatic coupling, the transported quantity $n$ has the statistics of a passive scalar whose gradient drives the flow. The two fields $n$ and $\phi$ are largely uncorrelated, as the near-circular contours of the left frame show. The phase shifts of all $k_y$-modes peak at $\pi/2$, which yields maximum drive efficiency, and the distributions are equally wide. The scatter is indicative of strong, randomised turbulence on top of the main structures which cause most of the transport. The dominant effect in the pressure equation is the ExB advection, which is the same randomising phenomenon leading to the direct cascade of free energy to high-$k_\perp$. These forms reflect the absence of an adiabatic response, and form the opposite extreme to what one would find in nearly adiabatic drift wave turbulence. They are very similar to the form shown by these diagnostics above for the two-dimensional case.

The amplitude probability distributions are shown in figure 13.21. These are unremarkable, except for small deficit rather than excess in the large-event tails, compared with the Gaussian reference. The vorticity also does not show significant tails, being closer to Gaussian than to the power-law passive scalar distribution. This can be compared with figure 12.16 of Chapter 12, which shows these tails to a larger degree.

The spatial form of the disturbances in shown in figure 13.22 for the latest snapshot. The larger and seemingly more pronounced eddies and sinuous form of the vorticity sheets reflect a character closer to hydrodynamic turbulence than that of the drift Alfvén cases, which are exemplified by the form of the case with the same parameters in figure 12.17 from Chapter 12.

The three-dimensional morphology is shown in figure 13.23. The twisting of identifiable structures with the field lines is clearer than in the drift Alfvén case

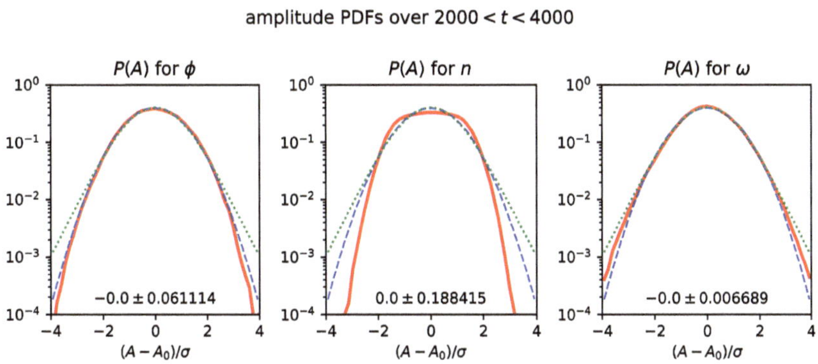


**Figure 13.21.** Amplitude probability distributions of $\phi$ (left), $n$ (centre), and $\varpi$ (right). The blue (dashed) and green (dotted) curves show Gaussian and passive scalar distributions for comparison. The absence of strong tails (excess over the Gaussian) at large amplitude is noteworthy, especially for $n$, where there is a tail deficit. Compare with figure 12.16 from Chapter 12.

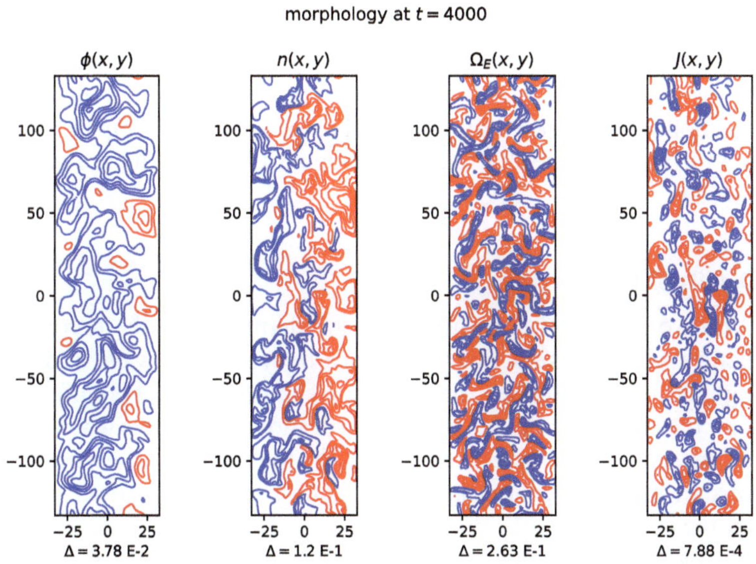


**Figure 13.22.** Spatial morphology at $t$ = 4000 of the state variables, potential, and current. Positive contours are in red; negative in blue. The structure is at larger scale and more coherent than the drift Alfvén cases (figure 12.17 from Chapter 12).

shown in figure 12.18 from Chapter 12. The form of the radially extended structures in the perturbed density field gave them the moniker 'streamer' in the literature. The streamers are visible because they are large and coherent enough to last for the time it takes Alfvén dynamics couples $\phi$ versus $J$ from plane to plane, and then $\phi$ drives $n$ through the background gradient while spatially aligned to the magnetic field. This is due to the weaker vorticity in this model compared with the drift Alfvén one.

In general, interchange and ballooning dynamics is dominated by the lower-$k_y$ range of the spectrum, and $\phi$ tends towards the steep spectrum of the extreme

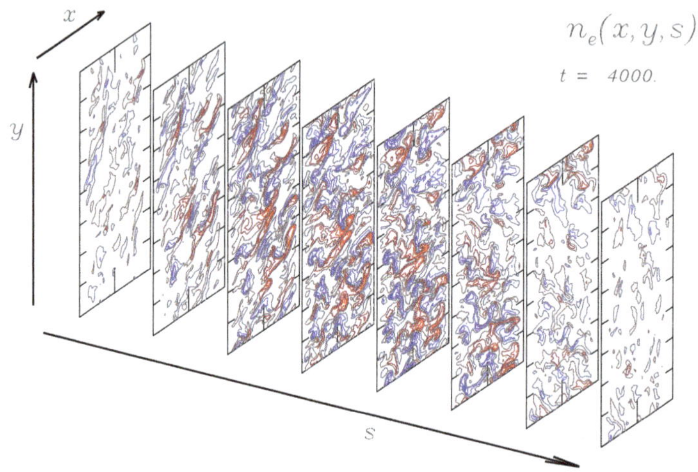


**Figure 13.23.** Three-dimensional structure of $n$ for the representative case. Of the 32 planes in $s$ in the computation, 8 are shown. In this case the radial 'streamer' structures are large enough to be visible as they are twisted by the shear of the magnetic field.

hydrodynamic limit. These diagnostics from the three-dimensional toroidal model of ballooning mode turbulence can be compared with the simpler 2D cases of interchange turbulence. The 2D and 3D mode structure results are comparable in most every way.

### 13.12.1 Details of the energetics

What will turn out to be the main interest in Chapter 14 is the ingredients of the mode structure which make a clear difference between drift wave and interchange (ballooning) turbulence. Besides the cross correlation and phase shift distributions already seen, the parallel envelope structure will show differences when the adiabatic response is present. The clearest contrast is in the energetics, especially the part which describes how free energy is shifted from $n$ to $\phi$ through curvature forcing rather than the current.

The isotropic spectra of the state variable energetics are shown in figure 13.24, which one can compare with figure 12.19 from Chapter 12, for the drift Alfvén case. The measurement technique is the same as given therein, with the addition of the curvature terms and the deletion of the adiabatic coupling channel for the thermal free energy. The cross-quantity energetics in detail are given by

$$-\phi\frac{\partial \varpi}{\partial t} = -\phi\big(-[\phi, \varpi] + \nabla_{\|}J - \mathcal{K}(n)\big), \tag{13.109}$$

$$n\frac{\partial n}{\partial t} = n\big(-[\phi, n] - \nabla_{\|}u + \omega_n n v_E^x + \mathcal{K}(\phi)\big), \tag{13.110}$$

for the ExB and thermal free energy pieces, respectively. The $\nabla_{\|}J$ term is absent in the density equation but now we have the two curvature terms which combine into a

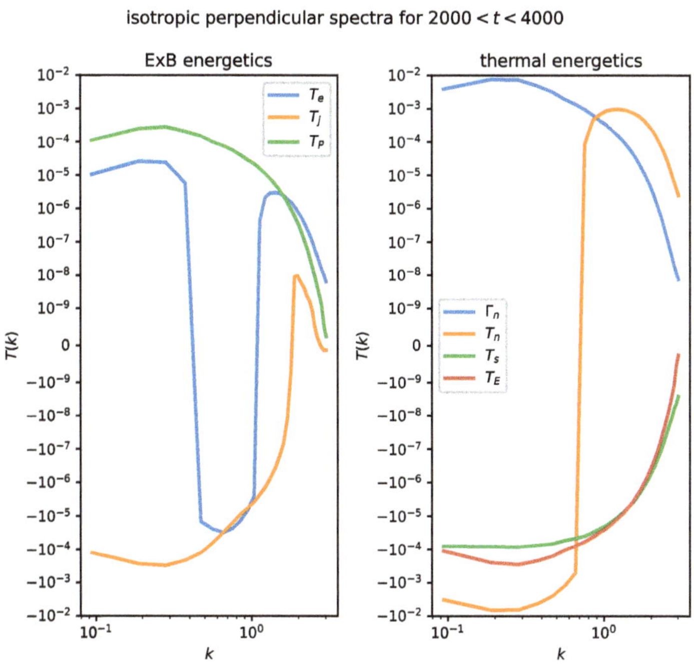


**Figure 13.24.** State variable energetics, for the ExB (left) and thermal free energy (right), for the toroidal RMHD case. Compare with figure 12.19 from Chapter 12, for the drift Alfvén case at the same parameters. The Alfvénic transfer ($T_j$) is out of the ExB energy and balances curvature forcing (through $T_P$) at all scales. In the thermal free energy the adiabatic transfer (through $n\nabla_{\parallel}J$) is absent, and the gradient drive ($\Gamma_n$) balances nonlinear ExB advective transfer ($T_n$) especially at long wavelength. These form the clearest contrast to the drift Alfvén case. The acoustic ($T_s$) and ExB compressional ($T_E$) channels are subdominant.

conservative coupling. The relevant curvature forcing term in the energetics is labelled ‘P’ in the ExB case and ‘E’ for ExB compression in the thermal free energy case. While in the drift Alfvén case the ExB energy had to come in from the current, it now comes in over the curvature and back out over the current in basic balance. Moreover, the ExB vorticity nonlinearity, in a controlling role in the drift Alfvén case, is now subdominant at essentially all wavelengths. That the curvature forcing has the same shape as the gradient drive in the thermal free energy follows from the same term $\partial n/\partial y$ times a geometric coefficient. In the thermal free energy the drive acts as a source and it is balanced by both the nonlinear ExB advection and the curvature in the scale range where most of the energy resides. Without adiabatic coupling, free energy can be passed from the thermal towards the ExB side solely over the curvature.

The isotropic spectra of the flux variable energetics are shown in figure 13.25, which one can compare with figure 12.20 from Chapter 12, for the drift Alfvén case. The labelling is as it was there. In the ballooning case the adiabatic transfer is absent, so that in the Ohmic energy the MHD Alfvén transfer is the only mechanism which can act as a source. It balances resistive dissipation at all scales. Maintaining the current at finite levels is why the Alfvénic transfer into the current now must be

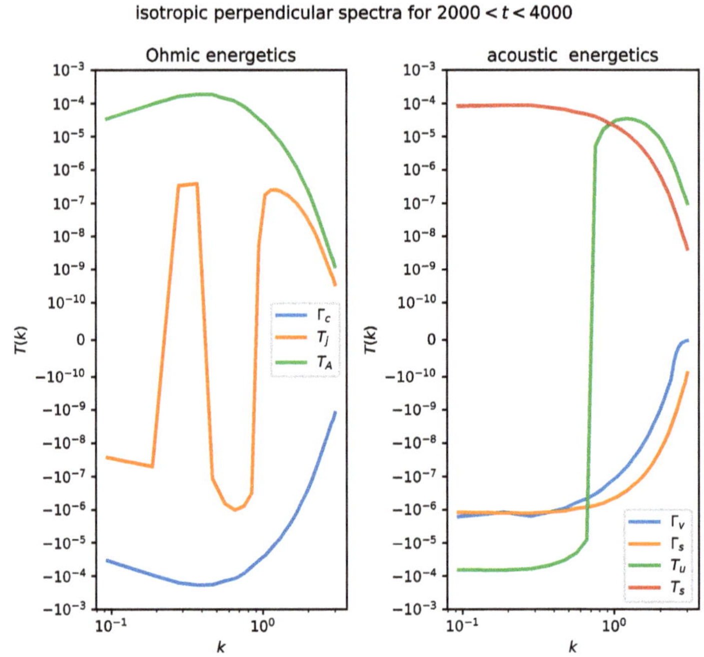


**Figure 13.25.** Flux variable energetics, for the electron (left) and ion channels (right), where positive (top) and negative (bottom). The electron channel includes the magnetic energy. In the RMHD case the adiabatic transfer (through $J\nabla_{\parallel}n$) is absent, and the Alfvénic transfer ($T_A$) incoming from the ExB energy balances resistive dissipation ($\Gamma_c$) at all scales. The latter is especially in contrast to the drift Alfvén case. Compare with figure 12.20 from Chapter 12.

positive everywhere. In the acoustic energy the incoming acoustic transfer, positive nearly everywhere, balances viscous dissipation. This is the energetic pattern one would more naïvely expect from magnetised fluid dynamics with a pressure gradient—basic drive through a forcing similar to convection, balanced by simple interchange coupling to maintain the flows, with the rest of the system in a subordinate, dissipative role, and with the polarisation current dynamics in a very subdominant role. It is in strong contrast to the drift wave pattern of circular energetics with bi-directional adiabatic coupling acting at both ends of the spectrum to balance the transfer through the two types of ExB nonlinearity. Control cases of this level of clarity are necessary for the diagnosis of physical mechanism in more complicated situations.

## 13.13 Electromagnetic and collisional effects

We now consider the range of values for $\hat{\beta}$ and $C$ which were done for the drift Alfvén cases of Chapter 12. The parameters controlling the electrons are

$$1 < \hat{\beta} < 50 \qquad \hat{\mu} = 2.72 \qquad 0.139 < C < 139, \tag{13.111}$$

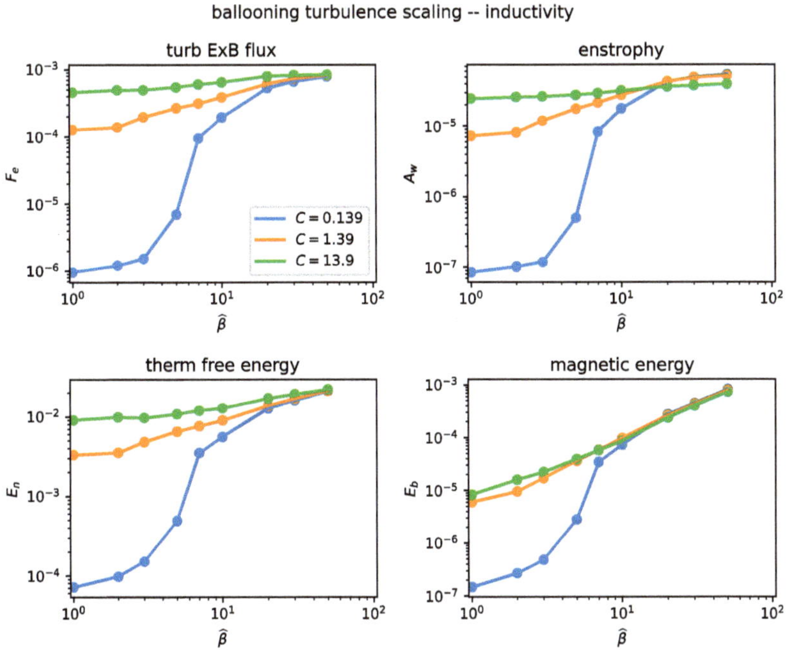


**Figure 13.26.** Scaling of the turbulent ExB transport flux ($F_e$), the enstrophy ($A_w$), the thermal free energy ($E_n$), and the Ohmic energy ($E_b$), all in volume average, against magnetic inductivity $\hat{\beta}$, for various $C = 0.139$, 1.39, and 13.9. For the ballooning cases the finite $\hat{\beta}$ is never stabilising. The linear magnetic flutter mechanism from the drift Alfvén cases ($[\phi, n_0]$ in the Ohm's law) is absent here. Only induction remains, and that is found to be destabilising. The run of parameters, even the abrupt rise between $5 < \hat{\beta} < 7$ for $C = 0.139$, does not follow the simple linear ballooning scaling (see text).

indicating the main parameter sweeps in the magnetic induction and the collisionality (these values arise from setting $\nu_e L_\perp / c_s = 0.1$ to 100).

The results of varying inductivity $\hat{\beta}$ are shown in figure 13.26, in which the measured quantities are the turbulent ExB transport flux across the field lines, $F_e = n v_E^x$, the enstrophy, $A_w = \varpi^2/2$, the thermal free energy, $E_n = n^2/2$, and the Ohmic energy (magnetic and parallel electron flow), $E_B = J(A + \hat{\mu} J)/2$, all in volume average. The benchmark level for these is $\delta_0^2 = 2.25 \times 10^{-4}$ for comparison. The higher collisionality cases all show robust turbulence. By contrast to the drift Alfvén cases, increasing inductivity always leads to stronger turbulence, which is understandable since the stabilising mechanism in the drift Alfvén case was magnetic flutter acting through $\nabla_\| n_0$, and that is absent in this model. It is interesting to note that for the lowest collisionality value, the turbulence jumps up to stronger values between $\hat{\beta} = 5$ and 7, which are values of $\alpha_M$ between 0.15 and 0.21 (since $\omega_B = 0.03$), and that is well below the nominal ideal MHD stability boundary. That boundary of $\alpha_M = 0.66$ (since $S = 1$) falls at $\hat{\beta} = 22$, and in the ballooning model that is indeed where the instability appears. However, with finite $C$ also present, the nominal threshold is seen not to play a strong role. That is the effect of all of (1) field line connection, (2) finite resistivity, and (3) nonlinearity. We see the importance of testing the equations and numerics within the simple ballooning model, in which it indeed produced the threshold where existing theoretical results said it should. When the adiabatic response is restored, the longest wavelengths we see here in the turbulent cases are subdued. The MHD ballooning threshold then returns, since the long wavelengths require it to escape the constraint of the adiabatic response.

The next test, shown in figure 13.27, takes the $\hat{\beta} = 5$ case and varies the resistivity (collisionality) over three orders of magnitude. This value of $\hat{\beta}$ is just below where the $C = 0.139$ cases showed the abrupt rise. So it is not surprising to see a jump appear for $C$ within this range, and indeed one is found between $0.139 < C < 0.278$, after which the turbulence rises weakly with increasing resistivity. Unlike $\hat{\beta}$, which understandably has its strongest effect on the magnetic energy, all four curves rise with increasing $C$ essentially together, suggesting no strong variation in character or mode structure with resistivity. The tendency to flatten off for extreme $C$ follows from the eventual suppression of parallel dynamics for high enough $C$, returning to an essentially 2D dynamics. One can see this from the form of Eq. (13.105), imagining an extra factor of $\gamma/C$ in front of the term on the left-hand side. But this does not enter unless $C$ takes extreme values which are not attained in realistic cases. The dynamics remains three-dimensional.

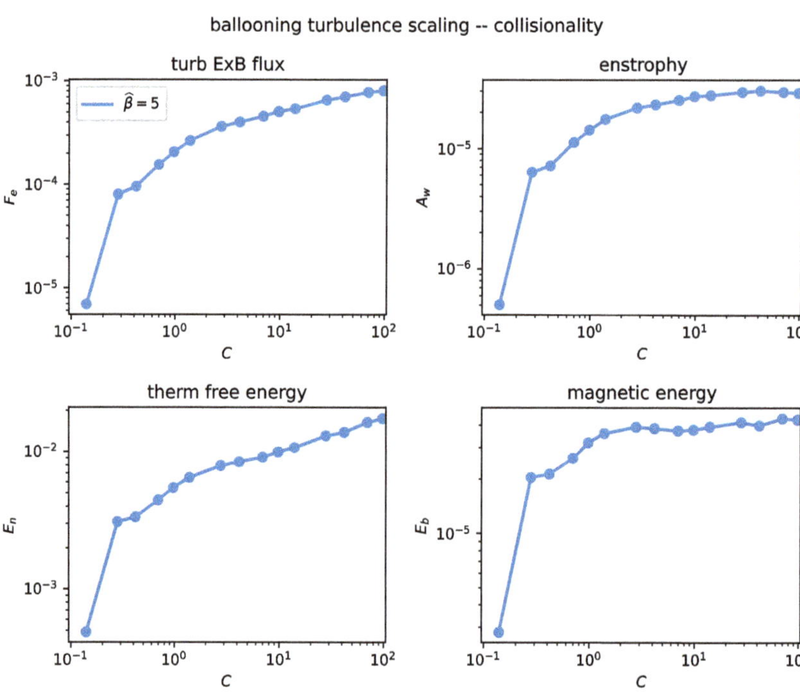


**Figure 13.27.** Scaling of the turbulent ExB transport flux ($F_e$), the enstrophy ($A_w$), the thermal free energy ($E_n$), and the Ohmic energy ($E_b$), all in volume average, against collisionality $C$, for $\hat{\beta} = 5$. As this value is just below the rise in the $C = 0.139$ cases with $\hat{\beta}$ in figure 13.26, one finds a corresponding rise for $\hat{\beta} = 5$ for $0.139 < C < 0.278$. The main regime of validity for this model is for these larger $C$ values. We will find this result changing again when drift wave and ballooning effects are merged in Chapter 14.

## 13.14 Summary

Interchange turbulence and its ballooning mode counterpart in three dimensions have a simpler physical basis and mode structure than drift wave turbulence. The quantity whose gradient is the free energy source develops disturbances via ExB advection in the gradient, and then nonlinear advection leads to the turbulence. The dynamics of the ExB flows arises from interchange forcing, which has the same physical role as Rayleigh–Taylor forcing on a neutral fluid, even though in this case the mechanism is a charge separation and does not constitute a potential energy source. The magnetic curvature which gives rise to the interchange forcing in fluid drift dynamics is a catalyst which allows free energy in the pressure disturbances to reach the electrostatic potential, and hence the ExB eddies. There is no strict cross coupling to consider, however, because the back reaction of the pressure on the ExB flows is smaller than the gradient drive level, by contrast to the situation with adiabatic coupling. Relatively simple analytical modelling is therefore possible, comparable to what has been done in astrophysical fluid dynamics.

Interchange flows are primarily streamers or radial plumes, rather than more or less isotropic eddies. In a computation they are sensitive to the aspect ratio, and it is important to leave enough room in the purely periodic coordinate so as to avoid a spurious conclusion involving secondary instabilities of these flows—if the aspect ratio is too low, the widest possible radial flow is Kelvin–Helmholtz unstable and therefore the radial flows do not develop properly. Interchange turbulence has a wider spectrum than drift wave turbulence, and in a situation where absolute transport levels are paramount one will require more resolution than was necessary here to investigate the basic physical properties.

The three-dimensional interchange variant becomes a ballooning mode in a torus, because the interchange forcing is positive only on the outboard midplane region of the configuration. The disturbances still tend to build flute modes, but they localise to the region of unfavourable curvature because of the magnetic shear. The associated turbulence produces a state with large amplitude in the pressure disturbances but low amplitude in the electrostatic potential, since the latter is now alone in being held down by the shear Alfvén dynamics. The ExB vorticity is fairly small, so that the linear mode structure largely remains. The result violates the basic assumption that pressure forces on the electrons along the magnetic field can be neglected, even for extreme collisionality. So we have to restore the adiabatic response and re-investigate the findings with both interchange and drift Alfvén mechanisms present. We turn to this in Chapter 14.

## Further Reading

The ideal interchange and ballooning instabilities are treated in the texts on ideal MHD listed in Chapter 6. MHD interchange models in astrophysics are treated in the text by E N Parker 1979 *Cosmical Magnetic Fields* (Oxford: Clarendon Press). For geophysical fluid dynamics see the text by J Pedlosky 1992 *Geophysical Fluid Dynamics* (Berlin: Springer).

Ideal MHD stability and the energy principle provide a topic with a vast literature. I was helped by the following texts: J P Freidberg 1987 *Ideal Magnetohydrodynamics* (New York: Plenum); W M Mannheimer and C N Lashmore-Davies 1989 *MHD and Microinstabilities in Confined Plasma* (Boca Raton, FL: CRC Press); R J Goldston and P H Rutherford 1995 *Introduction to Plasma Physics* (Boca Raton, FL: CRC Press); R B White 2001 *The Theory of Toroidally Confined Plasmas* (Singapore: World Scientific); and a lecture by White from Princeton Plasma Physics Laboratory, from 2012, which gives his derivation of the delta-W principle and the ballooning equation. These texts also cover the Grad–Shafranov equation for axisymmetric equilibria.

The resistive MHD paradigm for toroidally confined plasmas grows out of the reduced MHD model introduced by H R Strauss 'Dynamics of high beta tokamaks' *Phys. Plasmas* **20** (1977) 1354; and was advanced for fluid-like turbulence in toroidal magnetically confined plasmas by B A Carreras and P H Diamond *Phys. Fluids B* **1** (1989) 1011.

Balloning modes were identified in experiments by A M M Todd, M S Chance, J M Greene, R C Grimm, J L Johnson and J Manickam *Phys. Rev. Lett.* **38** (1977) 826. They were calculated for toroidal magnetic fields by B Coppi *Phys. Rev. Lett.* **39** (1977) 939; and by J W Connor, R J Hastie, J B Taylor 'Shear, periodicity, and plasma ballooning modes' *Phys. Rev. Lett.* **40** (1978) 396; and by R L Dewar and A H Glasser 'Ballooning mode spectrum in general toroidal systems' *Phys. Fluids* **26** (1983) 3038.

Resistive ballooning modes were introduced by H R Strauss 'Resistive ballooning modes *Phys. Fluids* **24** (1981) 2004.

**IOP** Publishing

Turbulence and Instabilities in Magnetised Plasmas, Volume 1
Fluid drift turbulence
**Bruce Scott**

# Chapter 14

# Toroidal drift Alfvén turbulence

## 14.1 Introduction

A toroidal magnetic field is the central concept for a confined plasma for reasons discussed in Chapter 13. In this context 'confined' means holdable by statistical equilibrium for arbitrary time, which precludes inertial confinement or systems with open ends through which the plasma flows. The only practical means for this is confinement due to a magnetic field. In this context, turbulence is a major theme because even if you solve the problem of equilibrium and confined orbits (i.e. no leakage), turbulence leads to transport (averaged, or net flows, even if unsteady) of any quantity across magnetic field lines, by extension closed magnetic flux surfaces.

We treated the axisymmetric toroidal magnetic field in Chapter 13 as a relevant setting for interchange dynamics due to magnetic curvature forcing. Herein we address the general dynamics of drift Alfvén turbulence in this geometry. We now have two different mechanisms for coupling the ExB turbulence to the gradient drive, adiabatic response or magnetic curvature, and they are in competition. Having established the mode structure of 2D interchange and 3D ballooning mode turbulence previously, and before that the mode structure of 2D drift wave and 3D drift Alfvén turbulence, we have those two settings as control cases to ascertain the dynamics that we find in toroidal geometry. We treat this competition at the level of the four-field models because the differences are so minimal. Delete the curvature terms and we have pure drift Alfvén turbulence. Delete the adiabatic coupling and we have pure ballooning mode turbulence. This use of control cases and testing corresponding models against each other is necessary to understand the physics of the turbulence in its nonlinear, statistically saturated context.

After doing this, we have another round of comparing linear to nonlinear mode structure. It is a commonly held opinion that linear stability theory is the guide to understanding turbulence, not only whether or not one will find it but also what the physics of the scaling is in relationship to what the linear dispersion relation indicated on the parameter scaling of the linear growth rate. We have already

doi:10.1088/978-0-7503-2504-2ch14

seen the drift wave nonlinear instability, the existence of which should have disarmed the picture of turbulence as linear instabilities in nonlinear interaction. In turbulence, most of the free energy is in stable modes, merely because the nonlinear three-wave interactions are incoherent and transfer free energy from different spectral regions into another one, in a sense spraying the energy indiscriminately into all available degrees of freedom. When the turbulent vorticity is greater than any underlying linear growth rate, then the dynamics is too fast and incoherent for the linear mode structure to impose itself. We have already seen that the time taken for a given linear mode to have its structure crystallise from a general initial state is comparable to the time that turbulence takes to saturate. Indeed it happens that the correlation time, scaling against the RMS value of the turbulent vorticity, is much shorter than either. In this context the linear instabilities lose their dominance and indeed in some cases even their relevance.

Turbulence in a toroidal magnetic field with finite shear will encounter interchange dynamics as well as drift wave dynamics. The diagnostic package we have built, most especially the energetic analysis, is necessary to distinguish among the possible mechanisms of drive and saturation in fully developed turbulence. The situation of saturated turbulence is more generally relevant than any initial transient or instability. We will find that the situation with instabilities can be quite different from the same situation with turbulence. That is the most important lesson of what we will address in this Chapter.

## 14.2 The toroidal drift Alfvén model

The drift Alfvén equations contain both the curvature terms and the adiabatic response. The equations are for four dependent variables: the vorticity $\varpi$, the current $J$, the density $n$, and the parallel velocity $u$. These are normalised as Chapters 11 and 12. The equations are the same as listed in those Chapters, up to the addition of the curvature terms. As such, they form a superset of the drift wave and drift Alfvén models as given in 3D Cartesian slab geometry. In particular, the field-aligned coordinates of the sheared slab are used, synonymous with the toroidal setting as given for 3D ballooning mode turbulence in Chapter 13. As such, they also form a superset of the reduced MHD ballooning mode model listed therein.

### 14.2.1 Equations

The equations are

$$\frac{\partial \varpi}{\partial t} + [\phi, \varpi] = B\nabla_{\parallel}\frac{J}{B} - \mathcal{K}(n), \tag{14.1}$$

$$\frac{\partial}{\partial t}(A + \hat{\mu}J) + \hat{\mu}[\phi, J] = \nabla_{\parallel}(n_0 + n - \phi) - C\,J, \tag{14.2}$$

$$\frac{\partial n}{\partial t} + [\phi, n_0 + n] = B\nabla_{\parallel}\frac{J - u}{B} + \mathcal{K}(\phi - n), \tag{14.3}$$

$$\hat{\varepsilon}\frac{\partial u}{\partial t} + \hat{\varepsilon}[\phi, u] = -\nabla_{\parallel}(n_0 + n) - B\nabla_{\parallel}\frac{P}{B}, \tag{14.4}$$

for the dependent variable evolution equations, and

$$\varpi = \nabla_{\perp}^2\phi \quad J = -\hat{\beta}^{-1}\nabla_{\perp}^2 A \quad P = -X\nabla_{\parallel}u_{\parallel}, \tag{14.5}$$

for the auxiliaries and field equations, in which $P$ is the viscous correction to the pressure and $X$ is the parallel viscous diffusivity.

The one addition we make here is to make the distinction between the parallel divergence and the parallel gradient, which one finds in the placement of the factors of $B$. If $B$ has $s$-dependence, this procedure ensures that the unperturbed field $\mathbf{B}$ is treated as divergence free, since

$$\mathbf{B} \cdot \nabla f = \nabla \cdot f\mathbf{B} \qquad \mathbf{B} \cdot \nabla\frac{f}{B} = \nabla \cdot f\mathbf{b} \tag{14.6}$$

with arbitrary scalar $f$, and since the divergence operation contains the coordinate Jacobian or volume element, and due to field aligning $B^x = B^y = 0$ while $B^s$ remains finite, it follows that

$$\mathbf{B} \cdot \nabla = B^s\frac{\partial}{\partial s} \quad \nabla \cdot \mathbf{B} = \frac{1}{\sqrt{g}}\frac{\partial}{\partial s}\sqrt{g}B^s, \tag{14.7}$$

so that if $B^s$ is given $s$-dependence, the coordinates have to be modified accordingly. The nonlinear magnetic divergence is therefore

$$B\nabla_{\parallel} = \mathbf{B} \cdot \nabla + \nabla(-A_{\parallel}R) \times \nabla\varphi \cdot \nabla, \tag{14.8}$$

in physical units, with $R$ the general locational major radius, in which we simplify the nonlinear piece such that

$$B\nabla_{\parallel} = B^s\frac{\partial}{\partial s} + [\,-AR_0, \quad], \tag{14.9}$$

in keeping with a local fluxtube model. The units are normalised around values at $r = r_0$, where $r_0$ is the toroidal minor radius of the reference surface, about which the model is localised. The bracket in physical units with $x$ and $y$ in units of distance is

$$[a, b] = \frac{R_0}{B_0}\nabla\varphi \cdot \nabla a \times \nabla b = \frac{R_0/B_0}{\sqrt{g}}\left(\frac{\partial a}{\partial x}\frac{\partial b}{\partial y} - \frac{\partial b}{\partial x}\frac{\partial a}{\partial y}\right), \tag{14.10}$$

with arbitrary scalars $a$ and $b$, where $R_0$ is the toroidal major radius of the magnetic axis, assumed to be the centre of every flux surface, and $B_0R_0 = I_0$, all constant, gives the toroidal magnetic field covariant component. The bracket becomes

$$[a, b] = \frac{1}{\sqrt{g}}\left(\frac{\partial a}{\partial x}\frac{\partial b}{\partial y} - \frac{\partial b}{\partial x}\frac{\partial a}{\partial y}\right), \tag{14.11}$$

in normalised units. In this model the geometry is kept simple, and the coordinate metric, volume element are

$$g^{xx} = 1 \quad g^{xy} = S(s - s_k) \quad g^{yy} = 1 + S^2(s - s_k)^2 \quad \sqrt{g} = 1, \tag{14.12}$$

and the magnetic field strength and nonvanishing component are

$$B = 1 \quad B^s = 1 \quad b^s = 1, \tag{14.13}$$

in normalised units. The perpendicular Laplacian is therefore the 2D Cartesian one in the $xy$ plane. In the equations, the geometry is concentric cylindrical except for the existence of the curvature operator, given by

$$\mathcal{K} = \omega_B\left(\sin s\frac{\partial}{\partial x} + (\cos s + S(s - s_k)\sin s)\frac{\partial}{\partial y}\right), \tag{14.14}$$

with $\omega_B = 2L_\perp/R_0$ the curvature forcing parameter with its nominal value, as in Chapter 13. The perpendicular terms (all except $\partial/\partial s$) are evaluated at $s = s_k$ on each drift ($xy$) plane, and the parallel derivatives $\partial/\partial s$ are taken across drift planes each with its own coordinate system, with the method detailed in Chapter 12.

The relationship to the previous models, as stated in the Introduction, are such that the pure 3D drift Alfvén model is recovered by setting $\omega_B = 0$, while the pure 3D ballooning mode model is recovered by deleting $\nabla_\parallel n$ in Eq. (14.2) for the current and deleting $B\nabla_\parallel(J/B)$ and $\mathcal{K}(n)$ in Eq. (14.3) for the density.

### 14.2.2 Energetics

Since we allow in general for $s$-dependence in $\sqrt{g}$ and $B^s$ we detail how the energetics works in that case. Energetic consistency is maintained so long as $[,\ ]$ and $B\nabla_\parallel$ and $\mathcal{K}$ are all pure divergences. Multiply Eqs (14.1)–(14.4) by $-\phi$ and $J$ and $n$ and $u$, respectively, to obtain

$$\begin{aligned}\frac{1}{2}\frac{\partial}{\partial t}|\nabla_\perp\phi|^2 - \nabla\cdot\left(\phi\frac{\partial}{\partial t}\nabla_\perp\phi\right) + [\phi, -\phi\varpi] + B\nabla_\parallel\frac{\phi J}{B} - \mathcal{K}(\phi n)\\ = J\nabla_\parallel\phi - n\mathcal{K}(\phi),\end{aligned} \tag{14.15}$$

$$\begin{aligned}\frac{1}{2}\frac{\partial}{\partial t}\left(\hat{\beta}^{-1}|\nabla_\perp A|^2 + \hat{\mu}J^2\right) - \nabla\cdot\left(\hat{\beta}^{-1}\frac{\partial A}{\partial t}\nabla_\perp A\right)\\ + \frac{1}{2}\hat{\mu}[\phi, J^2] = J\nabla_\parallel(n_0 + n - \phi) - C\,J^2,\end{aligned} \tag{14.16}$$

$$\begin{aligned}\frac{1}{2}\frac{\partial}{\partial t}n^2 + \frac{1}{2}[\phi, n^2] + B\nabla_\parallel\frac{n(u - J)}{B} + \frac{1}{2}\mathcal{K}(n^2)\\ = (u - J)\nabla_\parallel n + n\mathcal{K}(\phi),\end{aligned} \tag{14.17}$$

$$\frac{1}{2}\frac{\partial}{\partial t}\hat{\epsilon}u^2 + \frac{1}{2}\hat{\epsilon}[\phi, u^2] + B\nabla_{\|}\frac{uP}{B} \\ = -u\nabla_{\|}(n_0 + n) - X(\nabla_{\|}u)^2. \tag{14.18}$$

This is mostly a set of free energy transport fluxes and a conservative network of transfer channels, as one finds by doing the volume integration

$$\frac{\partial}{\partial t}\mathscr{V}^{-1}\int d\mathscr{V}\ \frac{1}{2}\Big(|\nabla_{\perp}\phi|^2 + \hat{\beta}^{-1}\,|\nabla_{\perp}A|^2 + \hat{\mu}n^2 + \hat{\mu}J^2 + \hat{\epsilon}u^2\Big) \\ = \mathscr{V}^{-1}\int d\mathscr{V}\ \big(n[n_0, \phi] + (J - u)[n_0, A] - CJ^2 - X(\nabla_{\|}u)^2\big), \tag{14.19}$$

which leaves the ExB gradient and magnetic flutter gradient drive as sources, and the resistive and viscous dissipation as sinks. One sees that magnetic curvature does not represent drive but a conservative transfer directly between ExB energy ($\phi$) to/from thermal free energy ($n$), for $p\mathscr{K}(\phi)$, the ExB compression. The adiabatic compression $J\nabla_{\|}n$ transfers between thermal free energy to/from magnetic energy and electron kinetic energy ($A$, $J$). The Alfvén compression $J\nabla_{\|}\phi$ transfers between magnetic energy and electron kinetic energy and to/from ExB energy. Finally, the acoustic compression $u\nabla_{\|}n$ transfers between thermal free energy to/from ion kinetic energy ($u$). The diamagnetic compression $n\mathscr{K}(n)$ transfers energy within $n$ among degrees of freedom. The nonlinearities all act as previously in the pure drift Alfvén model. To the sinks is added subgrid dissipation $D$ on $\varpi$, $\hat{\mu}J$, $n$, and $\hat{\epsilon}u$, respectively. Computations are checked by measuring the error given by the difference between the total nonlinear growth rate $\Gamma_T = (1/2E_T)\partial E_T/\partial t$ and $\Gamma_+ - \Gamma_-$, where $E_T$ is the total free energy and $\Gamma_+$ and $\Gamma_-$ are the collective sources and sinks, respectively, as before.

## 14.3 Toroidal drift Alfvén turbulence

The realm of relevance for this model is the range of magnetised plasma situations in which either or all of the parameters controlling the adiabatic response, namely $\hat{\beta}$ and $\hat{\mu}$ and $C$, are larger than unity and the drift parameter $\delta_0 = \rho_s/L_{\perp}$ is in the range to make a moderate domain size reasonable. The effective size $L_{\perp}/\rho_s$ is no larger than $10^3$ and usually much less. For example, in an edge region of a confined plasma the scale length decreases towards the outer boundary much as in a stellar atmosphere, so that $L_{\perp}/\delta_0$ drops below 100 (in a linear decline of the temperature, $L_{\perp}$ falls with $T_e$ while $\rho_s \propto T_e^{1/2}$). In the core regions of moderate- to high-performance tokamaks (i.e. temperatures above 1 keV and densities above $10^{19}$ m$^{-3}$), the first requirement is not met, since $\hat{\beta}$ and $\hat{\mu}$ and $C$ are all small, mostly due to the moderate $R_0/L_{\perp} \sim 10$, but we still have $L_{\perp}/\rho_s < 200$. This is the realm of temperature gradient driven turbulence which is left to Volume Two. In the edge regions of these tokamaks, on the other hand, all of these conditions are satisfied. Indeed, $\hat{\mu} > 1$ can be taken as a definition of the edge region while still on closed magnetic flux surfaces. For practically any case we also have $\hat{\beta} > 1$, such that the long wavelengths are always electromagnetic, especially in a nonlinear setting where due to the statistical nature of the mode structure simple cancellations as seen for the linear drift Alfvén case do not enter. For high temperature edge cases $C$ is still usually larger than unity, since a smaller normalised $\nu_e L_\perp / c_s$ often combines with a larger $\hat{\mu}$ such that $C$ is still large enough to be what limits the parallel spectrum of the current. We will investigate the validity of the reduced MHD model in this parameter regime, which is the one that is expected to be functional in the first place. We have already seen it fail its basic assumption to neglect pressure forces in the electron forces along the magnetic field. Here, we investigate the situation while keeping both basic mode types in the dynamics.

The first task is to re-run all the 3D ballooning mode cases looked at in Chapter 13 with the toroidal drift Alfvén model. We consider the latter in its own right first. Then we do the explicit comparisons, especially energetics and mode structure. Finally, we take representative cases and start them deeply linear enough so as to compare linear and nonlinear energetics and mode structure.

As before, we start with the same initial state as in Eq. (12.59) of Chapter 12, with the initial RMS amplitude of $A_0 = 1$ to get a nonlinear start. The parameters controlling the electrons are

$$1 < \hat{\beta} < 50 \quad \hat{\mu} = 2.72 \quad 0.139 < C < 13.9, \tag{14.20}$$

and we will also consider lowering $\hat{\mu}$ to 0.1 to ascertain its effect.

The scale ratio and domain size parameters are

$$\hat{\varepsilon} = 10\,000 \quad \delta_0 = 0.015 \quad L_y/L_x = 4, \tag{14.21}$$

and we also set $X = 100$. The nominal grid is $64 \times 256 \times 32$ in $xys$, with perpendicular grid step sizes comparable to $\rho_s$ (i.e. $\delta_0^{-1} = 67$), and the time step is $\tau = 0.01$. The node count in $s$ is raised and the time step dropped if any of the cases show a need for it (the energetic error and Courant limit are the indicators). Dirichlet boundaries are used in $x$, set up as in Eq. (12.58), with the same coefficients.

## 14.4 The energetics of toroidal turbulence

In evaluating the energetic processes in the turbulence we have the same drive and sink terms as in Eqs (9.62)–(9.65), with Eq. (11.36) for the subgrid dissipation. These are the same terms as in the 3D slab cases, because the curvature terms collectively conserve energy as an ideal compression process. The only difference to the slab cases is the presence of $\mathcal{K}$ with $\omega_B$ having a nonzero value.

The total drive $\Gamma_+ = \Gamma_n + \Gamma_v$ is collectively positive, despite the indeterminate sign, since otherwise we cannot have driven and dissipative turbulence in statistical equilibration. This gradient drive is split as before into the ExB advection and magnetic flutter channels,

$$\Gamma_n = \frac{1}{2E_T}\mathcal{V}^{-1}\int d\mathcal{V}\ \omega_n\, n v_E^x \quad \Gamma_v = \frac{1}{2E_T}\mathcal{V}^{-1}\int d\mathcal{V}\ \omega_n (u - J) b^x, \tag{14.22}$$

where $E_T$ is the total free energy and $v_E^x = [\phi, \quad]$ and $b^x = [-A, \quad]$ are the radial components of the ExB velocity and the magnetic direction, as before. Of the magnetic flutter there are two instances, one from the Ohm's law through $J$, and one through the sound waves through $u$, together giving the electron parallel velocity flowing down the gradient. The one due to $J$ integrates away, due to the dependence of both $J$ and $b^x$ upon $A$.

The sink effects are the same as they were in the drift wave and ballooning mode models,

$$\Gamma_c = \frac{1}{2E_T}\mathcal{V}^{-1}\int d\mathcal{V}\ C\,J^2 \qquad \Gamma_s = \frac{1}{2E_T}\mathcal{V}^{-1}\int d\mathcal{V}\ X(\nabla_\| u)^2, \tag{14.23}$$

for resistive and viscous dissipation, respectively. To this is added the subgrid dissipation,

$$\Gamma_w = \frac{1}{2E_T}\mathcal{V}^{-1}\int d\mathcal{V}\ (-\phi\ D(\varpi) + n\ D(n) + \hat{\mu}J\ D(J) + \hat{\epsilon}u\ D(u)), \tag{14.24}$$

with $D$ given as in Eq. (12.58) from Chapter 12. The total sink is then given by $\Gamma_- = \Gamma_c + \Gamma_s + \Gamma_w$. The total growth rate and error are given by

$$\Gamma_T = \frac{1}{2E_T}\frac{\partial E_T}{\partial t} \qquad \Gamma_E = \Gamma_T - (\Gamma_+ - \Gamma_-). \tag{14.25}$$

The Γ's are normalised as rates which can be compared with the linear growth rate of the underlying instability (should there be one). The error is used to check the integrity of the numerics.

Then we also have the energetic transfer processes within the conservative part of the system. These are the parallel dynamics and the curvature terms, as well as the nonlinearities. We have seen the relevance of the $\phi \leftrightarrow J \leftrightarrow n$ set through the parallel dynamics before, not least the direction of it, namely, from $\phi$ to $n$ or vice versa. The same enters for the curvature terms, which go $\phi \leftrightarrow n$ directly through the ExB compression. Whether one or both of these are functional is now to be tested within the self-consistent turbulence. The Alfvén transfer between $\phi$ and $J$ is given by

$$T_A = \frac{1}{2E_T}\mathcal{V}^{-1}\int d\mathcal{V}\ J\nabla_\|\phi, \tag{14.26}$$

normalised as a rate, signed positive for transfer from the magnetic/parallel (both $JA$ and $J^2$ are involved) to the ExB energy. The adiabatic transfer between $n$ and $J$ is given by

$$T_a = \frac{1}{2E_T}\mathcal{V}^{-1}\int d\mathcal{V}\ J\nabla_\| n, \tag{14.27}$$

signed the same way, so if $T_A$ and $T_a$ are both positive, the transfer proceeds preferentially from the thermal free energy to the ExB flow energy.

The transfer over the curvature coupling is more direct, passing directly between state variables through the ExB compression

$$T_E = \frac{1}{2E_T}\mathcal{V}^{-1}\int d\mathcal{V}\ n\mathcal{K}(\phi), \tag{14.28}$$

which in this form gives the transfer from the ExB energy to the thermal free energy resident in the density component (later on, we will separate this among the densities and temperatures). A linear instability driven by curvature would have this negative, indicating energy passing through compression from $n$ to $\phi$ having been driven in $n$ by the background gradient.

For any linear instability, because the only drive is in the thermal free energy, one of these transfer channels must show energy passing into the ExB energy, indicating the instability mechanism. As we have seen, it is interesting when the nonlinear dynamics, transferring from short to long wavelength in the ExB energy and opposite in the thermal free energy, is able to upset the balances between linear effects, showing that the energetics can only be nonlinear. But if the curvature forcing dominates, then the transfer should be into the ExB side through the curvature terms, and therefore perhaps back out through the parallel dynamics or simply to short wavelength through the nonlinearities. This is how it went in the ballooning mode cases of Chapter 13. Our task is to document which way these go and to know the statistics as well as possible.

## 14.5 The mode structure of toroidal turbulence

The distribution among the sources and sinks and the energy transfer channels is what is most relevant to the physics of what determines the mode structure. We measure these effects in the same way as done for the pure drift wave and ballooning mode turbulence cases. The case with $\hat{\beta} = 5$ and $C = 13.9$ is referred to as nominal and is showcased. The setup and grid is the same as before, with the only difference being the presence of both the curvature forcing and the two-fluid processes—adiabatic response and diamagnetic compression, the latter being another curvature effect which will have a larger role when we consider flows later on. As before we start nonlinear with an RMS amplitude of $A_0 = 1.0$ and run to $t = 4000$, measuring mode structure and other averaged quantities over the $2000 < t < 4000$ range. We will consider the transition from linear instability to turbulence again afterward. In the following all mention of other cases refer to the drift wave and ballooning mode turbulence cases of Chapters 12 and 13 for these same parameters.

With the nonlinear start, the turbulence saturates into statistical equilibrium as rapidly as in the other cases which produce robust turbulence. The time traces of amplitude and energetics are shown in figure 14.1. These are unremarkable compared with the other cases. The drive $\Gamma_n$ especially is similar for all three cases. However, the resistive dissipation, relatively weak in the ballooning mode case, is robust here, of similar magnitude to the subgrid dissipation. The error $\Gamma_E$ is measured at $(9.21 \pm 1.75) \times 10^{-6}$, versus $\Gamma_n = (2.32 \pm 0.20) \times 10^{-2}$ for the drive. The dotted curve dropping to and then hovering near zero is $\Gamma_T$, measured at $(0.04 \pm 1.49) \times 10^{-3}$, indicating effective saturation.

The isotropic perpendicular spectra are shown in figure 14.2, for amplitudes (top left), energetics (bottom left), and the four ExB nonlinear transfer channels (right, top where positive; bottom, negative). The spectra are weighted towards larger scale, as in the ballooning mode cases, while the ExB vorticity is very robust and resident

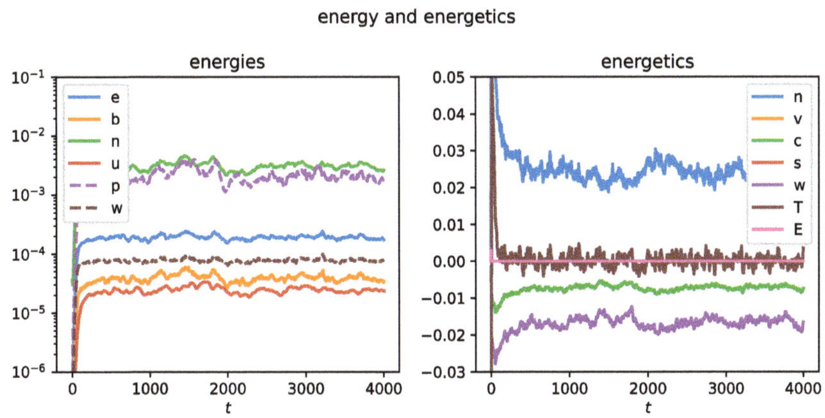


**Figure 14.1.** Time traces of (left) energy pieces and amplitudes of $\phi$ and $\varpi$, and (right) energetic drive and dissipation, nonlinear growth rate, and error. The amplitudes of $n$ and $\phi$, measured through $E_n$ ('n') and $A_p$ ('p'), respectively, are now comparable. The enstrophy ($A_w$, 'w') is still smaller than the ExB ($E_e$, 'e') or thermal free energy $E_n$, but is now larger than the Ohmic energy ($E_b$, 'b') and is a larger fraction of $E_n$. The total drive rate is about the same, with the gradient drive ($\Gamma_n$), but now the resistive dissipation ($\Gamma_c$) is a larger fraction of the balance. Compare with the ballooning mode and slab drift Alfvén turbulence cases at the same parameters, given in figure 13.16 from Chapter 13.

in the region around $k_\perp\rho_s = 1$, as in the drift wave cases. The nonlinear transfer spectra are more as in the drift wave cases, due to the robust ExB energy content and attendant vorticity nonlinearity. A short description of this is that the scales are more like the ballooning case but the mode structure is more like the drift wave case.

The parallel wavenumber spectra are shown in figure 14.3, for amplitudes (top left), energetics (bottom left), and the four ExB nonlinear transfer channels (right, top where positive; bottom, negative). The standout feature is the much wider vorticity spectrum for this and the drift wave cases, where for the ballooning case it does not differ in shape from the density spectrum. The same is true for the nonlinear ExB vorticity transfer into the flute mode ($k_\parallel = 0$) region. Both features are therefore the result of the adiabatic coupling. The inverse transfer in perpendicular scale for this nonlinearity is common in signature to all three models, so the nonlinear transfer into the flute mode in $k_\parallel$-space is not a simple consequence of the transfer direction in $k_\perp$-space. It is noteworthy in both perpendicular and parallel spectra that the vorticity and resistive dissipation are more like the slab case, while the drive is more towards larger scale as in the ballooning case.

The parallel envelopes are shown in figure 14.4, for the amplitudes (left) and the transport flux (right). The curves are ballooned (localised towards $s = 0$) to a finite extent, but more moderately than in the ballooning mode case. It is significant that in this model the amplitudes of $n$ and $\phi$ are comparable, with both larger than that of the nonadiabatic response, $h = n - \phi$, even though $h$ remains robust. Due to the adiabatic coupling, the Alfvén dynamics constrains $h$ and $J$ against each other, not just $\phi$ and $J$, since the pressure forces on the electrons are not neglected. The turbulent ExB transport flux is ballooned, but only by about two to one, not one to zero as a simplified ballooning model might suggest. The actual ballooning mode case is also

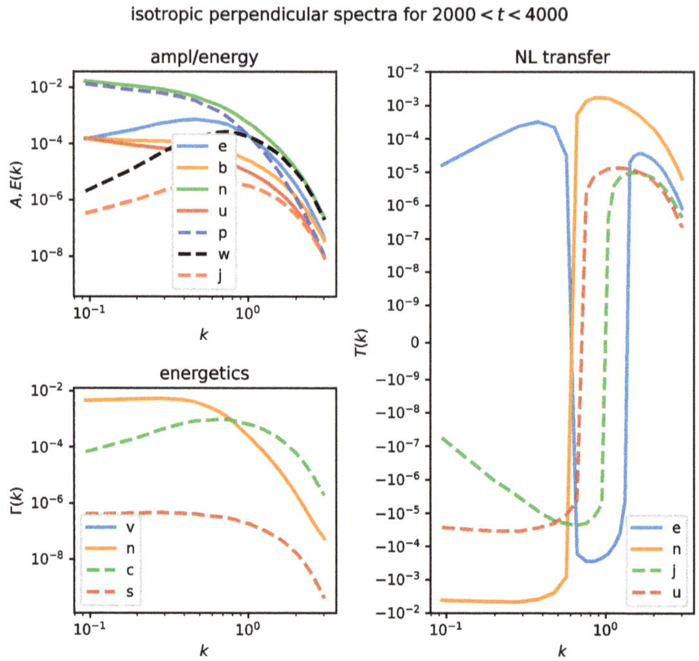


**Figure 14.2.** Isotropic perpendicular spectra for the nominal case during $2000 < t < 4000$, in the saturated phase. See the contrast against figure 13.17 from Chapter 13. The spectra are still weighted towards larger scale, with the vorticity (top left, 'w') still near $k_\perp \rho_s \sim 1$ and very robust. The nonlinear transfer (right) has the same directionality but the ExB vorticity nonlinearity (blue) is much larger, more like the pure drift wave case from figure 12.12 from the Chapter 12. The scales are more like the ballooning case but the mode structure is more like the drift wave case.

not so extreme. It is noteworthy that if only the density and transport flux spectra and parallel envelopes were available (i.e. no information on $\phi$ directly), one would have difficulty distinguishing this case from the ballooning mode one.

The cross coherence and phase shift distributions are shown in figure 14.5, which one can compare with figures 12.15 and 13.20 from Chapters 12 and 13. We find that the mode structure indicated by these diagnostics reverts to something close to the result in the drift wave case in the slab geometry without curvature forcing rather than the ballooning case. The phase shifts are positive, as they must be to reflect a positive gradient drive, but the width of the distributions is narrower, the peak values (mean phase shifts) are close to lining up across the $k_y$-spectrum, and most of the activity is reasonably close to zero. This indicates strong adiabatic coupling. In addition the cross coherence indicates that transported quantity $n$ is visibly correlated to the transporting flow potential $\phi$, which is opposite to the passive scalar or hydrodynamic character.

The amplitude probability distributions are shown in figure 14.6, which can be compared with figure 12.16 for the drift wave case and figure 13.21 for the ballooning case. The main observation is that the more pronounced tails in the vorticity distribution have returned, showing the strength of its nonlinearity (polarisation current, ExB vorticity self advection).

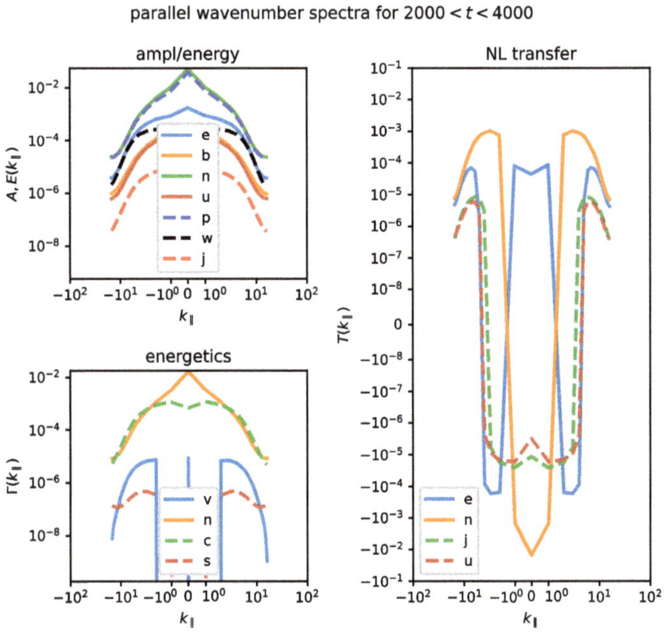


**Figure 14.3.** Parallel wavenumber spectra spectra for the nominal case during $2000 < t < 4000$, in the saturated phase. Compare with figure 13.18 from Chapter 13, for the ballooning mode case and with figure 12.13 from the Chapter 12, for the drift wave case. The spectra are wider again (left side) and the flute mode component is again driven by the ExB vorticity nonlinearity (left side, 'e'). The vorticity spectrum is again the widest of any of the dependent variables.

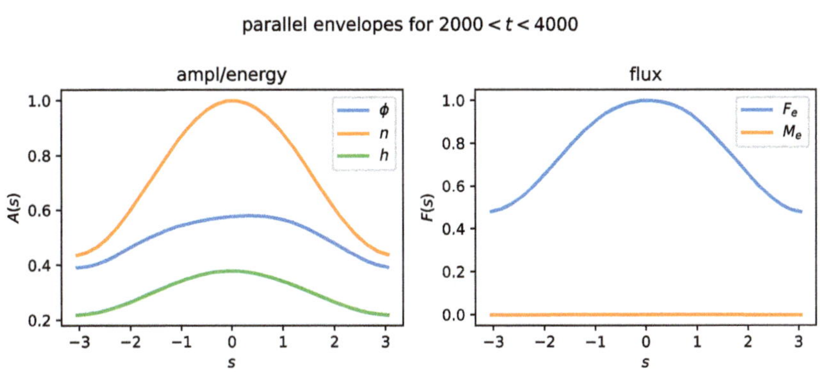


**Figure 14.4.** Parallel envelopes of the squared amplitudes (left) and turbulent ExB transport flux (right), which are averages over $x$ and $y$ and time $2000 < t < 4000$ of each quantity. Compare with figure 13.19 from Chapter 13. The curves are more moderately ballooned, and the amplitudes of the density (left, 'n') and the potential ('p'), are not so disparate. The nonadiabatic response $h = n - \phi$ is robust but smaller than either of $\phi$ or $n$. This is a result of the Alfvén dynamics constraining $h$ rather than $\phi$ alone. That the transport flux is ballooned is not by itself a revealing indicator.

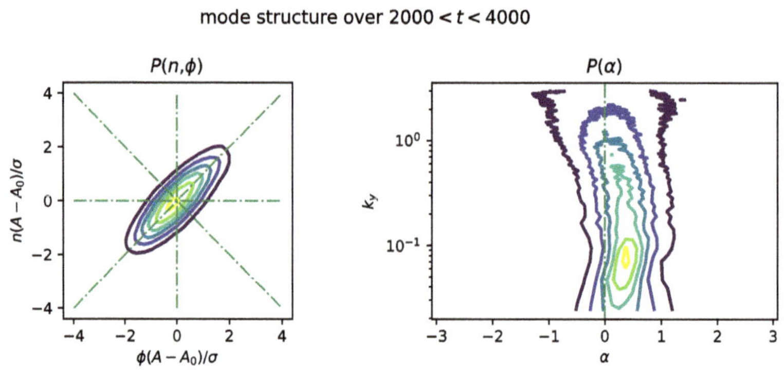


**Figure 14.5.** Cross coherence (left), and phase shift distributions (right) between $n$ and $\phi$, which one can compare with figure 13.20 from Chapter 13, for the ballooning mode case and figure 12.15 from Chapter 12, for the drift wave case. Both diagnostics show drift wave mode structure, with significant correlation between $n$ and $\phi$ and narrower phase shift distributions centred not on $\pi/2$ but much closer to zero, despite the robustness of the dynamics and $h = n - \phi$, the nonadiabatic response.

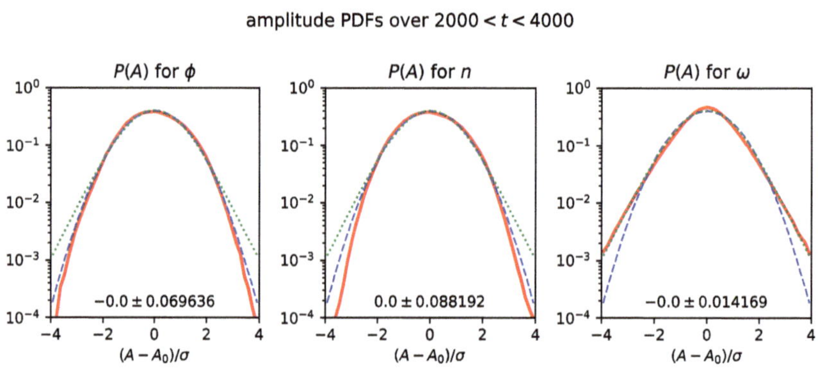


**Figure 14.6.** Amplitude probability distributions of $\phi$ (left), $n$ (centre), and $\varpi$ (right). The blue (dashed) and green (dotted) curves show Gaussian and passive scalar distributions for comparison. The vorticity is returned to a distribution more of the passive scalar type, with stronger large-event tails than the Gaussian. Compare with figure 13.21 from Chapter 13, for the ballooning mode case, and figure 12.16 from Chapter 12, for the drift wave case.

Taken together with the parallel envelopes, these diagnostics indicate that what we have is better described as drift wave turbulence forced by the curvature effects rather than ballooning turbulence with additional diamagnetic frequency effects. We will see more of this in the energetics and in the transition to nonlinearity, below.

The spatial morphology in the central drift plane ($s = 0$) is shown in figure 14.7, at $t = 4000$ for this case. One can compare with figure 13.22 from Chapter 13, for the ballooning mode case and figure 12.17 from Chapter 12, for the drift wave case. The structure is at intermediate scale and coherence between the two control cases, with both large and small components visible. Noteworthy is the similarity between $\phi$ and $n$, especially at large scales, and the still-visible sheet structures in the vorticity. This is the ExB vorticity, $\Omega_E = \delta_0^{-1}\varpi$ in normalised units, with an RMS amplitude greater

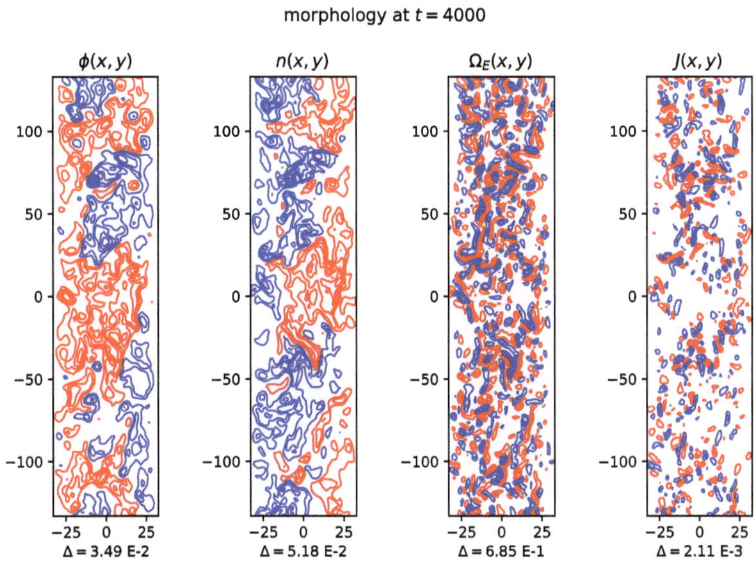


**Figure 14.7.** Spatial morphology at $t = 4000$ of the state variables, potential, and current. Positive contours are in red; negative in blue. Compare with figure 13.22 from Chapter 13, for the ballooning mode case, and figure 12.17 from Chapter 12, for the drift wave case. The structure is at intermediate scale and coherence between the two control cases, with both large and small components visible. Note the similarity between $\phi$ and $n$, especially at large scales, and the still-visible sheet structures and large amplitude in the ExB vorticity.

than unity, much larger than in the ballooning mode case. The adiabatic response is necessary to get $\phi$ and $n$ to track each other like this.

The three-dimensional morphology of the density fluctuations is shown in a snapshot at $t = 4000$ in figure 14.8 for this case. Now that a larger-scale component is present, one can identify individual structures which twist with the shear of the magnetic field lines. These last two figures illustrate the basic look and feel of the manifestation of drift wave mode structure in the morphology, but by themselves they do not discern the physical processes, for which we need the other diagnostics, and especially the energetics.

### 14.5.1 Details of the energetics

The main interest is now how the turbulence maintains and processes itself energetically. The details on how free energy is shifted from $n$ to $\phi$ and how the back reaction works tell us more than any of the morphology about how these processes work and which ones are in a controlling role in the fully developed turbulence. We now have the superset of the drift wave and ballooning physical mechanisms, following from the equations themselves. Explaining any of this cannot be done in a heuristic fashion. Any conclusion is something that must be measured. The cross-quantity energetics in detail are given by

$$-\phi \frac{\partial \varpi}{\partial t} = -\phi\big(-[\phi, \varpi] + \nabla_{\parallel} J - \mathscr{K}(n)\big), \tag{14.29}$$

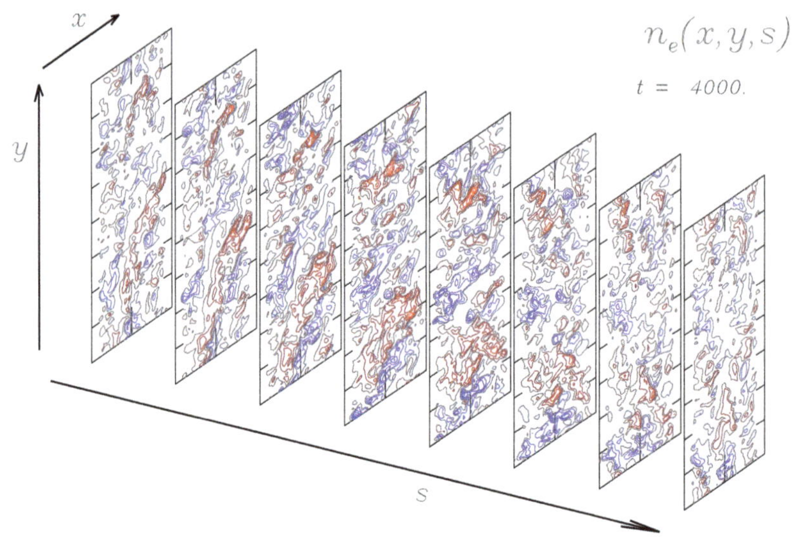


**Figure 14.8.** Three-dimensional structure of $n$ for the representative case. Of the 32 planes in $s$ in the computation, 8 are shown. Some twist with the shear of the magnetic field is visible in the largest structures.

$$n\frac{\partial n}{\partial t} = n\big(-[\phi, n] + \nabla_\parallel J - \nabla_\parallel u + \omega_n n v_E^x + \mathcal{K}(\phi) - \mathcal{K}(n)\big), \tag{14.30}$$

for the ExB and thermal free energy pieces, respectively. Note that $n\mathcal{K}(n)$ is neither cross-quantity nor nonlinear, so it will not be shown. For the Ohmic and acoustic pieces, we have Eqs (12.67, 12.68) from Chapter 12, respectively. The labelling of each of the pieces is as in the slab and ballooning cases from Chapters 12 and 13, respectively. The measurement technique is the same as given therein, now with all terms, including the adiabatic coupling and curvature forcing together.

The isotropic spectra of the state variable energetics are shown in figure 14.9, which one can compare with figure 12.19 for the drift wave case, and to figure 13.24 for the ballooning case, from the previous two Chapters. We now find that the ExB vorticity nonlinearity transferring energy into large scales from small ($T_e$) is a large enough effect to overcome the MHD interchange forcing ($T_P$), to balance the Alfvénic coupling ($T_j$) except at the largest scales. Comparing this to the ballooning case, we find that this drift wave physics (specifically, polarisation) assumes control of the balance among the currents in the fully developed turbulence. In the thermal free energy the gradient drive ($\Gamma_n$) acts as a source which is balanced by the nonlinear ExB advection ($T_n$) at large scale and the adiabatic compression ($T_a$) at small. Now that the adiabatic coupling is restored, it remains the larger transfer mechanism sending energy towards the ExB side. This appears as $T_j$ in the ExB flow energetics, and it is what compensates the nonlinear transfer in the vorticity nonlinearity.

The isotropic spectra of the flux variable energetics are shown in figure 14.10, which one can compare with figure 12.20 for the drift wave case, and to figure 13.25 for the ballooning case, from the previous two Chapters. In the ballooning case the adiabatic transfer was absent, so that the Ohmic energy was maintained by MHD

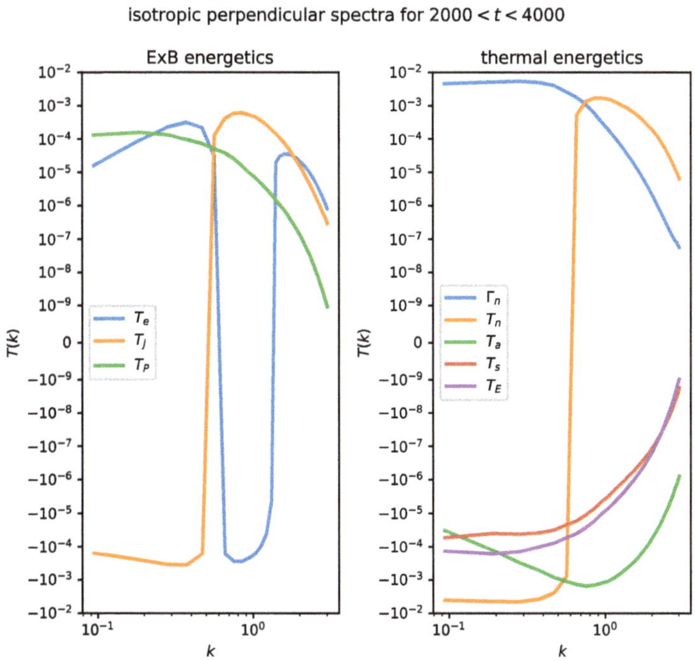


**Figure 14.9.** State variable energetics, for the ExB (left) and thermal free energy (right), where positive (top) and negative (bottom). Compare with figure 13.24 from Chapter 13, for the ballooning mode case, and figure 12.19 from Chapter 12, for the drift wave case. Here, although the curvature coupling adds to the energetic transfer, the incoming Alfvénic transfer ('j') is larger at small scale and the nonlinear ExB transfer is the principal source effect at large scale. This is due to the strength of the nonlinearity in this system. Note that the diamagnetic transfer through curvature coupling ('p'), becomes negative at large scale, in the clearest contrast to the ballooning mode case. In the thermal free energy the adiabatic transfer out to the Ohmic energy is dominant at small scale ('a'), and the gradient drive ('n') balances nonlinear ExB advective transfer ('P') at long wavelength. For this component the curvature coupling is subdominant at all scales.

Alfvén transfer ($T_A$) from the ExB energy at all scales. Here, the adiabatic transfer ($T_a$) from the thermal free energy is the principal maintainer of finite currents, while the MHD process is negative at most scales and small at large scale. Hence most of the Ohmic energy comes from the thermal side, not the ExB side. It is balanced by resistivity ($\Gamma_c$). In the acoustic energy the nonlinear ExB advective transfer ($T_u$) overcomes the viscosity ($\Gamma_s$) as a sink at the larger scales and is the main source at smaller scales, as in the drift wave case.

In general, restoration of the adiabatic coupling maintains enough vorticity at smaller scale that its nonlinear transfer dynamics can reassume control of the dynamics of the system, and provides the main source of Ohmic energy by adiabatic compression directly from the thermal free energy. The curvature forcing is present, helping the turbulence along particularly at medium collisionality and inductivity, especially in giving it a longer wavelength component. But the drift wave physics is in control of the underlying dynamics.

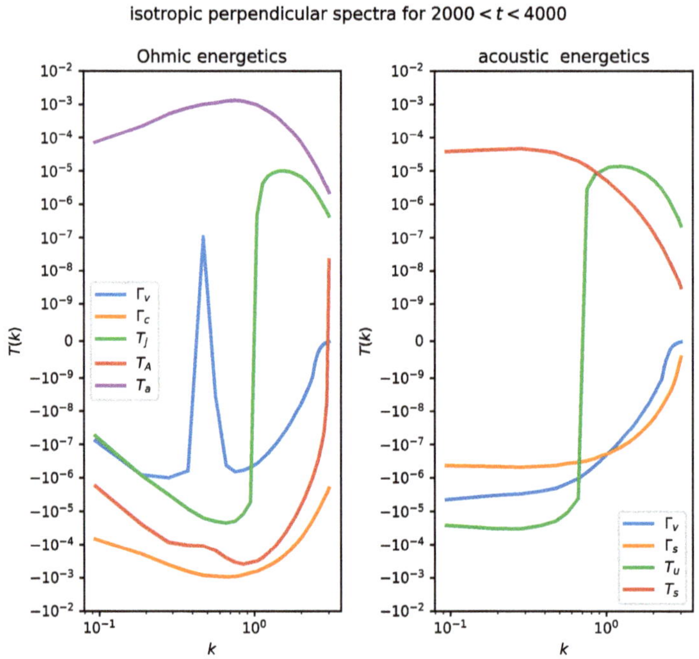


**Figure 14.10.** Flux variable energetics, for the electron (left) and ion channels (right), where positive (top) and negative (bottom). The electron channel includes the magnetic energy. Compare with figure 13.25 from Chapter 13, for the ballooning mode case, and figure 12.20 from Chapter 12, for the drift wave case. As in the drift wave case, the current balance is between adiabatic transfer in ('a'), Alfvénic transfer out ('j') and resistive dissipation ('c'). At this collisionality the dissipation is stronger. In the ballooning case the Alfvénic transfer was in and in the absence of adiabatic coupling provided the balance. The acoustic energy balance feels the strength of the nonlinearity, with the balance between advective mixing ('u') and viscous dissipation ('s', solid line) at small scale, as in the drift wave case.

## 14.6 From the linear stage to turbulence

Further insight into the nature of this turbulence, and how the nonlinearities act to set up the mode structure, is provided by giving representative cases a start in the linear regime. Then, examining the mode structure during the linear growth phase after the instabilities have attained their eigenmode structure, we learn the similarities and differences between the instabilities and the turbulence at the same parameters. Rather than studying linear cases in a stylised ballooning model, we do these in the same model as the turbulence. The only difference is the amplitude of the disturbances. No changes are made to the equations, and the same grid is used in both regimes. We simply measure during the growth phase, during the transition to nonlinearity, and then again in the later phase of fully developed turbulence.

We use a low beta case for this, with $\hat{\beta} = 1$, $C = 13.9$, as the comparison is between resistive ballooning, which carries the instability, and the drift wave mode structure in the turbulence. The grid is $256 \times 256 \times 32$ in the same domain of $L_x \times L_y \times 2\pi$ with $L_x = L_\perp$ and $L_y = 4L_x$, noting $L_\perp = 1$ in normalised units.

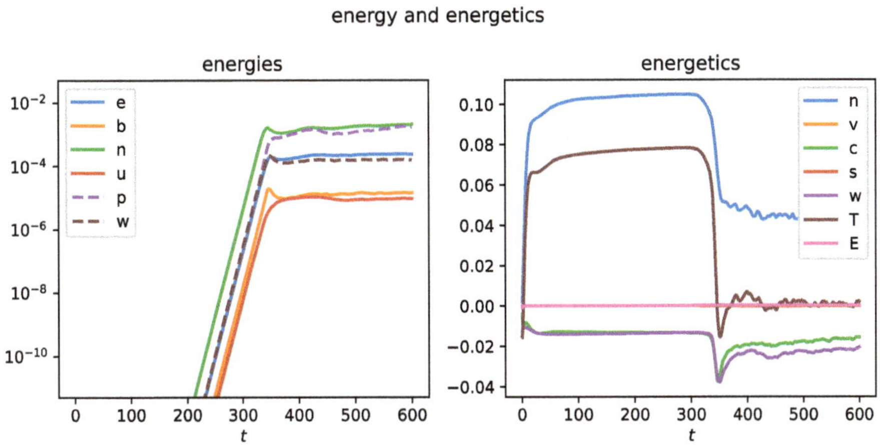


**Figure 14.11.** Evolution from a linear start with amplitude $A_0 = 10^{-10}$, for the $\hat{\beta} = 1$ and $C = 13.9$ case (see text). As in figure 14.1, these are the time traces of (left) energy pieces and amplitudes of $\phi$ and $\varpi$, and (right) energetic drive and dissipation, nonlinear growth rate, and error. All quantities rise at the same rate, as the linear instability forms its structure before $t = 100$ and then the time and space dependences are separate. The drive is robust, and the growth rate is comparable to those of resistive ballooning modes at similar parameters. Saturation occurs rapidly after $t = 300$, as the growth rate falls through and then oscillates around zero. The drive rate in turbulence is much less than during growth, and the dissipation rises. The error $\Gamma_E$ remains small, never larger than the $10^{-5}$ level.

The $N_x = 256$ in the $x$ direction follows the practice used for the linear modes to have $h_y/h_x = 4$ at the grid scale. The timestep is reduced to $\tau = 0.002$.

A linear instability scan is done first to confirm the validity of this choice. For linear-valid only runs the aspect ratio and $k_y$-spectrum considerations are suspended, and one can go to arbitrary resolution in the $x$ direction. The collisionality is set very low, with $C = 1.39 \times 10^{-3}$, to stress the grid and leave field line connection to function on the electrons by itself. The $x$-resolution parameter $h_y/h_x$ is set to powers of two from 1 to 16, with $N_y = 64$ and $L_y = L_x/4$, setting the grid scale near $h_y = \rho_s/4$. The $x$-resolution node count varies between 256 and 4096 for these cases. The same initial state as in a nonlinear case is used, except for the initial amplitude of $A_0 = 10^{-10}$, which gives a long enough linear growth phase for the eigenmodes to settle. We find measured maximum linear growth rates (before nonlinearity starts) of 9.30, 8.45, and 8.61, 8.61, 8.61, all times $10^{-2}$, for the resolution ratio $h_y/h_x$ from 1 to 16. These converge at $h_y/h_x = 4$, confirming that as the critical value to resolve the field line connection.

For the linear starts on the full grid we keep the $A_0 = 10^{-10}$ start and otherwise carry on as with a nominal nonlinear case. The time traces are shown in figure 14.11. Contrasting to the nonlinear start shown in figure 14.1, we see the linear growth phase taking finite time to settle, with the maximum growth rate attained near $t = 300$ just before nonlinearity starts to change the curves. This is due to the fact that in the unstable range there are several modes with similar growth rates. Any isolation (i.e. filtering a particular mode) further to this can be said to lose relevance, since one never has a plasma so quiescent that the density fluctuation level is less than $10^{-10}$, and 300 of these time units is far longer than any relevant time correlation measurement in the turbulence. We can see this visibly in the time traces when saturation occurs. The other significant finding is that the linear growth rates are far larger than the nonlinear drive rate in the fully developed turbulence. The total growth rate first falls through zero, and then as the mode structure adjusts it finds a statistical zero. The amplitudes are shown on a logarithmic scale, showing each piece growing at the same rate in the linear phase, which indicates a well-formed eigenmode structure. Only the initial $t < 600$ is shown, while the run itself carries to 4000 as in the nonlinear starts. This allows a convergence check on the results. The nonlinear start is the same case, with $\tau = 0.01$ and $h_y = h_x$ reverting to nominal. For the linear start case versus the nonlinear start case, averaged over $2000 < t < 4000$, we find

$$
\begin{array}{rclcll}
F_e & = & 1.70 \pm 0.14 & \text{versus} & 1.64 \pm 0.19 & \times 10^{-4} \\
A_w & = & 1.53 \pm 0.081 & \text{versus} & 0.915 \pm 0.068 & \times 10^{-4} \\
E_n & = & 1.96 \pm 0.26 & \text{versus} & 2.29 \pm 0.44 & \times 10^{-3} \\
E_b & = & 1.37 \pm 0.13 & \text{versus} & 1.21 \pm 0.17 & \times 10^{-5} \\
\Gamma_n & = & 3.87 \pm 0.23 & \text{versus} & 3.29 \pm 0.27 & \times 10^{-2} \\
\Gamma_c & = & 1.65 \pm 0.11 & \text{versus} & 1.08 \pm 0.10 & \times 10^{-2} \\
\Gamma_T & = & 0.004 \pm 1.43 & \text{versus} & 0.008 \pm 1.59 & \times 10^{-3} \\
\Gamma_E & = & 0.706 \pm 0.181 & \text{versus} & 1.69 \pm 0.21 & \times 10^{-5}
\end{array}
\qquad (14.31)
$$

This is essential convergence, with more vorticity for $h_x = h_y/4$. Standing out is the larger error for the nominal case, while still small. The $\Gamma_T$ results show the statistical zero in saturation mentioned above.

The rest of the mode structure is the same as given above. Here, though, we compare the various temporal phases *en route* to nonlinearity. The amplitude spectra for $\phi$, $n$, and $\varpi$ and the spatial morphology of $\phi(x, y)$ at $s = 0$ are shown in figure 14.12, for three times in the linear, transitional, and turbulence phases. The peak in the growing spectra feeds the other wavelengths, the linear streamer forms begin to move and break up, and then in turbulence we see the spectra properly spread out and a bath of flow eddies in the turbulent phase. For space, only one quarter of the $y$-domain with $L_y/L_x = 4$ is shown.

The nonlinear transfer spectra are shown for the same three times in figure 14.13, for the ExB nonlinearity in each of the four equations. The blue curve shows vorticity for the ExB flow energy, and the green curve shows advection for the thermal free energy. The nonlinear transfer is out of the unstable range into all the others for all four nonlinearities. Only when saturation is reached do the advection nonlinearities change direction at large scale and the usual pattern is set forever after.

The parallel envelopes are shown in figure 14.14, for the amplitudes (top row) and the transport flux (bottom), for short averages over the same three phases as above. These can be compared with figure 14.4 for the later-time saturated phase in the

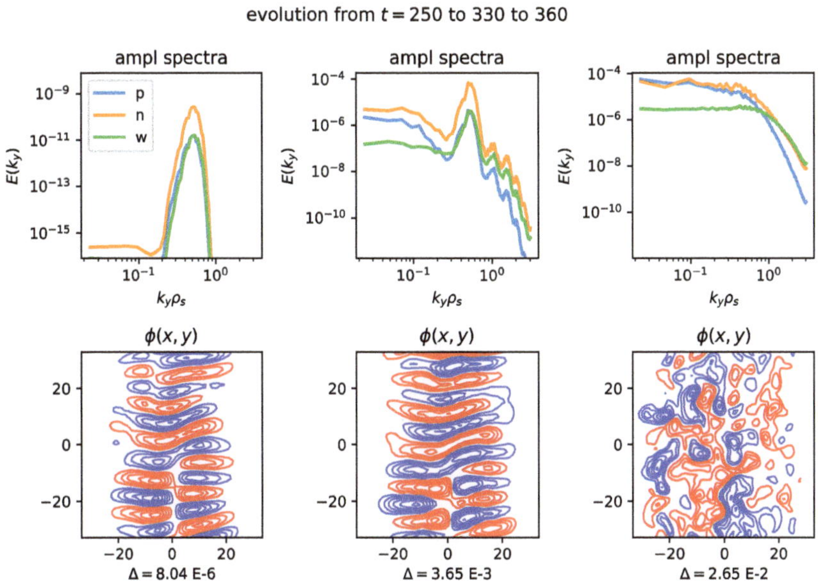


**Figure 14.12.** Transition out of the drift resistive ballooning mode instability to nonlinearity and into drift Alfvén turbulence. Compare with figure 13.14 from Chapter 13. Only 1/4 of the $y$-domain is shown. Drift direction wavenumber ($k_y$) spectra of the amplitudes of $\phi$ (top, 'p'), $n$ ('n'), and $\varpi$ ('w'), and spatial morphology of $\phi$ in the drift plane (bottom), for $t = 250$ in the linear growth phase, $t = 330$ in the transition to nonlinearity, and $t = 360$ at the beginning of the turbulence phase in which the spectra are close to their eventual statistical equilibrated forms. The transition is very rapid. First the linear structures begin to move, and then they are wiped out by the turbulence.

similar $\hat{\beta} = 5$ nominal case. In the linear growth phase once the eigenmode structure has settled in form, one finds the same structure as for a resistive ballooning mode. The envelope of $n$ is strongly ballooned, and $\phi$ is lower and flatter. Here we also show $h = n - \phi$, as its fate is very indicative. In the linear phase we have $h$ close to $n$ since $\phi$ is smaller. During the transition into turbulence, the envelope curves for $\phi$ and $h$ switch relative positions, so that in the turbulence we have the $n > \phi > h$ relationship as in drift wave cases. The curvature forcing still gives $n$ a ballooning character, which carries the rest of the quantities with it. The signature form is that $h$ is flatter and lower than either of the other two by themselves. The change away from the structure of the linear eigenmodes into turbulence is evident. By contrast, the flux envelopes by themselves are not very indicative.

The phase shift distributions are shown in figure 14.15, for short averages over the same three phases as above. These can be compared with figure 14.5 for the later-time saturated phase in the similar $\hat{\beta} = 5$ nominal case. During the linear growth phase the mode is a linear eigenmode, and the phase shift is part of the overall dispersion relation ($\omega$ and other quantities versus $k_y$). One finds different phase shifts for different modes, and the maximum phase shift is found for the most unstable modes just longer of the $k_y\rho_s = 1$ line. As the transition to nonlinearity is in progress (recall the moment of initial saturation is at $t = 330$ where $\Gamma_T$ first falls

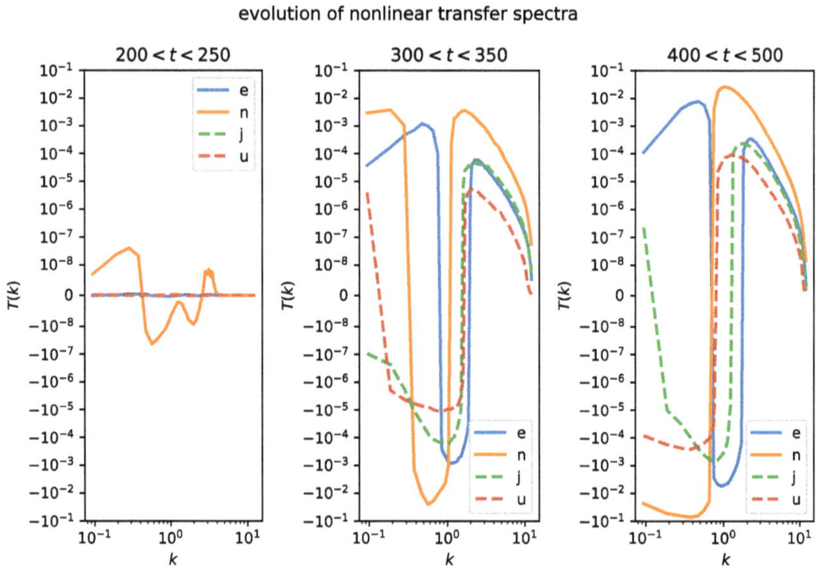


**Figure 14.13.** Transition out of the drift resistive ballooning mode instability to nonlinearity and into drift Alfvén turbulence. Compare with figure 13.15 from Chapter 13. Isotropic perpendicular spectra of the ExB energy transfer through the vorticity ('e') and advection of density ('n'), current ('j') and parallel ion velocity ('u'), drawn separately where positive (top row) and negative (bottom row), for $t = 250$ in the linear growth phase, $t = 330$ in the transition to nonlinearity, and $t = 360$ at the beginning of the turbulence phase in which the spectra are close to their eventual statistical equilibrated forms. Before saturation the transfer is out of the instability into all other ranges, with the longer wavelengths the strongest. The direct cascade in all channels except the vorticity only emerges in saturated turbulence.

through zero), the distributions have broadened, but the linear relation is still identifiable. As the dynamics first enters statistically equilibrated turbulence ($t < 500$) the phase shift distributions change completely and the structure assumes the forms in drift wave mode structure. Only once the linear eigenmode has been eliminated do the parallel envelope structure and the phase shift distributions assume their eventual forms.

The key identification in all of this evolution is found in the next set of measurements. These are the isotropic spectra of the ExB energetics shown in figure 14.16, for short averages over the same three phases as above. They can be compared with figure 14.9 for the later-time saturated phase in the similar $\hat{\beta} = 5$ nominal case. During the linear growth phase, the vorticity nonlinearity is negligible and the dominant modes are driven by curvature, which for $k_\perp \rho_s < 1$ is the only drive of ExB energy. This, more than any other measurement, identifies the instability as a drift resistive ballooning mode, namely, a resistive ballooning mode with some diamagnetic drift modifications (e.g. propagation with some fraction of $v_D$ in the positive $y$ direction). The ultimate source is the thermal gradient, and the transfer path by which the free energy gets into the ExB part of the system is dominantly the curvature forcing. However, this begins to change as the vorticity nonlinearity takes over the energetics, and into saturation the nonlinearity

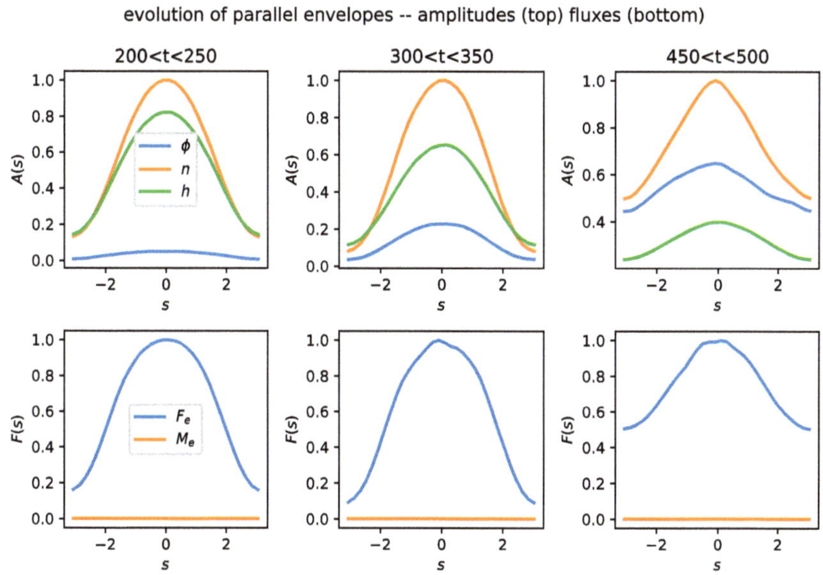


**Figure 14.14.** Parallel envelopes of the squared amplitudes (top) and turbulent ExB transport flux (bottom), which are averages of each quantity over $x$, $y$ and time, for $200 < t < 250$ in the linear growth phase (left), for $300 < t < 350$ as nonlinearity enters (centre), and for $400 < t < 500$ after the turbulence has saturated (right), which one can compare with the long-term statistical equilibration in figure 14.4. The linear phase shows a resistive ballooning mode structure, with $h = n - \phi$ satisfying $n \sim h > \phi$. During the transition, $h$ and $\phi$ switch, so that in turbulence one finds a drift wave mode structure, with $n \sim \phi > h$ and $h$ less ballooned than $n$ itself. The degree of ballooning of the flux envelope reduces somewhat but by itself would not yield information of this mode structure change.

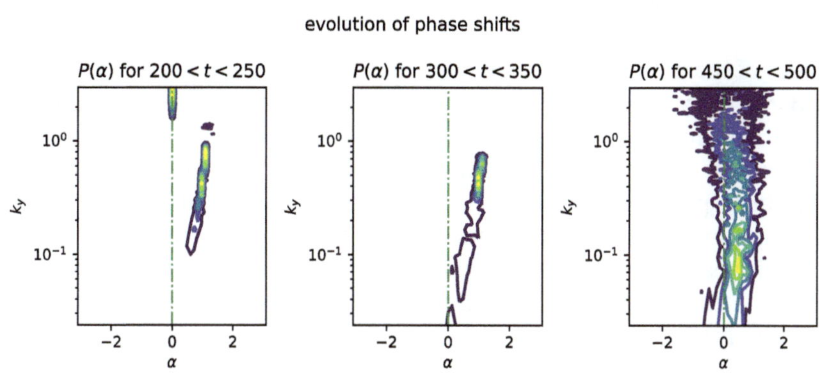


**Figure 14.15.** Phase shift distributions between $n$ and $\phi$, for $200 < t < 250$ in the linear growth phase (left), for $300 < t < 350$ as nonlinearity enters (centre), and for $400 < t < 500$ after the turbulence has saturated (right), which one can compare with the long-term statistical equilibration in figure 14.5. This shows the change from interchange mode structure during the linear phase to drift wave mode structure in turbulence.

is the dominant drive for ExB energy at all scales except the largest, the latter which is not where the gradient drive peaks.

Hence, the instability growing in the linear phase is identified as a drift resistive ballooning mode by these diagnostics. Then, when the nonlinear phase is entered,

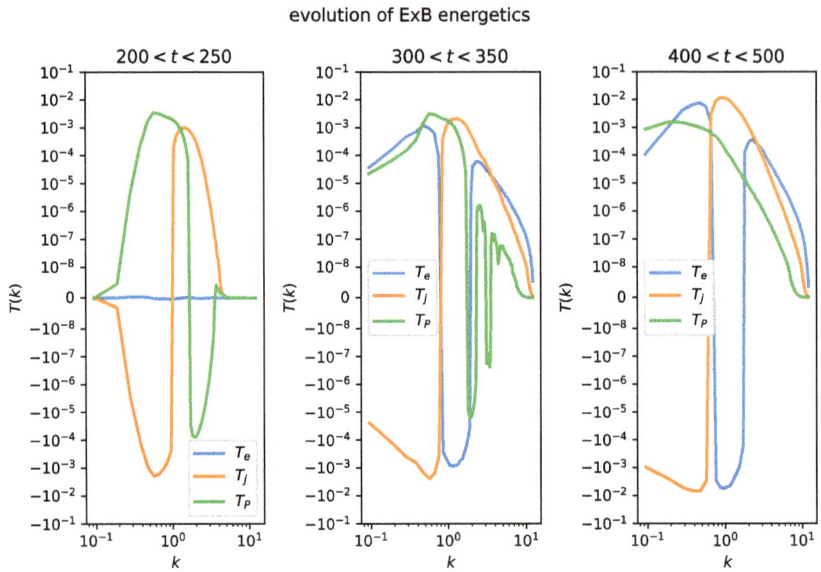


**Figure 14.16.** Transition out of the drift resistive ballooning mode instability to nonlinearity and into drift Alfvén turbulence. Isotropic perpendicular spectra of the ExB energetics, given by the polarisation ('$T_e$'), the parallel ('$T_J$'), and the diamagnetic ('$T_P$') currents. These reflect the ExB vorticity nonlinearity, Alfvénic transfer in through the current, and curvature forcing, respectively. They are averaged over $200 < t < 250$ in the linear growth phase, $300 < t < 350$ in the transition to nonlinearity, and $400 < t < 500$ at the beginning of the turbulence phase in which the spectra are close to their eventual statistical equilibrated forms. In the linear phase the curvature forcing maintains the ExB energy, while the onset of turbulence the vorticity nonlinearity replaces it in the range of strongest gradient drive. These are the energetic processes by which drift wave mode structure establishes itself in the turbulence.

the physics of the mode begins to change, and it loses its identity as a mode. Finally, the ExB vorticity nonlinearity is dominant, and the turbulence is drift wave turbulence forced upon by the curvature. The nonlinear polarisation current assumes control of the overall current balances, and one can no longer maintain that the finite parallel current is maintained and balanced by the diamagnetic current. The polarisation current first gets in the way and then overwhelms that linear physics balance, and at later times there is no longer any memory of how the turbulence started; much as with the drift wave nonlinear instability in the cases with no curvature forcing present.

## 14.7 Electromagnetic and collisional effects

The dependence of the turbulence amplitudes and transport fluxes on parameters, especially inductivity and collisionality, was covered in the pure drift wave and resistive ballooning models, and we repeat those investigations for the toroidal case in a sheared and curved magnetic field. We also take a set of control cases, setting the electron inertia artificially to 1/20 its nominal value, to separate the effects of the electron mass as part of the resistivity in the Ohm's law against the electron inertia itself. As in Chapter 13 the parameters controlling the electrons are

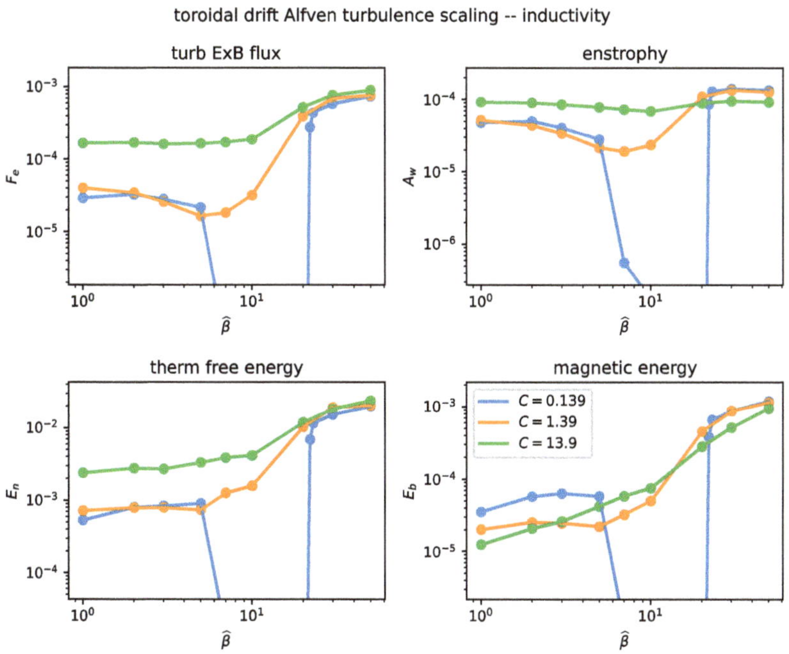


**Figure 14.17.** Scaling of the turbulent ExB transport flux ($F_e$), the enstrophy ($A_w$), the thermal free energy ($E_n$), and the magnetic energy ($E_b$), all in volume average, against magnetic inductivity $\hat{\beta}$, for various $C = 0.139$, 1.39, and 13.9. The stabilising tendency of finite $\hat{\beta}$ returns since it is due to the linear magnetic flutter mechanism from the drift wave cases ($[\phi, n_0]$ in the Ohm's law). It becomes a weaker effect at larger $C$. The nominal MHD ballooning transition at $\hat{\beta} = 22$ (since $\omega_B = 0.03$) is made more gradual by the resistive dissipation.

$$1 < \hat{\beta} < 50 \quad \hat{\mu} = 2.72 \quad 0.139 < C < 13.9, \tag{14.32}$$

indicating the main parameter sweep in the magnetic induction and varying collisionality (these values arise from setting $\nu_e L_\perp / c_s$ between 0.1 and 10).

The results of varying inductivity $\hat{\beta}$ are shown in figure 14.17, in which the measured quantities are the turbulent ExB transport flux across the field lines, $F_e = n v_E^x$, the enstrophy, $A_w = \varpi^2/2$, the thermal free energy, $E_n = n^2/2$, and the Ohmic energy (magnetic and parallel electron flow), $E_b = J(A + \hat{\mu}J)/2$, all in volume average. The nominal turbulence level is $\delta_0^2 = 2.25 \times 10^{-4}$ for comparison. The higher collisionality cases all show robust turbulence. As in the drift wave cases but in contrast to the ballooning mode cases, increasing inductivity reduces the turbulence at first, but then at larger $\hat{\beta}$ values the curvature forcing, remaining the same in these units, begins to overcome the adiabatic constraining, which becomes slower with increasing $\hat{\beta}$, so that the turbulence strength rises again. We do see a clear break at $\hat{\beta} = 22$, the nominal ballooning stability boundary, but only at low $C = 0.139$ for which $\hat{\beta}$ has stabilised the drift wave turbulence. To check this thoroughly, extra runs are done at $\hat{\beta}$ values around this point. Indeed we find that the low collisionality $C = 0.139$ case is stabilised for $\hat{\beta} > 1$, remains stable up to

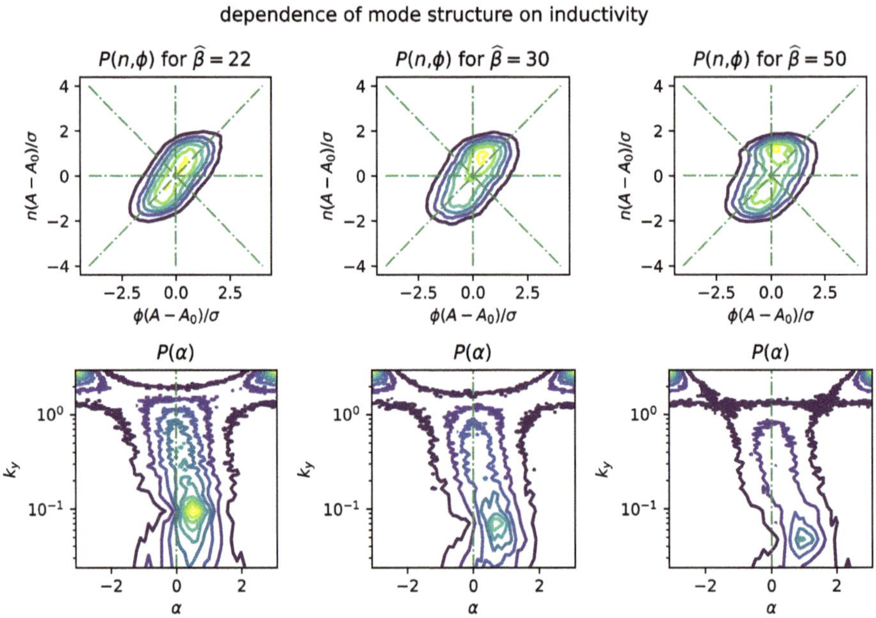


**Figure 14.18.** Phase shift distributions between $n$ and $\phi$, for $C = 0.139$ and $\hat{\beta} = 22$ (left) 30 (centre), and 50 (right), which one can compare with the lower-$\hat{\beta}$ nominal case in figure 14.5. This shows the survival of drift wave mode structure for $\alpha_M$ at the critical value, which for $S = 1$ and $\omega_B = 0.03$ is $\hat{\beta} = 22$. The transition to ballooning turbulence is gradual, becoming more so for larger $C$.

$\hat{\beta} = 20$ and 21, but for $\hat{\beta} = 22$ and 23 the turbulence re-emerges sharply, which is consistent with $\alpha_M = 0.66\ S$, with $S = 1$ and $\alpha_M = \hat{\beta}\omega_B$ with $\omega_B = 0.03$, namely the nominal MHD stability boundary. For larger $C > 1$, this MHD ballooning transition is washed out by the resistive dissipation.

If we check the mode structure for the high-$\hat{\beta}$ cases we indeed find the change away from drift wave signatures. However, it is very gradual and is even not seen for these $\hat{\beta} = 22$ and 23 cases just referred to. The cross coherence and phase shift distributions for the onset value of $\hat{\beta} = 22$ and the two cases $\hat{\beta} = 30$ and 50 at $C = 0.139$ are shown in figure 14.18. The linear instability itself grows slowly at the rather long wavelength of $k_y\rho_s = 0.15$, yielding 'fat' streamers in the spatial structure. The mode shows as a single peak in the phase shifts nevertheless at a value of $\alpha$ near zero. The parallel envelope is very strongly ballooned in $n$ while low and flat in $\phi$, which looks like the interchange cases of Chapter 13. The energetics done as in figure 14.9 show the dominant curvature transfer into the ExB energy from the thermal free energy in balance with the Alfvén transfer back out of it into the Ohmic energy, in the same midrange spectral region $k_\perp\rho_s \sim 0.2$ where the gradient drive peaks. As in the linear stage in figure 14.16, this is interchange energetics in a growing eigenmode.

The transition to turbulence takes place near $t = 2000$ and the statistical dynamics becomes well saturated for $t > 2500$. The measurement for $2000 < t < 4000$ then shows the drift wave mode structure seen in figure 14.18, which can be considered to be 'widened' by the curvature forcing. What is absent is the memory of the linear mode.

The cross coherence and phase shift distributions for the three values of $C$ at $\hat{\beta} = 30$ are shown in figure 14.19. Comparison to the cases at lower $\hat{\beta}$ in figure 14.5 to these cases shows the change from drift wave towards interchange mode structure, indicating that MHD ballooning has become a large enough part of the dynamics to take over control of it. This is especially obvious for the $C = 0.139$ case which without the curvature does not produce turbulence at all. Indeed, the nominal MHD ballooning result in the simplified linear model is done under ideal MHD and therefore nonexistent collisionality. So the more general toroidal model covering both limits indeed captures this result where we should expect it to do. But it is the case with $\hat{\beta} = 30$ and $C = 13.9$ which shows the clearest version of ballooning mode structure. The other diagnostics that go with this are the parallel envelopes, showing $n$ more ballooned with $\phi$ lower and flatter, and the state variable energetics, which shows the spectrum of the curvature forcing larger than the vorticity nonlinearity for about $k_{\perp}\rho_s < 0.4$; however, with the transfer through the current the dominant source for ExB energy for smaller scales where most of the vorticity is. The curvature controls the large-scale drive while the adiabatic coupling and the vorticity control the saturation and maintain robust decorrelation and mixing of any turbulent structures.

The dependence on collisionality $C$ is shown in figure 14.20, for the same four quantities as above, all in volume average. The robustness of the turbulence remains at this value of $\hat{\beta} = 5$ for $C$ down to the lowest value of 1.39 on this graph (recall this

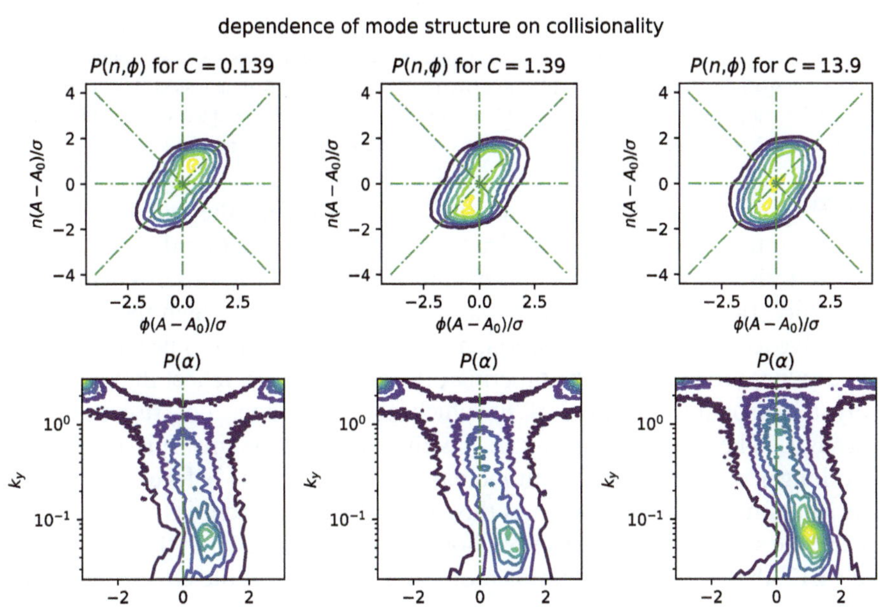


**Figure 14.19.** Phase shift distributions between $n$ and $\phi$, for $\hat{\beta} = 30$ and $C = 0.139$ (left) $C = 1.39$ (centre), and $C = 13.9$ (right), which one can compare with the lower-$\hat{\beta}$ nominal case in figure 14.5. This shows the partial survival of the drift wave mode structure for these cases even for $\alpha_M$ above the critical value, which for $S = 1$ and $\omega_B = 0.03$ is $\hat{\beta} = 22$. A change towards an interchange mode structure is apparent but also incomplete.

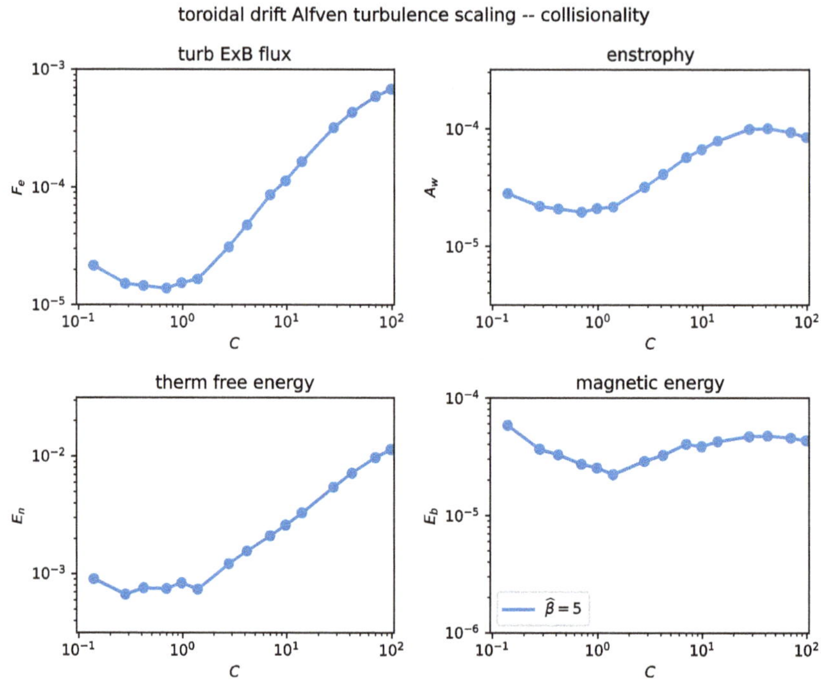


**Figure 14.20.** Scaling of the turbulent ExB transport flux ($F_e$), the enstrophy ($A_w$), the thermal free energy ($E_n$), and the magnetic energy ($E_b$), all in volume average, against collisionality $C$, for $\hat{\beta} = 5$. The turbulence is more robust towards small $C$ and tapers off at large $C$ since either limit leads to a weaker nonadiabatic response—it is $h = n - \phi$ rather than $\phi$ itself that is constrained, and the parallel dynamics becomes a smaller fraction of the whole (see especially the run of enstrophy) at extreme $C$ values. These scaling results change from the drift wave and ballooning mode cases of Chapters 12 and 13, and they will change again with the introduction of the temperature physics. The underlying mechanisms of energetics, however, remain.

was the highest $\hat{\beta}$ value still robust for $C = 0.139$ before the stabilisation due to magnetic flutter set in). The turbulence strength rises continually with $C$ as the enhanced resistivity leads to a more active adiabatic response. Then, for high collisionality, the parallel dynamics weakens to the point at which the turbulence is less sensitive to it. A pure 2D situation, however, is not reached unless unrealistically extreme values are reached. The tapering off does, however, reflect that $h = n - \phi$ is already a reasonable fraction of $n$, which therefore does not rise as fast with increasing $C$. The run of enstrophy ($A_w$) with $C$ also reflects this. It is already robust and no longer is strengthened with increasing $C$ as much as before.

The run of mode structure reflects this. While linear instabilities in this parameter range are the same strong resistive ballooning modes as shown in the linear stages in figure 14.12 above, the turbulence is more of a hybrid with the curvature forcing at large $\hat{\beta}$ or $C$ in a concurrent but not dominant role. The run of cross coherence and phase shift distributions with $C$ is shown for $\hat{\beta} = 5$ in figure 14.21. The basic structure is the same as in figure 14.5, with moderate hydrodynamic corrections in

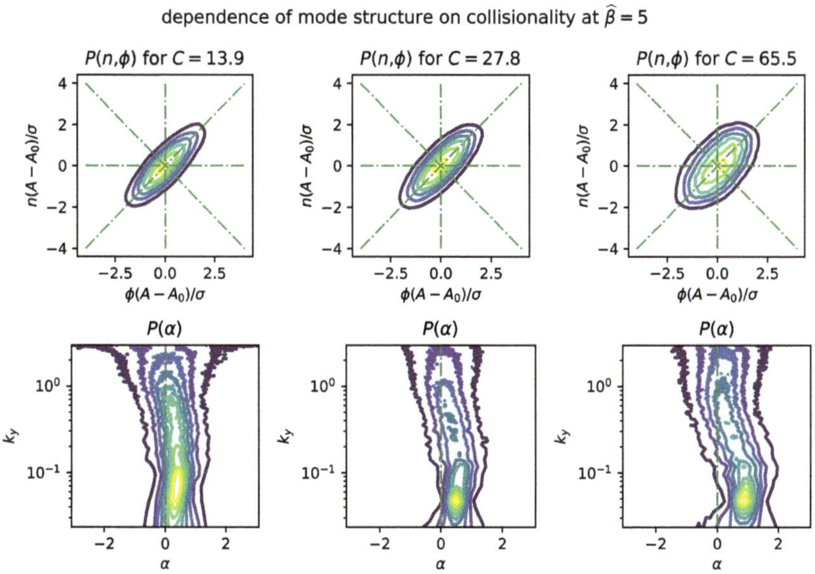


**Figure 14.21.** Phase shift distributions between $n$ and $\phi$, for $\hat{\beta} = 5$ and $C = 13.9$ (left) $C = 27.8$ (centre), and $C = 65.5$ (right), which one can compare with the nominal case in figure 14.5. Only for very large values are changes seen, and they are moderate hydrodynamics corrections to the basic drift wave mode structure.

the positive shift of phases in the longer wavelengths and the broadening of the cross coherences.

Finally, we provide control against electron inertia by maintaining the value of $C$ while reducing $\hat{\mu}$ by 20 to 0.139. These are shown in figure 14.22. Two sets of $\hat{\beta}$ values from 1 to 7 with $C = 1.39$ and 13.9 are carried, with the dependences upon $\hat{\beta}$ of the four quantities as done above are compared. We find there is indeed a role for electron inertia in the Ohm's law for the lower-$C$ cases, with the inertia providing some of the departure from force balance in the adiabatic response thereby making the turbulence stronger. But for robust collisionality there is little effect, as we could have surmised.

These scaling results with $\hat{\beta}$ and $C$ change from the drift wave and ballooning mode cases of Chapters 12 and 13, and they will change again with the introduction of the temperature physics. The underlying mechanisms of energetics, however, especially in anything involving the vorticity, remain the same as they are in these isothermal cases.

## 14.8 Warm ion effects

We now address the issue of finite ion temperature, which we have not yet done. Even without the temperature gradient and fluctuation physics, the ion temperature is significant in effectively doubling the relative strength of the curvature and acoustic coupling by the factor $(1 + \tau_i)$, which is two if the temperatures are equal. This is signified by the parameter $\tau_i = T_i/T_e$ giving their ratio. The more important

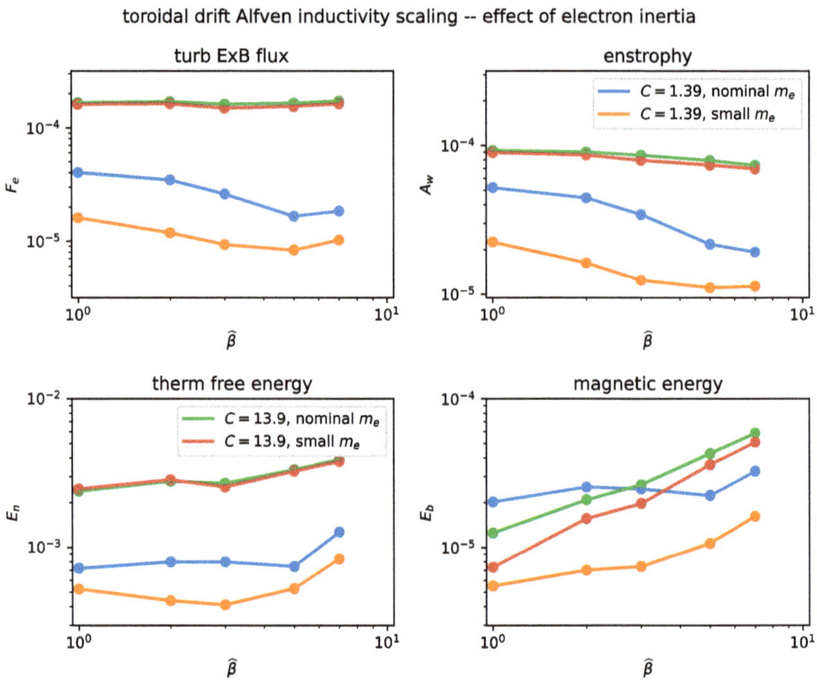


**Figure 14.22.** Effect of the finite electron inertia. Scaling of the turbulent ExB transport flux ($F_e$), the enstrophy ($A_w$), the thermal free energy ($E_n$), and the magnetic energy ($E_b$), all in volume average, against magnetic inductivity $\hat{\beta}$, for various $C = 1.39$ and 13.9, with nominal and small $m_e$ values for each. The small-$m_e$ cases reduce the electron inertia by a factor of 20 while preserving the value of $C$. This shows that the electron inertia assists robustness for moderate collisionality (nominally $\nu_e = c_s/L_\perp$) and as expected has no effect for strong collisionality (nominally $\nu_e = 10\ c_s/L_\perp$).

effect is on the vorticity nonlinearity, due to the gyroviscosity, in the divergence of the nonlinear polarisation current. This changes the Eulerian fluid form

$$-\nabla \cdot \mathbf{J}_p = \frac{\partial \varpi}{\partial t} + [\phi, \varpi], \tag{14.33}$$

in normalised units, which we now recognise as the cold-ion form, to

$$-\nabla \cdot \mathbf{J}_p = \frac{\partial \varpi}{\partial t} + \nabla \cdot \Big([\phi, \quad]\nabla_\perp W_i\Big), \tag{14.34}$$

where in both cases

$$\varpi = \nabla_\perp^2 W_i \qquad W_i = \phi + \tau_i n \quad \rightarrow \quad \phi \quad \text{if} \quad \tau_i = 0. \tag{14.35}$$

Obviously, the second form reverts to the first if $\tau_i$ is zero. If $\tau_i$ is of order unity, we refer to warm ions, and if it is much larger, then to hot ions, as briefly discussed at the end of Section 6.10 from Chapter 6.

The reason this matters so much is that the warm-ion case breaks the structure according to which the vorticity nonlinearity conserves enstrophy and thereby sets up the inverse ExB energy cascade. In our case this nonlinearity is responsible for converting a resistive ballooning instability into drift wave turbulence (more generally, drift Alfvén turbulence). So if the nonlinear dynamics is disrupted by the extra nonlinearity between $\nabla_\perp\phi$ and $\nabla_\perp p_i$, where $p_i = \tau_i n$, it might result in a qualitative change in the nature of the turbulence. We will briefly check this for relevant cases.

The first set is a re-run of the $\hat{\beta} = 1$ to 50 cases with $C = 1.39$ and 13.9, with $\tau_i = 1$ giving the warm ions. The cold-ion cases from figures 14.17 and 14.20 give the comparison. The dependence of the same four quantities as in those cases upon $\hat{\beta}$ for the two values of $C$ is given in figure 14.23. Although the turbulence is stronger this time we find similar behaviour to the cold-ion cases, with the weakening for intermediate $\hat{\beta}$ at the lower $C$ and the reduction of this effect with more resistivity. When evaluating the $\hat{\beta}$ dependence one should recall that the definition of $\alpha_M$ is in terms of the total pressure, so that for this model we now have

$$\alpha_M = (1 + \tau_i)\hat{\beta}\omega_B, \tag{14.36}$$

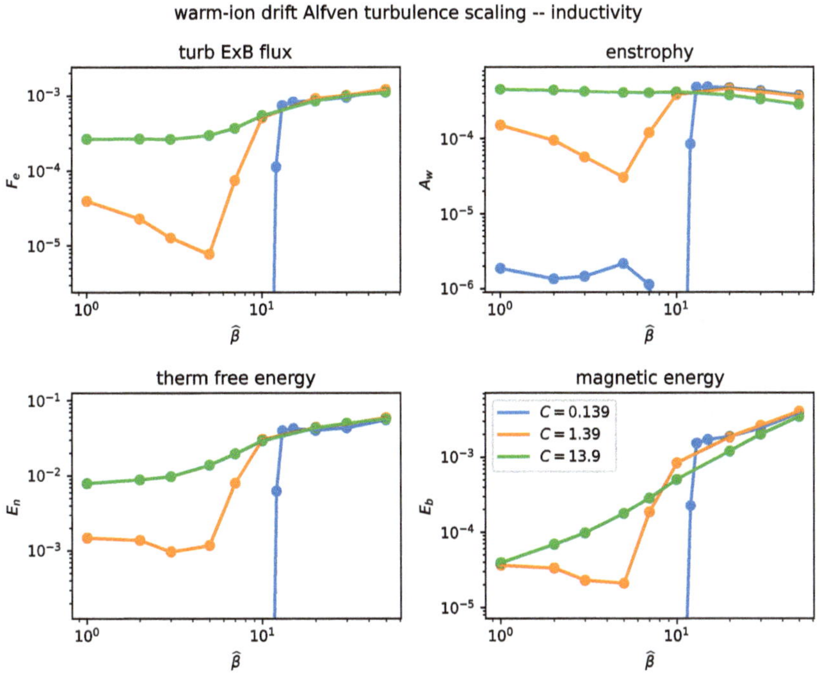


**Figure 14.23.** Scaling of the turbulent ExB transport flux ($F_e$), the enstrophy ($A_w$), the thermal free energy ($E_n$), and the magnetic energy ($E_b$), all in volume average, against magnetic inductivity $\hat{\beta}$, for the warm-ion cases with $\tau_i = 1$ for various $C = 0.139$, 1.39, and 13.9. Compare with the cold-ion cases in figure 14.17. The trends are very similar. The transition into the ballooning unstable regime ($\hat{\beta} > 11$ here) is very sharp for low $C$, and the values for $\hat{\beta} = 20$ and above are very close for all three $C$ values. The high-$\hat{\beta}$ result is indicative of MHD interchange dynamics controlling the turbulence, and it is borne out by the mode structure diagnostics.

for the isothermal model in general. Since we keep $S = 1$ and $\omega_B = 0.03$ the critical value to get $\alpha_M = 0.66\ S$ is now $\hat{\beta} = 11$ for these cases. For the smallest $C$ the transition to MHD ballooning is very abrupt and is indeed at the expected critical value. That this is the physics is also indicated by the nearly nonexistent dependence upon $C$ for the last three $\hat{\beta}$ values.

The mode structure is of interest. For $C = 13.9$ the transition to ballooning mode structure with $\hat{\beta}$ is as expected. The resistivity and inductivity work together to delay the adiabatic response sufficiently for strong curvature forcing to move the peak of the turbulence to larger scale, towards separation away from most of the vorticity which is at smaller scale. For $C = 1.39$ the collisional effect is weaker, leaving the inductivity as most of the cause of nonadiabatic responses. A sharp transition between $\hat{\beta} = 5$ and 7 is seen in all the diagnostics. The cross correlation and phase shift distributions are shown in figure 14.24. For $\hat{\beta} = 1$ the drift wave mode structure signature is as in the previous cases. For the other two values we see the switch of the low-$k_y$ component to $\alpha = \pi/2$, namely the interchange signature. The high-$k_y$ component remains in the nearly adiabatic relationship. The main difference at short wavelength is that for $k_y\rho_s > 1$ the phase shifts abruptly switch to $\alpha = -\pi$, that is, $n$ and $\phi$ become anti-phased. This is the warm-ion polarisation effect, signifying the effect of the diamagnetic contributions to the momentum stress, which is what gyroviscosity actually is. The cross correlations show the same switch at the same $\hat{\beta}$,

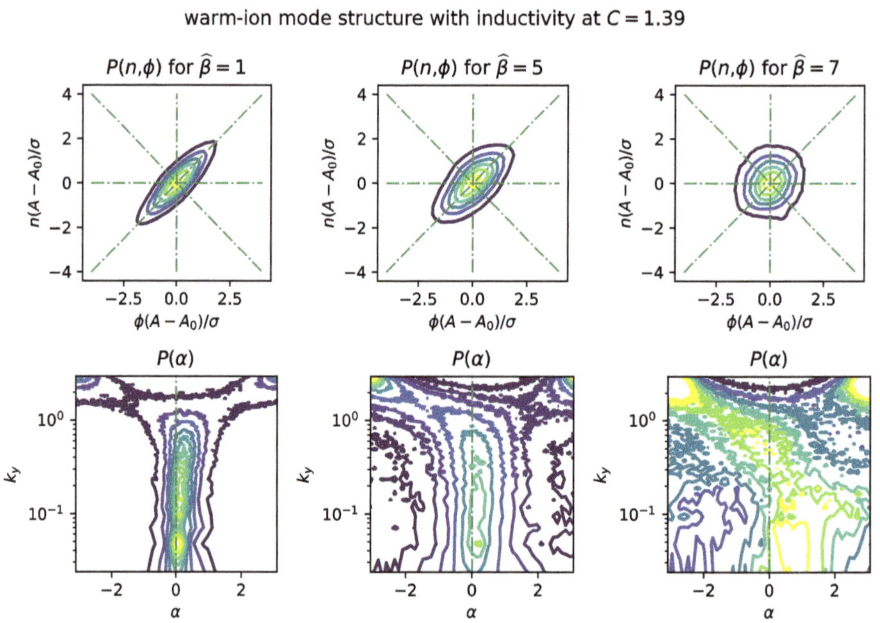


**Figure 14.24.** Phase shift distributions between $n$ and $\phi$, for $C = 1.39$ and $\hat{\beta} = 1$ (left) $\hat{\beta} = 5$ (centre), and $\hat{\beta} = 7$ (right), which one can compare with the cold-ion cases in figure 14.21. This shows a clearer transition towards a hydrodynamic relationship at long wavelength while leaving the shorter wavelengths closer to adiabatic, as the low-$k_y$ interchange drive becomes stronger. This is below the nominal MHD threshold, which is at $\hat{\beta} = 11$ due to $\tau_i = 1$.

away from the adiabatic correlation toward the incoherent hydrodynamic turbulence response seen in the pure interchange and ballooning cases. The parallel envelopes of the state variables are shown in figure 14.25. For $\hat{\beta} = 1$ the drift wave signature, with $h = n - \phi$ lower and flatter than the others, is found. At $\hat{\beta} = 5$ the transition to a ballooning mode structure begins and already at the next value it is complete. Now, $n$ is larger than $\phi$ and more strongly ballooned.

Since much has been made in the past of the question whether this type of turbulence rises or falls with increasing $\hat{\beta}$ or $\alpha_M$ for conditions commensurate with those in the edge of a tokamak, we adjust the parameters to fit that situation and then do the sweep in $\hat{\beta}$ again. The major and minor radii are set to 1.65 and 0.50 m, respectively, and the perpendicular gradient scale length is set to 3 cm. The field line pitch parameter is set to $q = 3.5$. At $T_e = T_i = 100$ eV and $n_e = n_i = 3 \times 10^{19}$ m$^{-3}$ and $B_0 = 2.5$ T with deuterium ions, we have

$$\frac{L_\perp}{\rho_s} = 52 \qquad \frac{R_0}{L_\perp} = 55 \qquad \frac{qR_0}{L_\perp} = 192, \tag{14.37}$$

for the scale ratios, yielding

$$\hat{\mu} = 10 \quad \hat{\beta} = 3.6 \quad C = 5.1 \quad \omega_B = 0.03 \quad \hat{\epsilon} = 36\,700, \tag{14.38}$$

for the normalised parameters. The collision frequency in normalised units is rather low, $\nu_e L_\perp / c_s = 1$, but coupled with the large value of $\hat{\mu}$ the value of $C$ is still moderate.

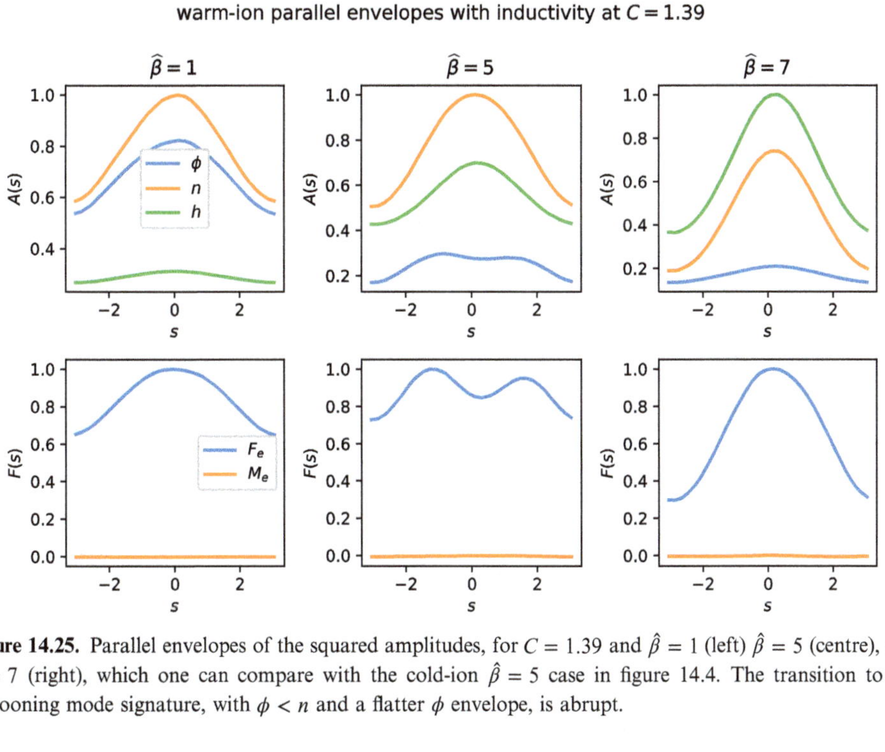


**Figure 14.25.** Parallel envelopes of the squared amplitudes, for $C = 1.39$ and $\hat{\beta} = 1$ (left) $\hat{\beta} = 5$ (centre), and $\hat{\beta} = 7$ (right), which one can compare with the cold-ion $\hat{\beta} = 5$ case in figure 14.4. The transition to the ballooning mode signature, with $\phi < n$ and a flatter $\phi$ envelope, is abrupt.

We have already seen that this is significant to the dynamics. It is important to note that in a real situation a steep gradient setting in a laboratory plasma always yields a rather small value for $L_\perp/\rho_s$, practically always below 100. This means that the dynamics at the largest scales remains in energetic contact with the range around $k_\perp\rho_s \sim 1$, so that the prospect of drift wave vorticity dynamics always enters. The dependence of the same four quantities as before upon $\hat{\beta}$ for this setting is given in figure 14.26, which can be compared with figure 14.23. The scaling is similar to the $C = 13.9$ cases in the cold ion parameter scan but not to the $C = 1.39$ cases. On the other hand, the mode structure transition is the same as that found in figures 14.24 and 14.25. This is why the main concentration here is on the physics, rather than on the scaling. The latter is sensitive to particular situations, which must be done with the correct scale ratios as well as the correct parameters. But the physics is robust and therefore a more reasonable focus of attention.

In general, the transition to ballooning mode structure for $\hat{\beta}$ within the vicinity of the critical value is enhanced by the warm ions, and not only because the critical value in terms of $\hat{\beta}$ is a factor of two lower. It is also because of the change in the nonlinear vorticity dynamics. If circumstances allow the longest wavelengths to energetically separate from the component at small scale where most of the vorticity

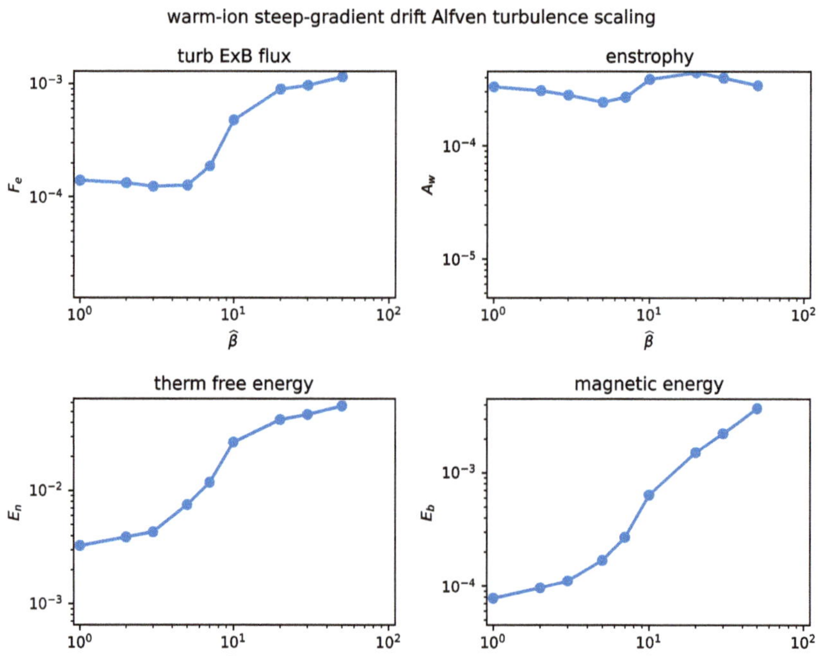


**Figure 14.26.** Scaling of the turbulent ExB transport flux ($F_e$), the enstrophy ($A_w$), the thermal free energy ($E_n$), and the magnetic energy ($E_b$), all in volume average, against magnetic inductivity $\hat{\beta}$, for the warm-ion cases with $\tau_i = 1$ for collisionality and scale ratio values reflecting conditions in the edge of a tokamak. Note that the steep drop of the turbulence strength with $\hat{\beta}$ below the MHD critical value is absent. The parameter situation is discussed in the text.

resides, then one can indeed find a change away from drift wave mode structure. However, the diagnostics show that the adiabatic response is still active across a large portion of the spectrum, and that the vorticity is still collected at scales near $\rho_s$. Hence, it is necessary to keep both of these in a computation.

## 14.9 Comparison to the control cases

We have the opportunity to summarise the result of drift Alfvén turbulence in toroidal geometry by comparison to the simpler pure drift wave and pure interchange cases in two dimensions and the pure drift wave and pure ballooning mode cases in three dimensions. The mode structure is already revealing, and the energetic analysis adds physical content. Finally, we emphasise the drift scale $\rho_s$ as a result of constraining toward electron force balance (adiabatic response), and the resultant scaling of the polarisation current in terms of the electron charge density and the parallel current in terms of a velocity. It is a result of the nonlinear vorticity dynamics, not an eigenmode of an instability or oscillation.

### 14.9.1 Drift wave and ballooning mode turbulence as controls

Drift wave turbulence in either of the 2D and 3D models occurs if the parallel dynamics either does (3D) or is assumed to do (2D) the parallel coupling that ties the electrostatic potential to the electron pressure through the adiabatic response, namely the tendency to restore electron parallel force balance. The action of field line connection is essential, as otherwise the $k_{\|} = 0$ flute mode takes control of the dynamics. With field line connection guaranteeing finite $k_{\|}$ for all available degrees of freedom, the adiabatic response ties the system together. As a result, the nonlinear transfer dynamics acting in *toto* in both directions maintains the ExB vorticity amplitude as the small scale thermal free energy transfers back into the ExB channel. By contrast, in a pure-MHD situation the ExB amplitude is reduced by the Alfvén relaxation.

In the drift wave cases the resulting small scale vorticity has a larger RMS level than the instability growth rate, so that the latter is eliminated. At the same time, the large scale ExB amplitude is held down by the adiabatic coupling, so that the amplitudes of the state variables are held together throughout the spectrum.

The strength of the vorticity is exemplified by comparison of the time traces for thermal free energy ($E_n$), enstrophy ($A_w$), and ExB turbulent thermal flux ($F_e$), for the nominal case, $\hat{\beta} = 5$ and $C = 13.9$, in all three models. These are shown in figure 14.27. The labels are 'slab', 'rmhd', and 'dalf' for the sheared-slab drift Alfvén case, the ballooning mode turbulence case, and the toroidal drift Alfvén case, respectively. The enstrophy is given by the dashed curves on the left side, and is the largest in the 'dalf' case, not only relative to $E_n$ but also in absolute terms. The thermal free energy and flux are the largest in the ballooning mode case. This can serve as a reminder that it is not always true that the model giving the largest flux or strongest turbulence for a given parameter set is the right one. Note as well that at early times the two toroidal cases track each other closely, before separating after about $t = 200$ into their respective nonlinear phases. Although the drive is strong in

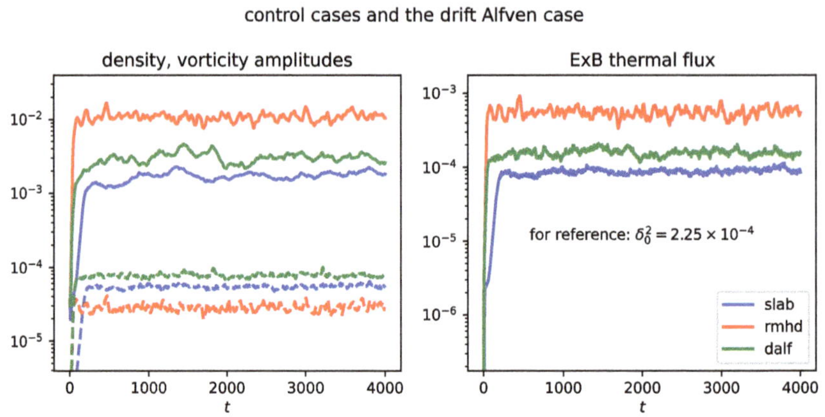


**Figure 14.27.** Time traces of the thermal free energy (solid) and enstrophy (dashed), on the left side, and turbulent ExB thermal flux, on the right, for the three models at the same parameter set, $\hat{\beta} = 5$ and $C = 13.9$, of the nominal case. The labels are 'slab', 'rmhd', and 'dalf' for the sheared-slab drift Alfvén case, the ballooning mode turbulence case, and the toroidal drift Alfvén case, respectively. Note especially the strongest turbulence for 'rmhd', the strongest vorticity for 'dalf', and the closeness of initial growth in the two toroidal cases before they separate widely into their respective nonlinear phases.

toroidal cases, the vorticity is decisive in mediating saturation, so much so that the levels of the turbulence and associated flux are brought down.

### 14.9.2 Mode structure and energetics

The comparisons and contrasts between the mode structure diagnostics of the toroidal case and the straight-field drift wave or toroidal ballooning mode models are the most revealing in deciding among the physical mechanisms involved in sustaining the turbulence. By comparison to the drift wave cases, we see the same vorticity dynamics, and the same cross correlation and phase shift distributions. Above all, we see the same energetics involving how the current, $J_\|$ or $J$ in normalised units, mediates the transfer from $n$, where the source is, to $\phi$, where the eddy flows are.

The pure interchange 2D and pure ballooning 3D models do not show these signatures in the same way. The phase shifts are at $\pi/2$ and there is little cross correlation. The thermal free energy can only feed the turbulent eddies through the curvature (interchange) forcing. The linear phase of a growing instability in the 3D toroidal drift Alfvén case indeed exhibits a similar form of energetics and mode structure to the interchange and ballooning cases. The phase shift is the largest at short wavelength (high $k_y$), and the distribution at each $k_y$ is narrow. In other words, the phase shift is part of the eigenmode structure, which stays the same for a growing eigenmode, once it has had time to form up. As the instability passes into turbulence, the mode structure and energetics is seen to change in response to the emergence into finite amplitude of the ExB vorticity nonlinearity transfer dynamics (as a nonlinear mechanism dependent upon the amplitude, it is negligible in the linear phase). Once the nonlinear mode structure becomes established, the direction of free energy transfer around the dynamical system is more like that in the pure drift wave cases in either 2D or 3D, with the balance between the vorticity nonlinearity and the parallel current pre-eminent in the ExB energetics, and the balance between the advective nonlinearity and the parallel current, as part of the parallel electron flow, in the electron thermal energetics. In the Ohmic energy, the essential balance is between the two linear terms given by the two static forces, pressure and electrostatic. In an MHD model the electron pressure force is assumed negligible in the Ohm's law, so these balances become impossible. It must be noted that the MHD assumptions behind the ballooning mode model, namely that with $k_\parallel$ given by geometry that $n \ll \phi$, is always violated by the results. The MHD model run with reasonable collisionality ($C$) and even inductivity ($\hat{\beta}$), almost always shows $n > \phi$ or at least $n \sim \phi$, so that the MHD assumption loses sense. The current cannot be ordered small against the flow velocity, and the pressure gradient cannot be ordered small against the components of the electric field.

### 14.9.3 Electron force balance and the drift scale

With electron pressure acting along the magnetic field and hence the adiabatic response present, we first see the definition of $T_e/e$ as the scale for the electrostatic potential. Using this scaling in the polarisation current divergence in units of a charge density divided by time, motivated by expressing the parallel current as a velocity, we first find $\rho_s$ as a relevant physical scale. An alternative way to derive this is through the same procedure as the ideas behind Hall MHD turbulence in Chapter 3. There, we assume an Alfvénic system and simultaneously expressing the current in terms of the electron charge density and the Alfvén velocity or of the magnetic field divided by a scale. Solving for this scale we find $v_A/\Omega_i$ as its reference value. Here, assuming low frequency perpendicular force balance, we replace $v_A$ by the acoustic velocity $c_s$ and do the same deliberation to find $c_s/\Omega_i$ as the reference scale. This is another way to derive $\rho_s$, before having to use the expression for the polarisation current.

It is of greatest importance to realise that the role of the $\rho_s$ scale in the dynamics does not result from the presence of a dominant instability at that scale. Resistive ballooning is unstable at all wavelengths, even though at reasonable $C$ the fastest growing mode is at relatively short wavelength. We find the turbulence at relatively long wavelength, either in resistive ballooning cases or in strongly collisional toroidal drift Alfvén cases, especially at finite ion temperature. But most of the vorticity is concentrated with a factor of two of $\rho_s$, and it is the dynamics of this vorticity which produces the energetics and mode structure that are found. It therefore follows that this drift scale is always relevant, whether linear or ballooning analysis indicates so or not. In a computation, it must always be resolved. The MHD picture of macroscopic fluid dynamics, perhaps with a few kinetic corrections to treat dissipation, therefore fails the basic test of applicability to magnetised plasma turbulence.

## 14.10 Summary

In general situations, turbulence in a magnetised plasma has several mechanisms of instability, drive, and also saturation available to it. The look and feel of the turbulence, even its correlation properties, spectra, and spatial morphology are fairly general and only in extreme cases indicative of the underlying energetic processes. This limitation is especially clear when one has information only of the electron density among the fluctuating variables. The significance in the real world is that it is often only technically possible to measure the frequency and wavenumber spectrum, along with auto-correlation information, of this variable. The temperature and electrostatic fluctuations are often not accessible, let alone their cross correlation information vis-a-vis each other or the density. The physical mechanisms governing the turbulence follow from the way it passes energy indirectly along the way between source and sink, or between thermal fluctuations and the ExB eddy flows. These mechanisms can only be directly measured if one can measure the direct relationship among the variables, scale by scale, which is the input to the quantities representing each of the energy transfer pathways. As a result, the necessary data to decide physical process are not even accessible, let alone available. These processes can only be determined in a computation by directly measuring the quantities that define them. Only in that case can one have falsifiable statements on the description of the physics of the turbulence.

In this case, drift Alfvén turbulence in toroidal geometry, we have elucidated the methods by which one can support these statements about the energetic process. The possible drive avenues were parallel flows down fluctuating field lines or down-gradient ExB advection of thermal fluctuations. The possible sink destinations were parallel flow viscosity or electrical resistivity (electron-ion frictional drag), or dissipation at arbitrarily small scale by collisional diffusion of thermal energy, here represented by the subgrid dissipation. Of greater significance are the pathways of free energy transfer from the drive to the ExB energy representing the vorticity of the turbulence, and also the transfer out of the ExB energy into the thermal free energy for subgrid dissipation, into the Ohmic energy (parallel currents) for resistive dissipation, or into the acoustic energy (parallel flows) for viscous dissipation. One could not have predicted in advance which of any of these were dominant, concurrent, or of secondary effect. In this case, we found the ExB advection of the gradient as the main source, the transfer between scales within the ExB energy (to larger) and the thermal free energy (to smaller), the transfer between these quantities mostly over the currents but also over the curvature, and finally the subgrid dissipation, then the resistivity, and then the parallel viscosity, in that order, as the active sinks.

Getting this information was complicated, but doing so, providing control cases along the way, was most productive in the establishment of scientific conclusions one can draw about the physical processes governing the turbulence. Drift Alfvén turbulence in toroidal geometry is just one example of a general problem, but the general methodology developed herein should be a useful guide on how to proceed on other physical problems.

## Further Reading

Drift Alvén turbulence in toroidal geometry was treated in the following papers, sometimes under other nomenclature. This set is a representative sample of the first efforts. Many alternative explanations of the physics have been given. Readers can sift among them to find how well documented they are. The references are: B Scott 'Three-dimensional computation of drift Alfvén turbulence' *Plasma Phys. Contr. Fusion* **39** (1997) 1635; and B N Rogers and J F Drake 'Enhancement of turbulence in tokamaks by magnetic fluctuations' *Phys. Rev. Lett.* **79** (1997) 229; and B N Rogers, J F Drake and A Zeiler 'Phase space of tokamak edge turbulence, the L-H transition, and the formation of the edge pedestal' *Phys. Rev. Lett.* **81** (1998) 4396; and B Scott 'Warm-ion drift Alfvén turbulence and the L-H transition' *Plasma Phys. Contr. Fusion* **40** (1998) 823; and X Q Xu and R H Cohen 'Scrape-off layer turbulence theory and simulations' *Contrib. Plasma Phys.* **38** (1998) 158; and B Scott 'Computation of warm-ion drift Alfvén turbulence' *Contrib. Plasma Phys.* **38** (1998) 171; and X Q Xu, R H Cohen, T D Rognlien and J R Myra 'Low-to-high confinement transition simulations in divertor geometry' *Phys. Plasmas* **7** (2000) 1951; and B Scott 'The nonlinear drift wave instability and its role in tokamak edge turbulence' *New J. Phys.* (2002) **4** 52; and X Q Xu, W M Nevins, R H Cohen, J R Myra and P B Snyder 'Dynamical simulations of boundary plasma turbulence in divertor geometry' *New J. Phys.* (2002) **4** 53.

Papers emphasising flows are referenced in the next Chapter. Papers specific to ion temperature gradient dynamics or gyrofluid equations will be referenced in Volume Two in the appropriate Chapters.

**IOP** Publishing

Turbulence and Instabilities in Magnetised Plasmas, Volume 1

Fluid drift turbulence

**Bruce Scott**

# Chapter 15

# Turbulence on open field lines

## 15.1 Introduction—open field line geometry

The geometry in many cases with open field lines is similar to or the same as in the closed flux surface cases we have been considering. Open field lines means the same thing in this context as open flux surfaces. Each surface is extended along the magnetic field until it opens to space or intersects a material surface. We will mostly consider the case with the field lines opening into space, and apply exit flux boundary conditions at some specified point along the geometric configuration. For the simplified slab and toroidal cases, this is a matter of declaring the system open at some value of the parallel coordinate $s$ and applying outgoing flux, or exit flux boundary conditions there. The equations are separated into pairs of state variables and flux variables; for example, vorticity and current for one pair, and pressure and parallel velocity for the other pair, which are the relevant state and flux variable pairs for the two-fluid dynamics in the model we studied in Chapters 12–14. Then, the characteristics are found, and the mode whose characteristic velocity would transport it into the flux tube from the open side of the boundary is set to zero. This is called an 'exit flux boundary' condition and is the one we will apply. Then, we develop the same type of coordinate system and model as for the simplified toroidal case of Chapter 14 for a dipole magnetic field. It is of interest to explore the turbulence one finds in such a system which is not really confined. One assumes a source of energy at the midplane of the dipole field of localised extent, around which one can build a flux tube which then extends along the purely poloidal dipole field to a point where the outflow is set, which eventually balances the source after the dynamics reached statistical equilibration.

### 15.1.1 Speeds, characteristics, exit fluxes

In the slab case we take the model without the curvature terms, set up as in a periodic slab, with or without magnetic shear. This is the model used in Chapter 12. Then we set the exit flux boundary conditions at $s = \pm\pi$ such that the length of the

doi:10.1088/978-0-7503-2504-2ch15

system is $L_{\|}$ with $K_{\|} = 2\pi/L_{\|}$ and $s$ the correspondingly normalised angle-like coordinate. These are referred to as the exit locations, or in current parlance, the ends of the field lines. The main effect of the open boundary is to break the constraint set by field line connection in a periodic system. The flute mode with zero parallel wavenumber is now allowed to exist.

To do this we assume $\beta \ll 1$ and $m_e/m_i \ll 1$ so that the kinetic shear Alfvén waves and sound waves separate. In this limit we don't have to solve a fourth-rank system. Instead we have two second-rank systems whose characteristics we find separately.

The first one is the kinetic shear Alfvén system. We did these already in Chapter 8, but it is useful here to start it in the context of a method used for analysis of global Alfvén modes cited in the Section Further Reading, below (Vlad 1990). The most basic ideal reduced MHD linearised equation set is

$$\frac{\partial \varpi}{\partial t} = B\nabla_{\|}\frac{\tilde{J}_{\|}}{B} \qquad \varpi = \nabla \cdot \frac{\rho}{B^2}\nabla_{\perp}\tilde{\phi}, \tag{15.1}$$

$$\frac{\partial \tilde{A}_{\|}}{\partial t} = -\nabla_{\|}\tilde{\phi} \qquad \mu_0 \tilde{J}_{\|} = -\nabla_{\perp}^2 \tilde{A}_{\|}, \tag{15.2}$$

in which we do not allow $\rho$ or $B$ or $k_{\|}$ to commute past $\nabla\cdot$ or $\nabla_{\perp}$. Taking $\mu_0\partial/\partial t$ of the first equation and $\nabla_{\perp}^2$ of the second, and substituting $\mu_0\partial\tilde{J}_{\|}/\partial t$ from the second into the first, we obtain

$$\nabla \cdot \frac{1}{v_A^2}\frac{\partial^2}{\partial t^2}\nabla_{\perp}\tilde{\phi} = B\nabla_{\|}\frac{1}{B}\nabla_{\perp}^2\nabla_{\|}\tilde{\phi}. \tag{15.3}$$

Fourier transforming $\partial/\partial t \to -i\omega$ and $\nabla_{\|} \to ik_{\|}$ produces

$$\nabla \cdot \frac{\omega^2}{v_A^2}\nabla_{\perp}\tilde{\phi} = Bk_{\|}\frac{1}{B}\nabla_{\perp}^2 k_{\|}\tilde{\phi}. \tag{15.4}$$

If $v_A$ and $k_{\|}$ are constants, then the $\nabla_{\perp}^2$ factors cancel and we recover the simple result from Chapter 8 without the adiabatic coupling. This is $\omega^2 = k_{\|}^2 v_A^2$. The point of a global mode analysis is that even in a cylindrical plasma with $B = B_0$ constant $k_{\|} = 1/qR_0$ has radial structure due to $q = q(r)$, namely the shear. Applying an identity

$$a\nabla_{\perp}^2(ab) = \nabla \cdot a^2\nabla_{\perp}b + (a\nabla_{\perp}^2 a)b, \tag{15.5}$$

and taking $B = B_0$ this can be recast into

$$\nabla \cdot \left(\frac{\omega^2}{v_A^2} - k_{\|}^2\right)\nabla_{\perp}\tilde{\phi} = \left(k_{\|}\nabla_{\perp}^2 k_{\|}\right)\tilde{\phi}. \tag{15.6}$$

This is the form with which global mode analysis starts. If variation of the profile $q(r)$ is neglected, then one has indeterminate spatial structure with $\omega^2 = k_{\|}^2 v_A^2$ everywhere. With the variation in $q(r)$ one can find global modes $\phi_{mn}(r)$, which, after the angles $\theta\varphi$ are Fourier decomposed, are functions of $r$ and are determined by the structure of the $q(r)$ profile. In a toroidal system, one has the curvature terms couple the $m$ mode with $m \pm 1$ to produce toroidal modes. If we reduce to the dynamics on one field line by neglecting radial variation, this produces the ballooning mode analysis referred to in Chapter 13.

For our purposes this analysis serves to establish the characteristic velocity as $\pm v_A$ without assuming local conditions. If we take the corresponding two-fluid system

$$\frac{\partial \varpi}{\partial t} = B\nabla_{\|}\frac{\tilde{J}_{\|}}{B} \quad \varpi = \nabla \cdot \frac{\rho}{B^2}\nabla_{\perp}\tilde{\phi}, \tag{15.7}$$

$$\frac{\partial \tilde{A}_{\|}}{\partial t} = \frac{T_e}{n_e e}\nabla_{\|}\tilde{n}_e - \nabla_{\|}\tilde{\phi} \quad \mu_0\tilde{J}_{\|} = -\nabla_{\perp}^2\tilde{A}_{\|}, \tag{15.8}$$

$$e\frac{\partial \tilde{n}_e}{\partial t} = B\nabla_{\|}\frac{\tilde{J}_{\|}}{B}, \tag{15.9}$$

with the adiabatic response $\nabla_{\|}\tilde{n}_e$ in the Ohm's law, the same analysis can be used to find

$$\nabla \cdot \frac{1}{v_A^2}\frac{\partial^2}{\partial t^2}\nabla_{\perp}\tilde{\phi} = B\nabla_{\|}\frac{1}{B}\nabla_{\perp}^2\left(\nabla_{\|}\tilde{\phi} - \frac{T_e}{n_e e}\nabla_{\|}\tilde{n}_e\right), \tag{15.10}$$

and then doing the same operation on $\partial\tilde{n}_e/\partial t$ as on $\partial\tilde{\phi}/\partial t$ we have

$$\nabla \cdot \frac{1}{v_A^2}\frac{\partial^2}{\partial t^2}\left(\nabla_{\perp}\tilde{\phi} - \frac{T_e}{n_e e}\nabla_{\perp}\tilde{n}_e\right) = \left(1 - \nabla \cdot \rho_s^2\nabla_{\perp}\right)B\nabla_{\|}\frac{1}{B}\nabla_{\perp}^2\left(\nabla_{\|}\tilde{\phi} - \frac{T_e}{n_e e}\nabla_{\|}\tilde{n}_e\right), \tag{15.11}$$

where again the drift scale $\rho_s$ is defined by

$$\rho_s^2 = \frac{m_i T_e}{e^2 B^2} \quad \text{with} \quad m_i = \frac{\rho}{n_e}. \tag{15.12}$$

This gives the correction due to the adiabatic response for the case where the parameters can have radial variation which does not commute past $\nabla_{\perp}^2$. We do not pursue the question of global modes here. The purpose of doing this is to establish that the characteristic velocities of the dynamical system do not depend on the local assumption. Therefore, the results of using characteristics can be applied generally. Here we see the correction $1 + K^2$ with $K = k_{\perp}\rho_s$ from a global analysis. One can include the electron inertia in the Ohm's law,

$$\frac{\partial \tilde{A}_{\|}}{\partial t} + \frac{m_e}{n_e e^2}\frac{\partial \tilde{J}_{\|}}{\partial t} = \frac{T_e}{n_e e}\nabla_{\|}\tilde{n}_e - \nabla_{\|}\tilde{\phi}, \tag{15.13}$$

to produce

$$\nabla \cdot \frac{1}{v_A^2}\frac{\partial^2}{\partial t^2}\left(\nabla_\perp\tilde{\phi} - \frac{T_e}{n_e e}\nabla_\perp\tilde{n}_e\right) = \left(1 - \nabla \cdot \rho_s^2\nabla_\perp\right)B\nabla_\parallel\frac{1}{B}\nabla_\perp^2 \times$$
$$\times\left(1 - d_e^2\nabla_\perp^2\right)^{-1}\left(\nabla_\parallel\tilde{\phi} - \frac{T_e}{n_e e}\nabla_\parallel\tilde{n}_e\right), \tag{15.14}$$

where again $d_e$ is the skin depth given by $d_e^2 = m_e/\mu_0 n_e e^2$ which represents the correction due to the inertia of the electrons. Taking the radially local approximations, neglecting radial dependence of parameters or $q$ and setting $\rho_s^2\nabla_\perp^2 \to -K^2$, then reproduces the velocity $V_a$ in Eq. (8.15) of Chapter 8.

Given a pair of equations, one for state variables and one for flux variables,

$$a\frac{\partial A}{\partial t} + b\nabla_\parallel B = 0 \quad c\frac{\partial B}{\partial t} + d\nabla_\parallel A = 0, \tag{15.15}$$

with $A$ the state variable and $B$ the flux variable and *abcd* constants, one can find the associated wave equation by the same means as above,

$$ac\frac{\partial^2 A}{\partial t^2} = bd\nabla_\parallel^2 A, \tag{15.16}$$

with the same equation also for $B$, to produce $V^2 = bd/ac$ where $V$ is the characteristic velocity. Then forming

$$u_\pm = A \pm \frac{b/a}{V}B, \tag{15.17}$$

the original set produces

$$\left(\frac{\partial}{\partial t} \pm V\nabla_\parallel\right)u_\pm = 0. \tag{15.18}$$

This identifies $\pm V$ as the characteristic velocity for the characteristic variable $u_\pm$ (minus for $u_-$ and plus for $u_+$).

Now, what is an exit flux? It is the characteristic variable for the characteristic velocity that would flow *out* of the computational domain along the field lines. The oppositely signed one is the entry flux, which would flow *into* the computational domain and which we now set to zero. Hence, at the positive end of the flux tube ($s = \pi$ in simple cases), setting the entry flux $u_-$ to zero, we find

$$B = \frac{a}{b}V\,A \quad \frac{\partial A}{\partial s} = 0 \quad \text{at} \quad s = \pi, \tag{15.19}$$

at this end of the flux tube. The state variable $A$ is given a Neumann boundary condition as indicated. At the negative end of the flux tube ($s = -\pi$) the opposite is done, setting the entry flux, now $u_+$, to zero, so that

$$B = -\frac{a}{b}V\,A \quad \frac{\partial A}{\partial s} = 0 \quad \text{at} \quad s = -\pi. \tag{15.20}$$

Each condition is applied at the point in $s$ indicated, which in our case falls between grid points. So with $s_k$ the last point within the domain on the positive side, we set

$$A' = A_{k+1} = A_k \quad B' = B(A') \quad B_{k+1} = 2B' - B_k, \tag{15.21}$$

where $B'$ is the value found by evaluating Eq. (15.19) at $A = A' = A_k$, and then the next point, the first outside the domain, is loaded with the $k + 1$ values as indicated. In a fourth-order scheme we need the next point still, so if we load these we would by applying the Neumann boundary condition in an even-reflecting way, for $A$, and the Dirichlet boundary condition in an odd-reflecting way, for $B$. This is

$$A_{k+2} = A_{k-1} \quad B_{k+2} = 2B' - B_{k-1}, \tag{15.22}$$

and now we have the variables in this pair at the necessary two guard cells outside the domain. This is the procedure for Eq. (15.19), and we now apply the same at the negative end $s = -\pi$, with $k$ now the first point in the domain on the negative side, with the signs changed according to

$$A' = A_{k-1} = A_k \quad B' = B(A') \quad B_{k-1} = -2B' - B_k, \tag{15.23}$$

$$A_{k-2} = A_{k+1} \quad B_{k-2} = -2B' - B_{k+1}, \tag{15.24}$$

which is the procedure for Eq. (15.20). Note how we set the incoming entry flux on each side to zero, so that we have the plus sign at $s = \pi$ and the minus sign at $s = -\pi$ for $B'$ which is always evaluated for positive quantities.

In a three-dimensional (3D) model this is how the exit flux boundary conditions are applied. If we want to restrict to a 2D model, then the exit flux boundary condition becomes an exit flux model. The variable $B$ is not advanced as a dependent variable but it effectively implemented in the form of a loss term according to

$$\frac{\partial A}{\partial t} = -\nabla_{\|} B \quad \text{becomes} \; -2\frac{V}{L_{\|}}A', \tag{15.25}$$

where $A'$ is the state variable combination associated with the flux variable $B$, not necessarily $A$ itself. We see this in the kinetic shear Alfvén response which ties $\tilde{J}_{\|}$ not to $\varpi$ or $\tilde{\phi}$ or $\tilde{n}_e$ but to $\tilde{\phi}-(T_e/n_e e)\tilde{n}_e$ as a combination. This is what the $A'$ notation represents. To see how this works we turn to the model characteristics in the drift Alfvén physics.

## 15.2 Model characteristics

We take the toroidal drift Alfvén system from Chapter 14, expressed in physical units. The dependent variables are $\varpi$, $\tilde{J}_{\|}$, $\tilde{n}_e$ and $\tilde{u}_{\|}$ and are advanced according to

$$\frac{d\varpi}{dt} = B\nabla_{\|}\frac{\tilde{J}_{\|}}{B} - \mathcal{K}(T_e\tilde{n}_e), \tag{15.26}$$

$$\frac{\partial \tilde{A}_{\parallel}}{\partial t} + \frac{m_e}{n_e e^2}\frac{d\tilde{J}_{\parallel}}{dt} = \frac{T_e}{n_e e}\nabla_{\parallel}(n_0 + \tilde{n}_e) - \nabla_{\parallel}\tilde{\phi} - \eta_{\parallel}\tilde{J}_{\parallel}, \tag{15.27}$$

$$e\frac{d}{dt}(n_0 + \tilde{n}_e) = B\nabla_{\parallel}\frac{\tilde{J}_{\parallel} - n_e e\tilde{u}_{\parallel}}{B} + \mathcal{K}(n_e e\tilde{\phi} - T_e\tilde{n}_e) + e\,Q, \tag{15.28}$$

$$\rho\frac{d\tilde{u}_{\parallel}}{dt} = -T_e\nabla_{\parallel}(n_0 + \tilde{n}_e) + B\nabla_{\parallel}B^{-1}\mu_{\parallel}\nabla_{\parallel}\tilde{u}_{\parallel}, \tag{15.29}$$

in which the background gradient is represented by $n_0 = -\omega_n(x/L_\perp)n_e$. If $n_0$ is not present then the source $Q(\mathbf{x})$ is expected to maintain the dynamics and the dependent variable $\tilde{n}_e$ also contains the profile gradient as the component with variation only in $x$. The field equations are

$$\frac{\rho}{B^2}\nabla_\perp^2\tilde{\phi} = \varpi \qquad -\nabla_\perp^2\tilde{A}_{\parallel} = \mu_0\tilde{J}_{\parallel}, \tag{15.30}$$

and the nonlinear operators are

$$\frac{d}{dt} = \frac{\partial}{\partial t} + [\tilde{\phi},\ \ ] \qquad \nabla_{\parallel} = \frac{B^s}{B}\frac{\partial}{\partial s} + \left[-\tilde{A}_{\parallel},\ \ \right], \tag{15.31}$$

where the spatial bracket

$$[a, b] = \frac{1}{B_0}\left(\frac{\partial a}{\partial x}\frac{\partial b}{\partial y} - \frac{\partial b}{\partial x}\frac{\partial a}{\partial y}\right), \tag{15.32}$$

includes the units as indicated. The $xy$ coordinates are in units of distance and $s$ is a normalised angle. The parameters $\rho$, $n_e$, $T_e$, $\eta_{\parallel}$, $\mu_{\parallel}$, and $B_0$ are constants and $B$ can depend on $s$. The metric coefficients in $\nabla_\perp^2$ and $B^s$ as well as the curvature operator will vary depending on the model as in the treatments laid out in Chapter 13.

To get the characteristics involving parallel dynamics for this model we dispense with the nonlinearities and the curvature, and set $B \to B_0$. Combining the vorticity and density equations into one for the adiabatic response $\tilde{h}_e = \tilde{n}_e - (n_e e/T_e)\tilde{\phi}$ produces Eq. (15.14), identifying $V_a$ with

$$V_a^2 = v_A^2\frac{1 + k_\perp^2\rho_s^2}{1 + k_\perp^2 d_e^2}, \tag{15.33}$$

as the velocity, provided we can Fourier transform the $xy$ coordinates. We can then follow the procedure leading up to Eq. (15.18), and find

$$\left(\frac{\partial}{\partial t} \pm V_a\nabla_{\parallel}\right)\left[\frac{\rho}{B^2}k_\perp^2\left(\tilde{\phi} - \frac{T_e}{n_e e}\tilde{n}_e\right) \pm \left(1 + k_\perp^2\rho_s^2\right)\frac{\tilde{J}_{\parallel}}{V_a}\right] = 0, \tag{15.34}$$

as the equation identifying the characteristics for the kinetic shear Alfvén part of the system. Hence, not only the wave velocity but the combination of the variables, for the dynamics in the parallel direction, is dependent on the scale of motion in the perpendicular plane.

We do the same for the acoustic wave part of the system, again closely following the operations in Chapter 8. We substitute the ion polarisation in for $\nabla_\parallel \tilde{J}_\parallel$ in the density equation and assume adiabatic electrons so that $\tilde{\phi}$ is replaced by $(T_e/n_e e)\tilde{n}_e$ under $\nabla_\perp^2$, and combine the polarisation term with the time derivative of the density. Then, combining the density and parallel flow equations into one, we find

$$\left(1 - \rho_s^2 \nabla_\perp^2\right) \frac{\partial^2 \tilde{n}_e}{\partial t^2} = c_s^2 \nabla_\parallel^2 \tilde{n}_e, \tag{15.35}$$

since $n_e T_e / \rho = c_s^2$. This identifies $V_s$ with

$$V_s^2 = c_s^2 \left(1 + k_\perp^2 \rho_s^2\right)^{-1}, \tag{15.36}$$

as the characteristic velocity for the acoustic dynamics, as in Eq. (8.23) of Chapter 8 (with $\Gamma = 1$). We then proceed as per the characteristic procedure leading up to Eq. (15.18), to find

$$\left(\frac{\partial}{\partial t} \pm V_s \nabla_\parallel\right)\left[\left(1 + k_\perp^2 \rho_s^2\right)\tilde{n}_e \pm n_e \frac{\tilde{u}_\parallel}{V_s}\right] = 0, \tag{15.37}$$

as the equation identifying the characteristics for the acoustic part of the system. In this case as well, the wave velocity and the combination are dependent on the scale of motion in the perpendicular plane.

Especially for a global model, but often even when applying Dirichlet boundary conditions in the radial coordinate $x$, it is impossible or undesirable to apply Fourier transforms, especially in $x$, to get these velocities and boundary conditions. This is more obvious as well, in the case that the radial domain includes both closed and open flux surface regions. Then it is not clear which $k_\perp$ is to be used and/or what the decomposition should mean. Conversely, if Fourier transforms are not or cannot be used, it is difficult to treat the square root of the operators involved in the formulae for the wave speeds,

$$V_a^2 = v_A^2\left(1 - \nabla \cdot \rho_s^2 \nabla_\perp\right)\left(1 - d_e^2 \nabla_\perp^2\right)^{-1} \quad V_s^2 = c_s^2\left(1 - \nabla \cdot \rho_s^2 \nabla_\perp\right)^{-1}, \tag{15.38}$$

which come from ion polarisation ($\rho_s$) and electron inertia ($d_e$).

In lieu of the full square root operations, we note that the turbulence is more able to compete with the exit flux losses at small scale, where most of the vorticity is. So we can take forms which give the correct limit at long wavelength, essentially by neglecting the $K^2$ corrections in both forms This reduces to

$$J' = -\nabla \cdot n_e e v_A \rho_s^2 \nabla_\perp^2 \left(\frac{e\tilde{\phi}}{T_e} - \frac{\tilde{n}_e}{n_e}\right) \quad u' = c_s \frac{\tilde{n}_e}{n_e}, \tag{15.39}$$

as the $B'$ expressions in the boundary conditions. The characteristic exit flux speeds are then $v_A$ and $c_s$, following the fact that these are due to Alfvén and acoustic dynamics, respectively. In a 3D model these are the forms to insert into the procedure in Eqs (15.21) and (15.22). In a 2D model, following Eq. (15.25), we would have

$$\frac{d\varpi}{dt} = 2\frac{J'}{L_\parallel} - \mathcal{K}(T_e\tilde{n}_e), \tag{15.40}$$

$$e\frac{d}{dt}(n_0 + \tilde{n}_e) = 2\frac{J' - n_e e u'}{L_\parallel} + \mathcal{K}(n_e e\tilde{\phi} - T_e\tilde{n}_e) + e\,Q, \tag{15.41}$$

which constitutes a 2D exit flux model with interchange forcing, with either the source $Q$ as an input or background gradient $n_0(x)$ as a drive. The two important things to note are, firstly, the relevant outflow velocity for vorticity is the Alfvén speed, and secondly, the adiabatic response is still active, so that $J'$ appears in both equations and effectively couples $\tilde{\phi}$ and $\tilde{n}_e$ together. The phenomenology is different from resistive dissipative coupling as in Chapter 4, since it acts with relatively equal strength at all scales, due to the appearance of $\nabla_\perp^2$ in $J'$, rather than dominantly at large scale. We will find in the 3D models that this coupling is very effective and, although it is an ideal effect intrinsically, it dissipates extrinsically due to the open boundaries through which the exit flux flows out.

## 15.3 Effects on the turbulence

We use the three-dimensional Eqs (15.26)–(15.30) as we did before. The normalised equations are given in Eqs (14.1)–(14.5), and the geometric coefficients are given in Eq. (14.12), from Chapter 14. The same warm plasma cases as therein are used here to isolate the effects of the open field line geometry. The gradient drive term with $\omega_n = 1$ is used. Exit flux boundary conditions are placed at $s = \pm\pi$ in lieu of the field line connection on closed flux surfaces.

### 15.3.1 The slab and toroidal cases

The slab cases use the gradient drive and no source, and set the curvature terms $\mathcal{K}$ to zero (this compares with the 3D results in Chapter 12). As in those cases these are taken to $t = 4000$ with mode structure averaged over $2000 < t < 4000$ and time traces over $2000 < t < 4000$. All manner of set up and initialisation is the same, while only the boundary conditions at the ends of the field lines are changed. The difference physically is that the exit flux conditions replace field line connection in determining the effect of parallel dynamics on the turbulence. The results show much stronger turbulence than with toroidal periodicity, and the transition to large-scale motion induced by finite induction ($\hat{\beta}$) occurs at much lower values. Moreover, the magnetic flutter is no longer stabilising, as the turbulence increases monotonically even at $C = 1.39$ with $\hat{\beta}$.

The toroidal cases use the gradient drive and no source, and set the curvature terms such that $\omega_B = 2L_\perp/R_0 = 0.03$ as before. This compares with the 3D results in Chapter 14. Again, all considerations are otherwise the same, isolating the effect of the change in the parallel boundary conditions. Even more so than in the slab cases, the turbulence is much stronger, and sees its transition to MHD motion at whichever largest perpendicular scale is allowed, with $\hat{\beta}$ values above unity.

Initially the effective coupling $\phi \leftrightarrow n$, given by the characteristics in Eq. (15.34). But this is only enforced at the ends of the field lines. The constraint propagates up and down the field lines with speed $V_a$, so if $\hat{\beta}$ is large enough the turbulence can overcome it. This occurs especially in the toroidal cases with ballooning curvature forcing. The destabilisation by curvature forcing is between the ends of the field lines, so strong turbulence there can limit the region of coupling to the vicinity of the field line ends in the parallel domain.

The phase shift distributions indicate the change in mode structure character caused by this slowing down of the parallel (shear Alfvén) propagation. For $\hat{\beta} = 1$ and less, the coupling is effective and the phase shifts are like drift wave cases. This is also reflected in the variation of the parallel envelope structure as $\hat{\beta}$ increases. These two sets of measurements are given for the three $\hat{\beta}$ values at $C = 1.39$ in figure 15.1, and for the three $C$ values at $\hat{\beta} = 1$ in figure 15.2. The transition away from drift wave mode structure to or towards a more hydrodynamic or MHD relationship is most apparent in the variation with $\hat{\beta}$, where the separation of the long wavelength component to a peak at $\pi/2$ is most apparent. For the collisionality variation the main effect is broadening of the phase shift distributions and a reduction of the cross correlations. But the parallel coupling in the adiabatic response remains effective.

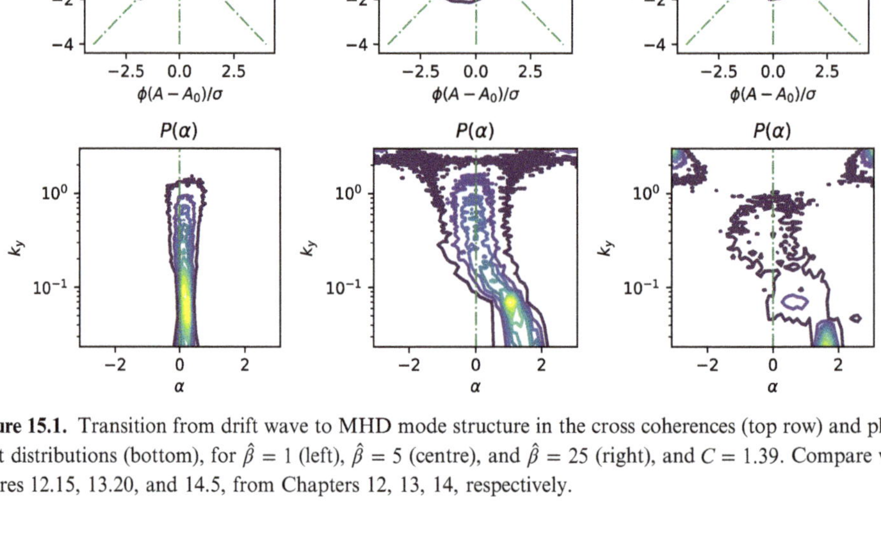


**Figure 15.1.** Transition from drift wave to MHD mode structure in the cross coherences (top row) and phase shift distributions (bottom), for $\hat{\beta} = 1$ (left), $\hat{\beta} = 5$ (centre), and $\hat{\beta} = 25$ (right), and $C = 1.39$. Compare with figures 12.15, 13.20, and 14.5, from Chapters 12, 13, 14, respectively.

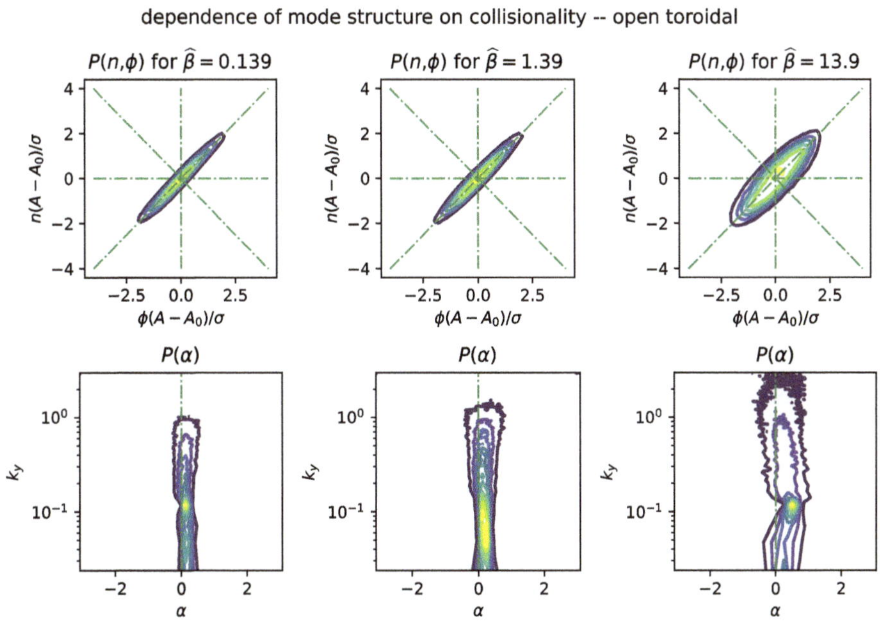


**Figure 15.2.** Transition from drift wave towards MHD mode structure in the cross coherences (top row) and phase shift distributions (bottom), for $C = 0.139$ (left), $C = 1.39$ (centre), and $C = 13.9$ (right), and $\hat{\beta} = 1$. In this case the stronger collisionality is not enough to break the adiabatic coupling and the long wavelength transition to MHD is not found. The larger $C$ is nevertheless felt in the broadening of the phase shift distributions and the weakening of the cross coherence. Compare this with figure 15.1.

We see this parallel coupling also in the state variable parallel envelopes for these cases, given as runs of $\hat{\beta}$ and $C$ in figure 15.3. In all but the one case with the largest $C$ the nonadiabatic density $h = n - \phi$ has a very flat envelope, more so than even the low-$\hat{\beta}$ cases under field line connection in toroidal geometry given in Chapter 14. The highest $\hat{\beta}$ case has a wave character to the envelope, resulting from the too-strong interaction with the open boundaries. When the dynamics is not localised, the situation at these boundaries becomes artificial, and the realism is lost. It is interesting to note that in the absence of field line connection, the envelope of $h$ is flat since the flute mode component is no longer disallowed, and the ballooning curvature and magnetic shear are not sufficient to localise the rest of the dynamics.

In the slab cases there is no curvature forcing to assist the gradient drive, and while some effect tilting the long wavelength side of the phase shift distributions towards larger values, the parallel coupling remains effective, with $h = n - \phi$ smaller than either $n$ or $\phi$ and with a very flat envelope despite the ballooning of especially $n$, caused in the absence of curvature forcing solely by the losses represented by the exit fluxes. Since the acoustic characteristics propagate more slowly than the shear Alfvén ones, it is natural that $n$ is the more strongly affected. The cross correlation and phase shift distributions are given for the slab cases in figures 15.4 and 15.5, with the same format as those just given for the toroidal cases. Only in the highest-$\hat{\beta}$ case is there any significant effect against the drift wave mode structure. The adiabatic coupling

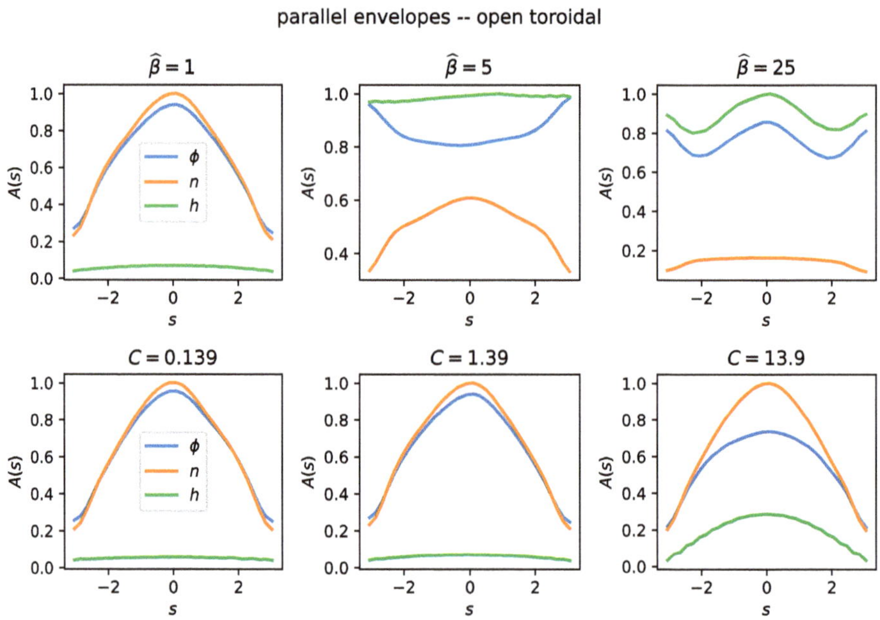


**Figure 15.3.** Parallel envelopes for the run of $\hat{\beta}$ (top row), and for the run of $C$ (bottom) for the driven toroidal cases, increasing values $\hat{\beta} = 1, 5, 25$, and $C = 0.139, 1.39, 13.9$, from left to right, respectively. One sees the lack of localisation in the parallel direction for the larger $\hat{\beta}$ cases. The only case with localisation in $h = n - \phi$ is $C = 13.9$ (lower right). The mode structure shows effective adiabatic coupling, but the lack of field line connection leads to a reduction of the tendency towards ballooning.

remains effective. On the other hand, the turbulence remains existent for high-$\hat{\beta}$ in the absence of field line connection. The parallel wavenumber spectra reflect the strong peak of the flute mode component, as in the three-dimensional slab cases of Chapter 11. This mode is not so strongly affected by the linear magnetic flutter component. The finite-beta stabilisation of this electromagnetic drift wave turbulence is found to be a result of the constraining provided by the field line connection.

### 15.3.2 Effective adiabatic coupling

The main lesson learned from these investigations is that the exit flux boundary conditions provide an effective field line connection that does not depend on the elimination of the flute mode by magnetic shear. Even for $k_{\|} = 0$ the coupling remains, since the absence of an inflow at the ends of the field lines where the boundary conditions are set results in effective line-tying. That is, not line-tying of the entire fluctuation set (towards zero at the ends of the field lines), but of $n$ and $\phi$ to each other. The nonadiabatic component $h$ is held down by this, only overcome if either $\hat{\beta}$ or $C$ is large enough. Hence, open field lines does not mean one can go back to an MHD model. That is because the characteristics of the parallel dynamics remain a two-fluid phenomenon. One finds a large amplitude for $h$, signifying $\tilde{\phi} > (T_e/n_e e)\tilde{n}_e$, only for large beta, or $\hat{\beta}$ larger than unity and $\beta_e$ larger than the mass ratio.

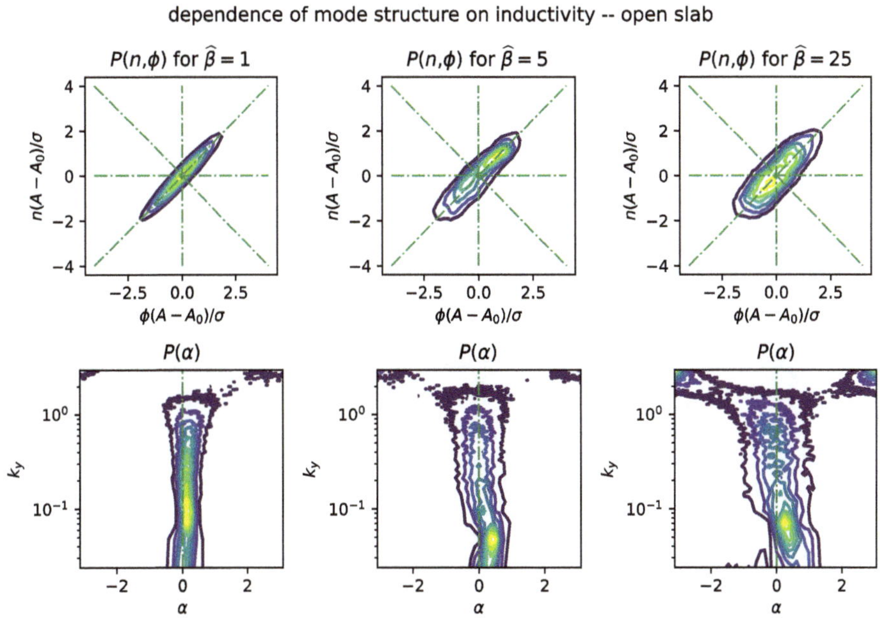


**Figure 15.4.** Transition from drift wave to MHD mode structure in the cross coherences (top row) and phase shift distributions (bottom), for $\hat{\beta} = 1$ (left), $\hat{\beta} = 5$ (centre), and $\hat{\beta} = 25$ (right), and $C = 1.39$, in the sheared slab cases with $\omega_B = 0$. Compare with figure 15.1.

The case of an open flux tube which is driven everywhere along the field lines by the gradient drive is most applicable in the case of a tokamak 'scrape-off' layer. The open flux surface region of the plasma is set by a material surface breaking the field lines. Outward from the high pressure regions in the plasma—which can still be thought of as a shaped cylinder wrapped around toroidally so that the flux surface can be closed—the turbulence spills over radially onto these open flux surfaces. Then, the plasma can flow out either end onto a material boundary. At these locations some sort of boundary condition is set. Either the situation is not very collisional, in which case the dynamics is not localised along the field lines, or it is collisional enough so that the path to the material surface can be regarded as a region in which fluxes exit the region of the strong dynamics. Then, an exit flux model is appropriate. In these regions the plasma beta is very small, essentially always satisfying $\beta_e \ll \mu_e$, such that the low-beta, high-collisionality case is the one which is relevant. For these cases the effective adiabatic coupling through the exit fluxes is a significant effect and should be included in any modelling effort.

In astrophysical applications it makes much less sense to have an external gradient drive. The plasma is sourced in some localised region in one way or another. Then, it tends to flow out along the field lines such that it only persists as long as the dynamical region is being fed. We consider a prototype for this next.

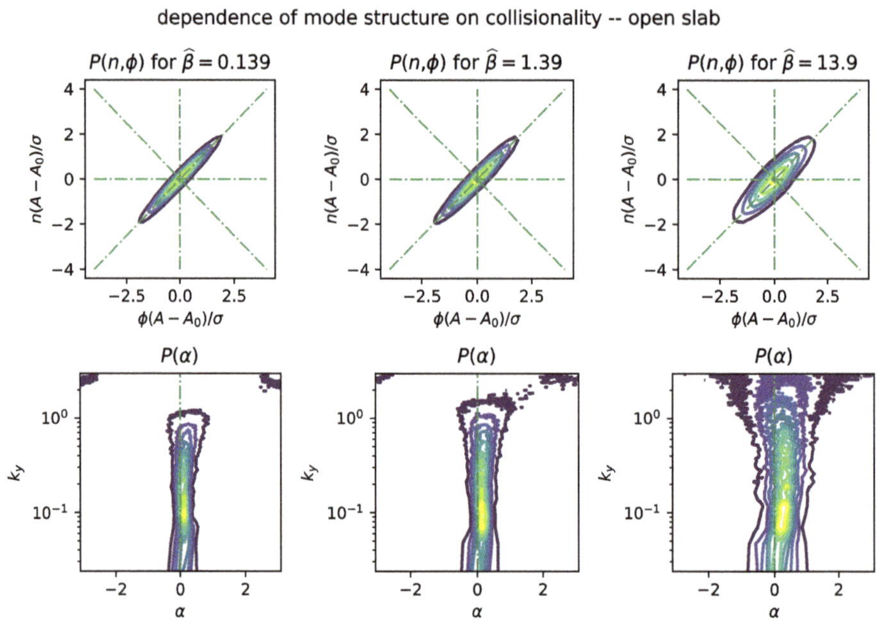


**Figure 15.5.** Transition from drift wave towards MHD mode structure in the cross coherences (top row) and phase shift distributions (bottom), for $C = 0.139$ (left), $C = 1.39$ (centre), and $C = 13.9$ (right), and $\hat{\beta} = 1$, in the sheared slab cases with $\omega_B = 0$. In this case the stronger collisionality is not enough to break the adiabatic coupling and the long wavelength transition to MHD is not found. The larger $C$ is nevertheless felt in the broadening of the phase shift distributions and the weakening of the cross coherence. Compare this with figure 15.2.

### 15.3.3 Self-consistent gradient drive

When the plasma is supported by a source, one no longer has an external gradient drive which is set, but a conservative energy transfer process between the profile and the turbulence. To analyse this we first have to define what is meant by a profile quantity. In the $xys$ coordinate system the field lines lie in the planes of constant $x$, which form the flux surfaces. We can have the following discussion with either closed (toroidally periodic) or open surfaces, as here. The only requirement is that $y$ represents a symmetry coordinate of the background and we can treat it as periodic. The domain in $s$ is either closed or open, and the domain in $x$ is bounded. In both $x$ and $s$ we assume we can take integrations by parts and neglect surface terms coming from integrations over divergences. When surface terms at the exit locations are kept, they produce loss terms. These are part of the parallel energetics in the unperturbed parallel gradient and divergence terms, not the nonlinear processes, and so they can be treated separately.

The profile of a quantity is its surface average over $y$ and $s$. Since we are working with a constant volume element (inverse of the Jacobian), the operations are straightforward. The surface average of any scalar quantity $f$ is given by

$$\langle f \rangle = \frac{1}{V'} \oint\oint \sqrt{g}\, dy\, ds\, f \quad \text{where} \quad V' = \oint\oint \sqrt{g}\, dy\, ds, \tag{15.42}$$

and $V'$ is the derivative of the volume with respect to $x$. In our case the factor of $\sqrt{g}$ cancels out. In the computations with the equidistant grid, this becomes a discrete average over grid points. This defines $\langle f \rangle$ as the profile of $f$. Then, the fluctuations are defined as

$$\tilde{f} = f - \langle f \rangle \quad \text{so that} \quad f = \langle f \rangle + \tilde{f}. \tag{15.43}$$

These have the property that their profile, or surface average, is zero.

The free energy can be split between the profile component and the fluctuation component, since it is a volume quantity. That is, a volume quantity is given by the radial integration (in $x$) of a profile quantity. This works since the profile of the cross term $\tilde{f}\langle f \rangle$ also vanishes. So we have

$$\int d\mathscr{V}\, f^2 = \int d\mathscr{V}\, \langle f \rangle^2 + \int d\mathscr{V}\, (\tilde{f})^2, \tag{15.44}$$

and we can treat their equations separately.

A nonlinear bracket enters as a composition of these pieces. With two scalar quantities $a$ and $b$, we have the following properties,

$$[\langle a \rangle, \langle b \rangle] = 0 \quad \langle[\tilde{a}, \langle b \rangle]\rangle = 0 \quad \langle[a, b]\rangle = \langle[\tilde{a}, \tilde{b}]\rangle, \tag{15.45}$$

all due to the fact that the bracket involves the $y$-derivative which vanishes for a profile quantity. Using the ExB advection of the density as an example, we have

$$[\phi, n] = [\langle \phi \rangle, \tilde{n}] + [\tilde{\phi}, \langle n \rangle] + [\tilde{\phi}, \tilde{n}]. \tag{15.46}$$

It follows that the two components of the pure ExB advection equation are

$$\frac{\partial}{\partial t}\langle n \rangle + \langle[\tilde{\phi}, \tilde{n}]\rangle = 0, \tag{15.47}$$

$$\frac{\partial \tilde{n}}{\partial t} + [\langle \phi \rangle, \tilde{n}] + [\tilde{\phi}, \langle n \rangle] + [\tilde{\phi}, \tilde{n}] - \langle[\tilde{\phi}, \tilde{n}]\rangle = 0. \tag{15.48}$$

Multiply the first by the profile and the second by the fluctuation, to find

$$\frac{1}{2}\frac{\partial}{\partial t}\langle n \rangle^2 + \langle n \rangle\langle[\tilde{\phi}, \tilde{n}]\rangle = 0, \tag{15.49}$$

$$\frac{1}{2}\frac{\partial}{\partial t}\tilde{n}^2 + \frac{1}{2}[\phi, \tilde{n}^2] + \tilde{n}[\tilde{\phi}, \langle n \rangle] - \tilde{n}\langle[\tilde{\phi}, \tilde{n}]\rangle = 0, \tag{15.50}$$

where in the second term $\langle \phi \rangle + \tilde{\phi} \rightarrow \phi$ has been combined back again. Then use the Leibnitz rule on the third term to recast this as

$$\frac{1}{2}\frac{\partial}{\partial t}\tilde{n}^2 + \frac{1}{2}[\phi, \tilde{n}^2] + [\tilde{\phi}, \tilde{n}\langle n \rangle] - \langle n \rangle[\tilde{\phi}, \tilde{n}] - \tilde{n}\langle[\tilde{\phi}, \tilde{n}]\rangle = 0. \tag{15.51}$$

The profile component is already a surface average, so now take the surface average of the fluctuation component to find

$$\frac{1}{2}\frac{\partial}{\partial t}\langle\tilde{n}^2\rangle + \frac{1}{2}\langle[\phi, \tilde{n}^2] + [\tilde{\phi}, \tilde{n}\langle n\rangle]\rangle - \langle n\rangle\langle[\tilde{\phi}, \tilde{n}]\rangle, \tag{15.52}$$

where the fifth term dropped since the average of $\tilde{n}$ vanishes. We already see that the product of the profiles of $n$ and the bracket appears equal and opposite in the profile and fluctuation energy components. Finally, do the integration over $x$, wherein the second and third terms in the fluctuation component vanish, to obtain

$$\int\frac{dx}{L_x}\frac{1}{2}\frac{\partial}{\partial t}\langle n\rangle^2 = -\int\frac{dx}{L_x}\langle n\rangle\langle[\tilde{\phi}, \tilde{n}]\rangle, \tag{15.53}$$

$$\int\frac{dx}{L_x}\frac{1}{2}\frac{\partial}{\partial t}\langle\tilde{n}^2\rangle = \int\frac{dx}{L_x}\langle n\rangle\langle[\tilde{\phi}, \tilde{n}]\rangle, \tag{15.54}$$

where the integration is over the entire domain (of width $L_x$). This remaining term is equal and opposite and so represents a conservative transfer between the profile and fluctuation components of the thermal free energy. Lastly, we can recast it by using the expression

$$\langle[\tilde{\phi}, \tilde{n}]\rangle = \frac{\partial}{\partial x}\langle\tilde{n}v_E^x\rangle \quad \text{where} \quad v_E^x = [\tilde{\phi}, x], \tag{15.55}$$

and then integrate this $x$-derivative by parts in both equations to find

$$\int\frac{dx}{L_x}\frac{1}{2}\frac{\partial}{\partial t}\langle n\rangle^2 = \int\frac{dx}{L_x}\langle\tilde{n}v_E^x\rangle\frac{\partial}{\partial x}\langle n\rangle = \int\frac{dx}{L_x}\langle F_e\rangle\langle\nabla n\rangle, \tag{15.56}$$

$$\int\frac{dx}{L_x}\frac{1}{2}\frac{\partial}{\partial t}\langle\tilde{n}^2\rangle = -\int\frac{dx}{L_x}\langle\tilde{n}v_E^x\rangle\frac{\partial}{\partial x}\langle n\rangle = -\int\frac{dx}{L_x}\langle F_e\rangle\langle\nabla n\rangle, \tag{15.57}$$

where

$$F_e = \tilde{n}v_E^x \quad \langle\nabla n\rangle = \frac{\partial}{\partial x}\langle n\rangle. \tag{15.58}$$

We already had this as the energy drive term in Chapter 9 in the discussion of phase shifts. Net turbulent flux down the gradient is the drive term. Here, the surface average flux times the surface average gradient is the transfer term between the profile and fluctuation components, so that we are discussing the same process just represented a different way. When the turbulence is maintained by a source, this becomes the way to understand how the turbulence is driven.

### 15.3.4 Cases with a source

In an alternative to the cases driven by a background gradient, which we looked at above, here we do not assume a confined plasma with background gradients, but a source supplying the domain near $x = s = 0$ which balances the loss terms arising from the exit flux boundary conditions. These are referred to as sourced cases as opposed to the driven cases. It is of course interesting to see whether one obtains turbulence at all, or will the plasma input by the source simply run down the field lines and out of the domain.

In these sourced cases, the geometry is set up exactly as before. Setting $\omega_n = 0$, the gradient drive is taken back out. Instead of it we specify a source $Q = Q_0 \exp(-ax^2 - bs^2)$ with $Q_0 = 10^{-2}$ and $a = 100/L_\perp^2$ and $b = 1$ as nominal choices. In this case we do not assume a confined plasma with background gradients, but a source supplying the domain near $x = s = 0$ which balances the loss terms arising from the exit flux boundary conditions. The source is a layer, whose extent in the $y$-direction is assumed to be long enough that it can be treated as independent of $y$ for these purposes. In another principal difference, we run these without magnetic shear, setting $S = 0$, so that only the localisation of the source and the ballooning of the curvature can contribute to localisation of the turbulence. The same exploratory cases as those above are taken: $\hat{\beta} = 1$ at $C = 0.139$ and 1.39 and 13.9, and then $\hat{\beta} = 5$ and 25 at $C = 1.39$, reflecting $\hat{\mu} = 2.72$ and $\nu_e L_\perp / c_s = 0.1$ and 1 and 10. The main interest is in the variation of the mode structure, not so much the scaling.

The collisionality variation at low beta, $\hat{\beta} = 1$, most closely reflect the expectations of localised dynamics. This is largely due to the fact that $h$ is effectively held down. The parallel envelopes are shown in figure 15.6, with cases with $\omega_B = 0.03$

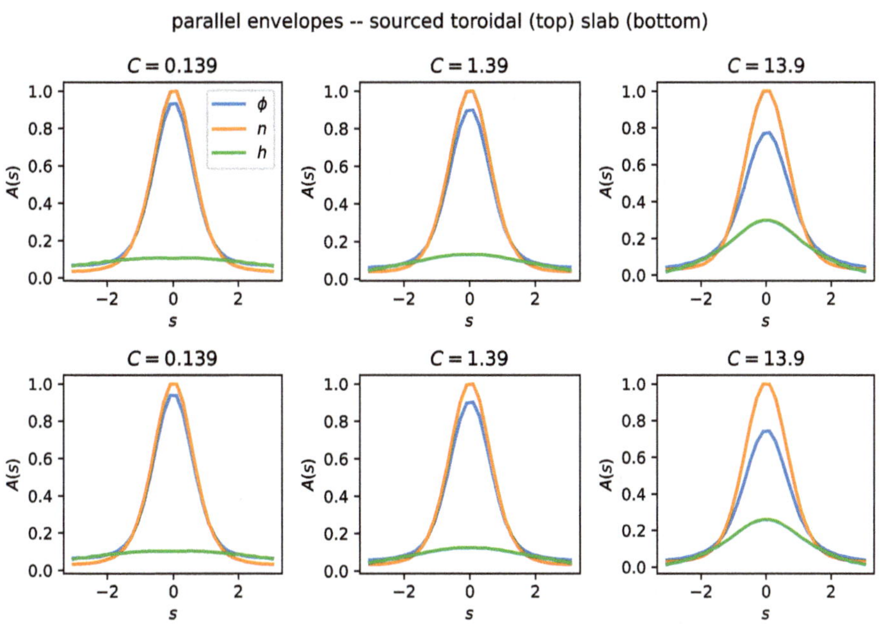


**Figure 15.6.** Parallel envelopes for the run of $C$ for the sourced toroidal (top row) and slab (bottom row) cases with $\hat{\beta} = 1$ and $S = 0$, with increasing values $C = 0.139$, 1.39, 13.9, from left to right, respectively. One sees the localisation in the parallel direction for all these cases. The differences between the slab and toroidal cases is moderate at most. This is classic drift wave mode structure, following the relatively low $\hat{\beta}$ value. With the turbulence effectively localised to the source, the scenario is not problematic.

and 0 shown together. Their essential similarity shows that the spatial localisation of the turbulence along the field lines is due more to the source than to the ballooning curvature forcing. The two cases with higher $\hat{\beta}$ cease to be localised. For $\hat{\beta} = 5$ the envelope of $h$ is still flat but especially $\phi$ is no longer localised along the magnetic field. For $\hat{\beta} = 25$ the envelope of $n$ remains localised but the amplitude of the Alfvén component is now large and not well localised.

The strength of the turbulence depends not on the drive but on how efficiently the free energy is transferred out of the $k_y = 0$ component into the $k_y \neq 0$ components of the thermal energy given by $T_e \tilde{n}_e$. In this case, with the source localised to $x = 0$, it makes a difference whether the turbulence is effectively driven on both $x < 0$ and $x > 0$ sides or not. For $\omega_B = 0$ only the drift wave mechanism is available. This, however, does not depend on the sign of the gradient, so with the $k_y = 0$ component maximum in the middle, the turbulence is driven on both sides, with the resulting flux driven away from $x = 0$ on both sides. So a full volume average yields small values. For the toroidal cases, the contrast between $\hat{\beta} = 1$ and $\hat{\beta} = 5$ at $C = 1.39$ shows this clearly. Figure 15.7 has this comparison. The top row of frames gives the profile components of $n$, the flux $F_e$, and then the profiles of $F_e$ and the gradient $\nabla n$ multiplied together. As analysed above, free energy is taken out of the gradient

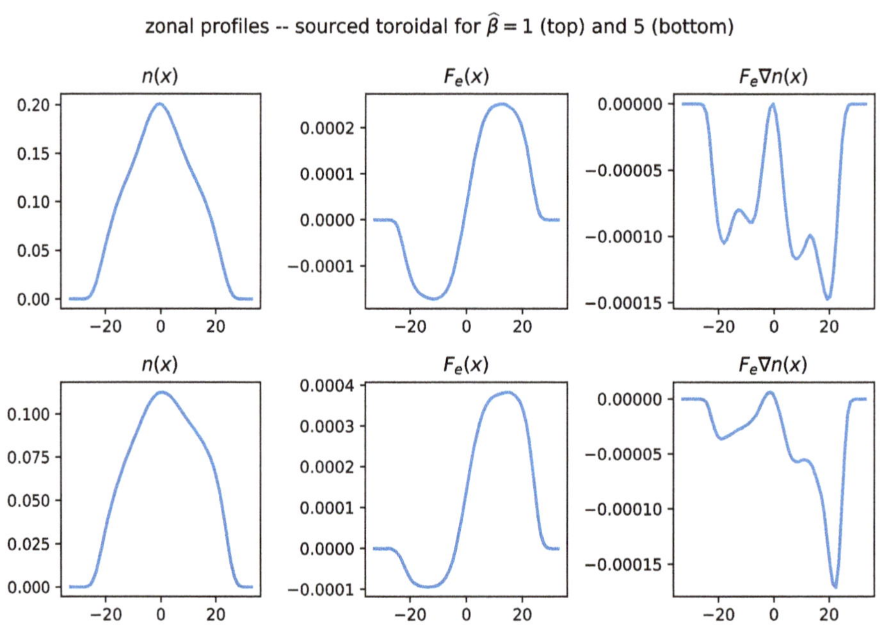


**Figure 15.7.** Profile components of the density (left), turbulent ExB flux in the $x$-direction (centre), and the product of the profile flux and the profile gradient (right), for the sourced toroidal ($\omega_B = 0.03$) cases with $\hat{\beta} = 1$ (top row) and $\hat{\beta} = 5$ (bottom) at $C = 1.39$. For low enough inductivity the dynamics is nearly symmetric. The flux is directed away from the central layer ($x = 0$) on both sides with nearly equal magnitude, and the self-consistent drive of the turbulence by the profile (third frame, negative values) is nearly equal on both sides. For $\hat{\beta} = 5$ the dynamics is more interchange-like, with most of the energetics on the right-hand side, where in the source region the curvature vector and gradient are aligned.

where this product is negative. At all locations on the profile a negative sign is found, indicating drive of the turbulence by the zonal component, which is maintained by the source. The flux is away from the central position on both sides, and nearly symmetric for $\hat{\beta} = 1$. For the higher value, however, the energetics is mostly on the $x > 0$ side, where the curvature forcing near $s = 0$ is effective for the interchange drive mechanism. The flux itself is comparable, but that is because the source is the same for both cases. The stronger interchange turbulence is now manifested in the self-consistent peak value of the profile, which is higher for the $\hat{\beta} = 1$ case, reflecting weaker turbulence at lower inductivity.

The axisymmetric spatial distributions of the density, flux, gradient, and flux gradient product are shown in figure 15.8, for the toroidal case with $\hat{\beta} = 1$ and $C = 1.39$. Following the $k_y = 0$ energetics, the flux and gradient are averaged before multiplication. These are averages over $y$ only, leaving the $k_y = 0$ component as a function of both $x$ (radial) and $s$ (parallel). This is done to sort out whether the maximum drive is at the source region or displaced up and down the field lines. The symmetry in the energetics is reflected in the rightmost frame, with a small enhancement on the $x > 0$ side reflecting the action of the curvature forcing. Essentially all of the energetics is in the vicinity of $s = 0$, which is also what the parallel envelopes show above.

We therefore compare averages over the fraction of the volume with $x > 0$, which we reckon as the right-hand side. Indeed, for all five cases the turbulent ExB

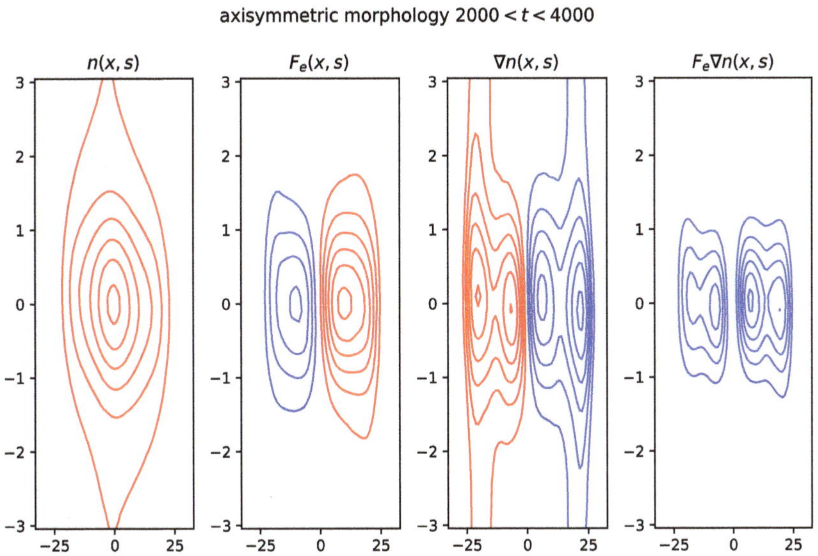


**Figure 15.8.** Axisymmetric components of the density, turbulent ExB flux in the $x$-direction (left), the gradient, and the energetic effect on the profile component (left to right), for the $\hat{\beta} = 1$ case from figure 15.7. The last is the product of the second and third, with negative values indicating self-consistent drive of the turbulence by the profile through the nonlinear advection. For this inductivity the dynamics is nearly symmetric. The same conclusions are drawn as in figure 15.7, with the additional information being the localisation of the energetics to the source region around $x = 0$.

transport flux on that side are only a factor of 1.5–2 larger with $\omega_B = 0.03$ than with $\omega_B = 0$. This also reflects the value of the source, so we also check the profile peak values for these cases. These range from 0.115 to 0.216 for all ten cases, signifying the strongest turbulence for the toroidal case with $\hat{\beta} = 5$ and weakest for the slab case with $\hat{\beta} = 1$. At $\hat{\beta} = 1$ there is very little sensitivity to collisionality and toroidicity, with all six values near 0.2. So we find similarly strong dynamics in all cases, with an enhancement of interchange drive for higher inductivity.

The spatial morphology in the central drift plane ($s = 0$) is shown in figure 15.9, for the toroidal case with $\hat{\beta} = 1$ and $C = 1.39$. One finds vigorous turbulence, which as indicated in figures 15.7 and 15.8 is present on both sides of the source region in $x$. All scales are present, not just the large ones. The look and feel tell us little about the energetics however, which is why we had to measure it. The vorticity shows a moderate tendency to become sheet-like but not as extreme as in pure hydrodynamic cases. This is understandable once we have seen the other diagnostics.

However, the results are different for stronger inductivity. The axisymmetric spatial distributions of the density, flux, gradient, and flux gradient product for the case with $\hat{\beta} = 5$ and $C = 1.39$ are shown in figure 15.10. The greater degree of radial asymmetry in the energetics found in figure 15.7 is reflected here, while the localisation around $s = 0$ along the field lines is still present. A large fraction of the energetics is found on the positive-$x$ side of the source location, resulting from the positive curvature forcing on that side. The spatial morphology for this case is shown in figure 15.11. The signature is more that of convection in the $x$-direction, with a few prominent large cells driving all the other scales. The vorticity layers have the

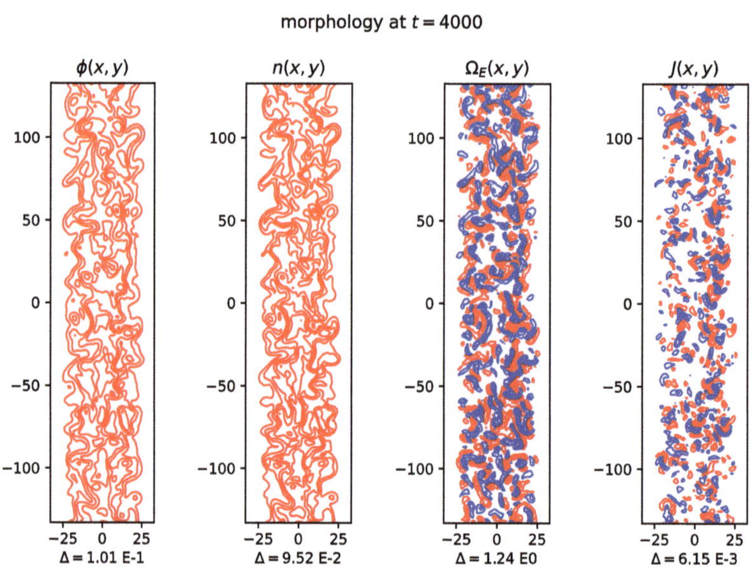


**Figure 15.9.** Spatial morphology of the turbulence in the midplane, for the $\hat{\beta} = 1$ case from figure 15.7. The dynamics involves all scales, with the largest available ones not dominant. The vorticity structures fill space, and even the current is only moderately intermittent.

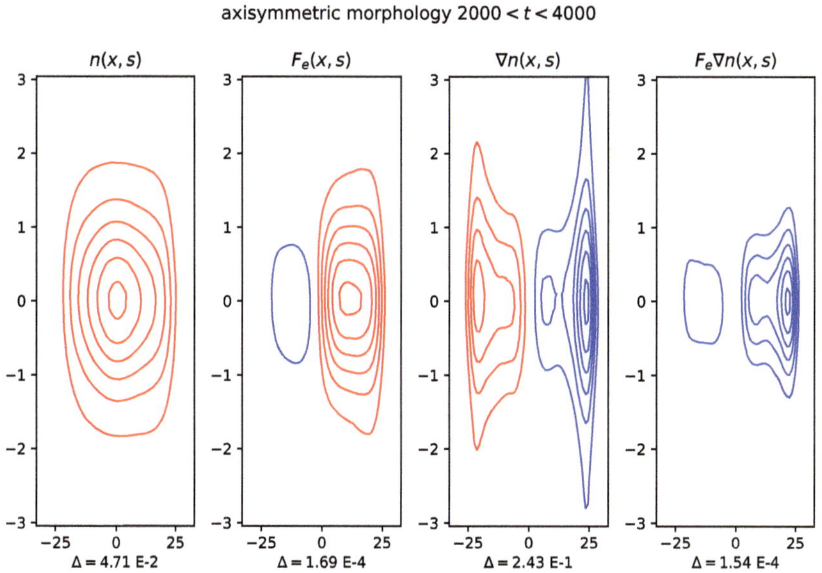


**Figure 15.10.** Axisymmetric components of the density, turbulent ExB flux in the $x$-direction (left), the gradient, and the energetic effect on the profile component (left to right), for the $\hat{\beta} = 5$ case from figure 15.7. The last is the product of the second and third, with negative values indicating self-consistent drive of the turbulence by the profile through the nonlinear advection. For this higher inductivity the dynamics is now skewed towards the right-hand side, due to the interchange forcing on the turbulence. The same conclusions are drawn as in figure 15.7, with the additional information being the localisation of the energetics to the source region around $x = 0$.

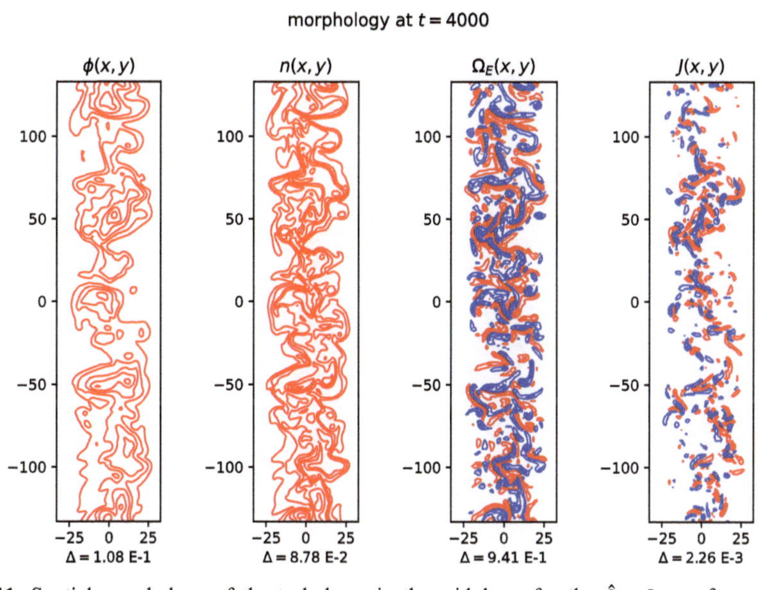


**Figure 15.11.** Spatial morphology of the turbulence in the midplane, for the $\hat{\beta} = 5$ case from figure 15.7. Contrast with figure 15.9. The dynamics is more prominently at the largest scales with a few centres of activity driving the rest. Sheet structure in the vorticity is more obvious, and the current is more intermittent.

hydrodynamic form of thin sheets. The current layers are more strongly localised. With strong interchange forcing, the turbulence shows more of a hydrodynamic character.

On the other hand, there is very little difference between the toroidal versions of the $\hat{\beta} = 5$ and even $\hat{\beta} = 25$ cases and the corresponding slab ones, except for the energetic signatures seen in figure 15.7. These are symmetric for the slab cases in contrast to the shift to the right seen in the energetic profile $\langle F_e \rangle \langle \nabla n \rangle$ for the toroidal cases. In all the other diagnostics, even the phase shifts and cross correlations, the slab and toroidal cases show identical structure. In the slab cases the spatial morphology is also inductive of a more hydrodynamic character. But there is no curvature forcing or interchange mechanism in the slab cases.

The three-dimensional structure is shown for $n$ in figure 15.12, for the toroidal case with $\hat{\beta} = 5$ and $C = 1.39$. Every fourth drift plane is shown, as there are 32 in total. When a pronounced large-scale structure emerges, one can see it following the field lines. The shapes are similar on all planes, even though the amplitude falls off strongly away from $s = 0$ on either side along the magnetic field. For this value of $\hat{\beta}$, the turbulence exists on both sides of the $x = 0$ line. For larger values, the one-sided nature becomes clearer. This is also seen in the $x$-dependence of the fluctuation envelope—to get this, the $k_y = 0$ component is stripped out and then $n^2$ is averaged over $y$ and $s$, leaving the envelope in $x$ as a profile. For $\hat{\beta} = 1$ the envelope is more symmetric, while for $\hat{\beta} = 5$ and 25 it is skewed towards the $x > 0$ side. This is due to the stronger interchange character, as seen in the profiles shown above.

So with all the control cases present we have to conclude that the character seen is a result of the differences in inductivity on the slab physics rather than an emergence of interchange signature, despite what is indicated in figure 15.7. The localisation in the parallel direction forms the main contrast to the driven cases in which the

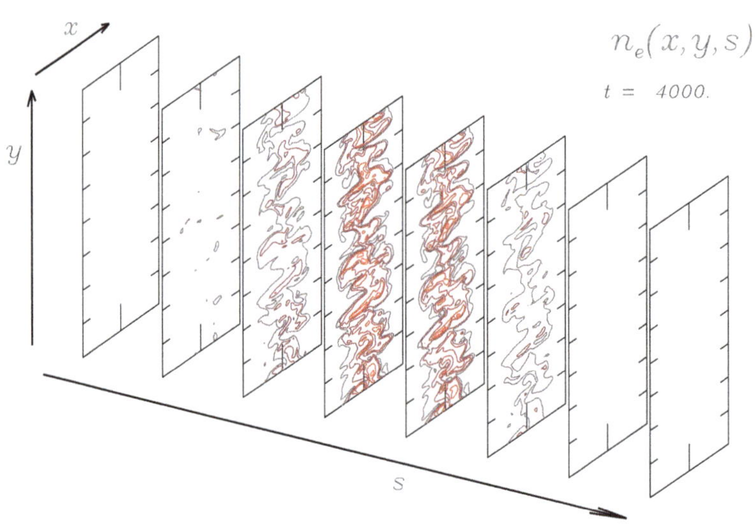


**Figure 15.12.** Three-dimensional morphology of the density, for the $\hat{\beta} = 5$ case from figure 15.11. Localisation along the field lines is visible, as is the tendency of the disturbance structures to follow the field lines. Every fourth drift plane is shown (8 of 32 in all).

gradient forcing is homogeneous in the parallel direction even if the curvature is not. This localisation, resulting mainly in the density for which the balance is largely between the source at the midplane and the spreading out via sound waves, occurs if the field lines are long enough relative to the gradient scale length, or conversely if the gradient is steep enough in an otherwise given configuration. If the field lines are too short then the dynamics is not localised and the parallel boundary conditions have to be different.

The localisation of the Alfvén dynamics itself is seen to break down for the $\hat{\beta} = 25$ cases, which causes several curiosities in the results. The main one is that weaker turbulence is found for these cases. The more hydrodynamic turbulence spreads the zone of activity in the $x$-direction so that the zone of finite energetics (in the $\langle F_e \rangle \langle \nabla n \rangle$ profile) is pushed to the radial boundary, in both directions for the slab case and mainly outward ($x > 0$) for the toroidal case, so that this zone is artificially compressed. It is apparent that physical realism is being lost in this case and a better model would be needed to treat it. It is always necessary to be aware of this type of limitation and to face it when it occurs.

The version of this turbulence in which we have a curved magnetic field with zero shear, the curvature forcing maximum at the $s = 0$ drift plane, and exit flux boundaries at some finite $s = \pm s_0$ has a natural analogue, to which we now turn.

## 15.4 Turbulence in a dipole magnetic field

Of interest in some areas is how turbulence functions in a magnetised plasma, when the background guide field is a dipole magnetic field. There are laboratory experiments with this geometry, and there is also relevant physics in a planetary magnetic field with this geometry. The dipole field, taken to be generated by loop currents near the origin of a spherical domain, is a textbook problem. We will treat it as a special case of the general axisymmetric magnetic MHD equilibrium, in which the region is close to vacuum so that a force-free field results. Consider it an alternative derivation that may be easier than the usual one. Then, we develop flux-tube geometry for the dipole field, which treats the severe volumetric changes that a flux tube encounters as the field lines are traced away from the equator towards the poles. This turns out to be another way to treat ballooning in the physics. Due to increase of the magnetic field strength in the parallel direction, the exit flux boundaries are placed where the dynamics is expected to be weaker. The lack of shear makes the problem tractable. Then, an introduction to the turbulence found in this geometry forms a useful example of how to approach this problem in general situations.

### 15.4.1 The dipole magnetic field

We consider the MHD equilibrium in the absence of pressure or currents for a purely poloidal magnetic field driven by currents sufficiently close to the origin of a spherical domain that only the outer vacuum region need be treated. We can use this as a model for the magnetic field provided $\beta = 2\mu_0 p/B^2$ is arbitrarily small. Our warm plasma case with $\beta \sim 10^{-3}$ should suffice.

The general axisymmetric magnetic field is once more given by

$$\mathbf{B} = I\nabla\varphi + \nabla\psi \times \nabla\varphi, \tag{15.59}$$

and with the field purely poloidal we have $I = 0$. The current is given by

$$\mu_0\mathbf{J} = -(\Delta^*\psi)\nabla\varphi = 0 \quad \text{hence} \quad \Delta^*\psi = R^2\nabla\cdot\frac{1}{R^2}\nabla\psi = 0. \tag{15.60}$$

The base coordinate system $r\theta\varphi$ is spherical, but $R$ is still the cylindrical radius, so we have

$$g^{rr} = 1 \quad g^{\theta\theta} = \frac{1}{r^2} \quad g^{\varphi\varphi} = \frac{1}{r^2\cos^2\theta} \quad \sqrt{g} = r^2\cos\theta \quad R = r\cos\theta. \tag{15.61}$$

Hence $R^2/\sqrt{g} = \cos\theta$ and with $\cos\theta$ commuting with $\partial/\partial r$ we have

$$r^2\Delta^*\psi = r^2\frac{\partial^2\psi}{\partial r^2} + \cos\theta\frac{\partial}{\partial\theta}\frac{1}{\cos\theta}\frac{\partial\psi}{\partial\theta} = 0. \tag{15.62}$$

This is separable, with the terms showing only $r$ and only $\theta$ dependence, respectively. The solution is then $r^n$ with either $n = 2$ or $n = -1$ for the first and $\cos^2\theta$ for the second. This yields

$$\psi = \frac{\cos^2\theta}{r}, \tag{15.63}$$

as the result, multiplied by a constant which can be understood as the normalisation of $\psi$. The choice $n = -1$ is due to the boundary condition that $\psi \to 0$ at radial infinity.

The magnetic field components we need are given by

$$B^2 = \frac{1 + 3\sin^2\theta}{r^6} \quad B^r = -2\frac{\sin\theta}{r^3} \quad B^\theta = \frac{\cos\theta}{r^4}. \tag{15.64}$$

To find these use $\partial\psi/\partial r = -\psi/r$ and $\partial\psi/\partial\theta = -2\psi\tan\theta$ and then $1 + \tan^2\theta = \cos^{-2}\theta$, etc, along with $\sqrt{g}$ above. As in the Clebsch case we are using the relationship between $B^\theta$ and $\sqrt{g}$.

### 15.4.2 Flux-tube geometry in a dipole field

Instead of $r$ as a coordinate we use $x = x(\psi)$ and therefore invert the relation between $r$ and $\psi$, so that $r = r(\psi, \theta)$, with

$$r = \frac{\cos^2\theta}{\psi}, \tag{15.65}$$

to be eventually substituted everywhere it occurs. In all of these forms, $\psi = 1$ and $r = 1$ at $\theta = 0$, in units of $r_0$, the reference radial location of the flux tube in the equator, with $\psi$ in units of $\psi_0$, the value of $\psi$ at $r = r_0$ in the equator. The magnetic field strength $B$ is $B_0$ at that location. These are used in the local normalisation.

The $r^{-3}$ dependence of many of the forms, native to the dipole field, eventually creates a very strong dependence of all metric coefficients with increasing $\theta$, which is to provide the same sort of localisation that the magnetic shear does in the ballooning theory. The structure in $\theta$ is exemplified by the magnetic field component

$$B^\theta = \nabla\psi \times \nabla\varphi \cdot \nabla\theta = \frac{\psi}{r}\nabla r \times \nabla\theta \cdot \nabla\varphi = \frac{\cos\theta}{r^4} = \frac{\psi^4}{\cos^7\theta}, \tag{15.66}$$

where in the last expression we are in a $(\psi, \theta)$ coordinate system, not an $(r, \theta)$ one. But at constant $\psi$ this is a very severe $\theta$-dependence indeed.

It is desirable to find an angle coordinate for the parallel direction for which $B^s$ is independent of $\theta$. Since we already have $B^\theta$ at constant-$\psi$, we can set up the definition of $s$ as a magnetic differential equation, $B^s = B_0/r_0$, or without the units,

$$B^s = B^\theta \frac{\partial s}{\partial\theta} = 1 \quad \frac{\partial s}{\partial\theta} = \cos^7\theta, \tag{15.67}$$

so that with the $\cos^2\theta$ dependence of $r$, we have at constant-$\psi$

$$s = \int_0^\theta d\theta' \cos^7\theta' = z - z^3 + \frac{3}{5}z^5 - \frac{1}{7}z^7 \quad z = \sin\theta, \tag{15.68}$$

with the choice of normalising such that $s = \theta$ at $\theta = 0$ on the midplane. For $-\pi/4 < \theta < \pi/4$ at constant $\psi = 1$, moving along the field lines, the radius drops from $r = 1$ to $r = 1/2$ in units of $r = r_0$ on the midplane, and $B/B_0$ rises from unity to more than ten, respectively, at these exit locations. It does seem reasonable to apply exit flux boundaries in this vicinity.

The formula for $s = s(\theta)$ at $\psi = \psi_0$ is then used to tabulate all of the above geometric coefficients as functions of $s$. The volume element in the new coordinate system is constant along the field line since it given by $B^s$ times a flux function. It remains to find the curvature terms. Instead of $\log R$ as in a tokamak-like configuration where $BR = I_0$ is assumed, we can use the vacuum value of $\log B$ in the evaluations of curvature since

$$\mathbf{b} \cdot \nabla\mathbf{b} = \nabla_\perp \log B \quad \text{for} \quad \beta = 0, \tag{15.69}$$

and in a kinetic treatment even $\nabla \times \mathbf{b} = \mathbf{b} \times \nabla \log B$ in the case of vanishing current. The curvature terms are found by evaluating

$$\nabla \cdot \mathbf{v}_E \equiv -\mathcal{K}(\phi) = \nabla\phi \cdot \mathbf{B} \times \nabla\frac{1}{B^2} = \mathbf{B} \cdot \nabla\frac{1}{B^2} \times \nabla\phi, \tag{15.70}$$

where the second form is used to show that only derivatives of $\phi$ with respect to $\varphi$ survive ($\mathbf{B}$ and $\nabla$ on any equilibrium quantity are both in the poloidal plane). Then we have

$$\mathcal{K} = -\mathbf{B} \cdot \left[\frac{\partial}{\partial r}\left(\frac{1}{B^2}\right)\nabla r \times \nabla\varphi + \frac{\partial}{\partial\theta}\left(\frac{1}{B^2}\right)\nabla\theta \times \nabla\varphi\right]\frac{\partial}{\partial\varphi}. \tag{15.71}$$

The unit vectors $\nabla r$ and $r\nabla\theta$ and $r\cos\theta\nabla\varphi$ are used to evaluate the cross products, then the components $B^r = \mathbf{B}\cdot\nabla r$ and $B^\theta = \mathbf{B}\cdot\nabla\theta$ and finally the derivatives of $B^{-2}$ are taken in the $r\theta\varphi$ coordinate system.

The coordinate transformation is

$$x = -\frac{r_0}{\psi_0}\psi \quad y = -r_0\varphi \quad s = s(\theta), \tag{15.72}$$

with the latter already given. These are normalised as unit Cartesian distances at $\theta = 0$ for $xy$ and $s$ as an angle equal to $\theta$ at the equator. The sense is to have $\nabla x \times \nabla y \cdot \nabla s$ positive. The gradients are outward, eastward, and northward, respectively.

Then, using $R = r\cos\theta$ and replacing $r = \cos^2\theta/\psi$ afterward, and normalising properly, we have the metric elements, the perpendicular Laplacian, the magnetic field components, and the curvature operator,

$$g^{xx} = \frac{1 + 3\sin^2\theta}{\cos^6\theta} \quad g^{yy} = \frac{1}{\cos^6\theta} \quad \sqrt{g} = r_0, \tag{15.73}$$

$$\nabla_\perp^2 = g^{xx}\frac{\partial^2}{\partial x^2} + g^{yy}\frac{\partial^2}{\partial y^2}, \tag{15.74}$$

$$B^2 = B_0^2\frac{1 + 3\sin^2\theta}{\cos^{12}\theta} \quad B^s = \frac{B_0}{r_0} \quad \mathbf{B} = B_0\nabla x \times \nabla y \quad \chi_x = B_0, \tag{15.75}$$

$$\mathcal{K} = \frac{6}{r_0 B_0}\left[\frac{1 + \sin^2\theta}{(1 + 3\sin^2\theta)^2}\cos^2\theta\right]\frac{\partial}{\partial y}. \tag{15.76}$$

The factor of $r_0$ in $B^s$ reflects $s$ being normalised as an angle, with $B^\theta = B^s = 1$ in units of $B_0/r_0$ at the equator. This means $K_\parallel = 1/r_0$. The field line length is still in the flute mode ordering regime so long as $L_\perp \ll r_0$ as the physical extent of the perpendicular domain at the midplane. The constancy of $\sqrt{g}$ means that in a flux-tube setting it drops out of the equations. Recalling that the geometric coefficients depend on $s$ only and the perpendicular derivatives are in the $xy$ drift plane only, one should not forget the strong factor of $B^{-2}$ in polarisation (Eq. 15.30).

Eqs (15.73)–(15.76) collect the geometric coefficients into one place. They are dropped into the drift Alfvén model as with the tokamak-like metric given in Chapters 13 and 14. The high powers of $\cos\theta$ experienced are due to the $r^{-3}$ factor in $B$. One can always replace them using $r = r_0\cos^2\theta$.

### 15.4.3 The turbulence

We use the three-dimensional Eqs (15.26)–(15.30) with the geometric coefficients given in Eqs (15.73)–(15.76). Highly arbitrary parameters are chosen, for comparison purposes the same ones as in the warm plasma cases above. Instead of gradient drive terms we specify a source $Q = Q_0\exp(-ax^2 - bs^2)$ with $Q_0 = 10^{-2}$ and $a = 100/L_\perp^2$ and $b = 100$ as nominal choices. Exit flux boundary conditions are placed at $s = \pm 0.44$ where $\theta = 0.72$ and $B/B_0 = 8.5$. Additionally, due to the pressure on the numerics, we set $X = 1$ so that $\mu_\| = c_s L_\perp$. The factors of $B$ in all the operators, especially any operation by $\nabla_\|$ but also in the polarisation involving $\rho/B^2$, have a large effect on the numerics, with the result that the parallel resolution is doubled ($N_s = 64$) and the time step is reduced to 0.0005 since the Alfvén speed is proportional to the magnitude $B$. In the characteristics used in the exit flux boundary conditions, the dependence upon $B$ is included in the factor of $v_A$ in Eq. (15.39).

In the absence of magnetic shear in a dipole magnetic field, one might expect the turbulence to show MHD or hydrodynamic character. Despite the variation of $B$ with $s$, the coordinate deformation does not involve any twisting, and the ratio between the two contravariant perpendicular metric elements, $1 + 3\sin^2\theta$, does not go through large changes. Indeed we find very little difference among the cases with differing $C$ at $\hat{\beta} = 1$. The decision on which way the mode structure result will go depends practically only on the product of $\hat{\beta}$ and $C$. However, the localisation of the source makes the choice of $\hat{\epsilon}$ significant, namely, the ratio of $K_\| c_s$ to the rate of the source given by $Q$. The losses are given by both the turbulence and the parallel streaming.

The domain in $s$ translates to much shorter field lines—a length of $0.88 r_0$ in this model compared with $2\pi qR$ in the other. With $K_\| = 1/r_0$ here, the difference is that between $2\pi$ and 0.88, nearly an order of magnitude. So we also run some cases with longer field lines for comparison. We take the same set of basic parameters with different choices for $L_\perp$ which gives the effective field line length, setting $(K_\| L_\perp)^{-1}$ to 100, 300 or 1000, giving the nominal parameters in the short model,

$$\hat{\epsilon} = 10^4 \quad \hat{\beta} = 5 \quad \hat{\mu} = 2.72 \quad C = 1.39 \quad \omega_B = 0.03, \tag{15.77}$$

while in the median model we use

$$\hat{\epsilon} = 10^5 \quad \hat{\beta} = 50 \quad \hat{\mu} = 27.2 \quad C = 13.9 \quad \omega_B = 0.01, \tag{15.78}$$

and in the long model,

$$\hat{\epsilon} = 10^6 \quad \hat{\beta} = 500 \quad \hat{\mu} = 272 \quad C = 139 \quad \omega_B = 0.003, \tag{15.79}$$

following $\delta_0$ and $\beta_e$ and $\mu_e$ and $\nu_e L_\perp / c_s$ being held fixed. The value of $\beta_e$ is moderate, larger than but not very large compared with the electron/ion mass ratio. All of the five toroidal cases from the sourced case study above were checked in the short and long models. Most of the short ones did not localise properly, while the long ones were all strongly hydrodynamic in character (i.e. the value of $\hat{\beta}$ matters more than $\hat{\epsilon}$ for this, while the latter determines parallel localisation). We highlight the three cases with parameters listed above, and refer to them as the short, median, and long cases.

Because of the differences in field line length (essentially: $\hat{\epsilon}^{1/2}$), the three cases are on different time scales. The time step values are $\tau = 5$, 20, and 50, all times $10^{-4}$, for the short, median, and long cases, respectively, and the run lengths are 2, 6, and 20, all times $10^3$, respectively. Averaged quantities such as spectra are measured over the last half of the run length in each case, or 101 snapshots of the variables.

The drift direction wavenumber spectra ($k_y$) are shown in figure 15.13. The first of these, for the short case, produces spectra that look like generic drift wave

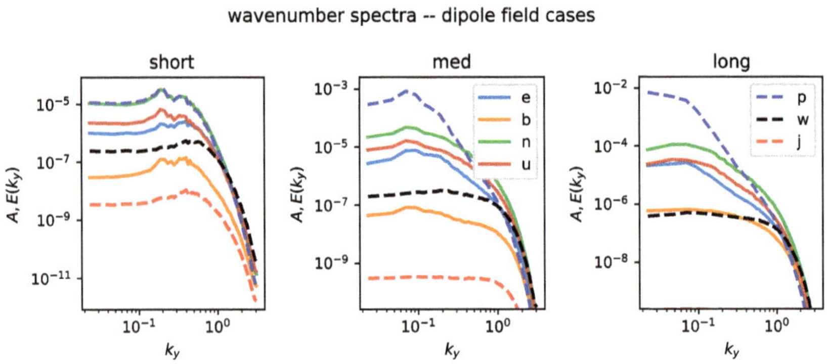


**Figure 15.13.** Drift direction wavenumber ($k_y$) spectra of the fluctuations, for the short, median, and long field line sourced dipole field cases. The transition from a drift wave signature (left), with the energy spread across the spectrum with $\phi \sim n$, to an MHD or hydrodynamic one (right), with the energy mostly at large scale where $\phi \gg n$, is the main finding. In this and following figures, the averaging is over the last half of the run in each case, as explained in the text.

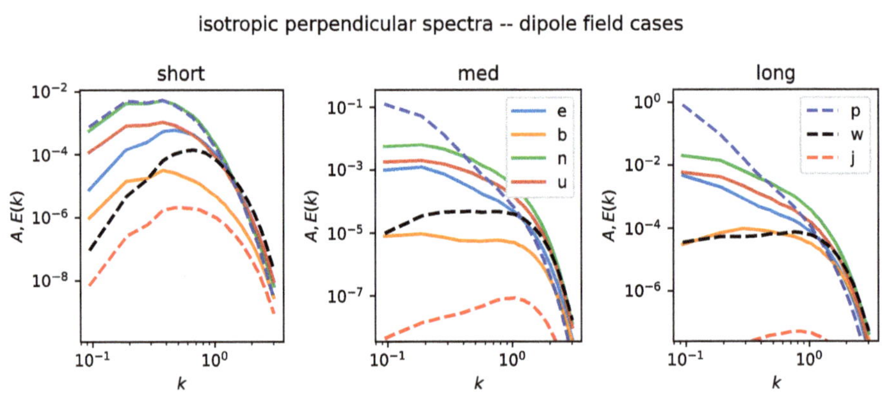


**Figure 15.14.** Isotropic perpendicular ($k_\perp$) spectra of the fluctuations, for the short, median, and long field-line sourced dipole field cases. The transition from a drift wave signature (left), with the energy spread across the spectrum with $\phi \sim n$, to an MHD or hydrodynamic one (right), with the energy mostly at large scale where $\phi \gg n$, is the main finding. The vorticity remains peaked at small scale, though the large scale component rises with field line length.

turbulence. All curves have moderately pronounced relative maxima in the 0.3 to 0.7 range in terms of $\rho_s$, and are relatively flat towards long wavelength. Moreover, $n$ and $\phi$ are very close in relative amplitude. In the median and long cases, the dynamics is more hydrodynamic and the spectra become more MHD-like. The amplitude of $\phi$ becomes by far the largest, and the ExB energy leads a general shift of the entire spectrum to long wavelength. One also notes that the vorticity drops, not only relative to $n$ or $\phi$ but also in absolute terms. The spectrum of the magnetic energy and current also drops accordingly.

The isotropic perpendicular spectra ($k_\perp$) are shown in figure 15.14. These trends are more clearly visible here, with the curves all peaking at smaller scale in the short case, while for the others the vorticity and magnetic energy drop away, the amplitude of $\phi$

separates from $n$, and the large-scale component is dominant in all channels of free energy. There is still small-scale vorticity, but its amplitude drops as one enters the hydrodynamic regime.

The parallel wavenumber spectra ($k_\parallel$) are shown in figure 15.15. What we see here is that the shape of the spectra in $n$ and $\varpi$ are the same in the short case, with only their amplitudes being different, and that $u$ has the widest spectrum. In the longer cases the vorticity spectrum broadens but $\phi$ is narrower and much more strongly peaked. The flute mode component at $k_\parallel = 0$ is already a visible peak for the short case but becomes strongly dominant in the others. It is interesting to note that the amplitudes of $\varpi$ and $n$ are not so disparate in the moderate wavelength range ($k_\parallel$ in the 3–10 range in terms of $K_\parallel$). This means that the disparities seen in the perpendicular spectra are due to the increasing role of the flute mode component into the hydrodynamic regime.

The parallel envelopes are shown in figure 15.16. The short case shows the density envelope with a peak given by the source at $s = 0$ and falling away slightly towards the exit flux locations at $s = \pm 0.44$. This case is not well localised, however. Localisation is stronger, with a hatted envelope shape, for the median case, and it is strongest in the long case. The shape is different from the sourced cases studies above, especially in the flatness of $h = n - \phi$, which is still strong in the long case. The nonadiabatic component reaches all the way from end to end despite the localisation in the state variables. This is a feature of the difference between the two models, with the effective wavelength increasing relative to the local drift scale, due to the factor of $B^2$ in the polarisation.

The cross correlations between $n$ and $\phi$ are shown in the top row of figure 15.17. These show the mode structure transition very clearly. In the short case two state variables are closely correlated, tied together by the adiabatic response. As the field lines relative to the perpendicular scale become longer, the inductivity breaks the static force balance and the variables become incoherent relative to each other, in a more hydrodynamic situation of a passively advected pressure field. The short case on the left indicates a drift wave case, while the other two indicate interchange dynamics.

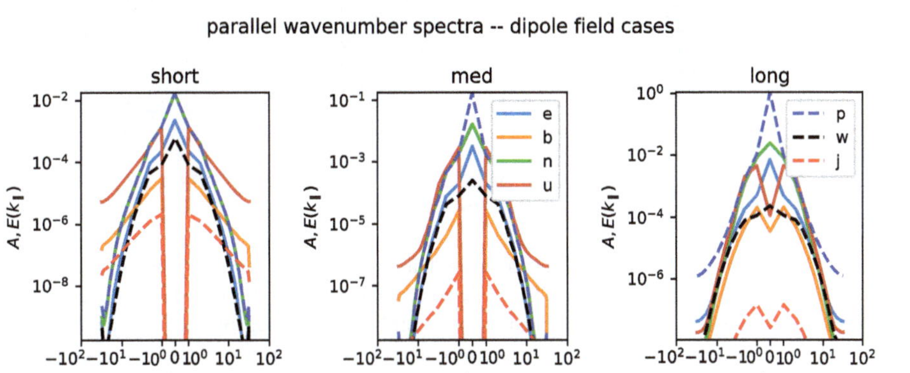


**Figure 15.15.** Parallel wavenumber ($k_\parallel$) spectra of the fluctuations, for the short, median, and long field-line sourced dipole field cases. Domination by the flute mode component at $k_\parallel = 0$ becomes more pronounced for stronger inductivity. In contrast to closed flux surface cases, the vorticity and the density fluctuations have the same shape in their spectra.

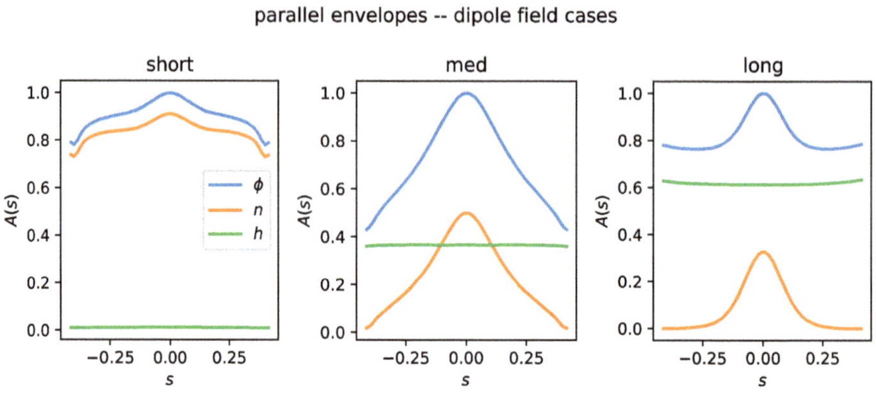


**Figure 15.16.** Parallel envelopes for the state variables, for the short, median, and long field-line sourced dipole field cases. The nonadiabatic component $h = n - \phi$ has a flat envelope for all three cases. For the short case the amplitudes of $n$ and $\phi$ are close. For stronger inductivity the source region is driven more strongly, first tending to overcome the parallel transport, and next rendering parallel dissipation small.

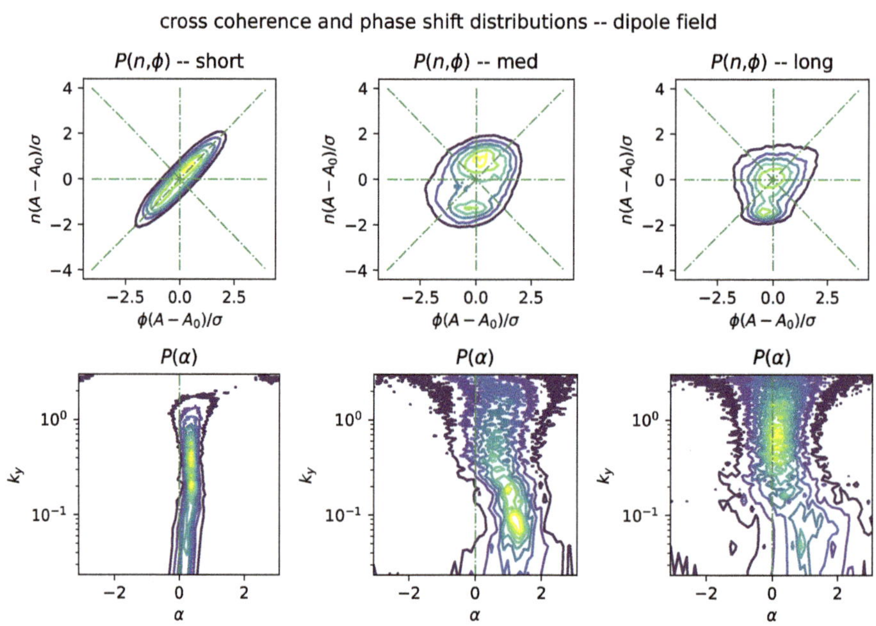


**Figure 15.17.** Cross coherence (top) and phase shifts (bottom) between $n$ and $\phi$, for the short, median, and long field-line sourced dipole field cases. The adiabatic coupling is effective for the lowest value, leaving the two electron state variables strongly correlated and the distributions lined up and narrowly peaked near zero. As the inductivity increases, the dynamics assumes a more MHD or hydrodynamic character reflective of interchange driven turbulence, the cross correlation weakens, and the phase distributions are wider and closer to $\pi/2$ for maximum drive by the self-consistent gradient.

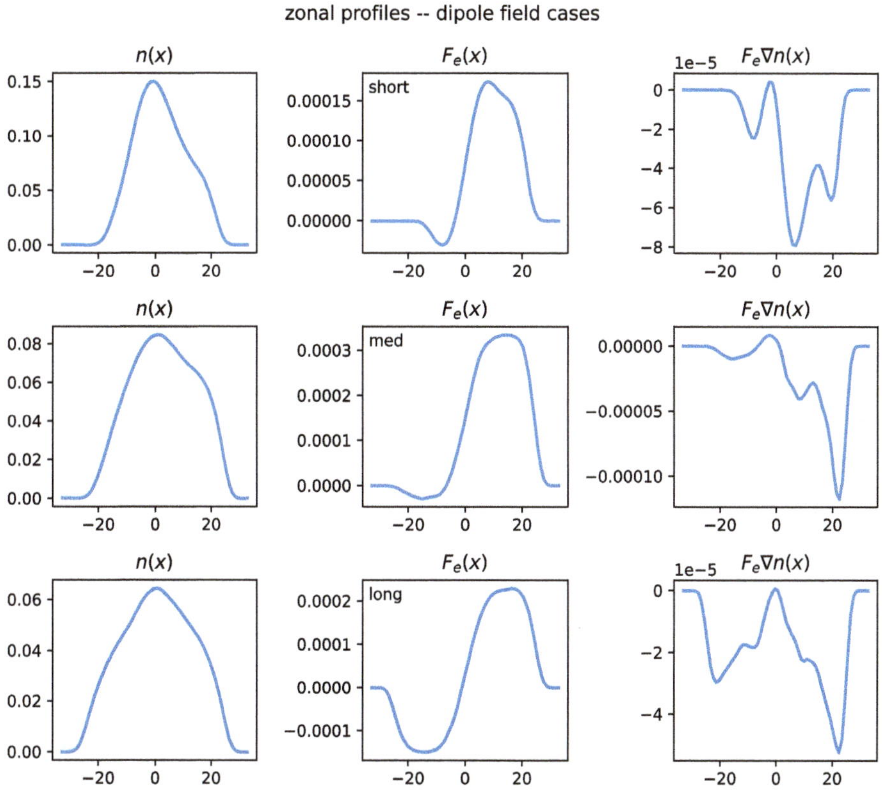


**Figure 15.18.** Profile components of the density, turbulent ExB flux in the $x$-direction, and the product of the profile flux and the profile gradient (left to right), for the short, median, and long field-line sourced dipole cases (top to bottom). For the long case the dynamics is nearly symmetric. The flux is directed away from the central layer ($x = 0$) on both sides, and the self-consistent drive of the turbulence by the profile (third frame, negative values) exists on both sides. For the short and median cases the dynamics is more interchange-like, with most of the energetics on the right-hand side, where in the source region the curvature vector and gradient are aligned.

The phase shift distributions between $n$ and $\phi$, in the bottom row of the same figure, show the same thing. In the short case the phase shifts are more strongly peaked at values that, while positive, are close to zero and mostly independent of $k_y$ in the spectrum. As the field lines become longer, the distributions are broader with the long wavelengths separating from the rest, moving the most probable values closer to $\pi/2$, where the ability of fluctuations to grow at the expense of the self-consistent gradient is maximum.

Profiles of the same quantities as in figure 15.7, for the sourced case above, are shown in figure 15.18 for the three dipole cases. These are the self-consistent profile $\langle n \rangle$, the turbulent ExB flux $\langle F_e \rangle$, and the product of the flux and the self-consistent gradient, $\langle F_e \rangle \langle \nabla n \rangle$, the quantity which determines exchange between profile and fluctuation thermal free energy. As analysed above, free energy is taken out of the gradient where this product is negative. The result is somewhat counterintuitive but can be understood by the way $\omega_B$ changes from the short to the long case. The one-sidedness with respect to radial position, referred to above, is most evident in the short case and reduced with the increasing field line length. However, the strength of the turbulence increases with the longer field lines. Although the parameters are held fixed, it is the scale ratios which determine the result. With the increasing product $\alpha_M = \hat{\beta}\omega_B$ from 0.15 in the short case to 1.5 in the long case, one might think the interchange character is increasing. Nevertheless, the actual decrease of the one-sidedness of the dynamics with longer field lines suggests that it is really the increasingly hydrodynamic character and not the interchange forcing which strengthens the turbulence in these cases.

The axisymmetric spatial distributions of the density, flux, gradient, and flux gradient product are shown in figure 15.19, for the median case, which can be compared with figure 15.8. The moderate localisation and strong radial one-sidedness are apparent. Most of the energetics is close to the source region near $s = 0$, so the spreading out along the field lines is the result of the relatively small combination $(0.44/\pi)^2\hat{\epsilon}$ affecting the sound wave dynamics.

It is interesting to note that the coupling $n \leftrightarrow \phi$ produces a localised vorticity layer in the profile, and this represents a flow layer. The flow layer is visible in the spatial morphology, shown for the density in figure 15.20, for the three cases. In the

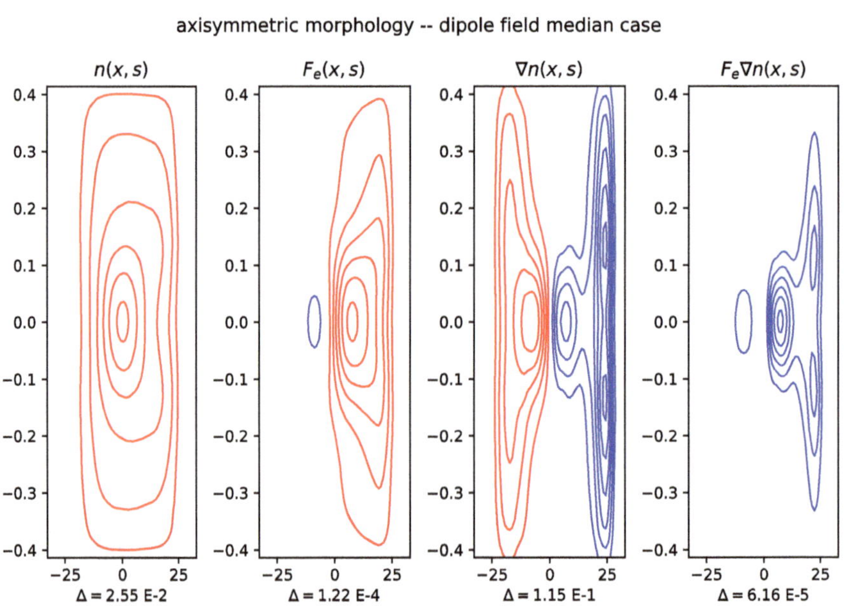


**Figure 15.19.** Axisymmetric components of the density, turbulent ExB flux in the $x$-direction (left), the gradient, and the energetic effect on the profile component (left to right), for the median dipole case from figure 15.18. The last is the product of the second and third, with negative values indicating self-consistent drive of the turbulence by the profile through the nonlinear advection. For this case the dynamics is strongly asymmetric, owing to the interchange drive through the curvature forcing. For the long dipole case this is much less pronounced despite the stronger turbulence, indicating strongly hydrodynamic rather than interchange turbulence.

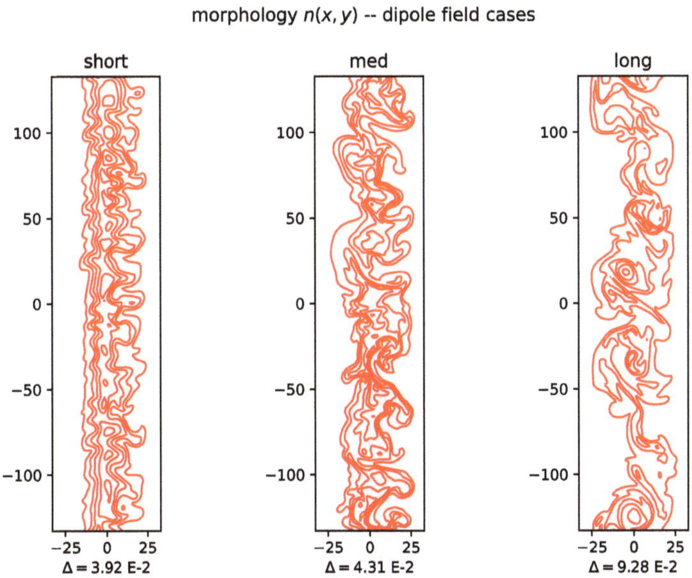


**Figure 15.20.** Spatial morphology of the density in the midplane, for the short, median, and long field-line sourced dipole field cases. Compare with figures 15.9 and 15.11. The turbulence is more interchange-like for the short field lines case, and becomes increasingly symmetric about $x = 0$ for longer field lines. By contrast, it is the latter cases which are more hydrodynamic, producing larger and more intermittent cells of dynamics. This is a case of the field line length (or steep gradient) effect overcoming the interchange effect, i.e. $\hat{\beta}$ is more relevant to this comparison than $\alpha_M = \hat{\beta}\omega_B$.

short and median cases, the interchange convective flow plumes are visible, with the flow layer oriented such that the vorticity is negative, with a $y$-direction flow positive on the left ($x < 0$) and negative on the right ($x > 0$). In the long case, this is less visible because the strength of the turbulence in producing the large-scale eddy cells overpowers it. Any such self-consistent flow dynamics is of interest, and it forms the basis for Chapter 16.

## 15.5 Summary

This study of magnetised plasma turbulence in an open magnetic field domain has served to emphasise the significance of the boundary conditions in the parallel direction upon the nature and mode structure of the resulting turbulence. By running the same cases in the slab and toroidal models on both closed magnetic surfaces with toroidal topology in the field lines and angle domains and open surfaces with open ends, we provide a control case on the effect of field line connection on the physics. The exit flux condition provides a sort of substitute adiabatic coupling since the incoming flux for which the current is proportional to the nonadiabatic response has to be set to zero. The current flowing out of the flux tube at both ends served to provide this effective clamp on the mode structure. If the field lines are too short, a strong sort of 2D result is found. If the field lines are long enough and the plasma is maintained by a source process in the middle, then the result is localised enough to make the boundary condition sensible. These differences are useful for introductory purposes in aiding those who are building models for particular situations of interest.

## Further Reading

Kinetic shear Alfvén waves were covered in Chapter 8 and the main references are given therein. The version of the Alfvén wave equation given in equation (15.6) is the starting point for global Alfvén modes in confined plasmas. A review was given in G Vlad, F Zonca and S Briguglio 'Dynamics of Alfvén waves in Tokamaks' *Riv. Nuovo Cim.* **22** (1999) 1.

The method of characteristics is standard in solving hyperbolic equations (of the advection type), and is treated in the text R Courant and D Hilbert 1962 *Methods of mathematical physics. Volume II: Partial differential equations* (New York: Wiley). It is behind the use of upwind methods for the numerical solution of this type of equation. See for example, W Hackbusch 'On a method of characteristics for solving a hyperbolic equation of second order' *Computing* **20** (1978) 47. A well known text is by R J LeVeque 1992 *Numerical Methods for Conservation Laws* (Basel: Birkhauser). This method as applied herein was used to set up a numerical scheme based on upwind methods for the kinetic shear Alfvén and acoustic wave systems in the gyrofluid version of these equations, see B Scott 'ExB shear flows and electromagnetic gyrofluid turbulence' *Phys. Plasmas* **7** (2000) 1845.

The magnetic coordinates we constructed by demanding that the contravariant component of the magnetic field vector be a flux function are due to S Hamada. He called them proper coordinates but since then they are known as Hamada coordinates, given originally in (Japanese) S Hamada 'The equilibrium configuration of a magnetic field' *Kakuyugo Kenkyu* **1** (1958) 524; and in (English) 'Hydromagnetic equilibria and their proper coordinates' *Nucl. Fusion* **2** (1962) 23.

A fusion experiment based on the properties of dipole magnetic fields was the 'Levitated Dipole Experiment (LDX)', based at the Massachusetts Institute of Technology. A description is given by D T Garnier, A K Hansen, J Kesner *et al.* 'Design and initial operation of the LDX facility' *Fusion Eng. Design* **81** (2006) 2371.

In tokamak physics there is a scrape-off layer wherein the magnetic field lines do not lie in closed surfaces but intersect material boundaries, where there is a region called a sheath in which the plasma begins to be affected. The distance of transition from plasma to field line ends is nominally a few Debye lengths, which can become several tens given the interSection angle of the magnetic field. This is outside of the present scope, but interested readers can consult the text J Wesson 1997 *Tokamaks* (Oxford: Oxford Univ Press). Treatments of the turbulence there include X Q Xu 'Fluid simulations of conducting-wall-driven turbulence in boundary plasmas' *Phys. Fluids B* **5** (1993) 3641; X Q Xu, W M Nevins, R H Cohen, J R Myra and P B Snyder 'Dynamical simulations of boundary plasma turbulence in divertor geometry' *New J. Phys.* **4** (2002) 53; T Ribeiro and B Scott 'Tokamak turbulence computation on open and closed magnetic surfaces' *Plasma Phys. Control. Fusion* **47** (2005) 1657.

**IOP** Publishing

Turbulence and Instabilities in Magnetised Plasmas, Volume 1
Fluid drift turbulence
**Bruce Scott**

# Chapter 16

# Drift wave turbulence and flows

## 16.1 Introduction—eddies and flows

In previous Chapters we have investigated the basic properties of the nonlinearities, parallel dynamics, magnetic curvature effects as energetic transfer processes in the turbulence, and the geometric properties of open and closed magnetic flux surfaces and how the relevant boundary conditions affect the resulting turbulence. In Chapter 15, we introduced the profile quantity as a surface average of the corresponding three-dimensional quantity. This can be referred to as a profile or a flux surface average, or simply, the profile of the given variable. For systems involving flows, the profile is also called the 'zonal' component, representing division of the system in the direction of gradient inhomogeneity into zones of finite extent, in the down-gradient or radial direction, encompassing the volume between two closed flux surfaces. A flow with this property is called a 'zonal flow', borrowing the concept from geophysical fluid dynamics of a flow in the direction of longitude, within a given band of latitudes; hence it is constant within arbitrarily thin zones (the term given to intervals of latitude, independent of longitude). The flow is zonal, or purely in the drift direction, because its stream function depends only on the radial direction, the only other dimension in the drift plane. In a three-dimensional setting, the zone becomes a thin volume between two flux surfaces, so that a zonal flow is rigid within the flux surface. This leads to the zonal component being equivalent to the flux surface average, or profile, component. The flux surface average is formally introduced in its own Section, below.

The zonal component has two properties which distinguish it from the turbulence: it does not participate in adiabatic coupling, and it does not participate in transport. However, it does have two energetic transfer connections to the rest of the system. Through the nonlinearities, especially the ExB advection, it is connected to the turbulence. Drive of the turbulence is a conservative transfer out of the zonal component, usually the pressure but in some cases the flow or magnetic energy. If represented by gradient drive terms, the maintenance of these by sources is the input.

doi:10.1088/978-0-7503-2504-2ch16  

If represented by self-consistent profile gradients, then the drive by source is either explicitly included or the profile is allowed to decay, as the free energy going to the turbulence lands in the sinks. In special circumstances the zonal component of an energetically subdominant quantity can be driven by the turbulence, a sort of reversal of the usual transfer direction of energy in turbulence. For zonal flows but also magnetic fields, the drive through the nonlinearities by turbulence is a possible transfer channel. Once energy lands in the zonal flow component, it is not subject to dissipation. Nonlinear transfer back to the turbulence or coupling through magnetic geometry to other degrees of freedom represent the only significant loss mechanisms.

The zonal dependent variable is the vorticity, not the flow, so that any flow it represents is sheared. Practically any significant background flow in a confined plasma is sheared to some extent, if only due to the boundary conditions. An equilibrium flow naturally has the same scale as the background gradient, especially in toroidal systems (for which see below). A zonal flow driven by the turbulence is on a scale of order of the profile scale or smaller, since the dynamics in the particular radial zone is mostly felt there. A zonal vorticity layer represents a flow in the drift direction which is radially sheared. The description 'ExB shear flow' is used as shorthand for this. In the same sense, a zonal current layer represents an addition to the shear in the magnetic field. Since the background field in a confined plasma is already strongly sheared, the amount of shear this adds is usually negligible. Not so for flows, however. Due to the smallness of the drift parameter, the shear in a zonal vorticity layer driven by the turbulence can easily be larger than the shear in an equilibrium flow.

If either the equilibrium flow or any self-generated zonal flow is sheared, then it can influence individual eddies and disturbances in the turbulence in a special way. Transport down the gradient represents a statistical wandering of the transported quantity in that direction, whether or not there is any displacement of the disturbances in the drift direction. If there is a strongly sheared zonal flow, a mean gradient in the kinetic energy of ExB flow is present. Transport of flow momentum down this gradient is called Reynolds stress, and if positive it leads to the well known Kelvin–Helmholtz instability. Under certain conditions and for certain scales of motion, this instability is eliminated: if the background vorticity is constant, there is no vorticity extremum and the Rayleigh criterion for instability is therefore not satisfied. Generally, the instability has a threshold expressed in terms of a ratio between the wavelength of a cross-flow disturbance and the scale of the flow shear; for a ratio well below unity the disturbances are stable. The special influence the sheared flow then has on disturbances maintained by other means (e.g. drift waves) is to literally rip them apart; both correlation times and lengths are reduced.

A sheared ExB flow can therefore either drive or suppress drift wave turbulence, depending on the situation. Furthermore, we have the issue of the origin of the sheared flow. Not only are externally applied flows important (i.e. those which are part of the background equilibrium), but the turbulence could itself actually drive or generate flows due to the intrinsic connection between electron pressure and electrostatic potential disturbances, and the tendency of the ExB flow eddies associated with those potential disturbances to transfer their energy towards ever larger scales. On energy conservation grounds alone, suppression of disturbances by flows implies a transfer of energy from disturbances to flows. The methods of diagnosis and control cases outlined in this Chapter serve to clarify the relationship between magnetised plasma turbulence and ExB flows and can be applied to any specific situation or model. Not only the parameters but also the geometry of the background magnetic field has a strong effect on the fate of the turbulence and the flows.

In a tokamak-like toroidal system (defined as a mostly toroidal magnetic field, curved mostly in the direction of decreasing toroidal major radius) there is another way for flows to arise, namely, as part of the equilibrium. In a global system the curvature forcing acts on the profile, due to the 'geodesic curvature'. The geodesic component of the curvature vector is the one which lies within the flux surface. The curvature in a tokamak is a quantity with structure in $\sin\theta$ or $\cos\theta$, namely mode numbers $m = \pm 1$ in the poloidal angle, but still axisymmetric. The components to which the zonal mode couples are called 'sidebands' and they have a mostly $m = \pm 1$ and $n = 0$ mode structure (the spectrum falls steeply with $m$). It was customary to neglect this component set in the 1970s and 1980s due to computational constraints. Then, in the late 1990s, study of the geodesic acoustic oscillations, first found in the 1960s, re-emerged after their detection in tokamak experiments.

The sideband modes also include the $m = \pm 1$ parallel current which relaxes into what in a tokamak is part of the MHD equilibrium. It is called the 'Pfirsch–Schlüter current.' Ultimately, these sideband modes are in every dependent variable, describing its part of the relaxation of a tokamak into equilibrium. All of currents, flows, heat fluxes, and finally the radial electrostatic potential representing the equilibrium ExB flow in force balance, are part of this equilibrium. The details of tokamak physics are outside of our scope, but the sideband analysis represents the same set of considerations on the perturbed equilibrium. All the zonal and sideband components of the local dependent variables, distinct from the background, are coupled to the turbulence. These sidebands become a strong mediator of the zonal flow energetic process. In saturated turbulence, every degree of freedom, including the zonal and sideband components, is in statistical balance against all transfer channels.

## 16.2 Kelvin–Helmholtz stability

We here review the basic textbook results concerning the stability of sheared flows. The first of these is the Rayleigh theorem, according to which there must be a vorticity extremum in order for there to be an instability. The second is the existence of a threshold, by which only disturbances of sufficiently long wavelength along the flow, but reflecting motion perpendicular to the flow, are unstable.

The classic Kelvin–Helmholtz problem consists of a two-dimensional Euler equation in the $xy$-plane, with a flow given by a stream function $\mathbf{v} = \hat{\mathbf{z}} \times \nabla\phi$. The mean flow in the $y$-direction has a profile in the $x$-direction, so that $\mathbf{v}_0 = \{0, v_0^y\}$ with $v_0^y = V_0(x)$. The vorticity profile is given by $\Omega_0(x) = \partial V_0/\partial x = V_0'$, and the vorticity gradient is $\Omega_0'(x) = V_0''$. The linearised equation for disturbances on a zonal flow is

$$\left(\frac{\partial}{\partial t} + V_0 \frac{\partial}{\partial y}\right)\varpi + v_x \Omega_0' = 0, \tag{16.1}$$

where the velocity and vorticity disturbances are given in terms of the perturbed stream function,

$$v_x = -\frac{\partial \tilde{\phi}}{\partial y} \qquad v_y = \frac{\partial \tilde{\phi}}{\partial x} \qquad \varpi = \frac{\partial^2 \tilde{\phi}}{\partial x^2} + \frac{\partial^2 \tilde{\phi}}{\partial y^2}. \tag{16.2}$$

As befits a linear problem with no $y$-dependence in the background, we may assume a dependence in $y$ and $t$ of the form $\tilde{\phi} \to \phi_{\mathbf{k}} \exp[i(ky - \omega t)]$, and obtain

$$\frac{\partial^2 \phi_{\mathbf{k}}}{\partial x^2} - k^2 \phi_{\mathbf{k}} = \frac{k\Omega_0'}{kV_0 - \omega}\phi_{\mathbf{k}}. \tag{16.3}$$

To find a useful criterion for stability we may multiply both sides by the complex conjugate $\phi_{\mathbf{k}}^*$, integrate over $x$, and take the imaginary part,

$$\gamma \int dx \, \frac{k\Omega_0'}{(\omega_R - kV_0)^2 + \gamma^2} \, |\phi_{\mathbf{k}}|^2 = 0, \tag{16.4}$$

where $\omega = \omega_R + i\gamma$ is complex, written in terms of two real components. This equation says that for general form of the disturbance $\phi_{\mathbf{k}}(x)$ the growth rate $\gamma$ must vanish unless there is a point within the domain (here, the entire $x$-line) in which $\Omega_0'$ vanishes, so that it can change sign. That is, there must be an inflection point in the velocity profile, or equivalently an extremum in the vorticity profile, in order for there to be an instability. This is the Rayleigh inflection point theorem for the stability of sheared flows.

If we take the real part we find

$$\int dx \, \frac{(\omega_R - kV_0)k\Omega_0'}{(\omega_R - kV_0)^2 + \gamma^2} \, |\phi_{\mathbf{k}}|^2 = \int dx \, \left|\frac{\partial \phi_{\mathbf{k}}}{\partial x}\right|^2 + \int dx \, k^2 \, |\phi_{\mathbf{k}}|^2. \tag{16.5}$$

Given Eq. (16.4) for finite $\gamma$ the quantity involving $\omega_R$ in the numerator vanishes, and we can combine the rest into

$$\int dx \, \left|\frac{\partial \phi_{\mathbf{k}}}{\partial x}\right|^2 + \int dx \, k^2 \left[1 + \frac{V_0 \Omega_0'}{(\omega_R - kV_0)^2 + \gamma^2}\right] |\phi_{\mathbf{k}}|^2 = 0. \tag{16.6}$$

This condition can be satisfied generally for localised vorticity layers whose product $V_0 \Omega_0'$ is negative. However, a lower bound on the growth rate can be found by noting that the size of the second term in the brackets must be larger than unity. Taking into account that we used $\gamma \neq 0$ to find this equation, we now have

$$\text{either} \quad \gamma = 0 \quad \text{or} \quad (\omega_R - kV_0)^2 + \gamma^2 < \left| V_0\Omega_0' \right| \sim \frac{V_0^2}{\Delta_V^2}, \tag{16.7}$$

where $\Delta_V$ can be taken to be the width of the flow. This implies in general that for $\gamma \neq 0$ the wavenumber $k$ cannot be too large; there is a threshold involving $k$ and $\Delta_V$ such that

$$\gamma \neq 0 \quad \text{only for} \quad k\Delta_V < O(1). \tag{16.8}$$

What this threshold exactly is depends on the detailed profile of $V_0$, but the existence of the threshold is true generally.

### 16.2.1 Aspect ratio effect

This is the explanation for the aspect ratio effects we found for the fluid and interchange cases (Chapter 13) and also for the 3D drift wave cases without magnetic shear (Chapter 11). If the computational domain is square, then a box mode can result, where in periodic cases the flow can jump back and forth from the $x$- to the $y$-direction. A flow through the domain is always at least twice as long as it is wide, so that with sinusoidal dependence enforced by the periodic boundaries in both directions, there is always the possibility for a Kelvin–Helmholtz wave to form on the flow, become secondarily unstable, and cause a new flow at right angles to the first. If one of the domain sides is short, a flow which goes the long way is unstable to the longest wavelengths, while a flow which goes the short way can always become wide enough for it to be stable, with the longest wavelength still shorter than the cutoff in Eq. (16.8). If the aspect ratio in a case with interchange forcing has $L_y < L_x$, a zonal flow will always result which takes up most or all of the energy. If $L_y > L_x$ then a radial flow will always result, which usually does not saturate. The radial flow becomes wide enough to remain stable and grows by interchange forcing without bound, unless the dissipation is quite large at long wavelength.

Physically realistic cases in almost any case will have $L_y \gg L_x$ by a reasonably large margin. For example, in a tokamak radial range equal to the profile scale length $L_\perp$ has

$$\frac{L_y}{L_x} = 2\pi \frac{r_0}{q_0 L_\perp}, \tag{16.9}$$

where $r_0$ is the reference minor radius at the flux-tube centre $x = 0$ and $q_0 = q(r_0)$, given by the minimum wavenumber in Eq. (13.68) allowed by the periodicity. For a case with $r_0/R_0 = 0.2$ and $q_0 = 2$ for the domain and field pitch and $L_\perp = r_0$ for the gradient scale length will have $L_y/L_x > 3$, which is a limiting case for the tokamak core. Indeed, if $L_\perp > r_0$ the flux-tube model becomes problematic. In outer core regions $r_0/L_\perp$ rises well above 3 and in the edge regions with $r_0 = a$, the limiting minor radius, well over 10. For a typical edge case with $a/L_\perp > 15$ and $q \sim 4$ we have $L_y/L_x \sim 20$, representing a thin-atmosphere geometry. In this case the radial flows are so strongly selected by the curvature forcing at $s = 0$ that it is imperative to make the flows turn around by imposing Dirichlet boundary conditions. In other words, the system should never be radially open with respect to flows. In the drift direction, flows are not artificially favoured, so that when a zonal flow actually does result in a case with $L_y/L_x = 4$ or greater, it can be considered to be physically realistic rather than spurious.

We already found this in the cases in Chapter 11 and later. The long-domain cases escaped the strong zonal flows found in the square-domain cases, showing that the latter were not long enough in the $y$-direction. Hence, especially in any modelling, one always needs $L_y/L_x$ to be large, unless one can actually show that the system size given by $2\pi r_0/q_0$ limits $L_y$ to smaller sizes.

## 16.3 Sheared flows and decorrelation

The scenario of suppression of turbulence by sheared ExB flows has received strong attention in plasma physics research. The original model for it, published in the early 1990s, is straightforward and easy to understand. It is important to recall the configuration assumptions upon which it is based. Firstly, the transported quantity is taken to be a passive scalar, with linear coupling back to the flow assumed to be subdominant. Secondly, the turbulence forcing rate, given by an instability, is set equal to the turbulence decorrelation rate, which in turn is the RMS vorticity. When evaluating the scenario to any situation, these should be checked, and it is not difficult to do so.

The results of Kelvin–Helmholtz stability of sheared flows concern direct interactions between disturbances and flows, for which the only free energy source is the gradient of the mean flow energy. But the main interest for magnetised plasma turbulence is what a sheared flow can do to turbulence caused by other means, usually the thermal gradient. The decorrelation model invokes a two-point correlation picture of turbulent diffusivity and leads to a result that the imposition of a large-scale sheared flow should interfere with the transport of passive scalar quantities by random scatter. Suppose we have a diffusive process, in which the average separation of two points grows with the square root of the elapsed time. In either the $x$- or $y$-directions, this distance will evolve as

$$\Delta_x = (2Dt)^{1/2} \qquad \Delta_y = (2Dt)^{1/2}, \tag{16.10}$$

where in the absence of other effects the diffusivity $D$ should be isotropic with respect to direction in the 2D perpendicular plane. In a sheared flow, the separation of two points in the direction parallel to the flow grows secularly, given an initial separation across the flow. If the separation in the $x$-direction is $\Delta_x$, then the separation in the $y$-direction will grow according to the flow shear $V_0'$ as

$$\Delta_y = V_0'\Delta_x t. \tag{16.11}$$

Expressing these infinitesimally, we have

$$\frac{\partial \Delta_x}{\partial t} = \frac{D}{\Delta_x} \qquad \frac{\partial \Delta_y}{\partial t} = \frac{D}{\Delta_y} + V_0'\Delta_x. \tag{16.12}$$

Given a definite correlation length $\lambda_y$ in the $y$-direction, the correlation time $\tau_c$ is reduced by the fact that with the flow shear the inter-point separation reaches $\lambda_y$ sooner than it would if only the diffusive process were present. As a result, we therefore find a smaller separation $\Delta_x$, due to the smaller time interval. This is the correlation length $\lambda_x$ in the $x$-direction. If the overall transport is a random walk process, its diffusion coefficient will scale as $\lambda_x^2/\tau_c$. If the flow shear is weak enough that $\lambda_y$ is reached by $\Delta_y$ comparably fast as $\lambda_x$ is reached by $\Delta_x$, then the flow shear has little effect. Beyond a threshold, however, roughly given by

$$V_0' > D/\lambda_\perp^2, \tag{16.13}$$

where $\lambda_\perp$ is the isotropic $\lambda_x$ or $\lambda_y$ in the absence of flows, the two-point correlation is limited by the flow shear rather than the diffusion. If the diffusivity is given by a linear instability within the linear mixing length model, the decorrelation rate is set equal to the growth rate $\gamma$ and the correlation length is set inverse to the $k_\perp$ of the eigenmode, and we have

$$D = \gamma/k_\perp^2 \sim \gamma\lambda_\perp^2 \qquad \text{hence} \quad V_0' > \gamma. \tag{16.14}$$

As a threshold value of the velocity shear, this gives the value beyond which the instability will be suppressed. For ExB flows in a magnetised plasma, this is the criterion for ExB flow shear suppression of the instability and by extension, of the resulting turbulence. It is implicitly assumed that the growth rate is comparable to the native vorticity of the turbulence.

The drive of disturbances by a background thermal gradient is given by the average transport times the gradient, as we have seen in previous Chapters. If the correlation implied in the transport average is reduced, so is therefore the drive, and so perhaps the amplitude of the disturbances. This is a simple heuristic argument, however, and even if it is dressed by more sophisticated theory, the assumptions into the character behind it (that the transport process should be diffusive, with no correlative effect of the transported quantity back to the eddy disturbances doing the transport) remain. Whether this process is actually active is something which must be checked by statistical diagnosis of a given situation, be it experimental data or computations.

## 16.4 ExB flow energetics

There is another way to think about suppression of turbulence by a sheared flow, which is how the imposed flow and the disturbances exchange energy. If the flow suppresses the turbulence directly, it follows that the turbulence loses energy to the flow. Conversely, if the flow were Kelvin–Helmholtz unstable, it should serve as a free energy source. Returning to Eq. (16.1) for disturbances on a zonal flow, we form the equation for the energy of the disturbances by multiplying by $-\tilde{\phi}$ and integrating over the entire plane to find

$$\frac{\partial}{\partial t}\frac{1}{2}\int dx\, |\nabla\tilde{\phi}|^2 = -\int dx\; V_0'(v_x v_y), \tag{16.15}$$

where the derivative on $\Omega_0 = V_0'$ is integrated by parts. This term on the right side multiplying $V_0'$ is the Reynolds stress. If $v_x v_y$ is positively correlated with the flow shear given by $V_0'$, the flow eddies are tilting with the flow, and they yield energy to the flow. In this situation, they are directly suppressed by the flow. In the opposite case, the eddies are tilting against the flow, taking energy from it. In this case they are directly driven by the flow, as in the case of a Kelvin–Helmholtz instability. We already know, however, that if there is no maximum in the mean vorticity or if the scale of variation of the flow shear is larger than the disturbances (i.e. in this case we imagine a constant background vorticity), then there is no direct drive. The flow either has no effect or it suppresses the disturbances. In the case of a complicated set of model equations, the measurement of the average product of the Reynolds stress and the background vorticity can tell us how vigorously the disturbances are driven or suppressed by the mean flow.

The mechanism of flow drive or suppression conserves energy: if the disturbances are driven, then the flow is locally reduced; if the disturbances are suppressed, then their energy goes into the flow. The equation for the effect of the disturbances on the mean flow is

$$\frac{\partial}{\partial t}\frac{1}{2}\int dx\ V_0^2 = -\int dx\ V_0 \mathbf{v} \cdot \nabla v_y = \int dx\ V_0'(v_x v_y), \tag{16.16}$$

using $\nabla \cdot \mathbf{v} = 0$ on the fluctuations in the integration by parts. Eqs (16.15) and (16.16) reflect the basic conservation of kinetic energy, as the Reynolds stress merely mediates transfer between the mean flow and the disturbances. We have already seen the tendency of two-dimensional flow dynamics to cascade energy inversely to the largest available scales. It follows that there is a natural tendency of flows in the absence of a Kelvin–Helmholtz instability to energetically suppress turbulence, as the turbulence transfers its energy into the flows.

The energetic analysis enables a simple cartoon picture of the stability of the flow against perturbation by a disturbance of a given wavelength along the flow. If the disturbances are of short wavelength, especially if they are radially extended structures like streamers or convective plumes, then they are tilted by the flow. Imagine the flow in the $xy$-plane, with $y$ up/down and $x$ left/right, such that $V_0' > 0$. A streamer flow eddy is like a left/right rod, tilted up to the right and down to the left by the flow. The resulting orientation then has $v_x v_y$ positive. The product of the Reynolds stress and the zonal vorticity is then positive. Energetically, it then loses to the flow and is suppressed. Now take an elongated wave disturbance like a von Karman vortex street (see the hydrodynamics texts cited in the Section Further Reading, below). The associated flow eddy is now more up/down, and the effect of the zonal flow upon it will be to roll it over, such that the top is displaced to the left and the bottom to the right. Now we have the energetic relationship given by a product of $v_x v_y < 0$ and $V_0' > 0$ to form something negative. This flow eddy gains energy from the flow and is driven by it, and hence is unstable. So the long wavelength disturbance is driven unstable, and the short wavelength disturbance is suppressed. This gives a heuristic interpretation of the wavelength cutoff in Eq. (16.8).

### 16.4.1 In a magnetised plasma

In a magnetised plasma with electromagnetic responses, there are two nonlinearities in the vorticity equation. Isolating these in a 2D model, as in the MHD cases of Chapter 3, we start with

$$\frac{\partial \varpi}{\partial t} + [\phi, \varpi] + [A_\parallel, J_\parallel] = 0, \tag{16.17}$$

where the bracket given by

$$[a, b] = \frac{1}{B_0}\left(\frac{\partial a}{\partial x}\frac{\partial b}{\partial y} - \frac{\partial b}{\partial x}\frac{\partial a}{\partial y}\right), \tag{16.18}$$

contains the factor of $B_0$ in the units. The fields are now given by

$$\varpi = \frac{\rho}{B^2}\nabla_\perp^2\phi \qquad \mu_0 J_\parallel = -\nabla_\perp^2 A_\parallel, \tag{16.19}$$

and the ExB flow and vorticity are given in terms of the electrostatic potential $\phi$ by

$$v_E^x = -\frac{1}{B}\frac{\partial\phi}{\partial y} \qquad v_E^y = \frac{1}{B}\frac{\partial\phi}{\partial x} \qquad \Omega_E = \frac{1}{B}\nabla_\perp^2\phi. \tag{16.20}$$

The disturbances in the magnetic field are given by

$$b^x = \frac{1}{B}\frac{\partial A_\parallel}{\partial y} \qquad b^y = -\frac{1}{B}\frac{\partial A_\parallel}{\partial x}. \tag{16.21}$$

In this analysis we keep the physical units, and drop the tilde symbols. The coefficients $\rho$ and $B = B_0$ are assumed to be constant parameters. Note the difference in units between the vorticity variable $\varpi$ as part of the polarisation current and the ExB vorticity $\Omega_E$ as the vorticity of the flow. The form is the same, but with different coefficients. With these definitions we do the energetics via the same analysis as for the self-consistent gradient drive in the thermal free energetics in Eqs (15.42) to (15.58) from the previous Chapter.

The zonal vorticity evolves according to

$$\frac{\partial}{\partial t}\langle\varpi\rangle + \langle[\tilde\phi, \tilde\varpi]\rangle + \left\langle[\tilde A_\parallel, \tilde J_\parallel]\right\rangle = 0. \tag{16.22}$$

The ExB energy exchange is found by multiplying by $-\phi_0$ to find

$$-\phi_0\frac{\partial}{\partial t}\langle\varpi\rangle = \phi_0\left(\langle[\phi, \varpi]\rangle + \left\langle[A_\parallel, J_\parallel]\right\rangle\right) = 0. \tag{16.23}$$

The zonal average facilitates processing of the brackets according to

$$\langle[\phi,\varpi]\rangle = \frac{\partial}{\partial x}\left\langle -\varpi\frac{1}{B}\frac{\partial\phi}{\partial y}\right\rangle = \frac{1}{B}\frac{\partial^2}{\partial x^2}\left\langle -\rho\frac{1}{B}\frac{\partial\phi}{\partial y}\frac{1}{B}\frac{\partial\phi}{\partial x}\right\rangle$$
$$\langle[\phi,\varpi]\rangle = \frac{1}{B}\frac{\partial^2}{\partial x^2}\langle\rho v_E^x v_E^y\rangle. \tag{16.24}$$

Similarly, we find

$$\big\langle[A_\parallel, J_\parallel]\big\rangle = -\frac{1}{B}\frac{\partial^2}{\partial x^2}\langle\rho v_A^2 b^x b^y\rangle, \tag{16.25}$$

where $v_A$ is built from the polarisability $\rho/B^2$ and from $\mu_0$ in the Ampère's law, and the sign also comes from the one in the Ampère's law. We then have

$$\frac{\partial}{\partial t}\langle\varpi\rangle + \frac{1}{B}\frac{\partial^2}{\partial x^2}\langle\rho v_E^x v_E^y - \rho v_A^2 b^x b^y\rangle = 0. \tag{16.26}$$

The two terms under the average are the ExB Reynolds stress and the magnetic flutter Maxwell stress. These stresses form the nonlinearities which can drive or deplete zonal vorticity, which is ExB zonal velocity shear in our language. The energetic effect is found by multiplying Eq. (16.26) by minus the zonal potential and averaging over the radial domain. This produces

$$\frac{\partial E_Z}{\partial t} = \int\frac{dx}{L_x}\langle\Omega_E\rangle\big(\rho\langle v_E^x v_E^y\rangle - \rho v_A^2\langle b^x b^y\rangle\big), \tag{16.27}$$

where $E_Z$ is the zonal ExB energy given by

$$E_Z = \int\frac{dx}{L_x}\frac{1}{2}\rho\left\langle\frac{1}{B}\frac{\partial\phi}{\partial x}\right\rangle^2 = \int\frac{dx}{L_x}\frac{1}{2}\big(-\langle\phi\rangle\langle\varpi\rangle\big). \tag{16.28}$$

This shows that flows are driven by the disturbances in the turbulence if the combination of the two stresses align with the zonal ExB vorticity, which in this context is assumed to be part of the self-consistent dependent variable, through $\varpi = (\rho/B)\Omega_E$. The products of the Reynolds stress and Maxwell stress with the zonal vorticity as energetic processes are labelled $R_E$ and $R_M$, respectively. As before, it is the sign of the product which is energetically relevant in each case, not the sign of the stress itself.

For the effect on the turbulence we first consider the effect of an externally applied zonal flow determined by a given $\phi_0(x)$ which also represents a zonal vorticity $\varpi_0(x)$ consistent with the field equation. In the nonlinearities of each equation we therefore add this piece as an additional background drive as already done for $n_0$ as the thermal gradient drive (recall the units of $T_e$ which here is constant). As for $n_0$ it is done not only in the ExB advection nonlinearity but also in the parallel gradient, so that the set of ExB flow shear drive terms is given by the profile quantities in

$$[\phi+\phi_0,\ ] \qquad [\phi,\varpi+\varpi_0] \qquad [-A_\parallel,\phi+\phi_0]. \tag{16.29}$$

The first is the advective ExB nonlinearity in every equation, an addition to $d/dt$. The second is the action upon $\varpi_0$ as part of the flow shear physics. The third is part of the parallel gradient in the parallel electric field. It is important not to forget this term in a computational study. It can be combined with the inductive electric field, so that $A_\parallel$ is locally advanced by the flow given by the $x$-derivative of $\phi_0$, which is in the $y$-direction. The flow is considered zonal because it is in the direction of the angle of symmetry in the background, namely, an advance of the entire magnetic surface within a radial zone. With radial variation, this implies radial shear of the dependent variable field as it evolves. The simple view is the ExB shear decorrelation model given above. But the interaction of this with a turbulent vorticity enhanced by nonlinearities is what changes things.

The effect of $\phi_0$ on the energetics of the turbulence is found by repeating the energetic analysis done in Chapters 9–12 and following the effect on these terms. We use the properties of the brackets as done with the zonal and fluctuating thermal free energy in Eqs (15.42) to (15.58) from the previous Chapter. Here, the dependent variable is kept together as a unit, with the external profile $\phi_0$ separate. We use

$$\phi[\phi, \varpi_0] = [\phi, \phi\varpi_0] \tag{16.30}$$

$$\phi[\phi_0, \varpi] = [\phi\phi_0, \varpi] - \phi_0[\phi, \varpi] \tag{16.31}$$

$$J_\parallel[A_\parallel, \phi_0] = [A_\parallel, \phi_0 J_\parallel] - \phi_0[A_\parallel, J_\parallel] \tag{16.32}$$

$$f[\phi_0, f] = \frac{1}{2}[\phi_0, f^2], \tag{16.33}$$

where $f$ is any dependent variable except $\varpi$ (i.e. multiplication by $\varpi$ does not occur in the energy theorem). The terms involving $\phi_0$ combine into a signed drive rate given by

$$\Gamma_V = \mathscr{V}^{-1}\int d\mathscr{V}\ \phi_0\big([\phi, \varpi] + [A_\parallel, J_\parallel]\big). \tag{16.34}$$

Now, with the volume average separating into a zonal and radial average, this is equivalent to

$$\Gamma_V = \int \frac{dx}{L_x}\ \phi_0\big\langle[\phi, \varpi] + [A_\parallel, J_\parallel]\big\rangle. \tag{16.35}$$

Processing these brackets as we did before (Eq. 16.24), we find

$$\Gamma_V = -\int \frac{dx}{L_x}\frac{1}{B_0}\frac{\partial^2\phi_0}{\partial x^2}\langle\rho v_E^x v_E^y - \rho v_A^2 b^x b^y\rangle, \tag{16.36}$$

where the form of $\phi_0$ to be considered depends on the model. Note that the multiplier of the Reynolds stress is the zonal ExB vorticity $\Omega_E$, not the same thing as $\varpi_0$, as noted above. In normalised units, $\varpi_0 = (\rho/B_0)\Omega_E$ differ by a factor of $\delta_0$, which is very significant to the results as we give them below.

This flow shear drive rate is the same as Eq. (16.27), with opposite sign. If the flow shear is an external drive term, $\Gamma_V$ gives its energetic effect. If instead of this a zonal flow has been created as part of the turbulence, $\Gamma_V$ gives the rate of the energy lost by the non-zonal component as part of the process, which we see conserves energy in an electromagnetic system in the same way as in the simple 2D Eulerian flow model in which the Kelvin–Helmholtz analysis is done. In a three-dimensional drift Alfvén model, this shear flow energetics combines with the rest of the analysis. As with the thermal gradient, the self-consistent zonal disturbance transfers free energy conservatively with the fluctuation components. As an external drive mechanism, it adds or depletes free energy according to a rate which is the same as the transfer rate in a system in which the profile physics is self consistent.

We see why we had to watch out for the effect of $\nabla_\| \phi_0$ in the Ohm's law. This piece gives rise to the Maxwell stress term in Eq. (16.36) so that the two expressions are the same. Instead of a transfer effect between turbulence and flows, with a background flow shear as a drive we have this as a drive term for the turbulence, with the background flow being maintained. The background ExB vorticity is now a simple input in terms of $c_s/L_\perp$, arranged among the background gradient drive parameters. The minus sign in the definition of $\Gamma_V$ reflects an effective sink if energy is transferred to the fixed background flow. We include $\Gamma_V$ within the energetics, with the rest given as in Chapter 14.

Whether a suppressive effect of a background flow on the turbulence proceeds via decorrelation or energetically is something that can be measured in a computation. Cross coherence of the flow eddies and the transported quantities (i.e. simply $\tilde{n}_e$ in the simplest drive wave model), correlation lengths, and Reynolds and Maxwell stress energetics are all directly accessible, and the measurement of these will tell us whether, and if so by which mechanism, the disturbances are affected by flows.

## 16.5 Effect of background flow shear

Whether a background shear flow drives or suppresses turbulence is something we can ascertain by measuring the zonal flow energy in Eq. (16.28) and the energetic process terms in Eq. (16.27). In normalised units with $\phi$ scaled with $T_e/e$ the ExB vorticity is

$$\Omega_E = \frac{1}{B}\frac{\partial^2 \tilde{\phi}}{\partial x^2} = \frac{T_e}{eB}\frac{\partial^2 \phi}{\partial x^2} = \frac{\rho_s}{L_\perp} L_\perp^2 \frac{\partial^2}{\partial x^2}\left(\frac{e\phi}{T_e}\right), \tag{16.37}$$

hence only one factor of $\delta_0$, so that in normalised units we have

$$\varpi = \delta_0 \Omega_E. \tag{16.38}$$

Hence, if $\phi_0 = V' x^2/2$ the ExB vorticity is only $\delta_0$ which is of order $10^{-2}$, that is, very small. So to parameterise the shear against $c_s/L_\perp$ we set

$$\phi_0 = V' \delta_0^{-1} \frac{x^2}{2}, \tag{16.39}$$

which follows the physical units given by

$$\frac{e\phi_0}{T_e} = V'(L_\perp c_s)\frac{eB_0}{T_e}\frac{1}{2}\frac{x^2}{L_\perp^2}. \tag{16.40}$$

That is, the factor $\delta_0^{-1}$ comes from the $L_\perp c_s$ and $eB_0/T_e$ factors. When we consider warm-ion cases in toroidal geometry, however, we will have to reconsider this scaling.

### 16.5.1 In toroidal geometry

The shear flow parameter $V'$ is swept from small to unrealistically large values for four cases, covering a low and high inductivity and a low and high collisionality,

$$\hat{\beta} = 1 \quad \text{or} \quad 5 \qquad C = 1.39 \quad \text{or} \quad 13.9 \tag{16.41}$$

The other parameters are set as they have been in previous Chapters, with $\hat{\epsilon} = 10^4$ and $\hat{\mu} = 2.72$ and $\delta_0 = 0.015$ and $X = 100$ and $S = \omega_n = 1$ and $\omega_B = 0.03$. The shear flow parameter is set to 0, and then 0.1 and 0.2 and 0.3 and 0.5 and 0.7, where we note that a value of 1.0 would imply a sonic level for the ExB velocity which is not realistic in a confined plasma. More on this below when we consider the effect of geodesic curvature forcing on flows. The results on the effect of ExB flow shear are shown in figure 16.1, for these toroidal cases. The four quantities measured are the turbulent ExB transport flux across the field lines, $F_e = nv_E^x$, the enstrophy, $A_w = \varpi^2/2$, the thermal free energy, $E_n = n^2/2$, and the Ohmic energy (magnetic and parallel electron flow), $E_B = J(A + \hat{\mu}J)/2$, all in volume average. The nominal turbulence level is $\delta_0^2 = 2.25 \times 10^{-4}$ for comparison.

The two values of $C$ reflect the effective range of validity for this model. As $V'$ is increased in the $C = 1.39$ cases, the turbulence is strongly sheared without apparent effect, and then it begins to strengthen for $V' > 0.3$, more so for stronger inductivity. By contrast, in the $C = 13.9$ cases the turbulence weakens slightly with $V'$ until for $V' > 0.3$ and higher it is sharply suppressed. This is an interesting result because the base $V' = 0$ case for the higher $C$ is much stronger than for the lower value. So the result that the more collisional case is eventually suppressed depends on mode structure and dynamical character rather than simply a competition between the strength of the turbulence and the flow shear.

### 16.5.2 In slab geometry

We have to control against the prospect just raised. If ballooning character is the decisive factor, a set of companion cases with $\omega_B = 0$, i.e. in the sheared slab case at the same parameters, should show that. If not, then we have to consider something else. For the same study in a sheared slab geometry we repeat the above cases while setting $\omega_B = 0$ so that we have the slab and toroidal geometry at otherwise identical parameters. This is a control case against the effects of magnetic curvature. The ballooning mechanism is now absent, but now we also acquire the native stabilisation due to increased inducticity ($\hat{\beta}$) at low collisionality ($C$). The results on the effect of ExB flow shear are shown in figure 16.2, for these slab cases. The measured quantities and baseline value $\delta_0^2$ are the same as for the toroidal cases above.

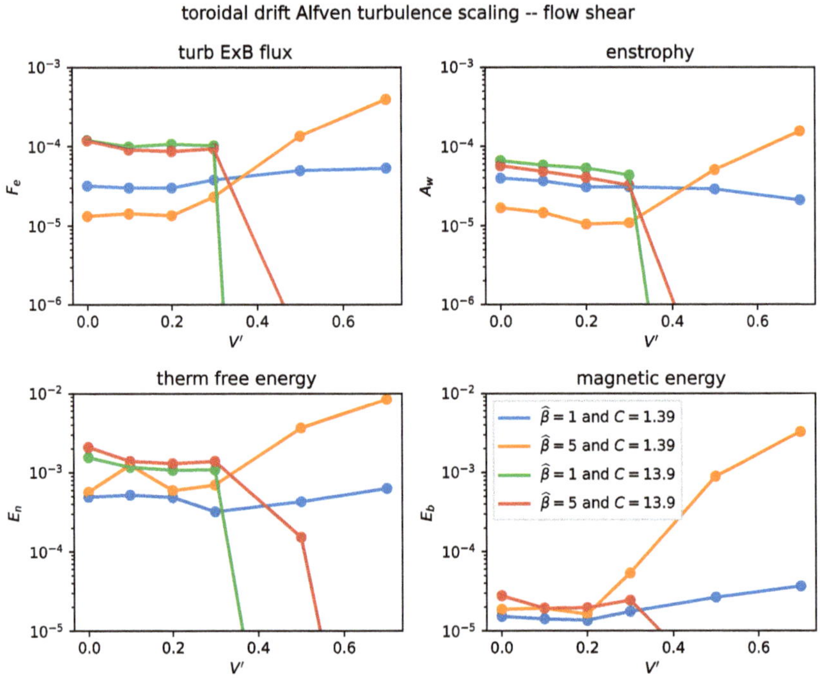


**Figure 16.1.** Scaling of the turbulent ExB transport flux ($F_e$), the enstrophy ($A_w$), the thermal free energy ($E_n$), and the magnetic energy ($E_b$), all in volume average, against imposed ExB flow shear $V'$ giving the background ExB vorticity in units of $c_s/L_\perp$. Four toroidal cases are considered, with $\hat{\beta}$, $C$ combinations of (1, 1.39), (5, 1.39), (1, 13.9), and (5, 13.9), colour coded as per the plot legend. For the lower $C$ the tendency for the flow shear is to mildly contribute to the strength of the turbulence, except for $\hat{\beta} = 5$ and $C = 1$ where the magnetic energy substantially increases for $V' > 0.2$. For the higher $C$ cases the flow shear leads to suppression of the turbulence for $V' = 0.5$ and 0.7.

We find the same basic flow shear parameter dependence as in the toroidal cases. The lower $C$ cases are not only not suppressed but are driven more strongly, at the higher $V'$ values. The high-$C$ cases show little or no effect until at $V' = 0.5$ the turbulence for them is suppressed. The $\hat{\beta} = 5$ case at $C = 1.39$ shows the same large increase in magnetic energy for $V' > 0.2$ as in the toroidal cases. This is also not explained by the level of the enstrophy versus the flow shear, as the case of the four with the highest enstrophy is $\hat{\beta} = 1$ and $C = 13.9$ at the control point $V' = 0$ (no flow shear). In fact, the case of the four with the lowest enstrophy ($\hat{\beta} = 5$ at $C = 1.39$) is the one which shows the largest increase in strength in all four quantities with higher flow shear. So the answer has to be something in the mode structure coming not from ballooning but from resistivity, in both the slab and toroidal cases.

### 16.5.3 On open field lines

Herein we repeat the toroidal study for cases with exit-flux boundary conditions on open field lines. This is another control case, here against the constraining by the field line connection property of the closed magnetic surfaces. The same four

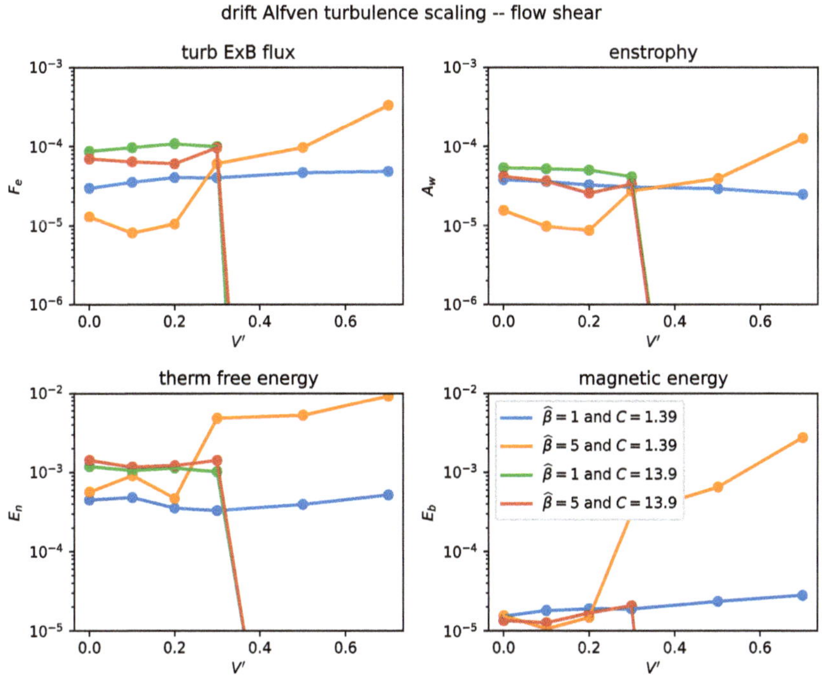


**Figure 16.2.** Scaling of the turbulent ExB transport flux ($F_e$), the enstrophy ($A_w$), the thermal free energy ($E_n$), and the magnetic energy ($E_b$), all in volume average, against imposed ExB flow shear $V'$ giving the background ExB vorticity in units of $c_s/L_\perp$. Four sheared slab cases are considered, with $\hat{\beta}$, $C$ combinations of (1, 1.39), (5, 1.39), (1, 13.9), and (5, 13.9), colour coded as per the plot legend. For the lower $C$ the tendency for the flow shear is to mildly contribute to the strength of the turbulence, except for $\hat{\beta} = 5$ and $C = 1$ where the magnetic energy substantially increases for $V' > 0.2$. For the higher $C$ cases the flow shear leads to suppression of the turbulence for $V' = 0.5$ and 0.7. This is the same basic effect of ExB shear as in the toroidal cases, indicating that ballooning or interchange drive effects are not the cause.

parameter sets as above are run again for the toroidal cases with the open boundary conditions using the exit-flux model of Chapter 15 instead of the field line connection ones of Chapters 13 and 14. The results on the effect of ExB flow shear are shown in figure 16.3 for these exit-flux toroidal cases. The measured quantities and baseline value $\delta_0^2$ are the same as for the toroidal cases above.

With open field lines with exit-flux boundary conditions, we find little or no effect of ExB shear on any of the representative quantities for any of the four cases. This makes it clear that field line connection, present in the other case sets but not here, is a necessary condition for the suppression of the turbulence by the flow shear in the more strongly collisional cases. Taken together, the results would not support a simple consideration that flute mode type turbulence is suppressed. The effect of the flow on the turbulence is found to depend on the presence of field line connection. It is also not a simple matter of flow shear preferentially overcoming weaker turbulence, since on closed field lines the more strongly collisional cases produce natively more robust turbulence.

Briefly, a variation of $C$ at fixed $\hat{\beta} = 5$ and $V' = 0.5$ in the closed-field line toroidal cases given above shows that resistivity is indeed the cause. Values of 1, 2, 3, 5, 7,

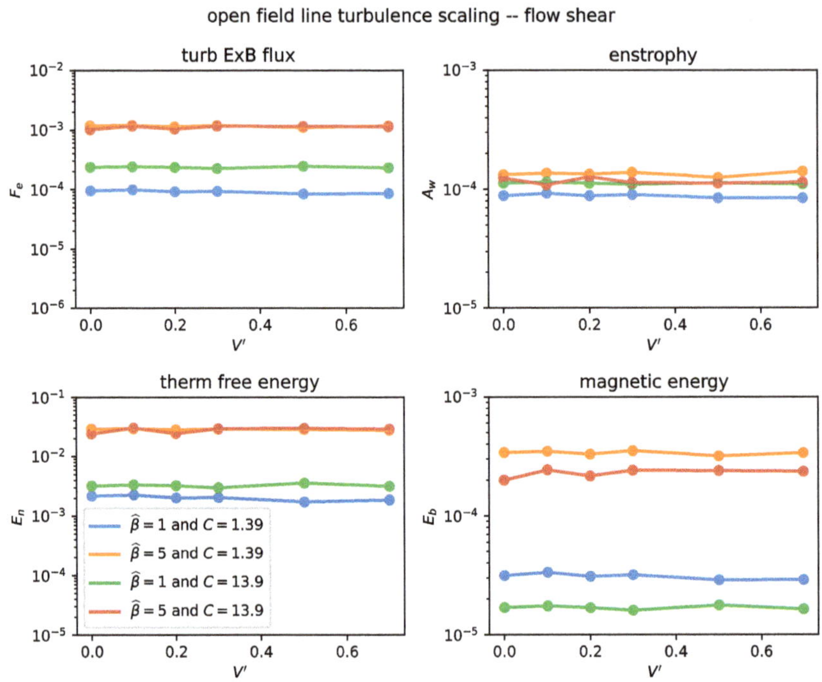


**Figure 16.3.** Scaling of the turbulent ExB transport flux ($F_e$), the enstrophy ($A_w$), the thermal free energy ($E_n$), and the magnetic energy ($E_b$), all in volume average, against imposed ExB flow shear $V'$ giving the background ExB vorticity in units of $c_s/L_\perp$. Four toroidal cases with exit-flux boundary conditions at the ends of open field lines are considered, with $\hat{\beta}$, $C$ combinations of (1, 1.39), (5, 1.39), (1, 13.9), and (5, 13.9), colour coded as per the plot legend. By contrast to figures 16.1 and 16.2, this time there is little or no effect of the flow shear on any of the channels.

and 10 times 1.39 are used for $C$, with the result that the turbulent flux $F_e$ begins to rise slightly, from 1.1 to 2.0 times $10^{-4}$ for the first two values, then 1.72 for $C = 4.17$ shows increasing effect of the flow shear. Afterward, the turbuent flux sharply drops, to $2.9 \times 10^{-5}$ for $C = 6.95$ and is then stabilised for higher $C$. The statistical variation is about 20 percent in all saturated cases. The values for $\Gamma_E$ are all below $10^{-5}$, with no tendency to increase for larger flow shear. So the result is real and it is due to the increasingly hydrodynamic nature of the turbulence.

## 16.6 Flow shear in warm-ion toroidal cases

The reduced MHD model arises from MHD ordering in which $n_e e \nabla \phi$ is large compared to $\nabla p$ for any species, so not only are the diamagnetic flows and adiabatic response neglected, so are the finite pressure corrections to gyroviscosity.

When the diamagnetic flows and adiabatic response are considered but the gyroviscosity is not, then the model becomes a cold-ion model assuming $\tau_i = T_i/T_e \ll 1$. This is because while keeping $\rho_s$ in the spectrum due to the vorticity dynamics, neglecting $\rho_i$ follows since $\rho_i \ll \rho_s$ implies $\tau_i \ll 1$, hence cold ions.

Under MHD ordering the pressure is regarded as the total pressure and there is no specific issue against the ion/electron temperature ratio, since both $\rho_s$ and $\rho_i$ are regarded as equally negligible. But when the adiabatic response is kept, then the gyroviscosity can only be neglected if the ions are considered cold. In many applications this assumption is not a reasonable one.

Gyroviscosity becomes a real issue with flows, since the nonlinearity breaks the enstrophy conservation upon which much of the basic phenomenology of ExB vorticity, Reynolds stress, and energy relies. With finite $\tau_i$ we are changing

$$\nabla \cdot \frac{d}{dt}\nabla_\perp \phi = \frac{\partial \varpi}{\partial t} + [\phi, \varpi], \tag{16.42}$$

in normalised units, to

$$\nabla \cdot \frac{d}{dt}\nabla_\perp W_i = \frac{\partial \varpi}{\partial t} + [\phi, \varpi] + [\nabla_\perp \phi; \nabla_\perp W_i], \tag{16.43}$$

with $W_i = \phi + \tau_i n$. In the cold-ion form, the two equivalent versions on the left- and right-hand side are used to prove energy and enstrophy conservation by multiplying with $-\phi$ and $\varpi$ and integrating, respectively. In the warm-ion form, the energy is still conserved by multiplying the form on the left-hand side by $-W_i$ and integrating. But now the form on the right has that third term which does not vanish, unless $\tau_i \to 0$ and hence $W_i \to \phi$ in the cold-ion limit. Energy is still conserved but enstrophy is not, since $\varpi[\nabla_\perp \phi; \nabla_\perp W_i]$ is a hanging term. This will have consequences on both the sensitivity to applied ExB flow shear, which we check here, and the self-generated zonal flow dynamics, which we check later on, below.

This diamagnetic contribution to the polarisation current leads to changes in the part of the flow energetics due to the Reynolds stress. We multiply equation (16.43) by $-W_i$ instead of $-\phi$ before integrating, and there is also a $\tau_i n_0$ term to go with $\phi_0$ under the gradient in $\nabla_\perp W_i$. The specific warm-ion additions are

$$W_i = \phi + \tau_i n \qquad W_0 = \phi_0 + \tau_i n_0 \qquad u_i^y = [W_i, y]. \tag{16.44}$$

Repeating the analysis leading to Eq. (16.36), we find

$$\mathscr{V}^{-1}\int d\mathscr{V}\, \frac{1}{2}\frac{\partial}{\partial t} \mid \delta_0 \nabla_\perp W_i \mid^2 = \mathscr{V}^{-1}\int d\mathscr{V} \left( -\delta_0 \frac{\partial^2 W_0}{\partial x^2} v_E^x u_i^y + \cdots \right), \tag{16.45}$$

with the other terms due to $\phi_0$ integrating out. We have put the factors of $\delta_0$ back in (three of them, two from $\varpi$ and one from the advection) to get the correct scaling. This is a mixed Reynolds stress term on the right-hand side, with one ExB flow component ($\partial\phi/\partial y$) and one total flow component ($\partial W_i/\partial x$, ExB plus diamagnetic), multiplied by the total flow vorticity profile. Now, in a self-consistent setting $\phi_0 + \tau_i n_0$ would actually come close to vanishing, as we indicate next. But for now we merely assume these two pieces are of comparable size, not exactly cancelling, and use this to evaluate the flow energetics.

The benchmark value for equilibrium vorticity in a toroidal magnetic field is warm-ion force balance under the curvature terms. This is given by the equilibrium profile condition

$$n_i e \mathcal{K}(\phi) + \mathcal{K}(p_i) = 0 \qquad \text{hence} \quad n_e e \frac{\partial \phi}{\partial x} + T_i \frac{\partial n_e}{\partial x} = 0, \tag{16.46}$$

in a plasma with $n_i = n_e$. Note how this goes with the ion temperature, not the electron one, since it is the ion diamagnetic flow which enters. Note how the curvature term translates to a geometric coefficient times the radial gradient. While interchange curvature is $\partial/\partial y$ here, this $\partial/\partial x$ piece arises from the geodesic curvature. It is due to the variation $\partial R/\partial \theta$ in a low-beta toroidal equilibrium, which arises according to Eqs (13.8), (13.56), (13.73), and (13.77) from Chapter 13. Note that the $\partial/\partial y$ piece in the curvature operator does not act on the axisymmetric component, for which $\partial/\partial y$ at constant $s$ is proportional to $\partial/\partial \varphi$ at constant $\theta$, so it does not matter if we think in terms of Eq. (13.73) or Eq. (13.77). There is more on the properties of geodesic curvature below, when we consider self-generated flows.

If the background profiles themselves have structure, the shear of the flow due to the zonal component of $\phi$ is determined by the scale of the profile and this condition. So we have the nominal level

$$\Omega_E \sim \Omega_E(\text{diamagnetic}) = \frac{v_D}{L_\perp}, \tag{16.47}$$

which is the 'diamagnetic shear' level. Note that it is not the diamagnetic flow that causes anything, just that the level of ExB shear is given by the diamagnetic velocity divided by the profile scale length, as a result of this force balance (more on the physical mechanism, below). We are not going to study tokamak equilibrium in detail herein, but we merely use this piece of it to motivate the concept and magnitude of equilibrium flow shear. Once we treat equilibrium flow shear in a warm-ion setting, we are no longer free to set its level, but have it scale as some multiple of $\delta$, given by the geometric shape of the profile.

For this sensitivity study we take the $\hat{\beta} = 5$ and $C = 13.9$ case from the warm-ion cases in Chapter 14 ($\delta_0 = 0.015$ and $\tau_i = 1$), and set $V'$ to values one to several times $v_D/L_\perp$. This sets a $\phi_0(x)$ given by

$$\phi_0 = V' \frac{x^2}{2}, \tag{16.48}$$

with $x$ in units of $L_\perp$. The ExB flow shear is then $\Omega_E = V'\delta_0$ while $\varpi_0$ is one factor of $\delta_0$ smaller. It is important to note this when considering flows in a tokamak, as this factor of $\delta_0$, being as small as it is, gets in the way of any model which considers arbitrarily strong flow shear. The resulting vorticity levels are always of order or smaller than $10^{-1}$ when compared to the drift frequency of $c_s/L_\perp$, even in steep gradient cases ($L_\perp/R_0$ such that $\hat{\mu} > 1$, but nevertheless $\delta_0 \sim 1/50$).

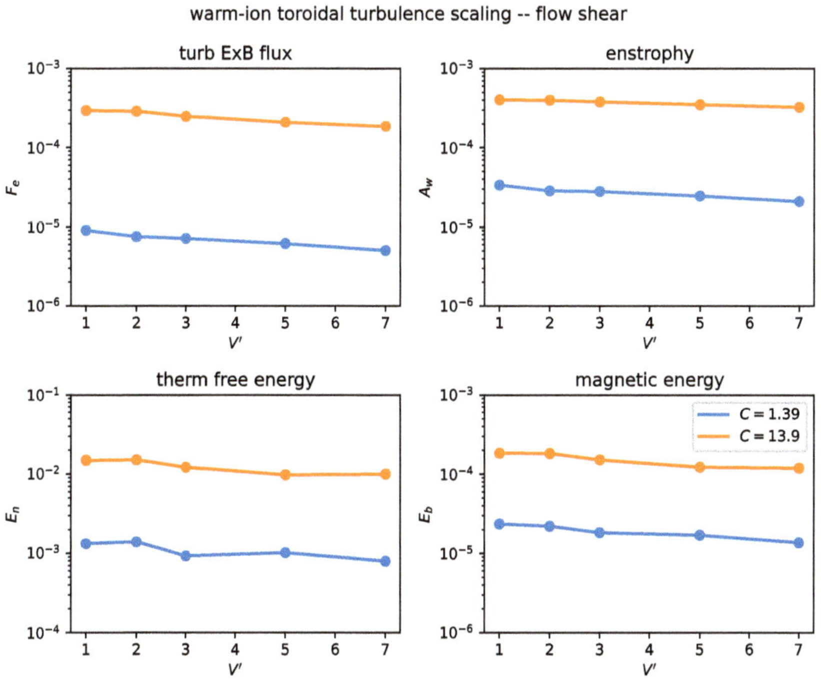


**Figure 16.4.** Scaling of the turbulent ExB transport flux ($F_e$), the enstrophy ($A_w$), the thermal free energy ($E_n$), and the magnetic energy ($E_b$), all in volume average, against imposed ExB flow shear $V'$ giving the background ExB vorticity in units of $v_D/L_\perp$, for $\hat{\beta} = 5$ and both $C = 1.39$ and $C = 13.9$. Note that $v_D/L_\perp$ is a factor of $\delta_0$, here 0.015, smaller than $c_s/L_\perp$. The flow shear at this level of toroidal force balance, several times $\delta_0$, has a weak but visible effect on the turbulence. The largest value, $7\delta_0$, corresponding to $V' = 0.105\ c_s/L_\perp$, is about as large as equilibrium flow shear can get.

The response of the drift Alfvén turbulence with warm ions to a flow shear given by several factors times the nominal $v_D/L_\perp$ is shown in figure 16.4, for a collisional, finite-beta case with $\hat{\beta} = 5$ and $C = 13.9$. The flow shear value is $V'$ times $v_D/L_\perp$, not times $c_s/L_\perp$ as in the cold-ion cases we considered before. The strongest case of $V' = 0.7$ has an ExB vorticity value of $\Omega_E = 0.105$. We find that this level of flow shear has a visible but (at most) moderating influence on any channel of the turbulence. The same scan is repeated with the lower value of $C = 1.39$, finding weaker turbulence as expected. The flow shear effect might be interpreted as slightly stronger, but is nevertheless a subdominant correction.

The energetics of the case with $V' = 0.7$, signifying a flow shear of $\Omega_E = 0.7\delta_0(c_s/L_\perp)$, is shown in figure 16.5. The positive trace is the gradient drive $\Gamma_n$. All the others are sinks. In descending order they are, the subgrid dissipation ($\Gamma_w$), the resistive dissipation ($\Gamma_c$), the flow shear transfer to background, ($\Gamma_V$, nearly all Reynolds stress), the viscous dissipation ($\Gamma_s$), and the two magnetic signals given by magnetic flutter gradient drive (negative) and the Maxwell stress part of the flow shear sink, both of which are not visible. The flow shear effect is a net sink, but it is not large enough to upset the balance between the other effects, namely, the overall energetics.

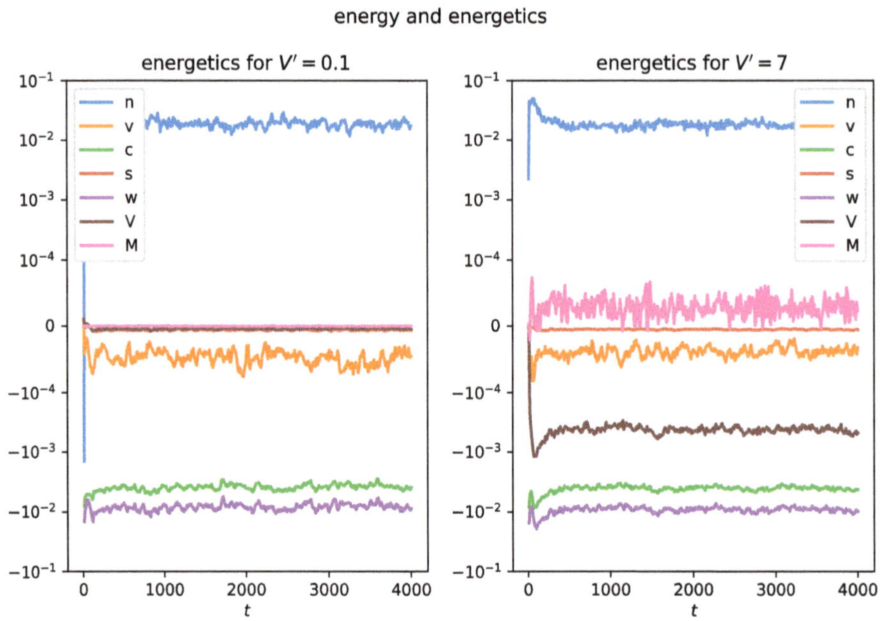


**Figure 16.5.** Time traces of the energetic quantities in the warm-ion cases with $\hat{\beta} = 5$ and $C = 13.9$, for negligible and strong shear flow ($V' = 0.1$ and 7 times $v_D/L_\perp$). The thermal gradient drive ($\Gamma_n$) is positive, while all the others are mostly sinks. The subgrid dissipation ($\Gamma_w$) is largest, then the resistive dissipation ($\Gamma_c$) and the effect of flow shear ($\Gamma_V < 0$). The others are much smaller: the parallel flow viscous sink ($\Gamma_s < 0$), the magnetic Reynolds stress transfer to background ($\Gamma_M > 0$) and the magnetic flutter gradient drive ($\Gamma_v < 0$). The energetic effect of the flow shear is stabilising but subdominant.

## 16.7 Properties of the flux surface average

The flux surface average is a special operation, not merely an average over angles. The distinction becomes relevant when there is spatial structure in the geometry, in the metric elements hence the volume element $\sqrt{g}$, the magnetic field strength $B$, or the contravariant components of the magnetic field $B^i = \mathbf{B} \cdot \nabla x^i$. There are two main properties. The flux surface average is the volume derivative of a volume integral, which means it is a volume integral over the infinitesimal domain defined by two neighbouring flux surfaces of infinitesimal separation. The flux surface average as an operator is the annihilator for the unperturbed parallel gradient, so that the flux surface average of a magnetic parallel divergence vanishes.

The volume $V$ is flux function, given by the volume enclosed by the flux surface. So if $x$ is a radial coordinate describing flux surface position, then flux functions satisfy $f = f(x)$, where we have $\psi = \psi(x)$ and $V = V(x)$ hence $V = V(\psi)$. In terms of the coordinates,

$$V = \oint_\psi \oint \oint \sqrt{g}\, dx\, d\theta\, d\zeta, \tag{16.49}$$

where $\theta$ and $\zeta$ are two arbitrary poloidal and toroidal angles. Due to the structure of the magnetic field

$$\mathbf{B} = I(\psi)\nabla\varphi + \nabla\psi \times \nabla\varphi, \tag{16.50}$$

with $\varphi$ the geometric toroidal angle periodic on $[-\pi, \pi]$, we have $\mathbf{B}\cdot\nabla\psi = 0$, hence $\mathbf{B}\cdot\nabla f = 0$ for any flux function. Flux surfaces are surfaces of constant $\psi$, which is why these quantities are called flux function and flux surface average. These details are for axisymmetric systems. Non-axisymmetry is a strong complication because it relaxes the requirement that a closed equilibrium is described by a nested sequence of closed magnetic flux surfaces. It is left to the references in the Section Further Reading.

The flux surface average defined with the first property is given by

$$\langle f\rangle = \frac{1}{V'}\frac{\partial}{\partial x}\oint_{\psi}\oint\oint \sqrt{g}\; dx\; d\theta\; d\zeta\; f, \tag{16.51}$$

where $f = f(x, \theta, \zeta)$ is a scalar quantity. The volume derivative with respect to $x$ is given by

$$V' = \frac{\partial V}{\partial x} = \frac{\partial}{\partial x}\oint_{\psi}\oint\oint \sqrt{g}\; dx\; d\theta\; d\zeta. \tag{16.52}$$

The differentiation strips the integration over $x$ so the integral is now over the surface of a given $x$,

$$V' = \frac{\partial V}{\partial x} = \oint\oint \sqrt{g}\; d\theta\; d\zeta. \tag{16.53}$$

This is now a volume-weighted integration over the angles covering the closed surface. So the volume derivative of the volume integral is the average over angles weighted by the volume element,

$$\langle f\rangle = \frac{1}{V'}\oint\oint \sqrt{g}\; d\theta\; d\zeta\; f. \tag{16.54}$$

This is the definition from which we now prove the second property.

The flux surface average of the parallel divergence is given by

$$\langle \mathbf{B}\cdot\nabla f\rangle = \frac{1}{V'}\oint\oint \sqrt{g}\; d\theta\; d\zeta \left(B^{\theta}\frac{\partial}{\partial\theta} + B^{\zeta}\frac{\partial}{\partial\zeta}\right) f. \tag{16.55}$$

As long as $\theta$ is held fixed, axisymmetry of the geometry is satisfied for any $\zeta(x, \theta, \varphi)$, since

$$c_0\frac{\partial f}{\partial\zeta}\bigg|_{x\theta} = \frac{\partial f}{\partial\varphi}\bigg|_{x\theta}, \tag{16.56}$$

with $c_0 = \partial\zeta/\partial\varphi$ constant. Hence, the $\zeta$-derivative may be pulled past the coefficient $\sqrt{g}B^{\zeta}$, and the integration over the $\zeta$-derivative vanishes. This leaves the poloidal derivative, for which we have

$$\sqrt{g}B^\theta = (\nabla x \times \nabla\theta \cdot \nabla\zeta)^{-1}(\nabla\psi \times \nabla\varphi \cdot \nabla\theta), \tag{16.57}$$

which is a constant times $\partial\psi/\partial x$. This is a flux function (as demanded by $\nabla \cdot \mathbf{B} = 0$) so it can be pulled out of the integration over angles. This leaves flux functions times

$$\oint\oint d\theta\, d\zeta\, \frac{\partial f}{\partial\theta} = 0, \tag{16.58}$$

which vanishes, so that

$$\langle \mathbf{B} \cdot \nabla f \rangle = 0, \tag{16.59}$$

proving the point. Hence, for example in the vorticity equation,

$$\left\langle \mathbf{B} \cdot \nabla \frac{J_\parallel}{B} \right\rangle = 0, \tag{16.60}$$

namely, the undisturbed parallel divergence of the current vanishes under the flux surface average. If disturbances are present we have the nonlinear bracket $[-A_\parallel, J_\parallel]$ leading to the zonal Maxwell stress.

Other properties follow. The flux surface average of a divergence is the volume derivative of the volume contravariant component,

$$\langle \nabla \cdot \mathbf{M} \rangle = \frac{1}{V'}\frac{\partial}{\partial x}V'\langle \mathbf{M} \cdot \nabla x \rangle, \tag{16.61}$$

where the factor of $V'$ cancels if $x$ is $V$. This is because the divergence is written as

$$\nabla \cdot \mathbf{M} = \frac{1}{\sqrt{g}}\left(\frac{\partial}{\partial x}\sqrt{g}M^x + \frac{\partial}{\partial\theta}\sqrt{g}M^\theta + \frac{\partial}{\partial\zeta}\sqrt{g}M^\zeta\right), \tag{16.62}$$

and the angle derivatives vanish under the integration. It follows that the flux surface average of the radial component of a divergence-free vector vanishes. We have

$$\nabla \cdot \mathbf{M} = 0 \qquad \text{hence} \quad \mathbf{M} = \nabla \times \mathbf{U}. \tag{16.63}$$

Then we have

$$\begin{aligned} \langle \mathbf{M} \cdot \nabla x \rangle &= \langle \nabla \times (\mathbf{U} \cdot \nabla x) \rangle = \langle \nabla \cdot (\mathbf{U} \times \nabla x) \rangle \\ &= \frac{1}{V'}\frac{\partial}{\partial x}V'\langle \nabla x \cdot (\mathbf{U} \times \nabla x) \rangle = 0, \end{aligned} \tag{16.64}$$

since $\nabla \times \nabla x$ vanishes. Hence, for example, the flux surface average of the radial component of the current,

$$\langle \mathbf{J} \cdot \nabla x \rangle = \frac{1}{\mu_0}\langle (\nabla \times \mathbf{B}) \cdot \nabla x \rangle = 0, \tag{16.65}$$

vanishes, and the net radial transport of electric charge is zero.

For flux-tube models, we have the following considerations. The angle coordinates $y$s are automatically normalised. Integrations by parts on angle derivatives leave no surface terms, since at the ends of the field line connection length, the phase shift in $y$ in a field-aligned coordinate system for $s$ is covered by the simultaneous integration over $y$. Another way to say this is that the integrals over the angles can be done sequentially. The average over $y$ strips the distinction between $y_k$ and $y$ so that no $s - s_k$ factors appear. This is why $s$ and $\theta$ are taken equivalently in the integration. The integration is over the full domain of the surfaces of constant $x$, namely the flux surfaces in that model. Double-periodicity, namely toroidal topology, is central to this. This operation is called the 'zonal average' even though it is taken at a singular radial location.

## 16.8 Zonal and equilibrium flows

In a slab geometry any zonal flow is an equilibrium flow, in the sense that if nonlinearities and gradient drive were switched off the flow would simply remain, except for a viscous dissipation acting on the vorticity profile but which is assumed to be arbitrarily small. If the flow is initialised and nonlinearities are allowed, but the gradient drive is switched off, the Kelvin–Helmholtz stability will determine whether it remains. In a magnetised plasma, the parallel dynamics couples ExB flow energy to the magnetic energy (and parallel electron kinetic, assuming $d_e$ is finite), so the disturbances are in the sort of 2D MHD system we saw in Chapters 3 and 5. With the adiabatic response also present we have the kinetic Alfvén and acoustic responses since the parallel current fluctuations now couple to the electron pressure. If the equilibrium flow is not maintained, it is simply a self-consistent zonal flow which has been initialised. The energetics of a situation like this is the same as seen before in Chapter 12, just with the energy initially in the flow and then coupled back to the rest of the system. The end state depends only on whether the flow is stable or not. If stable, the flow remains and there is no dynamics. If unstable, its energy is eventually processed and the dynamics at late times is as if the flow never existed. If the flow is maintained by a source, it can drive the rest of the system and a saturated balance is possible. If a zonal flow of width $\Delta_V$ is sourced in a system with $\Delta_V \ll L_y$, then the source for the flow becomes the source for the entire turbulent dynamics, since the flow is unstable. If the magnetic field is sheared, then the energetic fate depends on what the effect of the magnetic shear is on flow stability. With gradient drive present, drift wave or drift Alfvén dynamics becomes present and one has the interaction between gradient drive, flow source, and the turbulence sending the energy down to sinks. This is sufficient complexity to be decided only by well resolved (both $d_e$ and $\rho_s$) computations, not least due to the potential presence of the nonlinear drift wave instability.

### 16.8.1 Geodesic curvature and sidebands

In toroidal geometry, especially tokamak-like geometry, the situation is more interesting. This is because of the presence of the relaxation channels one has when geodesic curvature is present. A tokamak-like geometry is defined as a plasma column of minor radius $a$, curved around toroidally with nominal major radius $R_0$, with $a/R_0$ not too small, and a magnetic field line pitch $q$ larger than unity but not too large. Tokamak-like becomes tokamak when under these conditions it is also axisymmetric. A conventional tokamak is defined as one in which $\beta \ll 1$ and $(a/qR_0)^2 \ll 1$ in roughly the same measure. The model for it is the one set down in Chapter 13.

Geodesic curvature can be understood through the curvature operator

$$\mathcal{K} = \frac{2}{I_0}\left(\sin s\frac{\partial}{\partial x} + (\cos s + S(s - s_k)\sin s)\frac{\partial}{\partial y}\right), \tag{16.66}$$

with $I_0 = B_0 R_0$ constant, derived in Eq. (13.76) from Chapter 13. For axisymmetric quantities the $\partial/\partial y$ term vanishes so the magnetic shear doesn't enter. With finite radial structure the $\partial/\partial x$ term is present and does all the coupling. This term represents geodesic curvature, as its derivation depends on the poloidal variation of the magnetic field hence poloidal component of the curvature radius. In many other configurations, the magnetic field is mostly or only poloidal, so $\mathcal{K}$ only involves $\partial/\partial y$, as we saw for the dipole field. Even with a finite toroidal magnetic component, the smallness of it in some other configurations renders the geodesic curvature subdominant. But in tokamak and stellarator (3D equilibria with a 3D vacuum field set by coils with a finite helical component) cases, the magnetic field is dominantly toroidal, which is what small $a/qR_0$ means. Geodesic curvature is a strong effect in these configurations.

The sinusoidal dependence of $\mathcal{K}$ upon $s$ introduces the sideband components. In terms of poloidal Fourier mode numbers $m$, the curvature forcing couples neighbouring modes $m \pm 1$, which for the axisymmetric component $k_y = 0$ are the sidebands. With $y$-dependence filtered out, there is no distinction between $s$ and $\theta$ in a toroidal system and we can use them interchangeably. The sine and cosine sideband components of a variable $f$ are defined by

$$\langle f \sin\theta\rangle \qquad \langle f \cos\theta\rangle, \tag{16.67}$$

respectively. That is, first multiply by $\sin\theta$ or $\cos\theta$ and then take the flux surface average. These are the $m = \pm 1$ sidebands of the zonal component $\langle f\rangle$ which has both mode numbers at zero. The flux surface average over the linear terms in $B\nabla_\parallel$ vanishes but in general the average over $\nabla_\parallel$ itself does not. In our case, however, we will neglect the variation $B(s)$, so that extra factors of $B$ and $R$ as well as $\nabla_\parallel \log B$ are not carried around. In any of this analysis we assume appropriate flux coordinates, for which $B^s$ and $\sqrt{g}$ are flux functions, as in the dipole magnetic field analysis of Chapter 15.

For these axisymmetric components, only the average over $\theta$ enters the flux surface average, so that the parallel derivative, which can be considered as

$$\nabla_\parallel = K_\parallel \frac{\partial}{\partial\theta}, \tag{16.68}$$

can be done by parts under the integration. Hence, we also have

$$\left\langle \sin s \nabla_{\|} u_{\|} \right\rangle = -K_{\|} \left\langle u_{\|} \cos s \right\rangle \qquad \text{and} \qquad \left\langle \cos s \nabla_{\|} \phi \right\rangle = K_{\|} \langle \phi \sin s \rangle, \tag{16.69}$$

in the parallel divergence and gradient, respectively, since we are taking $B = B_0$ in the coefficients. Hence also, there are no parallel gradient or parallel divergence terms in the zonal components of the equations.

The second sidebands occur through the curvature forcing terms in the equations for the sine sidebands. We have

$$\langle \sin s \mathcal{K}(f) \rangle = \frac{2}{I_0} \frac{\partial}{\partial x} \langle f \sin^2 s \rangle = \frac{1}{I_0} \frac{\partial}{\partial x} \langle f (1 - \cos^2 2s) \rangle \rightarrow \frac{1}{I_0} \frac{\partial}{\partial x} \langle f \rangle, \tag{16.70}$$

with the factor of 1/2 arising from the average of $\sin^2 s$ over a cycle. This is where the second sidebands are neglected. The coupling is from the sine sideband ($m = \pm 1$) back to the zonal one ($m = 0$) and to the sine squared sideband ($m = \pm 2$), which is neglected as small. This stops the coupling chain and serves as a truncation operation which provides the analysis with a closed system of equations. These are energetically consistent, so that energy transfer mechanisms are net conservative.

### 16.8.2 Sideband analysis of the dynamics

We will give this here for the cold-ion case for clarity (avoiding $[\nabla_\perp \phi; \nabla_\perp W_i]$). The warm-ion changes are given later, when the warm-ion cases are considered.

With the geodesic coupling and sideband integrations in hand, the zonal average of the vorticity equation gives the zonal evolution as

$$\frac{\partial}{\partial t} \langle \varpi \rangle + \langle [\phi, \varpi] \rangle + \left\langle [A_{\|}, J_{\|}] \right\rangle = \frac{2}{I_0} \left\langle \frac{\partial p}{\partial x} \sin s \right\rangle \tag{16.71}$$

where the linear parallel divergence and the interchange curvature vanish under the zonal average. But the sine sideband of the pressure is now introduced (here, $p = T_e n$ with $T_e$ constant). The background gradient is not part of this because it is purely zonal. This new piece is part of the dependent variables and is part of their self-consistent system. The term on the right-hand side is the one introduced by geodesic curvature and is referred to as 'geodesic compression'. The nonlinearities couple the zonal flow back to the turbulence. The effect of geodesic curvature is to couple it to the sideband pressure.

The sine sideband component of the electron density then has its own equation, given by

$$\begin{aligned} &\frac{1}{n_e} \frac{\partial}{\partial t} \langle n \sin s \rangle + \langle [\phi, (n + n_0)] \sin s \rangle - \left\langle \left[ A_{\|}, \left( u_{\|} - \frac{J_{\|}}{n_e e} \right) \right] \sin s \right\rangle \\ &- K_{\|} \left\langle \left( u_{\|} - \frac{J_{\|}}{n_e e} \right) \cos s \right\rangle = \frac{1}{I_0} \left( \left\langle \frac{\partial \phi}{\partial x} \right\rangle - \frac{T_e}{n_e e} \left\langle \frac{\partial n}{\partial x} \right\rangle \right), \end{aligned} \tag{16.72}$$

in physical units, where $n$ is the dependent variable, $n_0$ the background gradient, and $n_e$ and $T_e/e$ are constant parameters. The top line contains the nonlinearities, with $n_0$

dropping out due to the averaging. The last term on the left-hand side is the parallel gradient, coupling not only to sound waves but also to the Alfvén system. This gives another potential drive or sink term for the zonal flow and sideband system, distinct from the Reynolds and Maxwell stresses in the zonal vorticity equation, and with different properties.

The two curvature terms on the right side involve the zonal electrostatic potential and density disturbances. The first represents the action of the curvature on the zonal flow and is called 'ExB compression'. It is present also in an MHD model. The second represents the action of the curvature on the zonal density disturbance, which represents the zonal diamagnetic flow due to the electrons. It is called 'diamagnetic compression', since it is the divergence of the diamagnetic flow. It is not present in an MHD model since the diamagnetic flow is different for electrons and ions. Any of these terms can have either sign, and it is only possible to get the results *a posteriori*.

In a flux-tube model the zonal density under this curvature term does not include the background gradient $n_0$. The background equilibrium in a toroidal setting includes the balance among all these zonal and sideband terms, with the zonal profiles set for the densities and temperatures, and all the rest of the terms determined by the balances. For our purposes we simply assume that the equilibrium is present in a relaxed balance, and so we only address the departures from equilibrium represented by the disturbances. The dependent variables do not include the self-consistent background gradient and equilibrium in a flux-tube model, but it is important to note that they always include the perturbed equilibrium, and therefore the physical processes that are involved in the relaxation to an MHD equilibrium are part of the overall dynamical system which governs the interaction of the turbulence with zonal flows.

### 16.8.3 Geodesic acoustic oscillations

The simplest model to understand this starts with the reduced MHD model without the adiabatic response or the diamagnetic current, as in Chapter 13. With more quantities present, we return to using the tilde symbols for the dependent variable quantities $\tilde{n}_e$, $\tilde{u}_\parallel$, $\tilde{J}_\parallel$, and the field variables. The state variables are $\varpi$ and $\tilde{n}_e$, and they are given the zonal and $\sin\theta$ sideband components. The flux variables are $\tilde{J}_\parallel$ and $\tilde{u}_\parallel$, they are given the $\cos\theta$ components only. The reason for the restriction to six quantities is the chain of coupling when starting with a zonal flow, or, for the case of tokamak equilibration, a zonal pressure profile. Due to the $\sin s$ nature of the geodesic curvature, a zonal flow couples to a sine sideband of the pressure, which then couples through the parallel dynamics to the cosine sideband of the parallel flow, and there it stops, since the second sidebands are neglected. In an MHD model these three quantities separate from the other ones given by a zonal pressure, a sine sideband of the vorticity (also ExB potential/flow), and then the cosine sideband of the current (also magnetic potential/field). The first three are the acoustic branch, and the second three are the Alfvén branch.

The first three produce what are known as 'geodesic acoustic oscillations' or modes. Given a Reynolds stress which can produce a zonal flow, or simply given an initialised zonal flow, the interactions are then governed by the equations for the zonal vorticity, the sine sideband of the density, and, as can be seen from Eq. (16.72), the cosine sideband of the parallel flow. In the MHD case the terms due to $J_{\|}$ and the diamagnetic compression are absent, and Eqs (16.71) and (16.72) without the nonlinearities are written as

$$\frac{\partial}{\partial t}\langle\varpi\rangle = \frac{2}{I_0}\left\langle\frac{\partial p}{\partial x}\sin s\right\rangle, \tag{16.73}$$

$$\frac{1}{n_e}\frac{\partial}{\partial t}\langle\tilde{n}_e \sin s\rangle - K_{\|}\left\langle\tilde{u}_{\|}\cos s\right\rangle = \frac{1}{I_0}\left\langle\frac{\partial\tilde{\phi}}{\partial x}\right\rangle, \tag{16.74}$$

$$\rho\frac{\partial}{\partial t}\left\langle\tilde{u}_{\|}\cos s\right\rangle + K_{\|}\langle p\sin s\rangle = -\mu_{\|}K_{\|}^2\left\langle\tilde{u}_{\|}\cos s\right\rangle, \tag{16.75}$$

where the pressure is $p = T_e\tilde{n}_e$ and $\{I_0, T_e, \rho, \mu_{\|}\}$ are constants. The field equation is

$$\frac{\rho}{B^2}\frac{\partial^2}{\partial x^2}\langle\tilde{\phi}\rangle=\langle\varpi\rangle, \tag{16.76}$$

with only the $x$-derivatives surviving the flux surface average.

To find the eigenmodes, the parallel viscosity term is neglected, and then we have a closed set of three equations for the three quantities. In the vorticity equation one of the $x$-derivatives is common and is dropped, and then the time derivative is taken. Then the time derivative of $\langle p \sin s\rangle$ is replaced by the latter's equation, and then the time derivative is taken again to then replace the time derivative of $\left\langle\tilde{u}_{\|}\sin s\right\rangle$ with the latter's equation. The result is a cubic equation for the frequency $\omega$, once the time derivative is replaced by $-i\omega$ using the Fourier transform. One of the factors of $\omega$ is common, and the other two give the oscillation. We find

$$\omega = 0 \qquad \text{or} \qquad \pm\,\omega_G \quad \text{with} \quad \omega_G^2 = 2\frac{c_s^2}{R_0^2} + K_{\|}^2 c_s^2, \tag{16.77}$$

using $I_0 = B_0R_0$ and $n_eT_e/\rho = c_s^2$. If as in a tokamak $K_{\|} = 1/qR_0$ then

$$\omega_G^2 = 2\frac{c_s^2}{R_0^2}\left(1 + \frac{1}{2q^2}\right), \tag{16.78}$$

which is the standard result. The zero frequency mode is the one in which there is no excitation of the pressure sine sideband since the parallel flow cosine sideband and the zonal flow balance, namely, the overall axisymmetric flow is divergence free. This mode represents a zonal flow in divergence balances. The eigenmodes with finite overall divergence produce the oscillations at frequency $\omega_G$, and these are the geodesic acoustic oscillations.

If a zonal ExB flow is initialised by itself, the divergence is not zero because there is ExB compression. This excites the pressure sideband and then the parallel flow sideband, and the oscillation results. The single factor of two represents $2/I_0^2$ in the curvature coupling compared to the simple square of $K_\parallel$ in the parallel coupling. The acoustic wave in the parallel responses (Chapter 8) becomes a geodesic acoustic oscillation in a tokamak-like configuration. The oscillation continues until the action of parallel viscosity relaxes the flow into divergence balance and the pressure sideband into force balance with the viscosity. In this simple model without thermal details, the relaxation of the flow goes all the way to zero since any finite $\tilde{n}_e$ drives dynamics in both the zonal flow and parallel sideband. The natural oscillation is with all three pieces, not just two. A general initialisation which includes the divergence-free piece will leave it undamped, so that it remains as a residual flow.

### 16.8.4 Geodesic Alfvén oscillations

The second set of variables in an MHD model are the ones which are involved in setting up a finite pressure toroidal equilibrium. The basic balance in the currents is found using the sine sideband of the vorticity equation. Again without the nonlinearities, we multiply by sin $s$ and then take the flux surface average. This gives

$$\frac{\partial}{\partial t}\langle \varpi \sin s\rangle - K_\parallel\left\langle \tilde{J}_\parallel \cos s\right\rangle = \frac{1}{I_0}\frac{\partial}{\partial x}\langle p\rangle, \tag{16.79}$$

where the second term is the parallel divergence, having integrated the $s$-derivative, part of $\nabla_\parallel$, by parts, and the term on the right is the curvature term. Now, instead of a zonal vorticity and a pressure sideband, we have the gradient in the zonal pressure and the vorticity sideband, and the parallel current sideband as well. If $p$ contains the entire gradient and $\tilde{J}_\parallel \to J_\parallel$ the entire toroidal current, the equilibrium of this equation is the same as the one for a tokamak. The pressure gradient implies a diamagnetic current, which requires a parallel current to balance the divergence. This current is called the Pfirsch–Schlüter current, after the authors who published it. In a flux-tube model $p$ does not carry the entire pressure as one is not solving the entire equilibrium but using it as a background. It does, however, carry the perturbed equilibrium, the one resulting from a zonal disturbance of the pressure, which is part of the model. The same physics of toroidal relaxation into equilibrium is therefore part of the model, in which disturbances in the zonal pressure drive currents.

When not in equilibrium, the parallel and diamagnetic currents do not balance in divergence, and the inertia term (vorticity time derivative) arises to keep the balance. This is the polarisation current, and maintaining the current divergence balance at all times is its role in the dynamics. If we start with a zonal pressure perturbation, i.e. a localised roll of zonal pressure adding to the equilibrium profile, then a sideband vorticity is excited. What happens next is determined by the cosine sideband of the Ohm's law. Neglecting the nonlinearities and the electron inertia, this is

$$\frac{\partial}{\partial t}\left\langle \tilde{A}_\parallel \cos s\right\rangle = -K_\parallel\langle\tilde{\phi}\sin s\rangle - \eta_\parallel\left\langle \tilde{J}_\parallel \cos s\right\rangle, \tag{16.80}$$

just the cosine sideband of the equation $E_\| = \eta_\| J_\|$, in which the disturbance in the current is the one which appears, since the equilibrium is given. The relevant field equations are

$$\frac{\rho}{B^2}\frac{\partial^2}{\partial x^2}\langle\phi\sin s\rangle = \langle\varpi\sin s\rangle \qquad \frac{\partial^2}{\partial x^2}\left\langle\tilde{A}_\|\cos s\right\rangle = -\mu_0\left\langle\tilde{J}_\|\cos s\right\rangle, \tag{16.81}$$

namely, the sine sideband of the polarisation equation and the cosine sideband of the induction equation (Ampère's law).

When the vorticity sideband is excited, the field equation gives the corresponding electrostatic potential sideband, which now drives the parallel current. In a flux-tube model the radial dependence of $K_\|$ is neglected (the magnetic shear appears only in the parallel boundary or field line connection condition for the disturbances), and we can take two $x$-derivatives and apply the field equations, to find

$$\frac{\partial}{\partial t}\left\langle\tilde{J}_\|\cos s\right\rangle = v_A^2 K_\|\langle\varpi\sin s\rangle - \frac{\eta_\|}{\mu_0}\frac{\partial^2}{\partial x^2}\left\langle\tilde{J}_\|\cos s\right\rangle, \tag{16.82}$$

using the definition $v_A^2 = B^2/\mu_0\rho$ with the polarisability $\rho/B^2$ and $\mu_0$ from the polarisation and induction equations, respectively. Hence, the vorticity and current sidebands are involved in the basic MHD shear Alfvén oscillation, here in the reduced MHD sideband description. The oscillation frequency is given by the coefficients as

$$\omega = \pm\omega_A \qquad \text{where} \quad \omega_A = K_\| v_A. \tag{16.83}$$

The resistivity term applied as a small correction dissipates this oscillation, leaving the currents in the divergence balance

$$\left\langle\tilde{J}_\|\cos s\right\rangle = -\frac{1/I_0}{K_\|}\frac{\partial}{\partial x}\langle p\rangle, \tag{16.84}$$

which describes the perturbed Pfirsch–Schlüter current, and a static parallel electric field given by the finite resistivity

$$-K_\|\langle\tilde{\phi}\sin s\rangle = \eta_\|\left\langle\tilde{J}_\|\cos s\right\rangle, \tag{16.85}$$

yielding the relaxed state of the textbook toroidal MHD problem, but in this case for the perturbed equilibrium.

The free energy for this comes from the zonal pressure, through the diamagnetic term in the sideband vorticity equation, above. Its equation, and the one that closes the set, is

$$\frac{1}{n_e}\frac{\partial}{\partial t}\langle\tilde{n}_e\rangle = \frac{2}{I_0}\frac{\partial}{\partial x}\langle\phi\sin s\rangle, \tag{16.86}$$

again with $p = T_e\tilde{n}_e$ and without the nonlinearities, noting that the parallel divergence vanishes under the zonal average. This says the zonal disturbance evolves according to ExB compression of the flow sideband. One usually does not keep flow compression in the MHD analysis. However, it yields a finite-beta correction given by the geodesic curvature coupling. The toroidal Alfvén oscillation proceeds much as for straight field lines, so long as the radial scale of the disturbance is large compared to $\rho_s$ and the two-fluid contributions can be neglected. However, the zonal pressure follows the oscillation and yields the last term in

$$\omega^2 = K_\parallel^2 v_A^2 + \omega_G^2. \tag{16.87}$$

As before the factors $1/I_0$ and $2/I_0$ and $p/\rho$ produce the geodesic frequency. This is a low-beta correction, since

$$\frac{\omega_G}{\omega_A} = 2q^2\frac{c_s^2}{v_A^2} = 2q^2\beta_e. \tag{16.88}$$

The result is the 'geodesic Alfvén oscillation', which is the other natural oscillation in a toroidal perturbed equilibrium. In the simple straight field line analysis of Chapter 8, we have the two pairs of waves given by the shear Alfvén and sound waves. In a toroidal equilibrium the axisymmetric versions of these become the geodesic Alfvén and acoustic oscillations. With all three equations present, the free energy is conserved up to the dissipation given by the finite resistivity.

### 16.8.5 Acoustic/MHD adiabatic coupling

When the adiabatic response is included, these two sets, the Alfvén and acoustic branches, are no longer separate. The parallel current cosine sideband and pressure sine sideband are now coupled and drive each other. This gives the $k_\perp^2\rho_s^2$ correction to the Alfvén frequency, and the restoration of the electron inertia gives the $k_\perp^2 d_e^2$ correction, restoring the same $k_\parallel V_a$ of the kinetic shear Alfvén waves in Chapter 8. Now the entire system relaxes into divergence and force balance between all six components. With $\beta_e$ and $\mu_e$ both small, and with $k_\perp\rho_s$ also small, there is little interaction between the two branches and they relax nearly independently. One can think of $k_\perp\rho_s \sim 0.2$ with a radial width of about $20\rho_s$ for the perturbed equilibrium layers.

On the other hand, the dynamics becomes more interesting when the turbulence continually forces the zonal/sideband perturbed equilibrium. An equilibrium flow in divergence balance is either initialised or caused by other means (see below, for the warm-ion corrections). It is coupled to the turbulence via Reynolds and Maxwell stresses as transfer mechanisms. At the same time, the zonal pressure is coupled to the turbulence via the zonally averaged nonlinearity. But the interesting part is the coupling from the zonal flow to the pressure sine sideband over the geodesic compression. Turbulence can put energy in or take it out of either piece. Moreover, the energetic coupling extends from the pressure to the current sidebands and therefore the connection from zonal flow to sidebands to resistive dissipation of the Alfvén system is now open. There are several places for zonal flow energy to go, should it be driven. And there are several places for the turbulence to react and become resilient to an applied equilibrium flow, which are absent in the simple decorrelation model and even in a reduced MHD model, which are more familiar.

The set of equations for this is the combination of the two sets above, plus the adiabatic response, diamagnetic compression, electron inertia, and the state variable nonlinearities which are the principal ones. We add the nonlinearities giving contact to the turbulence where energetically significant, i.e. in the zonal and sideband pressure and in the zonal vorticity only. In the same language as the above, with the sideband chains and the Reynolds and Maxwell stresses combined, we have

$$\frac{\partial}{\partial t}\langle\varpi\rangle + \frac{\rho}{B}\frac{\partial^2}{\partial x^2}(\langle v_E^x v_E^y\rangle - v_A^2\langle b^x b^y\rangle) = -T_e\frac{2}{I_0}\frac{\partial}{\partial x}\langle\tilde{n}_e \sin s\,\rangle, \tag{16.89}$$

$$\begin{aligned}\frac{\partial}{\partial t}\langle\tilde{n}_e \sin s\,\rangle &+ \left\langle\left(\tilde{n}_e v_E^x + \left(n_e\tilde{u}_\parallel - \frac{\tilde{J}_\parallel}{e}\right)b^x\right)\sin s\right\rangle\\ &= n_e K_\parallel\big\langle\tilde{u}_\parallel \cos s\,\big\rangle - \frac{1}{e}K_\parallel\big\langle\tilde{J}_\parallel \cos s\,\big\rangle\\ &\quad + \frac{1}{2}n_e\frac{2}{I_0}\frac{\partial}{\partial x}\langle\tilde{\phi}\rangle - \frac{1}{2}\frac{T_e}{e}\frac{2}{I_0}\frac{\partial}{\partial x}\langle\tilde{n}_e\rangle,\end{aligned} \tag{16.90}$$

$$\rho\frac{\partial}{\partial t}\big\langle\tilde{u}_\parallel \cos s\,\big\rangle = -T_e K_\parallel\langle\tilde{n}_e \sin s\,\rangle - \mu_\parallel K_\parallel^2\big\langle\tilde{u}_\parallel \cos s\,\big\rangle, \tag{16.91}$$

$$\frac{\partial}{\partial t}\big\langle\tilde{A}_\parallel \cos s\,\big\rangle = \frac{T_e}{n_e e}K_\parallel\langle\tilde{n}_e \sin s\,\rangle - K_\parallel\langle\tilde{\phi} \sin s\,\rangle - \eta_\parallel\big\langle\tilde{J}_\parallel \cos s\,\big\rangle, \tag{16.92}$$

$$\frac{\partial}{\partial t}\langle\varpi \sin s\,\rangle = -K_\parallel\big\langle\tilde{J}_\parallel \cos s\,\big\rangle - \frac{1}{2}T_e\frac{2}{I_0}\frac{\partial}{\partial x}\langle\tilde{n}_e\rangle, \tag{16.93}$$

$$\begin{aligned}\frac{\partial}{\partial t}\langle\tilde{n}_e\rangle &+ \left\langle\tilde{n}_e v_E^x + \left(n_e\tilde{u}_\parallel - \frac{\tilde{J}_\parallel}{e}\right)b^x\right\rangle\\ &= n_e\frac{2}{I_0}\frac{\partial}{\partial x}\langle\tilde{\phi} \sin s\,\rangle - \frac{T_e}{e}\frac{2}{I_0}\frac{\partial}{\partial x}\langle\tilde{n}_e \sin s\,\rangle.\end{aligned} \tag{16.94}$$

This shows how forcing can work. The nonlinearities can always go either way, but they are expected to drive the flow through Reynolds stress at the expense of the turbulence, and to return free energy to the turbulence through the perturbed gradient drive at the expense of the pressure sidebands. In electromagnetic cases the Maxwell stress can add or subtract to the Reynolds stress. There are no sources in the zonal vorticity equation, just forcing through Reynolds stress and geodesic compression. Energy can flow either in or out, and in saturated turbulence we expect a balance. The question remains, which way does the drive and depletion work? The answers will be more complicated than any of our simple pictures have led us to believe.

The energetic equation for the zonal flow is found by applying the steps leading up to Eqs (16.27) and (16.28) on the zonal vorticity Eq. (16.89). This yields

$$\frac{\partial E_Z}{\partial t} = \int \frac{dx}{L_x} \rho \langle \Omega_E \rangle (\langle v_E^x v_E^y \rangle - v_A^2 \langle b^x b^y \rangle) - \frac{2}{R_0} T_e \int \frac{dx}{L_x} \langle v_E^y \rangle \langle \tilde{n}_e \sin s \rangle, \quad (16.95)$$

where the radial integration acts on each of the terms in the whole equation, and the velocity and magnetic field components are given in Eqs (16.20) and (16.21). The extra factor of $B_0$ arranged to form $\Omega_E$ comes from the brackets, and we can write the equation in this way since we have taken $B \to B_0$. Without that we must remain with the more primitive forms such as the zonal bracket. The three terms on the right-hand side of Eq. (16.95) are Reynolds and Maxwell stress forcing and geodesic compression, respectively. Each of these are of indeterminate sign and therefore under circumstances could serve as drive or depletion of zonal flow energy. In a toroidal plasma this geodesic compression mediated by magnetic curvature can regulate the rest of the system by balancing whatever the turbulent stresses do. Or, alternatively, it can force them to act as flow depletion if the flow of energy originates somewhere else in the system.

Again, this is of sufficient complexity to be decided only by well resolved computations, since the direction of energy transfer in all the channels is not known *a priori*. Indeed, we have already seen nonlinear transfer reversing the direction of the flow of free energy between vorticity and pressure at medium wavelength in simpler models than even this one. We will want to measure PDF as well as mean average for any of these forcing/transfer effects in statistically equilibrated saturation.

## 16.9 Self-generated zonal flows

We re-examine the computations in toroidal geometry, from Chapter 14, and investigate the energetics of the entire zonal flow and sideband system. This will establish through which process the flows are driven, if at all, and if so, where their energy eventually goes or whether it stays in the flow, which of the driven should then increase steadily. If flows are driven and then saturate, it follows that all the energetic channels have to be in statistical balance.

The content of the zonal/sideband energy pieces for the more collisional warm-ion case, $\hat{\beta} = 5$ and $C = 13.9$, is shown in figure 16.6. The left side shows the turbulence, with thermal free energy larger than ExB energy in the fluctuations, $E_n > E_e$ as usual. The right side shows the zonal ExB flow energy ($E_Z$) and the sine sideband component of the thermal free energy ($E_{ps}$) and the ExB energy ($E_{es}$). The zonal flow energy is much less than $E_n$ or in this case even $E_e$, and is in equipartition with $E_{ps}$, with $E_{es}$ in the concurrent Alfvén dynamics rather less. The finiteness of $E_{es}$ nevertheless indicates robust Alfvén activity at a statistically equilibrated level. For the less collisional $C = 1.39$ case, the same diagnostic shows all curves lower, with $E_Z$ slightly larger than $E_{ps}$. The zonal thermal energy, not shown, is somewhat

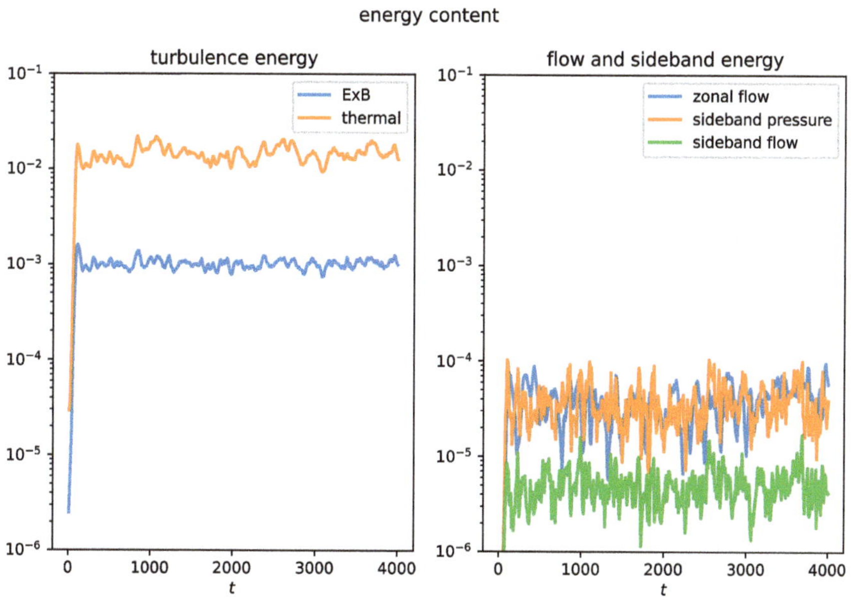


**Figure 16.6.** Time traces of the energetic content of the turbulence (left) and zonal and sideband channels (right), for the warm-ion case with $\hat{\beta} = 5$ and $C = 13.9$. The turbulence has most of the energy, with $E_n > E_e$ as previously. The zonal flow energy and sideband thermal free energy are in approximate statistical equilibration, with levels $E_Z = (4.17 \pm 1.79) \times 10^{-5}$ and $E_{ps} = (3.88 \pm 1.92) \times 10^{-5}$, respectively, averaged over $2000 < t < 4000$. This is the result of the geodesic curvature coupling, which apart from the associated oscillations, serves to keep the zonal flow energy at low levels in a toroidal system.

larger than $E_{ps}$. This is a result of the perturbed equilibrium—the gradient is slightly reduced in the radial region between the boundary dissipation layers, and this change is most of the signal. It does not have a large effect on this dynamics, however, as it is largely coupled to the Alfvén branch of the dynamics. But the main result is that one does not find the zonal flow energy rising to become concurrent to the energy of the turbulence.

### 16.9.1 Effect of geodesic curvature

The toroidal cases show that the zonal flow energy remains at low levels, more or less comparable to the flow energy in the turbulence, here defined as the $k_y \neq 0$ components. This energy level is much less than that in the thermal free energy, since with $e\tilde{\phi}/T_e$ comparable to $\tilde{n}_e/n_e$ in the turbulence, and with the $k_\perp \rho_s$ range mostly less than unity, it follows that

$$\rho \mid \tilde{\mathbf{v}}_E \mid^2 < n_e T_e \left( \frac{\tilde{n}_e}{n_e} \right)^2, \tag{16.96}$$

which is what we measure. Zonal flow shear only influences the dynamics if its vorticity is greater than that of the turbulence. And with the RMS vorticity in the range of $c_s/L_\perp$ an ExB energy less than the thermal free energy cannot produce a zonal vorticity larger than a factor of $\rho_s/\Delta_V$ times this, where $\Delta_V$ is the typical radial width of the zonal flow. That is because while $k_\perp\rho_s < 1$ for most of the turbulence, the vorticity itself is closer down to $\rho_s$, while the width of a typical zonal flow is some ten times larger. For zonal vorticity to take over, one must have the zonal flow energy itself rather larger than $E_e$, its counterpart in the turbulence.

### 16.9.2 Warm-ion zonal flow energetics

For warm ions the zonal flow sideband analysis is similar, just that we have to watch out for the structure of the polarisation. The total ion flow (ExB and diamagnetic) and the pressure, is given by $W_i$ and

$$u_i^x = -\delta_0\frac{\partial W_i}{\partial y} \qquad u_i^y = \delta_0\frac{\partial W_i}{\partial x} \qquad p = (1+\tau_i)n \qquad \varpi = \delta_0^2\nabla_\perp^2 W_i, \tag{16.97}$$

with the factors of $\delta_0$ explicitly written (it is necessary to count them in each term, in evaluating a computation). We make use of the re-casting of the nonadiabatic density $h = n - \phi$ as a nonadiabatic pressure $h = p - W_i$. This takes care of the ion diamagnetic polarisation due to $\tau_i n$ in the zonal flow energetics. In normalised units, the energetic equations for the zonal flow energy and the thermal sideband free energy are

$$\frac{\partial E_Z}{\partial t} = \left\{\delta_0^{-1}\langle\varpi\rangle\left(\langle v_E^x u_i^y\rangle - \hat{\beta}^{-1}\langle b^x b^y\rangle\right) - \omega_B\langle p\sin s\rangle\langle u_i^y\rangle\right\}, \tag{16.98}$$

$$\begin{aligned}\frac{\partial E_{ps}}{\partial t} = &\Big\{-2\langle p\sin s\rangle\left(\langle nv_E^x\rangle + \langle (u-J)b^x\rangle\right)\\ &+ 2\langle p\sin s\rangle\langle (u-J)\cos s\rangle + \omega_B\langle p\sin s\rangle\left(\langle u_i^y\rangle - \delta_0\left\langle\frac{\partial p}{\partial x}\right\rangle\right)\Big\},\end{aligned} \tag{16.99}$$

where the flux surface average is denoted as before, and the outer braces denote the radial average, in which several $\partial/\partial x$ are integrated by parts. In the second equation, note the two terms $\langle p\sin s\rangle$ with $\langle J\sin s\rangle$ in the parallel dynamics and $\langle p\sin s\rangle$ with $\langle\partial p/\partial x\rangle$ in the curvature ($\omega_B$). These represent adiabatic and diamagnetic compression, respectively. The significance here is that they give the zonal flow energy several channels to go to dissipation, after the geodesic compression through $\langle u_i^y\rangle$ and $\langle p\sin s\rangle$ sends it into the parallel dynamics. One sees that the structure is the same as in Eq. (16.95), with the replacement of $v_E^y$ by $u_i^y$, of $\Omega_E$ by $\delta_0^{-1}\varpi$, and of the electron pressure by the total.

The labels on these terms for computational purposes are according to

$$\frac{\partial E_Z}{\partial t} = R_E + R_M + T_G, \tag{16.100}$$

for the flows, such that a positive sign indicates net drive or net incoming energy transfer. For the sidebands they are

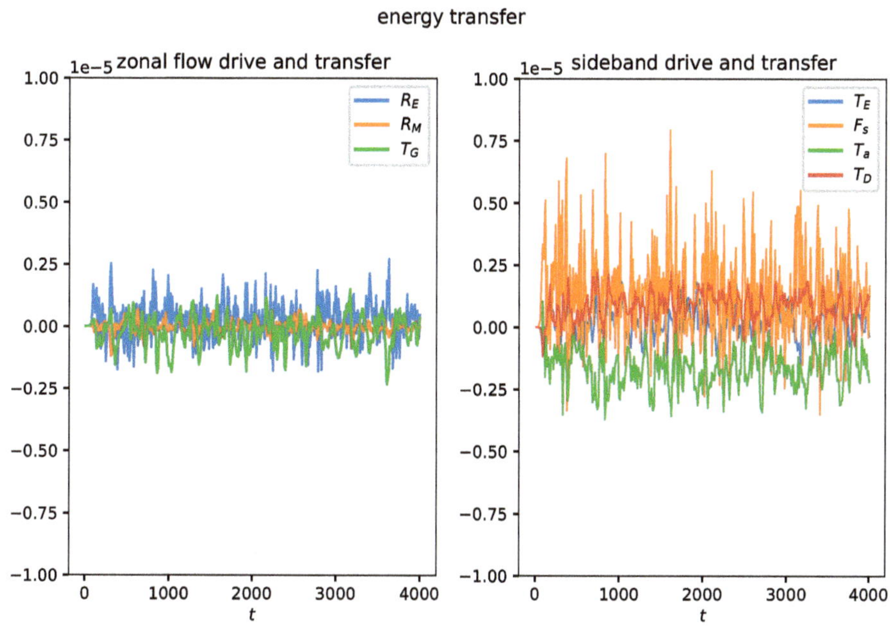


**Figure 16.7.** Time traces of the zonal flow and thermal sideband energy transfer channels, for the case with $\hat{\beta} = 3$ and $C = 13.9$ and $\tau_i = 1.0$. The zonal flow (left) and thermal sideband (right) energetic channels are shown separately for clarity. On the left, the curves are Reynolds ($R_E$) and Maxwell ($R_M$) stresses, and geodesic compression ($T_G$). On the right are the ExB compression ($T_E$), turbulent flux sideband ($F_s$), adiabatic compression ($T_a$), and diamagnetic compression ($T_D$). They represent drive of flows or sidebands where positive, and depletion where negative. Note the statistical character, in with prominent bursts. By construction, $T_E$ is the same as $-T_G$. See the text for the formulae.

$$\frac{\partial E_{ps}}{\partial t} = F_s + T_a + T_E + T_D, \tag{16.101}$$

with the positive sign again indicating net drive or net incoming energy transfer, but for this component. Note that $T_E = -T_G$ by construction.

The computational situation is complicated by the fact that the result depends on the run of three different parameters: the collisionality ($C$), inductivity ($\hat{\beta}$), and whether or not the ions are warm ($\tau_i$). The time traces of the terms in the zonal flow and thermal sideband energy equations are given in figure 16.7, for the case with $\hat{\beta} = 3$ and $C = 13.9$ for $\tau_i = 1$, warm ions. The key feature is the statistical character in the traces. The left side shows the zonal flow channels and the right side shows the sideband thermal free energy channels. In practically all cases the statistical scatter is substantially greater than the mean value, the only exception being the Reynolds stress in cold ions cases with relatively low $\hat{\beta}$ and $C$.

The sideband channels show that even when the turbulence drives the sideband through the sideband flux component ($F_s$), most of the energy stays in this channel and does not go to the zonal flow. Instead, it is mostly passed back out to the current through adiabatic compression ($T_a$, the effect of $\nabla_{\|}J$). There is enough scatter in the strength and statistics of these channels that the actual values vary by about ten to twenty percent. Nevertheless, a feel for them is adequately given. The numbers, averaged over $2000 < t < 4000$, are $T_E = (2.02 \pm 6.24) \times 10^{-7}$ for the ExB compression (incoming), $F_s = (12.2 \pm 18.2) \times 10^{-7}$ for the flux sideband (main drive), $T_a = (-15.7 \pm 77.9) \times 10^{-7}$ for the adiabatic compression (sink), and $T_D = (8.68 \pm 5.42) \times 10^{-7}$ for the diamagnetic compression (incoming). It is important to note that the drive of this component by the turbulence does not translate to a drive of the zonal flow itself. This is why we separate the zonal flow ($k_\| = 0$) and the thermal sideband ($k_\| = K_\|$) channels, as they represent different energetic processes.

In the zonal flow, it can be argued that all terms are statistically zero, but the small net positive component in the Reynolds stress ($R_E$) is to be noted. The numbers are $R_E = (2.15 \pm 7.83) \times 10^{-7}$ for the Reynolds stress drive, $R_M = (-1.80 \pm 2.18) \times 10^{-7}$ for the Maxwell stress drive (negative = sink), and $T_G = (-2.02 \pm 6.24) \times 10^{-7}$ for the geodesic compression, passing the energy out in rough net balance with the Reynolds stress.

The PDFs of the terms in the zonal flow energy and the sideband thermal free energy equation are given in figure 16.8, for the same case with $\hat{\beta} = 3$ and $C = 13.9$ for $\tau_i = 1$, warm ions. The samples are taken over $3000 < t < 4000$ and represent individual flux surfaces rather than their radial averages. Here we see more of the difference in statistical character, especially between $F_s$ and $T_a$. The signal in the latter is much more dominated by the tail (the significance of a PDF with two different power laws, the flatter one giving rise to the average). Large outliers are

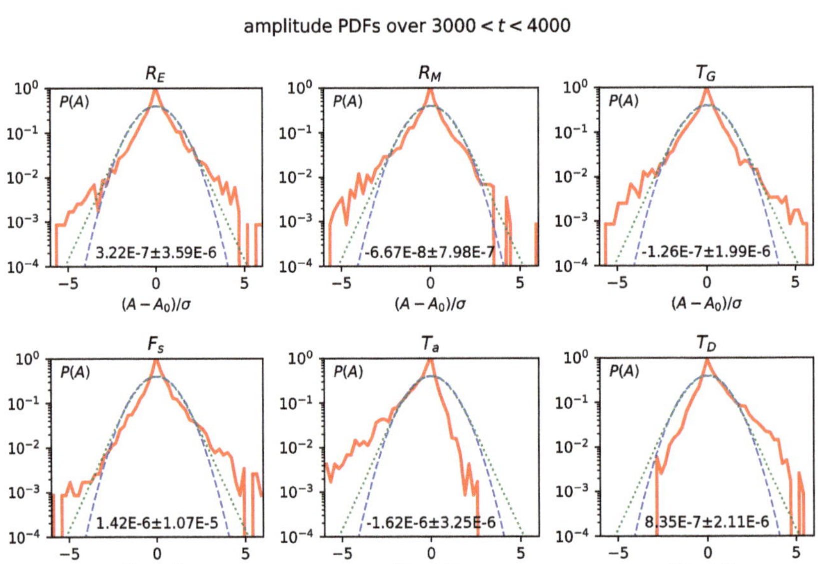


**Figure 16.8.** Probability distributions for the energy transfer channels, for the same case with $\hat{\beta} = 3$ and $C = 13.9$ and $\tau_i = 1.0$, in the zonal flow (top row), and the thermal sideband (bottom). The ExB Reynolds stress is net positive (left), the Maxwell stress is small (centre), and geodesic compression is net negative to balance $R_E$ (right). For the thermal sideband, the incoming energy is balanced by the flux drive (left), adiabatic compression (centre), and diamagnetic compression (right).

most of the signal. The flux channel giving transfer in from the turbulence is much more even.

The dependences of these transfer channels upon inductivity ($\hat{\beta}$), collisionality ($C$), and ion temperature ($\tau_i$) are given in figure 16.9. The more electrostatic and the less collisional, the more the Reynolds stress process based on pure ExB vorticity works towards flow generation in this case. On the other side of any of these, complications arise.

When more inductivity is present there is more of a tendency for the curvature term to change sign and allow an MHD plasmoid drifts process to function. With nonadiabaticity present at long wavelength, the clamps on sideband balance provided by the parallel dynamics are loosened, and the sideband responses become more dynamical. With the energy source in the pressure gradient, the direction of energy transfer is then into the flows (zonal to sideband pressure and then to flows, through the curvature coupling). The energy exit process starts to involve the Maxwell stress increasingly with rising inductivity.

When the ion temperature rises the short wavelength vorticity dynamics is affected. The diamagnetic vorticity breaks the enstrophy conservation, and the

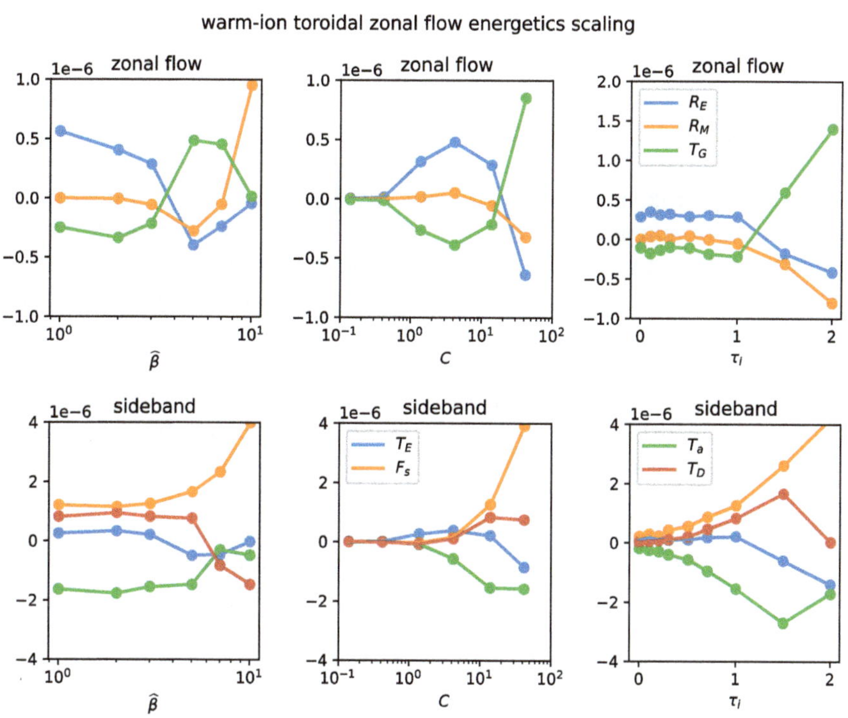


**Figure 16.9.** Dependences of the energy transfer channels for the zonal flow energy (top row) and the thermal sideband energy (bottom), upon inductivity ($\hat{\beta}$, left), collisionality ($C$, centre), and ion temperature ($\tau_i$, right). Labels for the top row are in upper right only, and for the bottom row they are split between the second and third frames, for clarity. Parameters are $\hat{\beta} = 3$, $C = 13.9$, and $\tau_i = 1$ unless varied in the scans. These complicated relations are discussed in the text.

Reynolds stress process is diminished. For finite inductivity at finite ion temperature, both ends of the wavelength spectrum are affected and the Reynolds stress picture breaks down. It is important to recall here, that the entire idea of the ExB shear and decorrelation picture is based on simple ideas of Reynolds stress in a 2D fluid. In a magnetised plasma at low inductivity, that survives. At low collisionality, the ion temperature has less effect because the responses, being closer to adiabatic, eliminate the extra diamagnetic nonlinearity when $[\nabla_\perp\phi; \nabla_\perp n]$ becomes closer to $[\nabla_\perp\phi; \nabla_\perp\phi]$, which vanishes. Then, the 2D fluid Reynolds stress process continues to work further into the finite inductivity regime.

The complications shown by these zonal flow energy diagnostics indicate that the situation in a toroidal magnetic field is much more subtle than a simple fluid analysis suggests. In terms of any modelling effort, one would have to examine this entire set anew for any particular situation to support any conclusion. The foregoing is intended as an introduction to the methods so that they can be used elsewhere.

## 16.10 Summary

Flows and turbulence are usually considered in rather simple terms, even for neutral fluids where the linear threshold problem is well known. Naïvely one expects a sheared flow to be Kelvin–Helholtz unstable, but in a magnetised plasma one considers the entire system, with the thermal gradient usually dominant. Equilibrium flow shear acts upon an already established turbulence, and self generated flows are part of the turbulence. The suppression by decorrelation model is a calculation with a simple diffusive model of the turbulence. In a magnetised plasma the underlying assumptions of the model are broken, with the vorticity in drift wave or drift Alfvén turbulence usually an order of magnitude or more larger than any instability growth rate present. In a physical setting, the flow of free energy among the separate components of the flow and fluctuating variables decides the eventual outcome of the interaction between turbulence and flows.

In a tokamak-like device, with a mostly axial magnetic field and a moderate aspect ratio, together with a low-beta plasma, the geodesic curvature enters to add transfer processes between turbulence and flows, mediated by sine and cosine sideband quantities in flows, pressure, and currents. There is a transfer pathway from flows to resistive dissipation, which receives a large fraction of the energy in return if the turbulence does set up to drive flows. The flow energy and ExB eddy energy in the turbulence tend to equipartition, with both smaller than the thermal free energy. The ratios descend from the fact that while $k_\perp\rho_s$ is finite, it is also smaller than unity, so that the turbulence always has more vorticity than either the equilibrium flow or any self-generated zonal flows. The result is that flows tend to regulate the turbulence but do not suppress it. Neither do they drive it appreciably. The flows are more or less passengers in the overall dynamical system.

## Further Reading

The criteria for Kelvin–Helmholtz stability are treated in most fluid dynamics textbooks, including S Chandrasekhar 1961 *Hydrodynamic and Hydromagnetic*

*Stability* (Oxford: Oxford Univ Press) (repr. New York: Dover, 1981); and P G Drazin and W H Reid 1981 *Hydrodynamic Stability* (Cambridge: Cambridge University Press). Extended criteria for flows in rotating or stratified systems are treated in most geophysical fluid dynamics textbooks, including A E Gill 1982 *Atmosphere-Ocean Dynamics* (New York: Academic).

The idea of the large-scale flows forming via a statistical self-organisation process was advanced by A Hasegawa 'Self-organization processes in continuous media' *Adv. Phys.* **34** (1985) 1 (see Chapter 11).

The scenario of shear flow suppression of plasma turbulence was introduced by H Biglari, P Diamond and P W Terry *Phys. Fluids B* **2** (1990) 1; and reviewed along with fluid counterparts by P W Terry *Phys. Plasmas* **7** (2000) 1653. The energetic signature of the background shear flow tilting the turbulence eddies is from B Scott *Plasma Phys. Control. Fusion* **34** (1992) 1977.

Geodesic acoustic oscillations were found with an MHD analysis by N Winsor, J Johnson and J Dawson *Phys. Fluids* **11** (1968) 2448. A two-fluid treatment of them as elements in a relaxation of a tokamak equilibrium to force balance was given by S V Novakovskii, C S Lui, R Z Sagdeev and M N Rosenbluth *Phys. Plasmas* **4** (1997) 4272.

Turbulence suppression by flows in a tokamak was expressed in terms of a growth rate threshold by T S Hahm and K H Burrell *Phys. Plasmas* **2** (1995) 1648. Zonal flow modification of low frequency turbulence was found by Z Lin, T S Hahm, W W Lee, W M Tang and R B White *Science* **281** (1998) 1835. Geodesic acoustic oscillations were brought into this picture by T S Hahm, M A Beer, Z Lin, G W Hammett, W W Lee and W M Tang 'Shearing rate of time-dependent ExB' flow *Phys. Plasmas* **6** (1999) 922.

Turbulence was found to drive geodesic acoustic oscillations by K Hallatschek and D Biskamp *Phys. Rev. Lett.* **86** (2001) 1223. Geodesic acoustic oscillations were found to act as an energetic sink for zonal flows in drift Alfvén turbulence by B Scott 'The geodesic transfer effect on zonal flows in tokamak edge turbulence' *Phys. Lett. A* **320** (2003) 53; and V Naulin 2003 Electromagnetic transport components and sheared flows in drift-Alfvén turbulence *Phys. Plasmas* **10** 4016; and B Scott 2005 Energetics of the interaction between electromagnetic ExB turbulence and zonal flows *New J. Phys.* **7** 92; and V Naulin, A Kendl, O E Garcia, A H Nielsen and J Juul Rasmussen 'Shear flow generation and energetics in electromagnetic turbulence' *Phys. Plasmas* **12** (2005) 052515.

Experimental detection of geodesic acoustic oscillations was reported by P Schoch *Bull. Am. Phys. Soc.* **44** (2002) RP1.011, and by G R McKee et al 'Experimental characterization of coherent, radially-sheared zonal flows in the DIII-D tokamak' *Phys. Plasmas* **10** (2003) 1712, in both cases resurrecting old data both groups had first obtained in the 1990s.

Pfirsch-Schlüter currents in the sideband analysis are part of the tokamak equilibrium including the lowest order aspect ratio corrections. The text by Goldston and Rutherford cited in Chapter 1 is a good place to look further.

Equilibrium flows in a tokamak are set up by force balance and ion collisional relaxation into a state such that further collisional transport is ambipolar (polarisation current drops to zero). See S P Hirshman 'The ambipolarity paradox in toroidal diffusion, revisited' *Nucl. Fusion* **18** (1978) 917. The scale of the flow shear under these conditions is a few $v_D/L_\perp$, as noted for the warm-ion cases above. The smallness of this was considered in the context of drift Alfvén turbulence in B Scott 'ExB shear flows and electromagnetic gyrofluid turbulence' *Phys. Plasmas* **7** (2000) 1845.

The flux surface average is given by R D Hazeltine, F L Hinton and M N Rosenbluth in 'Plasma transport in a torus of arbitrary aspect ratio' *Phys. Fluids* **16** (1973) 1645, which also contains a basic introduction to the geometry of a tokamak magnetic field as pertains to transport equations.

Flux surface geometry and coordinate systems are treated as part of by R D Hazeltine and J D Meiss 1992 *Plasma Confinement* (Reading, MA: Addison-Wesley).

Flux surface geometry and coordinate systems in general three dimensional plasma systems with closed flux surfaces are treated by W D D'haeseleer, W N G Hitchon, J D Callen and J L Shohet 1991 *Flux Coordinates and Magnetic Field Structure: A Guide to a Fundamental Tool of Plasma Theory* (Berlin: Springer). This has become a standard reference. Those interested in stellarator or heliotron or helias devices should look here.

**IOP** Publishing

Turbulence and Instabilities in Magnetised Plasmas, Volume 1
Fluid drift turbulence
**Bruce Scott**

# Chapter 17

## Interlude

This work has been split into two volumes due to its size, which was forced by the decision to include the three Chapter introduction with turbulence in fluid and MHD systems (Chapter 3), dissipative drift wave turbulence (Chapter 4), and absolute equilibrium ensembles (Chapter 5), with which the look and feel of turbulence and some of its energetic and consistency issues was introduced without the heavy theory one needs to go further. The transition from two- to three-dimensional concepts is also a large leap in a magnetised plasma, in many ways as large as for neutral fluids. Once these were passed through, it was possible to introduce the basic mode types of turbulence individually before combining them. One should be impressed by the level of complication and subtlety involved in the subject of flows versus turbulence or turbulence versus flows in a magnetised plasma. We have only scratched the surface of this up to now. Still remaining is the large set of further complications brought in by temperature gradients. In many applications one may consider the pressure gradient generically, as we have done in Volume One. Not only but especially in magnetically confined plasmas, the temperature gradient is dominant and brings in different physics. It remains to face some of these.

In Volume Two, the new issues brought in by both electron and ion temperature gradients and fluctuations are covered in the first two Chapters. In the case of the ions, we encounter another type of mode instability and mode structure which interacts with all the others: the ion temperature gradient mode, or as we call it, ITG turbulence. This class of dynamics puts pressure on the ion gyroradius scale in the spectrum, and we are then forced to entertain the idea of turbulence across and below this scale in the spectrum. And that requires kinetic theory.

Volume Two spends about half its time developing the special type of kinetic theory applicable to the situation of low frequency magnetised plasma dynamics, namely gyrokinetic theory. Up to now, we started with the standard kinetic theory which we just had to reference. Then, the low frequency approximations were taken on the fluid model. In the opposite case, the low frequency approximations are taken not only in the kinetic theory but directly on the Lagrangian which defines it. We only briefly touched upon classical field theory in Chapter 6, in the context of Maxwell's equation and plasma electrodynamics. But now this theory takes center stage because we need it to establish the gyrokinetic regime in a more rigorous manner to preserve exact energetic consistency all the way through. Once this is done, we then recover a set of fluid equations similar to the fluid drift equations but with a different underlying representation. Fluid equations for a plasma represent particles. In this case we will have gyrofluid equations, which represent gyrocenters in a low frequency system. To make a long story short, it is a matter of moving polarisation from a drift to a density. We still have a polarisation current but it emerges in a different way. Ultimately, we end up with a set of gyrofluid equations which are then shown to reduce to the fluid drift equations in the limit of small gyroradius and strong collisionality. Hence, we show we are solving the same problem ultimately with a better model. The complications of gyroviscosity are replaced with complications of finite gyroradius, but the latter are more computationally tractable and also result in the relaxing of any restriction on the gyroradius as compared to the scale of motion in the turbulence. The responses below the gyroradius take care of themselves, and we will find that the new types of turbulence in this regime can very well be covered with these gyrofluid equations. With an extension of the model, this is also the case with turbulence caused by trapped electrons, so named because they result from magnetic mirror trapping of particle orbits. We can treat the centers of these orbits as the basis of the representation, and then treat this dynamics as well with modifications of the gyrofluid equations. All that is presented in Volume Two. Afterward, there is some more on advanced topics in gyrokinetic theory, including the Hamiltonian functional bracket approach. Finally, some more MHD concepts and how they relate to turbulence driven by currents and flows are given, in order to show how these traditional MHD instability topics are captured by the gyrofluid equations, so long as the low frequency regime holds. This allows future study of global MHD phenomena in a turbulent plasma, or turbulence in the presence of and in interaction with the large scale MHD instabilities. And all that within the context of exact energetic consistency in the dynamics.

And that will then do it for this study of turbulence in magnetised plasmas. Other topics not covered will then be left to others.

# Index

All references are to Section numbers unless Chapter (Ch.) is indicated


absolute equilibrium ensembles (AEQ), Ch. 5 *ff.*
  dissipation and, 5.6
  energy, 5.2 *ff.*, 5.4.1
  equipartition, 5.2
  magnetic flux, 5.5.1
  MHD, 5.5
  vorticity, 5.2 *ff.*, 5.4.1
adiabatic, 4.2, 7.8, 9.1
  coupling, 4.2.3, 10.4, 11.3.1, 11.8, 14.5.1, 14.9.1, 14.9.3, 15.3.2, 16.8.5
  compression 8.2, 9.6.2, 10.4, 12.6.1, 12.7.1, 14.2.2, 14.5.1, 16.9.2
  electrons, 4.2.1, 4.3.2, 4.3.4, 8.3, 9.3, 9.4, 10.3.
  in energetics, 11.2, 12.7.1, 14.2.2, 14.5.1
  (*see* force balance)
  pressure dynamics 6.7
  response, 4.2.1 *ff.*, 4.2.2, 4.3.2, 4.3.3, 4.3.5, 4.4, 7.8, 8.1,8.4,8.5,8.6, 8.8,8.10,9.2,9.3,9.5.1,9.6.2,9.8,9.11
  start, 4.3, 5.4.1
  state (*see* electrons)
  (*see* energy transfer)
advection, 1.2, 2.4, 3.2.1, 4.2
  advective derivative, 6.2
    ExB, 4.2, 7.6, 11.2
    ion, 7.5
  Eulerian form, 6.2
  Lagrangian form, 6.2
  passive, 2.1, 4.2.3, 5.4.1, 5.5.1
Alfvén, 1.2
  (*see* energy transfer)
  Hannes, Ch. 1, Further Reading
  speed, 1.2
  waves, 1.2
    compressional, 6.3, 6.7
    quasineutrality in, 6.3
    shear, 6.7
aspect ratio, 11.4 *ff.*
  and flows, 13.6, 13.7.1, 16.2.1 *ff.*
  and gradient scale, 13.9
  parallel vs perpendicular wavenumbers, 6.7, 7.2, 9.3
    (*see* flute mode ordering)
  parallel vs perpendicular scale ratio, 1.2, 8.1, 9.5.2, 9.5.3, 11.7
  perpendicular domain, 11.4, 13.7
  grid cells, 12.5
  toroidal, 13.9, 16.8

ballooning, 13.10
  (*see* MHD)
Boussinesq approximation, 13.5
  (*see* delta-f, drift approximation)
  in fluid equations, convection, 13.5
brackets, 3.2.1 *ff.*
  advection, 3.2.1, 11.2.1, 12.3,
  definition, properties, 3.2.1
  and magnetic curvature, 13.9
  (*see* numerical schemes)
  and self consistent gradients, flows, 15.3.3, 16.4.1
  structure, 3.2.1
  in a toroidal model, 13.9, 14.2.1

Chapman-Enskog expansion, 6.10 *ff.*
  derivation of dissipative fluxes, 6.10
compression, 1.2
  (*see* Alfvén waves)
  (*see* force balance)
  regulation of, 2.1
    in hydrodynamics, 2.5
    in MHD, 2.6, 3.6 *ff.*
curvature, Ch. 13 *ff.*
  and drift waves, 14.1
  equilibrium, 13.2.1
  forcing, 13.2.2, 13.3
  geodesic, 16.1, 16.6, 16.8.1 *ff.*
  magnetic, 1.2


 

and MHD, 13.2.2
radius of, 13.2.1

delta-f, 7.7 *ff.*
(*see* drift approximation)
free energy versus energy, 7.7, 7.11 *ff.*
diamagnetic, 6.11
cancellations, 6.11, 7.6, 7.8
compression, 16.8.2
current, 7.3
drift, flow, 7.5
flows and Reynolds stress, 16.6
frequency, 9.2
and pressure, 13.2
flow shear level, 16.6
dipole magnetic field, 15.4 *ff.*
geometry, metric, fluxtube, 15.4.2
the turbulence, 15.4.3
dissipation, 5.1, 5.6 *ff.*
boundary, 11.5
collisional, resistive, 6.10, 7.8, 7.9, 9.6.2, 14.4
Landau damping, 6.10
parallel viscosity, 7.8, 7.9, 9.6.2, 14.4
subgrid, 3.2.2
introduction into turbulence, 3.2.2
in three dimensions, 11.2.3, 12.6
in two dimensions, 3.2.2
thermal conduction, 6.10
divergence, 2.4
current balance, 7.6, 7.9, 16.8.4
and drifts, 7.8
and energetics, 7.9
flow balance, 16.8.3
magnetic divergences, 7.3, 7.8
drift(s), Ch. 7 *ff.*
approximation, 7.2
(*see*
diamagnetic 7.3,
ExB 7.5,
polarisation 7.6)
direction, 9.2
(*see* drift Alfvén waves)
dynamics, 1.2, Ch. 7 *ff.*
fluid, 1.2, 7.5
operator, 7.3
ordering, 7.7
drift scale ('rho s'), 1.2, 4.2.1, 9.4, 14.9
definition, 1.2
and flows, 16.4.1, 16.5
and the ion gyroradius, 4.4, 16.6
and the vorticity spectrum, 12.7, 14.5
drift (Alfvén) waves, Ch. 9 *ff.*
definition, 9.3
instability, 9.6
numerical examples, 9.8
self consistent adiabatic response, 9.5

E-cross-B (ExB), 1.2
drift, 1.2, 7.5
energy, 4.3.1, 4.3.5, 7.9,
flow, 7.5, 9.2, 9.3, 9.6, 11.4, 12.7, 13.7,
nonlinearity, 11.2, 12.6.1, 12.7, 13.11.1, 13.12.1, 14.6
advective, 4.3.2, 10.9
vorticity, 4.3.2, 10.9
relation to force balance, 7.5
relation to Lorentz force, 6.6
(*see* flows and flow shear)
electrodynamics, 6.2
displacement current, 6.4
high frequency modes, 6.3
(*see* Maxwell equations)
plasma, 6.2, 6.3, 6.4
Elsässer variables, 2.6
EMHD, electron MHD, 3.4 *ff.*
current and electron flow, 3.4
energetics, 3.4.1
equations, 3.4
energetic consistency, Ch. 7 *ff.*
energetic error, 3.2.6, 14.4
and flows, Reynolds stress, 16.4
positive definite damping, 7.9
and turbulence, saturation, 4.3.5, 10.4
energy theorem, 6.8, 7.9
Alfvén waves, 8.4
drift waves, 9.6.2
2D turbulence ...

hydrodynamics, 3.2.6
MHD, 3.3.1
EMHD, 3.4.1
Hall MHD, 3.5.1
3D turbulence ...
drift (Alfvén) wave, 11.2.2
interchange forcing, 13.3
toroidal drift Alfvén, 14.2.2
energy transfer, 10.4 *ff.*
drift waves, 9.6.2
kinetic shear Alfvén waves, 8.4
sound waves, 8.4
in drift (Alfvén) wave turbulence ...
acoustic, 12.7.1, 14.5.1
adiabatic, 12.7.1, 14.5.1
Alfvénic, 12.7.1, 14.5.1
compressional (ExB, diamagnetic), 13.12.1, 14.4, 14.5.1
nonlinear (ExB advection, vorticity), 4.3.6, 10.4, 12.7.1, 14.5.1
enstrophy 2.5 *ff.*
and dual cascade, 2.5
generation by vortex tube stretching, 2.5
mean squared vorticity, 2.5
entropy, 6.11
gradient, 13.1
equilibration
(*see* saturation)
spectrum, 5.4.1, 14.6
equipartition, Ch. 5 *ff.*
energy, 5.2
magnetic flux, 5.5.1
spectrum (*see* AEQ)
enstrophy, 5.2, 5.5
Eulerian fluid equations, 3.1, 16.2
Euler-Lagrange equations, 6.2

field line bending, 1.2, 3.3.2, 5.5.1
field line connection, 11.8, 12.2 *ff.*
limit on parallel wavelengths, 12.2
periodicity, 11.3
field aligned coordinates, 12.5
slab, 12.2
toroidal, 13.9
rational surfaces, 12.2
spacing, 12.2
flows, Ch. 16 *ff.*
decorrelation, 16.3
suppression, 16.3
(*see* Kelvin-Helmholtz)
Maxwell stress, 2.6, 16.4.1
Reynolds stress, 2.6, 16.4
shear, 16.1
sheared flows, 16.1, 16.3
flute mode, 1.2
ordering, 7.2, 8.3, 9.5.3
in parallel spectrum, 11.3.1, 11.3.2
flux
exit, 15.1.1
magnetic, 3.3
conservation, 3.3.1
(*see* equipartition)
turbulent, 4.3, 10.2, 10.3, 10.4, 11.3.2, 12.8, 13.13, 14.7, 15.3.3, 16.5
flux function, 13.9
flux surfaces, 12.1, 12.2
averaging, 16.7 *ff.*
force balance, 1.2, 7.1, 8.1
parallel (electron), 4.2.2, 8.3, 14.9
perpendicular, 3.6, 7.3, 7.5
force potential, 3.3, 4.2, 7.8
electrons, 4.2
free energy, 4.3, 7.7, 7.9
conservation (*see* energy theorem)
versus energy, 7.9, 7.11
thermal, 4.3.6
frequencies ...
Alfvén, 8.3
curvature, interchange, 13.5
cyclotron (gyro), 1.2
diamagnetic, 9.2
drift frequency ordering, 1.2
geodesic acoustic, 16.8.3
geodesic Alfvén, 16.8.4
lower hybrid, 6.3
plasma, 6.3
sound wave, 8.3
upper hybrid, 6.3

geometry, 14.1
  coordinate, 3.2.3
  Jacobian, volume element, 3.2.1
  magnetic, 12.2, 12.5, 13.9, 14.2.1, 15.4, 16.7
  metric tensor, 3.2.1, 6.8, 7.8
geostrophic flow, 1.2, 2.5
  relation to drifts, Ch. 7, Further Reading
Grad-Shafranov equation, 13.9
guide field (magnetic), 1.2, 2.1, 2.8, 3.6, 7.4

Hall MHD, 3.5
  concept, 3.5
  energetics, 3.5.1
  equations, 3.5
Hasegawa-Wakatani (-Mima) equations, 4.2.3
helicity ...
  cross-, 2.6, 3.3.1,
  flow, Ch. 2, Further Reading
  magnetic, 2.6

interchange, Ch. 13 *ff.*
  dynamics, 13.1
  ideal growth rate, 13.5
  modes, 13.5
    ballooning, 13.10
    g-modes, 13.5
    convection, 13.1, 13.6, 13.7
  turbulence, 13.6

Kelvin-Helmholtz, 11.4,
  aspect ratio, 11.4, 16.2.1
  energetics, 16.4
  long wavelength cutoff, 16.2
  stability, 16.2
  vorticity extremum (Rayleigh theorem, criterium), 16.2

Lagrangian, 6.2
  density, 6.2
  field theory (*see* Volume Two)
Landau damping (*see* dissipation)

magnetic flutter, 3.3.1
  nonlinearity, 3.3, 7.4, 11.2
  energy transfer, 3.3.3
  parallel gradient ordering, 7.4
  stabilisation, 9.9, 12.8
Maxwell equations, 6.2
  derivation, 6.2
  energy, 6.8
  (*see* electrodynamics)
  gauge invariance, 6.4
  momentum, 6.8
  Poynting fluxes, 6.8
MHD, magnetohydrodynamics, 2.6, 3.3, 6.6
  ballooning, 13.10, 13.11
  conditions for validity, 6.6
  electron (*see* EMHD)
  energy theorem, 3.3.1, 7.9
  equations, 2.6, 3.3, 6.6
  equilibrium, 12.2, 13.2.1, 13.9
  full versus reduced MHD, 7.1
  (*see* field line bending)
  (*see* Hall MHD)
  momentum theorem, 7.9
  reduced MHD, 8.8, 13.4, 13.10, 13.11, Chs. 3 and 7, Further Reading
    comparison to drift Alfvén turbulence, 14.9.1
    and geodesic acoustic oscillations, 16.8.3
    subset of toroidal drift Alfvén equations, 14.2
mode structure, Ch. 10 *ff.*
  adiabatic, 4.3, 10.8
  correlation lengths, 10.5
  correlation times, 10.5
  cross correlation, coherence, 10.8
  drift wave, 10.8, 12.7, 14.5
  hydrodynamic, 4.3
  interchange, 13.8, 13.12
  phase shift distribution, 10.8
  parallel envelopes, 10.10, 13.12, 14.5, 15.3.1, 15.3.4, 15.4.3

probability distribution, 2.3, 4.3.6, 10.7, 16.9.2
(*see* spectrum)
moment hierarchy, 6.9
closure problem, 6.9
derivation of diamagnetic fluxes, 6.11

nonadiabatic electrons, 12.6.1
definition, 9.5.4
and energetics, 9.4, 9.6
nonlinearity, 2.1
advection, convection, 2.1, 3.2.1, 11.2, 11.4, 12.6.1, 13.7
one dimensional, 2.1
and ordering, 7.7
shocks, 7.1
numerical schemes, 3.2.3
Arakawa (space, bracket), 3.2.3
boundary conditions ...
bounded, Dirichlet, 11.5
and energetics, 3.2.4
periodic, 11.3
pseudo-periodic, 12.5
Karniadakis (time), 3.2.3
Laplacian, 3.2.3
shear, shifts, 12.5, 13.9

parallel and perpendicular, 1.2
directions, 6.2
divergences, 7.3, 7.4, 7.5, 7.8
dynamics, Ch. 7 *ff.*, 8.3, 9.2, 11.2.2, 12.6.1
metric, 12.5, 14.2.1, 15.4
(*see* guide field)
passive scalar, 2.1, 2.4, 4.2, 5.4.1, 10.7
equation, 2.4, 4.2
probability distribution, 10.7
periodicity, 11.3
and magnetic shear, 12.2
toroidal topology, 12.1, 12.2, 13.8
phase shift, 9.6
distributions, 10.8, 12.7, 13.8, 13.12, 14.5
and instability, 9.6.2
(*see* mode structure)
and transport, 9.6.1
polarisability, 1.2
plasma versus vacuum, 1.2, 7.12
(*see* quasineutrality)
polarisation, 1.2, 7.6, 9.4, 12.6,
drift, 1.2, 7.6
and drift ordering, 7.6
current, 7.3
and energetics, 12.7.1, 13.2.1, 14.5.1, 14.9.2
(*see* nonlinear instability)
vorticity nonlinearity, 12.6.1
Poynting cancellation, 6.8
Proudman-Taylor Theorem, 2.8

quadratic nonlinearity 2.4 *ff.*, 7.7
(*see* brackets)
quasineutrality, 1.2, 6.4, 6.5, 7.12
and charge density, 6.1, 6.4
and displacement current, 6.4
divergence free current, 6.3, 6.4
frequency limit, 6.3
and Lorentz force, 7.12
parameter limit, 1.2, 7.12
scale limit, 7.12
velocity limit, 7.12
vorticity and charge density, 7.12

ratio of specific heats, 8.2, 8.4
adiabatic (implied), 6.2
isothermal, 8.4
rational surface, 12.2, 13.9
definition, 12.2
(*see* field line connection)
spacing, 12.2
Rayleigh (inflection point) theorem, 16.2
Rayleigh Taylor instability, 13.5

saturation, 2.1
definition, 2.1, 2.3
of flows, 16.8.5
of turbulence, 10.6
(*see* turbulence scales)
space scales ...

Debye length, 1.1
drift scale, 'rho s', 1.2
field line connection length, 12.2, 12.5
flow shear length, 11.4, 16.2
fundamental wavenumber, 3.2.4, 5.2, 11.3, 12.2
(thermal) gradient, 1.2
gyroradius, 1.2
magnetic shear length, 11.6, 12.2
parallel periodicity, 8.4, 11.2.1, 11.6
shear ...
Alfvén waves (*see* Alfvén)
flow (*see* flows)
magnetic, 11.6, Ch. 12 *ff.*
spectrum (spectra), 2.1, 2.2, 2.5
(*see* absolute equilibrium ensemble)
(*see* turbulence)
energy, 3.2
isotropic perpendicular, 10.3
parallel wavenumber, 10.10
vorticity, 3.2
wavenumber, 10.3
sideband analysis, 16.8
flows in toroidal system, 16.8.2, 16.9.1, 16.9.2

thin atmosphere, 2.5, 11.4
(*see* aspect ratio)
three wave (triplets, or triads), 2.4, 2.5
definition, 2.4
energy transfer, 2.5
enstrophy transfer, 2.5
tokamak, 2.1, 13.9, 14.3, 14.8, 15.3.2, 16.6, 16.8
conventional ordering, 13.9, 16.8.1
equilibrium (*see* Grad-Shafranov equation)
sideband modes, 16.8.1
total-f, 7.1, 7.9
versus delta-f, 7.7, 7.11
energetics, 7.11
turbulence, Ch. 2 *ff.*
cascade, 2.2
dual, 2.5
dissipation, role of, 5.1, 5.6 *ff.*
hydrodynamic, 3.2
drift (Alfvén) wave, 11.3, 11.5, 11.6, 12.6
toroidal, 14.5, 14.6, 14.7, 14.8, 14.9
interchange, 13.6
ballooning, 13.11, 13.12
intermittency, 2.3
introduction, 2.1
MHD, 3.3
(*see* nonlinearity)
scales, 2.1
correlation length, 2.3, 10.5
correlation time, 2.3, 10.5
eddy turnover time, 2.2, 2.5, 10.2
RMS vorticity, 10.2
saturation time, 2.3
statistical nature of, 2.3
(*see* three wave)
two fluid, 5.5.2, 5.6, 8.2, 14.5
(*see* drift dynamics)
re electrodynamics, 6.5
versus MHD, 13.2.2, 13.10

vorticity, 1.2, 2.1, 2.2, 2.5
and charge density, 7.12
(*see* enstrophy)
(*see* flows, shear)
ExB, defintion, 1.2, 10.2
drift scale factor, 12.6, 14.5, 16.4.1, 16.6
and polarisation, 7.6
and quasineutrality, 7.12
nonlinearity, 4.3.1
in 3D drift wave turbulence, 11.6, 12.6.1
in toroidal drift Alfvén turbulence, 14.5, 14.5.1, 14.6, 14.7, 14.8, 14.9.2

www.ingramcontent.com/pod-product-compliance
Ingram Content Group UK Ltd.
Pitfield, Milton Keynes, MK11 3LW, UK
UKHW051910100926
8556IPUK00005B/836

9 780750 32502